NASA SP-4018

ASTRONAUTICS AND AERONAUTICS, 1973

Chronology of Science, Technology, and Policy

Text by
Science and Technology Division
Library of Congress

Sponsored by
NASA Historical Office

Scientific and Technical Information Office   1975
NATIONAL AERONAUTICS AND SPACE ADMINISTRATION
Washington, D.C.

Foreword

This volume in NASA's continuing series of annual chronologies records the events of 1973. The 15th year of the space agency's existence is memorable to the public as the Year of Skylab and the Year of Jupiter. The nation's first earth-orbital laboratory was launched. Beginning with a dramatic 10-day fix-it effort that restored Skylab's capability to generate electrical power, three three-man crews rendezvoused and docked with the space laboratory. Their total of 12 351 man-hours in space supported the feasibility of prolonged manned space flight. Even more impressive were the volume and quality of experiment results. The sophisticated battery of telescopes in the Apollo Telescope Mount produced 182 842 frames of film and 941 hours of manned viewing of the sun. The earth was not neglected either. Orienting the laboratory toward the earth, astronauts took a total of 40 286 frames of earth-resources photographs. Electric furnace experiments with zero-gravity manufacturing offered interesting possibilities of new industrial processes that cannot be achieved in earth gravity. Assimilation of the knowledge contained in the mass of data is still continuing.

As the year closed, *Pioneer 10* swept by the planet Jupiter after a 21-month flight that traversed a billion kilometers. Passing through the intense radiation surrounding the largest planet in our solar system, *Pioneer 10* sent back 300 closeup photographs of the planet and its inner moons, along with other measurements of the swirling atmosphere and its massive Red Spot and of the magnetic, radiation, and electrical environments. Even as it returned the data—much of which contradicted previous theories—*Pioneer 10* was heading outward toward the orbits of Saturn and Pluto. In the late 1980s it will become the first man-made object to leave the solar system and voyage into interstellar space. The events of Skylab and Pioneer were only the most visible successes of 1973, however; many other important events accomplished their objectives with less notice.

On a smaller scale, but particularly important to a few, was the announcement of a new heart pacemaker developed from electronic components and batteries originally designed for use in space. The new model was rechargeable, thus eliminating the need for periodic surgery to replace the earlier instrument.

Uses for the vast amount of data from the Earth Resources Technology Satellite *Erts 1* multiplied as it completed its first year in space in July 1973: planning land use, managing water resources, estimating crop yields, inventorying timber, exploring for minerals and petroleum. The practical benefits from this and similar programs seem at this writing almost limitless. Indeed, the principal problems are no longer the acquisition of such data, but rather its rapid and economical reduction into usable forms.

In aeronautics, work went forward on a number of programs: digital fly-by-wire, which will significantly improve aircraft control; the supercritical wing, which provides significant fuel savings; reduction of noise and pollution; and hydrogen injection to increase efficiency and reduce pollution from gasoline engines. Collectively these efforts promise safer flight, fuel economies, and less damage to our environment.

Thus on most counts NASA activities seemed to bear out the Administrator's statement in late 1973 that NASA was providing more emphasis on becoming "more like one of the service agencies of the government."

Altogether 1973 was a productive year, punctuated at times with excitement. And while real achievements were being realized, plans were being laid for subsequent programs, and this volume offers specific information about both.

Edwin C. Kilgore
*Deputy Associate Administrator
for Center Operations*

December 1974

Contents

	PAGE
Foreword	iii
Edwin C. Kilgore, Deputy Associate Administrator for Center Operations	
Preface	vii
January	1
February	35
March	61
April	101
May	131
June	171
July	203
August	235
September	255
October	279
November	307
December	335
Summary	357
Appendix A: Satellites, Space Probes, and Manned Space Flights, a Chronicle for 1973	363
Appendix B: Chronology of Major NASA Launches, 1973	395
Appendix C: Chronology of Manned Space Flight, 1973	399
Appendix D: Abbreviations of References	403
Index and List of Abbreviations and Acronyms	407

Preface

A brief, chronological account of key events of the year in space and in the atmosphere, *Astronautics and Aeronautics, 1973* records dates, actions, hardware, persons, scientific discoveries; plans, decisions, achievements, preliminary evaluations of results, and samples of public reaction and social impact. The volume is offered for reference use within the Federal Government and by the public. It should be of use to present and future scholars in a variety of disciplines.

The sources, identified by abbreviations that are explained in Appendix D, were those immediately available in NASA and other Government agencies, the Congress, and the professional societies, as well as the press. Contradictory accounts have been resolved and doubtful ones verified whenever possible by querying participants. Cross-references are given in the text, and the detailed index will aid in tracing related events through the year. The index also serves as a glossary of acronyms and abbreviations.

General editor of the volume was the Publications Manager of the NASA Historical Office, Frank W. Anderson, Jr., and the technical editor was Mrs. Carrie E. Karegeannes. Archivist Lee Saegesser collected current documentation. The Science and Technology Division of the Library of Congress, under an exchange of funds agreement, drafted monthly segments in comment edition form, which were circulated for corrections, additions, and use. At the end of the year, the entire manuscript was reworked to include comments received and additional information that was not available when the monthly segments were prepared. At the Library, Patricia D. Davis, Nancy L. Brun, Shirley M. Singleton, and May Faye Johnson carried principal responsibility. Arthur G. Renstrom of the Library prepared preliminary indexes for the first six months; Informatics TISCO, Inc., prepared the detailed index for the annual volume.

Appendix A, "Satellites, Space Probes, and Manned Space Flights, 1973," Appendix B, "Chronology of Major NASA Launches, 1973," and Appendix C, "Chronology of Manned Space Flight, 1973," were prepared by Leonard C. Bruno of the Library of Congress. Appendix D, "Abbreviations of References," was prepared by Mrs. Karegeannes.

Without the assistance of many persons throughout NASA and other Federal agencies, the content of this volume would be less reliable and complete. Comments, additions, and criticisms are always welcomed by the NASA Historical Office.

Monte D. Wright
Director, NASA Historical Office

January 1973

January 1: NASA had scheduled 15 spacecraft launches in 1973 from Kennedy Space Center, Eastern Test Range, and Western Test Range, KSC announced. Launches would include four Skylab missions, three of them manned; Intelsat-IV F-6, F-7, and F-8 for Communications Satellite Corp.; Pioneer-G to fly by Jupiter; Canada's Telesat-B comsat; RAE-B Radio Astronomy Explorer; United Kingdom's Skynet II-A comsat; Mariner mission to fly by Venus and Mercury; ITOS-E Improved Tiros Operational Satellite; ERTS-B Earth Resources Technology Satellite; and AE-C Atmosphere Explorer. (KSC Release I-73)

- NASA and Soviet space officials had agreed to permit representatives of the U.S. and U.S.S.R. to observe and advise in Apollo and Soyuz flight control rooms during orbital phase of July 1975 joint mission, *Aviation Week & Space Technology* reported. No agreement had been reached to permit U.S. observers at the Soviet launch site during liftoff of Soyuz spacecraft. [See March 15–30.] (*Av Wk*, 1/1/73, 13)

- An *Aviation Week & Space Technology* editorial noted the aerospace industry forecast for 1973 was for continuing improvement "across the entire technical spectrum." The year 1971 had bottomed out the 1969–1970 recession and provided the "first tangible evidence of upswing." The year 1973 should continue the uptrend "on a modest but solid curve." Industry had been "under increasing pressure from Congress, the Pentagon and the taxpaying public for the last five years to improve its managerial efficiency and the reliability performance of its products. The pressure has now accumulated to the point where the industry must respond, both internally in cleaning its own house and mounting an aggressive campaign for government contracting reforms to match, or face the very real spectre of nationalization in the public mood that may well prevail four years hence." (Hotz, *Av Wk*, 1/1/73, 11)

- The "deeper meaning" of *Apollo 17* (Dec. 7–19, 1972), the last manned lunar landing mission in the NASA program, was discussed by author William I. Thompson in a *New York Times* article: The mission had represented a "sunset" of rocket technology. "Apollo 17 turned the night into day, but elsewhere smaller lights were going on as men began to discover relationships between consciousness and the growth of plants . . . , between enzyme change and faith healing . . . , and between mind and matter in psychokinesis. . . . The space program was an important scaffolding, but now that the building of the new human culture is up, we no longer need the scaffolding." To spend a fortune on rockets now "would be the same as spending a fortune on dirigibles in 1916. There are other forms of space travel to be invented, and these forms are more likely to spring from the new paradigms emerging in science than from the hardware of the old technology. The era of the rocket has climaxed in Apollo 17; if we push on and ignore the sense of an ending, we shall find only the bitterness and disappointment of the anticlimax." (*NYT*, 1/1/73, 3:13)

January 2: Dr. James C. Fletcher, NASA Administrator, expressed gratification in his monthly letter to the staff "with NASA's excellent performance in 1972, and with the successful transition . . . from the completed programs of the Sixties to the new programs and new challenges of the Seventies. In 1972, Phase I of the national space effort of the United States was completed and Phase II well begun."

NASA's first year with all launches and missions successful underlined that it was NASA's custom "to do the unprecedented and then quickly make it routine." Zero failures could now be the goal year after year. Dr. Fletcher also noted that of 18 space launches in 1972 half had been for other organizations or countries, with 8 being reimbursed.

On another "important front," the NASA-wide payload cost-reduction campaign begun in 1972 had gone well, with valuable suggestions from Centers and contractors. The effort "was not a routine cutback exercise and not a threat to anyone's security. It is, instead, a creative effort to expand our usefulness even though funds are limited. And you can be sure it has a lot to do with NASA's long-range survival as the instrument of the U.S. Government to handle most kinds of space and aeronautical R&D and other related assignments." (*NASA Activities*, 1/15/73, 2–3)

- Dr. Edward E. David, Jr., Presidential Science Adviser, resigned to return to private industry. The former communications systems scientist at Bell Telephone Laboratories, Inc., had served as Director of White House Office of Science and Technology for 28 mos and had instituted reorganization of the Federal science complex. The *New York Times* quoted White House sources as saying the resignation was prompted by disappointment that his advice had not been heeded by the Nixon Administration. Federal support for science and technology expressed as percentage of budget outlays for research and development had reached the lowest point since latter years of the Eisenhower Administration. The White House had declined comment on the resignation and had not released the text of Dr. David's letter to the President. (Lyons, *NYT*, 1/3/73, 1)

- Grumman American Aviation Corp. was formed by the merger of Grumman Corp.'s general-aviation interests and American Aviation Corp. Grumman's Gulfstream 1 and 2 and Ag-Cat program assets were transferred to American Aviation Corp. in exchange for shares in American. Corporate headquarters of the new company would be in Cleveland, Ohio, with Grumman President E. C. Towl as company Chairman and American President and Chief Executive Officer R. W. Meyer as President. (*Interavia*, 4/73, 301; Grumman Corp PIO)

January 3: President Nixon accepted the resignation of Dr. Edward E. David, Jr., as Science Adviser and Director of Office of Science and Technology [see Jan. 2]. (*PD*, 1/8/73, 10)

- NASA planned to spend up to $100 million during the next five years on the post-Apollo lunar science program to extract information from 385 kg (850 lbs) of lunar samples, 30 000 photos, and "miles" of magnetic tape from six Apollo lunar landing missions, the *New York Times* reported. Director Anthony J. Calio of Manned Spacecraft Center Office of Science and Applications had said in an interview that recommendations in an October 1972 report of the Lunar Science Institute were being followed in much of MSC's planning, but the MSC Lunar Receiving Laboratory would be closed on completion of *Apollo 17* sample processing. The sample collection might be divided for storage at another Government facility as a "precaution." The LRL report had said preservation of MSC curatorial facilities for lunar materials was "absolutely essential" to lunar science objectives. Calio said sample inventory and preparation of rock slivers for the thin sections library should take two years. Repository for Apollo geophysical material, including magnetic tapes, probably would be Goddard Space Flight Center. (Wilford, *NYT*, 1/3/73, 25)

- The Aviation Advisory Commission sent to the President a report of its two-year study of long-range U.S. aviation needs. Acting under Public Law 91–258, the Commission had concluded that the "U.S. is facing the greatest combination of threats to its position of world preeminence in aviation since it established that position in the late forties." Recommendations included establishment of Under Secretary for Civil Aviation in Dept. of Transportation to prepare and keep current a 10-yr plan for air services, airports, airways, air vehicles, and ground access. Civil-aviation research and development functions of NASA, the Federal Aviation Administration, and other Federal agencies should be placed under the USCA. (*CR*, 1/12/73, S608)

- Reaction of writers Norman Mailer and Katherine Anne Porter to space flight and to nighttime launch of *Apollo 17* were reported in the *Christian Science Monitor*. Authors had witnessed the Dec. 6, 1972, launch and participated in a space seminar during the "Voyage beyond Apollo" cruise of the S.S. *Statendam*, which left New York Dec. 4 for Cape Kennedy and the Caribbean.

    Mailer had said: "Moon colonies offer the possibility that we'll be able to discover definitively and for the first time what we are doing wrong. Whether we thrive on the moon or fail we're going to learn something that we're not going to learn by staying here on earth. . . ." Part of "the fundamental impasse of 20th-century man and woman is that they cannot find communities which express their philosophical ideas, their social ideas, and their private ideas. In other words, the one thing that we can't do any longer is verify what we think. We can have ideas. But we spend our lives just talking about them.

    "There are very few people who have a profound sense of commitment. One of the reasons that the astronauts have been so revered . . . is that they have this huge sense of commitment. . . . they are absolutely devoted to a living a life that they believe in utterly and are willing to go through great hardship for it. I think that until our society reaches the point where many more people live lives of such commitment it's going to remain essentially unhealthy. . . .

    "The decision to go to the moon is such a declaration of intent. If nothing else, one has to risk dying on it. One has to recognize that one is taking a major step, and the idea of taking a major step brings out in people at least for a short period . . . the best in them. Then, if there's something really alive in the venture, they're likely to keep changing their habits and go on to make more of themselves."

    Miss Porter had said that, in almost anything men undertook together here on the earth, "in no time at all there's a pull and haul for power. They're betraying and tricking each other, getting in one another's way. And yet when it comes to something they consider big enough and grand enough they do it in a grand way and they do it well, like the astronauts. . . . We've always built and saved a little more than we've been able to destroy, don't you think?" (*CSM*, 1/3/73)

- Rep. Alphonzo Bell (R–Calif.) introduced H.R. 32, National Science and Policy Priorities Act of 1973. The bill differed from S. 32, passed by the Senate in 1972 but never reported out by the House Committee on Science and Astronautics [see Jan. 4]. The new bill sought to protect the National Science Foundation's pure science functions by allocating at least 40% of the science budget to NSF. (CR, 1/3/73, H27; NASA LAR, XII/1)
- Discovery of a new comet near the constellation Orion had been reported to Harvard Univ. by 15-yr-old amateur astronomer David Sams of Columbus, Ohio, the *Washington Evening Star & Daily News* said. Sams had observed the comet through a $90 telescope. (W Star & News, 1/3/73, A3)
- Rep. Charles S. Gubser (R–Calif.) introduced H.R. 611, to limit and control more effectively the use of Government production equipment by NASA and Dept. of State contractors. (CR, 1/3/73, H76)
- The sky show "When Earth Became a Planet" opened at the American Museum–Hayden Planetarium to mark the 500th anniversary of the birth of Polish physician, economist, statesman, and astronomer Nicholas Copernicus. Copernicus, born in 1473, taught that the sun was the center of the solar system. The theory had paved the way for the work of Kepler, Galileo, Newton, and others who eventually made space age possible. (Am Mus–Hayden Planetarium Release)
- Grumman Corp. spokesman told the press in Bethpage, N.Y., that Grumman had accepted an $18-million, short-term loan from the Navy at the same time the firm had said it would not honor a contract to provide the Navy with F–14 jet fighter aircraft. (UPI, W Post, 1/4/73, D6)

January 4: President Nixon announced that Dr. James C. Fletcher, NASA Administrator, and Dr. George M. Low, Deputy Administrator, would remain in the NASA positions during the President's second term of office. President Nixon also accepted the resignation of Gerard C. Smith as Director of the Arms Control and Disarmament Agency, U.S. Ambassador, and Chief of the U.S. delegation to the Strategic Arms Limitation Talks (SALT). He announced the resignation of Secor D. Browne as Chairman and of Whitney N. Gillilland as Vice Chairman of the Civil Aeronautics Board. He submitted to the Senate the nominations of Elliot L. Richardson to be Secretary of Defense and of Secretary of Transportation John A. Volpe to be U.S. Ambassador to Italy. (PD, 1/8/73, 10, 12)

- The Federal Communications Commission authorized Western Union Telegraph Co. to build the first domestic satellite communications system in the U.S. Six other proposals were pending and FCC had indicated it would approve "all qualified applicants." (Jacobs, WSJ, 1/31/73, 1)
- The Senate adopted S.Res. 10, establishing the number of members of the Senate Committee on Aeronautical and Space Sciences at 13, and S.R. 12, which elected Chairman and majority members to the Committee. Majority members were Democrats Sen. Frank E. Moss (Utah), Chairman, succeeding Sen. Clinton P. Anderson (D–N. Mex.), who had retired; Sen. Warren G. Magnuson (Wash.); Sen. Stuart Symington (Mo.); Sen. John C. Stennis (Miss.); Sen. Howard W. Cannon (Nev.); Sen. James G. Abourezk (S. Dak.); and Sen. Floyd K. Haskell (Colo.). (CR, 1/4/73, D6; A&A 1972)
- Sen. Edward M. Kennedy (D–Mass.), with 30 cosponsors, introduced S. 32, National Science Policy and Priorities Act of 1973—reintroducing S. 32

from the preceding Congress. The bill, passed by the Senate but not the House, provided three-year authorization of $1.8 billion for "civilian research and engineering problems of our society." (*CR*, 1/4/73, S103-12)

- Sen. Edward M. Kennedy (D–Mass.) introduced S. 60 to authorize acquisition and maintenance of the Goddard rocket launching site in Auburn, Mass., by the National Park Service as tribute to late U.S. rocket pioneer. The first liquid-fueled rocket had been launched on the site by Dr. Robert H. Goddard March 16, 1926. (*CR*, 1/4/73, S135)
- The People's Republic of China was expanding communication lines to rest of world with the aid of American communications satellite technology, the *New York Times* reported. Communications Satellite Corp., RCA Global Communications, Inc., and Western Union International had provided ground stations that would be operational before the end of 1973, to link Peking and Shanghai via satellite to 63 nations that also had ground terminals. (Lyons, *NYT*, 1/5/73, 4)

January 5: NASA began program cutbacks to adjust to a lower spending level in compliance with President Nixon's $250-billion target for total Government spending in FY 1973. Space shuttle manpower buildup would be slowed, delaying the first orbital flight. Work on the High Energy Astronomy Observatory (HEAO) was being suspended at least one year and work on communications satellites would be phased out, with cancellation of ATS–G. Work on nuclear propulsion would be discontinued and work on nuclear power curtailed. Plum Brook Station of Lewis Research Center would be closed. In aeronautics, quiet, propulsive-lift, experimental, short takeoff and landing (QUESTOL) aircraft would be canceled, but technology would continue to be developed.

Skylab, space shuttle, Apollo-Soyuz Test Project (ASTP), Viking, Mariner Jupiter-Saturn mission, and many applications and aeronautics projects would be retained to continue essential elements for a balanced and productive space and aeronautics program within tight fiscal restraints. NASA also would continue development of a new front fan to reduce jet aircraft engine noise. NASA spokesmen told the press that the cutbacks would trim 700 civil servants from the NASA payrolls and would save the Government $200 million through June 30, 1973. (NASA Release 73–3; *W Post*, 1/6/73)

- Reductions in the development of space nuclear propulsion and space electric power were announced by the Atomic Energy Commission. The action paralleled NASA determination to focus research and technology programs on near-term development, AEC said. Programs to be terminated included nuclear rocket propulsion work at Los Alamos Scientific Laboratory and at Nuclear Rocket Development Station. Contractors affected included Pan American World Airways, EG&G, Inc., and Westinghouse Electric Corp. Cutbacks also would affect space reactor thermoelectric programs at Atomics International Div. of North American Rockwell's Aerospace & Systems Group and space reactor thermionic programs of Gulf General Atomic Div. of Gulf Oil Corp. AEC would continue programs in radioisotope thermoelectric generators, Navy's Transit navigation satellite program, and other unmanned space applications, and work on the isotope electric generator for NASA's Viking Mars landing program, Mariner Jupiter-Saturn mission, and the Dept. of Defense Lincoln space satellite. (AEC Release R–13)

January 5

- An *Apollo 17* post-mission press conference—the last press conference of the Apollo program—was held at Manned Spacecraft Center. Geologist-astronaut Dr. Harrison H. Schmitt said the orange soil discovered at Shorty Crater in the moon's Taurus-Littrow valley was "extremely young material," less than 10 million yrs old, that probably had been formed by "local volcanic vent." The discovery meant "that the moon is still active enough to produce volcanic rock." With data from *Apollo 17* and previous Apollo missions, "we now have the capability . . . to get a first order understanding of the evolutionary sequence of the Moon and . . . recent history of the Sun from the soil, and maybe an increased understanding of the Sun itself." Significance of Apollo lunar discoveries to mankind was that we could "hope to get insight . . . primarily for our children and their children" into the distribution of elements on the Earth and what processes effected that distribution for use in the future. Two revolutions in earth science during the decade had been "knowledge we've gained from the Moon" and "insight that terrestrial geologists have gotten into the formation of ocean basins. The effect of those two revolutions, particularly when they're integrated . . . is going to be profound." If man could "understand the evolutionary sequences through which Earth has progressed . . . we're going to be able to understand how to utilize the resources of that Earth." Astronaut Eugene A. Cernan said lunar study might "tell us something about the possibility of some ancient existence of some civilization . . . possibly within our own universe." Cernan and Astronaut Ronald E. Evans said they planned to stay in the space program. Schmitt said he would "stay in government service" but "I'm not sure what that service may be." (Transcript; *W Post*, 1/6/73)

- Sen. Frank E. Moss (D–Utah), who became Chairman of the Senate Committee on Aeronautical and Space Sciences Jan. 4, commented on his new post in a statement released to the press in Washington, D.C.: "Many people may have the impression that the space program is over now that the last Apollo moon landing has taken place. This is not the case. One important phase of our space exploration program has ended, but a new phase—a more productive phase in terms of benefits to man on earth—has already begun." Sen. Moss would "work diligently with my colleagues on the Committee and with NASA to develop a clear understanding of the program" and its relation to other national needs and priorities. Sen. Moss has been concerned with the "apparent high cost of the space program in relation to other important demands on the Nation's resources." He would "approach the question of current, present and future level of the nation's space program with an open mind." (Text)

- Petrographic and chemical description of *Apollo 16* lunar samples was given in a *Science* article by the *Apollo 16* Preliminary Examination Team. Preliminary characterization of samples had "substantiated the inference that the luna terra are commonly underlain by plagioclase-rich or anorthositic rocks. No evidence has been found for volcanic rocks underlying the regolith in the Apollo 16 region. In their place, we have found anorthositic rocks that are thoroughly modified by crushing and partial melting." Anorthositic rocks varied texturally and chemically. Occurrence of deep-seated or plutonic rocks in place of volcanic or pyroclastic materials suggested that "the inference from physiographic evidence that the latter materials are widespread in terra regions may be incorrect."

Other conclusions of the preliminary examination were: (1) Combination of data from the Descartes region and data from the orbital x-ray fluorescence experiment indicated that "some backside, highland regions are underlain by materials that consist of more than 80 percent plagioclase." (2) Soil or upper regolith between North Ray and South Ray had not been completely homogenized since formation of the craters. (3) Chemistry of the soil indicated that rocks rich in potassium, uranium, and thorium—similar to those at the Fra Mauro site—were relatively abundant in the Descartes region. (4) Potassium-uranium ratio of the lunar crust was similar to that of KREEP basalts. (5) The carbon content of the premare lunar crust was "even lower than that of the mare volcanic rock." (*Science*, 1/5/73, 23–24)

- The U.S.S.R. had fitted its Tu-144 supersonic transport with retractable wings on the nose for increased lift and control at lower speeds during landing and takeoff, the *London Daily Telegraph* reported. The aircraft had been lengthened to carry 140 passengers—32 more than the Anglo-French Concorde supersonic transport. Newly designed undercarriage had been installed, but engines had no reverse thrust to help braking after landing and relied on extra large tail parachutes. The U.S.S.R. still expected the Tu-144 to be operational in 1975. (Donaldson, *London Daily Telegraph*, 1/5/73)

- The Dept. of Defense had recommended expansion of the U.S. Minuteman intercontinental ballistic missile force to include the highly accurate triple warhead multiple independently targetable reentry vehicle (MIRV) on all 1000 missiles, the *Washington Post* reported. DOD had originally planned to put the MIRV on 550 Minuteman missiles and would recommend expansion of the program in the 1974 military budget request. (Getler, *W Post*, 1/5/73, A1)

- A *New York Times* editorial commented on the research recession caused by changing priorities of the Federal Government and insufficient budget appropriations in Federal research programs: "The brilliant record of American scientists . . . testifies to the quality of the national research enterprise. But how long can such top quality survive in the face of the present economic stringency and the threat of even worse times ahead?" (*NYT*, 1/5/73, 28)

January 7: Astronaut James A. Lovell, Jr., had announced his decision to leave NASA and retire from the Navy effective March 1 to accept a position with private industry, Manned Spacecraft Center announced. Lovell, who had been Deputy Director of Science and Applications at MSC since May 1971, would become Senior Executive Vice President of Bay Houston Towing Co. Lovell was a veteran of four space flights including *Gemini 7* (launched Dec. 4, 1965); *Gemini 12* (launched Nov. 11, 1966); *Apollo 8* (launched Dec. 21, 1968), the first mission to circle the moon; and *Apollo 13* (launched April 11, 1970), the mission scheduled to make the second lunar landing but aborted after an onboard explosion. He held the record for total time in space—715 hrs (30 days). (MSC Release 73-01)

- The skeleton crew at AEC–NASA Space Nuclear Systems Office at Jackass Flats, Nev., would be disbanded by late June, SNSO Nevada Operations Chief John P. Jewett was quoted as saying. The facility would be shut down June 30 following termination of the nuclear engine for rocket vehicle application (NERVA) program [see Jan. 5]. More than $1.2 billion had been spent on program since its inception in 1959. (UPI, *Orlando Sentinel*, 1/7/73)

January 8–23: The U.S.S.R. launched its *Luna 21* unmanned lunar probe from Baykonur Cosmodrome at 11:55 am local time (1:55 am EST). The "automatic station" was separated from the "satellite" in earth orbit and placed on a close to the planned trajectory for the moon. Tass said the purpose of the mission was to "further scientific studies of the moon and near-lunar space." During the flight to the moon all systems were reported functioning normally. On Jan. 12 *Luna 21* entered lunar orbit with 110-km (68.4-mi) apolune, 90-km (55.9-mi) perilune, 1-hr 58-min period, and 60° inclination. Orbital corrections were made Jan. 13 and 14 and on Jan. 14 the spacecraft was placed in an elliptical orbit with a low point of 16 km (10 mi) from the moon's surface.

Luna 21 softlanded on the eastern fringe of the Sea of Serenity Jan. 16 (Jan. 15 EST) and released the 840-kg (1850-lb) self-propelled *Lunokhod 2* vehicle to make scientific investigations of the moon's surface. *Lunokhod 2* resembled its predecessor, *Lunokhod 1* (placed on the moon Nov. 17, 1970, by *Luna 17*). It carried flags, pennants with Lenin's bas relief, state coat of arms, and the inscription "50th Anniversary of the U.S.S.R." and was equipped with a control system, radio and TV devices, and scientific instruments including a French-made laser reflector. The reflector was designed to measure the distance between the earth and the moon to within 20–30 cm (10–12 in).

Between Jan. 16 and Jan. 23 *Lunokhod 2*, powered by solar batteries and controlled from the ground by a special crew, photographed the lunar soil, rocks, and panoramic moonscape; determined chemical composition of the surface layer; measured the glow of zodiacal light near sun; and looked for cosmic dust. A highly skilled ground crew maneuvered vehicle over the shortest and safest routes, orienting it to a complex relief map of route. *Lunokhod 2* was turned to bypass obstacles without decreasing speed. On Jan. 23 *Lunokhod 2* was commanded to terminate operations in preparation for lunar night.

Luna 17 was the first unmanned spacecraft to land on the moon and deploy an automatic lunar explorer. The last mission in the series, the unmanned *Luna 20* (launched Feb. 14, 1972), had landed on the moon and returned to earth with lunar samples. (GSFC *SSR*, 1/31/73; Tass FBIS–Sov, 1/8/73, L1; 1/15–19/73, L1; 1/22–23/73, L1; 1/26/73, 4; *SBD*, 1/9/73, 38; Shabad, *NYT*, 1/17/73, C14)

January 8: Dr. Rocco A. Petrone, NASA Apollo-Soyuz Test Project (ASTP) Program Director, was quoted by *Aviation Week & Space Technology* as saying U.S.S.R. would use a U.S.-supplied very-high-frequency (VHF) transceiver on Soyuz spacecraft during rendezvous and docking maneuvers with Apollo spacecraft in July 1975 mission. Transceivers had been used in lunar modules for Apollo rendezvous and docking in lunar orbit. The U.S. also would build a receiver for Apollo spacecraft that could operate on Soviet communication frequencies and might lend to the U.S.S.R. equipment for use by cosmonauts and astronauts during joint portions of mission. Dr. Petrone had said NASA engineers, in discussions with Soviet engineers, had noted that necessity of concentrating on interpreter for exchange of technical information had made them more aware of technical problems than they would have

been in direct English conversation. March meeting of U.S. and Soviet ASTP planners would consider flammability problems. (*Av Wk*, 1/8/73, 17)
- The Air Force would consider use of aerostats—tethered balloons that operated in galelike winds—to provide air defense radar surveillance of the U.S. southern perimeter, *Aviation Week & Space Technology* reported. Aerostats, operating at altitude of 3000 m (10 000 ft) could cover area and be effective against low-altitude surprise attacks in Air Force project Veto. Balloons also were attracting military and civil interest as modest-cost, long-endurance airborne platforms for battlefield reconnaissance and electronic intelligence and for modest-cost, quickly deployable communications and television relay or scientific observation. (Klass, *Av Wk*, 1/8/73, 36, 39)
- The U.S.S.R. had decreed the death penalty for aircraft hijackers in a new law "to make air navigation safer and to improve protection of the lives and health of the passengers and crews," *Aviation Week & Space Technology* reported. (*Av Wk*, 1/8/73, 21)
- *Aviation Week & Space Technology* editorial praised Soviet aircraft designer Andrey N. Tupolev, who died Dec. 23, 1972: "His warmth, intense and lively professional curiosity and sense of humor did much to open channels of communication between the secrecy-shrouded and isolated Russian aerospace industry and the West." Until close to his death "he was still the gracious host to foreign technicians at his experimental facility outside Moscow." Tupolev had "escorted astronaut Neil [A.] Armstrong around his supersonic transport prototype in 1970." (Hotz, *Av Wk*, 1/8/73, 9)

January 8–10: The American Institute of Aeronautics and Astronautics held its 9th Annual Meeting and Technical Display in Washington, D.C. Langley Research Center Director Edgar M. Cortright was General Chairman. LaRC Director for Space Eugene S. Love delivered 1972 von Kármán Lecture, "Advanced Technology and the Space Shuttle." Lecture discussed shuttle development, need to maintain aggressive advanced research and technology programs independent of near-term needs of major systems, and future transport systems that would blend successful shuttle development technology with nonshuttle technology advancements that could be accomplished within the current decade.

Guest speaker Sen. Lowell P. Weicker, Jr. (R–Conn.), criticized Administration budget policies that had affected NASA programs and gone "far beyond the space program." Policies implied "cutback and stand-still philosophy." Sen. Weicker was "concerned over the recently announced White House directed NASA cuts but I'm much more concerned about the leadership of this nation . . . playing to our lowest, most immediate desires rather than our highest future potential." *Apollo 17* had been "15 years in the making and thousands of years in the future will benefit." Vietnam war "in 10 years has undone 200 years of building the American image and contributed not one year to the American dream."

Dr. Holt Ashley, Director of Exploratory Research and Problems Assessment in the Office of Research Applications of National Science Foundation, was installed as 1973 AIAA President.

AIAA presented the Daniel Guggenheim Medal posthumously to William C. Mentzer, Senior Vice President for Engineering and Maintenance with United Air Lines, Inc., for "many-fold accomplishments in airline engineering, maintenance, and economic disciplines, which accomplishments contributed significantly to the achievement of today's civil air transportation system."

The Elmer A. Sperry Award was presented to Leonard S. Hobbs and Perry W. Pratt, United Aircraft Corp. engineering executives, for "their contributions to the development of turbojet engines."

The History Manuscript Contest award of certificate and $500 was given to Dr. Richard K. Smith for *First Across—U.S. Navy Transatlantic Flight of 1919*. Section Special Event Award went to AIAA Houston Section for the public relations program "Down to Earth Space Program Applications."

The De Florez Training Award was presented to James W. Campbell, Special Assistant, General Aviation Accident Prevention Program, Flight Standards Service of Federal Aviation Administration, for outstanding contributions to improving the skill and knowledge of flight instructors throughout the general-aviation community. The John Jeffries Award was presented to Roger G. Ireland of the Office of the Chief of Naval Operations, Dept. of the Navy, for "outstanding efforts and achievements, over a long time span, in improving the flight safety of aerospace flight personnel, particularly in the areas of life support and rescue systems."

The Robert M. Losey Award was given to George H. Fichtl of Marshall Space Flight Center Aerospace Environment Div. for "original work in the definition and interpretation of atmospheric wind environments for the design and operation of aeronautical systems." The Sylvanus Albert Reed Award was presented to I. Edward Garrick, Chief Mathematical Scientist at LaRC, for "outstanding contributions in the areas of aeroelasticity, structural dynamics, noise, flutter, and landing dynamics of aerospace vehicles." (AIAA Release; *AIAA Bull*, 1/73; Program; *CR*, 1/18/73, S985)

January 8–12: The Civil Service Commission held an intensive equal employment opportunity (EEO) evaluation review at Marshall Space Flight Center to eliminate any discrimination practices and to ascertain that the Center had an affirmative action program. Similar reviews would be conducted at Ames, Kennedy, Langley, and Manned Spacecraft Centers. (*NASA Activities*, 2/15/73, 34)

January 9: The first Skylab space vehicle, Saturn IB with "boilerplate" spacecraft, was rolled out from the Vehicle Assembly Building to Complex 39, Pad B, at Kennedy Space Center for fit checks and fueling tests. The vehicle would remain at the pad for three weeks before return to VAB for removal of boilerplate spacecraft and mating with flight spacecraft. The launch of unmanned Skylab was scheduled for April 30, 1973. The Apollo spacecraft carrying the first three-man crew to the orbiting Skylab would be launched 24 hrs later. (KSC Release 6-73)

January 10: The American Institute of Aeronautics and Astronautics held its honors night banquet in conjunction with the Jan. 8–10 9th Annual Meeting and Technical Display and Jan. 10–12 11th Aerospace Sciences Meeting. AIAA Fellows inducted included Dr. James C. Fletcher, NASA Administrator; Director Bruce T. Lundin of Lewis Research Center (LeRC); and Harris M. Schurmeier, Manager of the proposed 1977 Jupiter-Saturn mission at Jet Propulsion Laboratory. AIAA presented the Louis W. Hill Space Transportation Award to Dr. Richard H.

Battin, Director of Apollo Mission Development, and David G. Hoag, Director of Apollo Guidance and Navigation, both of Charles Stark Draper Laboratory at the Massachusetts Institute of Technology. The citation was for "leadership in the hardware and software design of the Apollo spacecraft primary control, guidance and navigation system which first demonstrated the feasibility of on-board autonomous space navigation during the historic flight of Apollo 8."

The G. Edward Pendray Award was presented to Marcus F. Heidman, aerospace research engineer, and Dr. Richard Priem, Head of the Rocket Combustion Section of LeRC Chemical Research Div., for contributions to the literature in the field of liquid rocket combustion, particularly on performance and instability criteria for liquid motor design. The Lawrence Sperry Award was presented to Sheila E. Widnall, Professor of Aeronautics and Astronautics at MIT, for "contributions to the understanding of vortex flows in wing wakes, aerodynamic noise and lifting surface theory." The Spacecraft Design Award went to Grumman Aerospace Corp. Vice President Thomas J. Kelly for his contribution to the design, development, production, and operation of the lunar module.

The AIAA Space Science Award was presented to Dr. Paul W. Gast, Chief of Manned Spacecraft Center Planetary and Earth Sciences Div., for "significant scientific accomplishment resulting from his examination of material returned from the lunar surface. His precise measurements of lunar sample elements by the stable isotope dilution method has been one of the most outstanding scientific achievements in lunar science." The Goddard Award was given to Dr. Edward S. Taylor, Professor Emeritus of MIT, for continuous contributions to the advancement of air-breathing propulsion over 45 yrs, "as designer, inventor, researcher, teacher, advisor and as founder and leader of a major educational and research center of aircraft engine activity." The von Kármán Lecture Award was presented to Director for Space Eugene S. Love of Langley Research Center. (Program)

- Scientists at an unidentified Soviet laboratory had reported successful conversion of hydrogen gas into metal, the *New York Times* said. In letter to the *Soviet Journal of Experimental and Theoretical Physics*, scientists had said a cylindrical charge of a high explosive had driven a projectile into a compression chamber filled with liquid hydrogen. At a pressure 2.8 million times that of the atmosphere at sea level, the density of hydrogen had jumped from 1.08 to 1.30 g per cu cm (from 0.04 to 0.05 lb per cu in), indicating conversion to a metallic state. Hydrogen as metal, hitherto unknown on earth, was thought by scientists to be the chief constituent of the planets Jupiter and Saturn and was expected to have superconducting characteristics. (Sullivan, *NYT*, 1/10/73, 3:23)
- The Air Force announced issuance of a $1 358 695, firm-fixed-price contract to McDonnell Douglas Corp. for special tooling for F–4 jet fighter aircraft. (DOD Release 20–73)

January 10–12: The American Institute of Aeronautics and Astronautics held its 11th Aerospace Sciences Meeting in Washington, D.C. The Dryden Research Award winner for 1973, Dr. Herbert Friedman of the Naval Research Laboratory Space Sciences Div., delivered the Dryden Lecture "High Energy Astronomy." The lecture, which carried $1000 honorarium, had been named in 1967 in honor of the late Dr. Hugh L. Dryden, first NASA Deputy Administrator, who died in 1965. Dr. Friedman said high-energy astronomy was "almost exclusively the subject of space research." X-ray astronomy to date had revealed about 125 discrete sources and a "diffuse background which spans the energy range from 100 ev to tens of Mev." X-ray sources revealed included supernova remnants, pulsars, and binary systems in the galaxy. Extragalactic sources included radio galaxies, quasars, Seyfert galaxies, and clusters of galaxies. Gamma ray astronomy had shown "no evidence for antimatter within the universe." (Program; Text)

A committee established by AIAA to evaluate NASA's space shuttle program reported its conclusion that a shuttle system, including both manned and unmanned flight, ought to be built. The committee recommended a research and applications module system with interchangeable modules that could be removed in space to become part of an orbiting laboratory. Other nations should be encouraged to participate in shuttle development, the committee said. (*Tech Rev*, 5/73, 55)

Advanced applications of the shuttle were discussed by J. E. Blahnik of Science Applications, Inc., and D. R. Davis of ITT Research Institute. An exploratory study had concluded that use of the shuttle orbiter for advanced, manned, high-energy missions from low earth orbit should be competitive with other proposed concepts, if orbital and refueling requirements were satisfied by using normal shuttle missions to deliver propellant in addition to their payload. (Program)

January 10–13: The American Astronomical Society met in Las Cruces, N. Mex. Cal Tech astronomer Dr. Robert Howard said six-year observations through the 46-m (150-ft) solar telescope at Mt. Wilson Observatory had shown sections of the sun's outer atmosphere to be in motion with respect to time. Movements had extended to 259 000 sq km (100 000 sq mi) of photosphere—the sun's gassy surface—and had appeared to be east-west or west-east. Some motions had been calculated at 6257 km per hr (3888 mph); others, at 7300 km per hr (4536 mph). Cause and significance of the motion were unknown.

At a press interview Dr. John C. Brandt, chief of the Goddard Space Flight Center Solar Physics Laboratory, said Indians in New Mexico, Arizona, and California nine centuries ago might have recorded a star explosion. Drawings on walls and ceilings of a cave at Chaco Canyon National Monument, N. Mex., closely resembled an explosion of July 4, 1054, as recorded by the Chinese. Similar drawings had been found in Arizona and California. (Alexander, *LA Times*, 1/11/73; AP, *W Post*, 1/14/73, A13)

January 11: *Cosmos 543* was launched by the U.S.S.R. from Baykonur Cosmodrome into orbit with 309-km (192.0-mi) apogee, 202-km (125.5-mi) perigee, 89.6-min period, and 65.0° inclination. The satellite reentered Jan. 24. (GSFC *SSR*, 1/31/73; *SBD*, 1/12/73, 54)

• Jet Propulsion Laboratory scientists Dr. Roy E. Cameron and Frank A. Morelli had departed for a two-month research project in Antarctica to determine the degree of man's impact on the environment, JPL announced. The National Science Foundation-sponsored ecological measurements of Antarctica's primitive soils would also provide information for NASA's development of life-detection systems for Viking landers on Mars. (JPL Release 643)

• Soviet Novosty Press science writer Yury Marinin said in a letter to the *Washington Post* that the U.S.S.R. "abandonment" of manned lunar missions did not mean that nation was lagging behind the U.S. in space

exploration. "Soviet specialists think it more expedient to explore the moon by automatic devices, with the present level of world technology. Besides, manned flights to the moon involve tremendous expenses." Soviet automatic devices would be "steadily improved as they continue their planned studies of the moon." (*W Post*, 1/11/73, A19)
- Sen. Harry F. Byrd, Jr. (Ind.–Va.), introduced S. 328 to limit the amount payable in advance on any contract signed by NASA, the Army, Navy, Air Force, or Coast Guard to $20 million. (*CR*, 1/11/73, S427)
- National Park Service officials blamed sonic booms for damage to rock formations at Death Valley National Monument where Air Force jet pilots flew low over the desert floor to claim they had flown below sea level, United Press International reported. The Air Force had agreed in 1971 to avoid monuments or parks listed by Assistant Interior Secretary Nathaniel P. Reed as "fragile environment," but the agreement had not worked in Death Valley as fliers found the "urge to zoom near desert floor irresistible." (UPI, *W Post*, 1/11/73, A2)

January 12: *Oao 3* (*Copernicus*) Orbiting Astronomical Observatory (launched Aug. 21, 1972, from Eastern Test Range) was adjudged successful by NASA. The spacecraft had completed 2000 earth orbits and all systems had operated satisfactorily. *Oao 3* had obtained high-resolution spectra of stars in the ultraviolet range and investigated composition, density, and physical state of matter in interstellar space. (NASA prog off)
- The Senate adopted S.R. 16, appointing minority members to the Senate Committee on Aeronautical and Space Sciences: Republicans Sen. Barry M. Goldwater (Ariz.), Sen. Carl T. Curtis (Neb.), Sen. Lowell P. Weicker, Jr. (Conn.), Sen. Dewey F. Bartlett (Okla.), Sen. Jesse A. Helms (N.C.), and Sen. Pete V. Domenici (N. Mex.). (*CR*, 1/12/73, D22)
- NASA announced publication by Univ. of New Mexico's Technical Application Center (TAC) of *Noise Pollution Resource Compendium*. The compilation of references on noise pollution research, legislation, and technical problems was offered as service to Government, industry, and business officials with noise abatement responsibilities. TAC operated one of NASA's six Regional Dissemination Centers for NASA-developed technology. (NASA Release 73-8)
- *Apollo 17* Astronauts Eugene A. Cernan, Ronald E. Evans, and Dr. Harrison H. Schmitt began an 11-week schedule of public appearances throughout the U.S. with an appearance at the Super Bowl football game in Los Angeles. (*NASA Activities*, 5/15/73, 89)
- General Electric Co. announced it had signed an agreement with the U.S.S.R. for "broad scientific and technical cooperation in fields of mutual interest." The agreement emphasized cooperation in power generation, including steam and gas turbine and nuclear energy technology; exchange of specialist delegations, information, and production samples; exchange, acquisition, or transfer of licenses and "know-how"; and joint research and development programs and projects. (GE Release)
- The Air Force announced issuance of $282 743 000 firm-fixed-price contract to McDonnell Douglas Corp. for production of F–4E jet fighter aircraft, parts, aerospace ground equipment, and related data. (DOD Release 25-73)

January 13: A three-month study of fog dispersal techniques made by the Federal Aviation Administration at two Pacific Northwest airports had indicated seeding of cold fog (below 273.15 K [32°F]) significantly increased visibility but had no effect on warm fog (above 273.15 K [32°F]), FAA announced. Photos had indicated that jet aircraft movements on airfield taxiways and in traffic patterns around airfields contributed to dissipation of fog through heat of jet engine exhaust. (DOT Release 73-9)

- A *New York Times* editorial praised Federal Communications Commission policy in authorizing construction of communications satellites: "Fortunately, the F.C.C. has not approved one chosen entity or single system but has decided to throw open the airwaves to various competitive groups." There should be no corporate dominance of systems that depended on Federal funds. As FCC developed its open skies policy for satellites, "it should also consider the use of educational and public service broadcasts on a free or reduced-rate basis. Here is where the main 'profit' should accrue from quasi-public satellite communications."

A second *Times* editorial praised the agreement announced Dec. 5, 1972, that permitted the U.S. firm RCA Global Communications, Inc., to use channel of Canada's *Anik 2* satellite to provide domestic communications service in the U.S.: The agreement signaled "a beam of light in broadcasting and business communications across the international oceans of the world." In the near future Americans might receive some programs originating in Canada; later, from Europe. "Operating procedures and telecommunications systems must be synchronized; problems of copyright and aerial piracy must be overcome. And then international agreements would have to be hammered out as difficult as those on human and other national rights that have been dormant for years in the United Nations. Intelsat, a temporary arrangement between nations interested in orbital systems, will soon have the machinery for regular coordination with a secretariat in Washington." (*NYT*, 1/13/72, 28)

January 14: A building boom of nuclear power plants in 1972 was reported by the *New York Times*. The Atomic Industrial Forum, a group seeking to foster development of atomic energy for peaceful uses, had reported 33 plant commitments made, with a total capacity of 36.3 million kw. The Atomic Energy Commission had reported 29 nuclear power plants operable in the U.S. with a total capacity of 14.7 million kw, or 3.8% of total electric utility capacity. Fifty-five plants with a total capacity of 47.8 million kw were being built and 76 with 80-million-kw capacity were planned. AEC had predicted that in 1980 nuclear power capacity would be 23% of total U.S. electric capacity. (Smith, *NYT*, 1/14/73, 5:3)

January 14–17: The American Chemical Society held a regional meeting in Washington, D.C.

Dr. Cyril A. Ponnamperuma and Dr. Peter M. Molton, Univ. of Maryland chemists, presented results of studies in which a Jupiterlike atmosphere of methane and ammonia had been produced in a flask at the University's Laboratory of Chemical Evolution. Results had shown the chemicals necessary for establishing life probably existed in Jupiter's upper atmosphere. "Jupiter today may have the kind of atmosphere that the primitive earth had 4 billion or 4½ billion years ago," Dr. Ponnamperuma, former NASA specialist in life chemistry, said.

Further studies had shown that bombarding primitive chemicals with electrical discharges triggered complex chemical reactions forming aminonitriles—precursors of purines and pyrimidines, main components of DNA and RNA, chemicals dictating the genetic make-up of every living cell. Aminonitriles combined with water became amino acids, ingredients of all proteins.

Ponnamperuma and Molton said they would ask NASA to explore Jupiter's atmosphere during the planned 1977 flyby mission to that planet. If a scientific probe could find the same life building blocks as produced in the laboratory, life in Jupiter's clouds would be almost a certainty. (Kirkman, W *Star & News*, 1/17/73, A9; Cohn, *W Post*, 1/18/73, A14)

January 15: President Nixon had sent Dr. James C. Fletcher, NASA Administrator, a letter praising NASA's contribution to the Administration program to assist minorities in finding a profitable role in the U.S. economic system, *NASA Activities* reported. (*NASA Activities*, 1/15/73, 6)

- Seventy-seven Florida firms had received $182 633 000 in prime contracts awarded by Kennedy Space Center during FY 1972, KSC announced. KSC spending in 12 other states had amounted to $2 536 000. (KSC Release 9-73)

January 16: Appointment of Capt. Chester M. Lee (USN, Ret.) as Program Director of the Apollo-Soyuz Test Project was announced by NASA. Lee, Apollo Mission Director for *Apollo 12* through *17*, would be responsible for direction of the U.S. ASTP effort, including management of the spacecraft and docking module, flight and crew operations at Manned Spacecraft Center, launch vehicle operations at Marshall Space Flight Center, and launch operations at Kennedy Space Center. (NASA Release 73-5)

- Dr. James C. Fletcher, NASA Administrator, presented an award of $10 000 to Jet Propulsion Laboratory scientists Paul M. Muller and William L. Sjogren in a JPL ceremony. The award was for a 1968 discovery of mascons (mass concentrations of dense material beneath the lunar surface) and for imaginative interpretation of the discovery's results and implications significant to the success of the Apollo program and future studies of the earth and other planets. The mascon discovery had received international recognition. Muller and Sjogren had been awarded the 1971 Magellanic Gold Medal of American Philosophical Society and had been elected members of the Society of Sigma Xi and Fellows of Britain's Royal Astronomical Society. (NASA Release 73-6; NASA PAO)

- NASA launched three sounding rockets, two from Wallops Island and one from Andoeya, Norway.

A Nike-Apache launched from Wallops carried a Univ. of Illinois payload to a 157-km (97.5-mi) altitude. The launch was the second of three to measure ionospheric properties on three kinds of anomalous days: (1) "L," a day with very low electron densities in the 70- to 80-km (40- to 50-mi) region and a 10-day minimal storm activity preceding; (2) "H_1," a day with very high electron densities and the same altitude and magnetic storm prerequisites; and (3) "H_2," a day with very high electron densities in the same region with two or three days of major magnetic disturbance and magnetic storm aftereffects. The Nike-Apache was launched on an H_1 day and measured electron densities, temperatures, and collision frequencies. The rocket underperformed but the scientific objectives were satisfied. The first launch in the series had been Dec. 5, 1972.

A Super Arcas, launched from Wallops Island 34 min after the Nike-

Apache, carried a Pennsylvania State Univ. payload to a 95.1-km (59.1-mi) altitude to collect data on positive and negative ion conductivities. The rocket and instrumentation performed satisfactorily.

A Nike-Tomahawk sounding rocket was launched from Andoeya, Norway, carrying a National Oceanic and Atmospheric Administration experiment to a 280.0-km (174-mi) altitude to collect data on energetic fields and particles and to study the aurora. The rocket and instrumentation performed satisfactorily. (NASA Rpts SRL; GSFC proj off)

- Small business firms had received $12 436 000 in Kennedy Space Center procurement awards during FY 1972, KSC announced, an increase over $8 433 000 in the previous year. Overall procurement for KSC amounted to $192 882 000 in FY 1972 and $211 996 000 in FY 1971. (KSC Release 10–73, 1/16/73)

January 17: NASA Hq. held its first press briefing on *Mariner 9* (launched May 30, 1971) since spacecraft ceased transmission October 27, 1972, one year after it had gone into Martian orbit. Photos were released showing three kinds of channels on the Mars surface thought to be formed by flowing water sometime in Martian history. Harold Masursky of the U.S. Geological Survey said one kind seemed to have been formed by melting permafrost; the other two looked "as though it's necessary to have precipitation and a collection of that runoff into flowing water channels."

New Mexico Univ. geologist Dr. Bradford A. Smith said riverbeds on Mars probably had dried because of orbital changes. When the Mars orbit was more elliptical than at present, the perihelion would be much closer to the sun, perhaps producing long-term climate changes which might, "at certain times, permit liquid water to exist."

Dr. Conway B. Leovy, Univ. of Washington, said dust storms encountered by *Mariner 9* were "phenomena which will recur on Mars depending on the particular astronomical relationship between the perihelion and the solstice." Dust storms might have been responsible for "the remarkable laminated terrain which has been observed in both polar zones."

Scientists had determined from *Mariner 9* photos that the planet was a more varied and dynamic body than expected. Dr. John E. Naugle, NASA Associate Administrator for Space Sciences, said: "Mars is like three planets: the old cratered planet, the young volcanic active planet that you see, and then an entirely different situation in the polar regions in Mars." Mars as "planetary laboratory" of towering volcanoes, deep rift valleys, broad plateaus broken by fault lines, ancient planes with moonlike craters, sand dunes, clouds with water ice, and ever-changing polar caps would "substantially increase our basic knowledge of planets . . . and . . . when that history is written you will find that planetary exploration has contributed in a major way to the health and welfare of life on this planet." (Transcript)

- The U.S. and United Kingdom concluded an agreement on U.K. access to U.S. space launch capabilities. The U.K. would purchase boosters and launch services from NASA for satellite projects of the U.K. Dept. of Trade and Industry. Launches would occur at NASA launch sites in U.S. The first U.K. satellite launched under terms of agreement would be X–4 technology satellite scheduled for 1974 launch from Western Test Range. (NASA Release 73–11)

- Dr. James C. Fletcher, NASA Administrator, outlined Apollo program achievements in a speech before the National Security Industrial Assn. in Los Angeles. The Apollo program had reasserted U.S. technological leadership, exploded theory that only a controlled economy and Communist society could succeed with large scientific and technological undertakings such as space exploration, demonstrated "that we could respond effectively to a totalitarian challenge without resorting to totalitarian methods," and produced the kind of Government-industry cooperation on large-scale research and development programs needed to make the American system work and keep it competitive with any other systems in the world. Apollo had reinforced "Can Do attitude that has made America great," met national security requirements in 1960s for rapid progress in manned space flight technology and wide experience in the field, obtained valuable scientific knowledge obtainable no other way, "discovered" planet Earth on the way to the moon, forced creation of new technology "as no peacetime enterprise in history has done—or could do up to now," contributed spiritually to lives of all Americans, "and stayed within estimated cost bounds." The program had cost $25 billion, or $11 per person per year from 1961 through 1972. "Who or what could have produced more security, more economic benefits, more pride in achievement, more confidence in our future as a free people—for $11 per year?" (*NASA Activities*, 2/15/73, 29–30)
- The International Civil Aviation Organization legal committee, representing 58 countries, voted to defeat U.S.-backed proposals to take sanctions against countries which failed to act against air piracy. The committee approved a U.S.-backed proposal to allow signers of the proposed antihijacking treaty to make recommendations against nonsignatory nations that violated the treaty. (Shaw, *W Post*, 1/19/73, A3)
- The People's Republic of China was preparing to deploy the first group of 10 nuclear missiles capable of reaching U.S.S.R. targets, Joseph Alsop reported in *Washington Post*. U.S. intelligence agencies had observed missile sites carved into mountain sides and believed the new intermediate range missile had already been successfully tested. (*W Post*, 1/17/73, 17)

January 17–19: A Skylab simulation at Manned Spacecraft Center picked up the Skylab flight plan in the 10th day of the mission scheduled for April 30 launch. Flight controllers worked in shifts to tally the day's activities, prepare a summary flight plan for the day following, and complete a detailed flight plan for use the next day. Activities included all elements of flight between 8 am and 5 pm each day but were oriented toward experiments rather than flight operations. Skylab astronauts manned crew simulators, principal investigators manned support stations, and the Marshall Space Flight Center Operations Support Center for Skylab provided support for the Orbital Workshop. Prime contractors North American Rockwell Corp. and Martin Marietta Corp. supported the simulation. MSC controllers inserted simulated malfunctions to exercise procedures and mission rules for the flight and to observe the reaction of control teams. (MSC *Roundup*, 1/19/73, 2)

January 18: The European Space Research Organization Council met in Paris and unanimously authorized establishment of a project to develop the sortie laboratory to fly with NASA's space shuttle in the 1980s. Participating states would provide funding and would use ESRO facilities and assistance. The Federal Republic of Germany, Italy, Belgium, and

Spain would participate initially, with other ESRO member countries expected to join later. After definition studies of the research and applications module (RAM), the sortie lab final design would be agreed on by NASA and ESRO. ESRO would proceed to development phase, with delivery of the sortie lab to NASA scheduled for 1979. The laboratory would consist of a pressurized manned module to house equipment, experiments, data-processing equipment, electrical power and environmental control systems, and a six-scientist crew. An external, unpressurized instrument platform would house experiments and large instruments controlled remotely from the laboratory. Project cost was estimated at $250 million to $300 million. Negotiations at agency and governmental levels would begin shortly. (ESRO Release, 1/19/73; NASA Release 73-12)

- Advanced communications satellites were discussed at a joint meeting of the Cleveland Chapter of the American Institute of Aeronautics and Astronautics and the Institute of Electrical Engineers at Lewis Research Center. Joseph N. Sivo, head of LeRC's Advanced Communications Systems Study Office, said higher powered communications satellites offered "real potential in educational and health service areas." Communications satellites could beam educational TV to the people of India or Appalachia, link medical centers with patients in distant places, or speed mail service. Because of NASA program cutbacks, "industry will have to pick up where NASA left off." (LeRC Release 73-3)

- The Senate Committee on Appropriations assigned members to the Subcommittee on Dept. of Housing and Urban Development, Space Science, Veterans: Sen. William Proxmire (D-Wis.), Chairman; Sen. John O. Pastore (D-R.I.); Sen. John C. Stennis (D-Miss.); Sen. Mike Mansfield (D-Mont.); Sen. Daniel K. Inouye (D-Hawaii); Sen. Birch Bayh (D-Ind.); Sen. Lawton M. Chiles, Jr. (D-Fla); Sen. Charles M. Mathias, Jr. (R-Md.); Sen. Clifford P. Case (R-N.J.); Sen. Hiram L. Fong (R-Hawaii); Sen. Edward W. Brooke (D-Mass.); and Sen. Ted Stevens (R-Alaska). (*CR*, 1/18/73, D31)

- Rep. Charles E. Bennett (D-Fla.) introduced H.J.R. 198, to redesignate Cape Kennedy as Cape Canaveral. (*CR*, 1/18/73, H365)

- Selection of Fairchild Industries, Inc., for full-scale development of the Air Force A-X specialized close air support aircraft was announced by Dr. Robert C. Seamans, Jr., Secretary of the Air Force. The Fairchild A-10 had been selected after competitive evaluation with the Northrop A-9 aircraft. The Air Force would negotiate a cost-plus-incentive-fee contract with Fairchild for 10 preproduction aircraft. The full-production decision would be made late in 1975. (DOD Release 36-73)

- NASA released a list of personal mementoes carried by *Apollo 17* crew Eugene A. Cernan, Ronald E. Evans, and Dr. Harrison H. Schmitt on the Dec. 7-9, 1972, lunar mission. The crew had been permitted to carry 12 items each. Cernan had carried two *Apollo 17* pins, a lunar module tie pin, wedding ring, moon flag pin, personal jewelry piece, $20 gold piece, religious memento, three *Apollo 17* gold commemorative medals, and memento from *Gemini 9* and *Apollo 10*.

  Evans had carried a wrist watch, wedding ring set, gold brooch set, ring, three wedding diamonds, a moonstone gem, identification bracelet, and three *Apollo 17* gold commemorative medals.

  Schmitt had carried eight moon pins, turquoise beads, a family memento set, and two *Apollo 17* silver commemorative medals. (NASA PAO; *W Post*, 1/20/73, A7)

- The Senate confirmed the nomination of Claude S. Brinegar to be Secretary of Transportation. (*CR*, 1/18/73, S799)
- NASA launched an Aerobee 200 sounding rocket from White Sands Missile Range carrying a Lockheed Missiles & Space Co. solar physics experiment to a 245.3-km (152.4-mi) altitude. The rocket and instrumentation performed satisfactorily. (GSFC proj off)
- Rep. Robert McClory (R–Ill.) and 20 cosponsors introduced H.R. 2351 to provide for systematic U.S. conversion to the metric system over a 10-yr period. (*CR*, 1/18/73, H361)

January 19: Director Andrey Severny of the Crimea Astrophysical Observatory said that the Soviet lunar vehicle *Lunokhod 2* (deposited on the moon by *Luna 21*, launched Jan. 8) carried an astrophotometer, or electron telescope without lenses, to register, the glow in wide areas of the sky in visible and ultraviolet spectral bands. (*SF*, 4/73, 121)

- A study by the Aerospace Medical Research Laboratory at Wright-Patterson Air Force Base, Ohio, had confirmed that fighter pilots under high g forces would perform better if cockpit seats were tilted backward, the Air Force Systems Command announced. The tilted seat would relieve some discomfort caused by g force while also providing maximum visibility. (AFSC Release 141.72)
- A *Science* editorial commented on the resignation of President Nixon's Science Adviser, Dr. Edward A. David, Jr. [see Jan. 3]: The resignation and "likelihood that the office of the President's science adviser will be abolished come as a disappointing shock. David is a scientist-engineer of very broad competence who maintained intellectual integrity in the emotional heat of politics." Abolition of office was "one of many consequences of President Nixon's determination to reorganize the Executive Office of the President." It was probable that the President's Science Advisory Committee and the Office of Science and Technology "would vanish" in reorganization. Criticism of Presidential Science Advisers had been "that they often became so involved in the day-to-day fighting of political brush fires that they lost sight of the forest." The scientific establishment had fought against creation of a Dept. of Science. "It has been argued that science and technology permeate the activities of virtually all executive agencies and that centralization would not be wise. The argument had been driven home too well. In consequence, there is now danger that science, while being everywhere, will be nowhere." (Abelson, *Science*, 1/19/73, 9)

January 20: *Cosmos 544* was launched by the U.S.S.R. from Plesetsk into orbit with 547-km (339.9-mi) apogee, 510-km (316.9-mi) perigee, 95.1-min period, and 74.0° inclination. (GSFC *SSR*, 1/31/73; *SBD*, 1/23/73, 109)

- The training version of *Apollo 17* lunar roving vehicle (LRV), participating in President Nixon's inaugural parade, required replacement of two batteries to negotiate the distance from the Capitol to the White House, because batteries had only 30-min life. *Apollo 17* Astronauts Eugene A. Cernan, Ronald E. Evans, and Dr. Harrison H. Schmitt watched the battery change from an automobile following the LRV in the procession. (*Av Wk*, 1/29/73, 13)

January 21: NASA established the Office of Management Planning and Review with James R. Elliott as Director, under Associate Administrator for Organization and Management. The Office of Management Development was abolished and its personnel and responsibilities transferred to the new office. Elliott had joined NASA in February 1962 and had been a senior management consultant in Office of Management Development. (NASA Ann)

- Jet Propulsion Laboratory staff changes were announced by the Deputy Director, Gen. Charles H. Terhune, Jr. Geoffrey Robillard, Manager of the Applied Mechanics Div., had been appointed Deputy Assistant Laboratory Director for Technical Divisions. His successor was Glenn E. Vescelus, who had joined JPL in 1969. (JPL Release 654)

- A decision by de Havilland Aircraft of Canada Ltd. to build the world's first short takeoff and landing (STOL) passenger transport aircraft was reported by the *Washington Post*. The firm would shortly begin work on two prototypes of 48-passenger, turboprop DHC-7 aircraft to be test-flown within two years. The aircraft would be certifiable by 1975. The Canadian government had approved $80 million for development of the DHC-7, designed for routes under 800 km (500 mi). The aircraft would operate from 600-m (2000-ft) runways with minimal noise and pollution. (Botwright, *W Post*, 1/21/73, L2)

January 22: Former President Lyndon B. Johnson died of heart attack in Austin, Tex., at the age of 64. During his senatorial career preceding his election to the Vice Presidency in 1961, Johnson had served in 1957 as Chairman of the Prepardeness Investigating Subcommittee of the Senate Committee on Armed Services, where he had instituted an inquiry into the status of U.S. satellite and missile programs. Hearings led to establishment Feb. 6, 1958, of the Senate Special Committee on Space and Astronautics with Johnson as Chairman. The Special Committee prepared the National Aeronautics and Space Act of 1958 establishing NASA (effective Oct. 1, 1958). Before passage of the Space Act July 29, 1958, the Senate created the permanent Committee on Aeronautical and Space Sciences July 24. Johnson became first chairman—the position he held until 1961.

Johnson was Chairman of the National Aeronautics and Space Council 1961–1963 and was instrumental in persuading President John F. Kennedy to select attainment of a manned lunar landing within the decade as a national goal. Johnson had become 36th U.S. President on assassination of President Kennedy in November 1963 and had seen the first successful manned circumlunar flight, made by *Apollo 8* Dec. 21–27, 1968, before he left office in January 1969. (*W Post*, 1/23/73, A1; *CR*, 1/29/73, S1467)

- *Apollo 17* Astronauts Eugene A. Cernan, Ronald E. Evans, and Dr. Harrison H. Schmitt received the traditional greeting from Congress following their Dec. 7–19, 1972, mission. They were welcomed by the House of Representatives, since the Senate was not in session. Cernan told the House that *Apollo 11* Astronaut Neil A. Armstrong's placement of the first U.S. flag on the moon in 1968 had been "a symbol of the courage and the dedication and the effort and self-sacrifice of 200 million people in this country who made that effort possible." The act had "probably gained more pride and more respect throughout the entire world than any one thing that has happened . . . in the entire 200-year history of our country." With one step by Armstrong, "we are able to gain something that we had never really been able to grab hold of in my lifetime." Evans said his stay in the CM had made him realize that earth, like spacecraft, had limited consumable resources and "somehow the human survival requirement means that we have got to conserve these resources and man must learn to adapt to his environment." Schmitt said that, in obtaining knowledge through the Apollo program, man had evolved into entering universe. "Although the nature of that evolution was technological, I believe it will be marked a thousand years from now as a single unique event in human history" appearing "more distinctly even than history's record of our use of atomic energy. It is at times unseemly, at times shortsighted but always human pathway through time, mankind found that its reach could include the stars." (CR, 1/22/73, H372-4)

- The last Apollo splashdown party given by National Space Club in Washington, D.C., had the theme "Salute to Apollo: Man's Promise for the Future." The program included remarks and messages from President Nixon; Vice President Spiro T. Agnew; Sen. Frank E. Moss (D-Utah), Chairman of Senate Committee on Aeronautical and Space Sciences; Dr. James C. Fletcher, NASA Administrator; *Apollo 17* Astronauts Eugene A. Cernan and Dr. Harrison H. Schmitt; former NASA Administrator James E. Webb; North American Rockwell Corp. Vice President and President–Aerospace Group William B. Bergen; and Grumman Aerospace Corp. President Joseph G. Gavin, Jr.

    Former President Lyndon B. Johnson who had died during day, had sent a message commending space pioneers who had "made the Apollo miracle a living reality. It has been more . . . than an amazing adventure into the unexplored and the unknown. The Apollo program has been and will endure as a monument to many things—to the personal courage of some of the finest men our nation has produced—to the technological and managerial capability, which is the genius of our system—and to a successful cooperation among nations which has proved to all of us what can be done when we work together with our eyes on a glorious goal."

    Master of Ceremonies Peter S. Hackes of National Broadcasting Co. eulogized President Johnson as "the policy father of the space program." A message from Sen. Hubert H. Humphrey (D-Minn.), former Vice President and Chairman of the National Aeronautics and Space Council, said the Apollo program had been an "outstanding example of government, industry, and university cooperation and has really shown brilliance of management and technological competence." An Apollo film was narrated by Dr. Rocco A. Petrone, NASA Apollo Program Director.

    Dr. Petrone said the Apollo program had taken 12 yrs "and I don't look upon this as curtains at the end of a play. It's like the curtain at the end of the first act. It's just beginning." Dr. Wernher von Braun, Vice President for Engineering and Development with Fairchild Industries, Inc., and former NASA Deputy Associate Administrator, said President Johnson had been instrumental in achieving congressional support for the space program. *Apollo 8* Astronaut William A. Anders recalled President Johnson as "a very large man, very down to earth" who had the spirit of an astronaut, but was "a little too big" to be one. (Program; Texts; Smyth, *W Post*, 1/23/73, B2; RI PIO)

- NASA postponed the launches of the Skylab 1 Orbital Workshop and the Skylab 2 crew from April 30 and May 1 to indefinite dates in May. The postponement was necessitated by an unexpected accumulation of problems in launch preparations for vehicle modules and experiments and first-time Skylab testing at Kennedy Space Center. Skylab Program Director William C. Schneider set tentative launch dates of May 14–15 but would announce official dates in late March after continued testing and reviews of flight planning. (Schneider TWX to NASA Centers, 1/22/73)

- Dr. James C. Fletcher, NASA Administrator, issued a management instruction continuing in existence, and setting forth the charter for the NASA Space Program Advisory Council and its related standing committees under the Federal Advisory Committee Act of Oct. 6, 1972 (P.L. 92–463). The NASA Administrator had determined Jan. 4, 1973, that the Council and committees were in the public interest in the performance of duties imposed upon NASA by law, "and they are therefore continued for the period ending June 30, 1974." Council functions were to consult with and advise NASA, through the Deputy Administrator, on plans for, work in progress on, and accomplishments of NASA's programs. (NASA NMI 1156.20C)

- The Federal Republic of Germany's *Aeros* satellite (launched by NASA Dec. 16, 1972) was operating within expected limits. Five science experiments had been turned on and were operating satisfactorily. An anomaly had possibly degraded data from the U.S. neutral atmosphere temperature experiment; four other experiments were unaffected. (NASA prog off)

- NASA launched a Black Brant VC sounding rocket from White Sands Missile Range carrying a Goddard Space Flight Center payload to a 276.8-km (172-mi) altitude. The objective was to test the performance of the rocket configuration that would be used for a series of calibration rocket (CALROC) launches during the manned Skylab mission scheduled for May. The rocket and instrumentation performed satisfactorily.

    The CALROC launches would provide a reference for the calibration of equipment on Skylab's Apollo Telescope Mount (ATM). CALROC launches were also scheduled during the Skylab 3 and Skylab 4 missions, to acquire solar flux data on specific regions of the sun in conjunction with Skylab astronaut observations of the same regions with ATM experiments. Instrumentation on CALROC would be similar to but smaller than that on the ATM. Black Brant VC would be used for all launches. Taking part in the program, managed by Marshall Space Flight Center, were Ames Research Center, Naval Research Laboratory, Harvard College Observatory, and the White Sands Naval Ordnance Missile Test Facility. (GSFC proj off; MSFC Release 73–42)

- Measurements of ice, surface, and atmospheric conditions in the Bering Sea would be made by NASA and the U.S.S.R. Academy of Sciences, NASA announced. The experiment would be conducted Feb. 15–Mar. 7 on a Soviet weather ship and Il-18 aircraft and by a U.S. Coast Guard icebreaker and the instrumented NASA Convair 990. The project would be carried out under August 1971 recommendations of the Joint U.S.–U.S.S.R. Working Group on Satellite Meteorology for better understanding of interaction of sea ice and atmosphere on weather patterns in the Bering Sea area. (NASA Release 73–2)

- A Federal court in New York began hearing $2-million libel suit brought by Dr. W. Ross Adey, Univ. of California at Los Angeles brain research specialist, against the New York-based animal welfare organization United Action for Animals. Dr. Adey accused the organization of publishing "false, scandalous, defamatory statements" in an article in its bulletin on a NASA- and Air Force-funded *Biosatellite 3* experiment (launched June 28, 1969). The organization had criticized Dr. Adey's treatment of Bonny—an instrumented monkey that had been orbited to measure functions of central nervous, cardiovascular, and metabolic systems under weightlessness. Bonny had died shortly after its return to the earth. (Burks, *NYT*, 1/20/73, 27; 1/25/73, 45)

January 23: At a Marshall Space Flight Center Space Science Seminar, Dr. Maurice M. Shapiro, Naval Research Center chief scientist for cosmic ray physics, said studies had shown cosmic rays were fast atomic nuclei trapped in the Milky Way's magnetic fields. After acceleration, the rays diffused into the interstellar medium, where many underwent violent collisions and transformations. Surviving particles leaked out of the galaxy. (MSFC Release 73-3; NRL PIO)

- The Senate confirmed the nominations of William P. Clements to be Deputy Secretary of Defense and of Dr. James R. Schlesinger, former Atomic Energy Commission Administrator, to be Director of Central Intelligence. (*CR*, 1/23/73, D60)

January 24: *Cosmos 545* was launched by the U.S.S.R. from Plesetsk into orbit with a 492-km (305.7-mi) apogee, 268-km (166.5-mi) perigee, 92.1-min period, and 71.0° inclination. The satellite reentered July 31. (GSFC *SSR*, 1/31/73; 7/31/73; *SBD*, 1/26/73, 124)

- The U.S. National Scientific Balloon Facility launched a 20-m (64-ft) instrumented balloon for NASA from Oakey Field, Queensland, Australia, in Project Boomerang. The project was part of NASA's balloon program to study near-earth phenomena. [See March 22.] (NASA Release 73-48)

- Rep. John W. McFall (D-Calif.) introduced H.Res. 158, providing for 30 members of the House Committee on Science and Astronautics: Rep. Olin E. Teague (D-Tex.), Chairman, succeeding Rep. George P. Miller (D-Calif.), who had been defeated in a 1972 primary; Rep. Ken Hechler (D-W. Va.); Rep. John W. Davis (D-Ga.); Rep. Thomas N. Downing (D-Va.); Rep. Don Fuqua (D-Fla.); Rep. James W. Symington (D-Mo.); Rep. Richard T. Hanna (D-Calif.); Rep. Walter Flowers (D-Ala.); Rep. Robert A. Roe (D-N.J.); Rep. William R. Cotter (D-Conn.); Rep. Mike McCormack (D-Wash.); Rep. Bob S. Bergland (D-Minn.); Rep. J. J. Pickle (D-Tex.); Rep. George E. Brown, Jr. (D-Calif.); Rep. Dale Milford (D-Tex.); Rep. Ray Thornton (D-Ark.); Rep. Bill Gunter (D-Fla.); Rep. Charles A. Mosher (R-Ohio); Rep. Alphonzo Bell (R-Calif.); Rep. John W. Wydler (R-N.Y.); Rep. Larry Winn, Jr. (R-Kan.); Rep. Lou Frey, Jr. (R-Fla.); Rep. Barry M. Goldwater, Jr. (R-Calif.); Rep. Marvin L. Esch (R-Mich.); Rep. Lawrence Coughlin (R-Pa.); Rep. John N. Happy Camp (R-Okla.); Rep. John B. Conlan (R-Ariz.); Rep. Stanford E. Parris (R-Va.); Rep. Paul W. Cronin (R-Mass.); and Rep. James G. Martin (R-N.C.). (*CR*, 1/24/73, H442-H443; *A&A 1972*)

January 24-25: Symposium on recycling used glass containers was held in Albuquerque, N. Mex., by Univ. of New Mexico Technical Application Center (TAC), Albuquerque Environmental Health Dept., and Glass Containers Manufacturers Institute. Representatives of NASA's public technology program, glass researchers, and city and county officials discussed recycling glass into glassphalt road surfacing material, foams for building insulation, and other new products. TAC operated one of NASA's six Regional Dissemination Centers for space agency-developed technology. (NASA Release 73-7; NASA PAO)

January 25: The Global Atmospheric Research Program (GARP) Atlantic Tropical Experiment (GATE), largest and most complex international scientific experiment, would observe one third of the earth's tropical belt from June 15 to Sept. 30, 1974, the National Oceanic and Atmospheric Administration announced. A network of satellites, ships, aircraft including NASA's instrumented Convair 990, instrumented buoys, and land stations would collect meteorological data from the tropical Atlantic and adjacent land areas to improve forecast techniques for the equatorial belt. Brazil, Canada, France, Federal Republic of Germany, Mexico, Netherlands, Portugal, U.S.S.R., United Kingdom, and Venezuela, in addition to the U.S., would process data for compatibility. Data would be integrated by the GATE data center and distributed in 1976. U.S. participation would be coordinated by NOAA. (NOAA Release 73-5)

- Lunar laser ranging system at the Univ. of Texas McDonald Observatory located the French retroreflector array attached to the Soviet unmanned explorer *Lunokhod 2* (landed on the moon Jan. 16 by *Luna 21*). At the request of the French National Center for Space Studies (CNES), the 270-cm (107-in) McDonald telescope transmitted light pulses, which were observed after reflection from mirror array. Preliminary data indicated the surface location of *Lunokhod 2* was very near coordinates given by the Soviet Academy of Sciences. The lunar radius at that point appeared to be smaller than previous data had indicated. (NASA Release 73-13)

- Eruption of the Helgafell volcano on Heimaey Island off the Icelandic coast was recorded by a very-high-resolution radiometer on board *Noaa 2* National Oceanic and Atmospheric Administration meteorological satellite, launched by NASA Oct. 15, 1972. Data would be correlated with data from *Erts 1* (launched July 23, 1972) and an Air Force aircraft. (*Av Wk*, 2/19/73, 67)

January 26: *Cosmos 546* was launched by U.S.S.R. from Baykonur Cosmodrome into orbit with 613-km (380.9-mi) apogee, 574-km (356.7-mi) perigee, 96.5-min period, and 50.6° inclination. (GSFC *SRR*, 1/31/73; *SBD*, 1/29/73, 131)

- President Nixon transmitted to Congress Reorganization Plan No. 1 of 1973. The plan abolished the National Aeronautics and Space Council, including the position of Executive Secretary, and the Office of Science and Technology, including the posts of Director and Deputy Director. NASC had been created under the National Aeronautics and Space Act of 1958, enacted July 29, 1958. Provisions of the plan would take effect July 1 unless Congress acted against the plan.

    In his transmission message the President said NASC had "met a major need during the evolution of our nation's space program. Vice President Agnew has served with distinction as its chairman for the past four years. At my request, beginning in 1969, the Vice President also chaired a special Space Task Group charged with developing strategy alternatives for a balanced U.S. space program in the coming years. As a result of this work, basic policy coordination can now be achieved through the resources of the executive departments and agencies, such as the National Aeronautics and Space Administration, augmented by some of the former Council staff." In streamlining the Federal science establishment, the Administration was "firmly committed to a sustained, broad-based national effort in science and technology." The research and development capability of executive departments and agencies had been upgraded. The National Science Foundation had "broadened from its earlier concentration on basic research support to take on a significant role in applied research as well" and had "matured in its ability to play a coordinating and evaluative role within the Government and between the public and private sectors. I have therefore concluded that it is timely and appropriate to transfer to the Director of the National Science Foundation all functions presently vested in the Office of Science and Technology, and to abolish that office."

The President also announced his intention to ask Dr. H. Guyford Stever, NSF Director, to take on the additional post of Science Adviser to advise and assist the Administration "on matters where scientific and technological expertise is called for" and to act as the President's representative in selective cooperative programs in international science, "including chairing such joint bodies as the U.S.–U.S.S.R Joint Commission on Scientific and Technical Cooperation." (*PD*, 1/29/73, 75–8; NASA Leg Off)

- Hardware for Skylab inspace experiments proposed by U.S. high school students was shipped to Kennedy Space Center. Student winners in the Skylab Student Project national competition had worked with NASA specialists in designing the equipment, which was built and assembled at NASA's Marshall Space Flight Center. The experiments were in astronomy, biology, space medicine, and physics. (MSFC Release 73-8)

- *Erts 1* Earth Resources Technology Satellite (launched July 23, 1972) had returned valuable imagery and digital tapes from the spacecraft's multispectral scanner to the Brevard County, Fla., Planning Dept., Kennedy Space Center reported. John W. Hannah, Brevard Development Administrator and principal investigator in the Brevard ERTS experiment, had said interpretation of the images would yield information on beach erosion, street development patterns, landscape disturbances, identification of land use, and land drainage patterns. (KSC Release 14–73)

- Dr. Rocco A. Petrone, Apollo/ASTP Program Manager, became Marshall Space Flight Center Director on the retirement of Dr. Eberhard F. M. Rees. The appointment had been announced Dec. 22, 1972. (MSFC Org Ann 0101)

- Sen. Robert C. Byrd (D-W. Va.), for Sen. Lloyd M. Bentsen (D–Tex.), introduced S.J.R. 37 to designate Manned Spacecraft Center the Lyndon B. Johnson Space Center in honor of the late President. The resolution text said President Johnson had been "one of the first of our National leaders to recognize the long-range benefits of an intensive space exploration effort." During his presidency he had borne the "ultimate responsibility for the development of the Gemini and Apollo programs which resulted in man's first landing on the moon." (*CR*, 1/26/73, S1344)

- Manned Spacecraft Center announced the award of a $1 375 484 cost-no-fee contract to the Charles Stark Draper Laboratory at the Massachusetts Institute of Technology for technical support for guidance, navigation, and control in the space shuttle program. The Laboratory would develop and evaluate mission requirements, integration of guidance, navigation, and control with avionics and the shuttle, failure detection, navigation techniques, and guidance policies. (MSC Release 73-11)

January 27: NASA launched a Nike-Tomahawk sounding rocket from Andoeya, Norway, carrying a Norwegian Defense Research Establishment magnetospheric physics experiment. Preliminary data indicated that the rocket and instrumentation performed satisfactorily. (GSFC proj off)

January 28: Japan launched four-stage rocket from Uchinoura Space Center to test orbiting techniques. (AP, *NYT*, 1/29/73, 3)

- The U.S. National Scientific Balloon Facility launched a second balloon from Oakey Field, Queensland, Australia, for NASA's Project Boomerang [see Jan. 24 and March 22]. (NASA Release 73-48)
- Soviet cosmonaut Mrs. Valentina Nikolayeva-Tereshkova—on tour in Cacolla, India—told the press she was preparing for another space flight. "I am working for it and am keen on it because it is my work, but no date has yet been fixed for it." Mrs. Nikolayeva-Tereshkova, wife of Cosmonaut Andrian G. Nikolayev, had made her first space flight June 16–19, 1963, on *Vostok 6*. (AP, W *Star & News*, 1/29/73, A3)
- Novelist Joseph McElroy described his personal impression of the *Apollo 17* launch in a *New York Times* essay: "My main image was the Saturn rocket destroying itself stage by stage, separating behind the spacecraft, which was then alone, coasting a trajectory whose essence was to get away and not come back, escape the dialectical symmetries of up down, out back, growth and decay—and instead burst through into speeds and distances not just greater but other; yet also, yes, to come back changed, to save the earth, prove to Earthling ecologues it had been worth all that liquid oxygen and hydrogen in a few minutes of early burn." (*NYT, Bk Rev*, 1/28/73, 27)

January 29: President Nixon, in his message transmitting the FY 1974 U.S. budget to Congress, said the budget supported U.S. efforts to establish lasting peace by maintaining military strength while proposing a sound fiscal and monetary policy that "will contribute to prosperity and economic stability here and abroad." The U.S. had progressed substantially toward ending its part in war in Southeast Asia, concluded in the past four years "more significant agreements with the Soviet Union than in all previous years since World War II," and ended isolation of the U.S. from the People's Republic of China. Taken "with the success of the Nixon Doctrine, our substantial disengagement from Vietnam, and the increased effectiveness of newer weapons systems," defense outlays had been kept in line. When adjusted for pay and price increases, 1974 defense spending would be "about same as in 1973 and about one third below 1968." (*PD*, 2/5/73, 86–98)

- President Nixon sent a $268.7-billion FY 1974 national budget request to Congress—an increase of $18.9 billion over FY 1973. The budget projection for the succeeding fiscal year was included for first time, with the FY 1975 budget estimated at $288 billion. The national total for FY 1974 included an R&D budget of $17.4 billion (an increase of $300 million), with $3.1 billion requested for civilian and military space programs. Emphasis was on practical, immediate results from R&D rather than expanding scientific knowledge, with reduced priority for space research.

    The total request for NASA new obligational authority of $3.016 billion (1.2% of the total U.S. budget) was $391 million less than the FY 1973 NOA of $3.407 billion. FY 1973 funds of $91 million were to be carried over to FY 1974. NASA expenditures were budgeted to increase $74 million, from $3.062 billion in FY 1973 to $3.136 billion. The FY 1973 increase of $11 million had been largely for space shuttle engine and airframe design and development. Of the FY 1974 budget request, $2.288 billion would go for R&D, $112 million for construction of facilities, and $707 million for research and program management.

In manned space flight, elimination of Apollo funding reflected successful completion of the Apollo program with the *Apollo 17* mission. Manned space flight operations—including $233.8 million for Skylab; $90 million for the Apollo-Soyuz Test Project (ASTP); $220.2 million for development, test, and mission operations; $21 million for space life sciences; and $15.5 million for mission systems and integration—would decrease from $879 million in FY 1973 to $580.5 million. The decrease reflected removal of space shuttle funding from the manned spacecraft operations category to separate program status. The FY 1974 budget request allocated $475 million to the space shuttle program for the orbiter, main engine, solid-fueled rocket boosters, external tanks, technology and related development, and vehicle and engine definition. Advanced missions would receive $1.5 million, to bring the total for manned space flight (including the shuttle) to $1.057 billion, down $100 million from FY 1973.

Funding for space science programs would decrease $95.2 million, from $679.2 million in FY 1973 to $584 million in FY 1974. The decrease reflected suspension of the High Energy Astronomy Observatory (HEAO) program, in which two flights had originally been scheduled during 1976–1978. The physics and astronomy program would receive $95 million, down from $126.2 million; lunar and planetary exploration would receive $312 million, down from $331.9 million; and launch vehicle procurement would receive $177 million, down from $221 million.

NASA's space applications programs would receive $153 million (a decrease of $35.7 million), including $51 million for weather and climate research with satellites, meteorological soundings, experimental instruments and techniques, and for the Global Atmospheric Research Program (GARP); $13.9 million for pollution monitoring, including $9 million for Nimbus-G oceanographic and air pollution observation satellite; $42.6 million for earth resources survey; $10.7 million for earth and ocean physics; $3.1 million for space processing; $22.1 million for communications, including a multi-use communications experiment satellite (ATS–F) and a Canadian Cooperative Satellite (CAS–C); $3 million for a multidisciplinary Earth Observing Satellite (EOS); $4.5 million for shuttle experiment definition; and $2 million for application studies.

The total requested for NASA's aeronautics and space technology programs increased from $232.5 million in FY 1973 to $240 million. Aeronautical research and technology funding would increase from $150.6 million in FY 1973 to $171 million, including $22 million for experimental engine programs, $9.3 million for flight experiment programs, and $30 million for system technology programs. Space research and technology would receive $65 million, and nuclear power and propulsion $4 million (a drop from $17.1 million).

Tracking and data acquisition funding would rise from $248.3 million in FY 1973 to $250 million.

The Dept. of Defense FY 1974 budget of $85 billion was at an all-time high, $4.1 billion above the FY 1973 budget authority, but was 28.4% of total Federal outlays for FY 1974 (a percentage decrease from 29.0% in FY 1973 and the lowest proportion since 1950) and 6.0% of the gross national product (the lowest GNP percentage in 24 yrs, down from 6.2% in FY 1973). The DOD total included $8.7 billion for research, development, test, and evaluation (RDT&E), a $600 million increase over FY 1973. Of the requested RDT&E total, military astronautics—with major programs including military comsats, the first spaceborne ballistic missile early warning system, a prototype satellite to demonstrate precise navigation capabilities, continued flight experiment programs, and technology programs in advanced navigation, guidance, sensors, cooling, reentry, and propulsion—would receive $604.6 million (up $169.9 million). Military sciences would receive $509.8 million (up $13.7 million). Aircraft RDT&E would receive $1.787 billion (down $86 million), and missiles RDT&E $2.245 billion (up $143 million).

Major DOD increases were programmed for the Trident submarine-launched ballistic missile system ($1.712 billion, up $917 million), new Air Force F-4J jet fighter aircraft ($131 million), CVN-70 aircraft carrier ($657 million, up $358 million) and F-15 air superiority fighter ($1.148 billion, up $240 million). Decreases of $198 million in funding for the Safeguard antiballistic missile system, $208 million for the Poseidon missile, and $64 million for the short-range attack missile (SRAM) reflected a modernization program, as DOD continued the transition from combat support in Southeast Asia to strengthening baseline forces.

The Dept. of Transportation budget request of $9.025 billion was a decrease of $2.302 billion from FY 1973. DOT funding would continue a 10-yr, $10-billion program to improve urban mass transportation and reduce highway congestion. Federal Aviation Administration funding of $2.126 billion would continue efforts to reduce aircraft noise, ensure that advanced aircraft engines did not adversely affect the upper atmosphere, and design a safer and more productive airport system.

The Atomic Energy Commission FY 1974 budget request of $3.066 billion, up $382 million over FY 1973, would cover production of nuclear materials to make electricity and for use in nuclear weapons and as nuclear reactor fuel, research into availability of source materials, operation safety, and development of techniques for long-term storage of radioactive waste.

The National Science Foundation FY 1974 budget request of $582.6 million was down $63.1 million from FY 1973, but R&D funding would increase from $510.3 million in FY 1973 to $554.1 million. Programs would emphasize research into domestic problems, problems of long-range national interest, and effects of R&D on the economy. Priorities would include development of solar power as an energy source and of tunneling technology for urban use under the Research Applied to National Needs (RANN) program and development of a very-large-array radioastronomy facility.

The National Oceanic and Atmospheric Administration (in the Dept. of Commerce) budget request of $343.0 million, down from $366.7

million in FY 1973, would permit continued improvements in weather monitoring, prediction, and warning; continuation of its polar orbiting satellite system and implementation of a two-geostationary satellite system in the weather satellite program; major hurricane modification research; and expansion of research on marine ecosystems analysis to determine effects of ocean dumping.

The Environmental Protection Agency budget request of $148.7 million was down from $173.1 million in FY 1973. (OMB, *Budget of the US Govt,* FY 1974; *US Budget in Brief; Special Analysis;* NASA budget briefing transcript; DOD Release 44–73; AIAA Release 73–3; NSF Release 73–106)

- Dr. James C. Fletcher, NASA Administrator, released the statement given out at the NASA FY 1974 budget briefing held Jan. 27: "The budget the President has approved for NASA for FY 1974, like those of other agencies, has been reduced substantially below previous plans and expectations to hold total government spending to the targets set by the President. At the same time, the budget will permit NASA to carry forward a substantial and significant program in space and aeronautics."

The budget provided for continuing the major projects in the schedule announced Jan. 5 and for starting two new projects: Nimbus-G, "another in NASA's series of experimental earth oriented satellites, which will be directed at environmental pollution and oceanographic measurements, and LAGEOS, a new geodetic satellite for extremely accurate measurements of movements of the earth's surface."

In addition to reductions in programs announced Jan. 5, the budget required "that we stretch out schedules in a number of ongoing projects. The second Earth Resources Technology Satellite (ERTS-B) will be delayed two years, until 1976. This will allow it to carry an additional sensor. Tiros-N, the prototype of a new series of operational meteorological satellites for the National Weather Service, will be delayed by about one year to 1977.

"We will also undergo another belt-tightening in NASA. By the end of FY 1974 our civil service complement will be reduced 1880 positions below the FY 1973 end strength approved in the FY 1973 budget. This reduction includes the previously announced reductions related to the termination of our nuclear programs."

Despite a "sharp decrease from the level budget anticipated last year," Dr. Fletcher was "very pleased with the program content we were able to retain. Clearly from what has been said today and on January 5, we will have to curtail several activities that are of major importance to the nation." But "we have been able to keep NASA's main line program essentially intact and each of the programs curtailed will still be done, either at a later date than anticipated or in an alternate way to that planned last year. Although I, personally, would like to see a higher NASA budget, I believe we can capitalize on the belt-tightening required to produce even greater yield for the tax dollar in future years." NASA was "making substantial efforts to reduce future costs . . . while still giving the nation a strong program in space and aeronautics." Additionally, a prime object of the space shuttle program was to reduce costs of operating in space. NASA intended to "develop and introduce new design concepts and management approaches which, together with the relaxed weight and size constraints of the Space Shuttle, will result in substantially lower development and operating costs."

NASA Comptroller William E. Lilly discussed personnel cutbacks necessitated by the new budgetary limitations. The total NASA Civil Service reduction of 1880 positions would include a Marshall Space Flight Center cut of 650; Goddard Space Flight Center, 158; Kennedy Space Center, 100; Manned Spacecraft Center, 75; Wallops Station, 17; Lewis Research Center, 660; NASA Hq., 150; and Ames, Langley, and Flight Research Centers, a total of 70. From a peak Civil Service complement of more than 34 000 in 1967, NASA would be reduced to under 25 000 employees by June 1974—a drop of 30%. Total employees from all sources—contract and in-house—would drop from the 1966 peak of 420 000 to about 100 000. (Text; Transcript)

- U.S.S.R. ground control, during a communications session with *Lunokhod 2* to check onboard systems, found a temperature of 90 K ($-297°F$) at the end of a rod containing magnetometer sensors. It was the lowest temperature recorded near the lunar surface. *Lunokhod 2*, inactivated Jan. 23 for the lunar night after it had explored the lunar surface for seven days, had been launched aboard *Luna 21* Jan. 8 and landed on the moon Jan. 16. It would begin its second lunar day of exploration Feb. 9. (Tass, FBIS–Sov, 2/13/73, L1)

- Sen. Frank E. Moss (D-Utah), new Chairman of the Senate Committee on Aeronautical and Space Sciences, had promised an "in depth" inquiry into the FY 1974 NASA budget request at early date, *Aviation Week & Space Technology* reported. The Committee staff was being expanded and Sen. Moss would visit Manned Spacecraft Center in February. Sen. Barry M. Goldwater (R-Ariz.), the Committee's new ranking minority member, had urged that more attention be paid to aeronautics. Sen. Lowell P. Weicker, Jr. (R-Conn.), minority member, had said the Committee should be more forceful in pushing aerospace programs "because NASA seems reticent at times to speak for itself." (*Av Wk*, 1/29/73, 13)

- Library of Congress Congressional Research Service published *United States and Soviet Progress in Space: Summary Data through 1972 and a Forward Look*. The report—prepared by Dr. Charles S. Sheldon II, Chief of Science Policy Research Div.—was the latest in an annual review series and first to use the metric system in all quantitative measurements to conform to NASA and Soviet practice. The U.S. expected to have spent $70 billion by June 30 on combined civilian and military space programs. Physical evidence of Soviet space activity indicated the U.S.S.R. had committed a similar amount. There had been no manned Soviet launches in 1972. Moscow had been "filled with rumors" in late July of an impending "more ambitious Salyut-Soyuz manned mission," but press reports had indicated later that tracking ships which usually supported such missions were returning to port. NASA employment—peaked at 400 000 in 1966—had dropped to 150 000. Soviet space employment might be close to 600 000 because the "effort today seems to be at least equal to our 1966 peak" of a total 600 000 space employees in the U.S. The state of the Soviet effort toward a space shuttle was unknown but the "Soviet effort to maintain a position of leadership in space, and the continued high level of Soviet flight activity of more than double that of the United States" in-

dicated the U.S.S.R. "may have even more compelling reasons to develop a reusable shuttle as a cost saving device." (Text)
- Rep. Wright Patman (D-Tex.) introduced H.J.R. 255 to change the name of Manned Spacecraft Center to Lyndon B. Johnson Space Center in honor of the late President. He said: "Lyndon Johnson understood the importance of space exploration to the people of the United States and its ultimate benefits to all of mankind. He understood the necessity of this advance technology to our security. He understood the stimulation this new field would create in our educational systems. He understood the palpable gains to the average citizen from our weather, communications, and other applications satellites. He understood the challenge to excellence and achievement it posed to our scientists, engineers, and technicians. But most of all, he understood . . . the need of man to explore, to reach out, and to seek new ways to bend science and technology to our use." (CR, 1/29/73, S1467)
- The Senate confirmed the nomination of Elliot L. Richardson to be Secretary of Defense. (CR, 1/29/73, D51)
- The Senate Committee on Commerce announced appointments to its Subcommittee on Aviation: Sen. Howard W. Cannon (D-Nev.), Chairman; Sen. Warren G. Magnuson (D-Wash.); Sen. Philip A. Hart (D-Mich.); Sen. Vance Hartke (D-Ind.); Sen. Ernest F. Hollings (D-S.C.); Sen. Daniel K. Inouye (D-Hawaii); Sen. Frank E. Moss (D-Utah); Sen. John V. Tunney (D-Calif.); Sen. Adlai E. Stevenson, III (D-Ill.); Sen. Norris Cotton (R-N.H.); Sen. James B. Pearson (R-Kans.); Sen. Howard H. Baker, Jr. (R-Tenn.); Sen. Robert P. Griffin (R.-Mich.); Sen. Marlow W. Cook (R-Ky.); Sen. Ted Stevens (R-Alaska); and Sen. J. Glenn Beall, Jr. (R-Md.). (CR, 1/29/72, D52)
- NASA announced establishment of the Office of Safety and Reliability and Quality Assurance under George C. White, Director. The separate Safety Office and Reliability and Quality Assurance Office were disestablished and all responsibilities and personnel transferred to the new office. (NASA Ann)
- The Japanese government would spend $316 million on aerospace during Japanese fiscal year 1974, which began in April, *Aviation Week & Space Technology* reported. The figure was 13% less than FY 1973 aerospace funding. Japan's total defense budget would rise 16%, to $3.9 billion. (Av Wk, 1/29/73, 11)
- An *Aviation Week & Space Technology* editorial criticized the Jan. 3 recommendations of the Aviation Advisory Commission to President Nixon. Proposed creation of an under secretary of civil aviation in the Dept. of Transportation was a "rather old and oft-tried device that has an extremely poor track record wherever it has been attempted." The record of postwar civil aviation was clear. "No government-selected commercial transport has been able to sell a significant slice of the international market in the face of competitively developed aircraft." (Hotz, Av Wk, 1/29/73, 9)

January 30: The U.S. flight crew for the 1975 Apollo-Soyuz Test Project mission was announced by NASA. Prime crew: Thomas P. Stafford, commander; Vance D. Brand, command module pilot; Donald K. Slayton, docking module pilot. Backup crew: Alan L. Bean, Ronald E. Evans, and Jack R. Lousma. Support crew: Richard H. Truly, Robert F. Overmyer, Robert L. Crippen, and Karol J. Bobko. Joint crew training would begin in the summer of 1973, when Soviet cosmonauts

would visit the U.S. for several weeks. U.S. astronauts would spend an equal time in Russia in the fall of 1973. (NASA Release 73-15)
- A model for the interaction of the solar wind with very thin hydrogen gas in deep space, developed by National Oceanic and Atmospheric Administration geophysicist Dr. Thomas E. Holzer, would aid future space flight planning and provide data on the origins and fate of universe, NOAA announced. The presence of thin hydrogen gas in interplanetary space had been observed by *Ogo 5* (launched by NASA March 4, 1968). Interaction of the solar wind and interstellar hydrogen could become important at astronomical units from the sun. Neutral hydrogen could become charged and associated with the solar wind. Some solar wind energy would be converted to random motion or heat. At the same time, the velocity of the solar wind would be diminished. Neutral hydrogen could act as gentle, penetrable barrier to the solar wind. (NOAA Release 73-7)
- The last lunar sample containment bag from *Apollo 17*, the last Apollo mission to the moon (Dec. 7-19, 1972), was logged into the processing cabinets at the Lunar Receiving Laboratory at Johnson Spacecraft Center. (Apollo Sample Analysis Planning Team, *Science*, 8/17/73, 615)
- Effects of FY 1974 budget restrictions on Kennedy Space Center was conveyed to KSC employees in letter from KSC Director, Dr. Kurt H. Debus. The personnel ceiling for June 30, 1974, would be 2309—112 less than the current total onboard strength. The necessity for an average grade ceiling of 11.22 by June 30, 1973, would impose "further constraint in terms of available options." (Text)
- The Canadian National Research Council launched the first in a series of six sounding rockets, including two to be launched by NASA Feb. 2, to investigate growth and decay of auroras. A Black Brant IVB launched from Churchill Research Range carried a 84-kg (185-lb) payload to 714-km (443.7-mi) altitude to investigate the quiet prebreakup of the early evening aurora. Rocket and instrumentation performed satisfactorily. (NRC prog off)
- Cancellation of competition for procurement of a fixed-wing utility aircraft was announced by Secretary of the Army Robert F. Froehlke. Cancellation had been made because of "inconsistency between what was told the Congress and the eventual recommendation." Final decision on the award and contract had not been made, but details of proposals and their evaluation had been leaked by undetermined sources. A new solicitation would be based on a restated procurement objective, to be subject of new requests for proposals. (DOD Release 51-73)
- A People's Republic of China team of diplomatic and transportation officials arrived in Seattle, Wash., to discuss PRC purchase of 10 Boeing 707-320 transport aircraft. Boeing would train crews in Seattle and would assist with further crew training in Shanghai. (*Av Wk*, 2/5/73, 29)

January 31: Manned Spacecraft Center had named 10 investigators to receive the first *Apollo 17* lunar samples, NASA announced. Five U.S. and five foreign scientists would receive allocations that included tiny rock chips and polished thin sections cut from three large rocks. Rocks 70035 and 75055 were dark gray basalts typical of material underlying the Taurus-Littrow valley; rock 76055—lighter colored recrystallized breccia that Astronaut Harrison H. Schmitt had described as "anorthositic gabbro"—might have been part of the mountainside at one time. The first investigators had been recommended by the Lunar Sample Analysis Planning Team (LSAPT). Most would attempt to determine sample ages by rubidium-strontium and argon analysis. Other studies would analyze trace elements and mineralogical content. Early allocations were being made with the stipulation that research be conducted so that results could be reported at the 4th Annual Lunar Science Conference scheduled for March 5-8. Additional preliminary allocations would be recommended by LSAPT, including samples of orange soil found at Shorty Crater. (NASA Release 73-16)

- Fifteenth anniversary of the first U.S. satellite, *Explorer I*, launched by the Army Ballistic Missile Agency and Jet Propulsion Laboratory Jan. 31, 1958. The U.S. and the Western world had entered an age of space with the launch of the 14.0-kg (30.8-lb) stovepipe-shaped satellite. Dr. Kurt H. Debus, Director of Kennedy Space Center, recalled the mood of excitement surrounding the launch that had followed two successful U.S.S.R. launches and a failure in the first U.S. attempt to launch Vanguard. "I think we were all aware that this wasn't just another mission—that perhaps the entire world was watching this one."

  Explorer I had made significant scientific contribution with the discovery of the Van Allen Radiation Belt surrounding the earth. The satellite had transmitted data until May 23, 1958, when its small batteries were exhausted. It reentered the earth's atmosphere Mar. 31, 1970. (NASA Release 73-14; KSC Release 17-73)

- Pan American World Airways, Inc., and Trans World Airlines, Inc., announced a decision not to exercise their options to buy 13 Anglo-French Concorde supersonic transport aircraft. The action was expected to lead American Airlines, Inc., and Eastern Airlines, Inc., to drop their options, the *Washington Post* reported later. Concorde options would then stand at 30, but lack of competitive pressure from U.S. airlines was likely to cause additional cancellations. Nine firm orders for Concorde were from the state-owned British Overseas Airways Corp. and Air France. Iran Air and People's Republic of China had signed preliminary contracts for 5 Concordes but the *Post* said an order for only 14 aircraft could not sustain the Concorde production run. (Greer, Nossiter, *W Post*, 2/1/73, A1)

- The Federal Communications Commission notified U.S. companies bidding on the European Space Research Organization's aeronautical satellite system that participation was subject to FCC approval. (*Av Wk*, 2/12/73, 21; FCC PIO)

- The last Air Force C-5A aircraft rolled off Lockheed Aircraft Corp.'s assembly line. The C-5A, world's largest subsonic jet transport, had been plagued by problems during its development phase but had since proved its effectiveness as a materiel and troop carrier. (*Atlanta JC*, 2/1/73, 8)

- The Air Force had ordered the last F-111 swing-wing fighter bomber from General Dynamics Corp., United Press International reported. Plans to procure parts to maintain and modify existing models would require a $29-million budget item. (*W Post*, 1/31/73, A16)

During January: NASA issued *The Interplanetary Pioneers*, Vol. 1, *Summary* (NASA SP-278). The three-volume series by William R. Corliss traced NASA's Pioneer program from its May 1960 inception through *Pioneer 9* (launched Nov. 8, 1968). The program had begun with the launch of early lunar probes *Pioneer 1* through 5 by Ames Research

Center. Later Pioneers—*Pioneer 6* (launched Dec. 16, 1965), *Pioneer 7* (launched Aug. 17, 1966), *Pioneer 8* (launched Dec. 13, 1967), and *Pioneer 9*—had been the first spacecraft to orbit the sun systematically at widely separated points in space. They had collected information on space, solar winds, and cosmic radiation of both solar and galactic origin. The Pioneers had been "superbly reliable scientific explorers, sending back information far in excess of their design lifetimes over a period that covers much of the solar cycle."

Vol. II, *System Design and Development* (SP-279), and Vol. III, *Operations* (SP-280), had been issued in 1972. (Text)

- The National Science Foundation published *Scientists, Engineers, and Physicians from Abroad: Trends through Fiscal Year 1970* (NSF 72-312). A total of 13 300 immigrant scientists and engineers admitted to U.S. in 1970 was an increase of one third over 1969 and two and one half times the 1965 total. Natural scientists had increased from 10% of the 1950-1964 total to 20% of the total since 1965. Scientists and engineers from Asia had risen from 10% of a much smaller total in 1965 to 50% in 1970. Indian scientists and engineers admitted in 1970 numbered 2900, the largest group admitted from any country over the previous 20 yrs. Among 3800 born in one country but living elsewhere before entering the U.S. had been 740 born in the People's Republic of China and 620 born in India.

  The number of foreign science and engineering students had increased from 56 800 in 1967 to 72 100 in 1970, with Asia the source of 50%. The number of recipients of doctorates of science and engineering from U.S. universities had grown 222% between 1960 and 1970. (NSF 72-312)

February 1973

February 1: The U.S.S.R. launched *Cosmos 547* from Baykonur Cosmodrome into orbit with a 311-km (193.3-mi) apogee, 201-km (124.9-mi) perigee, 89.6-min period, and 65.0° inclination. The satellite reentered Feb. 13. (GSFC *SSR*, 2/28/73; *SBD*, 2/2/73, 183)

- NASA held an Apollo-Soyuz Test Project news briefing at Manned Spacecraft Center. U.S. Technical Director Glynn S. Lunney said the U.S.S.R. had agreed to lower the Soyuz cabin pressure to within 350 newtons per sq cm (5 lbs per sq in) of Apollo pressure during the July 1975 joint docking mission, to eliminate two-hour depressurization for each crew member before transfer to the other spacecraft. The change would permit time for all three astronauts and two cosmonauts to fly in the others' spacecraft. Previously, only two astronauts had been scheduled to make the transfer.

    Astronaut Thomas P. Stafford, commander of the U.S. crew, said "the mission is going to be one of the most difficult . . . because it involves a different country, different language, different operating techniques. . . . But it can certainly open the doors for future cooperation between the countries, can enhance our capabilities" for progress together. Astronaut Donald K. Slayton, who would fly his first mission since he was disqualified from Project Mercury's *Aurora 7* flight (launched May 24, 1962) because of an erratic heart rate, said, "I think I've been pretty fortunate. And I don't believe in looking back at the past too much except in relation to what benefit it will give me in the future."

    Astronaut Vance D. Brand, who would make his first space flight, said, "I am really happy to get on this mission. I think as we look back over the years, it'll be regarded as a very historic thing . . . having the doors open between the Soviet Union and the United States." (Transcript)

- NASA's *Explorer 48* Small Astronomy Satellite (launched Nov. 16, 1972, from Italy's San Marco Equatorial Range) successfully completed a two-month comprehensive study of celestial gamma rays. Its gamma ray detector had been turned on Nov. 19, 1972, and had observed gamma rays in the region of the galactic center, galactic plane, and various x-ray sources. A sky map of gamma rays would be prepared. (NASA Release 73-17)

- Tenth anniversary of the formation of Communications Satellite Corp. In 10 yrs, ComSatCorp had grown to its present size from the mandate in the Communications Satellite Act of 1962 stating the corporation should be formed to establish a global commercial communications satellite system in cooperation with other countries. The organization employed 1100 persons at 11 locations to operate satellites in a global system for the International Telecommunications Satellite Consortium (INTELSAT), 7 U.S. earth stations for international satellite communications, ComSatCorp laboratories, and a wide range of technical activities. (ComSatCorp Release)

- Sen. Henry M. Jackson (D-Wash.) introduced S. 734 for himself and Sen. Edward J. Gurney (R-Fla.) and Sen. Lawton M. Chiles, Jr. (D-Fla.), "to authorize the Secretary of the Interior to establish a Man in Space National Historic Site at Cape Kennedy." The bill would establish a site on the area comprising Launch Complexes 5 and 6, which had been used for the Redstone, Jupiter C, Juno I and II, and Mercury-Redstone launches, as well as Little Joe II in support of the Apollo program. (*CR*, 2/1/73, S1798; Draft KSC Hist Note, 2/72)

February 2: Dr. James C. Fletcher, NASA Administrator, discussed NASA's outlook for the future in a letter to the NASA staff: "I regret that we have not been able to stabilize NASA's budget at about the $3.4 billion level . . . and that our reduced program workload necessitates a further reduction in . . . NASA employees. I am pleased, though, that we can proceed with most of our mainline programs and schedules for this decade with little change." With the May launch of Skylab "we should be able to give the world a very impressive demonstration of the tremendous new capabilities we now have for using manned and unmanned spacecraft in Earth orbit. I believe that Skylab will awaken public interest throughout the world in the usefulness of scientific and practical observations to be made from Earth orbit, and in the 'space manufacturing' experiments. I am also confident that the reports from Pioneer 10 as it swings past Jupiter in December will stimulate strong public interest in expanded programs to explore the planets. Our joint flight with the Russians in 1975 will stress the potential of large-scale international cooperation in space. The Viking Lander on Mars may produce the first evidence of life in the universe beyond the confines of our home planet Earth; and it may suggest to Americans the real potential of space exploration as a continuing and expanding activity during our third century as an independent and democratic nation." Dr. Fletcher forecast a favorable NASA future: "With another year of excellent performance in 1973 like we had in 1972, and continued strong bi-partisan support in Congress, this forecast is bound to become a reality." (*NASA Activities*, 2/15/73, 26-8)

- The U.S.–U.S.S.R. Working Group on Interplanetary Exploration released a statement following a weeklong meeting in Moscow. Scientists had met to analyze data about the Mars surface and atmosphere for selection of future landing sites for unmanned spacecraft. They also had discussed new results from exploration of Venus by *Venus 8* (launched by U.S.S.R. March 27, 1972) and earth-based optical and radio measurements. Recommendations for cooperation would be announced when confirmed by both sides [see March 5]. The Group had been established under a NASA and Soviet Academy of Sciences agreement of January 1971. U.S. scientists were led by Dr. S. Ichtiaque Rasool, Deputy Director of Planetary Programs in NASA's Office of Space Science. Soviet participants were headed by Academician Georgy I. Petrov, Director of the Institute of Space Research. (NASA Release 73-20)

- The Senate Committee on Aeronautical and Space Sciences approved a favorable report on S.J.R. 37—to change the name of NASA's Manned Spacecraft Center at Houston, Tex., to Lyndon B. Johnson Space Cen-

ter—after hearing testimony from Dr. James C. Fletcher, NASA Administrator. (CR, 2/2/73, D71)
- A new satellite series, Intelsat IVA would begin service in mid-1975 to expand and improve communications capacity among nations, ComSatCorp announced. The new series would double capacity by reusing frequencies for the first time by beam separation. The series would replace the global system of Intelsat IV satellites jointly owned with International Telecommunications Satellite Consortium, in which ComSatCorp held the major interest. Since *Early Bird* (*Intelsat 1*, launched April 9, 1965), satellites had expanded and improved world communications until, to date, there were 251 pathways among 80 antennas at 65 stations in 49 countries. Two thirds of all long-distance communications were via satellite. (ComSatCorp Release 73–5)
- The European Launcher Development Organization (ELDO) Council, meeting in Paris, agreed to defer its final decision on the fate of the Europa II launcher until a further meeting, scheduled for March 30. France advocated continuing the project to provide a launch vehicle for communications satellites. West Germany said it desired to leave the Europa program. (SF, 4/73, 212)
- Four sounding rockets in a series of six [see Jan. 30 and Feb. 8] were launched in Canada to investigate growth and decay of aurora phenomena.

  NASA launched two Super Arcas sounding rockets from Churchill Research Range carrying a Univ. of Houston payload to supplement investigation of auroral zone disturbances, with emphasis on the explosive phase of auroral substorms, and to study auroral phenomena in the region of atmosphere above that investigated by balloons and below the altitude investigated by rockets. The first rocket reached 84-km (52.2-mi) altitude. Probable parachute-system malfunction caused rapid descent and rendered most data unusable. The second reached 81.2-km (50.5-mi) altitude; rocket and instrumentation performed satisfactorily.

  The Canadian National Research Council launched two Black Brants. A Black Brant IIIB launched from Gillam, Manitoba, carried a 63.5-kg (140.0-lb) payload to 191.7 km (119.1 mi) to investigate plasma properties of upper-atmosphere radio aurora plasma. Black Brant VB launched from Churchill Research Range carried a 175-kg (386-lb) Univ. of Belgium payload to 325 km (202 mi) to investigate the expansive phase of an aurora storm. Both rockets and instrumentation performed satisfactorily. The third Black Brant was postponed because of dissipation of the aurora substorm into which it was to be launched [see Feb. 8]. (NASA Rpts SRL; NRC prog off)
- *Erts 1* Earth Resources Technology Satellite (launched by NASA July 23, 1972) was "important tool" to facilitate better management of the earth "that has not received appropriate recognition," a *Science* editorial said. Many investigators, selected from 600 research proposals, were studying its images. (Of 310 proposals accepted, 100 were from scientists of foreign nations.) "Policy with respect to distribution of pictures is one of complete openness. Nationals of any country are free to purchase them at a nominal cost." Canada was operating its own *Erts 1* receivers; Brazil, Mexico, and Venezuela were moving toward establishing their earth stations. "The Brazilians are particularly enthusiastic about ERTS, for it is giving them a first look at much of

the Amazon valley. Their enthusiasm is likely to be contagious, and other developing countries will find ERTS a valuable source of many kinds of information." (Abelson, *Science*, 2/2/73, 431; NASA OA)

- The National Science Foundation released *An Analysis of Federal R&D Funding by Function: FY 1963-73* (NSF 72-313). Federal expenditures for research and development were expected to grow faster than total Federal outlays in FY 1973 for the first time since FY 1965. Between FY 1972 and 1973 R&D expenditures were expected to rise 3.6%, with total outlays up 2.6%. The Federal R&D percentage of the total was expected to be 7.3% in FY 1973. Of total outlays for space research and technology, 98% was expected to go toward R&D, but the R&D share of the total outlay for international affairs and finance, veterans benefits and services, general government, and income security would be less than 1%. Functions other than national defense and space R&T had increased emphasis on R&D significantly 1963-1973, from 10% to an expected 23%. The trend toward more civilian-oriented programs within the Federal R&D total was continuing, with an expected rise of 2% between 1972 and 1973. (Text)
- The Air Force announced the award of $13 974 000 firm-fixed-price-incentive contract to Boeing Co. for Minuteman force modernization for FY 1973. (DOD Release 61-73)

February 3: The U.S.S.R. launched *Molniya 1-23* communications satellite from Baykonur Cosmodrome into orbit with a 39 772-km (24 713.2-mi) apogee, 578-km (359.2-mi) perigee, 717.6-min period, and 65.0° inclination. The satellite was to help provide a system of long-range telephone and telegraph radio communications in the U.S.S.R. and transmit Soviet central TV programs to the Orbita network. (Tass, FBIS-Sov, 2/6/73, L1; *SBD*, 2/6/73, 198; GSFC *SSR*, 2/28/73)

- A suggestion that quasars, mysterious bright light sources and radio waves, might be part of galactic evolution was made by Dr. Jerome Kristian, Hale Observatory astronomer, in a telephone interview with Reuters News Agency. Using photos taken through a 508-cm (200-in) reflector, Dr. Kristian had located quasars at the heart of six galaxies by a complex series of calculations of brightness and darkness. "The observations are consistent with the hypothesis that all quasars occur in the nucleii of giant galaxies." If quasars were part of the process of the development of galaxies, "then it could mean they are the start of the galaxies." (Reuters, *B Sun*, 2/3/73, A3)

February 4: India launched a two-stage Indian-made Centaur sounding rocket from Thumba Equatorial Rocket Launching Station (TERLS) to an altitude of 148 km (92 mi). The rocket carried instrumentation to measure nighttime airglow emission. Rocket and instrumentation performed satisfactorily. (Delhi Domestic Service, FBIS-India, Bhutan, Sikkim, 2/5/73, 3)

- U.S.S.R. scientists were developing highly efficient equipment for undersea exploration, *Krasnaya Zvezda* [Red Star] reported. A newly designed bathyscaphe, designated Sever-2, included a high degree of comfort for crew members. It had been fitted with complex instruments and could move independently in any direction, remain at a designated depth for long time, or descend to the bottom. (FBIS-Sov, 2/12/73, L4)

February 5: The new status of the High Energy Astronomy Observatory (HEAO) planning was clarified by Dr. F. A. Speer, Manager of the HEAO Project at Marshall Space Flight Center. HEAO was expected to fly three missions between 1977 and 1979 but, because of budgetary constraints, would be smaller, carry fewer experiments, and use a smaller and more economical launch vehicle. Although HEAO funds of $15 million were much less than the original level, scientific objectives—to observe the galaxy and the universe in high-energy x-ray and gamma ray stellar sources and to investigate cosmic ray flux—would remain mostly unchanged. (MSFC Release 73–13)

- Manned Spacecraft Center announced selection of Technology, Inc., for $794 000, one-year, cost-plus-award-fee contract to provide operational and research support for MSC's Life Science Laboratories. The contract had become effective Feb. 1, 1973. (MSC Release 73–17)
- Secretary of Transportation Claude S. Brinegar announced establishment of the Office of Assistant Secretary of Transportation for Congressional and Intergovernmental Affairs to ensure "an effective program of communications and coordination with members of Congress, Congressional staff members, and Congressional committees" and "serve as our primary liaison with officials of other agencies. . . ." (DOT Release 7–73)
- Lewis Research Center Director Bruce T. Lundin announced appointment of Harold Ferguson as Equal Employment Opportunity Officer. Ferguson was a physicist with 17 yrs experience in fluid mechanics and electric propulsion at LeRC. (LeRC Release 73–6)
- The National Science Foundation published *Federal Funds for Research, Development and Other Scientific Activities, Fiscal Years 1971, 1972, and 1973* (NSF 72–317). The FY 1973 budget had indicated an upward trend since 1970 in Federal research and development support. The Federal R&D obligation total was expected to rise from $15.5 billion in FY 1971 to an estimated $16.8 billion in FY 1972 and to an all-time high of $17.8 billion in FY 1973. Of the $970-million increase in R&D obligations scheduled for FY 1973, the Dept. of Defense accounted for $400 million and NASA and the Atomic Energy Commission $67 million each. The three agencies made up more than 50% of the 1973 growth. (Text)

February 6: Marshall Space Flight Center announced establishment of the Large Space Telescope Task Force to direct planning and preliminary design of the LST to be launched by the space shuttle in the 1980s. The LST—to be capable of looking at galaxies 100 times farther than those seen by the most powerful ground-based telescope—would study energy processes that occurred in galactic nuclei, study early stellar and solar system stages, and observe supernova remnants and white dwarfs. The LST Task Force would be managed by James A. Downey III, Associate Director for Science in the Program Development Directorate. (MSFC Release 73–15)

- Sen. William Proxmire (D–Wis.), Chairman of the Senate Committee on Appropriations' Subcommittee on Housing and Urban Development, Space, Science, Veterans, proposed to the Senate a counterbudget that would cut $4 billion from the President's $268.7-billion FY 1974 U.S. budget request. The counterbudget would reduce the $3.1-billion space budget by $500 million. "The prime candidate for cutting is the space shuttle, which this year will cost us $400 million. . . . I think Skylab, where spending of $315 million is scheduled, could be stretched out and postponed. Additional savings could be made in a much more vigorous effort to substitute unmanned for manned space efforts. Scientists tell us that unmanned flights can be as productive as manned flights at roughly half the cost." (CR, 2/6/73, S2108-10)

- The Senate passed S.J.R. 37 with an amendment and sent it to the House. The bill would redesignate the Manned Spacecraft Center the Lyndon B. Johnson Space Center. The amendment substituted the word "space" for "science" in the resolution preamble. (CR, 2/6/73, S2229-30)
- The Federal Aviation Administration put into effect Phase II of the tightened airport security measures announced Dec. 5, 1972. An armed guard would be stationed at every departure gate of every U.S. air carrier airport; every passenger would be screened before boarding aircraft; and all baggage would be inspected before being carried aboard. (DOT Release)
- Sen. Lowell P. Weicker, Jr. (R-Conn.), told the press in Washington, D.C., that a hold had been ordered on the Air Force AX tactical fighter plane contract with Fairchild Industries, Inc. The Air Force had said Fairchild had won a $14-million contract to build 10 prototypes by outbidding Northrop Corp. but congressional critics had disagreed, saying Northrop could build the same aircraft more cheaply. The Air Force would conduct its own investigation. (AP, W *Star & News*, 2/7/73, A7)
- President Nixon designated Dr. Dixy Lee Ray Atomic Energy Commission Chairman, succeeding Dr. James R. Schlesinger. Dr. Schlesinger had become Director of the Central Intelligence Agency in January. Dr. Ray, former Univ. of Washington zoologist and Director of the Pacific Science Center, had become the first woman nominated as a member of AEC in July 1972. The nomination had been confirmed in October 1972. (PD, 7/17/72, 1148-9; 7/24/72, 1165; 2/12/73, 124)
- NASA launched a Nike-Tomahawk sounding rocket from Andoeya, Norway, carrying a Norwegian Defense Research Establishment magnetospheric physics experiment to a 254.3-km (158-mi) altitude. The rocket and instrumentation performed satisfactorily. (GSFC proj off)
- The Army announced award of a $1 969 269 firm-fixed-price contract to Industrial Contractors, Inc., for modification of the acoustic model engine test facility at Marshall Space Flight Center. (DOD Release 63-73)

February 7: The House passed and cleared for the President's signature S.J.R. 37, to designate the Manned Spacecraft Center the Lyndon B. Johnson Space Center in honor of the late President. Rep. Olin E. Teague (D-Tex.), for himself and Rep. Kenneth J. Gray (D-Ill.), introduced H.J.R. 328 for the same purpose. H.J.R. 328 was referred to the House Committee on Science and Astronautics. (CR, 2/7/73, H838.9, H877)

- Selection of John E. O'Brien as Assistant General Counsel for Procurement Matters was announced by NASA. O'Brien, who had been Chief Counsel of Kennedy Space Center since December 1970, would replace John A. Whitney, who would return to private law practice. (NASA Release 73-19)
- Rep. Olin E. Teague (D-Tex.), Chairman of the House Committee on Science and Astronautics, introduced NASA-related bills: H.R. 4115, to authorize coinage of 50-cent pieces to commemorate the *Apollo 11* lunar landing and to establish the Apollo Lunar Landing Commemorative Trust Fund; H.R. 4119, to amend the National Aeronautics and Space Act of 1958 to provide for additional reports to Congress; and

H.R. 4120, to authorize the NASA Administrator to convey certain lands in Brevard County, Fla. (*CR*, 2/7/73, H875)

- The Federal Aviation Administration announced award of a $3 485 861 contract to Westinghouse Electric Co.'s Aerospace and Electronic Systems Center to design, fabricate, install, and test one prototype air-route-surveillance radar, designated ARSR–3. The radar would include the latest technology to improve coverage area and target detection and to reduce clutter. (FAA Release 73–26)

- A *New York Times* editorial commented on the new role of science in President Nixon's Administration: The extent to which technical expertise was indispensable to governmental decision-making in fields of environmental pollution, defense, and medical policy made it "surprising that the President now has chosen to lessen the role of science and scientists in his Administration." Downgrading science and technology would have effect of a handicap "in getting first-class scientific advice quickly when he needs it. . . . The United States is powerful and prosperous today only because it has made such effective, large-scale use of scientific knowledge in all civilian and military fields." (*NYT*, 2/7/73, 34)

February 8: U.S.S.R. launched *Cosmos 548* from Plesetsk into orbit with a 284-km (176.5-mi) apogee, 209-km (129.9-mi) perigee, 89.4-min period, and 65.4° inclination. The satellite reentered Feb. 21. (GSFC *SSR*, 2/28/73; *SBD*, 2/9/73, 227)

- Rep. Kenneth J. Gray (D–Ill.) introduced H.R. 4225 for himself and Rep. Thomas P. O'Neill, Jr. (D–Mass.), Rep. John J. McFall (D–Calif.), and Rep. Charles M. Price (D–Ill.) Section 3 of the bill would repeal section of P.L. 92–520 that had changed the name of Jet Propulsion Laboratory to H. Allen Smith Jet Propulsion Laboratory. (*CR*, 2/8/73, H932; Sen Com on Public Works)

- French oceanographer Jacques-Yves Cousteau spoke via *Ats 3* satellite (launched Nov. 6, 1967) from his research vessel *Calypso* in the Antarctic to reporters in the U.S. and Europe, to describe his research. In experiments with Ames Research Center, Cousteau had made constant measurements of chlorophyll, temperatures, and water transparency, sending data to ARC for comparison with satellite data. The experiment sought to learn how to monitor biological productivity of oceans from outer space. His expedition had been in Antarctic waters for two months. It would head up the South American coast and terminate in Los Angeles in the summer. (Blakeslee, *NYT*, 2/9/73, 57; Transcript Cousteau Press Conf, 3/1/73)

- The Canadian National Research Council launched the sixth and final sounding rocket in a series to investigate growth and decay of auroras [see Jan. 30 and Feb. 2]. The series had included two NASA launches. A Black Brant IVB carried a 84-kg (185-lb) payload from Churchill Research Range to 729-km (453-mi) altitude to investigate the quiet pre-breakup of the early evening aurora. Rocket and instrumentation performed satisfactorily. (NRC prog off)

- President Nixon took his first ride in the new *Spirit of '76* presidential Boeing 707 jet aircraft. The aircraft, which replaced the 707 put into presidential service in 1962, had been flight-tested 100 hrs by the presidential pilot, Col. Ralph D. Albertazzi (USAF), before the President's flight from Washington, D.C., to San Clemente, Calif. (UPI, *W Post*, 2/9/73, 27; Boeing Co. PIO)

- A Washington *Evening Star and Daily News* editorial commented on the supersonic transport effort: Environmental and economic problems and order cancellations had created a bleak outlook for the French and British Concorde. The modest budget request for research into the environmental impact of the U.S. SST by the Nixon Administration was realistic and should be approved by Congress "for the nation's protection." But the Administration should be "resigned to its defeat of 1971 in the building of an SST prototype." (W *Star & News*, 2/8/73, A10)

February 9: Aspects of President Nixon's FY 1974 budget requests were discussed in a *Science* article: The space program seemed "alive and well as it makes the transition into the post-Apollo era, despite recent fears . . . that its activities might be cut back severely." NASA officials seemed confident the space shuttle would be built. The President had supported the project "although his new budget message contained no mention of the space program" and, "in Congress, it has survived handily all past attempts to kill it. The fact that the project helps sustain an aerospace industry that has suffered grievously from layoffs is a point lost on no one. And then too, NASA has going for it the fact that, both in Apollo and in the unmanned programs, it has generally met its goals and stayed within its budget." Evidence existed "that the President does indeed look to a possible revival" of the supersonic transport "but not until later in the 1970's. The new NASA budget contains $28 million—more than twice as much as last year's budget—for research and development on supersonic technology. The work focuses on problems of noise, pollution, and efficiency of configuration." (Carter, *Science*, 2/9/73, 551–2)

- The United Nations Working Group on Remote Sensing of Earth by Satellite reported the success of U.S. and Soviet programs in identifying soil types, mapping minerals, charting geological features, locating fish schools, and producing "highly accurate" maps in areas incompletely mapped by conventional methods. Data had come largely from *Erts 1* Earth Resources Technology Satellite launched by NASA July 23, 1972. The Group suggested satellites might eventually detect underground water and record snowfalls. Valuable but incomplete information had been obtained about oceans. The Group urged priority research to develop methods of mixing carbon dioxide and ozone in the atmosphere, so satellites could monitor air pollution, and proposed putting a sensing system on the moon to monitor the earth's atmosphere. (*LA Times*, 2/10/73; *Atlanta JC*, 2/22/73)

- Dr. Thomas O. Paine, Vice President and Group Executive of General Electric Co. and former NASA Administrator, cited technological advances in power systems by the space program in a speech on the national energy crisis before the Morgan Guaranty International Council in San Francisco: "As we meet here today five small isotope generators on the moon are powering unmanned research stations there. Such technological advances by the U.S. aerospace industry promise to have major future impacts on world power systems, from fuel cells and the safe handling of hydrogen in the Apollo program to high temperature military jet engine technology for gas turbines and semiconductor technology for high voltage DC transmission." (Text)

- NASA launched two sounding rockets, one from Sweden and one from White Sands Missile Range. A Nike-Tomahawk launched from Kiruna, Sweden, carried a Swedish Space Corp. aeronomy experiment to a

180.2-km (112-mi) altitude. The rocket and instrumentation performed satisfactorily.

A Black Brant VC was launched from White Sands Missile Range carrying a Univ. of California astronomy experiment to a 212.4-km (132-mi) altitude. The rocket and instrumentation performed satisfactorily but the payload was damaged in recovery. (GSFC proj off)

- The Federal Aviation Administration announced award of a $735 213 contract to Collins Radio Co. to develop new area aviation techniques to increase airport and airways capacity. The two-year study would emphasize two-dimensional (positional and altitude guidance) and three-dimensional (added time element) area navigation concepts. (FAA Release 73–30)

February 9–19: The U.S.S.R. ground control commanded the lunar explorer *Lunokhod 2* to unfold its solar battery panel and begin its second lunar day of exploration. The moon car explored its parking area—a small deep crater with outcroppings of bedrock at its edge, formed 3 million yrs ago by a large meteorite that hit the surface of congealed lava. *Lunokhod 2* made measurements of the magnetic field inside the crater by approaching it four times from different directions and making 120 precision maneuvers.

The vehicle moved southeasterly to the Taurus Mountains. En route it measured physicomechanical properties of the soil and tension of the local magnetic fields, took panoramic photos of surrounding terrain, and tested riding characteristics of the lunar vehicle in various modes of operation. A monolith one meter (three feet) long of unusual lunar material was discovered Feb. 13. The plate had a strong, smooth surface unlike surrounding pockmarked stones and appeared to be much younger.

During exploration of the giant crater *Lunokhod 2* skidded and swerved over 2.3 km (1.4 mi) of difficult terrain with slopes up to 25° and sank in dust to the hub of its wheels as it ascended the rim of the crater. At the top of the rim, TV cameras showed a complex surface with ejecta piled on the rims of lunar holes formed when three meteorites hit the area within millions of years. *Lunokhod 2* descended the slope of a 100-m (330-ft) crater, leaving deep ruts in loose rock. The difficult ascent was interrupted for 20 hrs to replenish electrical energy Feb. 19. *Lunokhod 2* had covered 9.6 km (6.0 mi) since landing on the moon Jan. 16. (Tass, FBIS–Sov, 2/14/73, L1; 2/15/73, L2–L3; 2/21/73, L2–L3)

February 10–March 10: Water impact and towing tests of the space shuttle's solid-propellant booster were conducted by the Navy at the Long Beach, Calif., Naval Shipyard. Tests, under Marshall Space Flight Center direction, gathered data on retrieval of reusable solid-fueled-motor casings jettisoned from the orbiter stage at 40-km (25-mi) altitude. Test hardware, a 77%-scale model of the shuttle booster, was dropped from heights of 0.3 to 12 m (1 to 40 ft) at angles of 10°, 20°, and 30° off vertical. The model was towed at speeds of 3.7 to 14.8 km per hour (2.3 to 9.2 mph).

Kennedy Space Center was responsible for development of the retrieval techniques; MSFC, for design and fabrication of solid-propellant reusable boosters. (MSFC Release 73–12; KSC Release 22–73; NASA prog off)

February 11: Rising costs of the Skylab missions were discussed by Thomas O'Toole in a *Washington Post* article: Nearly $400 million had been added to cost estimates since the Skylab program had originated seven years ago. Postponement of the first Skylab launch from April to May had added $5 million in overtime costs and NASA had conceded another two-week delay and $5 million in additional costs was possible. Another reason for skyrocketing costs was the post-Apollo switch of emphasis from the moon and planets to the earth. Skylab's original purpose had been to study the sun and addition of the earth studies program on Skylab had added greatly to costs.

Skylab was the flagship of the U.S. manned space program. It was the "most impressive machine ever built," weighing 90 metric tons (100 U.S. tons) and providing living quarters equal to a three-bedroom house. Scientists saw Skylab as the link between the first decade of space exploration and the 1980s, when the reusable shuttle would become operational. It would provide much information about effects on the human body of long periods in space. The last of three Skylab crews might stay in space for 70 days. "If they make it that long without any serious side effects, they will have done as much for space exploration as any of the pioneers before them." (O'Toole, *W Post*, A1)

February 12: A statement by Western Union International, Inc., Vice President Thomas S. Greenish that mainland China had no plans for establishment of a domestic satellite communications system was quoted by *Aviation Week & Space Technology*. Greenish had returned from two-month stay in People's Republic of China in connection with an earth terminal being built by WUI near Peking. (*Av Wk*, 2/12/73, 17)

- Definitive agreements drawn up to establish International Telecommunications Satellite Organization (INTELSAT) went into effect, replacing the interim arrangements in effect since August 1964. Under the agreements, INTELSAT would operate under a four-component structure of a Board of Governors, Assembly of Parties, Meeting of Signatories, and an Executive Organ. The Board would have primary responsibility for design, development, construction, establishment, operation, and management of the comsat system. The Assembly—composed of a representative of the government of each member nation—would meet every two years to consider matters of concern to the governments. The Meeting of Signatories—representatives of governments or telecommunications entities—would convene annually to consider operational and other matters of interest to investors. Each member of the Meeting would have one vote.

The Executive Organ would be headed by a Secretary General responsible to the Board of Governors for performance of financial and administrative services. Under the agreements, Communications Satellite Corp. would furnish technical and operating management services for INTELSAT for a six-year period ending Feb. 12, 1979. Services would be performed under policies set by the INTELSAT Board of Governors. At the end of the ComSatCorp contract, INTELSAT would continue to contract with "one or more competent entities" to provide technical and operational management functions "to the maximum extent practicable." Agreements called for election of a Director General by Dec. 31, 1976. The Board of Governors would hold its first meeting in Washington, D.C., March 14. (Text; ComSatCorp Release, 2/1/73)

- An *Aviation Week & Space Technology* cover photo showed orange and black particles brought back from the moon's Shorty Crater by *Apollo 17* astronauts during the Dec. 7-19, 1972, mission. Particles were described by the magazine as 20 to 45 μm in size—"about the size of terrestrial silt." They were "glass, but in a process of devitrification, that is transition from non-crystalline form to crystalline minerals." The orange soil was "rich in titanium—8%—and iron oxide—22%—but is also high in zinc." (*Av Wk*, 2/12/73, cover, 20)
- The U.S.S.R. had launched three times as many reconnaissance satellites as the U.S. during 1972 but had achieved only 25% more time in orbit because of the expanding U.S. use of long-endurance "Big Bird" spacecraft, *Aviation Week & Space Technology* reported. The U.S.S.R. had obtained 366 days in orbit for its Cosmos satellites; the U.S. total was 295 days.

   An improved design of U.S.S.R. reconnaissance satellites had been used for 80% of the launches, the magazine said. The new design provided for ejection of hardware—probably a vernier propulsion unit—one or two days before spacecraft recovery. The propulsion unit was used to change the spacecraft orbital path to bring it directly over the ground target. Spacecraft might also be equipped with high-resolution, narrow-field-of-view cameras for a close-look mission.

   Three of a total eight U.S. launches had been new-generation Big Bird satellites equipped with a giant high-resolution camera, the magazine said. The remaining satellites had been launched by a Titan IIIB-Agena booster, while the number of launches using the Thorad booster had continued to decline. The mission of the Titan IIIB-launched spacecraft was probably changing from mainly close-look to both close-look and search-and-find missions and used as gap fillers for the high-cost Big Birds. (*Av Wk*, 2/12/73, 50–51)

February 12-17: The NASA and Soviet Academy of Sciences Joint Working Group on the Natural Environment met in Moscow to review progress and plan continued cooperation in geology and geomorphology; vegetation, soils, and land use; water, snow, and glaciology; microwave techniques; and oceanology. Delegations were led by NASA Deputy Associate Administrator for Applications Leonard Jaffe and Soviet Academician Yu. K. Khodarev. (NASA Release 73–106)

February 13: Pioneer-G, the second Jupiter-mission spacecraft, arrived at Kennedy Space Center in preparation for the opening of the launch window on April 5. If successfully launched with an accurate trajectory in the first few days of window, the spacecraft, designated Pioneer 11, might be programmed to acquire sufficient speed from the Jupiter flyby to continue to Saturn. When the spacecraft reached the Jupiter vicinity in January 1975, Jupiter and Saturn would be nearly aligned. The spacecraft could, under favorable conditions, acquire part of Jupiter's orbital velocity during the flyby to speed it to Saturn. The predecessor *Pioneer 10* (launched March 2, 1972) had been the first spacecraft to undertake an outer planetary mission. It was due to arrive at Jupiter in December 1974. (KSC Release 27–73)

- NASA turned off *Oao 2* Orbiting Astronomical Observatory at 10:40 pm EST during the 22 000th earth orbit by the spacecraft launched Dec. 7, 1968. *Oao 2*, designed to operate for one year, had "far exceeded the fondest hopes" of NASA officials by operating more than four years. (NASA prog off; AP, *SD Union*, 2/14/73)

February 13

- Establishment of task team to focus on space tug activities was announced by Marshall Space Flight Center. The team, under Manager William Teir and Deputy Manager William G. Huber, would direct early planning and design of the tug—a vehicle to supplement the space shuttle's capabilities. The tug was to be shuttle payload and would become the shuttle 3rd stage after being deployed in earth orbit by the shuttle. Its rocket engine would propel payloads to different earth orbits or send payloads on planetary missions. (MSFC Release 73-19)
- The mission of *Erts 1* Earth Resources Technology Satellite—launched July 23, 1972, to determine feasibility of exploring earth from space—had exceeded all expectations, John N. Wilford said in a *New York Times* article. During seven months since the satellite had been put into a polar, sun-synchronous orbit, it had produced 125 000 photos, surveyed earth resources, and monitored crop growth, glacier advances, and spread of pollution and population.

    With scanners measuring reflected light in both visible and infrared bands, *Erts 1* had discovered previously unmapped fractures branching off the San Andreas fault, prepared a detailed map of cracks and faults in the Wind River Mountains, traced linear terrain features where India had slammed into Asia millions of years earlier, produced accurate maps of the U.S. underground water supply, discovered two hitherto unknown lakes in Iran, and provided better knowledge of the Amazon River basin. (Wilford, *NYT*, 2/13/73, 22)
- NASA launched an Arcas sounding rocket—first in a series of two—from Antigua, West Indies, carrying a Goddard Space Flight Center payload to a 58.1-km (36.1-mi) altitude. The objectives were to measure the ozone distribution in the upper atmosphere, monitor anomalous ultraviolet absorption for the evaluation of the *Nimbus 4* (meteorological satellite launched April 8, 1970) backscatter UV experiment, and extend the data base for a climatology of stratospheric ozone in the tropics. The rocket was launched during a *Nimbus 4* overpass. The rocket and instrumentation performed satisfactorily. (NASA Rpt SRL)
- U.S. Patent 3 715 962 was awarded to Dr. Edward F. Yost, Jr., Professor of Engineering at Long Island Univ., for a multispectral aerial photography system. The equipment, used on *Apollo 9* and *12* (launched March 3, 1969, and Nov. 14, 1969) and on *Erts 1* (launched July 23, 1972), was able to detect water, soil, and agricultural conditions with greater subtlety than previously possible. (Jones, *NYT*, 2/17/73, 41; Pat Off PIO)
- Secretary of Transportation Claude S. Brinegar announced completion of Phase I of the air-route-control-system automation program. Memphis had been the last of 20 cities to install IBM 9020 computers in a nationwide network system to transfer flight data from one facility to another. The network had 61 automated radar terminal systems (ARTS). (FAA Release 73-32)

February 14: Selection of four firms to negotiate contracts to study systems for the space tug (space shuttle 3rd stage) was announced by Marshall Space Flight Center. General Dynamics Corp. Convair Aerospace Div. and McDonnell Douglas Astronautics Co. each would perform a 10-mo study of a cryogenic tug using liquid-hydrogen and liquid-oxygen propellants.

Grumman Aerospace Corp. and Martin Marietta Corp. would perform parallel studies of tug using storable propellants. (NASA Release 73-24)

- The U.S.S.R. weather ship *Priboy* and special Il-18 aircraft arrived in the Bering Sea to take part in the first U.S.S.R.–U.S. research expedition [see Jan. 22]. *Priboy* and Il-18 would meet in the area of Saint Matthew Island with a U.S. Coast Guard icebreaker and NASA Convair 990 aircraft to begin the expedition Feb. 17. (Tass, FBIS–Sov, 2/16/73, L3)
- Cancellation of U.K.'s Hovertrain program was announced by Michael R. D. Heseltine, Minister of Aerospace and Shipping. The project was canceled after 4½ yrs and $12-million development cost because of the excessive expenditures necessary to bring the Hovertrain to a competitive status and because of a lack of a foreseeable market. (Arbose, *NYT*, 2/18/73, 76)

February 15: NASA's *Pioneer 10* probe, on its way to the planet Jupiter, safely completed its crossing of the Asteroid Belt, NASA announced at a Headquarters press conference. Launched March 2, 1972, the spacecraft had completed the 430-million-km (270-million-mi), seven-month trip through the belt with no damaging hits from asteroid particles. Dr. William H. Kinard, Langley Research Center scientist, said NASA was "firmly convinced that the Asteroid belt presents little hazard to future spacecraft going out to explore outer planets." *Pioneer 10* had encountered an average of one large particle of 0.1- to 1-mm (0.004- to 0.04-in) dia a day and had come within 1640 million km (400 million mi) of only one asteroid over 10 m (33 ft) in diameter. Particle encounters had been observed by measuring sunlight scatter of particles by telescopes and by pressurized cells that leaked when penetrated by a particle. *Pioneer 10* project scientist Dr. John H. Wolfe of Ames Research Center said the asteroid distribution was fairly uniform. *Pioneer 10* had made the most distant observations of solar atmosphere and solar wind to date and had found that the sun's atmosphere expanded with distance while the solar wind slowed but heated up with distance. The spacecraft had covered 70% of its 1-billion-km (620-million-mi) flight and would reach Jupiter Dec. 3. After exploring Jupiter's atmosphere it would escape the solar system. Pioneer-G, scheduled for April launch, would travel a similar course to Jupiter. (Transcript; NASA Release 73–27)

- The U.S.S.R. launched *Prognoz 3* from Baykonur Cosmodrome into orbit with a 200 000-km (124 274-mi) apogee, 590-km (367-mi) perigee, 96-hr 23-min period, and 65° inclination. Tass announced the primary objective of the 845-kg (1863-lb) spacecraft was to explore corpuscular, gamma, and x-ray solar radiation; solar plasma flow; and magnetic fields in near-earth space to determine effects of solar activity on the interplanetary medium and magnetosphere of the earth. All spacecraft systems and instrumentation functioned satisfactorily. (GSFC *SSR*, 2/28/73; *Spacewarn*, 3/6/73; FBIS–Sov, 2/15/73, L1)
- North American Rockwell Corp. shareholders voted at annual meeting in Los Angeles, Calif., to change the company name to Rockwell International Corp. The new name dropped all reference to North American Aviation, Inc., a household word in the U.S. and a contractor for the development of the Apollo spacecraft before its 1967 merger with the smaller Rockwell Standard Corp. The company's Aerospace Group would become North American Aerospace Group. Changes would be effective Feb. 16. (NR Release NR–6; *LA Times*, 2/16/73; Ertel, Morse, *The Apollo Spacecraft: A Chronology*, NASA SP–4009)

- Federal Communications Commission Chairman Dean Burch wrote Dr. James C. Fletcher, NASA Administrator, that FCC hoped NASA would continue satellite communications programs to "discharge its statutory responsibilities in the Communications Satellite Act of 1962" and "be a potent factor in maintaining the position of the United States at the very forefront of this new and vital technology." The letter was in response to a Feb. 2 notification to FCC that NASA planned to terminate its comsat programs because of FY 1974 budget restrictions [see Jan. 5]. Burch wrote that FCC had "over the years placed heavy reliance on the advice and technical expertise of your agency with respect to communications services via satellite." NASA also had been "invaluable" to FCC in other fields, including "detailed evaluation of the orbital assignments and space engineering of the various proposals before this Commission in the domestic satellite field." With the use of comsat technology spreading to aviation, maritime, and educational activities, "it would be particularly unfortunate if this Commission were not to be able to receive the same high level advice and assistance upon which it has relied over the past decade." (Text)
- An emergency team put together by NASA, the National Oceanic and Atmospheric Administration, and the Navy's Fleet Weather Facility was aiding Jacques-Yves Cousteau's damaged oceanographic research vessel *Calypso*. The *Calypso*, damaged by an iceberg Jan. 16 while exploring the Antarctic Ocean, would have to navigate the treacherous Drake Passage to the port of Ushuaia, Argentina, for repairs. The best conditions for the crossing passage would be determined from weather and ice floe information supplied by Nimbus and NOAA weather satellite photos and FWF sea ice experts. *Calypso* had been instrumented to receive weather and satellite communications via *Ats 3* (launched by NASA Nov. 6, 1967) by Goddard Space Flight Center for participation in an experiment with NASA [see Feb. 8]. (NASA Release 73-28)
- Dr. Harold A. Rosen, codeveloper with Dr. Donald D. Williams of the first synchronous communications satellite (*Syncom 1*, launched Feb. 14, 1963), was named the "Southern California Inventor of the Year" by the Patent Law Assn. of Los Angeles. Dr. Rosen, manager of the commercial satellite systems division of the Hughes Aircraft Co. Space and Communications Group, held 18 patents for inventions related to the synchronous satellite. He was cited for "revolutionizing the satellite industry." (LA *Her-Exam*, 2/16/73; *SBD*, 2/27/73, 301)
- Rapid detection of contaminated oxygen had been developed by researchers at the Air Force School of Aerospace Medicine, the Air Force Systems Command announced. Contaminated oxygen, which could cause serious illness or death for high-altitude pilots, previously had been detected by analysis of samples at a regional laboratory. The new portable infrared and gas chromatographic analyzers would provide reliable analysis on the spot in 15 min. (AFSC Release 014.73)
- President Nixon submitted to the Senate the nomination of Assistant Secretary of Labor for Labor-Management Willie J. Usery, Jr., to be Federal Mediation and Conciliation Director. Before joining the Dept. of Labor, Usery had been special representative of the International Assn. of Machinists at the Cape Canaveral Air Force Test Facility, IAM representative on the President's Missile Sites Labor Committee at Kennedy Space Center and at Marshall Space Flight Center, and coordinator for union activities at Manned Spacecraft Center. He had helped to form the Cape

Kennedy Labor-Management Relations Council in 1967 and had become its first Chairman in 1968. The nomination was confirmed March 29. (*PD*, 2/19/73, 154, 156; Usery Off)

February 16: Scientists at the State Univ. of New York had concluded the orange soil found on moon by *Apollo 17* Astronauts Eugene A. Cernan and Harrison H. Schmitt was billions of years older than first thought, the *New York Times* reported. The discovery had originally led the scientific community to believe the moon might have been volcanically active into geologically recent times, but Dr. Oliver A. Schaeffer, Professor of Earth and Space Sciences at State Univ., had said, "It can now be reasonably stated that volcanism . . . ended about three billion years ago." The team had tested the orange soil sample by potassium-argon dating. By measuring atomic properties, the soil had been found to be 3.71 billion yrs old. It was not known whether soil was of volcanic origin. (Rensberger, *NYT*, 2/16/73, 46)

- The Manned Spacecraft Center had awarded a shuttle contract to Charles S. Draper Laboratory at Massachusetts Institute of Technology, the MSC *Roundup* reported. The Laboratory had received a $1 375 484 cost-no-fee contract to provide technical support for guidance, navigation, and control subsystems in space shuttle program. (MSC *Roundup*, 2/16/73, 1)

- Plans to operate the new Trident nuclear-powered submarine with 24 missiles of 6500- to 9700-km (4000- to 6000-mi) range from the Pacific Ocean were announced by Secretary of the Navy John W. Warner. Facilities at the Bangor Annex of the Naval Torpedo Station Keyport near Bangor, Wash., would be expanded as base for Trident, to broaden the area of operations of deterrent submarines. (DOD Release 84–73; *NYT*, 2/17/73, 30)

- A *Science* editorial criticized the Nixon Administration's abolition of the post of Presidential Science Adviser and the Office of Science and Technology: The deed had been done in a way "not worthy of a great nation." The office had first been abolished, "then someone woke up to the fact that it served important functions. After scrambling around, someone had the inspiration to transfer the functions of the office to the National Science Foundation (NSF) and appoint [Dr. H.] Guyford Stever (head of the NSF) as science adviser. The solution has merit. However, if it is to represent more than a gesture, Stever and the NSF will be overloaded with conflicting responsibilities." (Abelson, *Science*, 2/16/73, 641)

February 17: The NASA–U.S.S.R. joint mission to measure ice, surface, and atmospheric conditions in the Bering Sea began, following arrival at the experiment site of the U.S. Coast Guard ship *Staten Island*, NASA instrumented Convair 990 *Galileo* aircraft, and Soviet weather ship *Priboy* and Il–18 aircraft. (FBIS–Sov, 3/6/73, L1)

- President Nixon signed S.J.R. 37 into Public Law 93–8, redesignating the Manned Spacecraft Center the Lyndon B. Johnson Space Center in honor of the late President, who died Jan. 22. The President said: "Few men in our time have better understood the value of space exploration than Lyndon Johnson." As senator, Johnson had written, introduced, and helped to enact legislation which created NASA. He had called NASA the "proudest legislative achievement" of his congressional career. As Vice President and Chairman of the National Aeronautics and Space Council, he had served in the "early years of exploration when the groundwork was laid, and the determination made to put a man on the moon." As

President, Johnson had overseen the first Apollo flights, "in a way that led people beyond the adventure and the pride to the deeper meaning and the deeper benefits of space exploration." President Johnson "by his vision and his work and his support" had drawn America up closer to the stars "and before he died he saw us reach the moon—the first great plateau along the way." (PD, 2/26/73, 160)

- The U.S.S.R. had fired nine intercontinental ballistic missiles within two weeks over its Asian territory north of the People's Republic of China, the *New York Times* reported. Nixon Administration officials had said this suggested a step-up in Soviet training of crews for missiles with potential targets in China. (Beecher, *NYT*, 2/17/73, 4)

February 19: NASA was measuring aerodynamic noise made by aircraft in unpowered flight, to solve noise problem in communities surrounding airports. Lockheed-California Co. studies made for NASA and the Navy had indicated that aerodynamic (nonengine) noise might be a significant portion of total aircraft noise. Knowledge of sources could aid in designing quieter aircraft. Tests had been made with twin-engine, propeller-driven aircraft at Flight Research Center. Flights with twin-engine jet aircraft were planned for late January; flights with four-engine jet aircraft were being considered. (NASA Release 73–22)

- Lockheed U–2 high-altitude reconnaissance aircraft equipped with sensors and data-transmission link had begun flight tests off the Southern California coast, *Aviation Week & Space Technology* reported. In one flight, an X-band commercial weather radar modified by Lockheed had detected a surfaced submarine at distance of 240 km (150 mi). (*Av Wk*, 2/19/73, 15)

- Five hundredth anniversary of the birth of Polish astronomer Nicolaus Copernicus, who was the first to propound the theory that the earth moved around the sun. A *New York Times* editorial said: "Today, after centuries of development of Copernican ideas we know that the earth and this solar system are a relatively unimportant byway in a universe with billions of stars and of similar solar systems. Statistically, the probability is high that there are other sentient forms of life in the universe, separated from us by uncountable light years and perhaps exceeding our species in understanding. Copernicus thus should be honored today both as the founder of modern science and as a pioneer of a more realistic understanding of man's place in the universe." (*NYT*, 2/19/73)

February 20: A full-scale mockup of the Skylab Orbital Workshop—shipped from McDonnell Douglas, Huntington Beach, Calif.—arrived at NASA's Marshall Space Flight Center. It would be combined with other Skylab hardware and used for system engineering and integration support to Skylab earth orbital missions scheduled to begin in May. (MSFC Release 73–14)

- The Office of Management and Budget released $2.53 million of NASA funds appropriated for the quiet, clean engine development in FY 1973 but impounded by OMB in 1972. OMB also released $2 million appropriated for NASA's quiet, experimental, short takeoff and landing aircraft (QUESTOL) in FY 1973. The project had since been terminated [see Jan. 5]. (OMB PIO)

- *Apollo 17* Astronauts Eugene A. Cernan, Ronald E. Evans, and Dr. Harrison H. Schmitt visited Kennedy Space Center to thank employees for their efforts in making manned exploration of the moon possible. Dr. Schmitt said man would explore Mars in the not too distant future. "The reason it is possible to talk this way and not be accused of talking science fiction is because there's no longer science fiction. It's now only science prediction." (UPI, LA *Her-Exam*, 2/21/73)

- The U.S.S.R. called for participation of the People's Republic of China and France in negotiations on an accord to end all nuclear weapon testing. During his opening statement at the first 1973 session of the Geneva Disarmament Conference, Soviet delegate Aleksey A. Roshchin said negotiations to end nuclear tests "require the participation of all nuclear states." Neither China nor France had shown any interest in joining the conference. Progress in the talks continued to be blocked by the issue of onsite inspection to detect nuclear explosions. The U.S.S.R. opposed international controls and the U.S. supported them. (*NYT*, 2/21/73, 7)

- NASA launched an Arcas sounding rocket—second in a series of two—from Antigua, West Indies, carrying a Goddard Space Flight Center payload to a 52.5-km (32.6-mi) altitude. The objectives were to measure the ozone distribution in the upper atmosphere, monitor anomalous ultraviolet absorption for the evaluation of the *Nimbus 4* (meteorological satellite launched April 8, 1970) backscatter UV experiment, and extend the data base for a climatology of stratospheric ozone in the tropics. The rocket was launched during a *Nimbus 4* overpass. The rocket and instrumentation performed satisfactorily. The first rocket in this series was launched Feb. 13. (NASA Rpt SRL)

February 21: The emergency team formed by NASA, the National Oceanic and Atmospheric Administration, and the Navy's Fleet Weather Facility to aid Jacques-Yves Cousteau's crippled research vessel *Calypso* [see Feb. 15] had been disbanded, NASA announced. The *Calypso* had safely navigated the dangerous Drake Passage between Antarctica and the Ushuaia port, Argentina, using weather and ice information supplied by *Ats 3* satellite. Satellite photos had been processed at Goddard Space Flight Center, turned over to FWF, and analyzed and information transmitted to Cousteau aboard the *Calypso*. (NASA Release 73-33)

- Rep. J. Herbert Burke (R–Fla.) introduced three resolutions: H.J.R. 358 to provide for a portion of the moon to be displayed in the Capitol; H.J.R. 359 to establish an Astronauts Memorial Commission to construct a memorial at Kennedy Space Center in honor of U.S. astronauts; and H.J.R. 362 to redesignate Cape Kennedy Cape Canaveral. (*CR*, 2/21/73, H1061)

February 22: Presidential Reorganization Plan No. 1, abolishing the Office of Science and Technology and the National Aeronautics and Space Council [see Jan. 26 and July 1], was reviewed by the Senate Committee on Government Operations' Subcommittee on Executive Reorganization and Government Research. Chairman Abraham A. Ribicoff (D–Conn.) said, "Historically, Congress has allowed the President wide latitude in organizing his own office and so far no opposition has developed to this plan," but added that questions had arisen as to whether the plan downgraded the voice of the scientific community within the Executive branch. Frederick V. Malek, Deputy Director of the Office of Management and Budget, said it was "no longer necessary to have a single office within the Executive Office specifically directed to science and technology." National Science Foundation Director H. Guyford Stever, newly designated Science Adviser to the President, said that NSF had "many inputs to the scientific community and they have many inputs to NSF" and that he expected the routes for advice to remain open. (*CR*, 2/22/73, D128; *Sci & Govt Rpt*, 3/1/73, 7)

- NASA announced expansion of its program to sponsor research at four-year colleges and universities with predominantly minority enrollments. One hundred institutions had been invited to consider whether and how they might participate by submitting proposals relevant to NASA's mission, contributing to solution of problems of concern to agency, and of technical merit. The expanded program had been developed in cooperation with the National Association for Equal Opportunity in Higher Education. (NASA Release 73-25)

- *Apollo 17* geologist-astronaut Dr. Harrison H. Schmitt was among 10 young Government employees named to receive the 25th annual Arthur S. Flemming Award for outstanding contributions to Federal management and science, the Washington *Star and News* reported. The Schmitt citation was for "total performance as a scientist and astronaut, including geological training of early Apollo flight crews, and as lunar module pilot for *Apollo 17*." Awards would be presented at a Washington, D.C., luncheon sponsored by the Downtown Jaycees Feb. 29. (*W Star & News*, 2/22/73; NASA Special Notice, 2/26/73)

- The Federal Women's Award program committee announced six winners of its 13th annual awards for outstanding careers in the U.S. Civil Service. Goddard Space Flight Center's Small Astronomy Satellite Project Manager, Mrs. Marjorie R. Townsend—the only woman to manage a U.S. satellite program—was among those who would receive the award at a March 6 banquet in Washington, D.C. *Explorer 42* (*Uhuru*), first satellite launched in the program (Dec. 12, 1970), had provided the first complete picture of the sky in x-rays, bringing a major advance in astronomy. (*W Post*, 2/22/73; NASA Hq *WB*; *A&A 1970*)

- The National Space Club held its Awards Luncheon in Washington, D.C. The Dr. Robert H. Goddard Historical Essay Award of $500—named for the U.S. rocket pioneer—was presented to Barton C. Hacker of Iowa State Univ. for his essay "From Space Station to Orbital Operations in Space Travel Thought, 1895–1951." The Dr. Hugh L. Dryden Memorial Fellowship award of $2000, presented annually to an "individual . . . pursuing the challenges of research, exploration, and administration in astronautics, so that he may continue to seek and maintain pre-eminence in outer space" for the U.S., was given to Dr. Palmer Dyal of NASA Ames Research Center Electrodynamics Branch. The award was named for the first NASA Deputy Administrator, who died in 1965. The Dr. Robert H. Goddard Scholarship of $2000 was presented to David O. Starr of Catholic Univ. (Program; NSC *News Letter*, 2/22/73)

- Analysis of *Apollo 15* and *16* data had shown that certain lunar craters, including Aristarchus, emitted gas puffs, the *New York Times* reported. Data from alpha particle experiments had indicated the events occurred in dark-floored or rimmed craters. Most events recorded had occurred when the moon was nearest earth orbit—the same time at which moonquakes occurred. (Sullivan, *NYT*, 2/22/73)

- An experimental connector smaller than a dime had been developed by Kennedy Space Center engineers to apply small electrical currents from an outside power source to move paralyzed human muscles, NASA announced. The connector—which would detach easily but provide good electrical contact while attached—was being tested at Rancho Los Amigos Hospital in Downey, Calif. The NASA Office of Technology Utilization was supporting the hospital's research on application of high-purity vitreous carbon—developed originally for rocket engine linings—to implants of tiny platinum wires attached to small pads on human nerve endings. Earlier connectors had damaged human skin during the process of detachment and had cost $1000 each. The KSC-developed connector detached readily and cost $50 to $100 each. (NASA Release 73-30)

February 22-23: Marshall Space Flight Center's Science and Engineering Directorate sponsored a Research and Technology Review. The purpose was to expand the familiarity of scientists and engineers in government, industry, and the academic community with activities and accomplishments at MSFC having significant applications in space projects. Topics discussed were space sciences, manufacturing in space, optical contamination, materials and manufacturing, propulsion, computation and simulation, control and atmospheric flight mechanics, electronics communications and instrumentation, and apparatus. (MSFC Release 73-16; 73-21)

February 23: Gamma ray spectrometers on *Apollo 15* and *16* had mapped the moon's radioactivity over 20% of the lunar surface, scientists from Jet Propulsion Laboratory, Goddard Space Flight Center, and Univ. of California at San Diego reported in *Science*. Highest levels of natural radioactivity had been found in Mare Imbrium and Oceanus Procellarum with contrastingly lower enhancements in eastern maria. The rate of potassium to uranium was higher on far side of moon than on near side, "although it is everywhere lower than commonly found on earth." (Metzger *et al.*; *Science*, 2/23/73, 800-3)

• Results of the *Mariner 9* ultraviolet spectrometer experiment to measure seasonal variation of ozone on Mars were reported in *Science* by Univ. of Colorado and Cal Tech scientists. Observations made between Nov. 14, 1971, and Oct. 27, 1972, had shown ozone to be present in polar regions of Mars and to have seasonal variation. In summer, the amount in the polar atmosphere was less than three micrometer-atmospheres. In autumn, ozone increased and was found in association with formation of the polar hood. In winter, the maximum amount of ozone was present, "57 micrometer-atmospheres over the polar hood and 16 over the polar cap. In spring, the amount over the polar cap decreases monotonically until by the beginning of summer the ozone disappears. Ozone is not observed in the equatorial region during any season." (Barth *et al.*, *Science*, 2/23/73, 795-6)

• Dr. Joseph V. Charyk, President of Communications Satellite Corp., announced organizational changes to strengthen new roles in domestic and international satellite communications.

Three new vice presidents were elected by ComSatCorp's Board of Directors: David C. Acheson, Senior Vice President and General Counsel; John A. Johnson, Senior Vice President; and George P. Sampson, Senior Vice President in charge of International Systems Division.

The subsidiary COMSAT General Corp. was formed to carry out ComSatCorp's U.S. domestic programs. It would be headed by John A. Johnson, President; Joseph H. O'Connor, Vice President and Treasurer; John L. Martin, Jr., Vice President; and Jerome W. Breslow, Secretary.

ComSatCorp had formed the International Systems Div. to consolidate all activities related to the International Telecommunications Satellite Organization's Global Satellite System. The division would be headed by George P. Sampson, Senior Vice President in charge of International Systems Div.; Richard R. Colino, Assistant Vice President for Technical Development and Support; and H. William Wood, Assistant Vice President for International Systems Operations. (ComSatCorp Release 73-9)

- The Philadelphia *Evening Bulletin* editorial commented on the name change of Manned Spacecraft Center to Lyndon B. Johnson Space Center: "While the U.S. is cutting back its spending in space, this nation and the world can never forget the adventure and exploration of the Sixties and the person largely responsible, after John F. Kennedy's death, for getting man to the moon." It was therefore "appropriate that Mr. Johnson be honored. . . . At one point Mr. Johnson was reported to have called the creation of NASA his 'proudest' legislative achievement as a congressman. The Lyndon B. Johnson Space Center will keep that achievement alive." (P *Bull*, 2/23/73)

- The U.S.S.R. had abandoned its women's cosmonaut training program, the Washington *Evening Star and Daily News* reported. Gen. Georgy T. Beregovoy, commander of the Soviet cosmonauts, had said in an interview with the Polish newspaper *Express Wieczerny* in Warsaw, "We train only young men, not women." (UPI, W *Star & News*, 2/23/73, A3)

February 24: A Baltimore *Sun* editorial commented on the Feb. 15 press conference on the *Pioneer 10* Jupiter probe (launched March 2, 1972): "There was reason to worry for Pioneer's wellbeing, as it traveled the section beyond Mars; so many chunks of rock float along there, on such erratic tracks, that our expensive new-model spacecraft could have inadvertently smacked into something a great deal bigger than itself. Even a sideswipe might have thrown it off course. Happily, it has registered contact only with the sort of fine-grained debris that litters space in general." *Pioneer 10* had indicated the existence of more asteroids between the earth and Mars than previously realized and fewer between Mars and Jupiter. "Or is it so new a phenomenon, that the closer to earth's big cities you come, the thicker the traffic gets?" (B *Sun*, 2/24/73)

February 25: The 18th U.S.S.R. Antarctic expedition of 100 geophysicists, aerometeorologists, glaciologists, and medical workers launched its first sounding rocket from Molodezhnaya Observatory to measure the intensity of corpuscular radiation. A computer would be set up on site to analyze the data. (*Izvestiya*, FBIS-Sov, 3/2/73, L1)

February 26: Representatives from 16 companies had visited Lewis Research Center's Plum Brook Station to interview personnel affected by the Station's closing, which would begin July 1973, LeRC announced. A special Outplacement Service Office, set up at LeRC, had prepared resumés, sought job opportunities, and arranged interviews for highly trained displaced engineering and technical personnel. (LeRC Release 73-10a)

- Johnson Space Center announced award of a one-year, $9 300 000 contract to Northrop Services, Inc., for operational and maintenance support services to JSC laboratories and test facilities. (JSC Release 73-21)

- Sen. John J. Sparkman (D-Ala.), for himself and Sen. James B. Allen (D-Ala.), introduced S.J.R. 69 to authorize the President to proclaim

July 20 of each year as National Space Day commemorating July 20, 1969, lunar landing of *Apollo 11*. (*CR*, 2/26/73, S3213)

- President Nixon announced the designation of Robert D. Timm Chairman of Civil Aeronautics Board, effective March 2. He would succeed retiring Secor D. Browne. The President also accepted the resignation of the first NASA Administrator, Dr. T. Keith Glennan, as U.S. Representative to International Atomic Energy Agency. He submitted to the Senate the nomination of Alexander P. Butterfield to be Federal Aviation Administrator. (*PD*, 2/26/73, 189, 244; 3/5/73, 192)

February 26–March 3: The NASA and Soviet Academy of Sciences Joint Working Group on Space Biology and Medicine met in Moscow to exchange information on preliminary biomedical results of the *Apollo 17* mission (Dec. 7–9, 1972) and Soviet research in weightlessness modeling. The Group agreed on common procedures to test body negative pressure, to make active orthostatic evaluations, and to make biochemical blood and urine studies, permitting comparison of pre- and postflight data on body functions. The Group also discussed biological experiments in space and countermeasures to weightlessness. The U.S. delegation—headed by Dr. Charles A. Berry, NASA Director of Life Sciences—visited Soviet medical research facilities and the Gagarin Center for Cosmonaut Training. They examined spacecraft trainers, medical equipment used for cosmonaut training and evaluation, and medical instrumentation used on Salyut spacecraft. The Soviet delegation was headed by Dr. R. N. Gurovsky. The next meeting would be held in the U.S. in late 1973 or early 1974. (NASA Release 73–79)

February 27: The House Committee on Science and Astronautics' Subcommittee on Manned Space Flight began hearings on H.R. 4567, House version of the $3.016-billion NASA FY 1974 authorization bill. Dale D. Myers, NASA Associate Administrator for Manned Space Flight, said the "historic accomplishments of the Apollo program" were "an open record and an enduring tribute to a basic belief in human progress. The Apollo flights, in three short years, gave us an order of magnitude increase in our knowledge of the solar system. Although the epic Apollo voyages have ended, the results of these missions will provide the scientific, the technical, and medical, and the managerial communities with a rich store of data that will be studied and analyzed for many years to come." NASA had completed the Apollo program and "moved from the era of learning how to live and work in space to a new plateau, where this Nation can utilize space and its unique capabilities for expanding its horizons in science and in applications, in defense, commercial activities and in international cooperation at reduced costs." Skylab—scheduled for May launch—would be the "first post-Apollo step into the intensive utilization of space" and toward reorienting manned space activities to earth orbit. The Apollo-Soyuz Test Project (ASTP), scheduled for July 1975 launch, would be "a major step forward in meaningful and beneficial international cooperation in manned space flight" and a significant step in space safety and in reducing costs by cooperation. Later, the space shuttle would provide "ample additional opportunities for meaningful technical and economic cooperation in space." One opportunity would be to carry sortie lab being developed by group of European nations using their own funds. Closing out of Apollo activity would include "very careful screening of available production tooling for applicability to future shuttle or other NASA require-

February 27

ments." Available flight articles included two Saturn V and two Saturn IB launch vehicles, one completed command and service module, two partially assembled CSMs, and a backup Skylab cluster. This hardware would be placed in storage for potential use. (Transcript)

- A satellite station for international communications being built near Belgrade, Yugoslavia, would be completed by Oct. 21, the Belgrade *Politika* newspaper reported. The station, containing $4-million equipment, would be able to make direct contracts with many countries without the expense of buying cable communications. Japan had designed and built the equipment and would train Yugoslav engineers to operate it. (FBIS–Sov, 3/9/73, I19)

- The Air Force announced award of a $59 000 000 firm-fixed-price contract to Boeing Co. for the first two 747B jet transport aircraft for the Advanced Airborne Command Post Program. If Congress appropriated funds, a total of seven 747Bs would replace EC–135 airborne command posts of the National Military Command System and Strategic Air Command. The seven would be first 747s used by Air Force. (DOD Release 96–73)

- The Brevard County, Fla., Department of Civil Defense had obtained a 1-million-gal water tank valued at $60 000 from Kennedy Space Center under a Federal program which offered excess Government equipment to state, county, and local government agencies free of cost, KSC announced. The tank had been used to pump water onto Launch Complex 34 and 37 pads after Saturn launches to protect them from flame damage. Brevard County would use the tank to hold an emergency water reserve. (KSC Release 36–73)

February 27–May 15: The U.S.S.R. launched a series of meteorological sounding rockets from the French island of Kergulen in the Indian Ocean. Twenty-four rockets reached altitudes between 71 and 76 km (44 and 47 mi) under a Soviet-French agreement for joint research in space meteorology and aeronomy signed in 1972. The flights were to record changes in temperature, atmospheric pressure, and wind at high altitudes to aid in weather forecasting. (Tass, FBIS–Sov, 5/17/73, L1)

February 28: The U.S.S.R. launched *Cosmos 549* from Plesetsk into orbit with 723-km (449.3-mi) apogee, 516-km (320.6-mi) perigee, 92.2-min period, and 73.9° inclination. (GSFC SSR, 2/28/73; SBD, 3/1/73, 7)

- Dr. James C. Fletcher, NASA Administrator, compared the NASA FY 1974 budget program with FY 1973 plans in testimony before the opening session of the Senate Committee on Aeronautical and Space Sciences hearings on the FY 1974 NASA authorization: "Last year Congress approved $3.4 billion, the full amount of the President's recommendations. In presenting the budget last year, I was able to state to the Congress that the planning associated with the $3.4 billion budget request was configured so as not to require an increase in future NASA budgets above the $3.4 billion level, . . . except as necessary to meet the effects of inflation, or unless new decisions were made to expand the program." NASA then had had "a balanced and productive program which did not include any 'built-in' commitments that would require a higher total level of NASA funding in future years."

NASA's FY 1974 budget represented a change in this plan. "To meet the necessary constraints on total Government spending in FY 1973 and FY 1974 we have had to reduce our budget plans below the $3.4 billion level by a total of almost $400 million for the 2-year period." But program objectives were generally the same, "even though some of them will necessarily be achieved at later dates. For this reason, NASA budgets in future years will have to increase. . . . This need is recognized in the preliminary projections for FY 1975." NASA was "handling the reductions in our FY 1973 and FY 1974 budget plans in a way that will not increase the levels of future funding required above those projected last year." Dr. Fletcher believed this to be minimum level for space and aeronautics program U.S. should support. The program's major elements—space shuttle, unmanned planetary and scientific missions, space applications and aeronautics program components, and others—"are ones that have been presented and approved in previous years." NASA had "squeezed and stretched the program about as far as we can. In some of the reductions we have made in FY 1973 and FY 1974 we may have gone too far. Any further significant reductions would mean a real loss in a key element of the program."

NASA was requesting FY 1974 funds for an advanced supersonic technology program to provide technology for future military supersonic aircraft, data to assess environmental and economic impacts on the U.S. of present and future foreign supersonic aircraft, and a "sound technical basis for any future consideration that may be given by the United States to the development of an environmentally acceptable and economically viable commercial supersonic transport." The request did not include funds for initiating SST development and did not commit the U.S. to development.

Dr. Fletcher was accompanied by Astronauts Thomas P. Stafford and Charles Conrad, Jr., as well as other NASA officials. (Transcript)

- Rep. Thaddeus J. Dulski (D–N.Y.) introduced H.R. 4927 to amend Title 5 of the U.S. Code to make Level V of the Executive Schedule applicable to three additional NASA positions. The action was taken following receipt by the House Speaker of a letter from Dr. James C. Fletcher, NASA Administrator, requesting legislation to double the number of NASA Level V positions. Dr. Fletcher requested authority to establish titles of Associate Administrators "in a manner consistent with evolving functional responsibilities as they may develop and be altered . . . ; and to add three such positions to provide salary equality among all Associate Administrators having comparable responsibilities and authorities. Legislation was required because reorganization of NASA offices had created six offices headed by Associate Administrators, rather than three. (CR, 2/28/73, H1244)

- The Air Force announced award of a $27-million fixed-price-incentive-fee contract to Philco-Ford Corp. to build and test two communications satellites for the North Atlantic Treaty Organization. The satellites, scheduled for 1975 launch, would be space segments of NATO's Phase III Space Communications Satellite System and would handle long-range communications requirements. (DOD Release 102–73)

- The House Committee on Science and Astronautics' Subcommittee on Manned Space Flight continued hearings on the FY 1974 NASA authorization [see Feb. 27]. Skylab Program Director William C. Schneider described progress of the Skylab project at Kennedy Space Center: "The flight hardware has undergone intensive check-out at the module level including docking tests to verify the critical interface between the Multiple Docking Adapter and the Command and Service Module." Stacking of the Saturn V and Saturn IB launch vehicles "has proceeded apace and end-to-end integrated systems tests of the orbital assembly have been completed." Activities supported the mid-May dual launch, with rollout to Pad 39B scheduled for mid-April. The Skylab pyramid test program, "starting at the component qualification level and building up through the levels of subsystem and module testing, through acceptance reviews and checkouts, to the countdown demonstration, provides the confidence needed to launch and operate America's first space station."

Deputy Associate Administrator for Management Harry H. Gorman in the Office of Manned Space Flight discussed the NASA personnel changes brought by completion of the Apollo program. The research, development, and operations contract support effort during the Apollo program had peaked in FY 1968 at 17 500 persons. The FY 1974 budget request would provide employment at a 9500 level—a 45% reduction. The Civil Service staff would decrease by 825 positions in FY 1974, to a total of 10 525. The reduction was also related to completion of Skylab, termination of communications lead center activity, and suspension of the High Energy Astronomy Observatory program. (Transcript)

- NASA-developed technology to protect nuclear rocket nozzles from vibration was being used to modify automobile engines to reduce exhaust fumes, at Lewis Research Center. The project had begun in 1970 when the Environmental Protection Agency asked NASA to study materials for an automotive thermal reactor—a chamber in which hot engine exhaust was mixed with air to complete combustion of unburned gasoline, converting carbon monoxide and hydrocarbons into carbon dioxide and water. LeRC engineers were testing brittle ceramic materials for a reactor-system combustion chamber lining by cushioning materials with thin corrugated metal springs similar to those used during nuclear rocket nozzle testing. (LeRC Release 73–11)

- Dr. Robert C. Seamans, Jr., Secretary of the Air Force, announced that the Air Force had received congressional approval to proceed with the FY 1973 production program of 30 F–15 air superiority fighter aircraft. Four test aircraft had flown 270 successful hours since July. Production of an additional 77 was planned for FY 1974, pending congressional approval. (DOD Release 101–73)

- NASA launched an Aerobee 150 sounding rocket from Churchill Research Range carrying a Univ. of Michigan aeronomy experiment to a 179.9-km (111.8-mi) altitude. The rocket and instrumentation performed satisfactorily. (GSFC proj off)

- The National Science Foundation published *NSF Forecasts Rise in Company-Funded Research and Development and R&D Employment* (NSF 73–301). Company-funded R&D expenditures were projected to increase by 22% between 1972 and 1975, to a record level of $14 billion annually by 1975. A moderate upswing in employment of scientists and engineers on company R&D programs was anticipated. Industrial firms had spent estimated $11.4 billion on R&D during 1972, 7% more than in 1971. Between 1970 and 1971 company-financed R&D rose 4%. Between 1970 and 1972 the number of scientists and engineers employed on company R&D programs rose slightly, to 240 000; an increase to 260 000 was anticipated by 1975. Aerospace firms foresaw future R&D growth below the rest of the industry. The downward trend of employment on company-supported basic research had stabilized; it was expected to rise 10% between 1972 and 1975. (NSF *Highlights,* 2/28/73)

During February: Instrument technicians from Flight Research Center removed transducers from the XB-70 experimental aircraft on static display at the Air Force Museum at Wright-Patterson Air Force Base, Ohio, for installation in the YF-12 triple-sonic aircraft shortly to participate in a flight research program. Use of XB-70 transducers in the high-temperature environment to be encountered by new YF-12 aircraft would save Government more than $950 000 in instrument development and production costs. (FRC Release 3-73)

- The S.S. *Hope,* hospital ship of the People-to-People Health Foundation sailed from Baltimore and arrived in Maceio, Brazil, after having been equipped for communicating via satellite. Under an agreement between Communications Satellite Corp. and the Foundation, the ship had been fitted with a small, parabolic antenna and a transmit-receive comsat terminal to assess the use of reliable long-distance communications with medical teams in remote areas. (ComSatCorp brochure)

- The National Academy of Sciences released *Biological Impact of Increased Intensities of Solar Ultraviolet Radiation.* The report by the NAS Environmental Studies Board concluded that routine use of supersonic aircraft would increase risk of skin cancer, harm many food crops, diminish biological productivity of oceans, and adversely affect many useful insects. Emission of nitrogen oxide from SST engine exhaust would lessen ozone content of the upper atmosphere and increase the amount of UV radiation reaching the earth. In addition to increased cancer risk, UV rays could also change deoxyribonucleic acid (DNA), genetic messenger of cells, causing harmful mutations in future generations. (NAS-NRC Release, 7/24/73; Schmeck, *NYT,* 2/12/73, C11)

- The National Academy of Sciences released *Plans for U.S. Clear-Air Turbulence Research in the Global Atmospheric Research Program.* The report of a panel of the U.S. Committee for the Global Atmospheric Research Program (GARP) recommended that increased Federal logistic support be provided for clear-air turbulence studies in planning, because of importance to aviation safety and to understanding effects of small-scale motions on large-scale dynamics of atmosphere. The Dept. of Defense should make available additional C-130 and RB57 aircraft for vertical-motion measurements in Project Wamflex (wave-induced momentum flux experiment) of the National Center for Atmospheric Research. The National Oceanic and Atmospheric Administration's continuous-wave radar should be modified to detect clear-air turbulence in the tropopause. And NASA should provide additional instrument support including scaning lidar. (Text)

- The Federal Aviation Administration and industry Area Navigation Task Force issued the report *Application of Area Navigation in the National Airspace System.* It recommended a 10-yr program to establish area navigation as the primary method of aircraft navigation in the U.S. The program called for a total overhaul of the airspace route structure above 5000 m (18 000 ft) and redesign of high- and medium-density terminal areas. Small airborne computers would enable pilots to fly designed courses by extrapolating signals from ground navigation aids, freeing them from the necessity of flying directly between ground stations and allowing greater flexibility in route selection. (FAA Release 73-45)

During February

- A new low-cost high-performance sounding rocket was being developed by Bristol Aerojet Ltd. and Instituto Nacional de Técnica Aeroespacial of Spain, *Spaceflight* announced. The two-stage rocket, INTA 300, would lift a 50-kg (110-lb) payload to 320 km (200 mi) with low dispersion. First test flights would be held in 1974. (*SF*, 2/73, 69–70)

March 1973

March 1: Cosmos 550 was launched by the U.S.S.R. into orbit from Plesetsk with a 313-km (194.5-mi) apogee, 204-km (126.8-mi) perigee, 89.7-min period, and 65.4° inclination. The satellite reentered March 11. (GSFC *SSR*, 3/31/73)

- Contractor project definition and cost (Phase B2) studies were begun in Europe on the European Space Research Organization (ESRO) Spacelab (sortie lab) for NASA's space shuttle. (NASA Release 73-191)
- The House Committee on Science and Astronautics' Subcommittee on Manned Space Flight resumed hearings on the FY 1974 NASA authorization.

    Dale D. Myers, NASA Associate Administrator for Manned Space Flight, testified that a thorough analysis of the space shuttle's possible environmental effects had been made in 1972 and a report had been filed with the Environmental Protection Agency. The shuttle's effect on atmosphere, water, and noise would be "minimal and below allowable limits." Safeguards to be added would further minimize any potential environmental impact. NASA and Dept. of Defense studies with space-experienced companies to determine the reduction in satellite costs by use of the shuttle for launch and retrieval had shown "payload costs reductions of 40 to 50 percent" from "present-day payload costs." Plans for FY 1974 called for "an orderly buildup" in prime and subcontractor manpower for the orbiter and space shuttle main engine development and award of contracts for all remaining major elements of the shuttle, including the external tank and solid rocket booster. FY 1974 funds, although less than requested, would provide for "an expanded scope of design, development, and testing activities and for continuation of subsystems and component development."

    Director Douglas R. Lord of NASA's Sortie Lab Task Force testified that the lab would be designed for installation in the space shuttle's cargo bay "to allow nonastronaut scientists to use nearly conventional laboratory equipment at orbital altitudes." The lab would consist of a closed, pressurized module for experimenters and their laboratory subsystems and apparatus, as well as a pallet section exposed to space when the payload bay doors were opened, for large sensors requiring space exposure, such as telescopes. The pallet could be used either with the pressurized module or separately, mounted in the payload bay and supported by the shuttle orbiter. The sortie lab would provide "an opportunity for international participation, . . . for the first time in the manned program, in the initial development of a completely new capability." (Transcript)

- NASA Associate Administrator for Applications Charles W. Mathews testified on NASA's FY 1974 applications program in authorization hearings before the House Committee on Science and Astronautics' Subcommittee on Space Science and Applications. With communications satellite services being provided throughout the world "on a highly useful, economic, and profitable basis," NASA had turned its emphasis to earth resources observations from satellites. "Perhaps the most significant event occurred last July [23] with the launch and successful operation of the first Earth resources technology satellite, . . . whose very special images of the Earth's surface are being provided to users and investigators here and throughout the world every day." More than 300 investigations were under way and "important uses are being defined in almost every area under study." The *Noaa 2* satellite launched by NASA for the National Oceanic and Atmospheric Administration Oct. 15, 1972, "afforded for the first time the measurement of temperatures at various heights in the atmosphere throughout the entire world on a truly operational basis." The information was being rapidly provided to the U.S. Weather Service for prediction programs. In addition, more precise images of global cloud cover were being obtained night and day. *Noaa 2*, in conjunction with the continuation of the experimental program through the launch of *Nimbus 5* in December, "bodes well for continued progress in meteorology." New efforts in earth observations to start under the FY 1974 budget were in pollution monitoring, oceanography, and earth and ocean physics.

In technology applications, NASA was supporting the Dept. of Housing and Urban Development on integrated utility systems to minimize energy consumption and waste output in housing units and the Dept. of Transportation in urban mass transportation. Special programs also included manufacturing in space.

In a new lead-center concept, responsibility for support in particular disciplines had been assigned to Centers: Johnson Space Center—Earth Resources Center; Goddard Space Flight Center—Weather and Climate Center; and Ames Research Center—applications aircraft program. The Applications Program Integration Board (APIB) had been established to evaluate major new proposals for applications or modifications to major ongoing efforts. (Transcript)

- NASA held a Hq. press conference with French oceanographer Jacques-Yves Cousteau on the joint NASA and National Oceanic and Atmospheric Administration project in which Cousteau and his oceanographic research ship *Calypso* had participated [see Feb. 8, 15, and 21]. Cousteau had gathered data on ocean temperature, color, and breeding areas which would be compared with data from satellites in the NASA Applications Satellite program and from NOAA's Environmental Satellite Service. Cousteau said the opportunity to contribute to the project had been "very thrilling, because we believe that one of the most urgent tasks is to monitor the earth's resources and the ocean resources, and also to control—later on—directly from outer space, the degree of pollution of the surface of the ocean." During the expedition in Antarctic waters Cousteau had received satellite photos from *Nimbus 2* satellite (launched by NASA May 15, 1966), *Essa 8* (launched Dec. 15, 1968), and *Noaa 2* (launched Oct. 15, 1972) to help him navigate the dangerous Drake passage between Antarctica and Ushuaia, Argentina. They were "the most valuable tool we had to plan our trip." (Transcript)
- Remote-sensing techniques developed by NASA for space and aircraft investigation of earth resources had been used to map land uses in the 18-county area surrounding Houston, Tex., Johnson Space Center announced. The experiment had been carried out with a specially equipped B-57 aircraft by the Earth Observations Div. of JSC to show the effec-

tiveness of remote sensing to prepare fast, accurate land use inventories for large areas. (JSC Release 73-15)

- NASA launched a Nike-Apache sounding rocket from Churchill Research Range, Canada, carrying a Univ. of Pittsburgh payload to 135.1-km (84-mi) altitude. The primary objective was to confirm an initial observation of nitric oxide in the auroral zone and collect data on processes producing it in certain auroral arcs. The rocket and recovery system performed satisfactorily. The telemetry signal varied from good with good scientific data on the upward trajectory to poor with marginal data on the down. (NASA Rpt SRL)

March 2: Communications Satellite Corp. would receive a $27 912 000 firm-fixed-price contract for lease of ultrahigh-frequency satellite communications service, the Navy announced [see Mar. 7]. (DOD Release 109-73)

- Dr. Robert C. Seamans, Jr., Secretary of the Air Force, announced the award of Air Force contracts to Fairchild Industries, Inc., and General Electric Co. to develop the A-X close air support aircraft.

  Fairchild Industries, Inc., would receive a $159 279 888 cost-plus-incentive-fee contract to test prototype aircraft and to develop and build 10 pre-production A-10 aircraft for flight-testing.

  General Electric Co. would receive a $27 666 900 fixed-price-incentive firm contract to develop and deliver 32 TF-34 engines to power the A-10 aircraft. (DOD Release 105-73)

- *Science* commented on the new congressional Office of Technology Assessment. OTA, now in the planning stage, was unlikely to receive any of its $5-million budget before July 1 and therefore would not be operating before the summer. The OTA board would be a joint committee of Congress including Sen. Edward M. Kennedy (D-Mass.), chairman; Sen. Hubert H. Humphrey (D-Minn.); Sen. Ernest F. Hollings (D-S.C.); Sen. Peter H. Dominick (R-Colo.); Sen. Clifford P. Case (R-N.J.); Sen. Richard S. Schweiker (R-Pa.); Rep. John W. Davis (D-Ga.); Rep. Olin E. Teague (D-Tex.); Rep. Morris K. Udall (D-Ariz.); Rep. Charles A. Mosher (R-Calif.); and Rep. James Harvey (R-Mich.).

  OTA would be a general consulting service to Congress to aid it in its decisions, in part by contracting out long-term studies to think tanks and universities. Short-term work could be done by a panel of outside experts and ad hoc panels with members from industry, science, engineering, labor unions, and public interest groups. OTA hoped to make all of its business open to the public. (Shapley, *Science*, 3/2/73, 875-877)

- The small, $240 000 guidance-control computers used in the Minuteman I nuclear missile force had been made available to public organizations by the Air Force, *Science* reported. For a $30 delivery fee, the general-purpose computers, one meter (three feet) in diameter, were being recycled to hospitals, laboratories, and universities to begin reprogramming for peaceful uses. The computers had become available because of the deployment of the Minuteman III multiple-warhead missiles. (Wade, *Science*, 3/2/73, 880)

March 3: President Nixon entertained *Apollo 17* Astronauts Eugene A. Cernan, Ronald E. Evans, and Dr. Harrison H. Schmitt at the White House. Members of Congress and the Secretaries of Labor, Transportation, and Commerce also were among the guests at an entertainment starring singer Sammy Davis, Jr. (*PD*, 3/5/73, 223; *W Post*, 3/5/73, B1)

- The People's Republic of China was reportedly developing an intercontinental ballistic missile 20% larger than the largest one in the U.S.S.R., the *New York Times* said. Administration officials had said the missile was a three-stage, liquid-fueled ICBM expected to have a range of 8000 to 11 000 km (5000 to 7000 mi). Underground silos were being built at a missile test center west of Peking, a location which would permit the missile to reach major cities and military targets in the U.S. and U.S.S.R. Following the first successful tests, it would take three years before a force of 10 to 30 missiles could be deployed. (Beecher, *NYT*, 3/4/73, 17)

March 4: Laser beams, computers, and satellites were being used by NASA to gather data over a five-year span for the National Oceanic and Atmospheric Administration and the U.S. Geological Survey to help predict earthquakes along California's 965-km (600-mi) San Andreas fault. Lasers at precise locations on either side of the fault were beamed to satellites *Explorer 27* (launched April 29, 1965), *Explorer 29* (*Geos 1*, launched Nov. 6, 1965), and *Explorer 36* (*Geos 2*, launched Jan. 11, 1968) and bounced off tiny mirrors. Reflections were recorded by computers and sent to Goddard Space Flight Center, where analysis gave measurements of the slightest variation at a laser's base. A mathematical model would eventually be constructed to assist in predicting quake behavior. (GSFC proj off; GSFC PAO; LA *Her-Exam*, 3/4/73)

- NASA's Lunar Exploration Office in the Apollo/ASTP Program Office was reassigned from the Office of Manned Space Flight (OMSF) to the Office of Space Science (OSS) under the title Lunar Programs Office. The office would be responsible for formulation of future lunar programs, continued operation of lunar science stations, and analysis and synthesis of lunar data. Capt. William T. O'Bryant would continue as Director and Dr. Noel W. Hinners as Deputy Director and Chief Scientist. (NASA Hq *WB*, 3/12/73)

March 5: U.S.-U.S.S.R. cochairmen of the NASA and Soviet Academy of Sciences Joint Working Group on Near-Earth Space, the Moon, and the Planets had approved a data exchange negotiated at the Jan. 29–Feb. 2 Moscow working session, NASA announced. The U.S. would provide the U.S.S.R. with maps and photos of two Mars landing regions and the available atmospheric model of Mars, Mars ephemerides from ground-based radar data for the first half of 1974, radar measurements of Mars, and results obtained during a 1974 Mariner Mercury-Venus flyby mission. The U.S.S.R. would provide the U.S. with information on Viking landing sites from *Mars 2* and *3* probes (launched May 19 and 28, 1971) and future missions, data on atmospheric parameters and surface of Mars, data from *Venus 8* (launched Mar. 27, 1972) on the Venus atmosphere and surface, and radar measurements of Venus. The data exchanges would take place before April 15. (NASA Release 73–38; *Av Wk*, 3/12/73, 17)

- The discovery that Saturn's rings appeared to be made of solid chunks rather than of gas, ice crystals, or dust had been made by Jet Propulsion Laboratory astronomers, NASA announced. Dr. Richard M. Goldstein and George A. Morris, Jr., had made the first successful radar probes of Saturn, using NASA's 64-m (210-ft) antenna at Goldstone Station, and had received much stronger bounceback signals than expected. "From our radar results," Dr. Goldstein had said, "the rings cannot be made up of tiny ice crystals, dust, or gas. Our echoes indicate rough jagged surfaces, with solid material 1 meter (3.3 feet) in diameter or larger. Possibly much larger." The rings could be a great hazard to any spacecraft sent into them. (NASA Release 73-37)

- Communications Satellite Corp. applied to the Federal Communications Commission for authority to construct a maritime multifrequency satellite system to provide communications to the Navy and the commercial shipping industry [see Mar. 2]. The Navy would use ultrahigh frequencies allocated for Government use, for communications between satellites and Navy-provided ship and shore terminals. It would also make available to commercial shipping high-quality communications of greater reliability and scope than available before. (ComSatCorp Release 73-11; ComSatCorp PIO)

- The U.S.S.R. Academy of Sciences opened its Annual General Meeting in Moscow. Academy President Mstislav V. Keldysh reviewed the Academy's role in the second year of its five-year plan: Scientists had been given responsibility for ensuring that science achieved its greatest economic effect by accelerating technical processes. Space successes had included: *Venus 8* softlanding (July 22, 1972) of a capsule on Venus that transmitted from the surface 50 min; *Mars 2* and *3* orbiting of Mars (entering orbits Nov. 27 and Dec. 2, 1971) and the first softlanding of a capsule on Mars, by *Mars 3* (Dec. 2, 1971); research of the moon with *Luna 20* and the *Lunokhod 2* roving vehicle, which continued to explore the lunar surface; advances in communications and earth studies by Molniya, Meteor, Cosmos, and Intercosmos satellites; and advancement of cooperative programs with socialist countries, France, and the U.S. (*Pravda*, FBIS–Sov, 3/15/73, L1; *A&A 1971*; *A&A 1972*)

- NASA launched two sounding rockets from White Sands Missile Range. A Nike-Apache carried a Univ. of Texas payload to a 221.9-mm (137.9-mi) altitude to make redundant measurements of the electron density profile and photoelectron spectrum at different altitudes. The rocket and instrumentation performed satisfactorily.

  An Aerobee 170 carried a Harvard College Observatory solar physics experiment to a 241.4-km (150-mi) altitude. The rocket and instrumentation performed satisfactorily. (NASA Rpt SRL; GSFC proj off)

- The Air Force announced award of a $2 056 889 cost contract to Utah State Univ. to study disturbed and undisturbed infrared atmosphere. Rocketborne instruments would measure optical and infrared emissions, both natural and those stimulated by atmospheric dosing from a rocketborne electron source. (DOD Release 113-73)

March 5-8: NASA's Fourth Annual Lunar Science Conference, at Johnson Space Center, was attended by more than 750 scientists from the U.S. and 12 foreign countries. Visitors included scientists from the U.S.S.R. who presented papers on Apollo lunar samples received from the U.S.

At the opening session Dr. George M. Low, NASA Deputy Administrator said, "I can see at least a decade of fruitful analysis, synthesis, and integration of the information locked up in the material brought back and still coming back from the scientific stations left on the moon" in the Apollo program.

Dr. Gerald J. Wasserburg, California Institute of Technology scientist, described "lunar cataclysm." Lunar dust brought back by Apollo astronauts had been estimated to be as old as 4.5 billion yrs. But most lunar rocks were less than 4 billion yrs old and none were younger than 3 billion yrs. The best explanation was that 3.9 billion yrs ago the moon was battered and bombarded on an unprecedented scale, creating so much heat that old rocks were destroyed and a new generation of rocks was created. Since then the moon might have experienced minor volcanic eruptions but nothing like that period of cataclysm.

Some scientists believed the moon had cooled and gone into a period of hibernation. Dr. Oliver A. Schaeffer of the State Univ. of New York said even the orange soil found by *Apollo 17* astronauts had turned out to be 3.71 billion yrs old, an age disappointing to scientists who had believed the moon had been formed by more recent volcanic activity.

Dr. Keith A. Howard, U.S. Geological Survey geologist, compared the lunar landslide found near the *Apollo 17* landing site to a large, high-velocity avalanche on earth. Dr. Howard suggested that gas within the lunar soil had been released during the slide. *Apollo 17* samples would be searched for evidence of fluids.

One *Apollo 17* instrument placed on the lunar surface had made an atom-by-atom search for evidence of lunar atmosphere. Although the moon had no atmosphere, traces of argon, neon, and helium had been found. Dr. John H. Hoffman, associate professor of physics at the Univ. of Texas, said only argon appeared to have been generated from the moon's interior. Neon and helium apparently had been deposited by the solar wind.

Conference attendees unanimously voted to congratulate NASA on completion of the Apollo program by sending a letter to Dr. James C. Fletcher, NASA Administrator: The conception, design, and implementation of lunar exploration represented "an extraordinary human and technological achievement." To explore another planet, it had been necessary for man "to achieve many successive levels of technological capability and to surpass formidable barriers. This was accomplished brilliantly by the dedicated engineers and astronauts of NASA in conjunction with skilled management. On this technical base a new branch of science has been built." (Transcript; *Goddard News*, 3/73, 2; JSC *Roundup*, 3/16/73, 1; *NASA Activities*, 4/15/73, 67)

March 5–9: More than 1000 administrators, congressmen, and government, industry, and university scientists attended the *Erts 1* Earth Resources and Technology Satellite Symposium sponsored by Goddard Space Flight Center in New Carrollton, Md. Two hundred papers presented results from *Erts 1* (launched July 23, 1972) experiments in agriculture and forestry, mineral resources and geology, environmental surveys, land-use mapping, and marine surveys.

Dr. Marian Baumgardner, scientist at the Purdue Univ. Laboratory for the Applications of Remote Sensing, said results from *Erts 1* study of semi-arid regions suggested that the satellite could identify and map vegetative, species, and management differences in range lands; gross soil patterns and differences related to agricultural and land-use-management problems and practices; areas of surface water and changes related to ground water recharge; crop damage by hail and windstorms; and areas of bare soil and related problems of erosion and conservation.

Dr. William A. Fischer, U.S. Geological Survey scientist, said data from an *Erts 1* study of Alaska had reinforced recent plate tectonic theories of a continuing mobility in the earth's crust. The satellite had mapped fractures previously unknown or not put together as continuous structures in the earth's crust. The fractures indicated "there has been a fundamental movement, extending deep down into the crust."

Dr. John Miller of the Univ. of Alaska said seismic data and *Erts 1* identification of previously unknown fractures "has been an important input to the design of a bridge structure across the Yukon River . . . and possibly the pipeline." (Transcript)

- The Communications Subcommittee of the Intergovernmental Maritime Consultative Organization—a United Nations affiliate—voted 20-to-1 in London to convene an international conference of governments in October 1974 on the establishment of a global maritime satellite communications service. The one dissenting vote was cast by the U.S., which argued that formation of a new organization was premature. The U.S. favored operation of a maritime service by the 80-nation International Telecommunications Satellite Organization. (*Av Wk*, 3/26/73, 59–60)

March 6: The Air Force launched an unidentified reconnaissance satellite on an Atlas-Agena booster from Eastern Test Range at 4:30 am EST. Orbital parameters: 36 679-km (22 791.3-mi) apogee, 35 855-km (22 279.3-mi) perigee, 1435-min (23.9-hr) period, and 0.2° inclination. (UN Registry; *Today*, 3/7/73)

- *Cosmos 551* was launched by the U.S.S.R. from Baykonur Cosmodrome into orbit with a 292-km (181.4-mi) apogee, 202-km (125.5-mi) perigee, 89.4-min period, and 65.0° inclination. The satellite reentered March 20. (GSFC *SSR*, 3/31/73; *SBD*, 3/7/73, 35)

- Dr. James C. Fletcher, NASA Administrator, testified on long-term objectives during the Senate Committee on Aeronautical and Space Sciences hearings on the FY 1974 NASA authorization. Among effects of space and aeronautics on "all Americans and, to a degree, people throughout the world, in 1985," he foresaw a rise in air cargo revenues from the current $1.6 billion per year to $14 billion per year, with development of automated handling and compatible aircraft. Trade and tourism would flourish with the use of hypersonic jet aircraft, "which we will be thinking about in 1985 and maybe other nations will have developed." Such an aircraft could fly from Washington to Peking in 1½ hrs. Hydrogen fuel that produced water rather than pollutant fumes would replace present aviation fuels.

In space, Dr. Fletcher saw a proliferation of domestic satellites leading to economic use of long-distance telephone calls for "discussing almost anything." Large-scale digital computers—spinoffs of the space age—would be used educationally; "computer-aided instruction is the coming thing. This is the quickest way to learn if we can afford to build the system, and by 1985, we think we can." Other NASA objectives to be realized within the next 10 to 15 yrs were a satellite-operated search-and-rescue system that picked up signals from pocket-sized radios aboard life rafts, the use of space age sensors in preventive medicine and for rapid diagnosis, a global disaster-alert and relief-control system to monitor major disasters, biological space processing to separate viruses or bacteria in the production of vaccines, an international space laboratory, an international lunar expedition, and expanded planetary exploration.

Dr. Fletcher said NASA could be called an environmental agency. "It is not just that space is our environment, but it is rather that . . . virtually everything we do, manned or unmanned, science or applications, helps in some practical way to improve the environment of our planet and helps us understand the forces that affect it. Perhaps that is our essential task, to study and understand the earth and its environment." (Transcript)

- The House Committee on Science and Astronautics' Subcommittee on Aeronautics and Space Technology opened its hearings on the FY 1974 NASA authorization. Rep. Ken Hechler (D-W. Va.), Committee Chairman, said the Subcommittee would be particularly concerned with "the extent of progress toward the enunciation of a national aeronautics and aviation policy." The Subcommittee had been partially responsible, he said, for the reduction of emphasis on aeronautics because it was "not aggressive enough in its insistence that more emphasis be placed on aeronautics during . . . the late 1950's when the Committee was first established, and the early 1960's when the tremendous emphasis on space tended to push aeronautics . . . into the background."

Roy P. Jackson, NASA Associate Administrator for Aeronautics and Space Technology, testified that reorientation of NASA's space and aeronautical programs to meet changing national goals had been completed "without jeopardizing the continuing development of a broad base for new knowledge, and the potential for meeting future, but presently unseen, needs." In the OAST aeronautics program the research and technology base took priority over experimental programs. The R&T base was "work that is unparalleled in the U.S. It is the data base from which springs new ideas for development in the future." When faced with the necessity to cut expenditures, "we had to consider primarily a reduction in our work on flight experiments and experimental programs." Among examples of meeting future needs, hydrogen as a long-term aircraft fuel for the distant future was a "concept of potential major importance to the reduction of air pollution, while simultaneously conserving our increasingly scarce petroleum reserves." Hydrogen eliminated hydrocarbon pollution and was "a more efficient fuel, but the reduction of NO_x [oxides of nitrogen] may require additional effort." The "earliest likely commercial use of hydrogen as an aircraft fuel will be about 15 to 20 years hence." Hydrogen was "a low-density, cryogenic fuel which represents problems of relatively large, insulated tanks, the need for materials to keep their strength through wide temperature variations, very low temperature hardware including pumps, valves, seals and associated maintenance. In addition to cost, hydrogen is difficult to store and transport. . . . Our experience with the practical use of liquid hydrogen in NASA's space program suggests that there will be some complexities to solve but these should not be insurmountable."

Dr. Seymour C. Himmel, Deputy Associate Administrator for Technology in OAST, reported on the quiet, clean, short-haul experimental engine (QCSEE) program to integrate and demonstrate technology required for an environmentally acceptable and economical powered-lift propulsion system. "QCSEE will utilize a basic existing engine core, modified as required to mate with specially designed low-pressure spool fans and turbines to build up an experimental engine. This, together with appropriate flight-type acoustic nacelles and wing section installations, will be tested for noise and performance and operating characteristics. The design and test program will provide the technology base needed to help guide Government . . . environmental rulemaking and to provide industry with data to evaluate propulsion concept effectiveness and development risks."

George W. Cherry, Deputy Associate Administrator for Programs in OAST, described the status of aeronautics programs after revisions necessitated by FY 1974 budget restrictions. The General Electric Co. quiet engine contract had been completed and both engines delivered to NASA. Engines A and C were scheduled for extensive in-house research programs at Lewis Research Center. An in-house study underway at LeRC would determine the technical value of another contracted effort, for a second-generation quiet engine. "In the meantime, we are working to generate component noise design and performance data, so that a significant noise reduction below that demonstrated by the first Quiet Engine can be obtained in any future engine program. No important area in aeronautical propulsion has been terminated, although funding levels for certain portions of the aeronautics programs have been readjusted based on a review of priorities. The F-8 high-speed supercritical wing verification project was not cut back or terminated. During FY 1973, the project was completed according to plan. The results were very encouraging and gave confidence as to wind-tunnel test results at transonic speeds." A follow-on to the F-8 was under study. Depending upon results, a follow-on F-8 flight-test program might be undertaken in FY 1975. (Transcript)

- The Subcommittee on Manned Space Flight of the House Committee on Science and Astronautics continued hearings on NASA's FY 1974 authorization. Dale D. Myers, NASA Associate Administrator for Manned Space Flight, identified locations for space shuttle development: Kennedy Space Center and Vandenberg Air Force Base, Calif., as the launch and landing sites; NASA's Michoud Assembly Facility for the external tank production; and the Rockwell International Corp. facility at Downey, Calif., for orbiter and systems integration. "The major unknown, at the moment, is where the solid rocket boosters will be produced and tested. We expect that this location will be identified when the solid fueled rocket booster contract is awarded this November."

Philip E. Culbertson, Director of Mission and Payload Integration in OMSF, contrasted a typical operational communications satellite program using expendable launch vehicles with one using the space shuttle. Launched on the Titan IIIB-Centaur launch vehicle, each spacecraft was estimated at 467 kg (1030 lbs). To fulfill 1979–1990 requirements for reliable communications services, 26 spacecraft would be required. Use of the Titan IIIB-Centaur would require 26 new spacecraft and 26 new launch vehicles, at an average cost for each flight of $25.8 million.

"If the same program is carried out using the Shuttle and Tug, the cost changes significantly. First, with the Shuttle we can apply the low cost approaches to payload design. We are no longer constrained by weight limitations; therefore, we can ruggedize the spacecraft structure, simplify the design and verification test program, and thus reduce both development and unit spacecraft cost." The shuttle and tug combination could deploy and retrieve 2 satellites on a single flight, making 13 flights to deploy 26 spacecraft—10 new and 16 refurbished. On 8 of these flights 2 spacecraft would be deployed and retrieved. "The total cost for conducting the program would be $14.9 million per spacecraft flight, compared to the $25.8 million for the expendable approach—a savings of 42 percent."

M/G Robert H. Curtin (USAF, Ret.), NASA Director of Facilities, discussed disposition of facilities at Lewis Research Center's Plum Brook Station, which would be closed because of FY 1974 budget cuts. LeRC was discussing the station's availability and capability with potential user agencies—including the Electric Research Council, which planned to establish an electric power research institute, and the Business and Employment Council of the Governor of Ohio, which had recommended establishment of a state-funded Ohio Development Center. LeRC was studying possible use of the station's reactor for environmental research by the Environmental Protection Agency. The Plum Brook space power facility was being prepared to test the shroud for NASA's Viking spacecraft but "there are no additional programmatic needs for the use of this facility and it is anticipated that it will be placed in a standby mode by the end of Fiscal Year 1974." (Transcript)

- NASA Associate Administrator for Applications Charles W. Mathews continued testimony before the House Committee on Science and Astronautics' Subcommittee on Space Science and Applications during FY 1974 NASA authorization hearings: The ERTS-B launch was changed from November 1973 to the first quarter of 1976, permitting an instrument to measure the heat energy radiated from the land and the water surface to be incorporated. These data would be used to locate, map, and identify pollution in large lakes, bays and estuaries, as well as to obtain temperature information over land areas. NASA had shifted the emphasis of the earth observation aircraft program because of *Erts 1* (launched July 23, 1972) and Skylab (scheduled for May launch). "Heretofore the aircraft program was principally used to test instruments and to investigate techniques. It is now to be used in support of investigations associated with Earth observation spacecraft. The net effect is an increase . . . of aircraft flights, sensors and data processing. The aircraft program capability will be devoting approximately 80 percent of its time to this latter effort and the remainder will be used in experimental work on new instruments." (Transcript)
- The Thor-Delta launch vehicle and "its entire missile launch system" were being sold to Japan by McDonnell Douglas Corp., Director Andrew Biemiller of the American Federation of Labor–Congress of Industrial Organizations (AFL–CIO) Dept. of Legislation said in testimony before the House Committee on Finance's Subcommittee on International Trade. Biemiller told the Subcommittee that the U.S. aerospace industry, "where the U.S. has held technological supremacy," was "steadily being exported abroad." Dept. of State spokesman Charles W. Bray, III, later told the press in Washington that the sale had been approved under a 1969 agreement on space cooperation. Japan would be obligated to use the rocket system for peaceful purposes only. (*CR*, 3/6/73, S3977-9; Doder, *W Post*, 3/7/73, A10)
- The decision to publicize weather information from a secret Air Force satellite system was announced by Under Secretary of the Air Force John L. McLucas at a Washington press conference. Information would

be made available to the National Oceanic and Atmospheric Administration from "several" satellites orbiting at altitudes of 800 km (500 mi), providing more precise information on cloud cover than available from the higher altitude NOAA satellites. A USAF–NOAA meeting March 19 would arrange for transfer of information. (W *Star & News*, 3/7/73, A6)

- NASA and space shuttle contractors held the third quarterly Space Shuttle Review at Marshall Space Flight Center. Attended by 400, the review provided information on external tanks and solid rocket boosters. (*Marshall Star*, 3/7/73, 1)
- NASA and the Dept. of Defense were in disagreement over use of a high-frequency radio link between the earth and space, Thomas O'Toole said in the *Washington Post*. NASA wanted to use the frequency to transmit voice, data, and televised instructions to the ATS-F Applications Technology Satellite, to be launched in April 1974. The Air Force and the Army feared interference with their own radio signals. A move to lower the frequency for satellite communications would cost NASA as much as $4 million to install a new transponder and could delay the launch four months. (O'Toole, *W Post*, 3/6/73, 5)
- U.S.S.R. sounding rocket launchings, first step in joint Soviet-French meteorological studies, had begun from French Kerguelen Island, Tass reported. The first rocket had studied a vertical cross-section of the atmosphere over the island and reported temperature, pressure, and wind velocity in the upper atmosphere. (Tass, FBIS–Sov, 3/9/73, L1)

March 7: Elements of the space shuttle preliminary bioresearch laboratory simulator had been delivered to Marshall Space Flight Center, MSFC announced. Equipment would be installed aboard the payload carrier simulator for testing. The bioresearch laboratory model, the first shuttle-era payload to be delivered for testing and integration at MSFC, included a mass-measurement and microscopy unit, preparation unit, centrifuge, cryogenic freezer for storage of tissue, and an instrument for freeze-drying tissue. (MSFC Release 73–31)

- Dale D. Myers, NASA Associate Administrator for Manned Space Flight, testified on cooperation with the Dept. of Defense during Senate Committee on Aeronautical and Space Sciences hearings on the FY 1974 NASA authorization: NASA and DOD were evaluating alternative vehicles to the space tug and development approaches that could provide the capability to transport shuttle payloads to and from higher-than-earth orbits, or propel payloads to earth-escape velocity. A joint decision would be made in the fall as to which agency would conduct the development of "this very important propulsive stage." Air Force personnel had been assigned to the space shuttle program offices, and NASA personnel served in a liaison capacity at the Air Force Space and Missile Systems Organization (SAMSO). Civilian and military representatives of DOD served on source evaluation boards and working groups on aerodynamics, structure, propulsion, crew systems, electronics, and operations. DOD sponsored special industry and in-house studies such as payload requirements, security, and crew safety. DOD also provided results of research in solid rocket technology, lifting bodies, thermal protection, and similar topics. (Transcript)
- Apollo-Soyuz Test Project Program Director Chester M. Lee testified on the status of the joint 1975 U.S.-U.S.S.R. mission, during the House Committee on Science and Astronautics' Subcommittee on Manned

Space Flight hearings on the FY 1974 NASA authorization. Activities were being directed to identify and define candidate ASTP experiments to provide data for U.S. and Soviet scientists, provide for coinvestigators in both countries, require active cooperation by Soviet cosmonauts, and use existing hardware where possible. Experiments would be discussed at the March meeting of U.S.-U.S.S.R. working groups [see March 15-30]. No launch vehicle work specifically for ASTP had begun, but modification, testing, and checkout would begin during the coming year at Kennedy Space Center's Launch Complex 39 and other facilities. The program was "on schedule and moving well toward our current mission date of mid-July 1975." (Transcript)

- Roy P. Jackson, NASA Associate Administrator for Aeronautics and Space Technology, testified on aeronautical programs before the House Committee on Science and Astronautics' Subcommittee on Aeronautics and Space Technology during FY 1974 NASA authorization hearings: "NASA's advanced transport technology (ATT) program is aimed at expediting technology advances that could facilitate development of environmentally acceptable, and economically superior, next-generation U.S. air transports. The ATT system studies have shown that by the early 1980's a new subsonic/sonic transport—quieter, cleaner, safer, and more economical than the current wide-body jet transports—could be realized through technology advances which could be ready for commitment to application late in the 1970's. These advances would include supercritical aerodynamics and related configuration treatment, techniques for suppression of engine noise and emissions, composite structures, and active control systems." The studies would be extended in FY 1974 "by concentrating more heavily on the system improvement tradeoffs, and corresponding technology, involved in optimizing design for efficient terminal operations as well as cruise performance."

An analysis and wind-tunnel test program would investigate the sources and the reduction of airframe noise. "Recent studies and flight tests have indicated that as we continue to reduce engine source noise, it may also be necessary . . . to treat airframe noise." Further testing of airframe shielding and analyses would determine the operational and economic feasibility of new concepts like the yawed-wing, low-supersonic design being investigated at Ames Research Center. Propulsion technology programs underway would examine further noise reduction potential than that already being demonstrated in the conventional takeoff and landing aircraft experimental quiet engine program. When results were promising, plans for a follow-on CTOL experimental quiet engine, Mark II, would be developed.

The terminal-configured vehicles and avionics program, "a major new thrust in NASA research," was described by George W. Cherry, Deputy Associate Administrator for Programs in OAST: "While undertaken primarily to achieve increased capacity and quieter operations in the terminal area, this research will also aim at . . . improved accuracy, reliability, and automation of instrument flight in the approach, landing, and rollout process." Facilities at Wallops Station and at Langley Research Center would simulate an advanced complex air terminal through hard-wire connections to the Federal Aviation Administration's National Aviation Facilities Experimental Center. A specially equipped Boeing 737 would be flown to ensure compatibility of advanced airborne navigation, guidance, and control with the advanced landing aids and the air traffic control procedures under development by the FAA.

Dr. Seymour C. Himmel, Deputy Associate Administrator for Technology in OAST, reported NASA general-aviation programs were being directed to provide technology for designing future U.S. aircraft that are "safer, more productive, and clearly superior to the rapidly growing foreign competition." NASA's role encompassed "a broad technology effort . . . for a data base to improve the safety of all flight operations, while simultaneously pursuing the growing technological need for increased efficiency and performance" of general aviation. U.S. aviation products dominated world markets but this advantage was disappearing in the face of a rapidly emerging foreign production capacity. Cumulative imports of turbine-powered aircraft outnumbered exports, with a cumulative deficit of more than 200 aircraft and $200 million. (Transcript)

- The Anglo-French Concorde supersonic transport aircraft carried an 11 000-kg (24 000-lb) payload on a 6300-km (3900-mi), 3-hr 38-min flight-time journey from Toulouse, France, to the Madeira Islands and return. The distance flown equaled the distance between Washington, D.C., and Paris over which current scheduled flights took 7 hrs 42 min. (BAC-Aérospatiale Release 6C/73)
- President Nixon congratulated recipients of the 1973 Federal Women's Award at the White House. Among the recipients was Marjorie R. Townsend, Project Manager, Small Astronomy Satellite, Goddard Space Flight Center [see Feb. 22]. (PD, 3/12/73, 233)
- Rep. Lawrence R. Coughlin (R-Pa.) resigned from the House Committee on Science and Astronautics and was elected to the House Committee on the Judiciary. (NASA LAR XII/26)
- The Air Force announced the award of a $62 426 400 fixed-price-incentive contract to Boeing Co. for operational and maintenance ground equipment for the Minuteman intercontinental ballistic missile force for FY 1973. (DOD Release 116–73)

March 8: Communications Satellite Corp. sought authority from the Federal Communications Commission to construct advanced satellites and related ground control facilities for lease to American Telephone & Telegraph Co. to provide domestic satellite services to all 50 states and Puerto Rico. Each satellite would provide 14 400 two-way voice-grade circuits. Ground stations for tracking, telemetry, and command duties would be in Connecticut and California. The ComSatCorp filing was similar to one made in October 1970 and revised in March 1971, and would expand coverage to include Hawaii and Puerto Rico. (ComSatCorp Release 73–12)

- *Erts 1* Earth Resources Technology Satellite (launched July 23, 1972) photos showing pollution spreading across Lake Champlain would be used in court as evidence against a factory accused of discharging pollution, the *Washington Post* reported. The case, brought by the state of Vermont against the state of New York and the International Paper Corp., would mark the first use of a satellite photo in legal action. The photo, of Lake Champlain from 900-km (560-mi) altitude, showed a brown circle extending 0.6 km (0.4 mi) from the New York shore and originating from an International Paper Corp. mill. (O'Toole, *W Post*, 3/8/73, A3)

March 8

- The U.S. and the U.S.S.R. completed their joint 20-day Bering Sea mission, begun Feb. 17 to determine the sea's surface characteristics, ice cover, and currents originating in the Pacific. NASA's instrumented Convair 990 aircraft, a Soviet Il–18 aircraft, U.S. Coast Guard icebreaker *Staten Island*, and Soviet weathership *Priboy* had gathered data for processing in mainland laboratories. (Moscow Domestic Service, FBIS–Sov, 3/8/73, C2)

- The House Committee on Science and Astronautics' Subcommittee on Aeronautics and Space Technology continued hearings on the FY 1974 NASA authorization.

  George W. Cherry, Deputy Associate Administrator for Programs in the NASA Office of Aeronautics and Space Technology, described the Dept. of Transportation Climatic-Impact Assessment Program (CIAP), in which NASA was cooperating: The CIAP "undertakes to assess the potential impact on the atmosphere of large fleets of SSTs [supersonic transports] operating in the eighties and the biological consequences of their potential modification of the stratosphere. We are cooperating with DOT by contributing to the technology needs of CIAP, with particular emphasis on the levels to which pollution can be limited." Contradictory requirements for civil transport aircraft (low-fuel consumption engines required by SSTs for good range and payload characteristics, tended to be extremely noisy in terminal areas) might be best met by variable-cycle engines that varied their mode of operation in each condition of the flight profile. NASA would "intensively study variable cycle-engine concepts in FY 1974."

  Dr. Seymour C. Himmel, Deputy Associate Administrator for Technology in OAST, testified that hypersonic research for potential application to military and civil airbreathing cruise vehicles was directed at "the critical aspects of propulsion, structures, and aerodynamics for this flight regime where hydrogen is used both as the fuel and as a coolant." Hydrogen was the most likely candidate for hypersonic aircraft fuel because its high coolant capacity was essential for a propulsion system to survive in the high-temperature environment, and liquid hydrogen's coolant capacity could cool portions of the vehicle's structure away from the engine area. NASA's structures and materials program provided essential technology for designing lightweight, reliable, actively cooled structures and advanced the technology of actively cooling structures heated to high temperatures. The program also extended shuttle technology on reusable surface insulation toward hypersonic cruise application.

  Flight Research Center test pilot Gary E. Krier testified on NASA's digital fly-by-wire program to develop an all-electric flight-control system using intelligence inserted into the digital computer memory. "We did this by using refurbished Apollo equipment and two surplus [Navy] F–8 fighters. Using this approach, we estimate we flew about 2 years earlier than we otherwise could have." To date, NASA had flown 15 DFBW flights with no major problems. "Airliners that we'll ride in the 1980's could profit from fly-by-wire in the form of smooth flight at very high speeds and the near elimination of the response to turbulence by computer application of smoothing controls." The "best use of FBW" was to build a control-configured vehicle [CCV] with reduced drag, increased lift, and slower approach and landing speeds. "It has been estimated that landing speed could be cut by 25 percent and range increased 15 percent with no sacrifice in mission capability, just by modifying existing aircraft. A much larger improvement in performance could be gained by starting from scratch with FBW. We have been refining aircraft for years now, and the FBW/CCV combination gives us a chance to make a quantum jump in aircraft performance." (Transcript)

- NASA launched an Aerobee 170 sounding rocket from White Sands Missile Range carrying an American Science and Engineering, Inc., solar physics experiment to a 170.1-km (105.7-mi) altitude. The rocket and instrumentation performed satisfactorily. (GSFC proj off)
- Secretary of the Navy John W. Warner and Grumman Corp. President John C. Bierwirth signed an agreement for the Grumman Aerospace Corp. to build 48 additional F-14 jet fighter aircraft at current contract prices. The Navy would not exercise further options under the current contract and would limit its purchases to 134 aircraft. Future purchases would be dependent upon congressional authorization and funding. (Text)

March 9: The Air Force launched an unidentified satellite by Titan IIID launch vehicle from Vandenberg Air Force Base into orbit with 263-km (163.4-mi) apogee, 152-km (94.5-mi) perigee, 88.6-min period, and 95.7° inclination. The satellite reentered May 19. (GSFC *SSR*, 3/31/73; 5/31/73; *NYT*, 3/10/73, 31)

- Photo-mapping of Mars by *Mariner 9* (launched May 30, 1971) had revealed multiple circular features surrounding both poles which might indicate that the Martian north and south poles had moved, California Institute of Technology scientists reported in *Science*. Some of the concentric patterns were centered on points displaced from the present north and south poles of Mars, implying that these were ancient positions of the poles and testifying to a slow drift of the Martian spin axis. Such a drift of the earth axis could explain the evidence for radical climate changes in the past—fossil coral reefs in Greenland and fossil forests in the South Pole—and revive the debate with scientists who believed the climate changes were due to the constant moving of the earth's continents and oceans. (Murray, Malin, *Science*, 3/9/73, 997–999)
- The Subcommittee on Manned Space Flight transmitted *Space Shuttle—Skylab, 1973: Status Report* (dated January 1973) to its parent House Committee on Science and Astronautics. The report covered the transition from detailed design to development phase of NASA's shuttle program (in which the first manned flight was planned for 1978) and the updated cost, performance, and status of the Skylab program (to be completed in 1973). The Subcommittee concluded that technology and resources existed for successful development of the NASA configuration for an earth-orbital shuttle and that the shuttle design would permit the total development cost to stay within $5.15 billion and the per-flight operational cost within $10.5 million. Success in meeting this cost was "particularly sensitive" to the cost of the hydrogen-oxygen tanks and an acceptable recoverable and refurbishment cost of the solid-fueled rocket boosters. Space tug development was "of key importance to gaining full utility of the space shuttle." Development of a low-cost space shuttle system was essential "if the nation is to realize the full benefits of near space in . . . scientific exploration, practical application, and national security."

The Skylab program was within costs projected by NASA for FY 1973. Development schedules were being met "with some problems being encountered in procuring and integrating experiments." NASA was recommended to consider the possibility of flying Skylab in the 1974–1976 period and also possible Skylab revisits after the first three manned missions. (Text)

- The 16th Annual Goddard Memorial Dinner was held in Washington, D.C., with *Apollo 17* Astronauts Eugene A. Cernan, Ronald E. Evans, and Dr. Harrison H. Schmitt among the honored guests.

  The Dr. Robert H. Goddard Memorial Trophy was presented to Dr. George M. Low, NASA Deputy Administrator. The citation read: "To the Honorable George M. Low who has held key positions in the space program since its beginning, and has played a vital role in putting men into space and later, on the moon. Among Dr. Low's outstanding contributions was, as acting NASA Administrator in 1971, the negotiation of the space agreement with the Soviet Union which provided the foundation for the Apollo/Soyuz flight in 1975 and other joint space endeavors."

  The National Space Club Press Award was presented to Ralph Morse of *Life* magazine for "his ingenuity in photographically interpreting the dynamics of the United States Space Program and for his resourcefulness in portraying the human experience in its larger dimension." The Astronautics Engineer Award went to Wilfred E. Scull of Goddard Space Flight Center for "outstanding leadership—from decision to fruition—of the Earth Resources Technology Satellite Project." The Nelson P. Jackson Aerospace Award was presented to The General Electric Co. for "outstanding accomplishment" as prime contractor to NASA for the Earth Resources Technology Satellite and ground data handling system. (Program)

- Chairman Maurice M. Levy of the European Space Research Organization Council described steps to complete plans for European participation in the space shuttle program, during FY 1974 NASA authorization hearings before the Senate Committee on Aeronautical and Space Sciences: "The ESRO Council has already approved the text of an agreement laying down the arrangements under which the Spacelab [sortie lab] will be developed, first in ESRO and subsequently within the European Space Agency which, in conformity with the decisions taken in Brussels, is expected to be set up in 1974. This agreement is open for signature between March 1 and the end of July. It will probably be implemented very shortly." A technical agreement would have to be negotiated between NASA and ESRO as to Spacelab's procurement and its integration into the space shuttle. A second agreement between governments would be necessary to cover commitments for nonduplication of efforts, the possibility of transferring technology to other European programs, access to the whole shuttle system, and the availability of conventional launch vehicles.

  M/G Robert H. Curtin (USAF, Ret.), NASA Director of Facilities, testified that the $8.9-million budget request for space shuttle facility planning—in addition to $67.2 million for shuttle facility modification projects—included studies, engineering support, and preliminary engineering reports for upcoming projects and the final design for the pending FY 1975 facility needs. "This specific final design, when accomplished, will essentially complete facility design for the shuttle

ground test program, the early and initial phase of Solid Rocket Booster (SRB) production and test facilities," and the pads and mobile launchers at the Kennedy Space Center. (Transcript; Text)
- Microwave measurements of the atmosphere of Venus were described by Univ. of California scientists in *Science*. Two sets of passive radio observations—measurements of the spectrum of the disc temperature near the 1-cm wavelength and interferometric measurements of the planetary limb darkening at the 1.35-cm water vapor resonance—had shown no evidence of water vapor in the lower atmosphere of Venus. The upper limit for the mixing ratio of water vapor to dry air was substantially less than the amounts derived from the Venus probes. The amount of water vapor could not produce dense clouds nor contribute significantly to a greenhouse effect. (Janssen et al., *Science*, 3/9/73, 994–996)
- Problems of the European space program were described in a *Science* article by Dominique Verguèse of Paris *Le Monde:* Since 1964 the European Space Research Organization and the European Launcher Development Organization had spent $1 billion on space. One third had been used for the successful launchings of seven satellites but the remainder had been spent on an expensive and yet unsuccessful booster. Interest in the booster had dissipated despite the French desire to continue and the offer to pay for 60% to 70% of the costs.

  At two European conferences on space (December 1971 and December 1972) plans were made to merge ELDO and ESRO into a single agency and to allow member nations to cooperate with NASA on building a sortie lab for the space shuttle. France had threatened to leave ESRO unless more emphasis was placed on applications satellites and ESRO closed its plasma laboratory in Italy and its sounding rocket range in Sweden. France had also insisted that more use be made of national facilities and teams rather than ESRO's technical center.

  Final agreements and objectives on a European applications satellite could not be made because of the conflicting interests of each nation. Nor could it be decided what role national programs should play. But Europe had successfully launched 26 satellites and had acquired an industrial competence which could easily provide a firm basis for more ambitious projects. (*Science*, 3/9/73, 984–986)
- Experiments by international groups using the intersecting storage rings of the European Nuclear Research Center (CERN) on the Swiss-French border had shown that protons grow when accelerated to great energy, the *New York Times* reported. New questions were raised as to what would happen within the nucleus of an atom that was packed with protons and neutrons and subjected to high-energy acceleration. Experiments had indicated that the proton—once thought a homogenous building block of the atomic nucleus—harbored processes and forces yet to be discovered. (Sullivan, *NYT*, 3/9/73, 1)

March 11: The U.S.S.R.'s *Lunokhod 2* began its third working day on the moon. The roving vehicle was 400 m (1300 ft) above its landing site at the upper rim of the 3.5-billion-yr-old-crater. *Lunokhod 2*'s primary task during its third working day would be to study the crater thoroughly. (Tass, FBIS–Sov, 3/13/83, L1)
- The book *13: The Flight That Failed*—an account of the aborted *Apollo 13* mission (launched April 11, 1970) by Henry S. F. Cooper, Jr.— was reviewed by Joseph McElroy in the *New York Times Book Review:*

In a spacecraft so full of safety valves and designed redundancies "it had seemed incredible that two of three fuel cells from the two liquid oxygen tanks in the service module could go dead. . . . But to lose two fuel cells and then eventually two oxygen tanks? Not possible." Command module pilot John L. Swigert, Jr., had said after the mission that if anyone had thrown at the crew a simulation of what actually happened "they'd have thought it unrealistic." To have had less than this confidence, McElroy said, would have been unrealistic. The care and skill with which the many teams of NASA technicians work, plus the computers whose development may be "the main reason NASA could achieve a moon landing so soon" had created "an efficiency as unprecedented as the Apollo 13 disaster was implausible to the engineers at Houston Mission Control." Cooper's "plain, detailed picture of the teams at their consoles and in discussion and over the intercom 'loop' and under the adroit supervision of Eugene Krantz, a flight director who was unquestionably an executive hero of Apollo 13, gives their work a fascinating muscle and texture that might surprise those inclined to think this side of NASA boring." (*NYT Book Review*, 3/11/73, 4–5)

March 12: Dr. John E. Naugle, NASA Associate Administrator for Space Science, testified on the status of space science programs and plans during hearings on NASA's FY 1974 authorization before the Senate Committee on Aeronautical and Space Sciences: "The largest and most significant current project is Viking, a program to continue the exploration of Mars with two automated spacecraft during the 1975–76 opportunity." A Mariner spacecraft would be launched in October or November toward Venus. "As the Mariner passes close to Venus its path will be bent and it will be accelerated toward . . . Mercury. If all goes well, we will make our deepest penetration toward the sun and get our first view of the sun-scorched Mercury. Pioneer G is ready for shipment to Cape Kennedy. It will be launched in April 1973 and will follow Pioneer 10 [launched March 2, 1972] through the asteroid belt to explore Jupiter.

"We are proceeding with the development of the two Mariner spacecraft planned for launch to Jupiter and Saturn in 1977. The experiments and experiment teams have been selected. We are proceeding also with spacecraft design and will be purchasing some of the subsystems—those common to Viking—in Fiscal Year 1974. The use of common subsystems and leftover Viking hardware is helping us to minimize costs on this project. The major manned mission of 1973 is . . . Skylab, which will have many scientific experiments aboard. The most important and the largest . . . is the Apollo Telescope Mount (ATM), which contains five major experiments using eight scientific instruments, all devoted to observing and studying the sun. This complex assembly of solar instruments is designed to be man-operated in a manner similar to a ground-based observatory. The ATM is now assembled as part of Skylab, awaiting launch in May 1973. An international campaign of solar observations in 15 countries . . . will be run simultaneously with ATM; a number of sounding-rocket payloads will be launched to help calibrate the instruments on ATM and to make complementary observations."

Oso 7, in orbit, "made many unique observations during the August 1972 period of intense solar activity and made the first observations of gamma rays in solar flares. OSO-I is planned to be launched late in 1974. It will carry several major experiments to make detailed measurements of the process that transports energy from the sun's surface to its intensely hot corona. The broad international participation in OSO-I is exemplified by a major French instrument and by a group of over 40 guest investigators from the U.S., France, Germany, Sweden, the United Kingdom and the U.S.S.R."

Associate Administrator for Applications Charles W. Mathews testified on special applications programs: "The Earth and Ocean Physics Applications Program (EOPAP) is an interrelated series of projects directed to the solution of important problems in earth and ocean dynamics that can best be addressed by the application of space techniques. Major areas of EOPAP are (a) global monitoring, reporting and forecasting of ocean-surface conditions and (b) monitoring of motions of the whole earth and its crust to obtain a better understanding of the processes that produce earthquakes, with the expectation that this will lead to the development of techniques for predicting the occurrence, magnitude and location of earthquakes." The second special program underway, the space-processing applications program, was "organized to exploit the unique environment of space flight, weightlessness and unlimited vacuum, for research and technological developments in material sciences and processing with the specific goal of inventing, developing and commercializing new products and manufacturing techniques for use on earth or in space." NASA's role was "that of an initiator of new technology and, following a suitable 'incubation period,' a supplier of supporting services to a developing area of commercial endeavor." (Transcript)

- Dr. Philip Handler, National Academy of Sciences President and National Research Council Chairman, announced the first steps in an extensive reorganization of the NRC "to provide a closely coordinated structure whereby we may more effectively deploy the resources of the Academy and those of the nation's scientific, technical, and health communities." NRC would consist of assemblies concerned with studies and multidisciplinary commissions concerned with public-policy problems. The Assembly of Behavioral and Social Sciences—succeeding the NRC Div. of Behavioral Sciences—and the Commission on Natural Resources had already been established. A total of three assemblies and five commissions would become constituent parts of NRC to develop coherent programs and manage advisory activities of NAS. NAS also would strengthen procedures for selecting advisory committees to preserve the quality of scientific advice to the Government and ensure that committee membership was drawn from the entire national scientific community. (NAS–NRC–NAE *News Report*, 4/73, 1, 6)

- RCA Global Communications, Inc., and RCA Alaska Communications, Inc., applied to the Federal Communications Commission for authority to build and operate an interim communications satellite system by Aug. 1. The $7.4-million system would use Telesat Canada's satellite with five earth stations to provide voice, message, and TV traffic between both U.S. coasts and Alaska until RCA's domestic satellite system, proposed in 1971, was approved and began operation. (*W Post*, 3/13/73)

- The Senate confirmed Alexander P. Butterfield as Administrator of the Federal Aviation Administration. (*CR*, 3/12/73, D214)

- President Nixon submitted *National Science Foundation: Twenty-Second Annual Report, for Fiscal Year 1972* to Congress. In his transmittal message the President said the Foundation in 1972 had "increased its support for scientific research in all disciplines and further expanded its involvement in research focused on domestic problems." (PD, 3/19/73, 252)
- The 1972 British Gold Medal for Aeronautics had been awarded to Dr. G. S. Hislop of Westland Aircraft Ltd. for design and development of rotorcraft, *Aviation Week & Space Technology* reported. The Silver Medal for Aeronautics had been awarded to Mallinson Powley of Ferranti Ltd. for developmental work on electronic systems, particularly advanced weapon systems for British aircraft. (Av Wk, 2/12/73, 9)
- The U.S.S.R. would produce the passenger door of the Boeing 727 jet transport under license, *Aviation Week & Space Technology* reported. The transaction had been negotiated between Boeing Associated Products and Z/O Licensintorg, the U.S.S.R. licensing agency, and included the sale of tooling, parts drawings, and a small amount of hardware. (Av Wk, 3/12/73, 26)
- The Atomic Energy Commission announced it would propose issuance of a construction permit and operating license to the Univ. of Utah for a research reactor on its Salt Lake City campus. (AEC release R-101)

March 13: NASA Associate Administrator for Aeronautics and Space Technology Roy P. Jackson testified on the shuttle technology programs before the House Committee on Science and Astronautics' Subcommittee on Aeronautics and Space Technology during FY 1974 authorization hearings: With the shuttle under development, the program had been converted from preparing technology to assisting in development. With this transition, "we have reduced in fiscal year 1973, and will further reduce in fiscal year 1974, our rate of Shuttle-oriented resource investments," but wind tunnels and arc jets would play a key role in shuttle development. Aerodynamic and heat-transfer investigations at subsonic through hypersonic speeds in simulated shuttle ascent would focus on effects of the rocket plume and separation of rocket engines. Wind-tunnel and structural laboratory vibration tests of scale models would provide vehicle dynamic-response characteristics. Additional tests in high-temperature wind tunnels would determine structural behavior and integrity of thermal protection systems and supporting structure for simulated earth-atmosphere-entry environments. About 14 wind tunnels would be used in the program.

Dr. Seymour C. Himmel, Deputy Associate Administrator for Technology in OAST, testified on power production for long-distance space communications: "For future spacecraft requiring 500–2000 electrical watts, our work is focused on the Brayton dynamic conversion system. This system uses thermal energy from the isotopic sources to heat a working gas which expands through a turbogenerator to produce electricity at four to five times the approximately 6 percent efficiency of current RTG's [radioisotope thermoelectric generators]. Work in fiscal 1974 will be focused on the design and fabrication of the components of an experimental conversion system. Our technology goal is to demonstrate 50,000 hours of operation." (Transcript)

- NASA Associate Administrator for Applications Charles W. Mathews described NASA's Earth Observatory Satellite (EOS) program in continued testimony before the House Committee on Science and Astronautics'

Subcommittee on Space Science and Applications during FY 1974 authorization hearings. The satellite, a potential new start for 1974, was intended to provide a multidisciplinary earth observations capability to support the areas of meteorology, environmental quality, earth resources and, in some respects, the earth and ocean physics program. The system would be the first major shuttle-compatible spacecraft with modularity, giving ability to retrieve and repair modular elements of the spacecraft, and the spacecraft might also be modular in basic size. "We may be able to build up a different size and weight spacecraft, using these modular techniques." Elements were expected to serve the several disciplines without significant changes in components through the modular buildup.

George W. Cherry, Deputy Associate Administrator in OAST, testified that Lewis Research Center was providing technical management on several Environmental Protection Agency contracts. Turbines, heat exchangers, combustors, and fuel controls obtained from NASA's low-cost jet engine research would be evaluated for use in automobile turbine engines. A memorandum of understanding with EPA would expand LeRC responsibility for component and system development for advanced automotive power systems and would exploit LeRC test facilities. (Transcript)

- L/C Eduard Burchard, West German Air Force medical officer, had begun two-year tour of duty with the Flight Medicine Section of Johnson Space Center, JSC announced. Burchard would work with U.S. scientists on Skylab. (JSC Release 73-26)

March 14: Dr. George M. Low, NASA Deputy Administrator, cited examples of space technology spinoff in testimony before the Senate Committee on Aeronautical and Space Sciences during FY 1974 authorization hearings: A heat recovery system developed from an improved means for eliminating thermal gradients in spacecraft would soon be available for use as a home unit that would provide adequate warm air and hot water from a single flame source. A fast-scan infrared camera adapted from Marshall Space Flight Center research would project 60 frames per second, giving "the same flicker-free image you receive on your TV set." An irreversible warm-up indicator adapted from NASA-sponsored low-temperature balloon battery research could warn when frozen food had been exposed to too high a temperature.

A low-cost portable drug detector developed under Ames Research Center auspices would detect drugs in urine in microscopic amounts. ARC had cooperated with the NASA-sponsored Biochemical Application Team at Stanford Univ. to develop a consumable pill to telemeter body temperature to a remote display unit. The pill could determine the location of ulcers or tumors within the intestines. The need to monitor astronauts' sleep had led to the development of a sleep analyzer—a soft electrode cap—that required no scalp preparation for good electrode contact. The NASA-sponsored Biomedical Team at Southwest Research Institute had helped make Temper Foam available for prevention of bedsores and as an impact-absorbing material for artificial limbs. The foam had been developed under ARC contract. (Transcript)

- *Apollo 17* geologist-astronaut Dr. Harrison H. Schmitt testified on the legacy of the Apollo program before the House Committee on Science and Astronautics' Subcommittee on Manned Space Flight in hearings on NASA's FY 1974 authorization: As man emerged "from the scientific revolution brought about by Apollo on the Moon with the simultaneous revolution brought about by new insight into the origins of ocean basins and continents on the Earth, we may begin to understand the great stresses and strains within our crust as ocean floors grow and continents move. These stresses and strains profoundly affect the everyday lives of people living within belts of present earthquakes and volcanic activity. Within future understanding of the frozen ocean of basalt of [Oceanus] Procellarum and the vast ridge and volcanic system that splits it, may lie the simplification of thought about past events that leads to the expansion of thoughts about present events. I will be the first to admit that these last comments give free run to the imagination. However, knowledge never becomes a resource until it is married to imagination. It is thus . . . that the scientific legacy of Apollo will be realized."

Dr. Rocco A. Petrone, Marshall Space Flight Center Director and former Apollo Program Director, reported the status of the five scientific stations deployed on the moon during the Apollo program: The first station deployed by *Apollo 12*, in November 1969, "is still functioning over 3 years after its initial deployment in the very demanding environment of the Moon. . . . The fact that this equipment can stand these rigors of temperature and vacuum . . . and still function . . . demonstrates one facet of the technology we have developed." All five stations were still functioning, though some of the experiments had been lost. The first total station, on *Apollo 12*, had been designed for a one-year life. "And yet we have had it over 3 years . . . and the seismometers worked beautifully all 3 years." The superthermal ion detector was "working beautifully and had for 3 years, night and day." NASA had "five seismic units just giving us beautiful information on the seismic tremors of the Moon and meteorite impacts, so now we can locate them fairly accurately."

Dr. Charles A. Berry, Director of Life Sciences in OMSF, reported: "We are vastly wiser in many medical and technical areas for our 'decade in space.' Much has been learned about the effects of the space flight environment upon man. We have found that some body systems respond to space flight factors . . . with changes . . . observable in the period immediately following space flight. Occasionally changes have been noticeable during the inflight period of the mission. None of these changes has been of such a severity as to cause any real concern for man's safety in space, but all changes are . . . being watched closely lest they have implications for long duration flight." Cardiovascular responses which showed the greatest changes also appeared most amenable to regulation by inflight countermeasures, such as lower body negative pressure. Medication had not proved of value in treating either the cardiovascular deconditioning or the loss of exercise capacity. Countermeasures were also being investigated for the small bone density losses noted. A decrease in heart size had been noted in both U.S. and Soviet space crewmen and some Soviet crewmen had experienced postflight muscular difficulties. "Both astronauts and cosmonauts have also reported sensations resembling sea sickness during spaceflight. On the whole, however, U.S. astronauts do not appear to be excessively plagued by motion sickness symptoms. The growth of opportunistic microorganisms appears to be favored in the space environment." (Transcript)

- Alexander P. Butterfield was sworn in as Administrator of the Federal Aviation Administration by Secretary of Transportation Claude S. Brinegar. Butterfield succeeded John H. Shaffer, who had returned to private life after four years as FAA Administrator. (FAA Release, 3/22/73)
- The American Photographic Interpretation Award of the American Society of Photogrammetry was presented in Washington, D.C., to NASA's Earth Resources Technology Satellite Team for important contributions to ASP's interests. (*NASA Activities,* 4/15/73, 72)

March 15: NASA announced selection of Lockheed California Co. and Boeing Co. for negotiations leading to contracts for structural technology studies of an arrow-wing, supersonic, cruise aircraft configuration. Each contractor would apply a Langley Research Center concept for a four-engine SST with sharply swept-back wings to the structure of a hypothetical aircraft. The aircraft would have a 340 000-kg (750 000-lb) takeoff weight, 7800-km (4800-mi) design range, 22 000-kg (49 000-lb) payload, and 15-yr economic service life. Structural concepts would be developed under two $12\frac{1}{2}$-mo, $700 000, cost-plus-fixed-fee contracts. (NASA Release 73–43)

- The Aerospace Industries Association reported that aerospace exports continued to be a major factor in the U.S. balance of trade. In 1972, aerospace exports of $3.8 billion, offset by imports of $546 million, had netted $3.2 billion positive contribution to the U.S. position. The U.S. had registered an overall negative balance of $6.4 billion. Dept. of Commerce statistics had revealed that aerospace exports had included $2.9 billion in commercial sales and $900 million in military equipment. (AIAA *News,* 3/15/73)
- NASA Associate Administrator for Tracking and Data Acquisition Gerald M. Truszynski testified before the House Committee on Science and Astronautics' Subcommittee on Aeronautics and Space Technology during NASA FY 1974 authorization hearings: Support in 1972 had been provided to more than 40 ongoing missions as well as 18 new flight projects. New launches—which included *Apollo 16* and *17* and the first probe to Jupiter, *Pioneer 10*—"placed increased demands on the capacity and reliability of the tracking systems." During the two-year, 1 000 000 000-km (621 400 000-mi) journey of *Pioneer 10* to Jupiter's vicinity, the Deep Space Network would continuously monitor the spacecraft's condition and acquire data from its 13 experiments.

  Major changes in the Spaceflight Tracking and Data Network (STDN) had been the installation of a transportable station at St. John's, Newfoundland, to provide Skylab launch support and the closing of the station at Fort Myers, Fla. The closing was "the first of the station closures planned to take place over the next few years as the spaceflight tracking and data network moves toward a single integrated network in the post-Apollo-Soyuz test project time." (Transcript)
- Johnson Space Center announced the issuance of requests for proposals to build hydraulic actuators for space shuttle applications and to test them under simulated space conditions. The actuators were being considered for aerodynamics-surface and thrust-vector control on the shuttle. Proposals, leading to a firm-fixed-price research-and-development contract, were due by April 2. Work was to be concluded 12 mos after the contract award. (JSC Release 73–27)

March 15–30: Five U.S.-U.S.S.R. joint working groups met at Johnson Space Center and reached agreements on the July 1975 Apollo-Soyuz Test Project. Technical specialists from each country would be in the other country's control center during the mission and flight crews and specialists would be trained and made familiar with each country's spacecraft and systems. Under a flammability agreement, the U.S. would provide fire resistant material for the Soviet cosmonauts' flight clothing; the cosmonauts would use an Apollo headset in the Apollo spacecraft for communications to the Soyuz spacecraft over the Soviet communications link; the Soviet TV camera used in the Apollo spacecraft would be in a fireproof hermetically sealed container; and the still camera and film used by the cosmonauts in Apollo would be stored in a flameproof container. The U.S. would provide a 16 mm camera for Soviet use during docked operations in both Apollo and Soyuz spacecraft. Agreements were reached on the prescribed procedure if either country encountered anomalous results during test activities that concerned the other country and on mission planning for rendezvous.

The groups agreed that additional testing was required on the docking seals, that the U.S. would limit Apollo control system thrusting in the roll axis to two thrusters to preclude excessive structural loads on the Soyuz spacecraft, and that an additional docking test could be performed on the second day, with the Apollo systems passive and the Soyuz active. Both countries agreed to perform and exchange safety assessments in fire, pyrotechnic devices, cabin pressure, and propulsion and control systems.

The 47-man delegation of Soviet scientists and engineers included Cosmonauts Vladimir A. Shatalov and Aleksey S. Yeliseyev and the Soviet ASTP Technical Director, Academician Konstantin D. Bushuyev. At a March 19 press conference Bushuyev and U.S. ASTP Technical Director Glynn S. Lunney agreed that cordial, friendly, and cooperative progress was being made in planning the joint mission. Preliminary drawings of the two spacecraft had been exchanged so that mockups could be built. (C. M. Lee presentation to Aerospace Advisory Panel, 4/10/73; MSC Release 73-24; JSC *Roundup*, 3/16/73, 1; Wilford, *NYT*, 3/20-73, 32; JSC PIO)

March 16: Apollo 16 and 17 lunar samples, representing the widest variety of soil and rock from both missions, were presented at Johnson Space Center to Soviet Academy of Sciences representatives Vladimir Shcherbina and Lev Tarasov by Dr. Paul W. Gast, JSC Chief of Planetary and Earth Sciences Div. The presentation of three grams (one tenth ounce) each of rock and soil from the Descartes and Taurus-Littrow landing sites was part of the U.S.-U.S.S.R. agreement for exchange of lunar samples. (JSC Release 73-28)

• Some members of Congress were urging a second series of Skylab missions beginning in 1976, *Aviation Week & Space Technology* reported. The purpose would be to avoid the hiatus in manned space exploration between the joint U.S.-U.S.S.R. Apollo-Soyuz mission in 1975 and the first orbital flight of the space shuttle in 1978. Dr. James Fletcher, NASA Administrator, had estimated the cost of a second Skylab series at less than $1 billion if the same experiments were flown. Sen. Lowell P. Weicker, Jr. (R-Conn.), would try to add funds for a second Skylab to NASA's FY 1974 authorization. (*Av Wk*, 3/12/73, 16)

- Brazil was reported to be negotiating with several companies for a $100-million domestic communications satellite system to link its cities with remote areas not reached by ground systems. A special government commission would study proposals for the satellite program to relay thousands of simultaneous phone calls and multiple TV channels. If the program was authorized, Brazil would become the third country in the world with a nationwide domestic comsat system. (LATNS, *Today*, 3/16/73)
- A crew escape module was being tested by the Air Force at Holloman Air Force Base, N. Mex. The detachable crew compartment of a B–1 advanced bomber, designed by Air Force System Command's Aeronautical Systems Division and Rockwell International Corp., would be able to save four men from a disabled aircraft over land or water. The compartment could be blasted free of the aircraft by rocket engines; the fins and spoiler would deploy for stabilization until the parachute opened to bring the men down safely. (AFSC Release 15.73)
- The Subcommittee on Priorities and Economies in Government of the congressional Joint Economic Committee released the report *Federal Transportation Policy: The SST Again*. The British-French Concorde supersonic transport was a commercial failure because of high costs, limited range-payload capabilities, and unknown environmental effects, the report concluded. It recommended that no further attempts be made to provide funds for a U.S. SST. (Text)
- NASA launched a Nike-Tomahawk sounding rocket from Poker Flats, Alaska, carrying a Univ. of Minnesota and Univ. of New Hampshire auroral studies experiment. The rocket and instrumentation performed satisfactorily but the tracking radar lost the signal at 55 sec. (GSFC proj off)

March 17: Dr. James C. Fletcher, NASA Administrator, reported on NASA's contribution toward environmental improvement, in a speech before the 37th Annual Meeting of the National Wildlife Assn. in Washington, D.C.: "No part of the changing, moving face of the globe we inhabit is free of human influence or removed from human interest. We therefore can afford to leave no part unmonitored. From forest fires to hurricanes, from the slow erosion of granite hills to the short-lived volcanic eruption or avalanche, from the atmospheric particulates to crop diseases to oil spills—we need to know the condition of our environment, minute by minute." From this knowledge would flow the interrelated set of local, national, and international measures "that will make our planet what we want it to be." The key was "to know how they will be deployed and how they will be used. I personally believe that the greatest lag today is in the development of mature social and political institutions to collect, manage, and act on the whole new level of environmental information we are able to generate." The major impact of current activities, largely data collection by *Erts 1* Earth Resources Technology Satellite (launched by NASA July 23, 1972), would be "to force the development of a new level of maturity on society." NASA's jobs were to extract, display, and use the information its tools could provide. "We must develop and expand natural and predictive models of the new environment . . . to provide . . . real assessments so necessary in real-world decision making." The larger job was the "development of competent local, regional, national, and global institutions that will build and operate such systems." (Text)

- Infrared photos were made from a leased aircraft by Marshall Space Flight Center's Environmental Applications Office during an Alabama flood emergency. Photos—for analysis of flooded areas, water sources, flow patterns, and runoff routes—had been requested by the U.S. Geological Survey. They would be used also by civil defense organizations, the Tennessee Valley Authority, Redstone Arsenal engineers, and the North Alabama Regional Council of Governments. (MSFC Release 73-35)
- Selection of four companies to continue second-phase development of a microwave landing system was announced by Secretary of Transportation Claude S. Brinegar. International Telephone & Telegraph Corp. had been awarded a $4 765 760 contract; Hazeltine Corp., $4 401 000; Bendix Corp., $3 196 090; and Texas Instruments, Inc., $3 063 840. The MLS would provide Category III (zero visibility–zero ceiling) landing capabilities as well as low-cost ground facilities for use at low-density airports. (FAA Release 73-15)

March 18: Proposals to outfit Boeing 747 aircraft with instruments to monitor pollution in the upper atmosphere were being solicited for the Global Air Sampling Program, NASA announced. Manufacturers would submit bids covering design and fabrication of instruments to sample gaseous and particulate pollutants, installation on Boeing 747, and flight-testing. The contract would be awarded by July. (LeRC Release 73-15)

- A secret Department of Defense study and an independent Air Force study had shown that costs on the F-15 fighter aircraft could run $1.7 billion —21%—above estimates given to Congress, the Washington *Star and Daily News* reported. The engine developed for the F-15 had had 52 hitherto undisclosed hardware breakdowns. Secretary of the Air Force Dr. Robert C. Seamans, Jr., was convinced the F-15 would not encounter the cost problems plagued by other large aircraft programs but was concerned about the engine problems. Seamans had asked NASA to help solve the problem and find better ways to develop new aircraft engines, the newspaper said. (Kelly, W *Star & News*, 3/18/73, A1)

March 19: President Nixon submitted *Aeronautics and Space Report of the President: 1972 Activities* to Congress. He said in his transmittal message: "The Apollo program was successfully concluded with the flights of Apollo 16 and 17. These missions were designed to obtain maximum scientific return and provided almost half the lunar exploration time in the Apollo program. Though it is far too early to attempt a definitive assessment of the value of this program, it is clear that one result will be a quantum jump in both our scientific knowledge and our technical expertise. Our unmanned satellites include a variety of vehicles ranging from meteorological, navigational and communication satellites to a new experimental spacecraft providing information on our resources and environment. Increasing practical applications for satellite technology confirm the immediate value of our efforts in space, while observatory satellites and others carrying specialized scientific instruments provide accurate and dependable data never before available to scientists on earth. The conclusion of the Apollo programs marks only another step in this Nation's push into space."

The year 1972 had also seen advances in aeronautical research and development. "It should be emphasized that work in this field is particularly vital if America is to maintain its leadership in the development and production of civil and military aircraft and engines. Our efforts in aeronautics and space will allow us to meet demands in these and other important domestic and foreign areas."

The report said the year had been one "of significant, advanced accomplishments in lunar exploration, exploration of the planets and the universe, and in aeronautical concepts." A balanced program had been initiated to "continue exploration in space, space science studies, application of space for benefits on earth, and the development of new technology for space and aeronautics" by the 16 Federal agencies reporting. (Text)

- An on-the-job training program began at Johnson Space Center to train a team from Mexico's National Commission for Outer Space (Comision Nacional del Espacio Exterior) in remote sensing. The program would prepare the team to use a remote-sensing aircraft purchased by Mexico in December 1972. NASA's instrumented NP-3A aircraft would carry the team on a survey flight that also would provide information to six Mexican investigators using the earth resources experiment package (EREP) aboard Skylab. (JSC Release 73-43)

- Sen. Henry M. Jackson (D–Wash.), with 27 cosponsors, introduced S. 1283, to establish an interagency "Earth Management Project" for research, development, and demonstration in fuels and energy for the coordination of and financial supplementation of Federal energy research and development. The bill would include an assistant administrator from NASA. (CR, 3/19/73, S5021–37)

- Academician Anatoly A. Blagonravov discussed the role of automatic vehicles in a *Pravda* article: Automatic explorers could pave the way for manned exploration but both were necessary. Space science was an inalienable part of the national economy. "It is impossible to conceive of mankind's tomorrow without further penetration of outer space. This is not a mere curiosity but a necessity of the development of science and our civilization." (FBIS–Sov, 3/23/73, L1–L3)

March 19–21: The U.S.–U.S.S.R. Joint Commission on Scientific and Technical Cooperation met in Washington. The U.S.S.R. agreed to help guide and finance a U.S. deep sea drilling project in the Antarctic. The project's drill ship, *Glomar Challenger,* had discovered evidence of natural gas in the floor of the Ross Sea. Drilling had also shown the Antarctic ice sheet to be 15 million yrs older than originally thought and was expected to increase knowledge of ocean basins, their climates, and inhabitants. Participation of the U.S.S.R. with the Joint Oceanographic Institutions for Deep Earth Sampling, a consortium of five U.S. institutions, would open up the Black and Baltic seas.

Also agreed on were joint projects in pollution control and water resources management. (Sullivan, *NYT*, 3/23/73, 1)

March 19–29: U.S. and U.S.S.R. experts meeting in Moscow on air pollution exchanged information on purification of smoke gases at power plants and on reduction of harmful waste from diesel engines. Provision for the meetings had been made in a memorandum adopted September 1972 at the first session of the U.S.–U.S.S.R. Joint Commission on Cooperation in the Field of Environmental Protection, which had provided for joint participation in designing projects, exchange of models, and exchanges of technical groups for testing, participation in symposiums, and obtaining supplementary information. (Tass, FBIS–Sov, 3/30/73, G1)

March 20: The U.S.S.R. launched *Meteor 14* from Plesetsk to obtain meteorological information for weather forecasts. Orbital parameters: 892-km (554.3-mi) apogee, 872-km (541.8-mi) perigee, 102.5-min period, and 81.2° inclination. The satellite would study cloud and snow cover and gather data on reflected heat. (GSFC *SSR*, 3/31/73; Tass, FBIS–Sov, 3/21/73, L1; *SBD*, 3/22/73, 124)

- Dr. James C. Fletcher, NASA Administrator, with other NASA officials, testified on the FY 1974 NASA authorization request in hearings before the House Committee on Science and Astronautics.

    Dr. Fletcher pointed out that the NASA request of $3.016 billion was down $400 million from FY 1973 but said the agency program was significant. In 1972 NASA had completed the transition from manned exploration to a focus on practical benefits, unmanned exploration, and development of the space shuttle "as the means to make more practical and more economical all future uses of space, unmanned and manned, especially for applications of economic, commercial, international, or national security significance." Work on uses of the shuttle was "reinforcing our view that the Space Shuttle system will not only provide the launch capability needed for future space activities of all types but will revolutionize the ways in which space is used as well. The more the specialists in various fields look at the shuttle's capabilities, especially in the sortie mode, the more they see how its weight and volume capacity and flexible operations will permit them to carry out experiments and useful applications in a much simpler and economical manner."

    Assistant Administrator for Institutional Management Joseph F. Malaga testified that NASA Hq. would have 1581 permanent positions in FY 174 after agency reduction in force, or 6.3% of the total NASA Civil Service employment. "This is down 600 positions from the 1968 peak, when it represented 6.7 percent of the NASA total. Headquarters has thus kept its size consonant with agency size."

    NASA's policy on disposition of manpower and equipment necessitated by FY 1974 budget cuts was described by Richard C. McCurdy, Associate Administrator for Organization and Management, in a prepared statement: "We are already well into the organization and operation of a comprehensive program to identify specific employee preferences and abilities and corresponding suitable potential employers. We have set up training for staff members in gathering job information and in counseling employees, and have arranged for a coordinated flow of information about job openings throughout NASA, and other Federal agencies, state and local government units, educational institutions and private industry. . . . But even more basic . . . is our recognition that these people represent a virtually irreplaceable national resource." He referred not only to "this highly talented and trained professional-technical work force but also to the complex of sophisticated plant and equipment which must be partially idled. . . . It would be unconscionable to permit key elements of America's aeronautics and space capability to be disbanded or retired without making every effort to keep them available for further service to the nation."

    Dr. George M. Low, NASA Deputy Administrator, testified on "practical returns from our space investment," including the work of NASA Applications Teams in the technology utilization program. Teams assisted local, state and Federal Government agencies in biomedicine, air and water pollution, fire safety, transportation, housing, law enforcement, postal service, and mine safety. "Those teams work with the user to explore important technical problems in his specific field and then, often with the help of specialists at NASA field centers, search for aerospace technology which can be used in the solution of the problems." A technique used by NASA to inspect space vehicles had been adapted to improve the x-ray diagnosis of tumors and cardiovascular problems, with a tenfold reduction in radiation dosage. (Transcript)

- Cochairman of the U.S.–U.S.S.R. Joint Commission on Scientific and Technical Cooperation—the National Science Foundation Director, Dr. H. Guyford Stever, and Soviet Academician V. A. Trapeznikov—met with President Nixon at the White House. Dr. Stever told the president: "The signing of the agreement on science and technology, plus the other initiatives that you have taken, have triggered a great amount of cooperation in science and research, exchange of technical data with firms. And I think the progress is excellent." (PD, 3/26/73, 288)
- NASA launched a Nike-Tomahawk sounding rocket from Poker Flats, Alaska, carrying a Goddard Space Flight Center magnetospheric physics experiment to a 225.3-km (140-mi) altitude. The rocket and instrumentation performed satisfactorily. (GSFC proj off)
- A *New York Times* editorial commented on results of the March 5–8 Fourth Lunar Science Conference: The scientists had had "far more factual information . . . than has ever before been available" but current hypotheses about the moon's history "have an obvious weakness. They are based almost entirely upon material and observations obtained from the near side of the moon." The moon's far side remained "primarily luna incognita." The next major objective of lunar exploration "over the decades ahead must be to land instruments and/or men on that far side so that scientists will at long last have samples and other data permitting them to analyze the total moon, not—as now—primarily the side we see in the sky." (NYT, 3/20/73, 36)

March 21: NASA Associate Administrator for Aeronautics and Space Technology Roy P. Jackson testified on aircraft programs during hearings of the Senate Committee on Aeronautical and Space Sciences on NASA's FY 1974 authorization: "The objective of the engine refan program . . . at Lewis Research Center is to demonstrate the feasibility and cost of significantly reducing the noise of current civil fleet engines. The JT8D, which powers the Boeing 727 and 737 and the McDonnell Douglas DC–9, will continue in extensive civil fleet use well into the 1980's. The first JT8D refanned engine ground test is scheduled for January 1974." Flight tests would occur in FY 1975. Analysis of benefits from refanning indicated that approximately 50% greater community noise improvement could be gained from a fleet-wide JT8D refan retrofit than from a fleet-wide JT3D refan retrofit. The JT3D powered the four-engine Boeing 707 and Douglas DC–8. "This analysis resulted in the decision in January to concentrate, within available resources, on the JT8D and to defer work on the JT3D." NASA's experimental quiet engine program had achieved its major goals during FY 1973 and would be continued in FY 1974. "Altitude performance tests . . . will start soon in the christening run of the Lewis Propulsion Laboratory expansion authorized in 1967. After initial acoustic testing, a sonic inlet, which chokes off forward sound propagation, will be added to engine C." Technology was in hand to permit a new generation of civil aircraft engines, with

March 21

acceptable economics, whose noise footprint in terminal operations would be less than 50% of that of the generation of relatively quiet wide-body jet aircraft now coming on the line in the civil fleet. (Transcript)

- Confirmation of U.S.S.R. development of the SS–17 missile—a follow-on to the SS–11 with first strike capability against the U.S.—was given by Dept. of Defense spokesman Jerry W. Friedheim. The press had quoted U.S. military analysts as saying the U.S.S.R. had successfully tested the SS–17, which had an onboard computer to direct it to a predetermined impact point and to adjust for variables. Reports had said the SS–17 could attack U.S. Minuteman missile silos and advance Soviet development of accurate multiple independently targetable reentry vehicles (MIRVs) for its intercontinental missiles. (DOD PIO; Beecher, *NYT*, 3/21/73, 4; *Aero Daily*, 3/22/73; *Av Wk*, 3/26/73, 11)

- NASA launched a Nike-Tomahawk sounding rocket from Poker Flats, Alaska, carrying a Goddard Space Flight Center magnetospheric physics experiment to a 239.8-km (149-mi) altitude. The rocket and instrumentation performed satisfactorily. (GSFC proj off)

March 21–25: Skylab mission simulation and flight-readiness tests were completed at Kennedy Space Center, including experiment hardware and the control system. The first Skylab crew—Charles Conrad, Jr., Dr. Joseph P. Kerwin, and Paul J. Weitz—participated in the checkout. Decisions were made to replace three water accumulators and bladders in the airlock module and the multiple docking adapter coolant system. (NASA Release 73–55)

March 22: Cosmos 552 was launched from Plesetsk by the U.S.S.R. into orbit with 308-km (191.4-mi) apogee, 202-km (125.5-mi) perigee, 89.5-min period, and 72.8° inclination and reentered April 3. (GSFC SSR, 3/31/73; 4/30/73; *SBD*, 3/23/73, 32)

- Dr. James C. Fletcher, NASA Administrator, continued testimony before the House Committee on Science and Astronautics during hearings on the FY 1974 NASA authorization. Dr. Fletcher explained the decision to cancel the quiet, experimental, short takeoff and landing (QUESTOL) aircraft program because of FY 1974 budgetary restrictions. "The Air Force STOL program turned out to use some of the advanced concepts that we had developed in NASA and some of which we were planning to use on the QUESTOL." The main difficulty with the Air Force advanced medium STOL transport (AMST) "is that it doesn't have the proper engine—you couldn't operate a noisy transport like that for commercial uses, especially in a city—and we are proceeding post haste with the advancement of technology for an engine for missions such as that, in the quiet, clean, STOL experimental engine (QCSEE) program." It was not clear when the STOL would be phased into the current fleet. "The technology would be developed by and large for commercial aircraft, with quiet engines and at some later date either we or more probably a commercial company can put these technologies together and develop that STOL as the need requires." (Transcript)

- The success of Project Boomerang, first earth-orbital flight of a balloon carrying scientific instruments, was announced by NASA. The 19-m (64-ft) balloon (launched Jan. 24 for NASA by the U.S. National Scientific Balloon Facility from Oakey Airfield, Queensland, Australia) had circled the earth 36 days at 24 000-m (78 000-ft) altitude and made two orbits before being brought back. A second balloon, launched

Jan. 28, was still aloft but had been slowed by changes in the prevailing winds. It was expected to complete its second orbit by the month's end. Each balloon had carried about 40 kg (90 lbs) of passive experiments to study cosmic rays and their effect on corn seedlings and other vegetation and to collect micrometeorites. The first balloon's payload had parachuted to within 15 km (9 mi) of the launch site. Project Boomerang was primarily an engineering test in NASA's balloon program to study near-earth phenomena. The program's goal was to build superpressure balloons that would orbit at altitudes to 40 000 m (130 000 ft), stay aloft for six months or more, and carry up to 227 kg (500 lbs) of scientific instruments. (NASA Release 73-48; NASA PIO)

- The Senate Committee on Aeronautical and Space Sciences continued FY 1974 NASA authorization hearings with testimony by NASA Associate Administrator for Tracking and Data Acquisition Gerald M. Truszynski: The NASA Communications Network (NASCOM) "during the life of the Apollo program advanced the expansion of communications satellites throughout the world by relying solely on them for direct reliable communications from certain remote tracking stations where previously marginal communications were available." Innovations in data transmission had been introduced to meet Apollo program data requirements, "and many of these new techniques have been adopted by other Government agencies, industry, and the developing countries of the world." Examples included network technical control and monitor techniques, high-data-rate transmission at low error rates, and the use of a single wideband channel for voice, data, or teletype, or any combination of them. "World standards for data transmission which were unique for Apollo . . . are now becoming commonplace. The NASCOM . . . has not only successfully completed every task assigned to it in support of our complex program, but during the process . . . has advanced the level of performance in overall communications systems."

  Assistant Administrator for Institutional Management Joseph F. Malaga and Associate Administrator for Organization and Management Richard C. McCurdy testified on NASA manpower adjustments necessitated by FY 1974 budgetary restrictions. A prepared statement by McCurdy said that college recruiting figures showed junior professionals recruited by NASA had dropped from 965 in 1966 to 56 in FY 1972 and rose only to 118 for the first half of FY 1973. "The necessary reinvigoration of the organization through the introduction of new blood is thus not yet in sight, nor will it be until some stability in staffing and supporting funding has been assured."

  Assistant Administrator for International Affairs Arnold W. Frutkin testified on the 1975 U.S.-U.S.S.R. Apollo-Soyuz Test Project: "The cooperation achieved with the Soviet Union seems to offer very considerable promise. If United States and Soviet manned spacecraft can rendezvous, dock, and transfer crew members, both countries will have increased their chance of rescuing astronauts under stress without commensurate increases in the cost of standby rescue capabilities." If successful, the ASTP mission would "point the way to future joint activities which should help both countries gain more in space than they would from separate programs." (Transcript)

- Declining jobs at Kennedy Space Center would level off at about 9600 total for contractors and Civil Service in 1976 when the space shuttle buildup began, *Today* quoted KSC Director of Administration George

March 22

A. Van Staden as saying. The number would be a drop from the current 14 500 but would mean community stability in employment for the first time. Previously NASA had projected 6000 to 7000 jobs for shuttle operations. Employment would fall by 3000 at the end of 1973, with completion of Skylab, and would drop again after the mid-1975 Apollo-Soyuz mission. When the 9600 level was reached in 1976, jobs would gradually build up for two years, to about 10 800 in 1978 when shuttle testing began. (*Today*, 3/22/73, A1)

- American Telephone & Telegraph Corp. planned to sell its stock ownership in the Communications Satellite Corp. to the public, the Washington *Star and Daily News* reported. AT&T had purchased 2 895 750 shares, 29% of the total, in 1964 and would have to sell as a condition set by the Federal Communications Commission for participation in a domestic satellite system. (W *Star & News*, 3/22/73, A20)

- The Florida Senate Natural Resources Committee had unanimously endorsed a proposal to restore the name "Cape Canaveral" to Cape Kennedy, *Today* reported. If the bill—sponsored by State Sen. Henry Sayler (R–St. Petersburg), State Sen. John Vogt (D–Merritt Island), and State Sen. Lori Wilson (I–Merritt Island)—became law, "Cape Canaveral" would appear on all official state maps and publications. (*Today*, 3/22/72, B1)

- General Dynamics Corp. successfully completed static and failsafe tests of a composite aircraft fuselage structure that was 47% graphic-epoxy, 12% boron-epoxy, and 14% fiber-epoxy. The structure demonstrated the applicability of advanced composite materials to primary aircraft fuselage construction in a program to develop techniques that would make composite fabrication competitive with conventional metal construction. (*Av Wk*, 4/2/73, 24)

- The House passed H.J.R. 5 to designate the week of April 23 as Nicolaus Copernicus Week, marking the quinquicentennial of the pioneer astronomer's birth. (*CR*, 3/22/73, D286)

March 23: NASA and the Army Air Mobility Research and Development Laboratory had requested contract bids for design, construction, flight-test, and delivery by mid-1976 of two rotor system research aircraft (RSRA), NASA announced. The RSRA would be used in the joint NASA-Army advanced rotor concepts program to develop technology for increased performance, safety, reliability, and strength and for reduced noise, maintenance, and vibration of rotary wing aircraft. The RSRA would be based at Langley Research Center as flying laboratories to test new rotary wing concepts for civil aviation. (NASA Release 73–49)

- NASA launched an Aerobee 170A sounding rocket from White Sands Missile Range carrying a California Institute of Technology payload to a 199.6-km (124-mi) altitude to map extended x-ray sources using multi-wire proportional counters. The rocket and instrumentation performed satisfactorily. (GSFC proj off)

- The Air Force Aerospace Defense Command said in Colorado Springs, Colo., that 2897 man-made objects were orbiting the earth. The U.S. had orbited 1978 and the U.S.S.R., 843. (UPI, *NYT*, 3/24/73, 7)

- The Air Force announced award of a $2 233 952 cost-plus-incentive-fee contract to TRW Inc. for operational support of a space test program applicable to atmospheric research. (DOD Release 141–72)

March 25: Kennedy Space Center's Earth Resources Office would participate with Florida's Dept. of Natural Resources in a three-year experiment to determine whether Siberian white amur could check the growth of aquatic weeds choking Florida's inland waters, KSC announced. Four experimental ponds and lakes would be stocked with white amur. Specially equipped NASA-6 twin-engine Beechcraft aircraft would photograph the fish's progress using color infrared and standard photography. If the resulting vegetation maps indicated the white amur ate undesirable vegetation, it would be introduced to other areas, saving the Government $10 million a year in harvesting and research expenses. (KSC Release 50-73)

March 26: Dr. James C. Fletcher, NASA Administrator, testified before the House Committee on Appropriations' Subcommittee on Housing and Urban Development-Space-Science-Veterans during hearings on NASA FY 1974 appropriations: "I recognize that in the allocation of funds to support the multitude of agencies and programs in the overall budget, hard decisions must be made each year based upon overall considerations of national priorities and needs. And with respect to NASA, hard decisions have already been made—in the past in cutting NASA spending almost in half and again this year in cutting NASA about $400 million below the level approved for FY 1973. But when we consider the benefits of NASA's programs—in advancing scientific knowledge, in exploration, in the practical applications of aeronautics and space, in contributing to international cooperation, and perhaps most importantly in advancing technology—and when we consider the NASA portion of the overall Federal budget for FY 1974 amounts to less than 1.2 percent, down from 4.3 percent in 1965, a factor of more than three—I am convinced that NASA more than justifies its present place in the ranking of national priorities." (Transcript)

- Establishment of the lunar data analysis and synthesis program was announced by NASA. Scientists throughout the world would be invited to propose investigations drawing on a variety of lunar data and cutting across a number of scientific disciplines to build a detailed picture of the moon's origin, history, and characteristics. The knowledge would carry implications for other solar system planets, including the earth, and would provide the basis for future mission planning and broad dissemination of new knowledge to the public. The program would be conducted by the Lunar Programs Div., headed by William T. O'Bryant, in the Office of Space Science. (NASA Release 73-50)
- The U.S.S.R. Academy of Sciences and NASA had agreed to a joint exhibit of the Apollo-Soyuz Test Project at the 30th International Exposition at LeBourget Airport, Paris, May 24 to June 3, NASA announced. Each agency would exhibit actual size models of Apollo and Soyuz spacecraft in docked configuration. (NASA Release 73-28)
- The U.S.S.R. intended to purchase 10 Apollo extravehicular spacesuits of the kind used on the *Apollo 15, 16,* and *17* missions to use in a comparative study with its own suits, *Aviation Week & Space Technology* reported. ICX–International Computer Exchange, Inc., had been negotiating the terms of the purchase between ILC Industries, Inc., which manufactured the spacesuits, and the U.S.S.R. State Committee for Science and Technology and the Institute for Medical Biological Problems of the U.S.S.R. Academy of Sciences. A contract could be signed within 30 to 60 days and the suits delivered during the last quarter of 1973 if ICX could obtain an export license from the U.S. Department of State. (*Av Wk,* 3/26/73, 17-18)

March 26

- NASA selected Kentron Hawaii, Ltd., Continental Operations, for negotiation of a contract for engineering support services at Johnson Space Center. The contract would be awarded on a cost-plus-fixed-fee basis with provisions for an award fee. (JSC Release 73–30)
- Wallops Station announced contract awards: $349 291 to RCA for a laser tracking system, $154 749 to Dynalectron Corp. for photographic laboratory services, $106 714 to Metric Systems Corp. for a real-time meteorological data-processing system, and $205 000 to the Joseph S. Floyd Corp. for rehabilitation of a sounding rocket facility. (WS Release 73–1)
- National Academy of Engineering President Clarence H. Linder, on behalf of the NAE Council, issued a statement calling for incorporation of NAE, independent from its affiliation with the National Academy of Sciences, and establishment of an NAE foundation to collect and disburse funds. Linder said the action arose "from the existence of apparently irreconcilable differences in arranging for the joint governance of the National Research Council by the two Academies." (NAS–NRC–NAE *News Report*, 4/73, 1)
- The General Accounting Office released its study *Cost Growth in Major Weapons Systems* made at the request of the House Committee on Armed Services. Cost overruns on 45 major weapon systems in the U.S. arsenal had totaled $31.3 billion. Each of the 45 systems had incurred an overrun. (Text)

March 26–27: Adm. Thomas H. Moorer (USN), Chairman of the Joint Chiefs of Staff, delivered the report *United States Military Posture for FY 1974* to the Senate Committee on Appropriations' Subcommittee on Defense Appropriations: Strategic parity existed between the U.S. and the U.S.S.R. "While the Soviet Union has a substantial advantage, not only in the numbers of strategic missiles, but also in missile throw-weight, the United States still has a substantial advantage in other areas of key importance to the overall strategic balance—e.g., missile accuracy, MIRVs [multiple independently targetable reentry vehicles], submarine quietness, sonars, and numbers of bombers and bomber payload. And, while the Soviet Union has enormous advantage in air defense, without an extensive ABM [antiballistic missile] defense . . . the air defense could be substantially undercut by ballistic missile attack, particularly . . . a large number of MIRVs." The U.S. advantages were transitory, resting on a "technological lead that has been steadily narrowing over the past decade." With the momentum of the Soviet offensive research and development program, "we cannot preclude the possibility that our technological lead in this area may further diminish or disappear altogether during the current decade." Adm. Moorer believed "that the Soviet Union will soon succeed in its efforts to develop an effective MIRV system for its strategic missiles. Thus, the Soviet Union may be able to challenge our lead in both missile guidance and MIRVs . . . , and thereby pose a significant threat to our Minuteman force." (Text)

In a statement confirmed later by Dept. of Defense spokesman Jerry W. Friedheim, Adm. Moorer said the U.S.S.R. was actively testing three new intercontinental ballistic missiles and might deploy 60 improved multiple-warhead versions of an existing missile during the summer. He said the People's Republic of China was expanding production of weapon-grade nuclear materiel and slowly but steadily building a stock of strategic and tactical nuclear weapons. (Beecher, *NYT*, 3/28/73, 4; DOD PIO)

March 27: Dr. James C. Fletcher, NASA Administrator, and Dr. Fernando de Mendonca, Director General of the Brazilian Institute for Space Research (INPE), signed a Memorandum of Understanding to extend the NASA–INPE cooperative project in remote sensing using the *Erts 1* Earth Resources Technology Satellite launched by NASA July 23, 1972. The project's purpose was to advance applications of spacecraft and aircraft remote sensing to monitor environmental conditions. INPE would establish a data-acquisition station at Cuiaba, central Brazil, and a data-processing facility in Sao Paulo state. There would be no exchange of funds; both agencies would share all data and information necessary for the conduct of the program; and INPE would make available to NASA, cost free, copies of ERTS data it acquired and processed. The data would be available to the domestic and international community. (NASA Release 73-82)

- Langley Research Center had purchased a 737–100 aircraft from Boeing Commercial Airplane Co. for research on advanced flight systems for future air terminal environments, NASA announced. The twin-jet transport, the first 737 built, would be delivered to Langley in January 1974. The $2-million contract with Boeing included modification, refurbishment, and NASA pilot and ground crew training. Special equipment would permit study of efficient air paths, aircraft performance requirements, automatic systems, displays, and pilot workloads. (NASA Release 73-57)

- A Federal Aviation Administration rule protecting the public from sonic booms generated by civil supersonic aircraft was announced by Secretary of Transportation Claude S. Brinegar. The rule, prohibiting civil aircraft from exceeding mach 1 when flying over land mass or territorial waters of U.S., would become effective April 27, 1973. (FAA Release 73-57)

- Award by NASA contractor Informatics TISCO, Inc., of $410 000 in subcontracts to three minority business enterprises for technical and production activities at NASA's Scientific and Technical Information Facility in College Park, Md., was announced by NASA. Awards to Reliable Engineering Associates, Inc., Automated Typographics, Inc., and Plato Systems, Inc., brought total NASA minority contracts to $513 540, or 15% of the total $3-million contract value. (NASA Release 73-53)

- Federal Aviation Administration award of a six-month $125 082 contract to Boeing Co. to develop data to identify aircraft noise patterns in airport vicinities was announced by Secretary of Transportation Claude S. Brinegar. Boeing would collect and analyze noise and performance data on its four in-service commercial aircraft—Boeing 707, 727, 737, and 747. (FAA Release 73-55)

- Rep. Charles A. Vanik (D-Ohio) introduced H.R. 6194, Energy Development and Supply Act of 1973. The bill would establish the Energy Development and Supply Commission, a NASA-like agency to "coordinate the Nation's energy policies and develop new sources of clean, cheap energy." (*CR*, 3/27/73, H2189)

- NASA launched a Nike-Tomahawk sounding rocket from Poker Flats, Alaska, carrying a Goddard Space Flight Center magnetospheric physics experiment to a 233.2-km (144.9-mi) altitude. The rocket and instrumentation performed satisfactorily. (GSFC proj off)

- The Air Force announced award of a $2 000 000 cost-plus-incentive-fee contract to General Electric Co. for services and supplies for reentry system components and related assembly, checkout, and integration support for an advanced ballistic reentry system. (DOD Release 146–73)

March 28: Secretary of Defense Elliott L. Richardson testified on the Air Force advanced medium STOL (short takeoff and landing) transport (AMST) prototype development program before the Senate Committee on Armed Services in hearings on the FY 1974 Dept. of Defense budget and FY 1974–1978 program: The AMST program's purpose was "to determine the feasibility of developing an operationally useful STOL aircraft which could be procured in quantity at an average unit flyaway cost (in FY 1972 dollars) of about $7 million each, as an eventual replacement for the C–130 in the 1980s. A decision on engineering development and production of this aircraft will be made only after the prototypes have been evaluated in terms of both performance and cost." The FY 1974 budget included $67 million to continue prototype development of the AMST. (Testimony)

- Grumman Aerospace Corp. would lay off 1000 employees to reduce operating expenses, the *New York Times* reported. The expense paring had been ordered by Deputy Secretary of Defense William P. Clements to increase operating efficiency after the controversy over increasing costs of the Grumman-produced F–14 fighter aircraft. (Andelman, *NYT*, 3/28/73, P28)

March 28–29: Dr. John E. Naugle, NASA Associate Administrator for Space Science, testified before the House Committee on Science and Astronautics' Subcommittee on Space Science and Applications during NASA FY 1974 authorization hearings: The Office of Space Science had assumed total responsibility for collecting, handling, and analyzing data from the Apollo program, and "we plan to devote the next several years to their detailed analysis and interpretation. Our objective is a better understanding of the origin, history and present environment of the Moon, and the definition of goals and objectives of possible future flight programs. These studies will also use the data from the still-continuing flight program of the Soviet Union." The French-built laser reflector on the Soviet lunar roving vehicle *Lunokhod 2* deposited on the moon by *Luna 21* (launched Jan. 8), "in addition to the three carried by Apollos 11, 14, and 15, provides another control point for lunar mapping studies and other scientific objectives. The close cooperation between the French and Soviets in putting the device on the Moon and between the Americans and French in securing early observations is a striking example of the possibilities for effective scientific cooperation in space among the nations of the world."

Pioneer 10 (launched March 2, 1972) was "very much a pathfinder as it goes to Jupiter. We know Jupiter has intense radiation fields. We plan for *Pioneer 10* to pass by Jupiter about three-and-a-half Jupiter radii from the planet. We think that *Pioneer 10* at that distance . . . will be able to survive as it goes through those radiation belts. If by chance it does not, then we would probably retarget *Pioneer G* so that it stayed farther out." If *Pioneer 10* survived the radiation belts, "we would probably target *Pioneer G* closer so that we could get a better look at Jupiter and get more information about the intensity of the radiation belts as we get closer to Jupiter." (Transcript)

March 28–30: A symposium on federally sponsored university research in transportation noise was conducted at Stanford Univ. by the Dept. of Transportation in cooperation with NASA and the National Science Foundation. Papers presented by university researchers and representatives of Government agencies covered jet noise, combustion noise, noise from moving bodies, sound propagation through ducts, rotor noise, high-intensity sound, tire noise, and economic and community aspects of noise. (DOT Release 15–73; DOT PIO)

March 29: USNS *Vanguard*, 181-m (595-ft) NASA tracking ship, left Port Canaveral, Fla., for Mar Del Plata, Argentina, to support Skylab missions scheduled to begin in May. *Vanguard* would be a link in NASA's worldwide tracking network relaying two-way information between a spacecraft and mission control at Johnson Space Center in Houston via Goddard Space Flight Center. En route *Vanguard* would record tracking data for the Pioneer G launch scheduled for April 5. (GFSC Release; Skylab prog off)

- Four seven-year subcontracts totaling $140 million for the design and fabrication of major structural components for the space shuttle orbiter were awarded by Rockwell International Corp. Fairchild Hiller Corp. received a $13-million contract for the design and fabrication of the vertical tail. Grumman Aerospace Corp. received a $40-million contract to design and build the double delta orbiter wing. Convair Aerospace Div. of General Dynamics Corp. was awarded a $40-million contract to build the mid-fuselage that formed the payload bay section. McDonnell Douglas Astronautics Co. received a $50-million contract to design and build the orbital maneuvering system that would aid in orbital circularization and change and in rendezvous and deorbit maneuvers. (RI Release SP–17)

- President Nixon, in a nationwide TV and radio address, said: "Our defense budget today takes the lowest percentage of our gross national product that it has in 20 years." But the U.S. "must never forget that we would not have made the progress toward lasting peace that we have made in this past year unless we had had the military strength that commanded respect. This year we have begun new negotiations with the Soviet Union for further limitations on nuclear arms. . . . If prior to these negotiations we . . . unilaterally reduce our defense budget, or reduce our forces in Europe, any chance for successful negotiations for mutual reduction of forces or limitation of arms will be destroyed." (PD, 4/2/73, 311–15)

- NASA-developed spacecraft sterilization technology was being used by the Bird Co. to produce a breathing machine that could be sterilized entirely by dry heat to prevent transfer of infectious organisms, Jet Propulsion Laboratory announced. A JPL team had worked with industry to produce a sterilizable ventilator to aid surgical patients and sufferers from respiratory diseases. Ventilator prototypes would be field-tested in at least one Los Angeles area hospital. (JPL Release 654)

- All outstanding options on the Concorde supersonic transport had been dropped, The *Washington Post* reported. British Aircraft Corp-Aérospatiale combined British-French manufacturers, would take only direct orders. The action had followed the cancellation of 20 options by U.S. commercial airline companies and would end marketing "brinkmanship." (Egan, *W Post*, 3/29/73, C11)

March 30: Preparations for twin launches of the Skylab space station and its first three-man crew continued with few significant problems, NASA announced. Major elements of the space station had completed integrated mission simulation and flight-readiness testing. Seventy percent of stowed items had been placed in the Workshop section. Simulated loading of propellant into the Saturn IB launch vehicle had been completed. A leaking valve in the service module had caused replacement of a 244- by 91-cm (96- by 36-in) panel containing propellant tanks, plumbing, and engines. Excessive leakage in the oxidizer tank bladder was also being investigated. (NASA Release 73-61)

- A video tape recorder aboard *Erts 1* Earth Resources Technology Satellite (launched July 23, 1972) had developed sporadic bursts of noise that degraded recorded photographic images, NASA announced. Investigations were underway to determine the cause and full impact of the problem. Real-time photos were unaffected. (NASA Release 73-62)
- NASA launched two sounding rockets, one from Alaska and one from White Sands Missile Range. A Nike-Tomahawk from Poker Flats, Alaska, carried a Goddard Space Flight Center magnetospheric physics experiment to a 233.2-km (144.9-mi) altitude. The rocket and instrumentation performed satisfactorily.

  An Aerobee 170A from White Sands carried a Massachusetts Institute of Technology x-ray astronomy experiment to a 184.3-km (114.5-mi) altitude. The rocket and instrumentation performed satisfactorily. (GSFC proj off)
- Light pollution was a threat to astronomy, a *Science* article reported. The level of sky light caused by outdoor lighting systems was growing at a rate of 20% per year nationwide. Light pollution was already damaging to some astronomical programs and was "likely to become a major factor limiting progress in the next decade. Suitable sites in the United States for new dark sky observing facilities are very difficult to find." Observatories should establish programs to monitor sky brightness. "The astronomical community should establish a mechanism by which such programs can be supported and coordinated." (Riegel, *Science*, 3/30/73, 1285–91)

March 31: President Nixon accepted the resignation of V/A John M. Lee as Assistant Director of the United States Arms Control and Disarmament Agency, effective April 1. (PD, 4/9/73, 335)

- NASA launched a Nike-Tomahawk sounding rocket from Poker Flats, Alaska, carrying a Goddard Space Flight Center magnetospheric physics experiment to a 239.8-km (149-mi) altitude. The rocket and instrumentation performed satisfactorily. (GSFC proj off)

During March: A large comet from 16 to 160 km (10 to 100 mi) in diameter was discovered past Jupiter's orbit by Czechslovakian-born astronomer Dr. Lubos Kohoutek, using a telescope at Hamburg Observatory in Bergedorf, West Germany. Dr. Brian Marsden, Director of the Smithsonian Institution's Central Telegram Bureau, said later that Comet Kohoutek should reach perihelion, its closest point to the sun— 21 million km (13 million mi)—Dec. 27, "when it should shine perhaps as luminously as the star Sirius." Its nearest approach to the earth would be 121 million to 129 million km (75 million to 80 million mi) in mid-January 1974. The comet should be visible six weeks before, and northern hemisphere observers should enjoy a good view of it Jan. 10–15, 1974. (Alexander, *W Post*, 4/15/73, G7)

- The House Committee on Science and Astronautics published *The Federal Government and Energy R. & D.: Historical Background*. The report, by the Science Policy Research Div. of the Library of Congress Congressional Research Service, said energy research and development in the Federal Government had "followed a typically American pragmatic course. A need has appeared—necessary research has been conducted—development has occurred almost simultaneously." One exception was the National Science Foundation. "Its support of basic research includes energy R&D projects which may or may not be based on any demonstrated needs for practical results." To a much less extent, "basic research is conducted in other Federal agencies but their projects are usually mission-oriented." There was no comprehensive national energy R&D program or policy. (Text)
- Russian translation computers of the Foreign Translation Div. of Wright-Patterson Air Force Base, Ohio, would be used in the exchange of technical and scientific information necessary for the Apollo-Soyuz Test Project, the Air Force Systems Command *Newsreview* reported. The computers, called Systrain, translated Russian to English at the rate of 100 000 to 300 000 words per hour using 10 topical glossaries and 300 000 terms. Systrain could translate multiple-meaning Russian words, resolve problems of ambiguity, analyze sentence structure, and provide finished English translations of technical documents. (AFSC *Newsreview*, 3/73, 12)

April 1973

April 1: Kenneth E. Hodges joined NASA as Director, Aeronautical Operating Systems Office, in the Office of Aeronautics and Space Technology. He had recently finished a one-year tour of duty with the Aeronautics and Space Council and formerly had been a flight-test engineer with Lockheed-California Co. (NASA Hq *WB*, 4/9/73, 1)

- *Science & Government Report* commented on results of the March 19–21 Washington, D.C., meeting of the U.S.–U.S.S.R. Joint Commission on Scientific and Technical Cooperation: "What is clear is that after two postponements of the Commission's debut and a good deal of indecisiveness, at least on the Washington end, concerning how much R&D [research and development] cooperation to offer the Soviets, both parties are approaching the relationship with caution and no evidence of haste. We will no doubt be seeing more Soviet researchers than ever before, and access to labs in the USSR will be eased a bit for American visitors, but the politically inspired match of the two research communities should not be mistaken for a raging love affair." (*Sci & Govt Rpt*, 4/1/73, 3)

April 2: Dr. James C. Fletcher, NASA Administrator, testified on NASA's role in supersonic transport aircraft development before the House Committee on Appropriations' Subcommittee on Housing and Urban Development-Space-Science-Veterans during FY 1974 appropriation hearings: "I don't think we can turn our back on supersonic technology. This is a fact of life. Military planes are flying supersonically in this country; SSTs are flying in Europe and Russia. We can not just say we don't want to build a transport ourselves and therefore we should turn off all technology in this area. What we are doing in NASA is trying to address ourselves to the main problems with supersonic flight, both military and civilian. The main problems include the economic factors involved, the pollution of the stratosphere, and the noise." NASA was the only agency qualified to investigate these problems but "I do think that the Congress has responsibility to determine any future development of a supersonic transport in this country, at least as far as the Government's support of such a program, and it is not our intention to in any way change that decision." (Transcript)

- NASA released to industry a request for proposals for design, development, and production of the space shuttle external tank. Invited to bid were McDonnell Douglas Astronautics Co., Boeing Co., Chrysler Corp., and Martin Marietta Corp.'s Aerospace Div. (NASA Release 73-64)

- New York advertising art director Richard Kline had stumbled on a radically new concept in aerodynamics while designing paper airplanes for his son, *Time* magazine reported. Wind-tunnel tests had confirmed that Kline's wing would greatly resist stalling. Instead of being curved like most airfoils, the wing was flat on top and, from the leading edge, the cross-section gradually thickened into a wedge. The underside swept abruptly forward. Tests had shown that the wing could provide lift

even when tilted at 19°, when conventional wings lost their lift and caused stalls. (*Time*, 4/2/73)
- Sir Geoffrey E. Knight, Vice Chairman of British Aircraft Corp., and Henri Ziegler, President of Aérospatiale, had received the 1973 Tony Janus Award for "significant contribution in the field of commercial aviation" for developing the Anglo-French Concorde, *Aviation Week & Space Technology* reported. (*Av Wk*, 4/2/73, 11)
- The National Science Foundation released *Resources for Scientific Activities at Universities and Colleges, 1971* (NSF 72–315). Scientists in research and development declined 1% per year between 1969 and 1971 from an annual growth rate of 6% from 1965 to 1969. Teaching staff growth rate remained at about 10% per year 1965–1971. R&D scientists dropped from 26% of the total teaching staff in 1965 to 20% in 1971. The increase of graduate students working on R&D projects fell from an annual rate of 8% 1965–1969 to 2% 1969–1971. Science expenditures in 1970 totaled $7.9 billion—an increase of 6% per year over the $7.0 billion in 1968. The 1964–1968 annual growth rate was 15%. Federal financing of R&D leveled off to a 3% increase per year in 1968 to 1970, but recent data indicated a higher rate of increase in Federal obligations to colleges and universities for FY 1972 and FY 1973. (Text)

April 3–May 28: The U.S.S.R. launched unmanned *Salyut 2* scientific space station from Baykonur Cosmodrome into orbit with 260-km (161.6-mi) apogee, 215-km (133.6-mi) perigee, 89-min period, and 51.6° inclination. The Soviet news agency Tass announced objectives were to perfect design, onboard systems, and equipment and to conduct scientific and technical experiments in space flight. Tass said systems were functioning normally.

Western observers speculated that a manned spacecraft would be launched to dock with *Salyut 2* and establish a manned orbital workshop, resuming Soviet manned space flight after a two-year hiatus. The large Soviet tracking ships *Gagarin* and *Komarov* had been sighted in March moving from the Black Sea toward positions in the Atlantic Ocean.

On April 11 Tass confirmed earlier U.S. press reports that *Salyut 2* had been maneuvered into a higher orbit. April 4 and 8 maneuvers had raised it to a 296-km (183.9-mi) apogee and 261-km (162.2-mi) perigee, with an 89.8-min period and a 51.6° inclination. Stable radio communication was being maintained and the "improved design of onboard systems and instruments" was being tested. Tass said systems and instruments were functioning normally. The new orbit, out of normal range of a manned Soyuz spacecraft, indicated to some observers a holding orbit because of delay in launching cosmonauts to join the station.

Tracking data showed an increase in the number of fragments following the spacecraft after April 14. By the end of the month Goddard Space Flight Center listed 27 objects, of which 25 had reentered the atmosphere.

On April 18 Moscow sources said the U.S.S.R. had no plans for a manned flight to link with the space station and that *Salyut 2* was carrying out experiments in connection with the joint U.S.–U.S.S.R. Apollo-Soyuz mission planned for 1975. Earlier Maj. Gen. Vladimir A. Shatalov, chief of the cosmonaut training project, had said Salyut laboratories could gather information for weather forecasting, geology, transport, communications, agriculture, and environmental protection—without mentioning manned flight.

By April 25 U.S. observers suspected that an April 14 failure, possibly an explosion or a wildly firing thruster, had sent the station tumbling end over end, tearing off the four solar panels and damaging the compartment, making cosmonaut manning of the station impossible. Radio signals from *Salyut 2* were reported to have ceased. On April 28 Tass reported that *Salyut 2* had "concluded the programme of flight" and that the data obtained in the experiments confirmed "the correctness of design and structural decisions and the properties chosen for the main systems and on-board equipment of the station. These data will be used in building a new spacecraft." The spacecraft reentered May 28.

Salyut 1, the world's first experimental orbiting workshop, had been launched April 19, 1971. A three-man crew—launched aboard *Soyuz 10* April 23, 1971—had docked with *Salyut 1* but did not board, returning to earth after two days in space. Three more cosmonauts—launched on *Soyuz 11* June 6, 1971—docked, boarded the station, and conducted scientific experiments for 23 days. During their return to earth in *Soyuz 11* June 30, they died after a valve accidentally opened and evacuated the air from the spacecraft compartment. A second Salyut was believed by Western observers to have been launched in July 1972 without achieving orbit. (GSFC *SSR:* 4/30/73; 5/23/73. Un Gen Assembly Release A/AC.105/INF.272. FBIS–Sov: 4/4,12,30/73. *NYT:* Wilford, 4/4,11,25/73; 4/29/73. *W Post:* 4/16/73; O'Toole, 4/18,25/73, 5/2/73, 10/29/73; UPI, 4/19/73. *Av Wk:* 4/9/73, 21. *A&A 1971.*)

April 3: Dr. John S. Foster, Jr., Director of Defense Research and Engineering in the Dept. of Defense, testified that DOD space shuttle efforts were being expanded, during NASA FY 1974 authorization hearings before the Senate Committee on Aeronautical and Space Sciences. A program memorandum on DOD shuttle use, to be coordinated with NASA, would include plans for early entry into the shuttle program, identify key DOD planning milestones, and treat funding and management issues. DOD would "place increased emphasis on new concepts of payload design and operation . . . essential to achieving the Shuttle's full potential for more effective, less costly space operations." Goals of DOD planning and coordination were to ensure that NASA knew and understood DOD needs so that the shuttle would be of maximum utility to DOD, to provide data essential to integrated planning and support decisions on future shuttle use, to explore ways to benefit most from the shuttle's unique capabilities, and to coordinate DOD shuttle activities with NASA to the best possible use of both agencies. The DOD Shuttle-User Committee had been established as "a focal point for our broad and varied interests in the future military use of space." The Committee, under Air Force chairmanship, included representatives of other military departments and the Joint Chiefs of Staff; NASA would be invited to provide an observer. The Committee would guide studies of payloads and missions, explore the transition from launching DOD payloads with expendable vehicles to using the shuttle, and possibly would identify experimental DOD payloads to be carried on shuttle research and development flights for NASA.

April 3

DOD's FY 1974 space program budget request of $1.6 million, $205 million more than in FY 1973, would reduce the vulnerability of the Transit navigation satellite system; procure spacecraft and boosters for the Fleet Satellite Communications (FLTSATCOM) System and the Defense Satellite Communications System (DSCS); cover research and development costs of early-warning general support; and increase supporting R&D for the space test program, for advanced surveillance technology for earth-limb-measurement satellites, and for a joint-service navigation satellite experiment. FY 1974 funding for procurement of early-warning satellites and boosters had been reduced. (Transcript)

- NASA Deputy Associate Administrator for Space Science Vincent L. Johnson testified on NASA's balloon program as the House Committee on Science and Astronautics' Subcommittee on Space Science and Applications concluded FY 1974 authorization hearings: "Balloons are used for a variety of programs in both atmospheric science and astronomy. We, in many cases, launch with a balloon a prototype of an astronomical instrument which we may later fly on a sounding rocket or orbiting satellite." Payloads ranged from test instruments to an infrared experiment; "as you get above the water vapor in the atmosphere you can do very good infrared experiments, and you can keep the instrument pointed precisely for quite long periods . . . and do very valuable scientific work at nominal cost." Balloons were launched from Texas, South Dakota, Missouri, and Alabama in the U.S., and from Canada, Australia, and Argentina. "One of the beauties of the balloon program is that they can be launched from many areas, and we can recover the payloads. . . . We use the same payload many times for balloon observations, and it's not at all unusual to use the same payload for half a dozen flights with sometimes slightly different instrumentation each time." (Transcript)

- U.S. Geological Survey geologist Dr. Peter D. Rowley said in Denver, Colo., that results of a three-month field trip in Antarctica had backed the theory of a continental drift occurring over hundreds of millions of years. The field work, ending in January, had shown that mountains on the Antarctic peninsula's Lassiter Coast and adjacent areas were "missing pieces that can link the great mountain chains of the world." Rock samples taken by his team were 65 million to 135 million yrs old and had been pushed above the ocean's surface by upheavals that occurred when the earth's crust cooled. (UPI, LA *Her-Exam*, 4/4/73)

- NASA launched a Black Brant VC sounding rocket from White Sands Missile Range carrying a Naval Research Laboratory payload to a 245.3-km (152.4-mi) altitude to evaluate instrumentation for the calibration rocket (CALROC) series to be launched in support of the manned Skylab missions [see Jan. 22]. The rocket performed satisfactorily but minimum data were acquired from the instrumentation. (GSFC proj off)

April 4: Launching of Skylab, the Nation's first space station, was set for May 14 at 1:30 pm EDT from Kennedy Space Center. The launch was scheduled after completion of a comprehensive two-day review of prelaunch test results and remaining work by top Skylab program officials. (NASA Release 73–67)

- *Nimbus 5* (launched by NASA Dec. 11, 1972) was adjudged a success. It had satisfied the mission objectives of improving the capability for vertical sounding of temperature and moisture in the atmosphere for 10 wks and demonstrating improved thermal mapping of the earth.

Four of the six *Nimbus 5* experiments were functioning normally at the end of the 10-wk period ending Feb. 19, 1973. The infrared temperature profile radiometer was operating in a limited mode. The electrically scanning microwave radiometer had failed Jan. 4. (NASA prog off)

- The Apollo Honor Awards Ceremony at NASA Hq. celebrated the success of the Dec. 7-19, 1972, *Apollo 17* mission and honored 47 persons for contributions to the success of the entire program. Dr. James C. Fletcher, NASA Administrator, said in a foreword to the ceremony that Apollo's richest achievement might be the spiritual unity it provided "by enabling all Americans, and all the world watching our astronauts' progress moonward, to discover the uniqueness of our beautiful yet fragile Earth as viewed from the perspective of the Moon. Unique in the solar system, it is the only home of the Family of Man; and seen from the perspective of the Moon, the common humanity that we all share becomes more evident than the temporal differences that keep us apart."

  Dr. Fletcher and Dr. George M. Low, NASA Associate Administrator, presented awards, while *Apollo 17* Astronauts Eugene A. Cernan, Ronald E. Evans, and Dr. Harrison H. Schmitt congratulated recipients. The NASA Distinguished Service Award was presented to Dr. Charles A. Berry, Director of Life Sciences, and Director Chester M. Lee of the Apollo-Soyuz Test Program, who had been Mission Director for *Apollo 12* through *17*. Exceptional Service Medals were presented to 21 employees; the Exceptional Scientific Achievement Medal to 14 scientists from Government, industry, and universities; and the Group Achievement Awards to 10 teams from industry and Government.

  The Distinguished Public Service Medal, NASA's highest award to non-Government personnel, went to: Dr. B. Paul Blasingame, Delco Electronics Div., General Motors Corp.; Joseph F. Clayton, Aerospace Systems Div., Bendix Corp.; Clinton H. Grace, International Business Machines Corp.; Robert E. Greer, George W. Jeffs, and Joseph P. McNamara, Rockwell International Corp.; Thomas J. Kelly, Grumman Aerospace Corp.; H. Douglas Lowrey, Chrysler Corp. Space Div.; Richard H. Nelson, Boeing Co. Space Div. Aerospace Group; and Theodore D. Smith, McDonnell Douglas Astronautics Co. (Program; NASA Release 73-60; RI pro)

- Appointment of Dr. Myron S. Malkin, Deputy Assistant Secretary of Defense (Technical Evaluation), as Director of the State Shuttle Program in the NASA Office of Manned Space Flight effective April 9, was announced by NASA. Malkin would be responsible for planning and directing the design, development, and test of the space shuttle system. (NASA Release 73-68)

- Dr. Karl G. Harr, Jr., President of the Aerospace Industries Assn. of America, Inc., testified on the effects of the Apollo program before the Senate Committee on Aeronautical and Space Sciences during NASA FY 1974 authorization hearings: Apollo "did not just sweep across our landscape to successful achievement of its specific objectives. Its effects on Earth are permanent and pervasive. Not just the hundreds of thousands of jobs it provided, the schools and homes it caused to be built, and the economic impact of the program itself; not just the invaluable accretion of scientific knowledge; not just the restoration of faith in American technological leadership—but a radical and enduring upgrading of technological capability throughout American industry.

It is such advances that have enabled the United States to retain its preeminence in high technology production. And it is the continuation of such preeminence that is at stake when we consider the future of our space program." (Transcript)

April 4–26: Apollo 15 Astronaut James B. Irwin was hospitalized in Aurora, Colo., after a heart attack. He was released April 26 after three weeks treatment in the Fitzsimmons Army Medical Center in Aurora. A hospital spokesman said recovery ultimately would be complete, but "it will be considerable time before he can return to full action." Since leaving NASA, Irwin had worked for the Johns-Manville Corp. and had begun an evangelistic organization, High Flight, Inc. (B *Sun,* 4/27/73; *NYT,* 4/27/73, 43)

April 5–26: NASA's *Pioneer 11* (Pioneer-G), second Jupiter probe, was successfully launched into an excellent trajectory from Eastern Test Range at 9:11 pm EST by a three-stage Atlas-Centaur–TE–M–364–4 launch vehicle. Eighteen hundred persons watched the nighttime launch after touring the Kennedy Space Center Flight Training and Vehicle Assembly Buildings, where Skylab was being prepared for a mid-May launch.

Pioneer 11 sped away from earth at a velocity of 51 800 km per hr (32 200 mph), equaling the speed of *Pioneer 10* (launched March 2, 1972, and still heading toward Jupiter), which flew faster than any previous man-made object. The primary objective was to obtain precursory scientific information beyond the Mars orbit with emphasis on investigation of the interplanetary medium during the 609- to 825-day journey, the Asteroid Belt, and Jupiter and its environment. The secondary objective was to advance the technology for long-duration flights to the outer planets.

The 259-kg (570-lb) spin-stabilized spacecraft carried 12 scientific experiments to measure magnetic fields, plasma, cosmic rays and charged particles; the electromagnetic radiation in the ultraviolet, visible, and infrared ranges; and the asteroid-meteoroid population. During the encounter trajectory with Jupiter, the spacecraft would provide the best possible information on the radiation environment, provide good viewing conditions of Jupiter before periapses, and obtain a short occultation of the spacecraft by Jupiter.

The spacecraft also carried a pictorial plaque identical to the one carried on *Pioneer 10* for identification of its origin by any possible intelligent beings of another solar system.

Six hours following launch the large number of thruster pulses used to precess the spacecraft and point the antenna toward the sun provided the impulse to deploy fully one of the two radioisotope thermoelectric generators that had failed to deploy completely at first. The magnetometer boom was deployed and the spin rate stabilized at 5.4 rpm. On April 6 four science instruments—the helium vector magnetometer, charged particles experiment, geiger tube telescope, and meteoroid, detector—were turned on.

An April 11 midcourse correction moved the target point at Jupiter from slightly ahead of the planet as it moved on its orbit, and above the equatorial plane, to slightly behind the planet and below the equator. The adjusted course would allow several mission choices: to make an equatorial pass either close to or distant from the planet to inspect one or more of its moons; to pass over the flattened pole of the planet;

to fly on to Saturn in 1980; or, like *Pioneer 10*, to leave the solar system and escape into interstellar space. The option chosen would depend on *Pioneer 10* findings during its pass of Jupiter in December. Nine of the twelve experiments were turned on and were operating normally.

The control center for *Pioneer 11* was transferred April 16 from Jet Propulsion Laboratory to the Pioneer Missions Operations Center at Ames Research Center. On April 26 *Pioneer 11*, traveling at 9.34 km per sec (20 900 mph), was 14 350 000 km (8 920 000 mi) from the earth. A second midcourse correction was made. The operation of all spacecraft subsystems was normal and 11 of the 12 experiments were transmitting good data. Six to seven hits had been recorded by the meteoroid detector. The infrared radiometer remained to be turned on. The spacecraft was expected to enter the Asteroid Belt in August 1973 and reach Jupiter in December 1974.

The Pioneer program, begun in 1958, was directed by Ames Research Center. *Pioneer 6* and *7* (launched Dec. 16, 1965, and Aug. 17, 1966, to study the heliocentric space environment) and *Pioneer 8* and *9* (launched Dec. 13, 1967, and Nov. 8, 1968, to study interplanetary phenomena) continued to supply data from heliocentric orbit on solar plasma, magnetic and electric fields, and cosmic rays. (NASA prog off; NASA Releases 73-41, 73-72; ARC Release 73-43; KSC Release 61-73)

April 5: The U.S.S.R. launched *Molniya II-5* communications satellite from Baykonur Cosmodrome into orbit with 39 828-km (24 748.0-mi) apogee, 525-km (326.2-mi) perigee, 717.7-min period, and 65.2° inclination. The satellite would help provide a system of long-range telephone and telegraph radio communications in the U.S.S.R. and transmit Soviet central TV programs to the Orbita network. (GSFC *SSR*, 4/30/73; Tass, FBIS–Sov, 4/6/73, L1; *SBD*, 4/6/73, 205)

- Dr. James C. Fletcher, NASA Administrator, and *Apollo 17* Astronauts Eugene A. Cernan, Ronald E. Evans, and Dr. Harrison H. Schmitt attended an awards ceremony at Kennedy Space Center honoring KSC personnel and contractor employees for their contributions to the *Apollo 17* mission (launched Dec. 7, 1972). (KSC Release 54-73)

- NASA research to improve general-aviation-aircraft flight control systems and increase the use of new flight equipment was described by Flight Research Center and Langley Research Center engineers at the Society of Automotive Engineers Business Aircraft Meeting in Wichita, Kan. FRC flight tests of a new attitude-command-control system aboard an experimental aircraft had shown that a flight-path control system similar to an automobile's steering system was a great improvement over conventional control systems. Use of the new system, plus an improved flight display tested by FRC, could reduce the general-aviation workload throughout most flights. Astronautics Corp. of America was developing a low-cost flight director under NASA contract. The Univ. of Kansas would evaluate its economical, separate, surface-stability-augmentation flight control system on a modified commuter aircraft under a NASA grant. LaRC was developing structural design technology to provide the greatest chance of pilot and passenger survival in a crash. Full-scale aircraft crashes were simulated by using a 60-m (200-ft) pendulum to swing instrumented aircraft into the ground at controlled angles and speeds. Resulting data were used to design aircraft to absorb more of the crash forces without destroying the cabin areas. NASA stall and spin studies at LaRC used a vertical wind tunnel, unique in the U.S., to develop data for designing aircraft with improved spin recovery characteristics. (NASA Release 73-63; FRC Release 7-73)

- Lewis Research Center scientists and the U.S. Coast Guard were developing an ice information system to reduce multimillion-dollar annual losses from ice that blocked commercial shipping on the Great Lakes, NASA announced. The system would use aircraft, or eventually satellites, with sensors to gather information on ice thickness, type, and distribution. The information would be relayed in map form to ice breakers and cargo vessels. (NASA Release 73-66)

- Long-operating orbital stations with changing crews would mark the main trend of Soviet cosmonautic development in the near future, Maj. Gen. Vladimir A. Shatalov, director of cosmonaut training and veteran of three space flights, said in a Moscow press interview. Along with spacecraft to deliver scientists and technicians, instruments, fuel, and food in orbit, the stations opened unseen possibilities for global meteorology, observations of the world's oceans, long-distance communications, and geological survey. (Tass, FBIS–Sov, 4/5/73, L3)

- The Federal Aviation Administration released *Aviation Forecasts, Fiscal Years 1973–1984*. Airline passenger traffic was expected to increase 10% annually, passing the 500 million mark in 1984. The number of hours flown by general-aviation aircraft would increase from 27.7 million in FY 1973 to 46.6 million in FY 1984 and the general-aviation fleet from 136 000 to 217 000 aircraft. U.S. airline passenger enplanements would increase from 200.4 million to 524 million in FY 1984, with international passengers increasing from 21.3 million to 62.8 million and domestic from 179.1 million to 461.2 million. Revenue passenger-miles recorded by U.S. air carriers were expected to reflect a continuing long-term trend of increased average passenger trip length; the FY 1984 total of 500.5 billion was expected to more than triple the FY 1973 figure of 162.1 billion. The air carrier fleet would increase by 1000 additional aircraft, bringing the total fleet to 3600, of which 60% would be wide-body aircraft accounting for 80% of available seat-miles.

    Operations at airport control towers were expected to rise from 56.8 million to 112.5 million. Annual production of U.S. civil aircraft would jump from 10 005 in FY 1973 to 24 465 in FY 1984, with general aviation accounting for most of the gain. Annual air carrier aircraft production would increase from 205 to 265, while the general-aviation figure would soar from 9800 to 24 200. (FAA Release 73-60)

- The National Academy of Engineering announced election of 70 members, including Director Edgar M. Cortright, Jr., of Langley Research Center; Dr. Eberhard F. M. Rees, Marshall Space Flight Center Director; Ames Research Center engineer Robert T. Jones; Robert J. Parks, Assistant Jet Propulsion Laboratory Director for Flight Projects; and Communications Satellite Corp. President Joseph V. Charyk. (NAE Release, 4/5/73)

- Award of a $6 826 250 cost-plus-fixed-fee contract to Boeing Co. for advanced development of an integrated aircraft propulsion control system was announced by the Air Force. (DOD Release 165-73)

April 6: The U.S. and Brazil governments confirmed the Memorandum of Understanding signed by NASA and the Brazilian Institute for Space

Research March 27. The agreement called for extension of a cooperative project in earth resources remote sensing. (NASA Release 73–82)
- NASA announced the appointment of Edward Z. Gray, Assistant to the President of Grumman Aerospace Corp., as Assistant Administrator for Industry Affairs and Technology Utilization, effective April 16. Gray, an Associate Fellow of the American Institute of Aeronautics and Astronautics, had been with NASA's Office of Manned Space Flight for four years before joining Grumman in 1967. From 1940 to 1963 he was with Boeing Co. (NASA Release 73–54)
- The Senate confirmed the nomination of Col. Michael Collins (USAF, R.) to be a brigadier general. Gen. Collins, *Apollo 11* command module pilot, was Director of the National Air and Space Museum of the Smithsonian Institution. (*CR*, 4/6/73, S6858; *NASA Astronauts*)
- Discovery of the most distant quasar yet detected was reported in the British journal *Nature* by Univ. of Arizona astrophysicists. Quasar OH471 appeared to be receding from the Milky Way galaxy at a speed 91% of the speed of light. The previous most distant quasar had been calculated to be receding at an 89% speed. (Carswell, Strettmatter, *Nature*, 4/6/73, 394)
- NASA Associate Deputy Administrator Willis H. Shapley testified on improved opportunities and working conditions for black employees at NASA's tracking and data-acquisition station near Johannesburg, South Africa, before the House Committee on Foreign Affairs' Subcommittee on Africa: "We believe that since 1971 we have made reasonable advances in obtaining the cooperation of the South African authorities in efforts to improve conditions for blacks employed at the station. While the situation will never be satisfactory to us as long as South Africa's apartheid practices continue, we believe that within the constraints of the existing situation . . . some concrete gains have been achieved and others will follow." (Testimony)
- President Nixon submitted to the Senate the nomination of Dr. Fred C. Ickle to be Director of the U.S. Arms Control and Disarmament Agency. The nomination was confirmed June 30. Ickle, a RAND Corp. executive and former Massachusetts Institute of Technology professor, would succeed Gerard C. Smith, who had resigned. (*PD*, 4/16/73, 340, 378; Kilpatrick, *W Post*, 4/7/73, A2; *CR*, 6/30/73, S12726)
- The dismantling of the Federal fellowship program would liquidate some excesses but, "on balance, is a destructive move" at a time "when the need for some kinds of scientists and engineers is actually growing," a *Science* editorial said. "After Sputnik was launched, this nation engaged in a frantic effort to expand its scientific capabilities. For a number of years government funds available for research in the physical and biomedical sciences increased rapidly. At the same time, the Apollo program was implemented. These developments created a shortage of scientists and engineers." When Government support ceased to grow and unemployment arose, those in the Government who wanted to dismantle the fellowship program had had a useful excuse. But the unemployment argument was no longer valid. "The index of employment opportunities has climbed above 100. In some regions there already are shortages of engineers." In years ahead, "this nation will encounter many unexpected problems requiring the skills of scientists and engineers. We may well come to regret bitterly the fact that we have been unable to do better than follow destructive blow-hot, blow-cold educa-

tional policies. We should adopt the more realistic assumption that this nation must have good science, good medicine, and good engineering, and we should make it possible for the top students, regardless of financial ability, to participate." (Abelson, *Science*, 4/6/73, 13)

April 6-13: Two satellites, five aircraft, and ten surface vessels were used to study the New York Bight. The waters extending from Staten Island to more than 32 km (20 mi) east of Asbury Park, N.J.—used as a dump for sewer sludge, acid wastes, and dredging and excavation materials—were dyed green. Water circulation at and near the surface was mapped to plan balanced use of the coastal environment and to assess the impact of natural and man-made substances on the coastal zone. Data gathered by sensors on the aircraft and on *Erts 1* (launched July 23, 1972) and *Noaa 2* (launched Oct. 15, 1972) were checked for accuracy by measurements gathered by surface vessels. The program was managed by the National Oceanic and Atmospheric Administration with participation by NASA and the Navy, Environmental Protection Agency, Coast Guard, Federal Aviation Administration, and State of New Jersey. (NOAA Release 73-60)

April 8: The first closeup radar images of the moon's surface—obtained during the Dec. 7-19, 1972, *Apollo 17* mission—had been assembled by scientists and engineers from Jet Propulsion Laboratory, the Univ. of Utah, the Univ. of Michigan, and the Center for Astrogeology of the U.S. Geological Survey, NASA announced. The "pictures," products of the Apollo lunar sounder experiment, would eventually provide a geological cross-section of the moon, detailing subsurface structure to a depth of 1.3 km (0.8 mi). (NASA Release 73-65)

• The April 6 announcement of the discovery of a distant quasar had strengthened the growing theory among astronomers that they had seen the edge of the universe, Walter Sullivan said in the *New York Times*. The implication was that, as Einstein believed, the universe was finite. "Beyond its expanding volume, . . . nothing exists—not even space, because, in this concept, over such great distances space curves back upon itself." By determining the rate at which a quasar was flying away from the earth, astronomers could estimate its distance in terms of the time its light had traveled to the earth. The universe appeared to have originated in a "big bang" some 13 billion yrs ago, and the earliest quasars should be visible far enough away for their light to have been traveling earthward for that long. "But looking across vast distances—and hence, far back into time—man can see out only to 12 billion years (in the most distant quasars)." Hale Observatory astronomer Dr. Alan R. Sandage had said that in these quasars astronomers were seeing "the edge of the world." Astronomers had long suspected the existence of some sort of wall preventing them from seeing quasars in the region beyond 12 billion yrs. "Now it is beginning to look as though the 'wall' is real." (*NYT*, 4/8/73, 63)

April 8-23: *Lunokhod 2* (landed on the moon by *Luna 21* Jan. 16) began its fourth lunar day in the Sea of Serenity near the Le Monnier Crater with the activation of its solar battery. All onboard systems functioned normally after the third lunar night. During an April 9 session cosmic ray characteristics were researched, magnetic measurements made, and panoramic photos of the parking site were taken. *Lunokhod 2* was brought to an area in the eastern part of the crater where a layer of basalt lava was bisected by a fissure 16 km (10 mi) long and 100 m

(328 ft) deep in some places, to study the chemical composition and magnetic properties of the lunar rock in the fissure. Ledge rocks aided *Lunokhod 2*'s movement. The vehicle did not stick in the loose ground and its velocity could be increased.

On April 17 the lunar vehicle was halted to replenish electric power reserves before making a complex trip around the fissure. On April 23 *Lunokhod 2* was parked for the lunar night after covering 11 km (7 mi). (Moscow Domestic Service, FBIS–Sov, 4/9/73, L1; Tass, FBIS–Sov, 4/11/73, L1–L2; 4/16/73, L1; 4/18/73, L1; Reuters, *NYT*, 4/23/73, 43)

April 9: Appointments of Philip E. Culbertson as Director of the new Mission and Payload Integration Directorate in the Office of Manned Space Flight and of Capt. Robert F. Freitag (USN, Ret.) as Deputy Director of the Advanced Programs Directorate, OMSF, were announced by NASA. Culbertson, formerly Director of Advanced Manned Missions, would be responsible for planning, direction, and coordination of payload activity and for management of the interface between users and the space shuttle, sortie lab, and tug. Freitag would participate in planning future manned space flight systems and in early development work. (NASA Hq *WB*, 4/9/73, 6)

- Soviet test pilot Aleksander Fedotov had flown an advanced MIG–23 jet fighter aircraft at 2599 km per hr (1615 mph) in a closed-circuit flight of 100 km (62 mi), to beat his own October 1961 Class C, Group III aircraft record of 2401 km per hr (1492 mph), the U.S.S.R. announced. Tass said details of Fedotov's flight were being submitted to the International Aeronautical Federation in Paris. (*NYT*, 4/9/73, 49; NAA Record Book)

- Flight Research Center technologist Kenneth W. Iliff—a 32-year-old, wheelchair-bound victim of bulbar and paralytic polio in his youth—had been selected by NASA as the agency's Outstanding Handicapped Federal Employee of the Year, FRC announced. Dr. James C. Fletcher, NASA Administrator, had nominated Iliff, to receive the same award for the entire Federal Government. Iliff would receive his doctorate from the Univ. of Southern California at Los Angeles in June, had helped develop advanced analytical techniques for NASA, including an optimal parameter estimation. (FRC Release 9–73)

- Two European space consortiums were in competition to develop the manned spacelab to be flown on the U.S. space shuttle, *Aviation Week & Space Technology* reported. Groups led by ERNO–VFW–Fokker and Messerschmitt-Boeklow-Blohm GMBH were in Phase B studies to define Spacelab (sortie lab) costs and configurations. The studies would end in July to allow the European Space Research Organization to decide on supporting Spacelab development. ERNO planned to offer two Spacelab configurations. One would be certain of meeting the established cost criteria. The second would be more expensive but technically more advanced. The MBB system used a common support system of a single module housing all general-purpose subsystems, with separate individual payload systems inserted as needed. Scientific payloads could be prepared independently of launch cycle constraints. One set of payload experiments could return to earth and another set be slid into place within two weeks. (*Av Wk*, 4/9/73, 22)

- NASA launched an Aerobee 170 sounding rocket from White Sands Missile Range carrying a Naval Research Laboratory astronomy experiment to

April 9

    a 197.8-km (122.9-mi) altitude. The rocket and instrumentation performed satisfactorily. (GSFC proj off)

April 10: Skylab Astronauts Charles Conrad, Jr., Joseph P. Kerwin, and Paul J. Weitz completed training at Marshall Space Flight Center for a three-hour extravehicular activity scheduled for the Skylab 1-2 mission, set for May 14 and 15 launches. Although Conrad and Kerwin were to perform the EVA, all three crew members trained so that any combination of two could accomplish the activity. The training—begun in February 1972—took place in MSFC's Neutral Buoyancy Simulator, which held a full-scale mockup of a major portion of the Skylab cluster in a simulated weightless environment. (MSFC Release 73-55)

- *Apollo 17* astronaut-geologist Dr. Harrison H. Schmitt championed manned space flight in testimony before the Senate Committee on Aeronautical and Space Sciences during FY 1974 NASA authorization hearings: "Man's unique capabilities in space flight come into play in situations requiring special insight in dealing with complex processes or events occurring over relatively short timespans. This has been most obvious in space in the geological exploration of the Moon, for example, in the rapid and comprehensive collection of samples, photographs, and observational data in the valley of Taurus-Littrow last December. This capacity will also be clear in certain aspects of Earth and solar observation where short-term dynamic processes are involved and in many space laboratory situations where rapid adjustments to changing experimental conditions are necessary. The dynamic range of the human eye and the mind's capability for rapid visual integration of changing geometry, color, and lighting make up an extremely versatile sensor system. In addition, when combined with experience and training of the man, this system allows for the efficient selection of other sensors pertinent to the problem at hand. By such selection man can intelligently reduce the quantity of data which analytical systems on the ground will be required to sort and treat in detail. This capability for on-the-spot data selection is not a trivial aspect of man's presence at the data collection point. Probably one of man's most useful capabilities is the precision by which he can operate, adjust, modify, or repair equipment. There is as yet little competition for man in this area on Earth or in space flight." (Transcript)

- President Nixon issued Proclamation 4206 designating the week of April 23 as Nicolaus Copernicus Week in celebration of the 500th anniversary of the Polish astronomer's birth. He said, "This anniversary should also serve to remind us that the study of science is one of man's noblest pursuits." (*PD* 4/16/73, 356)

- President Nixon submitted to the Senate the nominations of Principal Deputy Assistant Secretary of Defense for Public Affairs Jerry W. Friedheim to be Assistant Secretary of Defense for Public Affairs and of John O. March to be Assistant Secretary of Defense for Legislative Affairs. Friedheim would succeed Daniel Z. Henkin, and March, Gardiner L. Tucker, both of whom had resigned. (*PD*, 4/16/73, 355, 365, 378)

- A "quiet but intense" debate was developing in the Dept. of Defense over which service would "dominate America's nuclear arsenal and win the major share of appropriations," the *New York Times* reported. In "the most serious challenge of Air Force primacy since the start of the nuclear era," the Navy was pressing for construction of 12 Trident submarines with multiple independently targetable reentry vehicles (MIRVs). As a weapon system the Trident would be "more mobile and less vulnerable than ICBM's [intercontinental ballistic missiles] and more lethal than bombers." The debate had remained "in the house" to date, but few believed it could be kept from Congress and the media. "The amounts involved are too large and Congressional hostility to what is considered excessive spending too pronounced." (Middleton, *NYT*, 4/10/73, 16)

- Aviation pioneer Col. Clarence M. Young (USA, Ret.), holder of U.S. Civil Air License No. 2, died in Cottonwood, Ariz., at the age of 84. He had been appointed Director of Aeronautics in the Dept. of Commerce in 1927 and helped develop the Federal airways system. From 1929 to 1933 he was Assistant Secretary of Commerce for Aeronautics. Young joined Pan American World Airways, Inc., in 1934 to pioneer the first flying service across the Pacific. He was named to the Civil Aeronautics Board in 1946 and rejoined Pan Am in 1950 to direct its Pacific-Alaska Div. until his retirement. He learned to fly as a bomber pilot in World War I, was a member of the Aviation Hall of Fame, and had been named an Elder Statesman by the National Aeronautic Assn. (AP, *NYT*, 4/12/73, 44)

April 10–12: The Aerospace Industries Assn. of America's International Committee held its Spring National Meeting in San Francisco. Richard J. H. Barnes, Director of International Planning & Programs in NASA's Office of International Affairs,, discussed U.S.-European cooperation on the space shuttle sortie lab (designated Spacelab by the European Space Research Organization). ESRO was sponsoring a Phase B design competition between European industrial consortiums. Germany, Italy, Belgium, and Spain were participating, with other European countries expected to join. Germany had pledged to fund at least 40% of the program; Italy 20%. The program's hardware phase was expected to start in early 1974. ESRO would design, develop, and deliver one Spacelab flight unit to NASA. NASA would provide assistance and would purchase one or more production units. Participating European countries would receive preference over nonparticipating countries in future use of the Spacelab.

Ludwig Boelkow, President of the West German consortium Messerschmitt-Boelkow-Blohm GMBH, described the status of European civil aircraft production: "The resources which only the U.S. had in the early fifties are now also available in Europe. . . . Europe today enjoys a slight edge in SST [supersonic transport] technology, as evidenced by the Concorde, and I believe equality in subsonic jets, as evidenced by the excellent progress in terms of cost, schedule, and test performance of the A300 B [European airbus]." (AIA Memorandum GEN 73–43)

April 11: The Air Force Systems Command announced it was developing the Air Force Satellite Communications (AFSATCOM) System to satisfy high-priority Air Force requirements for command and control of forces. AFSATCOM's space segment would consist of Air Force ultrahigh-frequency transponders on the Navy Fleet Satellite Communications (FLTSATCOM) System, the Air Force Satellite Data System (SDS), and the global backup capability on other Dept. of Defense host satellites. The production contract award would be made after an extensive test program. (AFSC Release 029.73)

- The Federal Communications Commission granted permission to Communications Satellite Corp. to contract for three satellites to be used for a maritime communications satellite system. A construction permit would be granted later. FCC stipulated that ComSatCorp must allow all carriers already providing maritime service to invest and participate in the entire system. (AP, *W Star-News*, 8/30/73, D6; FCC PIO)
- U.S., French, and British scientists would witness the eclipse of the sun on June 30 while flying aboard the Anglo-French Concorde supersonic transport aircraft, Concorde's French manufacturer Aérospatiale told the press in Paris. A special flight at twice the speed of sound along the 3000-km (1900-mi) eclipse path from Las Palmas, Canary Islands, to Fort Lamay, Chad, would extend observing time to 80 min—11 times what it would be on the ground and 7 times what it would be from a subsonic aircraft. (UPI, *NYT*, 4/12/73, 13)
- Developing technologies in short-haul air transportation were discussed by Rep. Silvio O. Conte (R-Mass.) in a speech before the Short-Haul Air Transportation–STOL Symposium in McLean, Va.: "Steep approaching, quiet, short-runway aircraft are now appearing on the drawing boards. The 45-passenger DeHavilland DHC-7, able to operate from short runways quietly and economically, is scheduled to appear by the mid-70's. The Air Force is going ahead with construction of two prototype jet STOL [short takeoff and landing] transports from which technical fallout is inevitable. The major airframe manufacturers are seriously studying the type of short-haul aircraft that can meet the environmental and market demands of the next decade." NASA was developing a clean, quiet experimental engine. "Of all things necessary for a practical and acceptable short-haul aircraft to meet our future transportation needs, a quiet engine is the most crucial." (*CR*, 4/19/73, E2581-2)
- A *Kansas City Times* editorial commented on the proximity of Soviet *Salyut 2* and U.S. Skylab launch dates: "A minor fuss has been made over the timing of Salyut's launch [April 3], which was seen by some as an attempt by the Soviets, after several misfortunes and embarrassments, to steal a bit of the luster from a U.S. manned flight. The point is irrelevant. These are not merely showpiece, national prestige projects. They are bricks-and-mortar foundation-laying for both countries' futures in space. In the race for the moon, being first counted for something. In . . . learning to function usefully in a long-term space environment, the test is not how soon you get up but what's accomplished after you're there." (*KC Times*, 4/11/73)

April 11–14: The 10th Annual Space Congress met in Cocoa Beach, Fla. Dr. Franco Fiori, scientific counselor at the Italian Embassy in Washington, D.C., said space cooperation had brought a "wide spectrum" of advances in space science. Col. William G. Bastido of the Air Force Bureau of International Scientific and Technological Affairs noted that European countries had allocated $300 million toward the space shuttle program. Capt. Robert F. Freitag (USN, Ret.), Deputy Director of Advanced Programs in the NASA Office of Manned Space Flight, said international space cooperation might not be an easy solution to international problems, but was the right approach. NASA Associate Administrator for Applications Charles W. Mathews chaired a panel discussion on practical applications in space. (*O Sen*, 4/12/73; *Today*, 4/14/73)

April 12: The U.S.S.R. launched *Cosmos 553* from Plesetsk into orbit with a 486-km (302.0-mi) apogee, 270-km (167.8-mi) perigee, 92-min period, and 70.9° inclination—on the 12th anniversary of *Vostok 1*, the first manned orbital space flight. The satellite reentered Nov. 11. National Cosmonaut Day honored the April 12, 1961, mission flown by Cosmonaut Yuri A. Gagarin, the first man in space. Some observers had expected the announcement of a new manned Soviet mission, but in an interview published by *Pravda* Cosmonaut Vladimir A. Shatalov indicated there was sufficient time left for a crew to be placed aboard the *Salyut 2* space station, launched April 3. (GSFC *SSR*, 4/30/73, 11/30/73; *SBD*, 4/16/73, 253; *LA Times*, 4/13/73; *A&A 1972*)

- NASA's instrumented Convair 990 aircraft *Galileo*, piloted by Ames Research Center pilot J. Patrick Riley, and a Navy P-3 Orion antisubmarine patrol aircraft crashed in flames after colliding over Moffett Field, Calif. All aboard the Convair—the four-man crew and seven NASA technicians—were killed. There was one survivor among the six-man Navy crew. The *Galileo*, known worldwide as a flying laboratory and a test bed for instruments, was on an earth resources survey test flight. [See April 13.] (UPI, *W Post*, 4/13/73, A3; *CR*, 4/13/73, S7383; *Langley Researcher*, 4/27/73, 3; AP, *W Post*, 4/14/73, A3)

- The Federal Communications Commission approved Communications Satellite Corp.'s request to provide communications satellite service to Navy and commercial shipping [see March 5]. (B *Sun*, 4/13/73, C11)

- The Astronomy Survey Committee of the National Academy of Sciences and the National Research Council issued *Astronomy and Astrophysics for the 1970's*, Vol. 2, *Report of the Panels*. The report confirmed the Committee's 1972 observation that the opening of the radio sky had suggested the presence of undiscovered physical laws and requirements for new observations and explanations. The observation had been made in Vol. 1, released June 1, 1972. The Panel on Radio Astronomy report said: "The processes taking place in the radiation-emitting region of pulsars are the most extreme we know of in the galaxy and far surpass in energy production any power source yet invented." The understanding of the generating mechanism through further detailed study would improve the understanding of physics and could lead to "major developments in energy generation here on earth." Suspicion was growing that the molecules already discovered in space were part of "a large group of molecules of ever-increasing complexity that may exist in space." Belief was growing "that interstellar clouds . . . and fast particles may form large organic molecules. Such molecules, which may be as complex as the amino acids and may be of equal importance in our understanding of life, may be discovered by observations made at radio frequencies."

    The Panel on Optical Astronomy noted that new technology for basing detectors in the upper atmosphere and in space had begun to make infrared astronomy "realistic." The Panel on Space Astronomy said, "It appears that much of the energy emitted from astronomical objects and from the universe itself probably resides in the infrared, and hence some of the most important astronomical discoveries will be made in this region." The Panel on Astrophysics and Relativity said that violent activity in the nuclei of galaxies was "probably commonplace" and not due to thermonuclear processes. The report concluded: "Man has landed on his first planet—the moon. But his mind and eye have traveled billions of light years into the past and in the next decade will penetrate unimagined new worlds." (NAS–NRC–NAE *News Report*, 4/73, 1; NAS PIO)

- The prototype of a miniature medical diagnostic system developed for NASA's use aboard manned space stations by the Atomic Energy Commission's Oak Ridge National Laboratory might have clinical uses, NASA announced. The analyzer, being tested at Johnson Space Center, provided fast, automated blood analysis by using one fiftieth the amount of blood required by existing analyzers. It could help a doctor perform rapid analysis in his office, with results within minutes. (NASA Release 73–71)
- The Smithsonian Institution held the first showing of its "Experimental Experimentarium," a prototype of the "Spacearium" theater planned for the National Air and Space Museum, to open in 1976. The Experimentarium—a small planetarium in the old Air and Space Museum—showed a panorama of a NASA launch. (W *Star & News*, 4/13/73, D8)
- The possibility that three Soviet cosmonauts had died in training accidents early in the Soviet space program was discussed by James E. Oberg of the Dept. of Defense Computer Institute in a speech before the American Astronautical Society in Washington, D.C. Oberg, a Soviet space program observer, said photos and training films showed four cosmonauts—identified only as Ivan, Vasily, Grigori, and Valentin—in 1960 and 1961, before the April 12, 1961, *Vostok 1* first manned orbital space flight. None of these cosmonauts had been seen or heard of since. Oberg said they might still be training for a Salyut space station or a lunar landing mission or they could have been killed in accidents before making their intended flights. Oberg said there was no apparent truth to stories that as many as 18 cosmonauts had died secretly. (O'Toole, W *Post*, 4/13/73, A3)

April 13: Preparations for the first Skylab mission, launch of the Skylab 1 Orbital Workshop scheduled for May 14, continued with only minor discrepancies, NASA announced. Onboard experiments and major spacecraft elements that had not flown before required extensive first-time testing. Lower than desired voltage had been detected in one cell in each of two important flight batteries. The batteries would be replaced and tested. Stowage of crew equipment and consumables in the Orbital Workshop had been completed. Testing of the Saturn IB and the command and service modules continued without major problems. (NASA Release 73–74)

- Australian Prime Minister E. G. Whitlam and Dr. James C. Fletcher, NASA Administrator, officially opened the largest space-tracking antenna erected in Australia for NASA. The 64-m-dia (210-ft-dia) antenna at the Tidbinbilla Deep Space Communications Complex near Canberra would be part of NASA's global Deep Space Network (DSN) of three stations. The first station was built at Goldstone, Calif., in 1966 and the third was nearing completion in Madrid, Spain. (NASA Release 73–69; NASA PAO)
- The April 12 loss of NASA's instrumented Convair 990 aircraft *Galileo* would have "very serious effects" on NASA's research programs, Dr. Hans Mark, Ames Research Center Director, told the press in Mountainview, Calif. The $5-million aircraft had been carrying more than $1-million worth of equipment when it and a Navy aircraft collided and crashed in flames. The *Galileo* had participated in the Feb. 17–March 8 U.S.–U.S.S.R. survey of the Bering Sea and was scheduled

to chart the patterns of whales and other sea mammals. "It was one of the major programs at Ames." Dr. Mark said. "We do not have another aircraft to carry on its work." It would be "impossible" to go ahead with the project. A joint NASA–Navy investigation would try to determine the cause of the collision. R/A Herbert S. Ainsworth (USN) told the press that human error was responsible. (*W Post*, 4/14/73, A3)

- *Apollo 13* Astronaut John L. Swigert, Jr., was named Executive Director of the House Committee on Science and Astronautics staff, succeeding Charles F. Ducander, who was retiring effective June 30. Committee Chairman, Rep. Olin E. Teague (D–Tex.), said Swigert's "broadly based skill and enthusiasm" would "aid in expanding the effort being made by the Committee to assure that our national space program and federal research and development will receive adequate support in the mid-1970s." (AP, *Houston Post*, 4/14/73; Com Off)

- A U.S. Navy and United Kingdom agreement to begin an eight-month study of an advanced vertical or short takeoff and landing (V/STOL) Harrier aircraft was announced by Secretary of the Navy John W. Warner. Rolls-Royce Ltd. and Hawker Siddeley Group Ltd., of the U.K., and McDonnell Douglas Corp. and United Aircraft Corp. Pratt & Whitney Div. would study the design of the Pegasus 15 engine, specifications of the new power plant, projected costs, identification of airframe modifications for the uprated engine and an advanced technology wing, test performance predictions, preparation of a preliminary aircraft specification, and definition of a possible full development program. (DOD Release 189–73)

- The Air Force had made the first major improvements in aircraft landing techniques in more than two decades with installation of solid-state instrument landing systems (ILS) at six Air Force bases, the Air Force Systems Command announced. The new systems automatically transmitted signals that appeared on cockpit instruments to indicate the aircraft's position in relation to the runway's centerline and the pilot's glide slope. They replaced failure-prone tube circuits and introduced longer system life, greater reliability, and easier maintenance. (AFSC Release 035.73)

- President Nixon submitted to the Senate the nomination of William E. Kriegsman to be a Commissioner of the Atomic Energy Commission succeeding Dr. James R. Schlesinger, who had become Director of the Central Intelligence Agency. The nomination was confirmed May 31. Kriegsman was manager of the Washington, D.C., office of Arthur D. Little, Inc., and a former White House staff assistant in environment, space, nuclear energy, and oceanography. (*PD*, 4/16/73, 368, 378; *CR*, 5/31/73, D603)

- The Atomic Energy Commission announced that space reactor technology declassification had been completed with the declassification of engineering scale information developed in the SNAP–50 (systems for auxiliary nuclear power) space reactor program and the lithium-cooled reactor experiment. Thermionic-converter-reactor and uranium-zirconium-hydride-reactor technology had been declassified in 1972 and nuclear-rocket-propulsion technology earlier in 1973. (AEC Release R–156)

April 14: Rep. Don Edwards (D–Calif.) urged in a letter to the Dept. of the Navy that Moffett Field, where NASA's instrumented Convair 990 *Galileo* and a Navy P–3 Orion crashed after colliding in mid-air April

12, be closed because of the "dangerous situation" and "nuisance" it caused. (UPI, *Virginia-Pilot*, 4/15/73)

- An exhibit honoring black pilots, including the 450 who flew combat missions in World War II, was opened at the Air Force Museum at Wright-Patterson Air Force Base, Ohio. (UPI, W *Star & News*, 4/8/73, 4)

April 15–June 16: Storm penetrations by high-performance jet aircraft enabled ground-based scientists to check radar returns against airborne measurements of in-cloud turbulence for the first time. A two-month experiment by the Air Force Systems Command's Cambridge Research Laboratories and the National Severe Storm Laboratories of the National Oceanic and Atmospheric Administration at Norman, Okla., fingerprinted the wind to gather faster, more accurate information to identify tornadoes and devastating windstorms. Hazardous air currents swirling inside storms registered a distinct impression on doppler radar. Radar signals were processed through an electronics package to permit observers to pinpoint and measure gusts or turbulence. With severe storm patterns identifiable, observers could forecast their threat to aircraft in flight and people and installations on the ground. (AFSC Release OIP 101. 73; AFSC PIO)

April 16: The Skylab Orbital Workshop atop its Saturn V launch vehicle was moved from the Vehicle Assembly Building at Kennedy Space Center to Launch Complex 39, Pad A, for May 14 launch. The move marked the first time that both Complex 39 launch pads were occupied simultaneously by space vehicles. The Saturn IB rocket and Apollo spacecraft to convey Skylab's first crew had been positioned on Pad B for May 15 launch. (KSC Release, 4/10/73; KSC PIO)

- A lightweight version of the space shuttle was being refined by Rockwell International Corp.'s North American Aerospace Group for the major systems requirement milestone, *Aviation Week & Space Technology* reported. The lightweight version, with a gross liftoff weight of 1.86 million kg (4.10 million lbs), was scaled down in size from the previous baseline configuration, which had a gross liftoff weight of 2.38 million kg (5.25 million lbs). It had the same maximum payload capability of 29 480 kg (65 000 lbs). The major weight reduction had been achieved by switching to a more efficient double delta wing on the orbitor. Also, a newly designed external fuel tank, which could be separated from the vehicle before orbital velocity was reached, would eliminate the need for a solid-fueled-rocket deorbit motor on the nose of the tank. The solid-fueled boosters were also redesigned to a lighter weight. (*Av Wk*, 4/16/73, 18–19)

- The European Space Research Organization had begun negotiations with Communications Satellite Corp. and RCA Global Communications, Inc., to select a corporate partner that would own the U.S. portion of its proposed experimental communications satellite system, *Aviation Week & Space Technology* reported. Five companies had submitted bids to the ESRO competition. A final choice would be made before June. (*Av Wk*, 4/16/73, 9)

- Textron Corp.'s Bell Helicopter Div. had been selected to design and build two tilt-rotor aircraft for a joint NASA-Army research project, the *Wall Street Journal* reported. The aircraft would use large rotors at its wingtips to take off like a helicopter. Airborne, the rotors would be tilted forward and would serve as propellors for cruising. They would be tilted back again for landing. NASA had said the concept combined the longer range and higher speed of conventional aircraft with helicopter utility, factors desirable for Army air-mobile missions and civilian short-haul transport. (*WSJ*, 4/16/73, 3)

- Dr. Dudley G. McConnell, Director of NASA's Scientific and Technical Information Office, became Assistant Administrator for Equal Opportunity Programs, responsible for NASA's internal equal opportunity program and contractor compliance program. Mrs. Ruth Bates Harris, Director of NASA Equal Employment Opportunity and Deputy Director of the Equal Employment Opportunity Office, became Deputy Assistant Administrator for Equal Opportunity Programs. Dr. McConnell also would serve *pro tem* as Acting Director of the Scientific and Technical Information Office. (NASA Ann; NASA Release 73-73)
- The Nixon Administration's proposed Criminal Code Reform Act of 1973 contained language that could make it a felony for aerospace industry representatives in Washington to notify their companies of potential business until the Government officially permitted them to do so. *Aviation Week & Space Technology* reported. The broadly worded bill would make it illegal "to knowingly communicate information relating to the national defense to a person not authorized to receive it." (*Av Wk*, 4/16/73, 11)
- Award of a $19 500 000 cost-plus-incentive-fee contract to McDonnell Douglas Astronautics Co. for design, construction, and testing of an electromagnetic pulse simulator for testing large aircraft systems was announced by the Air Force. (DOD Release 190-73)

April 16-17: A joint meeting on the space shuttle Sortie Lab/Spacelab Project at Johnson Space Center and Marshall Space Flight Center was attended by 20 representatives of the European Space Research Organization (ESRO), European Space Research and Technology Center (ESTEC), and European space consortiums headed by Messerschmitt-Boelkow-Blohm GmbH and ERNO Raumfahrttechnik GmbH. The group was briefed by NASA on the space shuttle and Skylab programs, concept verification testing, and the MSFC sortie lab activities. The consortiums presented a summary of European Spacelab studies. (MSFC Release 73-44)

April 16-20: The American Geophysical Union met in Washington, D.C. Cornell Univ. engineers Thomas R. McDonough and Neil M. Brice propounded the theory that Saturn was encompassed by a great ring of hydrogen gas that had escaped from Titan's gravity pull and drifted into space. The ring was invisible from earth because of atmospheric interference, but should be visible to satellites that could see it with ultraviolet light. Other solar system planets might also have undetected rings.

Dr. Bruce Murray, Cal Tech astronomer and principal TV investigator of the *Mariner 9* mission (launched May 30, 1971), disputed the theory that ancient floods on Mars had carved the huge canyons shown in the *Mariner 9* photos. "Mars' atmosphere, which, at present, is only one-hundredth as dense as the earth's, very probably never has been dense enough for water to run on the planet's surface." There might have been times "when Mars had virtually no atmosphere at all." (*NYT*, 4/22/73, 31; Swaim, Pasadena *Star-News*, 4/22/73)

April 17: NASA announced the signing of a definitive contract with Rockwell International Corp.'s Space Div. for design, development, and production of the orbiter vehicle and for the integration of all space shuttle system elements. The cost-reimbursement, fixed-fee, and award-fee contract would have a $477 400 000 initial increment. The contract superseded an Aug. 9, 1972, letter contract and would continue through Aug. 3, 1974. A second work increment—the balance of design, development, test, and evaluation, plus delivery of two orbiters—was planned to begin Aug. 4, 1974. Rockwell planned to subcontract to firms and suppliers in almost every state. (NASA Release 73-76)

- The appointment of Gerald D. Griffin—flight director on the *Apollo 12, 15,* and *17* missions—as NASA Assistant Administrator for Legislative Affairs effective April 23 was announced by Johnson Space Center. He would succeed H. Dale Grubbs, who was leaving after holding the post since 1970. Griffin had received NASA's Exceptional Service Medal for his work on *Apollo 12* and *15* and the Presidential Medal of Freedom Group Achievement Award for *Apollo 13*. He joined JSC (then the Manned Spacecraft Center) in 1964 and was named a flight director in 1968. (JSC Release 73-39)

- Dr. John S. Foster, Jr., Director of Defense Research and Engineering in the Dept. of Defense, testified before the Senate Committee on Armed Services on the Soviet thrust for military and technological superiority: "The United States reacted to one Soviet technological advance—the Sputnik—with a successful ten-year effort to restore our technological superiority in space. Now we are lagging again in some technology areas and it may well require another decade after some new jolt before the United States regains an acceptable technological posture." The U.S.S.R. was "at such levels of technical competence in broad areas of defense technology that technological breakthroughs of major significance are possible. They could seriously destabilize the balance of world power and weaken our strength for deterrence and negotiations. They could encourage military opportunism in any of a number of areas and raise the risk of war or armed confrontation." (Testimony)

- The exhibit "The Image of the Moon—Galileo to Apollo XI," from the art collection of Anthony Michaelis, opened at the Univ. of Houston in Clear Lake City, Tex., in cooperation with the Univ. of Houston Libraries. (Invitation)

- Cost-plus-fixed-fee contract awards for demonstrator programs for turbofan engine technology were announced by the Air Force: $6 080 000 to United Aircraft Corp. and $1 141 633 to Teledyne Industries, Inc. (DOD Release 193-73)

- Sen. Frank E. Moss (D-Utah), Chairman of the Senate Committee on Aeronautical and Space Sciences, introduced S. 1610 to amend the Federal Aviation Act of 1958 to require installation of airborne, cooperative collision avoidance systems on certain military and civilian aircraft. (*CR*, 4/17/73, S7546-7)

- Sen. Edward J. Gurney (R-Fla.) reintroduced S.J. R. 94, to redesignate Cape Kennedy Cape Canaveral. (*CR*, 4/17/73, S7554-5)

- A *New York Times* editorial criticized the inclusion in the FY 1974 Federal budget of $5.8 million for further studies of the supersonic transport's possible effect on the stratosphere: "While there may be no harm in pursuing a scientific study of the question, the Government's intent clearly goes beyond pure research or the problem raised by foreign and military supersonic planes. A high Administration official has bluntly stated that the study would be 'prefatory to possible introduction of an SST program at a later date.' The option 'is held open.' With an array of fiscal, political and scientific factors against it, the case for the SST at this late date would seem to rest solely on the advantage to a tiny handful of people crossing the oceans in three or four hours instead of six or seven—scarcely a reason for reviving a discredited project which is still being liquidated at public expense." (*NYT*, 4/17/73, 39)

April 18: NASA released a preliminary timeline of the Skylab missions scheduled to begin with the Skylab 1 launch of the orbiting laboratory at 1:30 pm EDT May 14 from Kennedy Space Center. The Skylab 2 mission, carrying the first three crewmen, would leave the pad at 1:00 pm EDT May 15 and dock with the orbiting laboratory at 8:40 pm EDT. At 8:30 am May 16 the tunnel would be pressurized and the astronauts would enter Skylab. On the 28th day, at 8:46 am June 12, the crew would reenter the command module, undock, and return for splashdown at 1:44 pm, 1300 km (800 mi) southwest of San Diego.

Skylab was to operate for eight months in earth orbit, occupied at intervals by three-men crews who would make scientific and technical investigations and observations related to such areas as earth resources, physiological effects of long-duration weightlessness, solar phenomena, and metal processing in zero-g conditions. (NASA Release 73-78)

- A series of calibration rocket (CALROC) launches at White Sands Missile Range during the manned Skylab 2 mission scheduled for May would provide a reference for the calibration of equipment used in two experiments on Skylab's Apollo Telescope Mount (ATM), Marshall Space Flight Center announced. CALROC launches were also scheduled during the Skylab 3 and Skylab 4 missions. The rockets would acquire solar flux data on specific regions of the sun in conjunction with Skylab astronaut observations of the same regions with ATM experiments. Instrumentation on CALROC would be similar to but smaller than that aboard the ATM. Black Brant VCs would be used for all the launches. Taking part in the program, managed by MSFC, were Goddard Space Flight Center, Ames Research Center, the Naval Research Laboratory, Harvard College Observatory, and White Sands Naval Ordnance Missile Test Facility. (MSFC Release 73-42)

- President Nixon issued Executive Order 11712 establishing a Special Committee on Energy and a National Energy Office in the Executive Office of the President. In a message to Congress on U.S. energy policy, he said: "If we are to be certain that the forward thrust of our economy will not be hampered by insufficient energy supplies or by energy supplies that are prohibitively expensive, then we must not continue to be dependent on conventional forms of energy. We must instead make every useful effort through research and development to provide both alternative sources of energy and new technologies for producing and utilizing this energy. For the short-term future, our . . . strategy will provide technologies to extract and utilize our existing fossil fuels in a manner most compatible with a healthy environment. . . . from 1985 to the beginning of the next century, we will have more sophisticated development of our fossil fuel resources and . . . full development of the Liquid Metal Fast Breeder Reactor. Our efforts for the distant future center on the development of technologies—such as nuclear fusion and solar power—that can provide us with a virtually limitless supply of clean energy." (*PD*, 4/23/73, 389-406; 412-3)

- Astronaut Fred W. Haise, Jr.—*Apollo 13* lunar module pilot, *Apollo 8* and *11* backup pilot, and *Apollo 16* backup commander—had been named Technical Assistant to the Manager, Orbiter Projects Office, at Johnson

Space Center, JSC announced. Haise, who had logged 142 hrs 54 min in space, would be responsible for assisting Manager Aaron Cohen in the overall management of the space shuttle orbiter development. (JSC Release 73-41)

- The appointment of William J. Hamon, Director of Program Budget and Control in the Office of Manned Space Flight's Sortie Lab Task Force, as NASA Resident Liaison Officer at the European Space Research Organization's Research and Technology Center (ESTEC) in Noordwijk, The Netherlands, was announced by NASA. He would be responsible for liaison between NASA Hq. and Marshall Space Flight Center and the ESTEC sortie lab project. (NASA Release 73-70)

- The Air Force Space and Missiles Organization (SAMSO) Air Force Satellite Facility (AFSF) received a Presidential Management Improvement Certificate for excellence in improving Government operations, from Secretary of Defense Elliot L. Richardson in Washington, D.C., ceremonies. The certificate was presented to Col. John J. Schmitt, Jr. (USAF), AFSF Commander. (AFSC *Newsreview*, 6/73, 4)

- Aviation pioneer Roy W. Hooe died in Martinsburg, W. Va., at the age of 78. He had been a member of the five-man crew of the *Question Mark*, a trimotor aircraft which set a 151-hr endurance record during pioneer tests of mid-air refueling in 1929. Hooe had retired as an Air Force master sergeant in 1950 after 30 yrs in the Air Force and its predecessor flying services. (W *Star & News*, 4/21/73, A12)

April 19: The U.S.S.R., with the cooperation of the Polish People's Republic, launched *Intercosmos Copernicus 500* from Kapustin Yar to mark the 500th anniversary of the birth of the Polish scientist Nicolaus Copernicus. The satellite entered orbit with a 1518-km (943.2-mi) apogee, 199-km (123.7-mi) perigee, 102.1-min period, and 48.4° inclination. *Copernicus 500*—built by Soviet and Polish scientists and carrying Soviet, Polish, and Czechoslovakian instruments—measured solar radiation and characteristics of the earth's ionosphere before reentering Oct. 15. (GSFC *SSR*, 4/30/73, 10/31/73; Tass, FBIS–Sov, 4/20/73, L1; Warsaw PAP, FBIS–Poland, 4/25/73, G4; *SBD*, 4/20/73, 238)

- *Cosmos 554* was launched by the U.S.S.R. from Plesetsk into orbit with a 329-km (204.4-mi) apogee, 171-km (106.3-mi) perigee, 89.4-min period, and 72.8° inclination. The spacecraft reentered May 27. Western observers believed it was a reconnaissance satellite and that it had exploded or been exploded during recovery. By the end of May more than 180 pieces had been recorded by tracking stations. (GSFC *SSR*, 4/30/73; 5/31/73; O'Toole, *W Post*, 5/19/73, A12; *SBD*, 4/20/73, 285; 5/21/73, 116)

- NASA and the Soviet Academy of Sciences had approved results of the Feb. 26–March 3 Moscow meeting of the U.S.–U.S.S.R. Joint Working Group on Space Biology and Medicine, NASA announced. The Group had agreed on common medical procedures to permit comparison of pre- and post-flight data on astronaut and cosmonaut body functions. (NASA Release 73-79)

- Johnson Space Center notified 45 Civil Service employees that they would be released because of NASA manpower reductions; an additional 38 were informed they would be reassigned or placed in lower-grade jobs. The action, to be completed by June 1, would reduce the JSC work force to 3727 Civil Service personnel. (JSC Release 73-40)

April 20–30: Canadian domestic communications satellite Telesat-B—named *Anik 2,* Eskimo for brother—was successfully launched at 6:47 pm EST from Eastern Test Range by a three-stage, long-tank, thrust-augumented Thor-Delta launch vehicle. *Anik 2* was placed in a highly elliptical transfer orbit with a 36 480-km (22 667.6-mi) apogee, 212-km (131.7-mi) perigee, and a 26.7° inclination. After checkout and reorientation of the spacecraft, the apogee motor was fired at 5:00 pm EST April 23, bringing the spacecraft into near synchronous circular orbit with a 35 709-km (22 188.5-mi) apogee, 35 604-km (22 123.3-mi) perigee, 1430.7-min period, and 0.1° inclination. The spacecraft was drifting 3° per day toward its final operational position off the west coast of South America.

Under NASA-Telesat Canada contract, the NASA mission objective was to place *Anik 2* into an orbit of sufficient accuracy to allow the spacecraft propulsion systems to place it in a stationary synchronous orbit while retaining sufficient stationkeeping propulsion to meet the mission lifetime requirements. The objective was met and the mission was adjudged a success April 30.

Anik 2—second in a series of Canadian domestic communications satellites—was built by Hughes Aircraft Co. under contract with Telesat Canada and designed to provide transmission of TV, voice, and data throughout Canada for seven years. The spacecraft system provided 12 channels of communications to relay 10 color TV channels or up to 9600 telephone circuits. Two channels would be protection channels for the 10 traffic-carrying channels.

The spacecraft was 1.8 m in diameter and 3.4 m high (6 by 11 ft). At launch it weighed 540 kg (1200 lbs) and its orbiting weight was 270 kg (600 lbs). The electronics system was powered by 23 000 solar cells with sufficient onboard battery capacity to maintain service during a sun eclipse of the solar cells.

Anik 1 had been launched Nov. 9, 1972. The two-satellite system would act as space repeaters to receive transmissions from earth stations and retransmitt them to earth stations in Canada. A third satellite was retained on the ground as protection against space failure of the first two spacecraft and could be launched in 1975 to expand the system if required.

NASA would be reimbursed by Telesat Canada for costs of the Thor-Delta launch vehicle and services. The project was managed for NASA's Office of Space Sciences by Goddard Space Flight Center. (NASA prog off; NASA Release 73–58; GSFC *SSR,* 4/30/73)

April 20: Checkouts of the first two Skylab launch vehicles and their spacecraft, in position on Pads A and B of Launch Complex 39 at Kennedy Space Center, continued without major problems. Top program officials from NASA Hq. and Centers had completed the final flight readiness review of all aspects of Skylab, including launch and mission operations, spacecraft and experimental hardware, safety, range readiness, recovery, and flight-crew readiness. (NASA Release 73–81)

- The appointment of Frederick A. Meister, Jr., as Deputy Associate Administrator of Plans for the Federal Aviation Administration was announced by FAA Administrator Alexander P. Butterfield. Meister would succeed Ronald W. Pulling, who had retired. (FAA Release 73–68)
- The year's biggest congressional defense debate would be whether to proceed with the Trident submarine project, the *Los Angeles Times* said.

April 20

The $1.7 billion requested in FY 1974 for the Trident was "the biggest single appropriation being sought for any weapons system." If the program went ahead, 10 missile-firing submarines would be built at a total $13.5-billion cost. Dept. of Defense officials said the Trident submarines were needed to compensate for the 62-41 edge in missile-firing submarines allowed the U.S.S.R. under the 1972 arms control agreement. Economy forces in Congress argued that an easier and less expensive way to guarantee continued invulnerability for U.S. sea-based deterrents would be to equip the existing Polaris-Poseidon fleet with a first-generation Trident with a range of more than 6000 km (4000 mi). (Conine, *LA Times*, 7/20/73, 7)

April 22: The development of Laserphoto—a system of news photo transmission using a laser beam receiver to deliver dry glossy prints, cut and stacked, to editors' desks—was announced in New York by Associated Press President J. Wes Gallagher. AP also announced plans for electronic dark rooms where photos would be stored in computers, edited on video screens, and transmitted at high speeds. Both systems had been developed at the Massachusetts Institute of Technology. (AP, B *Sun*, 4/23/73, A3; *WSJ*, 4/23/73, 1)

April 22–30: A Proton booster reportedly was launched by the U.S.S.R., failed, and fell into the Pacific Ocean during the final week of April. The *Washington Post* later quoted U.S. space observers as saying the booster probably had carried a Lunokhod moon vehicle. The *Post* said this failure and the failure of the *Salyut 2* (launched April 3) had thrown the 1975 U.S.-U.S.S.R. Apollo-Soyuz mission in doubt because both failures had involved hardware managed by the same Soviet space officials. (O'Toole, *W Post*, 5/4/73, A1)

April 23: Dr. Rocco A. Petrone, Marshall Space Flight Center Director, announced personnel actions to reduce the Center's total Civil Service personnel to 5214 by June 30. Separation notices had been given to 108 employees—87 at Huntsville, Ala., and 21 at MSFC installations elsewhere. A total of 67 MSFC employees would receive notices of change to lower grade, with 15 losing some salary. Reassignment of 57 would be made concurrently with the reduction in force. (MSFC Release 73-45)

• The European Space Research Organization had selected two new scientific satellite programs while rejecting participation in NASA's Venus orbiter program, *Aviation Week & Space Technology* reported. A joint International Sun-Earth Physics Satellite (ISEPS) program—to study solar winds and observe the discontinuities such as the magnetopause, bow shock, neutral sheet in the tail, and wave shocks—would place two satellites in orbit by a single booster in 1977. ESRO would develop one satellite at a cost of $28 million; NASA, the other. The second ESRO program, the Highly Eccentric Lunar Occultation Satellite (HELOS), would be launched in 1979 to define the position, map the spatial and spectral features, and monitor the time variabilities of x-ray sources. It would cost an estimated $75 million. (*Av Wk*, 4/23/73, 28)

• The U.S. Postal Service issued an eight-cent stamp commemorating the 500th anniversary of the birth of Polish astronomer Nicolaus Copernicus. First-day ceremonies, sponsored by the Smithsonian Institution and the National Science Foundation, were held at the National Museum of History and Technology. The first-day flight cover featured a photo of Orbiting Astronomy Satellite *Oao 3*, named *Copernicus* and launched by NASA Aug. 21, 1972. (USPS Philatelic Release 8; Flight cover)

- Exclusive astronaut writing contracts with *Life* magazine and other media had harmed the press, the public interest, the space program, and the astronauts themselves, Robert Sherrod said in a *Columbia Journalism Review* article released in advance of its May/June publication date. The contracts had brought relatively little money to all but a few astronauts and had distorted their individuality and diversity into a "deodorized, plasticized, and homogenized" image. Under the contract system, NASA could censor whatever the astronauts signed. "Even though this censorship was rarely invoked, it was inhibiting; so was the necessity of *Life*'s making the astronauts look good, and *vice versa.*" President Kennedy had originally opposed the contracts, but had been persuaded in their favor by Project Mercury Astronaut John H. Glenn. Former President Johnson had supported the contracts in a letter Sherrod received a few months before Johnson's Jan. 22 death. Sherrod said the contract system had had no adverse affects on the flow of space news. "Due in large part to the exclusion policy, many members of the working press deeply resented NASA." (Text)

April 23–25: A committee of 35 astronomers met at Marshall Space Flight Center to review proposed experiments for the Large Space Telescope (LST) to be launched by the space shuttle as a general-purpose facility for various astronomical instruments. The committee—representing 7 observatories, 10 U.S. universities, and NASA installations—would help NASA select science participants for LST definition studies. The LST would observe galaxies 100 times fainter than those seen by the most powerful ground-based telescopes and would provide long-term monitoring of atmospheric phenomena on Venus, Mars, Jupiter, and Saturn. (MSFC Release 73–46; *Marshall Star,* 4/25/73, 1)

- The National Academy of Sciences held its annual meeting and participated with the Smithsonian Institution in a joint program to honor the Copernican Quincentennial. Dr. Philip Handler, NAS President, announced that Dr. Robert C. Seamans, Jr., retiring Secretary of the Air Force, would become president of the National Academy of Engineering in May and that the NAE council had voted to sever NAE's affiliation with NAS. NAS elected 95 new members.

  During the program devoted to cosmological questions and humanistic aspects of scientific research, Dr. Stephen Toulmin, Provost of the Univ. of California at Santa Cruz, said he had detected a lack of purpose among younger scientists that might be leading to the end of the intellectual era inaugurated by Copernicus.

  Dr. John A. Wheeler, Princeton Univ. physicist, said the collapse of the universe into a single, great "black hole" with the annihilation of all matter and all physical laws seemed inevitable. Black holes—superdense objects in space—had been predicted as end products of the collapse of large stars when they no longer produced sufficient heat to counter the weight of their own material. Black holes could serve as examples of what would happen if the universe ceased its expansion and began to fall back upon itself. Since its expansion did not have sufficient momentum to continue indefinitely, the universe could reach its maximum size in 40 billion yrs. Dr. Alan R. Sandage of the Hale Observatory said a five percent decrease in the brightness of galaxies for each billion years of their lifetime would alter the calculation and indicate indefinite expansion of the universe. (NAS Memo to Press;

Sullivan, *NYT*, 4/26/73; 4/28/73, 48; *NYT*, 4/26/73, 74; 4/29/73, 47)

April 23–26: The American Physical Society held its annual meeting in Washington, D.C. Dr. Edward H. Teller, Univ. of California physicist and atomic scientist, urged legislation to declassify all Government scientific secrets after one year. He predicted the action would force the U.S.S.R. to take similar action "in maybe 10 years," because U.S. science and technology would advance faster without secrecy. Naval Research Laboratory scientist Dr. Herbert Friedman said most astronomers who believed that creation began with a "big bang" also believed that at least as much gas and dust had been left behind when the stars were formed as had been consumed in star formation. Recent findings had been that two clouds moved out in opposite directions from an exploding galaxy to form equally shaped spheres and that radio galaxies left wakes as they plowed through space. "In both cases you need great quantities of gas to put enough pressure on the clouds or the galaxies to slow them down. Galaxies moving through empty space leave no wakes or trains behind them." (*W Post*, 4/25/73, A3; *LA Times*, 4/26/73)

April 23–27: A symposium at Lewis Research Center's Plum Brook Station presented ideas of NASA engineers on new uses by business, industry, labor, other Government agencies, and universities of Plum Brook facilities available because of NASA's imminent shutdown of the station for budgetary reasons [see March 6]. (LeRC Release 73-21; LeRC PIO)

April 24: The first Skylab prime crew—Charles Conrad, Jr., Dr. Joseph P. Kerwin, and Paul J. Weitz—and backup crewmen Russell L. Schweickart, Dr. F. Story Musgrave, and Bruce McCandless, II, began the 21-day preflight isolation period at 8:30 am EST in preparation for May 15 launch. Isolation would continue for seven days following their return from Skylab. Crewmen would be restricted to specific areas and limited in the number of approved personal contacts allowed during the immediate preflight and postflight periods. During the isolation only Skylab food and water would be consumed by the crew members. This diet would also continue for 18 days following the return from the mission, to obtain baseline data for Skylab medical experiments in nutrition and musculoskeletal evaluation series. (NASA Releases 73-84 & unnumbered, 4/30/74)

- Richard W. Cook, Deputy Director, Management, at Marshall Space Flight Center, would retire in June, MSFC announced. He would be succeeded June 3 by John S. Potate, who was attending Massachusetts Institute of Technology, on leave from his position of Director of the Apollo Program Control Office in the Office of Manned Space Flight. Potate had joined the OMSF in 1969. (MSFC Releases 73-47, 73-77; MSFC Org Ann 0101)

- National Academy of Sciences' handling of the controversial supersonic transport was criticized by Philip M. Boffey, Managing Editor of the newsletter *Science & Government Report* and former *Science* magazine writer, in a report he released after a two-year "evaluation" study of NAS. The *Washington Post* later quoted the report as saying an NAS group in 1965 had "incredibly" urged the Government to conduct a campaign to persuade the U.S. public to accept the SST though it had become evident that the aircraft would produce harmful sonic booms. In 1968, NAS had downplayed the possibility of physical damage from the booms, though it later circulated a "clarifying" statement internally to its own members saying the SST would be unacceptable. (Cohn, *W Post*, 4/25/73, A2)

April 25: Cosmos 555 was launched by the U.S.S.R. from Plesetsk into orbit with a 230-km (142.9-mi) apogee, 216-km (134.2-mi) perigee, 88.9-min period, and 81° inclination. The satellite reentered May 7. (GSFC *SSR*, 4/30/73, 5/31/73; *SBD*, 4/30/73, 333)

- All three Skylab crews were briefed at Johnson Space Center by Robert A. Citron, Director of the Smithsonian Institution's Center for Short Lived Phenomena. The Smithsonian was under contract to NASA to report the major daily earth surface phenomena. During the Skylab missions, scheduled to begin May 14 and 15, the reports would be used to advise the crew on suitable targets for hand-held photography and to plan or modify earth resources experiment passes which required a major reorientation of the spacecraft. During the five months of Skylab activities, the crew might expect to see 10 major volcanic eruptions, 5 major cyclonic storms, and 5 earthquakes and 10 floods large enough to cause visible changes in topography. The Center would attempt to send research teams into areas where the Skylab scientists observed unusual changes. (JSC Release 73-47)
- The Air Force approved the use of $38 million for initial production of jet engines for its new F-15 fighter aircraft, despite failure of the prototype engine to satisfy fully key testing requirements called for in the contract with the Pratt & Whitney Div. of United Aircraft Corp. (Getler, *W Post*, 4/26/73, A2)

April 26: The 19 U.S. students whose scientific experiments had been selected for the Skylab missions would help analyze and compile the data and results from earth orbit, NASA announced. Henry B. Floyd, Manager of the Skylab Student Project at Marshall Space Flight Center, had said NASA would leave everything possible to the individual student. "We are not telling the student what to do." Each student experiment was being handled individually through principal investigators and science advisers who were expected to report results to NASA and the U.S. The students would be invited to report and exchange information at a symposium to be set up by NASA. (MSFC Release 73-49)

April 27: Countdown demonstration tests were underway at Kennedy Space Center for both the Saturn V that would launch Skylab on May 14 and the smaller Saturn IB that would boost the astronauts into orbit on May 15 to join it. The solar telescope flight film had been installed in the Skylab multiple docking adapter section and the hatch closed. Reassessment and further testing of a leakage problem with an oxidizer tank bladder in the command module reaction control system had determined that the bladder was not defective. A Mission Control Center network-validation simulation was conducted at the Johnson Space Center; an orbital operations simulation had been interrupted to take advantage of late software deliveries.

The Kennedy Space Center Public Information Office moved its operations to the Skylab News Center at Cape Canaveral, Fla. All KSC news activities would be conducted from there through two days after the launch of Skylab 2. (NASA Release 73-87; KSC Notice to Editors)

- Flight Research Center Director Lee R. Scherer had been named senior NASA representative on the Navy-NASA accident board investigating the April 12 mid-air collision of NASA's instrumented Convair 990 aircraft

April 27

and a Navy P-3 Orion aircraft, FRC's *X-Press* reported. FRC's 990 cockpit simulator had been set up to assist the board in determining the specific visibility available. (FRC *X-Press*, 4/27/73, 2)

- Dr. Paul A. Gast, Chief of Space and Planetary Sciences at Johnson Space Center, and Dr. Gerald J. Wasserburg of Cal Tech had been named to receive Columbia Univ.'s Kemp Medal for distinguished public service by a geologist, the Pasadena *Star-News* reported. The citation was for "performing a major role in the lunar projects of NASA, which have given the scientific world a better understanding of the moon." (Pasadena *Star-News*, 4/27/73)

- The role of commercial aircraft in monitoring meteorological parameters and atmospheric constituents was discussed in a *Science* article by Lewis Research Center scientist Robert Steinberg. Wide-body jet aircraft could supply global atmospheric and tropical meteorological data. "While scientists are not in total agreement on the magnitude of the effect of particulates and gases on the atmosphere, there is almost unanimous concurrence that we are severely limited in information, and that global baseline concentration must be established for particulates and gases in the troposphere and lower stratosphere as soon as possible." Commercial aircraft were flying 10 hrs a day on long-range flights, equipped with inertial navigation systems and central-air-data computers coupled to advanced data-storage systems capable of satellite interrogation. "This means that there is now a large amount of synoptic weather information which can be obtained with a minimum of effort and cost. Likewise, a start at obtaining measurements of atmospheric constituents on a global basis can be made now." (*Science*, 4/27/73, 375-80)

April 27-May 2: Goddard Space Flight Center sponsored three scientific meetings: the X-ray Astronomy Symposium and the Goddard Scientific Colloquium met April 27 and the International Symposium and Workshop on Gamma-ray Astrophysics met April 30-May 2.

Dr. Floyd W. Stecker, GSFC scientist, and Jean-Loup Puget of the Paris Observatory presented their theory before the gamma ray symposium that the galaxies had been formed when equal amounts of matter and antimatter in the universe coalesced into separate regions after the universe expanded and cooled from an extremely hot, dense state about 15 billion yrs ago. The regions could have grown to the mass of galaxy clusters by the time the universe had cooled enough to go from a plasma to an atomic gaseous state. The theory would account for the sizes, mean densities, and rotational speeds of galaxies and was consistent with Dr. Stecker's interpretation of recently observed cosmic gamma radiation as resulting from matter-antimatter annihilation. (GSFC Note to Editors; NASA Release 73-86)

April 28: Dr. Robert C. Seamans, Jr., retiring Secretary of the Air Force, was named by the National Geographic Society to receive its Gen. Thomas D. White Space Trophy. The citation said: "Under his leadership great strides were made in Air Force space programs, and new early warning and communications space systems were placed in operation." [See April 30.] (AP, W *Star & News*, 4/29/73, A6)

- U.S.S.R. scientists had named a newly discovered mineral "armstrongite" after the U.S. astronaut Neil A. Armstrong, first man to set foot on the moon, the Associated Press reported. The mineral, belonging to the zirconium-silicate compounds, had been found by a joint Soviet-

Mongolian expedition in the Gobi Desert. A sample of the mineral had been sent to Armstrong. (AP, *NYT*, 4/29/73, 53)

April 29: Los Angeles International Airport banned all landings and take-offs over populated areas between 11 pm and 6 am and became the first U.S. airport with regular two-way traffic in the same airspace. The measures had been taken in response to community protests over aircraft noise. Federal Aviation Administration air controllers would regulate the two-way traffic, called nose-to-nose operations, and had said there would be no safety problem. (Lindsey, *NYT*, 4/29/73, 58)

April 29–May 2: The First International Conference on Offshore Airport Technology was sponsored by the New York Section of the American Institute of Aeronautics and Astronautics and the Federal Aviation Administration in Bethesda, Md. Director W. D. Brinckloe of the Univ. of Pittsburgh's Graduate Center for Public Works spoke on the multi-purpose use potential of offshore airports: "Experience suggests that what should be planned is a *terminal*, mating all possible transportation modes, and complete with all the facilities and services needed to make the Terminal City thrive." Multiple use of the island airport "should be given equal attention with more conventional aspects of airport design and ocean construction. In the end, it may be the element that makes the project economically feasible." (AIAA *Release*, 3/1/73; Text)

April 30: The Thomas D. White Space Trophy Award of the National Geographic Society was presented in Washington, D.C., to Dr. Robert C. Seamans, Jr., Secretary of the Air Force [see April 28]. Secretary of Defense Elliot L. Richardson said at the award ceremony: "Bob Seamans is a distinguished scientist and administrator. He was formerly the Deputy Administrator of NASA and, for the past four years, has led the Air Force through a period marked by operational innovations and the initiation of a large number of essential modernization programs. Under his leadership a number of major advances have been made in the Department of Defense space programs, while the world has been mindful that United States operations in space do not pose a military threat to any country." Dr. Seamans had ensured "a close working relationship" between DOD and NASA, "including Defense support of NASA space operations and close cooperation in the design of the Space Shuttle." (DOD Release 213–73)

- Wallops Station reported on its cooperative research project with Virginia Polytechnic Institute's Agronomy Dept. to study the application of remote sensing to agriculture. Detailed investigations of circular soil ridges with depressed centers, common to Virginia's Eastern Shore, had been made by pedological and remote-sensing techniques. More than 150 ridges, called Carolina Bays, had been found in Accomack and Northampton Counties. The study included land management and microclimate aspects of the Bays. (Wallops Release 73–3)

- The National Science Foundation released *Federal Support to Universities, Colleges, and Selected Nonprofit Institutions, Fiscal Year 1971* (NSF 73–300). Total Federal obligations (exclusive of loans) to institutions of higher learning increased in current and constant dollars. Current-dollar volume rose $253 million, or 8%, to a record $3480 million. Constant-dollar increase was 2%, the first increase since 1967. The first 100 universities and colleges in Federal support accounted for more than $2.4 billion during 1971, or 69% of Federal funds to all academic institutions. Private universities and colleges receiving Federal funds outnumbered public institutions receiving Federal funds 1242 to 1126, but public institutions accounted for 61% of total Federal obligations, 65% of degrees awarded, and 75% of students enrolled.

In 1971 life sciences research and development funding was $741 million, or 48% of all Federal R&D obligations. Obligations for general support of NSF and National Institutes of Health science programs was $100 million, the same level as in 1970. The Atomic Energy Commission continued as the principal Federal agency funding academically associated R&D centers, with 64% of the $984 million Federal total. NASA and the Dept. of Defense awarded $312 million, or 32% of the total. (Text)

During April: *Apollo 17* Astronauts Eugene A. Cernan, Ronald E. Evans, and Dr. Harrison H. Schmitt concluded the largest and most successful schedule of postflight public appearances undertaken by an astronaut crew. The tour, begun Jan. 12 in Los Angeles, included 57 stops in 25 states and the District of Columbia in 11 weeks. The astronauts met with 12 state governors and addressed state legislatures in Kansas, New Mexico, Georgia, Arkansas, North Carolina, Oklahoma, Colorado, and Washington. Activities Bureau Chief Eugene A. Marianetti of NASA's Public Events Directorate said later that the astronauts had met no evidence of antispace sentiment, "only a lack of adequate information . . . on the actual amount spent on the space program." (*NASA Activities,* 5/15/73, 89)

- Iowa Geological Survey scientist James V. Taranik reported on the usefulness of a remote-sensing project to map temperatures of the Mississippi River, before the annual Iowa Science, Engineering and Humanities Symposium at the Univ. of Iowa. An infrared scanner aboard an aircraft flying at 2000 m (6000 ft) had provided data for a temperature map of waters beneath the site of a projected nuclear power plant. The map had shown temperatures above the standards set by the Environmental Protection Agency. If the plant had operated, the power company could have been penalized for a condition that had existed before it was opened. (Univ Iowa *Spectator,* 4/73, 8)

- A theory that the solar system had been visited by a space probe from another civilization was reported by Duncan A. Lunan, a graduate of Glasglow University, in *Spaceflight.* Lunan had found that certain long-delayed echoes of equally spaced radio signals transmitted from the earth could be interpreted in the form of a code. He had used data recorded in the 1920s by Norwegian, Dutch, and French experimenters who had noted that the delay times of the echoes varied from one signal to the next, but assumed a natural phenomenon. Lunan interpreted the pattern of an October 1928 signal as a star map identifying the probe's origin as the double star Epsilon Boötis and putting its arrival in the solar system at 13 000 yrs ago. (*SF,* 4/73, 122–31)

- Univ. of Wisconsin psychologist Dr. David A. Grant visited the U.S.S.R. as the first U.S. scientist to participate in an exchange agreement between the American Assn. for the Advancement of Science and its Soviet counterpart, All Union Znaniye [Knowledge] Society. Dr. Grant lectured in Moscow and Leningrad. (*AAAS Bulletin,* 6/73, 1)

May 1973

May 1: The joint NASA-Navy board investigating the April 12 mid-air collision of NASA's instrumented Convair 990 aircraft *Galileo* and a Navy P-3C Orion antisubmarine aircraft over Moffett Field, Calif., completed its investigation, forwarded its report to NASA and the Navy, and issued a statement: Both aircraft had been under visual flight rules (VFR) at the time of the accident; the P-3C was in the local landing pattern making touch-and-go landings and the Convair 990 was making a straight-in approach. "At initial contact with the control tower, the Convair 990 was cleared to continue for Runway 32 Right. At seven miles [11 km] the Convair 990 was again cleared to continue approach for the right runway. About three miles [5 km] out, the control tower operator cleared the Convair to land on Runway 32 Left. The Convair pilot acknowledged the new runway assignment. The control tower operator also cleared the P-3C to continue his approach to Runway 32 Left. It was the tower operator's intention to land the Convair 990 on Runway 32 Right, but he mistakenly called the wrong runway." There was no evidence that either pilot or the tower personnel were aware of the impending collision. (NASA Release 73-88)

- *Explorer 47* Interplanetary Monitoring Platform (launched Sept. 22, 1972) was adjudged a success. It had met prelaunch objectives of making detailed and continuous studies of the interplanetary environment and studying particle and field interactions in the distant magnetotail. Twelve of thirteen scientific instruments were operational and providing good data. Failure of the ultralow-energy telescope would reduce total scientific return from that experiment 10%. (NASA prog off)

- NASA's *RM* Radiation/Meteoroid Satellite—launched Nov. 9, 1970, with *Ofo* Orbiting Frog Otolith—was adjudged a success. It had substantially met mission objectives and the minimum mission duration of 90 days had been achieved by the time *RM* had reentered Feb. 7, 1971. The radiation experiment was partially successful, measuring the pulse-height distribution of electrons and of low-energy protons for spectral mapping of the South Atlantic magnetic anomaly. Above 50 mev the solid-state spectrometers malfunctioned. The meteoroid experiment had successfully measured meteoroid velocities and impacting flux despite damage to 40% of the front-plane and 10% of the rear-plane impact sensors during the first three weeks in orbit. (NASA prog off)

- The West German government announced it was withdrawing from the multinational European Launcher Development Organization (ELDO). ELDO had been building the Europa II launcher to launch a communications satellite into earth orbit Oct. 1, but the project collapsed with West Germany's defection. (Kent, LATNS, *M Her*, 6/1/73)

- U.S. Patent No. 3 730 287 was granted to NASA and to aerospace designer Edward Hryniewiecki of Comprehensive Designers, Inc., a Jet Propulsion Laboratory subcontractor, for a vehicle designed to explore planets with difficult terrain. Each of four legs with triangular wheels would have its own motive power for movement by walking, by turning wheels, or on tires revolving around the wheels like endless tracks. The vehicle had been planned for unmanned, remotely controlled exploration of Mars, but NASA was willing to license manufacture for oil and mineral exploration and rescue operations. (Pat Off PIO; Jones, *NYT*, 5/5/73, 45)

- Award of separate $10 000 contracts to nine aerospace companies for studies leading to advanced technology for highly maneuverable aircraft was announced by Flight Research Center. Boeing Co., Development Sciences, Inc., General Dynamics Corp., Grumman Aerospace Corp., Lockheed-California Co., LTV Aerospace Corp., Northrop Corp., Rockwell International Corp., and Teledyne Ryan Aeronautical Corp. would have three months to define design concepts and prepare a preliminary plan for developing design technology. Several contractors might then be selected for further contracts of $200 000 each for preliminary design of their concepts. (FRC Release 10–73; FRC PIO)

- A laser to transmit 1 billion bits of data per second between satellites was being developed for the Air Force by GTE Sylvania, Inc., United Press International reported. The system, using peripheral equipment with lighter weight and less power than present satellite radio networks, would send more information faster and would operate on a narrow bandwidth to reduce interference and interception. (*LA Times*, 5/5/73)

- National Oceanic and Atmospheric Administration scientists were using a new technique to map the sun's magnetic "weather"—constantly changing lines of magnetic-polarity reversal related to sunspots, solar flares, or storms, and configuration of the solar corona—NOAA announced. Developed by Patrick S. McIntosh of NOAA's Space Environmental Laboratory, the technique allowed observers to infer magnetic lines of force and magnetic polarities from hydrogen alpha photos of the sun made through a telescope with a light filter. The filter excluded all light except that in the red wavelength emitted by hydrogen. The inexpensive technique permitted even the smallst observatories to study magnetic fields. (NOAA Release 73–71)

- The proposal to redesignate Cape Kennedy Cape Canaveral was killed for the current session by the General Legislation Committee of the Florida State Legislature. The bill had cleared the Florida Senate March 22, but House opponents had said it would be disrespectful to Presidents Kennedy and Johnson. (*O Sen*, 5/2/73, 7A)

- *Science & Government Report* quoted the response of an unidentified alumnus of the President's Office of Science and Technology when asked if OST had participated in the Watergate episode: "No, they didn't trust us enough for anything as important as that. If we had been involved, do you think they'd have been using stone age electronics?" (referring to eavesdropping devices found in Democratic Party headquarters). (*Sci & Gov Rpt*, 5/1/73, 1)

May 2: The Skylab manned space laboratory to be launched May 14 would fly over 89% of the world's population and 65% of the earth's land areas and would be visible to the naked eye, NASA announced. Information would be distributed to enable people in most of the populated areas of the world to see Skylab in orbit. Skylab would be visible only in clear skies during the two hours before and after dusk when the viewer would be in the earth's shadow and the space station in

the sunlight. It would appear like the brightest star but would move fast enough to be distinguishable from stars. (NASA Release 73-90)
- U.S. space observers believed an accident to the Soviet *Salyut 2* space station launched April 3 had prevented the U.S.S.R. from carrying out a maximum one-month mission manned by a Soyuz crew, Thomas O'Toole reported in the *Washington Post*. Their assessment was that *Salyut 2* had begun tumbling in orbit when a maneuvering engine fired and then kept firing out of control. The firing had forced *Salyut 2* into cartwheel motion through space. Its solar panels had been torn loose, cutting the station off from almost all its electric power. The only electricity supplied to *Salyut 2* after the accident had come from batteries that might now be burned out. A few U.S. observers thought the Soviet tracking ship that instructed the station to make the fatal maneuver had received details of the accident by radio from *Salyut 2* before communications ceased. The ship had left its station off Newfoundland and had joined a sister ship in Curaçao, Dutch West Indies, where both had been refueled and refitted. The ships were now in Cuba, which had led some space observers to think they might put to sea again to await a further Soviet attempt to launch a Salyut station. (*W Post*, 5/2/73, A20)
- Thomas J. Lee had been named Manager of the Marshall Space Flight Center Sortie Lab Task Force by Dr. Rocco A. Petrone, MSFC Director, the *Marshall Star* announced. Lee succeeded Jack Trott, who had retired April 30. (*Marshall Star*, 2/5/73, 1)
- Johnson Space Center announced the award of a $1 947 000 cost-plus-fixed-fee contract to Chrysler Corp. to distribute and document wind-tunnel data for space shuttle development. (JSC Release 73-50)
- The Space Science Board of the National Research Council released *HZE-Particle Effects in Manned Spaceflight*. The report, prepared for NASA by the Board's Radiobiological Advisory Panel to the Committee on Space Biology and Medicine, concluded that the high-energy, heavy-ion irradiation encountered by astronauts outside the earth's magnetosphere or in high-inclination earth orbits would have negligible biological effects for periods of less than two years in space. (NRC Release)
- Former astronaut Frank Borman, now an Eastern Air Lines, Inc., senior vice president, left Moscow after a visit to the U.S.S.R. at the invitation of the Union of Soviet Societies for Friendship and Cultural Relations with Foreign Countries. During the visit, which began April 28, Borman laid a bouquet at the Kremlin wall where Yuri Gagarin and other cosmonauts were buried and visited the cosmonaut training center and a Salyut orbital station mockup at Star City. In an interview he said that he believed the 1975 U.S.–U.S.S.R. Apollo-Soyuz space flight was "the forerunner, the prototype to exactly the type of mission that we'll eventually see going to Mars. It is very difficult to project the pace at which space exploration will go forward, because it doesn't depend on technology so much as it does on economic and political action. But I think that by the end of this century we will have a manned earth mission to Mars." (Tass, FBIS–Sov, 4/30/73, L2; 5/3/73, G2; *Moscow News*, 5/19-25/73)
- Sen. Peter H. Dominick (R-Colo.) introduced S. 1686 to authorize the National Science Foundation to facilitate the application of science and technology to civilian needs and to assist in establishing civilian research and development priorities. The bill would seek to place a science

and technology adviser in each governor's office and enable states and localities to participate at the Federal level in establishing R&D priorities. (CR, 5/2/73, S8159)
- President Nixon submitted nominations to the Senate: Howard A. Callaway to be Secretary of the Army succeeding Robert F. Froehlke, whose resignation he had accepted May 1; Amrom H. Katz to be Assistant Director for Science and Technology of the U.S. Arms Control and Disarmament Agency; and Robert M. Behr to be USACDA's Assistant Director for Weapons Evaluation and Control. (PD, 5/7/73, 441, 444, 454)
- The Air Force announced the award of a $1 788 000 firm-fixed-price contract for component parts of the multimode radar system for the C-5A aircraft. (DOD Release 219-73)

May 3: NASA announced a delay in the launch of Intelsat-IV F-6 scheduled for May 4. Communications Satellite Corp. had said that a minor variation in the receiver gain had indicated the need for additional tests to determine possible causes of the variation. (NASA Note to Editors, 5/3/73)
- Dr. James C. Fletcher, NASA Administrator, issued a memorandum stating that private communications with astronaut crew members of the Skylab Orbital Workshop (to be launched May 14) would be permitted for morale, operational, and medical reasons. Private calls to families would be permitted once a week, without monitoring or public announcement. Private operational calls could be requested for "extreme operational emergency" and would be announced; paraphrases would be released to the public as appropriate. Private medical conversations would be scheduled daily and not announced. A daily medical bulletin on crew health would be issued to the public. (NASA Release 73-110, attaching memo)
- President Nixon sent to Congress *United States Foreign Policy for the 1970's: Shaping a Durable Peace.* The report recalled agreements made during the May 1972 Moscow Summit Meetings, including the agreement on a joint rendezvous and docking of Apollo and Soviet Soyuz spacecraft in 1975. "Since the summit, all of the agreements have been carried out as expected. Our space agencies have conducted preliminary tests of models of the spacecraft docking system and crew training will begin this summer. The Joint Committee on Environmental Protection met in Moscow in September 1972 and planned 30 collaborative projects on a variety of subjects. . . . The Joint Commission on Science and Technology met in Washington in March 1973 and agreed to carry out some 25 projects in . . . energy, chemistry, biology, and agricultural research. American and Soviet naval officers will meet this year to review the agreement on reducing incidents between ships and aircraft. This process of cooperation has begun to engage an ever widening circle of people in various professions and government bureaus in both countries. Direct contact, exchanges of information and experience, and joint participation in specific projects will develop a fabric of relationships supplementing those at the higher levels of political leadership. Both sides have incentives to find additional areas for contact and cooperation, and I anticipate further agreements patterned on those already concluded." (PD, 5/14/73, 455-653)
- The House Committee on Science and Astronautics began hearings on short-term energy shortages. Acting Director Darrell M. Trent of the Office of Emergency Preparedness testified that the U.S. "economy's

demand for all types of energy continues to increase at an amazing rate. However, our ability to find and produce domestic supplies of energy is not keeping pace with rising demand." The U.S. was becoming increasingly dependent on foreign supplies. "Except for Alaskan discoveries, additions to domestic petroleum reserves have fallen behind production since 1967. Gas is already in short supply. . . . At current growth rates, demand could double within the next 12 years over what it was in 1970. Because of the growing imbalance between domestic supply and demand, we may have to import as much as 60 percent of all the oil and gas we need in 1985. Such reliance on foreign sources of supply could pose serious problems for our national security. . . . The cost of our petroleum products alone could rise from about $6 billion in 1972 to as high as $45 billion by 1985, with grim implications for our troubled balance of payments." (Transcript)

- Western Union Telegraph Co. President Earl D. Hilburn said the company had received letters from seven unidentified companies of intent to order $3 million worth of communications satellite transmission services. Identification would be withheld until all applications had been received. Western Union's WESTAR satellite system was the first U.S. system to be authorized by the Federal Communications Commission. Inauguration of services was set for mid-1974. (Western Union Release)

- The Mackay Trophy for 1972 was presented by the Air Force and the National Aeronautical Assn. to Air Force Vietnam war aces Capt. Richard S. Ritchie, Capt. Charles B. DeBellevue, and Capt. Jeffrey S. Feinstein at a Dept. of Defense ceremony. The citation was for disregarding their own personal safety in providing protection for allied forces attacking high-priority targets deep in hostile enemy territory. Also honored was Col. John A. Macready (USAF, Ret.), the only three-time winner of the Mackay Trophy, on the 50th anniversary of the first nonstop flight across the U.S., which he piloted with Lt. Oakley G. Kelly in a Fokker T–2 aircraft May 3, 1923. (NAA *News*, 6/73, 1)

- The United Kingdom and the U.S.S.R. were negotiating for introduction of supersonic transport services in both countries, British Airways Chairman David Nicolson told a Washington, D.C., meeting of the Aviation/Space Writers Assn. The U.K. also would begin talks with the People's Republic of China toward inauguration of London-Peking air service before the year's end. (AP, *P Inq*, 5/4/73)

- The retirement of G. Merritt Preston as Manager of the Kennedy Space Center Shuttle Projects Office effective June 30 was announced by Dr. Kurt H. Debus, KSC Director. Preston, during 34 years of Government service, had been Assistant Chief of Operations for Project Mercury, Deputy Director of Launch Operations at KSC, KSC Director of Design Engineering, and KSC Director of Center Planning. He had received NASA's Outstanding Leadership Medal in 1963 for his Mercury work, two NASA Exceptional Service Medals for Apollo work, and the Spirit of St. Louis Award of the American Society of Mechanical Engineers in 1969 for meritorious service in the advancement of aeronautics. (KSC Release 85–73)

- Award of a $33 982 080 contract to Rockwell International Corp. for lease or purchase of 11 75A Sabreliner aircraft with electronic systems, to flight-test more than 7000 ground navigational aids, was announced by the Federal Aviation Administration. The contract was part of an FAA program to replace 47 DC-3s and T-29s with 21 modern, light, twin-turbine-powered aircraft for flight inspection. (FAA Release 73-78)

- French hydrogen bomb testing in the Pacific was "a moral outrage and a diplomatic embarrassment," a *Washington Post* editorial said, "but it is also a political fact, and one likely to become more rather than less important as the 1970s unfold." As a strategic weapon the French nuclear striking force was relatively elementary, lacking warheads to fit into long-range missiles, multiple warheads, long-range missiles, and missile-carrying submarines. "As a political force, however, the force de frappe is quite real. As [the late President Charles] de Gaulle intended, it is virtually certain to give France a larger voice in the evolution of European affairs than it otherwise would have." (*W Post*, 3/5/73, 26)

May 3-4: European Space Research Organization (ESRO) representatives met with U.S. space officials in Washington, D.C., to negotiate a draft Intergovernmental Agreement and draft a NASA-ESRO Memorandum of Understanding. Both agreements would cover U.S.-European cooperation in sharing costs of space shuttle development. The European countries, through ESRO, would fund and develop the $300- to $400-million Spacelab, a supporting system important to realization of the space shuttle's full potential. The Spacelab would facilitate joint-use programs with U.S. and foreign astronauts. (NASA Release 73-191)

May 4: Major countdown demonstration testing of the first two Skylab launch vehicles and their spacecraft on Kennedy Space Center Pads 39A and B, with crew participation, had been completed, NASA announced. Both launches remained on schedule for May 14 and 15. The three astronauts earlier in the week had been pronounced physically fit for the 28-day mission, after completing medical examinations at Johnson Space Center. (NASA Release 73-92)

- Dr. George M. Low, NASA Deputy Administrator, received the National Civil Service League's 1973 Career Service Award for Sustained Excellence at the League's 19th annual awards banquet in Washington, D.C. Dr. Low was cited as the individual most responsible for the success of the Apollo program because of his significant contributions toward solving reentry problems. Also honored was Paul G. Dembling, General Counsel for the General Accounting Office, who as NASA General Counsel had been a principal drafter of the National Aeronautics and Space Act of 1958 and a major participant in developing the legal framework governing peaceful uses of outer space. (NASA Hq *WB*, 4/16/73; *W Post*, 5/5/73, B4)

- Findings that the Aug. 4, 1972, solar storm—the largest ever recorded—had caused a small but measurable increase in the length of a day on earth were reported in *Nature* by Goddard Institute for Space Studies research associate Stephen Plagemann and John Gribben, an astronomer employed by *Nature*. Examination of U.S. Naval Observatory data had shown a daily increase in the day's length 10 times greater than usual between Aug. 4 and Aug. 8, indicating that the earth's rotation had slowed down fractionally. After a few days, the rotation rate had returned to its normal level. (*Nature*, 5/4/73, 26)

- Patents for X2048, a new aluminum alloy that could replace the two alloys used in the construction of supersonic aircraft, had been applied for by Reynolds Metals Co., a Reynolds spokesman said in Richmond, Va. X2048 provided up to 50% more fracture toughness than one alloy and was stronger and lighter than the other. (AP, W *Star & News*, 5/7/73)

- Magnetite from the Orgueil C1 meteorite was only 2.0 to 2.4 million yrs older than the next oldest meteorite, Karoonda C4, a Univ. of Chicago and Univ. of California at Berkeley team reported in *Science*. The oxide-xenon dating method had tied the primitive C1 meteorite to the chronology of the normal meteorites. If Karoonda and Orgueil had been formed from the same material, the age difference was an upper limit of the formation time of these meteorites and, by customary extension, of the solar system. (Herzog et al., *Science*, 5/4/73, 489)
- Award by Lewis Research Center of a $2 833 780 contract to General Electric Co.'s Gas Turbine Products Div. for two gas turbine assemblies was reported by *Lewis News*. The turbines were for a new test facility to bench-test advanced turbines and combustors for future aircraft. (*Lewis News*, 5/4/73, 4)

May 5: The U.S.S.R. launched *Cosmos 556* from Plesetsk into orbit with a 233-km (144.8-mi) apogee, 207-km (128.6-mi) perigee, 88.9-min period, and an 81.4° inclination. The satellite reentered May 14. (GSFC *SSR*, 5/31/73; *SBD*, 5/9/73, 49)

- Twelfth anniversary of the May 5, 1961, launch of *Freedom 7* by the Mercury-Redstone 3 launch vehicle, on the suborbital mission which carried Astronaut Alan B. Shepard, Jr., as the first U.S. man in space. (*Marshall Star*, 5/2/73, 1)

May 6: Langley Research Center Director Edgar M. Cortright received the degree of Doctor of Science *honoris causa* at the 152nd Annual Commencement of George Washington Univ. in Washington, D.C., and gave the commencement address to the University's School of Engineering and Applied Science. He said it was perhaps "a blessing in disguise" that the U.S. was running out of fossil fuels. In seeking new energy sources, "you will develop clean energy. And you will develop it within our own resources, thus alleviating the balance of trade problem." Three sources appeared particularly promising: nuclear fission, nuclear fusion, and solar energy. "In addition, you may even tap the heat of the Earth's core." (Newport News, Va., *Daily Press*, 5/10/73; *Langley Researcher*, 5/11/73, 1)

- The Federal Aviation Administration had set 1975 as a target date for phase-out of its seven DC–3 aircraft, United Press International reported. FAA was still using the aircraft—known as the workhorse that carved out much of U.S. aviation history—to check aerial navigation aids on the West Coast. Manufactured by the Douglas Aircraft Corp., the DC–3 had carried 95% of U.S. airline traffic by 1938. A 1966 survey had shown that one third of the world's transport planes were still DC–3s. The aircraft, nicknamed "The Gooney," had been designated the C–47 in World War II and the Korean war. By 1945 10 000 DC–3s were in military service. Built to carry 29 passengers in civilian configuration, the DC–3 flown by Gen. James H. Doolittle from China after his 1942 Tokyo raid had carried 72 persons. The DC–3 had been redesignated the AC–47 and armed with heavy machineguns for service in the Vietnam war. Scores of the aircraft were still flying in the U.S., in private hands and with Government agencies. (Clifford, UPI, *W Post*, 5/6/73, F15)

May 6–8: The Commission on Education of the National Academy of Engineering held a Washington, D.C., symposium to examine the problems of recruiting minority students in engineering to alleviate an impending engineer shortage. Only one percent of engineering students were members of an ethnic minority. (NAE Release 5/2/73, NAE PIO)

May 7: Selection of the Chryse valley near the mouth of the Martian Grand Canyon and of Cydonia in the Mare Acidalium, 1600 km (1000 mi) northeast, as landing sites for NASA's two unmanned Viking spacecraft in July and August 1976 was announced at an Hq. press conference. The two spacecraft were scheduled for launch on their year-long, 740-million-km (460-million-mi) journey toward Mars in the summer 1975. The sites had been selected by prominent scientists after a one-year evaluation of 22 potential areas for scientific interest and the probability of a successful landing. The possibility of finding water at these sites increased the chances of finding evidence of life. Backup sites were Tritonis Lacus and Alba.

Arriving at Mars, the spacecraft would enter a highly elliptical orbit and then separate into two parts, an orbiter and a lander. Each lander would carry a miniature chemical laboratory to analyze Martian soil for signs of life. A 3-m (10-ft) retractable claw would scoop soil samples for analysis. Other instruments would analyze the atmosphere and measure pressure, temperature, wind velocity, and quake activity. The orbiter would perform visual, thermal, and water-vapor mapping. (Transcript; NASA Release 73-91)

- NASA announced establishment of the General Aviation Technology Office within the Office of Aeronautics and Space Technology to develop the technology base for design and development of safer, more productive, and superior U.S. general-aviation aircraft. Roger L. Winblade had been named Manager, General Aviation Technology Office. (NASA Release 73-93)

- A Boeing 707 transport aircraft flew six flights over Dulles International Airport in Chantilly, Va., in Government-sponsored tests to demonstrate the results of Boeing Co.'s $7-million research project to reduce the 707 noise level to that of the newer Boeing 747 and McDonnell Douglas Co. DC-10. The tests proved the modified 707 was 10 to 15 db quieter than the original 707s. (Lindsey, *NYT*, 5/8/73, 78)

- World Airways, Inc., dedicated its George P. Miller World Air Center Hangar, one of the world's largest aviation buildings, at Oakland (Calif.) International Airport. The $14-million, 18 600-sq-m (200 000-sq-ft) facility had been named in honor of Rep. George P. Miller (D-Calif.), former Chairman of the House Committee on Science and Astronautics, who had been defeated in a 1972 primary. (NAA *News*, 6/73/73, 1-2; *A&A 1972*)

May 7-11: A 200-member U.S.-West German working group met at Kennedy Space Center to plan the Helios Project to launch two West-German-built spacecraft toward the sun with the greatest speed ever imparted to a man-made object. Helios' spacecraft velocity at insertion into trajectory would approximate 58 000 km per hr (36 000 mph). *Pioneer 10* and *11* (launched March 2, 1972, and April 6, 1973) had required a 51 000-km-per-hr (32 000-mph) velocity. Helios-A—to be launched by NASA from KSC by a Titan-Centaur rocket augmented by a solid-fueled 3rd stage, in September 1974—was to approach to within 45 million km (28 million mi) of the sun. Helios-B's target point would be determined by Helios-A results. Objectives of the project, named for the Greek god of the sun, would be to study solar physics by penetrating the outer solar corona, where charged particles received their final acceleration, and to demonstrate West German space technological capability. Equally funded by West Germany and the U.S., the project would be directed, after initial NASA liftoff control, by West Germany's control center at Oberpfaffenhofen. Cochairmen of the Helios Working Group were Gilbert W. Ousley, Chief of Goddard Space Flight Center's International Projects Office, and Ants Kutzer of the West German aerospace corporation Gesellschaft für Weltraumforschung. (KSC Release 86–73; KSC PIO)

May 8: An earthlike atmosphere, 70% oxygen and 30% nitrogen, would be used for the first time in a U.S. manned spacecraft on the Skylab mission, Marshall Space Flight Center announced. Pure oxygen had been an acceptable atmosphere for relatively short U.S. flights, but long-term breathing of pure oxygen might cause the red blood cells to become fragile. Three tons of breathing oxygen and three fourths ton of nitrogen would be launched aboard Skylab and combined into atmospheric distribution, circulation, and control systems at a nominal pressure of 3.45 newtons per sq cm (5 psia). (MSFC Release 73–57)

- The crew of Skylab 2—Charles Conrad, Jr., Dr. Joseph P. Kerwin, and Paul J. Weitz—participated in a simulated launch-to-rendezvous training session at Johnson Space Center. (AP, B *Sun*, 5/9/73, A10)

- *Lunokhod 2*, the U.S.S.R. automatic vehicle landed on the moon Jan. 16 by *Luna 21* (launched Jan. 8), began its fifth lunar day. The panel of the solar battery was opened and the onboard systems checked. The vehicle left the large tectonic fracture explored the previous lunar day and started northeastward toward the shore cusp of the Taurus Massif to continue exploration of the Le Monnier Crater. Lunokhod's Rifma radiation analyzer was sending data on the solar x-ray spectra and making comprehensive readings of a broad band of solar radiation. (Tass, FBIS–Sov, 5/11/73, L1; *Av Wk*, 5/21/73, 20)

- Award of a 12-mo, $234 788, cost-plus-fixed-fee contract to General Electric Co.'s Valley Forge Space Center to study future earth resources systems was announced by NASA. GE analysts would assist Johnson Space Center's Earth Resources Program Office in planning projects to be started at the decade's end. Guidelines would include unique space shuttle contributions to earth resources surveying. The study would focus on hardware and procedures to be developed for use from 1978 to 1982. (NASA Release 73–94)

- Completion of nine test flights in the Air Force Systems Command Flight Dynamics Laboratory's control-configured vehicle (CCV) program to produce a superstable aircraft was announced by AFSC. The flights, to validate the Laboratory's active-ride-control system design, were completed under part of a $4.1-million contract with Boeing Co. A second series of flights was expected to end in October. Tentative plans called for tests on fighter aircraft later. (AFSC Release 051.73)

- U.S. and U.S.S.R. delegates met in Vienna to continue Strategic Arms Limitation Talks (SALT). (Tass, FBIS–Sov, 5/9/73, H1)

May 9: The House Committee on Science and Astronautics favorably reported H.R. 7528, a $3.074-billion NASA FY 1974 authorization bill replacing the original $3.016-billion H.R. 4567. Increases in research and development funding included $25 million for the space shuttle, to total $500 million; $12 million for space applications, to total $159 million; $34 million for aeronautical research and technology, to total $180 million; $10 million for space and nuclear research and technology, to total $11 million; and $500 000 for technology utilization, to total $4.5 million. Overall R&D decreases were: $7 million in space flight operations, to $548.5 million; $8 million in space science programs, to $37 million; and $10 million in tracking and data acquisition, to $240 million. The budgets for construction of facilities and research and program management remained at $112 million and $707 million.

The Committee said its $10-million reduction in Skylab funding, to $224 million (within space flight operations), reflected a "success postured program" without major problems and expressed confidence in NASA Skylab management. Space shuttle funding was increased $25 million to provide more effective program implementation and hold total program funding "at or below the current projection." Total reductions of $8 million in funding for orbiting explorers, physics and astronomy, and lunar and planetary exploration supporting research and technology would be applied to the Earth Resources Survey Satellite program to permit immediate reinstatement of the ERTS-B project, on which the Committee placed "the highest priority." In the aeronautics program, a $14-million increase—from $18 million to $32 million—was recommended "to investigate and demonstrate noise reduction modifications to current narrow-body jet aircraft" and $20 million was added to reinstate the quiet, experimental, short takeoff and landing (QUESTOL) aircraft program terminated in January. The $10-million increase in space nuclear research and technology funding was "to maintain a viable, long-range capability in advanced nuclear power and propulsion research." It was probable "that any sensible energy research program undertaken by this nation would involve a significant part of this continued program." (CR, 5/1/73, D452; H Rpt 93–171; Text)

- Dr. Karl G. Harr, Jr., Aerospace Industries Assn. of America, Inc., President testified before the House Committee on Science and Astronautics' Subcommittee on Science, Research, and Development that the aerospace industry fully supported U.S. conversion to the metric system. He said the conversion was necessary if the U.S. was to continue to hold a prominent position in precision engineering in international trade, but urged that the conversion be timely, planned, orderly, and not necessarily exclusive of other measurement systems. (AIA Release 73–11)

- The keel for the Navy's first patrol hydrofoil missile ship (PHMS) was laid during a ceremony at Boeing's Seattle facility. One of two missile-carrying, fast hydrofoil patrol ships to be delivered to the Navy by Boeing in the summer of 1975 under a $46.5-million contract, the PHMS was propelled by waterjet systems and would cruise faster than 74 km per hr (40 knots). (Boeing Release A–0423)

- Eleven conservation organizations had petitioned President Nixon to halt plans for the Atomic Energy Commission's May 17 underground nuclear explosion in Colorado, the *Washington Post* reported. Conservationists had said the blast could contaminate the water supply for 27 million persons. The explosion was intended to free trapped supplies of natural gas. (UPI, *W Post*, 5/9/73, 3)

- President Nixon sent to the Senate the nomination of Dr. Gerald F. Tape to be the U.S. Representative to the International Atomic Energy Agency with the rank of Ambassador. Tape would succeed Dr. T. Keith Glennan, first NASA Administrator, who had resigned. The nomination was confirmed by the Senate June 7. (PD, 5/14/73, 658, 669; Tape Off)

- It was "puzzling," a Washington *Star and Daily News* editorial said, "that Soviet leaders remain so secretive about their space problems, when the general nature of a mishap cannot be hidden from modern tracking equipment and more openness could lead to a sharing of corrective knowledge. The Russian announcement that the last Salyut had merely completed its mission . . . fooled no one. Greater frankness, certainly, will be needed in the joint mission [U.S.–U.S.S.R. Apollo-Soyuz Test Project] requiring the confidence of each country's experts in the other's equipment and ability." (W *Star & News*, 5/9/73, A23)

May 10: Skylab 1, scheduled for launch May 14, would carry 950 kg (2100 lbs) of food, including lobster newburg, prime ribs of beef, filet mignon, asparagus, strawberries, and after dinner mints, the *Washington Post* quoted Dr. Malcolm C. Smith, Chief of Food and Nutrition at Johnson Space Center, as saying. Each Skylab mission would carry more than 70 different foods prepared by an automatic meal-reconstitution module. Semisolid food could be eaten out of the can with a fork and spoon and prevented from flying around in the weightless environment by a polyethylene membrane. Beverages could be consumed one swallow at a time from a multichambered bellowslike canister. Previously, astronauts had to drink the entire beverage at one time once the valve was opened. Fruit juices were fortified with potassium gluconate to replace the potassium lost in weightlessness. The use of alcoholic beverages had been rejected because of medical drawbacks and public opinion. The astronauts arranged their own meals with the correct combinations of nutrients and calories. Frozen foods would be used the first time in the U.S. space program. The cost of the Skylab food was $26.35 per man per day. Twenty percent more food than required would be carried. There would also be a rescue food supply of 16 days for three men. (Ross, *W Post*, 5/10/73, E1)

- The Senate passed S. 70, the Energy Policy Act of 1973, establishing a Council on Energy Policy to coordinate energy activities and recommend policy.

  The Senate also passed S. 373, to limit presidential power to impound appropriated funds. (*CR*, 5/10/73, S8809–10, 8828–55)

- Western Union Telegraph Co. announced it had received Federal Communications Commission authorization to construct the first of five earth stations for the $70-million WESTAR domestic communications satellite system [see May 3]. The first station would be built in Vernon Township, N.J. (Western Union Release)

- President Nixon's intention to nominate Dr. James R. Schlesinger, Director of the Central Intelligence Agency and former Atomic Energy Commission Chairman, to be Secretary of Defense succeeding Elliot L. Richardson, was announced by the White House Press Secretary. William E. Colby would succeed Dr. Schlesinger. The nominations were submitted to the Senate May 24 and confirmed June 28. Richardson would become Attorney General May 25. (*PD*, 5/14/73, 661–2; 5/28/73, 709, 714; *CR*, 6/28/73, S12394)

- Sen. Lawton M. Chiles (D-Fla.) introduced S.J. Res. 107 to redesignate Cape Kennedy Cape Canaveral. (*CR*, 5/10/73, S8740)

May 11: The U.S.S.R. launched *Cosmos 557* from Baykonur Cosmodrome into orbit with a 249-km (154.7-mi) apogee, 213-km (132.4-mi) perigee, 89.1-min period, and 51.6° inclination. Western observers noted the satellite, reported by some sources to be equipped with remote-controlled photographic equipment, was put into an orbit almost identical to that of the crippled *Salyut 2*, launched April 3. Observers later speculated it had been a Salyut-class spacecraft launched as a docking target for a manned Soyuz [see May 19]. The satellite reentered May 22 without maneuvering to extend its lifetime or an apparent attempt to recover it. (GSFC *SSR*, 5/31/73; *SBD*, 5/15/73, 81; O'Toole, *W Post*, 5/15/73, A14; 5/19/73, A12; *SF*, 1/74, 39–40; Lyons, *NYT*, 9/26/73, 33)

- Skylab 2 Astronauts Charles Conrad, Jr., Dr. Joseph P. Kerwin, and Paul J. Weitz completed their last medical checkups at Johnson Space Center and were declared ready for the Skylab launch, as countdown proceeded for May 15. Data were also gathered for inflight comparison of the astronauts' physical conditions. (Wilford, *NYT*, 5/12/73, 58)
- Objectives of Skylab's scientific and engineering experiments were outlined in a Kennedy Space Center press briefing. The 57 major experiment components aboard the space station scheduled for May 14 launch would provide data for 270 investigations in earth resources, solar astronomy, and materials science by 702 investigators and associates—200 of them foreign. Dr. Edward J. McLaughlin, NASA Director for Space Medicine, said 20 life sciences experiments would provide "a comprehensive picture of the interactions of man with his environment and of various body systems that are influenced . . . by the space environment." They would deal primarily with the metabolic and cardiovascular systems, exercise capacity, and nutrition.

    Dr. Goetz K. Oertel, NASA Chief of Solar Physics, said the objective of Skylab astronomy was "the study of the stars, including the sun, and other objects in space such as planets, comets, objects and libration points near the moon, and x-ray objects in the sky." Emphasis would be placed on the study of solar activity known to influence general weather patterns and on solar flares, responsible for radio communication blackouts and interference. Skylab was unique in that it would be manned. The astronaut would be able to align the instruments and select targets of interest, taking advantage of phenomena as it occurred.

    Dr. Robert A. Parker, astronaut and Skylab program scientist, said that corollary experiments would study gas and particles surrounding the Skylab and their effect on sensitive instruments, collect micrometeoroids, study man-made disturbances in Skylab and their effect on instrument alignment, and study flammable materials and manufacturing techniques in the zero-g environment. (Transcript)
- Dept. of Defense efforts to support Skylab 1 and 2, scheduled for launch May 14 and 15, were described by the Air Force Eastern Test Range newspaper the *Missileer*. More than 3250 DOD personnel members worldwide—including the crews of 3 ships and 53 aircraft—would assist NASA with Skylab communications, recovery, bioastronautics, weather, and public affairs. The DOD Manned Space Flight Support Office would coordinate NASA's requirements with DOD forces; AFETR, the Air Force Space and Missile Test Center in California, and Air Force advanced range instrumentation aircraft would provide tracking and instrumentation facilities; Air Force and Navy units would provide voice and teletype circuits for communications and recovery forces to assist NASA at the launch site and primary and secondary recovery zones; and Army, Navy, and Air Force personnel would contribute medical support at launch site and recovery areas. The Air Force hospital at Patrick

Air Force Base, Fla., had been designated the Skylab launch site medical facility. (*Missileer*, 5/11/73, 1)
- Flight tests of NASA's supercritical wing, developed by Dr. Richard T. Whitcomb in Langley Research Center wind tunnels, were being completed at Flight Research Center, NASA announced. Construction of the first wing for use on a commercial aircraft was planned for later in 1973 by LearAvia Corp. William P. Lear, developer of the Learjet executive transport aircraft, had said the wing could increase the Learjet's cruising speed by almost 10% and its range by 20%, without an increase in power or gross weight. Other aircraft companies were interested in the new airfoil shape, which had demonstrated its ability to increase aircraft efficiency 15% during more than 75 successful flights at FRC since 1971. (NASA Release 73-96)
- Employment assistance officers at Goddard Space Flight Center, Lewis Research Center, and Marshall Space Flight Center were providing detailed resumés of employees affected by NASA personnel reductions, NASA announced. Among affected employees were 200 engineers and scientists and 180 technicians at LeRC; 100 space scientists, engineers, and technicians at MSFC; and 106 GSFC specialists, including electronic, civil, mechanical, and aerospace engineers; physicists; chemists; mathematicians; and electronic and computer technicians. (NASA Release 73-99)
- Award of a $20 543 500 fixed-price-incentive contract for equipping two Boeing 747-200B advanced airborne command post aircraft with electronic equipment was announced by the Air Force. (DOD Release 240-73)

May 12: A press briefing on the earth resources experiment package (EREP) to be launched on the Skylab Workshop May 14 was held at Kennedy Space Center. NASA Associate Administrator for Applications Charles W. Mathews said the 150 EREP investigators included 50 from foreign countries. EREP included six sensors, two camera systems, a spectrometer, a multispectral scanner, a radiometer, a scatterometer, an altimeter, and a thermal data channel. EREP was similar in function to *Erts 1* (launched July 23, 1972) and could discriminate colors more completely, compare electronically and photographically derived imagery, and put up microwave instruments to operate in the different regions of the frequency spectrum, which would allow a look through clouds and under the surface of the earth.

Thomas L. Fischetti, EREP Project Manager at NASA Hq., said that if EREP failed the crew would have equipment to make corrections, modifications, or replacements. The flight plan could also be modified at the last moment for unfavorable weather conditions or new phenomena occurring on the earth. During the three missions EREP would return 40 000 photos and 25 magnetic tapes of data.

Dr. Verl R. Wilmarth, EREP project scientist at Johnson Space Center, said EREP had 9 major investigative areas with 47 subtasks, including stock and crop inventory, surveys of crop damage from insects, land-use classification, and water-resource and geologic mapping. A major effort would be made in sea state analysis. (Transcript)

May 13: NASA held a Skylab prelaunch press conference at Kennedy Space Center. NASA Skylab Program Director William C. Schneider said that Skylab hardware was "in good shape and ready to go." The countdowns for Skylab 1 and 2 launches were on schedule for May 14 and 15. The power transfer tests and fuel cell pressurization of Skylab 2 had been completed and the crew "is in good shape." Schneider said that the network, operational, and recovery forces were "in place and ready to go" and the weather was satisfactory.

Col. Alan R. Vette, U.S. Air Force recovery forces, said that the training of the recovery forces had been completed and they were "prepared for any type of landing." (Transcript)

- Skylab 2 Astronauts Charles Conrad, Jr., Dr. Joseph P. Kerwin, and Paul J. Weitz piloted their T–38 jet training aircraft from Johnson Space Center to Patrick Air Force Base near Kennedy Space Center to watch the May 14 launch of Skylab 1. KSC technicians meanwhile tested the onboard batteries and emergency detection system on the Saturn V scheduled to boost the Skylab 1 Workshop into orbit. The Skylab 2 launch crew applied the protective covers that would shield the manned Apollo spacecraft during Saturn IB launch vehicle liftoff May 15. (UPI, W Star & News, 5/13/73, 6; Wilford, NYT, 5/14/73, 24)

- Jet Propulsion Laboratory astronomers—using the 610-mm (24-in) telescope and coudé spectrograph at Table Mountain Observatory, Calif.—had found that a 20% fluctuation in carbon dioxide absorption was a recurring four-day phenomenon in the thick Venus cloud bank, NASA announced. The top of the cloud deck might move up and down more than one kilometer (two thirds mile) in a constant wave motion. The astronomers had said the observation was "a fundamental feature of atmospheric dynamics that is not explained by current theories of atmospheric circulation on Venus." (NASA Release 73-97)

- The U.S.S.R. had started production of the An-30 reconnaissance aircraft designed especially for aerial photography, the *New York Times* reported. The twin-turboprop aircraft, designed by Oleg Antonov, had possible applications in military reconnaissance and in civilian survey of the environment. (Shabad, *NYT*, 5/13/73, 8)

- A *New York Times* editorial commented on the cost-benefit ratio of the U.S. Skylab missions. "Is the advantage gained by putting men into space worth the high price that must be paid?" Although bold, imaginative, and potentially useful, Skylab came at a time when Americans were "oppressively aware of the limits of the nation's resources and of the consequent restraints and constraints on national policy." Skylab would be most important "in providing practical evidence in determining the relative value of manned activity in utilizing earth's immediate neighborhood to meet man's needs on this planet." (*NYT*, 5/13/73, 4:14)

May 14–June 22: The *Skylab 1* Orbital Workshop and—11 days later—the *Skylab 2* Apollo spacecraft carrying a three-man crew were launched into near-earth orbit to establish the first U.S. manned orbital laboratory, in the four-mission Skylab program. *Skylab 1*'s meteoroid shield was torn off and the solar array system damaged during launch, cutting power and raising temperature to threaten the mission, although the spacecraft achieved satisfactory orbit. *Skylab 2*, scheduled for May 15 launch, was delayed while damage was assessed and the flight plan modified. On May 25, *Skylab 2* carried three astronauts to rendezvous and dock with the earth-orbiting Workshop. The crew boarded the Workshop, repaired spacecraft damage, and conducted medical experiments and studies in solar astronomy and earth resources for a record 28 days before undocking for a safe return to earth June 22.

May 14–24: Skylab 1 (SL-1) was launched on time at 1:30 pm EDT from Kennedy Space Center's Launch Complex 39, Pad A, by a two-stage Saturn V launch vehicle after a nominal countdown. Liftoff was witnessed by U.S. and foreign TV viewers and by an estimated 500 000 persons at KSC, including U.S. and foreign dignitaries and 26 former prisoners of war invited by NASA. The Skylab 2 crew—Astronauts Charles Conrad, Jr. (commander), Paul J. Weitz (pilot), and Dr. Joseph P. Kerwin (science pilot)—watched the launch from quarantine.

Skylab's Saturn Workshop (SWS) cluster launched toward earth orbit included the Orbital Workshop (OWS) with its Apollo Telescope Mount (ATM), airlock module (AM), multiple docking adapter (MDA), and instrument unit (IU).

At 63 sec into the mission, indication was received that the meteoroid shield had deployed prematurely. The Workshop was placed in nominal, circular orbit with 438.2-km (272.2-mi) altitude, 93-min period, 50° inclination, and 7649.4-m-per-sec (25 096.6-fps) orbital velocity. The payload shroud was jettisoned on time at 15 min and 25 sec after launch and the ATM deployed at 21 min 34 sec with normal deployment of the ATM solar arrays at 26 min 38 sec, but ground controllers received no indication that the Workshop solar array system (SAS), which was to have been released by an onboard computer, had been deployed.

At 41 min, a ground command was sent to deploy the two SAS wings. They did not respond. The command was sent again 30 min later and a third time near the end of the first orbit, without result. During the first postlaunch press briefing, Director of Launch Operations Walter J. Kapryan said that if the arrays did not deploy "the mission will be seriously degraded." The crew was not set up to perform a fix-it-yourself extravehicular activity (EVA) and there were no controls for deployment inside the Workshop.

Anxious officials and flight controllers began analyzing radioed launch data. Analysis and later inspection revealed that the meteoroid shield had been torn off by vibration of the vehicle just after passing through mach 1 speed, near the time of highest dynamic pressure. Some early measurements indicated that the shield separation straps were still satisfactory, but later measurements indicated failure. The failure had released the securing mechanism on SAS Wing 2 and the wing ripped off. A piece of shield had wrapped around SAS Wing 1 and kept it from deploying. Loss of the wings, which were to provide power to the Workshop, reduced power to 50% of its average 8000-w output; the remaining 50% would be supplied by the four ATM solar panels.

At 11:00 pm EDT Skylab Program Director William C. Schneider announced the Skylab 2 launch had been rescheduled to May 20 to allow assessment of damage and development of a new flight plan to maximize scientific return from Skylab's experiments with limited power. Fuel cells of the Skylab 2 command and service module (CSM) would be used to produce electricity, after joining the Workshop, but the CSM oxygen and hydrogen for the cells would last only 16 to 21 days. It was too early to assess the effect on Skylab 3 and 4 missions, planned for 56 days each.

On May 15 Skylab faced a new threat as temperatures inside the Workshop reached above 311 K (100°F). The lost meteoroid shield had also been designed to reflect sunlight and help cool the Workshop. "We think we can live with the electrical power shortage," Schneider said, "but if we can't solve this thermal problem we have a serious situation." Schneider feared foods and film might be spoiled, plastic parts might begin to leak poisonous gases, and the laboratory might become uninhabitable.

The entire Skylab team was mobilized—at Johnson Space Center, Marshall Space Flight Center, Martin Marietta Corp., McDonnell Douglas Corp., TRW Inc., and Garrett AiResearch Co.—in an all-out attempt to overcome the loss of the shield and solar arrays. Schneider said NASA was studying the possibility of the Skylab 2 astronauts' putting "some kind of a thermal blanket around the spacecraft," but that would require a delay of the Skylab 2 launch beyond May 20. Astronauts would also inspect the solar wings, but whether they could make repairs was uncertain.

The Skylab 2 crew returned to JSC May 15 to help flight planners salvage the mission.

Special attitude maneuvers were run to obtain temperature data on the Skylab cluster and to limit Workshop temperatures, as internal temperature soared as high as 325 K (125°F). Temperature was reduced by pitching the Workshop with the longitudinal axis pointed at the sun for one orbit and, on May 16, the Workshop was tilted 55° away from the sun, with sufficient sunlight on the working solar panels to charge the batteries but moving the space station out of the sun's full glare.

Teams continued to compute options to overcome the loss of the shield: astronauts could spray-paint *Skylab 1*'s side to alter radiation properties of the gold foil coating the Workshop shell, could wallpaper the critical area, or could deploy a solar sunshade. Choice of the third option was announced May 16. The astronauts were trained in three possible methods of deploying the sunshade and in procedures to deploy the jammed solar panel. If they failed with the panel, a near-normal 16- to 21-day mission could still be realized by using the CSM power and curtailing some experiments. After that experiments would have to be severely cut back to reserve power for the CSM's trip home.

On May 17 NASA announced postponement of the Skylab 2 launch until May 25 to allow the astronauts to practice techniques for deploying the sunshade and mission engineers to design, fabricate, and test hardware. Astronauts Conrad, Kerwin, and Weitz flew to MSFC to practice installing the sun shield in the Neutral Buoyancy Simulator.

On May 20 the Workshop was purged of any gases from the overheated insulation material and turned slightly to allow the sun to warm the airlock module and prevent water stored there from freezing and causing breakage. Components of the astronauts' repair kit were shipped to MSFC for final testing before being installed in Skylab 2's Apollo CSM.

On May 22 the Skylab 2 astronaut crew flew to KSC to prepare for the May 25 launch. Three thermal shields were decided on at a May 23 review: a parasol deployable from the Workshop through the solar scientific airlock, a backup twin-pole shield deployable by the crew during EVA, and a standup-EVA (SEVA) sail deployable by the crew from the CSM if both other methods failed.

The Skylab 2 countdown resumed at 5:30 am EDT May 23 at KSC. It had been halted eight hours after *Skylab 1* liftoff May 14. On May 24 the three emergency sunshades and necessary tools were stored in the cm. Significant experiments—SO15, effects of zero g on single human cells; SO20, x-ray/solar photography; TO25, coronagraph contamination measurements; M555, gallium arsenide crystal growth; equipment for SO63, ultraviolet (UV) airglow horizon photography; and spare equipment for TV coverage of the ATM EVA—were removed to accommodate the additional 180 kg (400 lbs) of equipment. The crew would also carry an extra medical kit to replace medicines spoiled by high temperatures in *Skylab 1*.

Lightning struck the Skylab 2 mobile service structure at 5:24 pm EDT May 24, but retests of the vehicle indicated no damage.

At a May 24 press conference Program Director Schneider said there would be limited activity by the crew to deploy the partially opened solar panel. "We're not too optimistic that we'll be able to do too much, although we will give Capt. Conrad the option to try it if it looks like a reasonable job." NASA was "very confident that . . . we will be able to deploy a [thermal] shield, we will get the spacecraft temperatures back under control, and we will have a good 28-day mission."

May 25: Skylab 2 (SL-2) was launched from KSC Launch Complex 39, Pad B, on a two-stage Saturn IB launch vehicle into heavy dark clouds on time at 9:00 am EDT. As Astronauts Conrad, Kerwin, and Weitz headed toward the *Skylab 1* Workshop they were to attempt to repair in orbit and man for 28 days, Commander Conrad told Mission Control, "We fix anything."

The CSM/S-IVB combination was placed in a phasing orbit with 224-km (139.2-mi) apogee and 150-km (93.2-mi) perigee 9 min 56 sec after launch. Six minutes later the CSM separated. The S-IVB stage deorbit maneuver was performed at 2:23 pm EDT and the stage splashed into the Pacific Ocean.

The CSM rendezvoused with the Workshop at 4:30 pm EDT over the Guam tracking station on the fifth orbit. During 15 min of TV coverage the CSM flew around the Workshop, within 1.5 m (5 ft) of it. The astronauts confirmed that the SAS Wing 2 was missing and that the SAS Wing 1, deployed only 15°, was restrained by a piece of the micrometeoroid shield wrapped around the SAS beam fairing. Portions of the gold foil laminated to the Workshop skin for thermal protection had been torn and heat and UV radiation from the sun had scorched the skin.

The astronauts soft-docked the CSM with the Workshop, using only three capture latches; ate dinner; and undocked to attempt to deploy the jammed solar panel. Weitz stood in the open hatch of the CSM with Kerwin holding his knees to keep him from drifting into space. He tried to free the solar panel with a shepherd's crook and a pole with pruning shears attached without success. Conrad told Mission Control that the panel appeared to be "hooked in there like it's nailed on." The repair attempt was ended at 8:12 pm EDT.

The crew made four unsuccessful attempts to redock with the Workshop. Working in bulky spacesuits and gloves, the astronauts evacuated the air from the Apollo cabin and hooked up an electric cable to override a suspected short in the docking probe. "Yea. We got a hard dock out of it [a docking using all 12 sealing latches to seal the two spacecraft together]," Conrad said at 11:50 pm EDT. The crew found one of the capture latches had stuck in the retracted position, but troubleshooting procedures recommended by Mission Control restored the latch.

May 26–June 21: The *Skylab 2* crew entered the multiple docking adapter on the Skylab Workshop at 12:45 pm EDT May 26 and performed activation and Workshop preentry tasks. A chemical detection device inserted into the passage indicated no lethal gas was present. At 2:30 pm EDT Astronaut Weitz entered the Workshop wearing a gas mask, gloves, and soft shoes to protect him from the heat. Weitz told ground control, "The OWS appears to be in good shape. It feels a little bit warm as you might expect, like 90 to 100 degrees [F; 305 to 310 K] in the desert. I felt heat radiating from everywhere but I never felt uncomfortable. And nothing I touched felt hot to me. It's a dry heat." Kerwin and Conrad followed Weitz. After a quick inspection, the crew closed the hatch.

At 4:30 pm EDT the crew began to deploy the solar parasol—a mylar shade folded against a telescopic pole—through the solar airlock. As they readied the parasol, the astronauts looked out of the windows and identified landmarks across the U.S. Weitz recognized Seattle, Wash., and Conrad spotted both launch pads used for the Skylab launches at KSC.

Conrad and Weitz popped open the shiny aluminum and orange canopy at 8:30 pm EDT May 26 (Mission Day 2) while Kerwin took TV photos from the attached Apollo CM. Significant external temperature decreases were noted almost immediately after deployment. Internal temperatures decreased more slowly. The Workshop was headed back to solar inertial attitude.

By May 27 the Workshop temperature had dropped to 309 K (97°F). The crew activated systems and experiments while their back-and-forth excursions through the Workshop were televised to the earth. The *Washington Post* later said they resembled "skin divers snorkeling through a cave in the Caribbean" or a scene from the film "2001" with astronauts "in orange flight suits and white T-shirts swimming like fish through the big workshop, floating backwards and tumbling like gymnasts."

On May 28 the astronauts began the first medical tests, taking and processing blood samples and checking the conditions of their circulatory and metabolic systems. Checkout and preparation of the (ATM) was begun. During a seven-minute televised space-to-ground news conference beginning at 1:09 pm EDT, Conrad said the crew had adapted rapidly. "None of us has had any motion sickness. . . . It seems to be like the simulator except with the absence of gravity." Kerwin concluded that the brain and eyes must override the body's vestibule system in the inner ear. "You say to your brain, 'Brain, I want that way to be up.' And your brain says, 'OK, then that way is up.' . . . Your brain will follow you." Conrad agreed, "We're in good shape for 28 days."

Dr. Willard R. Hawkins, JSC Deputy Director of Life Sciences for Medical Operations, said later in the mission that it was "rather amazing" that the crew had shown no signs of the expected motion sickness. Data received were preliminary, but prelaunch exercises in a whirling chair might have helped avoid nausea.

Temperatures continued to drop slowly, reaching 304 K (88°F). At 8:50 pm EDT a 69-sec trim burn put the Workshop into a repeating ground track about 96 km (60 mi) west of normal, to acquire earth resources data.

On May 29 checkout of the ATM was completed and four passes were made. Kerwin trained an array of telescopes to observe x-ray emissions and examine the structure of the sun's atmosphere. He photographed three regions of the sun considered active by solar physicists and made detailed observations of the full solar disk and the corona. Conrad and Weitz prepared the earth-oriented cameras for earth resources experiment package (EREP) passes. Two biomedical runs were made. During a private conversation with JSC, Conrad reported that the internal temperature had stabilized near 304 K (88°F) and that, while this would be tolerable for the rest of the mission, he did not think the crew could carry out the full protocol on the bicycle ergometer (exercise device).

On May 30 the solar telescopes were aimed at the active regions of the sun while scientists on the earth received excellent TV photos of the solar corona and the solar disk in the UV spectrum. The first EREP photo pass was made over a 3200-km (2000-mi) strip from Oregon to the Gulf of Mexico. Good pictures were obtained of the Great Salt Lake, gypsum beds in New Mexico, and the soil salinity of the Rio Grande Valley, as well as microwave measurements of the Gulf of Mexico.

After the EREP pass, the batteries—at a reduced level of 45% to 50% capacity—tripped off line and ceased to produce electricity. They had been designed to trip off at 20% capacity. One battery also lost the regulator that allowed the solar panel to recharge it, decreasing the total Workshop power by 6%, or 250 w, leaving 4200 w. The Workshop required 3600 w for maintenance, leaving only 600 w for experiments. The second EREP pass was canceled for a power evaluation. Sixteen batteries were brought back on line but two remained off line and were useless. Heat, fans, and lights were powered down.

Workshop power continued reduced May 31 and EREP experiments were canceled. The crew continued medical experiments, each taking a turn in a rotating chair to provide data on motion sickness and maintaining balance in zero g. Weitz repaired the mirror system on the UV stellar photography experiment. Temperatures had dropped to 301 K (82°F).

The Skylab astronauts took their first day off June 1 (Mission Day 8) and their first shower in space. On a 15-min color telecast they performed housekeeping chores and—to the music from the film "2001"—did handstands, somersaults, and cartwheels in the zero-g atmosphere. Conrad led the crew in a run around the watertanks. Centrifugal force overcame weightlessness, so the men could keep their feet on the track. On the ground Skylab officials announced the decision to make a second attempt to free the jammed solar panel, while engineers and the backup astronauts, led by Russell L. Schweickart, tested EVA procedures.

On June 2, Conrad and Weitz made a 10-min EREP photo pass from California to Central Mexico to gather data on earthquakes, volcanoes, pollution sources, mineral resources, and land-use patterns. On the ground Schweickart successfully tested three ways of freeing the jammed solar panel using a bolt cutter, a prying rod, and a surgeon's bone saw. The astronauts completed the third EREP pass over North America and the fifth ATM pass on June 3. The nuclear emulsion experiment, the transuranic cosmic ray detector, and the neutron analysis detector were deployed.

Astronaut Conrad set a new world record for time in space at 1:17 am EDT, surpassing the 715-hr 5-min mark set by Astronaut James A. Lovell, Jr. The first ball game in space was held in the Workshop. Conrad said a ball thrown in weightlessness "goes straight as an arrow."

The fourth EREP pass was made June 4, as well as five hours of ATM solar viewing. Dr. John Zeiglschmid, Skylab Flight Surgeon, told a JSC press briefing that the physical condition of the Skylab crew was excellent. They were consuming the proper amount of food and getting adequate rest. Conrad and Kerwin had lost only 0.5 kg (1 lb); Weitz had lost 1.8 kg (4 lbs). Top Skylab officials at MSFC agreed that solar wing deployment techniques developed by Astronaut Schweickart were feasible, without unusual safety hazards.

On June 6 preparations were made for the next day's EVA. Conrad said, "I guess we have a fifty-fifty chance of pulling it off." The 16 Workshop batteries were producing 4200 w—only 600 w more than the minimum necessary for the astronauts' survival. One battery was down to half charge and others were believed to have been damaged by the heat. The astronauts photographed hurricane Ava during an EREP pass, made an ATM pass, and photographed UV spectra of stars.

Conrad and Kerwin opened the AM EVA hatch at 11:23 am EDT June 7 and extended their umbilicals and a rigged-up 7.6-m (25-ft) pole fitted with cutters for releasing the jammed solar wing, as Weitz took TV photos from the Skylab and the CSM. Outside Kerwin had difficulty keeping a firm footing, with only makeshift handholds and foot restraints, as he secured the end of the pole to antenna mount. His heart rate rose to 150 beats per minute and he was burning energy at 2000 to 2500 BTUs an hour. Finally he extended the pole to the beam holding the wing and—with Conrad repeatedly telling him to "take it easy" and "cool it"—managed after several tries to fit the cutters around the small bolt that restrained the wing. Conrad moved hand over hand along the pole to the wing. Tangling of the 10.6- and 18.3-m (35- and 60-ft) flexible umbilicals linking the astronauts to spacecraft life support systems complicated their movements.

Kerwin pulled a lanyard to snap the cutters and sever the bolt and then extended the wing using a rope attached to the upper surface. Conrad pulled a rope to break a frozen actuator. Two of the sections making up the wing were 40% deployed and the other about 30%.

The two astronauts also changed a film magazine in a telescope camera and latched open a balky door on an x-ray telescope before reentering the Workshop and closing the AM EVA hatch at 2:53 pm EDT, concluding 3 hrs 30 min successful EVA.

The Workshop was pitched up to expose the panels to more direct sunlight and unfreeze the hydraulic system that deployed them. After a few hours one section was 90% deployed, another 40%, and the third 29%. By the next day the sections were fully deployed. Additional electric power of 3000 w was obtained and all eight batteries supplied by the solar panels were charged and in good working order—paving the way for a full Skylab mission.

From Mission Control Schweickart said, "Everybody's shaking hands down here and we just wish we could reach up and shake yours." President Nixon sent a message: "On behalf of the American people I congratulate and commend you and your crew on your successful effort to repair the world's first true space station. In the two weeks since you left earth, you have more than fulfilled the prophecy of your parting words, 'We can fix anything.' All of us now have new courage that man can work in space to control his environment, improve his circumstances and exert his will even as he does on earth."

On June 12 the ninth EREP pass was made. Weitz said he could see pollution flowing into Lake Erie and he couldn't find Washington, D.C., because of haze. Conrad melted an aluminum alloy to study the behavior of molten metal in weightlessness and used an electron beam for welding inside a small vacuum chamber.

On June 13 Dr. Hawkins told a JSC medical review that Kerwin and Weitz were no longer able to perform at preflight levels on the bicycle ergometer, indicating their cardiovascular systems had been weakened by weightlessness. The condition was not at dangerous levels and was not interfering with their normal day-to-day performance.

Weitz photographed a massive explosion on the sun with the ATM solar telescopes June 15. The flare had erupted from the center of the sun's side facing the earth and covered an area 40 200 km (25 000 mi) wide. Conrad checked out the Apollo CSM for the return trip to earth, while Kerwin initiated the growth of bacteria in a culture medium and photographed its behavior in the space environment.

On June 17 a trim burn corrected a cross-track error and positioned the vehicle for the start of the Skylab 3 mission. President Nixon telephoned the astronauts from Key Biscayne, Fla.: "I guess the way I could summarize this project is that it proves that man still matters. With all the technical machines . . . that you had to work with, it proved that when there were difficulties the ingenuity of men in space was what really mattered, and you have made us all very proud with the way you handled some difficult problems in this project." The President invited the three Skylab astronauts to visit him at San Clemente, Calif., after splashdown.

At 3:22 am EDT June 18 (MD–25) astronauts Conrad, Kerwin, and Weitz surpassed the previous space endurance record held by the *Soyuz 11* cosmonauts (launched June 6, 1971). Conrad asked that a message be relayed to Soviet cosmonauts giving "our respects at this point in our flight to them and their comrades. Wish them good luck from us in the future."

EVA–2 began at 6:53 am EDT June 19. Conrad and Weitz moved through space to retrieve and replace the ATM film cassettes while Kerwin photographed the EVA for TV from inside the Workshop. Conrad also freed a stuck relay by striking it with a hammer. One of two dead batteries began charging immediately. From the ground Schweickart said, "It worked. Thank you very much, gentlemen. You've done it again." The cameras for the white light coronagraph and x-ray spectroheliographic experiments were replaced by the astronauts to permit photography during the unmanned period between *Skylab 2* and 3. EVA operations ended at 8:29 am EDT, after 1 hr 36 min.

During a June 20 press conference from space Kerwin said the astronauts' physical condition was "a pleasant big surprise. I'm tremendously encouraged about the future of long-duration flight."

June 22: The astronauts undocked the CSM from the Skylab Workshop at 4:55 am EDT on Mission Day 29. As the CSM pulled away and circled the Workshop, TV and photographic coverage were obtained by the crew. CM separation from the SM at 5:40 am EDT was followed by initial service propulsion system firing at 6:05 am EDT and a final deorbit maneuver at 9:11 am EDT.

The CM splashed down at 9:50 am EDT 1340 km (830 mi) southwest of San Diego. A recovery helicopter dropped swimmers who installed a flotation collar and a sea anchor. The recovery ship *Ticonderoga* was maneuvered beside the CM and a crane lifted it aboard with the astronauts inside. Medical examinations on board the recovery ship found some dizziness and ill effects from weightlessness. The crew would travel by ship to San Diego and by air to Ellington Air Force Base, Tex., June 24.

The *Skylab 1–2* mission achieved its primary objectives of establishing the Skylab Orbital Assembly in earth orbit, obtaining medical data on the crew for use in extending the duration of manned space flight, and performing inflight experiments. It contributed significantly toward program objectives of determining man's ability to live and work in space for extended periods, extending the science of solar astronomy beyond the limits of earth-based observations, and developing improved techniques for surveying earth resources from space. Despite mission anomalies, 80% of the solar data planned was obtained; 12 of the 15 earth resources data runs were accomplished; all 16 medical experiments were conducted as required and a time history of man's adaptations to zero-g environment was obtained for the first time; and data were taken on all 43 experiments scheduled for *Skylab 2* except those precluded by the use of the solar parasol and weight and power limitations. The performance of the flight crew was good and the mission was judged officially on July 14 to be a success.

Major records set to date included the longest-duration manned flight, 28 days 50 min; the longest total time in space for one man, 49 days for Conrad (veteran of *Gemini 5* and *11* and *Apollo 12* missions); the longest inflight EVA, 3 hrs 30 min by Conrad and Kerwin June 7; and the longest distance in orbit for a manned flight, 18 531 559 km (11 514 967 mi).

Skylab 1–2 was the first U.S. manned orbital workshop. Skylab 3 was scheduled to be launched in August and Skylab 4 in November. The Workshop would be operated unmanned between manned missions. The Skylab program was directed by NASA's Office of Manned Space Flight. MSFC developed and integrated the major program components including the OWS, AM, MDA, ATM, payload shroud, and most experiments and was responsible for the Saturn IB and Saturn V launch vehicles. KSC managed launch operations. JSC was responsible for flight operations, recovery, crew selection and training, and development of the modified CSM and the spacecraft launch adapter. Tracking and data acquisition was managed by Goddard Space Flight Center under the overall direction of the Office of Tracking and Data Acquisition. (NASA prog off; NASA PAO press briefing transcripts, 5/14/73–6/20/73; NASA Mission Rpt MR-3; *PD*, 6/25/73, 787, 827; *NYT*, 5/17/73–6/22/73; *W Post*, 5/16/73–6/22/73; W *Star & News*, 5/23/73–6/22/73; B *Sun*, 5/15/73–6/12/73; M *Her*, 5/14/73, 19A; H *Post*, 6/7/73)

May 14: Unsolicited cash contributions to NASA from persons in the U.S. and foreign countries since 1959 would be used to finance the 19 student scientific experiments on Skylab [see Jan. 26], NASA announced. The $5548 fund included individual gifts ranging from $1500 to the $0.35 contribution of an eight-year-old boy sent in 1970 because he had heard that NASA "didn't have enough money." A retired Navy chief petty officer had sent one day's pay after each manned flight in memory of his son who had been killed in the Korean war. A naturalized citizen from Eastern Europe had sent $100 in 1964 because he wished to be a part of the space program that would mean the survival of a free U.S. (NASA Release 73–98)

- Man's ability to survive in space was discussed in *New York Times* and *Washington Post* articles. Of 33 astronauts who had made Apollo flights, 11 had suffered from motion sickness for the first few days in space. *Apollo 13* (April 11–17, 1970), the aborted moon landing mission, had indicated that men in danger would have a tough time surviving in space. Astronaut James A. Lovell, Jr., had lost 6 kg (14 lbs) during the six-day mission. John L. Swigert, Jr., had been so overcome by fatigue near the end of the flight that he made several errors in checking reentry procedures. Fatigue might also have contributed to the deaths of the Soviet *Soyuz 11* (June 6–30, 1971) cosmonauts.

  Calcium loss in weightlessness, the most serious of potential medical problems, could lead to thinning of the bones and subsequent bone breaks after long periods in space. Other health problems included increased heart rates, weight loss, decreased oxygen consumption, reduced blood pressure, and the reduced flow of blood to the brain and kidneys.

  Psychological effects, as well as these problems, would be studied on Skylab missions. Were windows necessary? Should doors be doorshaped or would circular holes be sufficient? Would the crew tolerate sleeping upright in sleeping bags attached to the walls? (Schmeck, *NYT*, 5/14/73, 24; O'Toole, *W Post*, 5/14/73, A12)

- President Nixon signed Executive Order 11718 designating the International Telecommunications Satellite Organization (INTELSAT) as an international organization entitled to enjoy certain privileges, exemptions, and immunities. (*PD*, 5/21/73, 673–4)

- President Nixon announced the appointment of the National Science Foundation Director, Dr. H. Guyford Stever, as Acting Chairman of the Federal Council for Science and Technology. (*PD*, 5/21/73, 681)

- The U.S. Supreme Court ruled 5 to 4 that municipalities could not enact local curfews on jet flights at their airports to control noise. Justice William O. Douglas said in the majority opinion: "We are not at liberty to diffuse the powers given by Congress to F.A.A. [the Federal Aviation Administration] and E.P.A. [the Environmental Protection Agency] by letting the states or municipalities in on the planning." (*NYT*, 5/15/73, 11)

- An account of the successful Apollo program was part of an Air Force "cram" course in recent U.S. history being given to returned U.S. prisoners of war at Maxwell Air Force Base, Ala., the *Detroit News* reported. (Fleming, *D News*, 5/15/73, 25)

May 15: Sen. James S. Abourezk (D–S. Dak.), member of the Senate Committee on Aeronautical and Space Sciences, called for a full-scale congressional investigation of the malfunction of the *Skylab 1* Workshop [see May 14–June 22], during a Washington, D.C., press interview. (Lyons, *NYT*, 5/16/73)

- The resignation of Dr. Robert C. Seamans, Jr., as Secretary of the Air Force was accepted by President Nixon with deep regret and apprecia-

May 15

tion for his years of outstanding public service to the Air Force and to NASA. He had served NASA as Associate Administrator 1960–1965 and as Deputy Administrator 1965–1968. Dr. Seamans became President of the National Academy of Engineering, succeeding Clarence H. Linder, NAE president since 1970. (*PD*, 5/21/73, 682; NAE Release 5/16/73; NASA Biog; *A&A 1968*)

- NASA participation in the National Science Foundation's Research Applied to National Needs (RANN) program was described by Dr. Alfred J. Eggers, Jr., NSF Assistant Director for Research Applications and former NASA Assistant Administrator for Policy. Dr. Eggers testified before the House Committee on Science and Astronautics' Subcommittee on Energy during a hearing on energy research and development: "Particularly close ties are developing between NSF/RANN and NASA in the field of Energy Research and Technology." The major focus was on terrestrial applications of solar energy. Lewis Research Center was especially active in joint work on requests for proposals on solar thermal areas, photovoltaics (use of the generation of electromotive force when radiant energy fell between dissimilar substances), flat plate collection, and heating and cooling of buildings. NASA had made a preliminary inventory of its test facilities applicable to NSF's solar energy research program. NSF had requested NASA to review the status of solar insolation data from the National Oceanic and Atmospheric Administration and, working with NOAA, to examine the need for better data and methods. Jet Propulsion Laboratory and the Cal Tech Environmental Quality Laboratory were jointly conducting workshops and studies in water heating and other solar technology applications. A NASA Wind Energy Workshop at LERC would pursue plans for a wind energy research program. Other areas of NASA participation included research into bio-conversion of organic materials to fuels, thermal batteries and superconducting transmission technology, earthquake engineering, regional environmental systems, trace contaminants, and fire. (Transcript)

- Communications Satellite Corp. held its 10th Annual Meeting of Shareholders in Washington, D.C. Board Chairman Joseph H. McConnell reported that its global system, through the fourth generation of Intelsat satellites, provided more than 260 communications pathways among earth station facilities in 49 countries. "Along these pathways flows a major portion of all international long distance communications—telephone, television, teletype, high speed data, and facsimile. Almost 100 countries, territories, and possessions are leasing satellite services on a full-time basis. High-quality telephone calls can be made to more than 20 countries that cannot be reached by cable. . . . Over 900 million people on six continents can see an important event on TV as it happens." Dr. Joseph V. Charyk, ComSatCorp President, said, "At the end of 1965 . . . we were leasing 66 voice grade half circuits on a full-time basis. Today we are leasing 3,146." (ComSatCorp Rpt to Shareholders)

- The "spreading Watergate scandal"—arising from investigations of alleged White House implication in the burglarizing and electronic surveillance of Democratic National Committee Headquarters by members of the Committee to Reelect the President—had slowed down Administration research and development decision making, a *Science & Government Report* editorial said. The subsequent disintegration of presidential staff had created "a decision-making paralysis that is subtly but steadily spreading to agencies far remote from current headline events." The editorial quoted an Office of Science and Technology staff member as saying that "just about every thing that the President set out in his Science and Technology message 14 months ago requires some sort of approval at the White House level now and then, and there's no one to talk to there." The newly created White House National Energy Office had "already received some shock waves from the scandal through the resignation of presidential assistant John D. Ehrlichman," who had been appointed by President Nixon to a three-man Special Committee on Energy to "give political weight" to NEO decisions.

Difficulties flowing from Watergate were aggravated by the impending transition in management of the science-Government relationship. Many OST responsibilities would shift to the National Science Foundation Director, Dr. H. Guyford Stever, when the OST was abolished June 30. "Whether for good or ill, the change is a change, and necessarily involves the dissolution of longstanding relationships, some of them dating back to early post-Sputnik days." (*Sci & Gov Rpt*, 5/15/73, 7)

- Rep. Leslie Aspin (D–Wis.) in a House statement requested the General Accounting Office to investigate a "secret agreement" between the Air Force and the Pratt & Whitney Div. of United Aircraft Corp. under which, he said, the Air Force had certified the F–15 jet fighter aircraft engine for production without ensuring that it met contract standards. As a result, Rep. Aspin said, "tens of millions of dollars in developmental costs have been passed on from Pratt & Whitney to the taxpayer." (*CR*, 5/15/73, H3678)
- Russian-born atomic scientist Dr. Eugene Rabinowitch, senior chemist on the Manhattan Project that led to development of the first atomic bomb and founder and editor of the *Bulletin of the Atomic Scientists*, died in Washington, D.C., at age 71. An outspoken critic of the nuclear arms race, he had joined other workers on the Manhattan Project in 1945 in drafting the Franck Report, which cautioned the U.S. against use of the atomic bomb. At the time of his death, Dr. Rabinowitch had been working as a Woodrow Wilson Fellow at the Smithsonian Institution. His project was scientific revolution and its social implication. (Hailey, *W Post*, 5/16/73, B6)

May 16: The Air Force launched an unidentified satellite from Vandenberg Air Force Base by a Titan IIIB-Agena booster into orbit with a 397-km (246.7-mi) apogee, 134-km (83.3-mi) perigee, 89.8-min period, and 110.4° inclination. The satellite reentered June 13 (GSFC *SSR*, 5/31/73; 6/30/73; *SBD*, 5/18/73, 108)

- Sen. Frank E. Moss (D–Utah), Senate Aeronautical and Space Sciences Committee Chairman, urged in a letter to Dr. James C. Fletcher that the NASA Administrator "promptly appoint an investigative board to make a detailed inquiry" into the malfunction of the Skylab 1 Orbital Workshop—as ground crews at Marshall Space Flight Center studied options for overcoming the loss of the spacecraft's protective shield, torn away during its May 14 launch, and of augmenting its electrical power. (UPI, *NYT*, 5/17/73; *Marshall Star*, 6/13/73, 1)
- The Department of Defense had furnished NASA photos of the crippled *Skylab 1* Orbital Workshop after it was placed in orbit May 14, to help determine the status of the spacecraft, NASA told the press. *Aviation Week & Space Technology* later reported that Air Force tracking cameras in New Mexico had photographed the Workshop. The photos were not released publicly. (NASA PAO; *Av Wk*, 5/21/73, 12)

- Recent problems with the Soviet *Salyut* spacecraft and with *Skylab* emphasized "the great difficulties and substantial expense involved in the creation of even a small station in space, let alone the large manned space laboratories many scientists are looking forward to," a *New York Times* editorial said. "Under these conditions it becomes more evident than ever that the joining together of the Soviet and American programs, expanded into a truly international effort, would benefit everybody. If manned vehicles are to keep earth permanently under surveillance from space orbits, the interests of peace would be best served if those vehicles flew the flag of the United Nations, not of any one country." (*NYT*, 5/16/73, 46)
- The West German newspaper Ludwigshafen *Rheinpfalz* commented on the difficulties of NASA's *Skylab 1* Workshop, launched May 14: "The trouble of the Skylab experiment is apt to dampen even more enthusiasm of the U.S. public for costly spaceflight experiments. . . . Combatting cancer is the great project which the United States wants to tackle in this decade. The U.S. Government plans to set aside billions for this research, whereas space flights will continue only on a small scale. Since the Russians now likewise suffered setbacks in their space projects, this would be justifiable politically." (FBIS–W Germany, 5/18/73, U4)
- Appointment of R. Tenney Johnson, Civil Aeronautics Board General Counsel since 1970, as NASA's General Counsel effective May 21, was announced by NASA. He would replace Spencer M. Beresford, who had resigned to become counsel to the House Select Committee on Committees. (NASA Release 73–101)
- NASA and the Univ. of Southern California sponsored a Patent Licensing Conference at the Western Research Application Center at USC to aid manufacturers in using NASA-developed technology to produce new commercial products. (NASA Release 73–95)
- The Aerospace Industries Assn. of America, Inc., released *Monopsony: A Fundamental Problem in Government Procurement*. The report, prepared by the Orkand Corp., said the problems of a monopsony—a market dominated by a single buyer—were no longer restricted to defense and space suppliers as Government agencies took increasingly active roles in mass transit, education, health, and environmental protection. Results of Government monopsony power were higher final costs and lowered or nonexistent profits that compounded industry's difficulty in attracting capital. In the end, "what is in jeopardy is not one or two industries but rather our capability, as a nation, to undertake programs in which the resources of government and industry must be committed to the development and implementation of solutions to society's problems. The study recommended: establishment of a Government Procurement Practices Board (GPPB) to limit governmental monopsony power; a continuing review by GPPB of procurement policies, regulations, and practices; a "free market test" criterion to govern the procurement process; and formulation by GPPB of procurement principles that could be submitted to Congress and enacted into law. (Text)

May 17: The U.S.S.R. launched *Cosmos 558* from Plesetsk into orbit with a 491-km (305.1-mi) apogee, 267-km (165.9-mi) perigee, 92.1-min period, and 70.9° inclination. The satellite reentered Dec. 22. (GSFC SSR, 5/31/73; 12/31/73; SBD, 5/18/73, 108)

- A Baltimore *Sun* editorial commented on problems of *Skylab 1*, launched May 14: "We hope that through astute improvisations much of the

Skylab I mission can be salvaged. Yet if it has to be scrubbed, then NASA should be encouraged to go ahead with backup plans to send a second space laboratory into earth orbit next year." NASA could find the money "if it is prepared to cut back or postpone other activities, including the controversial $14 billion space shuttle program. . . . While space shuttles can operate independently, they will approach their full potential only when they can rendezvous with orbital space stations or laboratories." (B *Sun*, 5/17/73, A18)

- A 30-kiloton underground nuclear-gas-stimulation test, the third in the Plowshare program to stimulate recovery of natural gas from a thick geologic section of unconnected gas sand strata, was conducted by the Atomic Energy Commission in the Fawn Creek area, 48 km (30 mi) southwest of Meeker, Colo. Dr. Dixy Lee Ray, AEC Chairman, announced 30 min later that the experiment had proceeded "perfectly." Opponents of the blast had contended that seismic shocks could pour salt water into the drinking water supply throughout the West and Mexico and could make underground mining dangerous and that gas produced by the blasts could contaminate the region and its waters for centuries. (AEC Release R-186; Sterba, *NYT*, 5/18/73, 1)

- Grumman Corp. had submitted a proposal to the Navy for Dept. of Defense purchase of further lots of the F-14 jet fighter aircraft, Grumman Corp. President John C. Bierwirth told the annual shareholders meeting in Bethpage, N.Y. (Andelman, *NYT*, 5/18/73, 49)

- Dr. Paul W. Gast, Chief of Johnson Space Center's Planetary and Earth Sciences Div., died in Houston at age 43. He had contracted a fatal illness in 1971. Dr. Gast had been on leave from Columbia Univ.'s Lamont-Doherty Geological Observatory and, since 1970, had directed the lunar science team studying lunar samples at JSC. He had received the NASA Medal for Exceptional Service in 1970 and the NASA Distinguished Service Medal in 1972. He had been awarded the first Victor Goldschmidt Award of the Geochemical Society in 1972 and had been named to receive the James Furman Kemp Medal of Columbia Univ. in 1973 for directing the lunar rock study. (*NYT*, 5/18/73, 34; *SF*, 10/73, 389)

May 18: Newspapers commented on the *Skylab 1* mission as engineers at Johnson Space Center and Marshall Space Flight Center prepared sunshades to cool down the overheated space station:

New York Times: The "partial recovery from earlier disaster has set the stage for an unprecedented improvisation in space. Present plans are for the Skylab crew to blast off a week from today and install a sunshade or awning to replace the missing thermal shield. If that can be done, the temperature problem will be solved and the astronauts will be able to begin living in the space station and to carry out at least some of the projected experiments. All these plans transcend past space feats, and if successful, will mark a major advance in the nation's space capabilities. But the task will not be easy and a happy ending is by no means guaranteed." (*NYT*, 5/18/73, 32)

Chicago Daily News: "The space station experiments not only look outward toward the sun and stars, but also down at the Earth. From 270 miles [435 kilometers] up, men can learn much about the planet, its resources and its weather that they can't see on the ground. Much of the knowledge gained from Skylab can be expected to have practical application toward improving the quality of life on Earth." (*C Daily News*, 5/18/73)

May 18
- The U.S.S.R. launched *Cosmos 559* from Plesetsk into orbit with a 324-km (201.3-mi) apogee, 204-km (126.8-mi) perigee, 89.8-min period, and 65.4° inclination. The satellite reentered May 23. (GSFC *SSR*, 5/31/73; *SBD*, 5/21/73, 116)
- NASA launched a Black Brant VC sounding rocket from White Sands Missile Range carrying a Univ. of Colorado and Marshall Space Flight Center payload to a 267.0-km (165.9-mi) altitude. The primary objective was to check out the reliability and performance of subsystems of the CALROC rocket configuration which would be used during Skylab missions after a manned crew joined the *Skylab 1* Orbital Workshop launched May 14 [see April 18]. The rocket and instrumentation performed satisfactorily. (GSFC proj off)
- A sun-powered laser was being developed by the Air Force Avionics Laboratory at Wright-Patterson Air Force Base, Ohio, as part of an Air Force program to determine the feasibility of satellite optical communications. The device collected and focused solar rays through lenses and mirrors, stimulating material in the laser to produce light beams to transmit data, TV, voice, and other communications. The laser's power output would equal that of solid-state lasers. A demonstrable brassboard working model of a sun pump laser was hoped for by the end of the fiscal year. (AFSC Release 055.73)
- The first of six student volunteers in a Univ. of California at Davis experiment to determine if astronauts could survive social isolation was released after 105 days' isolation in two small rooms. Dr. Don A. Rockwell, project manager, said the experiment had shown that astronauts should be able to adjust psychologically to long periods in space. (UPI, *W Post*, 5/30/73, A4)
- Aerospace Industries Assn. of America, Inc., released *Aerospace Facts and Figures, 1973/74*. Aerospace sales in 1972 had increased for the first time since 1968, from $22.2 billion in 1971 to $22.3 billion. Profits increased from 1.8% in 1971 to 2.4% in 1972. Average employment declined in 1972 to 922 000 workers, from 969 000 in 1971. Aerospace exports declined from $4.2 billion in 1971 to $3.8 billion in 1972; but, with a net positive aerospace trade balance of $3.3 billion, the industry remained a principal supporter of the U.S. international trade position. (Text)
- French aviation pioneer Col. Dieudonné Costes died in Paris at age 80. He had piloted the first westward transatlantic flight from Paris to New York Sept. 1–2, 1930, with Maurice Bellonte in a Breguet biplane. Flying time had been 37 hrs 18 min 30 sec. Costes had established many records as chief test pilot for the Breguet Co. (UPI, *NYT*, 5/20/73, 64)

May 19: The *Economist* commented on Skylab mission plans as mission engineers worked to purge the *Skylab 1* Workshop of gases that might have emerged from overheated insulating material and the Skylab 2 astronauts trained for their May 25 launch: "The American genius with a spanner has never been more in evidence than when the astronauts have shown themselves capable of dealing with anything from a computer on the blink to the catastrophic break-up of *Apollo 13* in midflight. But what Nasa has to weigh now is that this is not an emergency —the men are safely on the ground—so is it really justified in taking the risk of sending them up to an already damaged and overheated Skylab? If the Americans decide to try emergency repairs, the nature of the Skylab mission will take on a new dimension." It had been a trial of man's ability to endure in space. "Now it could become a trial of his ability to work there and carry out improvised engineering. Not according to the book, but a good deal more interesting." (*Economist*, 5/19/73, 22)

- The U.S.S.R. had canceled the launch of two cosmonauts because an unmanned Salyut spacecraft with which they would have docked had failed in orbit, Western observers believed. The Salyut, identified as *Cosmos 557*, was launched May 11 into an orbit similar to that of the crippled *Salyut 2* space station, launched April 3. Observers believed the satellite would be a docking target for a manned Soyuz but its maneuvering engine apparently failed; it made an early reentry May 22. The tracking ships that had steamed to stations off Newfoundland and the Gulf of Guinea were returning to their home port in the Black Sea. (O'Toole, *W Post*, 5/19/73, A12; *SF*, 1/74, 39–40; Lyons, *NYT*, 9/26/73, 33)

- The U.S.S.R. Tu-144 supersonic transport had reached a speed of 2100 km per hr (1300 mph) during a test flight, Tass reported. It had carried Soviet journalists to an altitude of 17 km (11 mi) on a 90-min flight between Moscow and Saralov. The designer, Gen. Aleksey A. Tupolev, had said new design changes included extendable wings on the front that improved aerodynamic performance. The engine and landing gear had also been improved. (Tass, FBIS–Sov, 6/1/73, L4)

- Creation and detection of antitritium—the counterpart of heavy, radioactive, hydrogen isotope tritium—was reported in the Soviet press by a Soviet scientific team using the U.S.S.R.'s largest nuclear particle accelerator at Serpukhov. The *New York Times* later said the discovery further strengthened the hypothesis that the universe was made up symmetrically of ordinary matter and antimatter. In a joint project of the Soviet Institute of High Energy Physics and the Institute of Nuclear Research, nuclear physicists Valentin I. Petrukhin and Vladimir I. Rykalin and team had examined 400 billion particles over several months before identifying four antitritium nuclei. It was the second identification of an antinucleus at the Soviet accelerator. In February 1970, Soviet scientists had reported the creation of antihelium. The search for antimatter was being spurred by the theory that the mutual annihilation of matter and antimatter might ultimately yield a new form of useful energy. (Shabad, *NYT*, 5/20/73, 41)

May 21: French satellites Castor and Pollux, jointly launched from Kourou Space Center, failed to enter orbit and fell into the sea when the Diamant-B launch vehicle failed to produce sufficient thrust for the separation of the 2nd and 3rd stages. It was the second straight failure for the French space program. On Dec. 5, 1971, the Polaire satellite had failed to achieve orbit when the 2nd stage of the Diamant-B booster malfunctioned. (AP, *M Her*, 5/22/73; Tass, FBIS–W Europe, 5/21/73, F3)

- The People's Republic of China was preparing to launch weather satellites, Deputy Director Chi Sheng-ying of the Central Meteorological Institute in Peking told an international conference on weather satellites in Paris. (Agence France-Presse, *NYT*, 5/22/73, 21)

- The International Court of Justice at The Hague, Netherlands, began a public hearing on demands by Australia and New Zealand that France stop nuclear weapon testing in the Pacific. Australian Attorney General

Lionel Murphy asked for a temporary injunction banning the tests, pending the Court's final ruling. He said fallout from the tests endangered Australians and the tests violated Australian territorial sovereignty and freedom of the seas. France was not present at the hearings. (*W Post*, 5/22/73, A20)

- The Air Force announced the award of a $1 058 685 contract to Philco-Ford Corp. for feasibility demonstration of an aerial observation designator system for remotely piloted vehicles. (DOD Release 258–73)

May 22: Dr. James C. Fletcher, NASA Administrator, established the *Skylab 1* Investigation Board to determine why the micrometeoroid shield was lost following May 14 launch of the Orbital Workshop and why the solar panels failed to deploy. Dr. Fletcher named Director Bruce T. Lundin of Lewis Research Center Chairman of the Board. Lundin appointed NASA Deputy Associate Administrator for Space Science Vincent L. Johnson Vice-chairman. (NASA Notice 1154; NASA Release 73–112)

- The pointing direction of *Pioneer 11*, launched by NASA toward Jupiter April 5, was changed by NASA controllers to aim the spacecraft antenna more precisely toward earth. Ames Research Center engineers fired 26 short pulses from *Pioneer 11*'s thrusters to turn the spacecraft's spin axis four degrees. The change had allowed a shift to *Pioneer 11*'s narrow-beam, high-power antenna and a fourfold increase in the rate of data returned. *Pioneer 11* was 32 million km (20 million mi) from the earth and traveling at 127 000 km per hr (79 000 mph). It was scheduled to arrive at Jupiter in December 1974. (NASA Release 73–102)

- Dr. Homer E. Newell, NASA Associate Administrator, testified on NASA's role in the U.S. energy research program in joint hearings of the Subcommittee on Space Science and Applications and the Subcommittee on Energy of the House Committee on Science and Astronautics. NASA's energy work was primarily in development of aeronautics and space systems but "these systems and the NASA capabilities, that is, our highly skilled people, our facilities, and the technologies we develop, can be put to work on terrestrial problems of generation, conversion, and conservation of energy." Not only NASA technologies but "the systems engineering, systems analysis, and management experience which we have gained in our conduct of large, complex programs and projects can be brought to bear on terrestrial energy research programs." (Transcript)

May 23: Dr. James C. Fletcher, NASA Administrator, testified on the status of the Skylab missions before the Senate Committee on Aeronautics and Space Sciences hearing on the difficulties that had postponed Skylab 2, as the countdown resumed at Kennedy Space Center for May 25 launch: The micrometeoroid shield around the *Skylab 1* Orbital Workshop had been torn off about one minute after liftoff on May 14, during the time of maximum aerodynamic pressure. "At the same time, some of the solar panels used to generate electricity for the laboratory were apparently also damaged, jammed, or torn away. We don't know which. The full impact of this malfunction was not known until several hours after launch; we still do not know the specifics of the actual damage. The net result was that Skylab was in a good orbit, but had only approximately one-half of its power-generating capability in operation and also was overheating badly." The overheating occurred because the damaged micrometeoroid shield was also to have provided thermal balance. It was painted "to reflect the proper amount of sunlight so that the laboratory would stay cool." The Skylab 2 astronauts hoped to repair the damage and then carry out a 28-day mission. "If the planned repairs are successful, we will also be able to accomplish most of the activities scheduled for the two subsequent missions, each lasting 56 days." Skylab 2 would carry several sunshades. "We don't know whether the micrometeoroid shield is completely severed or whether parts of it are obstructing some of the areas where the sunshade might be fixed, but . . . by carrying several different shades, we should at least have one which will be suitable."

Dr. Fletcher described steps to be taken to deploy the solar panels. "If it looks like an easy thing to do, if one of the panels . . . is partially deployed and is hung up on something and all it needs is a little tug, we will put a loop of twine around the end of it, play it out so that we are well away from the panel in case it starts to deploy, and give a little tug on the twine. If it deploys, fine; but if it does not, we will go on without it. But we are not going to spend too much time on that fix. There are too many uncertainties and we do not really need the power." (Transcript)

- The U.S.S.R. launched *Cosmos 560* from Plesetsk into orbit with a 325-km (201.9-mi) apogee, 178-km (110.6-mi) perigee, 89.5-min period, and 72.8° inclination. The satellite reentered June 5. (GSFC *SSR*, 5/31/73; 6/30/73; *SBD*, 5/30/73, 162)
- The House passed H.R. 7528, the $3.074-billion NASA FY 1974 authorization bill [see May 9], by a vote of 322 to 73. Rejected were a motion to recommit the bill to the Committee on Science and Astronautics, an amendment by Rep. Bella S. Abzug (D–N.Y.) to delete $567 million for space shuttle programs, and an amendment by Rep. John Conyers, Jr. (D–Mich.), to prohibit use of tracking and data acquisition funds in South Africa because of South Africa's apartheid policy. (*CR*, 5/23/73, H3921–51)
- Success of an Ames Research Center project to provide the California Div. of Forestry a daily fire index measurement via *Erts 1* Earth Resources Technology Satellite (launched July 23, 1972)—from an unmanned remote station in a Sunol, Calif., fire area—announced by NASA. The project had enabled foresters to pinpoint probably sources of fires and had eliminated the necessity of using scarce fire-fighting personnel for monitoring. The system was being considered for monitoring air pollution. (NASA Release 73–104)
- Rep. Olin E. Teague (D–Tex.), Chairman of the House Committee on Science and Astronautics and Rep. Don Fuqua (D–Fla.), Committee member, introduced H. Con. Res. 223 requesting the President to proclaim July 16–22 of each year as "U.S. Space Week." The resolution was referred to the Committee on the Judiciary. (*CR*, 5/23/73, H3976)
- NASA launched an Aerobee 150A sounding rocket from White Sands Missile Range carrying a National Center for Atmospheric Research aeronomy payload to a 111.4-km (69.2-mi) altitude. The rocket and strumentation performed satisfactorily. (GSFC proj off)
- Preliminary data from the May 17 Plowshare underground nuclear blast in western Colorado indicated the blast had not damaged vast oil shale deposits in the area, Atomic Energy Commission spokesman Dr. Charles Williams told the press in Rio Blanco, Colo. Oil industry officials had claimed the blast could damage the shale, a future energy source. (UPI, *W Post*, 5/25/73, A3)

May 23–June 6: The 16th Annual Meeting of the Committee on Space Research (COSPAR) was held in Constance, West Germany. Data from the *Erts 1* Earth Resources Technology Satellite launched by NASA July 23, 1972, was presented in a three-day symposium on applications of space techniques to monitoring environmental problems. Recently analyzed data from the *Apollo 17* mission (Dec. 7–19, 1972) necessitating changes in existing lunar models were presented during working group meetings. Other topics included space biology, astronomical experiments conducted by space vehicles, and data from planetary explorations.

The National Academy of Sciences–National Research Council's Space Science Board submitted its report *United States Space Science Program,* including progress in NASA's geodetic satellite program. The program continued to provide essential ingredients in support of the emerging earth and ocean physics applications program (EOPAP), which would make geodetic satellite measurements for a global geometric network of reference points to a 10-m (33-ft) accuracy, refine knowledge of the gravity field for determination of geoid features of 1000 km (620 mi) accurately to about 10 m (33 ft), and evaluate the geodetic instruments used.

The U.S.S.R. presented its report *Space Research Performed in the USSR in 1972,* including data from *Luna 19* (launched Sept. 28, 1971, to orbit the moon). Meteoric particles near the moon were found not to exceed their density in interplanetary space. The moon's perturbing effect on the earth's magnetic loop and on interplanetary magnetic fields of solar origin were confirmed. Analysis of the sample collected by the Feb. 14–25, 1972, *Luna 20* mission showed that the primary rocks of continental regolith were chiefly anorthosite rocks which sharply distinguished "continents" from "maria." The sample contained finely dispersed metal iron of very high corrosion resistance, just as found in material from *Luna 16* and *Apollo 11* and *12.* (NASA Release, 3/23/73; NASA PIO; US text; JPRS 59778, trans of USSR rpt)

May 24: President Nixon sent a message to the first Skylab crew on the eve of its Skylab 2 launch to join the *Skylab 1* Orbital Workshop, launched May 14: "Already the courage and ability which you and your colleagues have shown in coping with the initial difficulties of the mission have given us a great sense of confidence in the Skylab team. Through those efforts you have shown the world that man is prepared to meet and overcome the challenges of space." (O'Toole, *W Post,* 5/26/73, A1)

- Soviet news agency Tass announced the Soviet crews for the July 1975 U.S.–U.S.S.R. Apollo-Soyuz rendezvous and docking mission. The first crew would be Cosmonauts Aleksey A. Leonov, the first man to walk in space (during the March 18, 1965, *Voskhod 2* mission), and Valery N. Kubasov, flight enginer on the *Soyuz 6* mission (Oct. 11–18, 1969). The second crew would be Anatoly V. Filipchenko, pilot on *Soyuz 7* (Oct. 12–17, 1969), and Nikolay N. Rukavishnikov, flight engineer on *Soyuz 10* (April 23–25, 1971). Third crew would be Cosmonauts Vladimir Dzhanibekov and Boris Andreyev and fourth would be Yuri Romanenko and Aleksander Ivanchenko—none of whom had been in space before. (FBIS–Sov, 5/25/73, L1)

- Kennedy Space Center's Launch Complex 39—site of the July 16, 1969, launch of *Apollo 11,* the first manned spacecraft to land on the moon—

was entered in the National Register of Historic Places of the Dept. of the Interior's National Park Service. The complex, built between November 1962 and October 1968, was first used for the Nov. 9, 1967, launch of the unmanned *Apollo 4,* the first Apollo launch, and was still in use for Skylab launches. (KSC Hist Off)

- A Washington *Evening Star and Daily News* editorial commented on the U.S. stake in the Skylab missions, endangered by damage to the *Skylab 1* Workshop during launch May 14: "Even if the present Skylab must be written off, the nation should not dismantle its unique capacity for manned space exploration. We have the equipment, including the Saturn V rocket, to mount another attempt in 15 months or so. More importantly, NASA currently has the brains and knowhow to operate in space with amazing powers of foresight and improvisation, which are being put to a severe test this week. The dispersal of this talent would be a loss to mankind in ways that can only be imagined." (W *Star & News,* 5/24/73, A18)

May 24–June 3: The 30th and largest Paris Air Show at Le Bourget Airport, marred by tragedy at its end, featured the first joint U.S.–U.S.S.R. pavilion—displaying models of Apollo and Soyuz spacecraft to be used in the July 1975 joint mission—and a record display of aircraft, subsystems, and equipment from manufacturers in 19 countries. The aircraft included the Anglo-French Concorde and Soviet production model Tu-144 supersonic transports, U.S. prototype Northrop F–5E international fighter and Grumman Navy F–14 variable-geometry air-superiority fighter, Soviet Il-62-M medium-range transport and Yak general-aviation aircraft, and the first European A–300B Airbus.

NASA Associate Deputy Administrator Willis H. Shapley said at the opening of the U.S.–U.S.S.R. exhibit that NASA was "pleased to have the opportunity to demonstrate . . . the rapid progress our joint working groups have made. . . . During the past year, American and Soviet engineers have worked closely together in an atmosphere of good will to solve many complicated and technical problems. We are confident that the target date for a joint mission in Earth orbit in 1975 can be met. This joint mission will be an exciting adventure as well as a highly visible expression of our nation's belief in the great value of exploring and using space in cooperation with other nations."

On May 30 the *Apollo 17* Astronauts Eugene A. Cernan, Ronald E. Evans, and Dr. Harrison H. Schmitt met with Soviet Cosmonauts Aleksey A. Leonov, Valery N. Kubasov, and Anatoly V. Filipchenko—of the U.S.S.R. prime and backup crew for the Apollo-Soyuz mission—and Dr. Aleksey S. Yeliseyev at the U.S.–U.S.S.R. Pavilion. They inspected a model of the Soviet *Lunokhod 2* lunar roving vehicle and Apollo and Soyuz spacecraft models and were invited aboard the Soviet Tu-144 supersonic transport by its principal designer Gen. Aleksey A. Tupolev.

On June 3 the Tu-144 crashed into the French village of Goussainville, near Le Bourget Airport, while finishing a series of demonstration flights. All six crewmen, including pilot Mikhail Kozlov, and seven persons on the ground were killed and 28 injured. The aircraft was destroyed. A team of French and Soviet officials was appointed immediately to investigate the accident, which observers said might have resulted from attempted aerobatics. Witnesses said the Tu-144's performance had been erratic just before it crashed. (*Av Wk,* 5/28/73,

6/4/73, 6/11/73; Shapley text; Tass, FIBS–Sov, 6/1/73, F5; 6/4/73, F3)

May 25: Skylab 2 was launched from Kennedy Space Center Complex 39, Pad A, on a two-stage Saturn IB launch vehicle to carry Astronauts Charles Conrad, Jr., Dr. Joseph P. Kerwin, and Paul J. Weitz to join the *Skylab 1* Orbital Workshop for repairs and a 28-day mission [see May 14–June 22 *Skylab 1-2* mission]. (NASA prog off)

- The U.S.S.R. launched *Cosmos 561* from Plesetsk into orbit with a 290-km (180.2-mi) apogee, 205-km (127.4-mi) perigee, 89.4-min period, and 65.4° inclination. The satellite reentered June 6. (GSFC *SSR*, 5/31/73, 6/30/73; *SBD*, 5/30/73, 162)

- A *New York Times* editorial commented on the manned *Skylab 2* launch: "The possibility of repairing a damaged spaceship, now brought to the fore by the last two weeks' developments, points up one of the advantages of man's presence in space. If something goes awry with an unmanned spaceship, and there are no arrangements for bringing repair men to it, it is more probable than not that little can be done to save the mission. Now the world is faced with a major effort to overcome unexpected difficulties by assigning additional duties to the astronauts originally involved. The success or failure of these repair efforts in the period immediately ahead will provide important evidence for the future debates between the advocates of manned and unmanned efforts to explore space." (*NYT*, 5/25/73, 32)

- Approval by both countries of results of the Feb. 12–17 Moscow meeting of the U.S.–U.S.S.R. Joint Working Group on the Natural Environment was announced by NASA. NASA and the Soviet Academy of Sciences had agreed to a cooperative project for detecting features linked to changes in structure of the earth's crust, for recognizing them on space images to develop a classification system, and for studying the relationship between the space imagery features and previously known subsurface geologic structures. The Group identified as a possible project development of remote-sensing methodology to determine vegetation productivity and agreed to develop a plan for a joint project to measure soil moisture content using microwave techniques. The U.S. would study a Soviet proposal to coordinate satellite water-temperature measurements with surface measurements obtained by a Soviet ship and to compare temperature, plankton, and color measurements by Soviet ships with U.S. satellite observations. The U.S.S.R. would study U.S. proposals to test a hydrological simulation model using remote-sensing data and environmental factors to infer the presence of shallow aquifers—areas of water underground and close to the surface. (NASA Release 73–106)

- Neutron-activation analysis of 27 *Apollo 16* lunar samples for 17 trace elements was reported in *Science* by Enrico Fermi Institute chemists. Several breccias had been found to be strikingly rich in volatile elements. Similar enrichments had been found in smaller quantities in all lunar highland soils. "It appears that volcanic processes took place in the lunar highlands, involving the release of volatiles, including water." Measurements suggested that the moon's original water content could not have exceeded the equivalent of a layer 22 m (72 ft) deep. "The cataclastic anorthosites at the Apollo 16 site may represent deep ejecta from the Nectaris basin." (Krahenbuhl *et al.*, *Science*, 5/25/73, 858–61)

- NASA launched an Aerobee 170 sounding rocket from White Sands Missile Range carrying a Princeton Univ. galactic astronomy payload to a 178.0-km (110.6-mi) altitude. The primary objective was to observe stellar ultraviolet spectra using an Echelle spectrograph and a TV camera. The rocket and instrumentation performed satisfactorily. (GSFC proj off)
- Establishment of the impact origin of Lonar Crater in Maharashtra, India, at less than 50 000 yrs ago was reported in *Science* by a team of U.S. and Indian geologists. Fission-track dating of shock-melted glass had also shown the preservation state to be in accord with such comparative youthfulness. As the only known terrestrial impact crater in basalt, Lonar Crater provided unique opportunities for comparison with lunar craters. Microbreccias and glass spherules from Lonar Crater had close analogs among Apollo lunar specimens. (Frederiksson *et al.*, *Science*, 5/25/73, 862–4)
- Installation of 150 additional Minuteman III multiple-warhead missiles at Grand Forks, N. Dak., had brought the Air Force more than halfway toward completion of its 550-missile program, the Air Force announced. (UPI, *LA Times*, 5/26/73)

May 27: Secrecy surrounding the possible use of lasers to produce nuclear fusion for energy was criticized by Univ. of California atomic scientist Dr. Edward H. Teller in a *New York Times* article: "Right now, the most hopeful model for a future fusion reactor, the Tokamak, has a Russian origin and a Russian name. If the energy problem should be solved in this promising manner, we shall certainly owe a debt of gratitude to our colleagues in the Soviet Union. Why not similarly open up the possibilities with the laser?" The subject had engaged the attention of world scientists and had been discussed openly at a Montreal conference in 1972; "in considerable part, however, the work had been carried on in secrecy." (*NYT*, 5/27/73)

- Mikhail D. Millionshchikov, Vice President of the Soviet Academy of Sciences and a principal spokesman in Soviet scientific contacts with the West, died in Moscow at age 60. He had contributed significantly to the studies of the turbulent flow of liquids and gases, nuclear energy, and new techniques for direct energy conversion. On April 28, 1970, Academician Millionshchikov—as head of a 10-man Soviet delegation to a conference sponsored in Washington, D.C., by the Fund for Peace—had suggested U.S.–U.S.S.R. cooperation in space exploration. This proposal had been "the first public suggestion by a Soviet spokesman of this hope," the *New York Times* reported later. *Apollo 8* Astronaut Frank Borman, also a conference participant, had proposed the next day that the two countries agree to exchange space engineers as a first step toward joint space cooperation. (*NYT*, 5/29/73, 32)

May 28: A *New York Times* editorial commented on the repair of *Skylab 1*, damaged on launch May 14: "Many people deserve credit for this remarkable turn of events. First there were the Houston ground controllers who maneuvered the crippled Skylab into an optimal position.... A small but varied army of technicians hurriedly contrived not one but three possible ways of creating a parasol in space so as to replace the missing heat shield that had created the most trouble. Finally, the Skylab astronauts mastered their new tasks in an incredibily short time, reached orbit in the neighborhood of Skylab, surveyed the damage, and made the most essential 'fix,' the sunshade, which yesterday brought the station's temperature down to tolerable levels." It had been a grueling test for all concerned. "But for the moment, at least, the Skylab project has been truly resurrected from near-death." (*NYT*, 5/28/73, 14)

- The U.S. aerospace industry had "weathered the economic storms of recent years" to emerge "a leaner and tougher competitor," an *Aviation Week & Space Technology* editorial said. For the first time in years, the industry was seeing "some of the government shackles that restricted its efforts abroad loosened and even active support forthcoming in certain key areas. The sad story of the balance-of-trade deficits and the decline of the dollar have finally shaken the U.S. government out of its economic isolation, and the aerospace industry will benefit accordingly in its increasing export campaign." (Hotz, *Av Wk*, 5/28/73, 9)

May 29: The U.S.S.R. launched a *Meteor 15* weather satellite from Plesetsk to obtain "meteorological information for the use of the operational weather service." Orbital parameters: 896-km (556.7-mi) apogee, 852-km (529.4-mi) perigee, 102.4-min period, and 81.2° inclination. The satellite carried equipment to receive day and night pictures of clouds and snow cover and data on the heat energy reflected and emitted by the earth and the atmosphere. (GSFC *SSR*, 5/31/73; Tass, FBIS-Sov, 5/30/73, L1; *SBD*, 5/30/73, 162)

- Newspapers commented on the Skylab missions as *Skylab 2* astronauts activated the Apollo Telescope Mount aboard the *Skylab 1* Workshop, following repairs to the damaged spacecraft they had joined May 25 [see May 14–June 23].

Miami News editorial: "To us, the conquest of the moon, and now the building of a space station, demonstrated that this nation can accomplish just about anything it sets its mind on. The secret lies in the magic word, commitment. With the kind of commitment we pledged to the space program, we could indeed surmount the housing crisis, the crime crisis, the energy crisis, or any other crisis." (*M News*, 5/29/73)

Wall Street Journal article: *Skylab 1*'s successful in-orbit repair had wide implications for the U.S. space program's future. "When the three Skylab astronauts erected a heat shield over their damaged space station, they both saved their mission and paved the way for future launching of two other missions in the $2.5 billion space-station program." They also had provided NASA with "a powerful argument for new manned-flight projects." NASA could contend that "the value of man in orbit has been proven in practice as well as in theory." While budget shortages would preclude any attempt to capitalize on Skylab's success before the space shuttle evolved, it had armed NASA against congressional critics of manned space flight. (Spivak, *WSJ*, 5/29/73)

New York Times columnist Russell Baker: "The space program until now has seemed less malevolent than most new ventures in technology, but to people watching those incessant countdowns at Cape Kennedy it has also seemed pointless." Lunar exploration had been "wonderful" but "it didn't really open any horizons for most of us, and it was certainly hard to see how it was going to improve man's lot." Skylab, by contrast, was wonderful. "There they are way up there—270 miles [435 kilometers] in the sky. They are attached to 100 tons [90 700 kilograms] of machinery which stretches out through space a distance of 118 feet [36 meters], about the length of an exciting pass play in the National Football League. And what are they doing? Repair work. They are fixing some of the machinery which doesn't work right." This was space with a purpose. "We ought to be asking why NASA didn't make this approach from the beginning instead of going after moon dust. The repair industry in this country could very well be rejuvenated by this trip, and not be a moment too soon, either." (*NYT*, 5/29/73, 31)

- Florida's Gov. Reubin Askew signed into law a March 22 proposal to restore the name "Cape Canaveral" to Cape Kennedy on Florida state maps and documents. The designation of Kennedy Space Center would remain unchanged. (AP, W *Star & News*, 5/30/73, A2; KSC PIO, Fact Sheet, 1/74)
- The Soviet research ship *Professor Vize* sailed from Leningrad with 60 scientists from Soviet weather service centers on an expedition to collect data on ionizing radiation of the sun and its influence on the upper stratosphere. During the two-month project the scientists would observe the June 30 solar eclipse, launch a series of high-altitude research rockets with onboard measuring instruments and photo equipment, and take deep measurements of the Atlantic Ocean. (Tass, FBIS–Sov, 6/1/73, L7)

May 30: The Senate Committee on Aeronautical and Space Sciences favorably reported H.R. 7528, a $3.046-billion FY 1974 NASA authorization bill, $28 million lower than the $3.074-billion House bill passed May 23. The Senate authorization—$30 million above NASA's requested $3.016 billion—would be NASA's lowest since FY 1962 and the lowest percentage of the total Federal budget—less than 1.2%—in 12 yrs. It would provide $2.231 billion for research and development, $34 million more than NASA's request but $24 million less than the House bill; $110 million for construction of facilities, as requested but $2 million less than the House bill; and $705 million for research and program management, as requested but down $2 million from the House bill.

The Senate Committee would retain Skylab funding at NASA's requested $233 million and space shuttle funding at $475 million. It concurred with the House addition of $7 million for Earth Resources Technology Satellite program funding and added $5 million to space applications funding to replace NASA's instrumented Convair 990 aircraft, lost in an April 12 crash. (*CR*, 5/30/73, S9866; S Rpt 93–179)

- Successful applications of earth resources surveying by satellite were described by representatives of agriculture and the lumber, oil, and aerospace industries at a Johnson Space Center press briefing sponsored by the American Institute of Aeronautics and Astronautics. Lockheed Electronics Co. earth resources coordinator Gordon R. Heath said surveys from Earth Resources Satellites would provide lumbermen "a continuous flow of up-to-date information which has never before been available to them," a new management tool for timber lands. "And better management will balance supply and demand . . . and help stabilize lumber prices." Lumber company executive James B. Webster said the development of earth resources surveying was "one of the most important events in the history of man." Incipient damage to trees from insects and forest fires had previously been impossible to detect. "And what we've got here is . . . a system which will permit us to know what's happening all over the world, with all the world resources at any given time."

Dr. Richard Phelps, technical information director for an agricultural firm, said a photo taken over a northern California area by *Erts 1*

May 30

(launched by NASA July 23, 1972) had been "a tremendous help" in complementing information from ground sources on farm crops. "It's still very difficult for us to realize that we can obtain so much detail from an electronically reconstituted print transmitted from a satellite as far away from the earth as Wichita, Kansas, is from Houston." (Transcript)

- A fire retardant suit that filled with foam at impact, designed by a NASA contractor, saved the life of racing driver Swede Savage when his car crashed into a retaining wall during the Indianapolis 500 automobile race. Associated Press said later that it was a "miracle" Savage had not burned to death. "The fire was so intense it seemed to illuminate the overcast skies." (*Washington Post*, 5/31/73, D6)

- Newspaper editorials commented on the repair of the Skylab Workshop by *Skylab 2* astronauts as the mission, begun May 14, continued to pursue its objectives of photographing and studying the solar atmosphere.

    The *Philadelphia Inquirer* found "vast excitement in the Skylab adventure beyond potential dividends and the high drama of the suspense itself." The astronauts' repair of the Workshop "tells us, all jokes aside, that Americans have managed an accomplishment of delicacy, complexity and force that would have been beyond the imagination of all but a few visionaries a decade ago. That tells us, all doubts aside, that America can do wonders when it sets its mind to it." (*P Inq*, 5/30/73)

    New York News: "We hope the lesson learned will not be lost on those who decry the 'expense' of manned space exploration. If we had relied on push-button robots instead of human ingenuity, it's a good bet Skylab would have ended up a complete washout." (*NY News*, 5/30/73)

    Christian Science Monitor: "Much of the technology spurred by space has applications nearer home. For example, modest solar-energy devices have already arrived on the market for home and commercial use after solar-energy pioneering in space. But even without such applications the Skylab venture would represent a remarkable manifestation of the problem-solving talent especially needed now that Americans are waking up to problems they hardly knew they had." (*CSM*, 5/30/73)

    San Francisco Chronicle: "Who could feel anything short of total admiration for the undeniable courage of the astronauts for venturing into an oven of 125 degrees Fahrenheit [325 kelvins] to carry out their endangered mission? The coolness and confidence of men who reckon the limits of what they can do and then run the risks of approaching those limits while their defective vehicle is speeding in space at 17,500 [28 200 kilometers] an hour are something to celebrate." (*SF Chron*, 5/30/73)

- President Nixon submitted to the Senate the nomination of John W. Barnum to be Under Secretary of Transportation succeeding Egil Krogh, Jr., who had resigned. Krogh, former assistant to President Nixon's Chief Domestic Adviser John D. Ehrlichman, had resigned while under investigation for his role in the Watergate activities. (*PD*, 6/4/73, 717, 727; *Time*, 5/14/73, 22)

- Associated Press quoted military sources in Brazil as saying the Brazilian Air Force had decided to buy $100-million worth of U.S. jet fighter aircraft. The order—for 48 Northrop F–5Es—was about double Brazil's two most recent orders for arms from France, 16 Mirage IIIs at $3 million each in 1970 and $2-million worth of French missiles more recently. (*NYT*, 5/31/73, 17)

- A Reuters report that the British government was secretly developing "death ray" weapons using powerful laser beams was published by the *Christian Science Monitor*. A Ministry of Defense spokesman had confirmed that work had been in progress for the navy, Royal Air Force, and army for some time and "we also have a close exchange of information with the United States on this subject, which is highly classified." (*CSM*, 5/30/73, 2)

May 31: Results from the Skylab Apollo Telescope Mount (ATM) launched on the *Skylab 1* Orbital Workshop May 14 were presented by principal investigators at a Johnson Space Center press briefing. Naval Research Laboratory astronomer Dr. Richard Tousey reported the detection by the extreme ultraviolet monitor—a TV system that showed what the sun looked like in extreme UV radiation of a broad band—of a "hole" in the solar corona. "And this coronal hole is believed to be a region where the corona is at a lower temperature and a lower density, and there's some reason to believe that the solar wind comes out of coronal holes to a greater extent than from any where else on the Sun." The first coronal hole had been recorded in 1966 but the significance of such holes had not been recognized until they were observed from NASA's unmanned Orbiting Solar Observatories. A major objective of the Skylab astronomy experiments was to determine how the solar corona was heated. Skylab instruments were observing at a wide range of wavelengths generated by gases of different temperatures at different elevations in the solar atmosphere. Sharper images than so far received were hoped for. The images, thus far seen only through a TV system with about one tenth the anticipated detail, "whet your appetite for what the instruments will bring back," Dr. Tousey said. (Transcript; Sullivan, *NYT*, 6/1/73, 6)

- A London *Daily Telegraph* editorial commented on the repair of the *Skylab 1* Workshop, while the *Skylab 2* astronauts proceeded with their tasks in space: "Hundreds of miles above the melee of the Watergate affair, the three American astronauts, having hazardously patched up the damaged Skylab and activated the equipment and workshop, have now settled down to a routine of planned experiments designed, above all, to test man's capacity to live in the weightlessness of space. Whatever the troubles in their political world, there is no question of the Americans' superb capability in scientific and industrial technology." (London *Daily Telegraph*, 5/31/73)

- Receipt of proposals for design, development, test, and evaluation of the space shuttle external tank from McDonnell Douglas Astronautics Co., Boeing Co., Martin Marietta Aerospace Corp.'s Aerospace Div., and Chrysler Corp.'s Space Div. was announced by Marshall Space Flight Center. The contract, calling for three ground-test tanks and six developmental flight tanks in the development phase, was expected to be awarded by Aug. 1. (MSFC Release 73-60)

- A Solar Observing Optical Network (SOON) to survey solar activity and warn of space events disruptive to satellite, radar, and communication operations was being acquired by the Electronic Systems Div. of the Air Force Systems Command, AFSC announced. The Air Force Weather Service would use the system to pinpoint the location and magnitude of solar disturbances. The system when completed would consist of five optical telescopes and a computerized data-processing and communications package. (AFSC Release 075.73)

During May: *Erts 1* Earth Resources Technology Satellite—launched by NASA July 23, 1972—had completed more than 4200 orbits, or 16 18-day whole-earth coverage cycles, and acquired 190 scenes per day for a 57 893 total. System performance had exceeded expectations; registration of multispectral scanner images in different spectral bands was within 55 m (180 ft), or one third, of prelaunch prediction. *Erts 1* had produced high-quality images of every major land mass; North America had been covered 10 times. Pictures had been made available to all citizens of the world. (NASA prog off; *Goddard News*, 6/73, 2)

- M/G Vernon R. Turner, Director of Aircraft and Missiles in the Office of the Assistant Secretary of Defense (Installations and Logistics), had been named Chief of Staff for the Air Force Systems Command, AFSC *Newsreview* reported. He would succeed M/G Lew Allen, Jr., who had been assigned to the Central Intelligence Agency. (AFSC *Newsreview*, 5/73, 2)

- Soviet Academicians Valery I. Popkov and Valentin A. Shteinberg visited the U.S. as the first Soviet participants in an exchange agreement between the American Assn. for the Advancement of Science and its Soviet counterpart, All Union Znaniye [Knowledge] Society. The first U.S. scientist had visited the U.S.S.R. under the agreement during April. (*AAAS Bulletin*, 6/73, 1)

- Abandonment of technology, a "source of man's strength," would be a senseless alternative to its further development and use, Dr. Jerome B. Wiesner, Massachusetts Institute of Technology President, warned in a *Technology Review* article. Man's exploitation of and dependency on the environment could not be changed through greater understanding of science and its application. Even a relevant technology alone would not suffice. Man needed "to develop the ability to estimate rationally and to choose among alternate courses of action, particularly when new technology is concerned. Above all, we need the humility to admit that we will not find any absolute answers or permanent solutions." But the fallacy of the approaching doomsday argument was "that it ignores the considerable evidence we already have that man can in fact modify his behavior fast enough to avoid the catastrophic disasters predicted by the doomsday-sayers." The evidence included the fact that scarcely a decade after Rachel Carson's warning against pesticides (in her book *Silent Spring*) "those chemicals are severely controlled—possibly too much so—and biodegradable equivalents are on the verge of being introduced." Lake Michigan's marine life was nearing extinction 10 yrs ago; "by vigorous ecological controls it has been restored."

    Development and control of the rapidly changing synthetic environment had become as important as contending with nature. Those who saw only evil in technology "fail to recognize that our situation would be much worse if the search for new technological solutions was stopped." The challenge was "to move on from where we are to technologies and ways of employing them that will avoid uncontrollable effects in the future. Stopping science will shut off new knowledge and weaken our efforts to reverse the present situation. Technology alone is not the answer, but without technological developments few answers are likely to be found." More than 30 yrs ago Dr. Vannevar Bush had called science the endless frontier; "it remains that today. The range of exciting research now exceeds that which was imaginable when Dr. Bush coined the title." (*Tech Rev*, 5/73, 10–13)

June 1973

June 1: Skylab 3's target launch date had been changed to July 27, NASA announced at a Johnson Space Center press briefing. SL-3, to carry the second crew to the Orbital Workshop launched May 14, previously had been scheduled for Aug. 8. The new target date was set because unexpected use of Skylab hardware during the *Skylab 1-2* mission had exposed electronics, batteries, and systems to unusual environments. Deputy Skylab Program Director John H. Disher said a July 27 launch would give greater solar power input, because the relation of the sun to orbit plane at that date would give the spacecraft a greater percentage of time in the sun. The precise date would be set after the end of the *Skylab 2* mission. (Transcript; NASA Release 73-111)

- Dr. James C. Fletcher, NASA Administrator, released a statement on private communications of crews of the Skylab Orbital Workshop (launched May 14), after meeting with the Deputy Administrator, Dr. George M. Low; Associate Administrator for Manned Space Flight Dale D. Myers; and Assistant Administrator for Public Affairs John P. Donnelly. A May 3 memorandum had authorized private communications for morale purposes (weekly calls from Skylab astronauts to their families), for "extreme" operational emergencies, and for medical reasons (to permit private doctor-patient relations). Following objections from some newsmen to the May 29 Charles Conrad, Jr., private conversation on operational matters, Dr. Fletcher had decided to instruct Mission Control's Capsule Communicator to query the seriousness of a situation before permitting a private conversation from Skylab.

    Dr. Fletcher said in his new statement that he would rescind this instruction because "we believe it would be imprudent to discourage requests by the crew for private operational communications." Paraphrases of the communications would be released to the press. "We will continue to leave it entirely up to the flight crew as to when to call for such a conversation, and will send no new instructions or clarification of instructions. This decision was reached so that we would in no conceivable way inhibit them from calling for a private conversation when they believe a need exists, and so that we will not jeopardize the safety of the mission." (NASA Release 73-110)

- A *Chicago Tribune* editorial hailed the *Skylab 2* astronauts (launched May 25 to man the crippled Workshop orbited May 14) as "tinkerers." The men who overcame the senseless stubbornness of inanimate objects had been "superlatively trained and admirably persistent and are the products of a highly scientific preparation. But the achievement also reflects the fine American heritage of tinkering. This was the quality that enabled the Pilgrim fathers to make do with strange, native materials when they landed on our alien shores." It was also "the practical genius that has enabled four generations of young Americans to compel model Ts to live and snort again when it was obvious that they were clinically dead. Its magic has guided these astronauts and their ground advisers to a thrilling kind of space achievement that is typically American." (*C Trib*, 6/1/73)

- The General Accounting Office released *Comptroller General's Report to the Congress: Analysis of Cost Estimates for the Space Shuttle and Two Alternate Programs*. The review, made by GAO at the request of Sen. Walter F. Mondale (D-Minn.), said GAO was "not convinced that the choice of a launch system should be based principally on cost comparisons" and "not certain that the Space Shuttle is economically justified (is less costly when the time value of money is considered), even though NASA's calculations show that it is." Payload cost was the major element in program cost estimates for both the shuttle and alternate use of existing expendable launch vehicles. NASA had estimated $30.2 billion for the shuttle and $50.1 billion for the expendable systems. The $19.9-billion difference was due to estimated low-cost design of shuttle payloads which could be recovered, refurbished, and reused. GAO recommended that Congress "consider the future space missions used in NASA's economic analysis . . . to determine whether these missions are a reasonable basis for space program planning at this time" and "review the estimates for the Space Shuttle annually, giving due consideration to the appropriateness of the missions used in making those estimates." (*CR*, 6/8/73, S10703-4)

- The Air Force announced a $3 547 890 firm-fixed-price contract award to Textron, Inc., Dalmo Victor Div. The contract was for aerospace ground equipment test sets applicable to the radar warning system for the F-105G, OV-1, and F-4E aircraft. (DOD Release 277-73)

- Jeanne M. Holm, age 51, became the first two-star female general officer in U.S. military history with her promotion to Air Force major general. She had begun her military career as a truck driver in the Women's Army Auxiliary Corps in 1942 and was now Director of the Air Force Personnel Council. M/G Holm was among seven women who had achieved the rank of general or admiral and the only woman to hold a command outside a service nurses' corps or women's branch. (UPI, *W Star & News*, 6/2/73, 2)

June 2: NASA launched an Aerobee 170A sounding rocket from White Sands Missile Range carrying an American Science and Engineering, Inc., payload to a 175.4-km (109.0-mi) altitude to collect data on the solar spectra. The rocket and instrumentation performed satisfactorily. (GSFC proj off)

- Soviet journalist Victor Louis discussed U.S.S.R. plans for the Tu-144 supersonic transport aircraft in a *New York Times* article. The aircraft was scheduled to go into regular passenger service in 1975, when it was expected to begin carrying 120 million passengers annually. Soviet territory was so sparsely populated that "fewer people will be troubled by noise than in Europe or the United States." Soviet statistics showed 3500 internal airfields in the U.S.S.R. Soviet aircraft flew to 63 major cities throughout the world—operating between Moscow and Santiago, Chile, and eastward between Moscow and Jakarta. SST watchers had calculated that the U.S.S.R. would not sell more than 20 Tu-144s initially. "Ten are estimated to be sufficient to satisfy [Soviet national airline] Aeroflot's demands for its internal and international European, African and Asian routes, and perhaps a further ten would be taken up by the Czechs . . . and other countries of Eastern Europe." Gen. Aleksey A. Tupolev—who had taken over Tu-144 planning after the death of his father, Tu-144 designer Andrey N. Tupolev—had predicted 50 times more air passengers by the year 2000 and talked of aircraft that would cruise in the stratosphere and carry 1000 persons at speeds of 10 060 km (6250 mi) per hr. (*NYT*, 6/2/73, 31)

June 3: The Soviet Tu-144 supersonic transport aircraft exploded and disintegrated in mid-air over the French village of Goussainville near Le Bourget Airport on the final day of the Paris Air Show, killing 13 persons [see May 24–June 3]. (Randal, *W Post*, 6/4/73, A1)

June 4: NASA was "fighting for its political life," a *Detroit News* editorial said as the *Skylab 2* astronauts considered ways to release a stuck solar wing on the *Skylab 1* Workshop launched May 14: "Its funds have been cut drastically since the moon landings were accomplished. Skylab itself is a vehicle costing $295 million. If it has to be abandoned the people will demand more expenditures be cut. So NASA is trying to prove that man really is boss of the machines he produces, that money is not being thrown away. There is a limit, however, to what NASA can devise and the astronauts accomplish. That limit must not be overstepped." (*D News*, 6/4/73)

- The U.S.S.R. lunar vehicle *Lunokhod 2* had completed its mission, Tass announced. The *New York Times* later said there were indications that *Lunokhod 2* had had an accident shortly after the start of its fifth lunar day, while heading toward the uplands east of Le Monnier Crater and that the vehicle's operation had been terminated unexpectedly. *Lunokhod 2* (landed on the lunar surface Jan. 16 by *Luna 21*, launched Jan. 8) had covered 37 km (23 mi) of complex terrain and withstood sharp temperature changes. It had taken the first measurements from the lunar surface of the luminosity of the lunar sky and had established that the moon was surrounded by a layer of dust particles in which the visible solar light and the reflected earth light were dispersed. *Lunokhod 2* had taken 86 relief panoramas and 80 000 TV photos of the lunar surface during 60 communications sessions. It had also measured physical, mechanical, and chemical properties of the upper layer of lunar rock and had studied the characteristics of the corpuscular emission of solar and galactic origin. (Tass, FBIS–Sov, 6/6/73, L5; *NYT*, 6/5/73, 18)

- A Japan Airlines spokesman said in Tokyo that the June 3 crash of the Soviet Tu-144 supersonic transport aircraft would compel JAL to delay further its decision on buying three Anglo-French Concorde SSTs. (*W Post*, 6/4/73, A10)

- President Nixon accepted the resignation of Grant L. Hansen as Assistant Secretary of the Air Force for Research and Development. (*PD*, 6/11/73, 751)

- NASA launched a Black Brant VC sounding rocket from White Sands Missile Range carrying a Harvard College Observatory payload to a 131.6-km (81.8-mi) altitude. The rocket was part of the CALROC series [see April 18] to provide a reference, using solar flux data, for the calibration of the Observatory's experiment on the Apollo Telescope Mount launched on the Skylab Workshop May 14. The experiment failed and the rocket was destroyed at 30.2 sec. (GSFC proj off)

June 5: The U.S.S.R. launched *Cosmos 562* from Plesetsk into orbit with a 473-km (293.9-mi) apogee, 270-km (167.8-mi) perigee, 91.9-min period, and 70.9° inclination. The satellite reentered Jan. 7, 1974. (GSFC *SSR*, 6/30/73; *SBD*, 6/6/73, 208)

June 5

- NASA's *Pioneer 11* probe, launched April 5, was traveling at 120 000 km per hr (74 000 mph) and had covered 48 million km (30 million mi), one fifth of its 20-mo journey to the planet Jupiter. All experiments and spacecraft systems were functioning well. Instruments were sampling the sun's field and wind, measuring solar and galactic cosmic ray particles, measuring hydrogen and helium in interstellar space, and making sky maps of the zodiacal light (sunlight reflected from cosmic dust between the earth and Jupiter).
  Meanwhile, *Pioneer 10* (launched March 3, 1972) had covered 560 million km (350 million mi) of its 1-billion-km (620-million-mi) flight path and was due to arrive at Jupiter in December. Good data were being returned from all scientific instruments. *Pioneer 10* was defining for the first time the interplanetary medium far beyond the orbit of Mars and the Asteroid Belt. (NASA Release 73-107)
- President Nixon accepted the resignation of Dr. John S. Foster, Jr., as Director of Defense Research and Engineering, and announced he would nominate as Dr. Foster's successor Dr. Malcolm R. Currie, Vice President for Research and Engineering of Beckman Instruments, Inc., in Fullerton, Calif. Dr. Currie had been with Hughes Aircraft Co. 1954–1969, serving as Vice President from 1964 until he went to Beckman in 1969. The nomination was submitted to the Senate June 6 and confirmed June 15. (*PD*, 6/11/73, 736, 752, 753; *CR*, 6/15/73, D695)
- An *Atlanta Constitution* editorial commented on Skylab as the *Skylab 2* astronauts planned a spacewalk to repair the power unit on *Skylab 1* (launched May 14): "The Skylab is a significant step in the developing technology of space exploration. Three men are there now, brave men. Their sheer competence makes it all seem routine, but they are risking their lives daily in doing things that literally no men have ever done before." (*Atlanta JC*, 6/5/73)
- NASA launched an Arcas sounding rocket from Antigua, West Indies, carrying a Goddard Space Flight Center payload to a 56.3-km (34.97-mi) altitude. The primary objective was to measure the ozone distribution in the upper atmosphere, monitor anomalous ultraviolet absorption, and to extend the data base for a climatology of stratospheric ozone in the tropics. The launch would collect data for comparison with data from a flyby of an Air Force RB–57 aircraft and the overpass of the *Nimbus 4* satellite (launched April 8, 1970). The payload was ejected near apogee and descended on a parachute. The rocket and the instrumentation performed satisfactorily and the experiment was successful. (NASA Rpt SRL)
- The appointment of French engineer Robert L. Mory as European Space Research Organization (ESRO) Sortie Lab liaison officer at Marshall Space Flight Center was announced by MSFC. Mory had been ESRO launcher coordinator for NASA launches of ESRO scientific satellites, a member of the commission to create ESRO in February 1963, and an ESRO member since April 1964. (MSFC Release 73–72)
- Former NASA scientist-astronaut Dr. Brian T. O'Leary, Hampshire College astronomer, criticized manned space flight spending in a *New York Times* article: "The spectacular problems of the $2.5-billion Skylab mission and the failure of the first three Soviet Salyut space stations may have dealt a death blow to manned space flight. The enormous expense, the high risk, the much-ballyhooed and grossly exaggerated claims about the pertinence of manned earth orbital flights to the qual-

ity of life on earth are creating an ever-widening credibility gap between the public interest and a vested interest inherited from the space of the nineteen-sixties. The result is the squandering of public funds." He suggested "indefinite postponement of the space shuttle program, a reduction in excessive NASA management costs and the establishment of a moderate unmanned space program emphasizing space science and applications" on a less than $2-billion budget. (*NYT*, 6/5/73, 39)

- U.S. patent No. 3 737 119 was granted to Princeton Univ. professor of aeronautical engineering Dr. Sin-i Cheng for a device to reduce supersonic aircraft sonic boom and improve SST flight performance. The device would create an antiboom jet with the same pressure as the flow around the aircraft and direct the jet toward the shock wave below. (Jones, *NYT*, 6/9/73; *Who's Who*; Pat Off PIO)

- A *Washington Post* editorial commented on the June 3 crash of the Soviet Tu-144 during the May 24–June 3 Paris Air Show: "With each new blow incurred by the cause of supersonic air transport . . . we wonder what it is beyond a kind of mindless national prestige contest that keeps the cause alive. Fairly or not, the flight risks of the new planes have now being sensationally dramatized." Spending billions to fly across an ocean or continent in three hours rather than six while "lacing the globe below with noise and God knows what other forms of ecological damage" was "one of the looniest notions of the 20th century—a triumph of technological and national zeal over good sense." (*W Post*, 6/5/73, A20)

June 6: *Cosmos 563* was launched by the U.S.S.R. from Plesetsk into orbit with a 306-km (190.1-mi) apogee, 190-km (118.1-mi) perigee, 89.5-min period, and 65.2° inclination. The satellite reentered June 18. (GSFC *SSR*, 6/30/73; *SBD*, 6/7/73, 211)

- The Pacific's first storm of the season, hurricane Ava, was probed by scientists and sensors aboard a National Oceanic and Atmospheric Administration C–130 flying laboratory and NASA's Skylab Orbital Workshop (*Skylab 1*, launched May 14). The storm was intercepted by the aircraft—equipped with microwave systems similar to those aboard *Skylab 1*—about 480 km (300 mi) southwest of Acapulco, Mex., as *Skylab 1* passed overhead. A series of tracks was made through the typhoon while the first deep-ocean-wave measurements ever made under a hurricane were taken with the laser altimeter. Maximum wave heights were 12 m (40 ft) at about 160 km (100 mi) from the eyewall. Winds of 110 to 240 km per hr (70 to 150 mph) were measured by an onboard inertial navigation system. The penetration was one of a series of NASA-funded Skylab underflights to prove the effectiveness of measuring large-scale processes in and over the ocean surface using microwave and other sensors aboard spacecraft. (NOAA Release 73–147)

- Deputy Secretary of Defense William P. Clements, Jr., announced that DOD would halt production of F–111 fighter-bomber aircraft when the current production run ended late in 1974 with the 543rd F–111. In 1962 DOD had planned to build 1726 F–111s for $3.4 million each, United Press said; the aircraft would wind up costing $14.6 million each, a cost overrun of 430%. (UPI, *W Post*, 6/7/73, C8)

- The U.K., West Germany, and Belgium agreed to buy U.S. Lance surface-to-surface missiles on the final day of a meeting in Brussels of Eurogroup, defense ministers from 10 European member countries of the

North Atlantic Treaty Organization (NATO). Italy had previously concluded a separate contract to buy the Lance. (Middleton, *NYT*, 6/7/73, 7)

- The Air Force announced the award of a $39.6-million contract to the Raytheon Corp. for construction of a large, new phased-array radar station, code-named Cobra Dane, at Semya Air Force Station in the Aleutian Islands. Cobra Dane would be part of the Air Force worldwide tracking program to trace space objects. An Aerospace Defense Command spokesman said that the total cost of the installation would be about $60 million. It would be operational in 1976. (Kelly, *W Star & News*, 6/7/73, A15)
- It would be "easy flying for the new Air Force bomber, the B-1, during the 93rd Congress," the *Congressional Quarterly* predicted. The FY 1974 Air Force request for $473.5 million to complete construction of the first prototype "probably will pass with relative ease." The prototype was scheduled for rollout at the Rockwell International Corp. factory in January 1974, with the first test flight scheduled for April. Two additional prototypes would be built to test the aircraft's structure and avionics, with test flights in January and April 1975. Final cost of the three aircraft would be $2.7 billion. B-1 supporters had said the real battle would come when Congress had to decide whether to begin mass production. (*CQ*, 6/16/73)
- The Coast Guard announced it would phase out its last two air cushion vehicles (ACVs) in July. A Coast Guard spokesman told the press in San Francisco that ACVs had met expectations but "we've just run out of money to operate them and they're too old now." (UPI, *NYT*, 6/7/73, 19)
- Award of a $9 098 952 firm-fixed-price contract to Boeing Co. for ground support for the Minuteman weapon system was announced by the Air Force. (DOD Release 286-73)

June 7: President Nixon was expected to bring Soviet Communist Party General Secretary Leonid I. Brezhnev to Johnson Space Center during his forthcoming visit for U.S. summit meetings, the *Houston Post* said. A JSC spokesman had said June 22, Skylab 2 splashdown day, was being considered and a survey team of Dept. of State and U.S. Secret Service personnel had visited JSC for preliminary planning. [See June 14.] (Maloney, *H Post*, 6/7/73)

- Deputy Secretary of Defense William P. Clements, Jr., directed the Navy to produce preliminary plans for a $250-million prototype development plan for a jet-fighter aircraft costing less than the controversial F-14 Tomcat missile-armed fighter. He also authorized the Navy to negotiate with Grumman Aerospace Corp. for an additional 50 F-14s armed with Phoenix missiles. The 50 aircraft would bring the F-14 total to 184 on order, with a possible 50 being ordered in FY 1974. The F-14 price was expected to average $25.8 million per aircraft. (Corddry, *B Sun*, 6/8/73, A8)
- Tass commented on the cause of the June 3 explosion and disintegration of the Soviet Tu-144 supersonic transport aircraft over Goussainville, France, during the May 24–June 3 Paris Air Show, as U.S.S.R. Council of Ministers Vice Chairman Leonid Smirnov and a Soviet government commission arrived in Paris to participate in the accident investigation: It had been hoped the Tu-144's black box, or flight recorder, would provide a clue to the cause. "But the box has been found open

in the garden of a house and the film in it exposed. The magnetic tape . . . is missing." Investigators were studying films taken by TV and amateur cameramen at the Air Show. "The last spurt of the plane before it hit the ground suggests that the pilot, powerless to prevent the disaster, fought to the last to take the plane away from the densely populated center of Goussainville." (FBIS-Sov, 6/8/73, F1)

June 7, 12: The House Committee on Science and Astronautics' Subcommittee on Energy held hearings on solar energy for heating and cooling. Dr. George O. Löf, Colorado State Univ. professor of civil engineering, testified June 7 that heating and cooling buildings by solar energy was technically feasible and closely approaching economic viability. "With adequate funding, the dual involvement of industry and university . . . can be expected to move solar heating and cooling into public use within a very few years." Electric power from solar energy, however, would cost several times the present cost of conventional power and the best method was not even known. A major study funded by the National Science Foundation was beginning at Colorado State Univ., with Westinghouse Electric Co. collaboration, of the best system and conditions for generating solar power. (Transcript)

June 7–15: The fifth meeting of the joint U.S.–U.S.S.R. editorial board for the preparation and publication of a review of space biology and medicine was held in Moscow. NASA and the Soviet Academy of Sciences announced June 18 that the meeting had developed further plans and schedules for the early publication of the three-volume work. (NASA Release 73-117)

June 8: The U.S.S.R. launched eight Cosmos satellites—believed by Western specialists to be a military communications data-relay system—on a single booster from Plesetsk. The satellites entered earth orbit:

Cosmos 564, with a 1482-km (920.9-mi) apogee, 1397-km (868.1-mi) perigee, 114.6-min period, and 74.0° inclination.

Cosmos 565, with a 1493-km (927.7-mi) apogee, 1448-km (899.7-mi) perigee, 115.3-min period, and 74.0° inclination.

Cosmos 566, with a 1484-km (922.1-mi) apogee, 1435-km (891.7-mi) perigee, 115.0-min period, and 74.0° inclination.

Cosmos 567, with a 1486-km (923.4-mi) apogee, 1413-km (878.0-mi) perigee, 114.8-min period, and 74.0° inclination.

Cosmos 568, with a 1483-km (921.5-mi) apogee, 1377-km (855.6-mi) perigee, 114.4-min period, and 74.0° inclination.

Cosmos 569, with a 1480-km (919.6-mi) apogee, 1361-km (845.7-mi) perigee, 114.2-min period, and 74.0° inclination.

Cosmos 570, with a 1482-km (920.9-mi) apogee, 1340-km (832.6-mi) perigee, 114.0-min period, and 74.0° inclination.

Cosmos 571, with a 1480-km (919.6-mi) apogee, 1322-km (821.5-mi) perigee, 113.7-min period, and 74.0° inclination.

(GSFC *SSR,* 6/30/73; *SBD,* 6/12/73, 233; Shabad, *NYT,* 6/10/73, 41)

• Sen. Frank E. Moss (D-Utah), Chairman of the Senate Committee on Aeronautical and Space Sciences, telegraphed congressional congratulations to Dr. James C. Fletcher, NASA Administrator, for the *Skylab 2* astronauts' achievements in repairing the solar array on the Skylab Workshop (launched May 14): "The magnificence of this accomplishment and the significance of this achievement is nearly impossible to put into perspective." Congress sent congratulations to the astronauts, NASA Associate Administrator for Manned Space Flight Dale D. Myers and his team, and "all of your group who work in complete anonymity but under the intensive examination of history." (CR, 6/8/73, S10691)

- President Nixon praised U.S. space technology in a commencement address at Florida Technological Univ. in Orlando: "The genius that could send men to the moon, the genius that could produce the Skylab, the genius that built America into the strongest and most productive nation in the world, the science, scientists, the technicians, all the engineers, all of those who could do that, certainly they can find the way to clean the air, and clean the water, and do the other things that will build a better environment in America." (PD, 6/11/73, 747)

- Life in weightlessness in the Skylab Workshop, orbited May 14, was described by a *Wall Street Journal* article as the *Skylab 2* crew began its 15th day on board the space station. Commander Charles Conrad, Jr., had reported, after attempting to eat canned tomatoes, that he was "flinging tomatoes all over the place." He had warned, "You've got to eat the whole tomato." Dr. Joseph P. Kerwin had told Mission Control: "I don't use the salt much. It tends to bounce off the food a lot." Later he had reported a minor kitchen disaster with a bag of dehydrated corn. "The bag failed at its seam. . . . We've got corn powder all over everything." Kerwin said taking a shower was pleasant, but the water didn't "fly through the air, it sticks to whatever is there, mostly you and partly the walls. It takes forever to dry both one's self and the walls . . . even using that inadequate little vacuum cleaner that we've got." The crew recommended allotting an hour for taking a shower on future missions. The astronauts were being awakened by the noise of the suction fan when the toilet was used at night. They found themselves growing hoarse from shouting at each other across the roomy Workshop during the day. Kerwin had found himself unable to whistle in zero g, but Astronaut Paul J. Weitz had assured him he could learn after a few days' practice. "You've got to hold your lips a little farther apart." (Bishop, *WSJ*, 6/8/73, 1)

- An *Atlanta Journal* editorial commented on space program spinoff as the *Skylab 2* astronauts began routine business after having freed the stuck solar array wing on *Skylab 1* (launched May 14): "The ambitious and expensive Skylab program is in trouble and daring measures are being taken to stabilize the planned programs it was to introduce. Before the screams of those who would dissolve the space program reach the astronauts aboard Skylab, it might be well to consider some spinoffs from the space program that preserve and lengthen the lives of those who cry for its cessation." Spinoffs included noise abatement, smog detection, medical devices, and fabric fireproofing. "The value of the space program is not up in the air; an awful lot of it is down to earth." (*Atlanta JC*, 6/8/73)

- The launch of Explorer 49 Radio Astronomy Explorer, scheduled for June 9, was postponed until June 10 following discovery of a sensitivity loss in 16 frequency channels of redundant burst receivers in the dipole experiment. A recalibration indicated the anomaly would not affect the mission seriously; the V antennas, when deployed, would provide data with higher sensitivity. (NASA prog off)

- Wallops Station announced it would negotiate with Computer Sciences Corp. for a two-year, cost-plus-fixed-fee contract worth approximately $1.9 million for engineering support and related services. The contract, effective July 1, would include support in range and airport instrumentation, spacecraft-system and sounding-rocket-payload-system design and development, development of aeronautical and earth resources programs, development of reliability and quality assurance plans, and environmental testing procedures, safety engineering, and data analysis and computation. (Wallops Release 73-4)

- The Federal Communications Commission authorized American Telephone & Telegraph Co. to lease 396 additional satellite circuits from Communications Satellite Corp. for service to Europe, Africa, and the Middle East. The circuits brought to 1450 the number leased from ComSatCorp by AT&T, at an annual $48.6-million charge. (DJ, W *Star & News*, 6/19/73; FCC PIO)

- A Washington *Star and Daily News* editorial commented on the June 3 crash of the Soviet Tu-144 supersonic transport aircraft during the May 24–June 3 Paris Air Show: "Coming on the heels of a series of Soviet air disasters which have killed more than 600 people in the last 18 months and many failures of its space program, the image of Soviet technology at this point is not so reassuring. . . . The crash of the Russian SST is certain to affect the commercial prospects of the Concorde as well." (W *Star & News*, 6/8/73, A18)

June 9: The *Skylab 2* astronauts' June 7 repair of the *Skylab 1* Orbital Workshop's solar wing [see May 14–June 22] had proved "that a manned vehicle outside the earth's atmosphere is not completely helpless when unforeseen contingencies arise," a *New York Times* editorial said. Skylab's future "now looks brighter than at any time since the initial launch. Valuable information remains to be collected on the reaction of human beings to prolonged stay in space, and on the usefulness of space stations for scientific and industrial activity. A promising start has already been made by this first Skylab crew; now it seems reasonable to expect that major contributions will also be made by the two other crews that are scheduled to occupy and work in Skylab later this year." (*NYT*, 6/9/73, 32)

June 10–20: NASA launched *Explorer 49* (RAE-B; RAE 2) Radio Astronomy Explorer by a long-tank, thrust-augmented, Thor-Delta launch vehicle from Eastern Test Range at 10:14 am EDT into a transfer trajectory for the moon. The second of two approved RAE missions and the last scheduled U.S. mission to the moon, *Explorer 49* was to measure galactic and solar radio noise at frequencies below ionospheric cutoffs and outside terrestrial background interference, using the moon for occultation, focusing, or aperture-blocking for increased resolution and discrimination.

At 17 min after launch, the spacecraft separated from the launch vehicle's 3rd stage. One midcourse correction was made at 11:28 am EDT June 11; the planned second correction was not required. During the transfer trajectory, data acquisition from telemetry systems was excellent. The lunar orbit insertion motor was fired at 3:21 am EDT June 15, putting the spacecraft into lunar orbit with 1334-km (828.9-mi) apolune, 1123-km (697.8-mi) perilune, 241-min period, and 61.3° inclination. On June 18 the lunar insertion motor was jettisoned and the hydrazine-fueled velocity-control propulsion system placed *Explorer 49* in a lower, near-circular orbit with a 1063-km (660.5-mi) apolune, 1052-km (653.7-mi) perilune, 221.2-min period, and 38.7° inclination.

On June 20 a 37-m (120-ft) dipole antenna was extended by ground control and the dipole experiment turned on. After a two-week calibration, the dipole would be retracted and the orbit adjusted as needed. The two main V antennas, joined in an X shape, would then be deployed to 183 m (600 ft). After another week, the V antennas would be deployed to their full 229-m (750-ft) length to achieve the mission's gravity-gradient, three-axis-stabilization mode.

The 200-kg (442-lb) spacecraft was the first known to use boron filament for the inertial booms that helped stabilize it during the lunar trajectory.

Explorer 38 (RAE 1)—the first successful radio astronomy satellite, launched July 4, 1968—had measured galactic radio sources and traced the movement of sporadic radio outbursts from the sun beyond the orbit of Mercury for four years. *Explorer 38*'s receivers, however, had been saturated 25%–40% of the observing time by interference from the earth's radio noise. *Explorer 49*, in lunar orbit, would be shielded from terrestrial noise as it orbited the moon's far side.

The RAE project was managed by Goddard Space Flight Center under the direction of the NASA Office of Space Science. GSFC designed, constructed, and tested the spacecraft and provided the scientific instrumentation. GSFC was also responsible for the Thor-Delta launch vehicle and the operation of the worldwide Spaceflight Tracking and Data Network (STDN). Launch services were provided by KSC. (NASA prog off; NASA Release 73–105; GSFC *SSR*, 6/30/73)

June 10: *Cosmos 572* was launched by the U.S.S.R. from Baykonur Cosmodrome into orbit with a 286-km (177.7-mi) apogee, 199-km (123.7-mi) perigee, 89.4-min period, and 51.8° inclination. The satellite reentered June 23. (GSFC *SSR*, 6/30/73; *SBD*, 6/13/73, 242)

• A *St. Louis Post Dispatch* editorial commented on the repair of the *Skylab 1* Workshop [see May 14–June 22] as its *Skylab 2* astronaut crew surveyed the U.S. heartland from space: "Now that Skylab is operating under something approaching full power, scientists will be able to judge how well man can exist for prolonged periods of weightlessness and whether experiments they carry out cannot be duplicated or outperformed by machines. The Skylab astronauts have managed to establish conditions under which the mission may receive a fair test and that is of paramount importance, for had they failed, the manned space program, for budgetary reasons, might have come to a halt." (*St. Louis P-D*, 6/10/73)

June 11: Skylab 3's Saturn IB launch vehicle and Apollo command module were moved from the Vehicle Assembly Building at Kennedy Space Center to Launch Complex 39, Pad B, for the July 27 launch of the second Skylab crew—Alan L. Bean, Dr. Owen K. Garriott, and Jack R. Lousma. The astronauts would join the Skylab Workshop (orbited May 14) for 56 days of experiments and observations. (KSC Release 124–73; *W Post*, 6/12/73, A13)

• A *Chicago Tribune* editorial commented on "the brave men of Skylab" as the *Skylab 2* astronauts (launched May 25 as the first crew to man the *Skylab 1* Workshop launched May 14) continued to conduct experiments in space: In repairing the Workshop's protective shield they had "behaved with such matter-of-fact competence, tinged with homely humor and flashes of exasperation when things refused to go right, that one almost overlooked the fact that this was an act of incredible sustained bravery." (*C Trib*, 6/11/73)

- NASA announced issuance of a request for proposals for an aircraft to replace the instrumented Convair 990 jet aircraft *Galileo*, lost April 12 in a mid-air collision over Moffett Field, Calif. The request called for a lease-purchase arrangement, for one year with three annual option periods, for an aircraft to make earth and sky surveys with infrared radiometers, microwave radiometers, cameras, telescopes, and other instruments for research and earth resource studies. NASA also was considering modification of any other Government aircraft. The replacement aircraft would be operated by Ames Research Center as a national scientific research facility. (NASA Release 73–116)
- Selection of 36 scientists, representing 27 organizations and 4 countries, to define experiments for the Large Space Telescope to be launched by the space shuttle in the 1980s was announced by NASA. The instrument evaluation teams had been selected after evaluation of 118 proposals submitted to NASA's Space Science and Applications Steering Committee. (NASA Release 73–109)
- Sen. Edward M. Kennedy (D-Mass.), Chairman of the Special Subcommittee on the National Science Foundation of the Senate Committee on Labor and Public Welfare, introduced S. 1977, $704 million FY 1974 NSF authorization bill. The bill would establish an Energy Research Div. and would authorize not less than $50 million for energy research programs. (CR, 6/11/73, S10810–12)
- NASA launched an Aerobee 200A sounding rocket from White Sands Missile Range carrying a Lockheed Missiles & Space Co. solar physics experiment to a 220.6-km (137.1-mi) altitude. The rocket and instrumentation performed satisfactorily. (GSFC proj off)

June 11–13: A Workshop on Wind Energy Conversion Systems, sponsored by the National Science Foundation and organized by Lewis Research Center, was held in Washington, D.C. The meetings, attended by scientists and industry and Government representatives from the U.S. and foreign countries, discussed past developments, wind characteristics and siting problems, rotor characteristics, energy conversion systems, small wind-power systems for remote and individual applications, wind-power systems for large-scale applications, and tower structures. The object of NSF's wind energy conversion program was to provide a reliable energy supply, in usable form, at a cost competitive with alternative systems. (NASA Release 73–114)

June 11–15: The Hugh O'Brian Youth Foundation sponsored a seminar at Kennedy Space Center in cooperation with NASA, the U.S. Junior Chambers of Commerce, and the National Assn. of Student Councils. Participants—67 high school sophomores from the U.S. and 6 foreign countries—toured KSC and were briefed on Skylab, the space shuttle, and astronaut training. (KSC Release 127–73)

June 12: The Air Force launched an unidentified satellite from Eastern Test Range by a Titan IIIC booster into orbit with a 35 787.0-km (22 237-mi) apogee, 35 778.9-km (22 232-mi) perigee, 1436.0-min (23-hr 56-min) period, and 0.3° inclination. The press quoted sources as saying the spacecraft was an early warning satellite expected to be stationed over the Indian Ocean to give split-second notice of missile launchings in the U.S.S.R. (Pres Rpt 74; AP, B *Sun*, 6/13/73, A16; *Av Wk*, 6/18/73, 17)
- NASA held a Skylab earth resources experiment package press briefing at Johnson Space Center. Dr. Verl R. Wilmarth, EREP project scientist at

JSC, said that the *Skylab 2* astronauts—Charles Conrad, Jr., Dr. Joseph P. Kerwin, and Paul J. Weitz, launched May 25 to crew the Orbital Workshop—had completed their ninth EREP pass. The sensors had been turned on over 190 areas designated by the principal investigators. A unique collection of laser propholometer data had been obtained over hurricane Ava. Data collected over Maryland and the Chesapeake Bay area would update topographic and geologic maps. (Transcript)

- A specially equipped United Air Lines, Inc., 727-200 jet transport aircraft was testing two-segment landing approach procedures to reduce noise levels near airport landing patterns in an Ames Research Center and Federal Aviation Administration cooperative program. The aircraft first descended at a steeper than usual glide-path angle. As the aircraft neared the runway and intersected the normal, shallow-approach flight path, the glide-path angle was decreased and the final landing made normally. A 67% reduction in the area disturbed by objectionable landing noise was expected to be achieved. (NASA Release 73–113)

- Sen. Frank E. Moss (D-Utah), Chairman of the Senate Committee on Aeronautical and Space Sciences, told the Philadelphia *Evening Bulletin* in an interview that he had been in the U.S.S.R. when the Soviet *Salyut 2* space station (launched April 3) disintegrated. "We knew that their Skylab had come apart, but it was classified at the time and didn't become public knowledge until after we had returned to the United States." Moss, in Philadelphia to address a meeting of the American Institute of Aeronautics and Astronautics, said that even if it experienced more space failures, the U.S.S.R. would do its utmost to make the joint U.S.–U.S.S.R. Apollo-Soyuz mission scheduled for July 1975. (P *Bull*, 6/13/73, 11)

- The International Telephone & Telegraph Corp. and the U.S.S.R. signed a five-year agreement in Moscow to cooperate on scientific and technical information in telecommunications, electronics and electromechanical components, and consumer products and on publication of scientific and technical data. (ITT Release, 6/14/73)

- Radio glow, a new phenomenon observed by U.S.S.R. scientists, was described in *Pravda* by Soviet radioastronomer Vsevolod Troitsky. The phenomenon was caused by charged particle flux from radiation belts in the lower layers of the earth's ionosphere and was thought to be generated at altitudes up to 100 km (60 mi) and related to changes in the area of solar spots and chromospheric flares. Observations had been made simultaneously in the far eastern, northern, and southern parts of the U.S.S.R. Irregular radiation on superhigh frequencies would permit study of dynamic processes in the ionomagnetic sphere of the earth. (Tass, FBIS–Sov, 6/12/73, L1)

June 13: Deputy Secretary of Defense William P. Clements, Jr., told a Washington, D.C., press conference that the U.S.S.R. was pressing ahead with "new families of missiles" in a quest for military superiority over the U.S. "It is hard information. We know they are testing these missiles. We know what size they are. We know what their capabilities are." (Corddry, B *Sun*, 6/14/73, A1)

- NASA launched a Black Brant VC sounding rocket from White Sands Missile Range carrying a Naval Research Laboratory payload to a 240.9-km (149.7-mi) altitude. The rocket was one of the CALROC series [see April 18] to provide a reference, using solar flux data, for the calibration of NRL's instrumentation aboard the *Skylab 1* Apollo Telescope

Mount. The rocket and instrumentation performed satisfactorily and good data were obtained. (GSFC proj off)

- The Robert J. Collier Trophy for 1972 was presented in Washington, D.C., by the National Aeronautic Assn. to the 7th and 8th Air Forces and Task Force 77 of the Navy for their "demonstrated expert and precisely integrated use of advance[d] aerospace technology" in the conduct of Operation Line Backer II, the 11-day air campaign in December 1972 that "led to the return of the United States prisoners of war" in Southeast Asia. Adm. Thomas H. Moorer (USN), Chairman of the Joint Chiefs of Staff, received the trophy on behalf of the servicemen at a luncheon hosted by the National Aviation Club and the National Aeronautic Assn. Guests included Dr. James C. Fletcher, NASA Administrator. (Invitation; NAC Off)

- Data from drifting buoys equipped with temperature and pressure sensors and satellite communications antennas that reported to French satellite *Eole* (launched by NASA Aug. 16, 1971) had shown that currents in the western Atlantic ocean were more erratic and unpredictable than earlier imagined, the National Oceanic and Atmospheric Administration announced. Although in continuous motion, some buoys had ended up approximately where they had started after two or three months of monitoring; others had "moved in unpredicted directions, changed directions frequently, circled, or doubled back." The buoys were part of a NOAA project to identify large-scale mean surface flows. (NOAA Release 73–118)

- Kennedy Space Center announced the award of a $98 000 contract to the Environmental Research Institute of Michigan for a feasibility study of the use of ultrahigh-frequency and microwave radars aboard aircraft to provide information on mosquito breeding grounds, determine water drainage patterns, and provide data for water resources management and land use planning in Brevard County, Fla. (KSC Release 132–73)

- The Air Force announced the award of a $1 924 913 fixed-price-incentive contract to McDonnell Douglas Corp. for incorporation of engineering change proposals on the F–4 fighter aircraft production contract. A $1 294 776 fixed-price-incentive contract to General Dynamics Corp. for additional funding for inspection and correction of F–111 landing gear pin deficiencies also had been awarded. (DOD Release 296–73)

June 14: Tentative selection of 22 scientists—including one each from France and Germany—from among 72 scientists who responded to NASA's August 1972 invitation to provide experiments for one of two Pioneer spacecraft to probe Venus in 1978 was announced by NASA. Mission objectives were to gather information on the Venusian atmosphere and clouds. Participating scientists would define experiments studying composition and structure of the atmosphere down to the surface, the nature and composition of the clouds, the circulation pattern of the atmosphere, and the radiation field in the lower atmosphere.

The mission—comprising a bus, a large probe, and three small probes —would use a spin-stabilized, solar-powered spacecraft. The trip from earth to Venus would take 125 days. The probes would be separated from the bus 10 to 20 days before entry into the Venusian atmosphere. The large probe would descend through the atmosphere in 1½ hrs; the small probes would fall free to the planet's surface about 75 min after entry. The bus would be targeted to enter the Venus atmosphere at a shallow angle and transmit data until it was destroyed by atmospheric-friction heat.

The atmosphere probe was scheduled for launch in May 1978, about three months before a Venus orbiter was to be launched. Both spacecraft would arrive in the vicinity of Venus in December 1978. By comparing the atmospheres of Venus, Mars, and the earth, scientists hoped to construct an efficient model of the earth's atmosphere to predict long-term climate changes and short-term effects of environmental pollution. Experiments for the orbiter mission would be selected in early 1974. (NASA Release 73–115)

- Soviet tests and investigations had shown that "supersonic civilian aircraft have a good future," Deputy Chairman Leonid Smirnov of the U.S.S.R. Council of Ministers, heading the Soviet commission investigating the June 3 crash of the Tu-144 [see May 24–June 3], said in a Paris TV interview. The Soviet airline Aeroflot was "preparing to use the supersonic Tu-144."

  Gen. Aleksey A. Tupolev, chief designer of the Tu-144, was quoted by Associated Press as saying in Paris that development of the aircraft would continue but would be delayed. It had been scheduled for service in late 1974. (FBIS–Sov, 6/15/73, F1; *NYT*, 6/15/73, 2)

- Dr. Henry A. Kissinger, Assistant to the President for National Security Affairs, discussed plans for the June 18–23 summit meeting to be held by President Nixon and Soviet Communist Party General Secretary Leonid I. Brezhnev in Washington, D.C., and San Clemente, Calif., at a Washington, D.C., press briefing. Asked why Brezhnev would not be visiting Johnson Space Center as had been reported [see June 7], Dr. Kissinger said, "There were various ideas for various possible stops, but given the fact that a maximum amount of time was always intended to be devoted to discussion . . . it simply proved impossible." (*PD*, 6/18/73, 771–8)

- Western Union Telegraph Co. announced it had received authorization from the Federal Communications Commission to construct four more earth stations for the terrestrial segment of its WESTAR domestic communications satellite system. It would build on sites near Atlanta, Ga.; Chicago, Ill.; Dallas, Tex.; and Los Angeles, Calif. The first WESTAR earth station was under construction at Vernon Township, N.J. (Western Union Release, 6/14/73)

June 15: *Cosmos 573* was launched by the U.S.S.R. from Baykonur Cosmodrome into orbit with a 401-km (249.2-mi) apogee, 195-km (121.2-mi) perigee, 90.5-min period, and 51.6° inclination. The satellite reentered June 17. The Western press later quoted a Soviet source as identifying the mission as a test of an unmanned Soyuz spacecraft in preparation for an upcoming manned launch. Western tracking stations reported the orbit was identical to one used during manned Soviet flights. (GSFC *SSR*, 6/30/73; *SBD*, 6/18/73, 265; B *Sun*, 7/6/73)

- NASA held two *Skylab 2* press briefings at Johnson Space Center as the astronaut crew, orbiting the earth in the Orbital Workshop launched May 14, carried out experiments and solar photography.

  At an Apollo Telescope Mount (ATM) briefing Dr. Robert Noyes of Harvard College Observatory said the experiment to look into the outer solar atmosphere "exceeds our wildest expectations. Every object that we've tried to look at, we've done successfully and we have seen . . . entirely new and different phenomena." For the first time there had been enough spatial resolution to see the fine structure. Data received by viewing the sun at different heights in the corona had established the existence of two fundamental structures—points of emission 12 900 km (8000 mi) in diameter and a cloud of relatively cool material emersed in the very hot corona. The structures should provide information on heat sources and physical properties of the corona. Dr. Robert MacQueen, High Altitude Observatory astronomer, said that a high point of the ATM experiments had been detailed examination of the corona for 45 consecutive hours, identifying the material making up the solar wind passing into the corona.

At a solar-flare activity briefing Dr. Guiseppe Vaiana, American Science and Engineering, Inc., scientist and a principal investigator on Skylab, described the TV coverage of the solar flare photographed by Astronaut Paul J. Weitz: "What you are seeing is the great event of the day." The flare had a double structure in the active region. A portion of it outlined the magnetic field of one polarity on one side and the other polarity on the other side, with a center neutral line. Two bright spots were seen on both sides of the neutral line. The flare extended 100 000 km (62 000 mi) up from the solar surface and covered 2½ trillion sq km (1 trillion sq mi). It had peaked in intensity 3 min after it had been discovered by Weitz and decayed during the following 30 min. (Transcript)

- Dr. James C. Fletcher, NASA Administrator, toured Ames Research Center and was briefed on the status of the Pioneer Venus mission. The mission was in the conceptual design state and, if approved in FY 1975, would lead to Venus orbiter and multiprobe launches in 1978. Dr. Fletcher said NASA intended to proceed with the dual Venus mission if possible and intended to make it one of NASA's most cost-effective programs. (ARC *Astrogram*, 6/21/73, 1)

- Boeing Co. and General Electric Co. Aircraft Engine Group had signed a $3-million subcontract under which GE would supply main propulsion engines for the first two ships in a new line of multinational, missile-carrying, hydrofoil patrol craft (PHMS), Boeing announced. The Navy had ordered two of the hydrofoils to be delivered in 1975. (Boeing Release 061573)

June 17: *Skylab 1* Orbital Workshop (launched by NASA May 14) had become Space Object 1973-027A—one of 584 man-made earth satellites and 41 deep space probes constituting "the increasingly heavy and varied traffic in space," the *New York Times* reported. The latest earth-orbital traffic count catalogued by the North American Air Defense Command's Space Detection and Tracking System showed 341 U.S. earth satellites, 204 Soviet, 9 French, 6 Canadian, 5 British, 4 Japanese, 4 European Space Research Organization (ESRO), 3 West German, 3 Italian, 2 North Atlantic Treaty Organization (NATO), 2 Chinese, and 1 Australian. The oldest was *Vanguard 1* (launched by NASA March 17, 1958). The heaviest traffic lay 160–965 km (100–600 mi) from the earth, where U.S. and Soviet scientific and military satellites orbited. Most remote was *Pioneer 10* (launched by NASA March 2, 1972), which was 547 million km (340 million mi) from the earth. Nearly 2300 pieces of space "junk"—used rockets, heat shields, and disposable spacecraft parts—floated in earth orbit but no serious collision or traffic jam had occurred in space to date. Two Navy experimental satellites had come close enough to tip antenna booms in 1965, "a possible warning of problems to come." (Wilford, *NYT*, 6/17/73, 4:4)

June 18: The first relay of a telecast across the U.S. by a domestic communications satellite was made. A cable TV program featuring House Speaker Carl B. Albert (D–Okla.) was transmitted from Germantown, Md., by American Satellite Corp. antenna via Canadian *Anik 2* comsat to Anaheim, Calif., received by new mobile ground station, and delivered to cable TV systems. (*SBD*, 6/2/73, 292; Am Sat Corp PIO)

- President Nixon welcomed Soviet Communist Party General Secretary Leonid Brezhnev on his first visit to the U.S. Brezhnev had arrived June 17 and was greeted on the White House lawn by President Nixon June 18 at the beginning of U.S. summit meetings. President Nixon recalled agreements signed during the May 22–29, 1972, Moscow summit meetings between the U.S. and U.S.S.R.: "What has happened since those agreements . . . lead me to conclude that this year at the summit in Washington we will not only build on the foundation that we laid last year, but that we have the opportunity to make even greater progress than we made last year toward the goals that we share in common—the goals of better relations between our two governments, a better life for our people, . . . and above all, the goal that goes beyond our two countries, but to the whole world—the goal of lifting the burden of armaments from the world and building a structure of peace." Brezhnev responded: "The distances between our countries are shrinking, not only because we travel aboard modern aircraft following a well-charted route, but also because we share one great goal, which is to ensure a lasting peace for the peoples of our countries, and to strengthen security on our planet." (*PD*, 6/25/73, 787–8)

- U.S.S. *Wasp*, World War II aircraft carrier and NASA prime recovery vessel, made her last voyage—to the ship breakers' yards at Kearny, N.J. The 30-yr-old ship had been purchased for scrap metal by Union Mineral Alloys Corp. for $505 250. She had served as the prime recovery vessel for NASA Gemini-Titan manned space flights: *Gemini 4* (June 3–7, 1965), *Gemini 6* (Dec. 15–16, 1965), *Gemini 7* (Dec. 4–18, 1965), *Gemini 9* (June 3–6, 1966), and *Gemini 12* (Nov. 11–15, 1966). (Phalon, *NYT*, 6/19/73, 39; NASA Hist Off)

- Rockwell International Corp. Space Div. announced the award of a $30-million-plus subcontract to LTV Aerospace Corp. Vought Systems Div. to design and develop the leading-edge structural subsystem of the space shuttle orbiter's thermal protection system. Final details were being negotiated. (RI Release SP–45)

- *Apollo 17* Astronauts Eugene A. Cernan, Ronald E. Evans, and Dr. Harrison H. Schmitt arrived in Pakistan to discuss the U.S. space program with Pakistani President Zulfikar Ali Bhutto and present a Pakistani flag they had taken to the moon during their Dec. 7–19, 1972, mission. (UPI, *W Post*, 6/20/73, A18)

- Dr. Carl E. Sagan, Cornell Univ. astronomer and NASA consultant, arrived in Paris to receive an International Galabert Prize for space research. Other recipients would be Soviet Academician M. Y. Marov and French astronomer Adouin Dollfus. (FBIS–Sov, 6/21/73, G15)

June 19: Soviet cosmonauts had telegraphed congratulations for new achievements in the conquest of space to the *Skylab 2* crew orbiting the earth in the *Skylab 1* Workshop, the Moscow Domestic Service reported. The astronauts, launched May 25, had surpassed the world manned-space-flight endurance record June 18 after repairing the Workshop. Maj. Gen. Vladimir N. Shatalov, director of cosmonaut training, wired on the cosmonauts' behalf, wishing the astronauts a successful conclusion to their program and a safe return to the earth. (FBIS–Sov, 6/20/73, G4)

- Rep. Edward P. Boland (D-Mass.), Chairman of the House Committee on Appropriations Subcommittee on Housing and Urban Development-Space-Science-Veterans, introduced H.R. 8825, FY 1974 HUD-Space-Science-Veterans appropriation bill that included a $2.989-billion NASA appropriation—$2.194 billion for research and development, $87.8 million for construction of facilities, and $707 million for research and program management. The bill would cut NASA's requested $2.197 billion for R&D by $3 million and the $112 million requested for construction of facilities by $24.2 million. Research and program management funding would remain at NASA's requested $707 million.

  The Subcommittee had reduced NASA's requested $28 million for supersonic flight research to $11.7 million; strongly urged NASA to continue advanced nuclear technology development "within the total funds provided in this appropriation"; urged NASA to reprogram funds to launch ERTS–B as early as possible if the malfunction in *Erts 1* Earth Resources Technology Satellite (launched July 23, 1972) continued to degrade performance; urged continued development of the quiet, experimental, short takeoff and landing (QUESTOL) aircraft "to insure American competitiveness in this largely underdeveloped area of commercial aviation"; and allocated $2.5 million to replace NASA's instrumented Convair 990 aircraft (lost in an April 12 mid-air collision). Reductions in construction of facilities funding would defer amounts for space shuttle facilities construction. (*CR*, 6/19/73, H4994, H Rpt 93–296)

- The Senate, by a vote of 90 to 5, passed H.R. 7528, $3.046-billion FY 1974 NASA authorization bill [see May 30], after rejecting an amendment proposed by Sen. J. William Fulbright (D–Ark.) that would have reduced the authorization by 4% ($122 million). (*CR*, 6/19/73, S11476–503)

- Secretary of State William P. Rogers and Soviet Foreign Minister Andrey A. Gromyko signed new U.S.–U.S.S.R. agreements on cooperation in transportation and ocean studies and a General Agreement on Contacts, Exchanges, and Cooperation in the Fields of Science, Technology, Education, and Culture, during State Dept. ceremonies attended by President Nixon and visiting Soviet Communist Party General Secretary Leonid I. Brezhnev. The General Agreement reaffirmed commitments for cooperation in environmental protection, medicine and public health, space exploration and the use of outer space for peaceful purposes, and science and technology signed during the May 22–29, 1972, U.S.–U.S.S.R. summit meetings in Moscow. (*PD*, 6/25/73, 791–800)

- Deputy Chairman Vladimir A. Kirillin of the U.S.S.R. Council of Ministers discussed U.S.–U.S.S.R. cooperation in science and technology in a Moscow TV interview while President Nixon and Soviet Communist Party General Secretary Leonid I. Brezhnev met for summit talks in the U.S. The broadening of cooperation in "the solution of the energy problem, space research, the study of the world's oceans, the struggle against disease, environmental protection . . . would mean not only a tremendous saving in national resources but would also go far beyond the confines of relations between the two states in its significance. This

June 19

cooperation in fact is becoming a necessary condition for the solution of a number of complex problems facing mankind in the last quarter of the 20th century." (Tass, FBIS–Sov, 6/21/73, G3–7)

- A delegation from the National Academy of Sciences—headed by Dr. Philip Handler, NAS President—arrived in Moscow to develop contacts between NAS and the Soviet Academy of Sciences. Dr. Handler told Soviet newsmen on arrival, "We have no doubt that these talks will promote development of Soviet-American scientific and technical cooperation." (Tass, FBIS–Sov, 6/20/73, G4)

- Cornell Univ. astronomer Dr. Thomas Gold replied to former astronaut Dr. Brian T. O'Leary's June 5 *New York Times* article in a letter published by the *Times:* "NASA believes that it is only with the public interest in manned flight that it can continue to collect the funds needed to feed its organization. Is it not time to evaluate the national priorities and to inform NASA of the result? The space successes of the next decade or more will be in the fields of greatly expanded communication systems, better weather prediction and earth surveys and in the deeper understanding of the earth and the solar system that space science can bring. It is true that this looks like hard work and not much glamour, but it is a program the country could stand behind, even in the face of occasional failures." NASA could and would carry out such a program if public support was evident, "and many persons inside NASA would be much happier if the facade and the make-believe of the present era could be abandoned." (*NYT*, 6/19/73, 36)

- The National Science Foundation released *National Patterns of R&D Resources: Funds & Manpower in the United States 1953–1973* (NSF 73–303). Total U.S. research and development spending was expected to reach $30.1 billion in 1973, 3% above the $29.2-billion 1972 level. R&D was expected to account for 2.4% of the gross national product (GNP) in 1973, down from 2.5% in 1972 and 3.0% in peak year 1964. The Federal Government would support 53% of U.S. R&D in 1973, with about 70% concentrated in defense and space programs. Almost 40% of R&D spending in the U.S. was for research, with 1973 basic research expenditures estimated at $4.5 billion and applied research at $6.8 billion. Development spending was estimated at $18.8 billion. Colleges and universities would perform nearly three fifths of U.S. basic research, with industry leading in applied research and development. An estimated 525 000 R&D scientists and engineers were employed in the U.S. in 1972, 1% fewer than in 1971. Of these, nearly 70% were employed by industry. (Text)

June 20: *Cosmos 574* was launched by the U.S.S.R. from Plesetsk into orbit with a 1014-km (630.1-mi) apogee, 985-km (612.1-mi) perigee, 105.0-min period, and 82.9° inclination. (GSFC *SSR*, 6/30/73; *SBD*, 6/21/73, 289)

- Johnson Space Center had signed a $700 000 contract with United Aircraft Corp. Hamilton Standard Div. to design, build, and install a laboratory model of a modular, integrated utility system (MIUS), NASA announced. The contract was part of a NASA project for the Dept. of Housing and Urban Development to use space technology to conserve natural resources, lessen air and water pollution, and reduce energy consumption. MIUS was a single, combined source of electric power, water management, solid-waste treatment, and heating and air conditioning. Communities could be served with networks of modules arranged to serve larger population concentrations. NASA and HUD would test the integrated concepts on a low-cost, small scale before full-scale demonstrations. Other agencies participating were the Atomic Energy Commission, National Bureau of Standards, and the Environmental Protection Agency. The contract called for completion of testbed installations at JSC by spring 1974. (NASA Release 73-119)

- Rep. Larry Winn, Jr. (R-Kan.), with cosponsors from the House Committee on Science and Astronautics, introduced H.R. 8871 to authorize the NASA Administrator to conduct research and development programs to increase knowledge of tornadoes, hurricanes, large thunderstorms, and other short-lived weather phenomena and to develop methods for predicting, detecting, and monitoring such atmospheric behavior. (CR, 6/20/73, H5064)
- A Navy F-14 jet fighter aircraft shot itself down over the Pacific with a Sparrow air-to-air missile launched erratically from its mount beneath the aircraft. The two Grumman Aerospace Corp. crewmen ejected and were rescued unharmed by a Navy helicopter. The F-14 had been flight-testing the missile's ability to drop clear of its mount before firing, the Navy explained in a statement issued June 21, but the missile failed to clear, struck the fuselage, and exploded. It was the third loss of an F-14 since the swing-wing aircraft's December 1970 maiden test flight. (Witkin, NYT, 6/22/73, 8)
- The House Committee on Armed Services voted in closed session to appropriate $172.7 million to continue production of F-111s at a one-a-month rate through 1975 [see June 6]. (Kelly, W Star & News, 6/21/73)
- The Aerospace Industries Assn. released *AIA Survey of Aerospace Employment*, its semiannual aerospace employment survey. Aerospace employment had increased from 924 000 to 944 000 in 1972 despite an anticipated decline, largely because of accelerated aircraft deliveries. A reduction during 1973 to approximately 913 000 was anticipated. From December 1972 to December 1973, production workers were expected to decline by 3.4%, from 473 000 to 457 000. Scientists and engineers were expected to decline 4.2%, to 161 000. Peak scientist and engineer employment of 235 000 was reached in 1967. Technicians were expected to remain at 1972 levels, with all other aerospace employees declining 3.4%. Aircraft production and research and development employment was expected to decline from 556 000 to 552 000 between December 1972 and December 1973, while employment in missiles and space was expected to be reduced by nearly 10%, from 217 000 to 196 000. (AIA Release 73-15; Text)

June 20–July 4: The American Assn. for the Advancement of Science celebrated its 125th anniversary with an inter-American meeting, "Science and Man in the Americas," in Mexico City. The meeting, cosponsored by the Consejo Nacional de Ciencia y Technologia of Mexico, explored Western Hemisphere problems: aridity, populations, education, earthquakes, nutrition, and the sea. (AAAS *Bulletin*, 4/73, AAAS PIO)

June 21: Cosmos 575 was launched by the U.S.S.R. from Plesetsk into orbit with a 267-km (165.9-mi) apogee, 204-km (126.8-mi) perigee, 89.1-min period, and 65.3° inclination. The satellite reentered July 3. (GSFC SSR, 6/30/73; SBD, 6/22/73, 304)

- President Nixon and visiting Soviet Communist Party General Secretary Leonid I. Brezhnev signed, during White House ceremonies, Basic

Principles of Negotiations on the Further Limitation of Strategic Offensive Arms and an agreement for scientific and technical cooperation in peaceful uses of atomic energy. The arms limitation document specified agreement on continued active negotiations to work out a permanent agreement to be signed in 1974, new agreements to be guided by recognition of each side's equal security interests, application of weapon limitations to quantity as well as quality, subjection of arms limitations to verification by national technical means, modernization and replacement of weapons under conditions to be formulated, formulation of agreements on separate measures to supplement the existing Interim Agreement of May 26, 1972, and to continue measures to prevent accidental or unauthorized use of nuclear weapons. The atomic energy agreement would establish a U.S.–U.S.S.R. Joint Committee on Cooperation in the Peaceful Uses of Atomic Energy.

At an evening Soviet Embassy dinner honoring President Nixon, General Secretary Brezhnev said: "The cause of developing Soviet-American relations is, indeed, moving forward. In 2 years, Soviet and American astronauts will fly into outer space to carry out the first major joint experiment in man's history. Now they know that from up there in space, our planet looks even more beautiful, though small. It is big enough for us to live in peace, but too small to be subjected to the threat of nuclear war." Brezhnev proposed a toast to President and Mrs. Nixon's health and "to the further success of the great cause which we have succeeded in advancing during our present meeting, to the docking, on earth as well as in outer space, of man's efforts and talents for the good of the peoples, to peace, friendship, and cooperation. . . ." (*PD*, 6/25/73, 810–21)

- West Germany's *Aeros* Aeronomy Satellite (launched by NASA Dec. 16, 1972) was adjudged a success by NASA. Prelauch objectives—to measure the main aeronomic parameters of the upper atmosphere and the solar ultraviolet radiation in the wavelength band of main absorption—had been satisfied. Four of the five scientific instruments had performed satisfactorily. (NASA prog off)

- NASA and the American Institute of Aeronautics and Astronautics held a joint press briefing at Johnson Space Center on the commercial benefits from Skylab electronics technology. Robert H. Webster of Ampex Corp. said that the 28-track, high-reliability earth resources experiment package tape recorder developed for the Skylab missions was being tested for use by auto and truck manufacturers. Airlines and heavy equipment manufacturers were using it because of its environmental strength.

    Thomas H. Kenton, Westinghouse Regional Vice President for the Southwest, said that Westinghouse had manufactured two sensitive, low-light-level TV cameras for Skylab. The cameras could operate at light intensities 500 million times less than typical studio lighting conditions and equal to the light level of a dark moonless night. The cameras could be used for commercial TV camera systems. (Transcript)

- A Washington *Evening Star and News* editorial commented on the Skylab mission as the *Skylab 2* astronauts (launched May 25) prepared for their June 22 return to earth: "It is discouraging to some advocates of the space program that the present mission has been observed so casually by the American people, over-shadowed as it has been in the headlines by Watergate, the Brezhnev summit and the price freeze. This comparative lack of attention is one of the costs of America's long string of successes in space. We tend to forget about the courage and skill required of the men we send into orbit and beyond, the ingenuity of their monitors on the ground and the billions we all have invested in the effort. These factors are worth reflecting upon as the country reassesses its future ambitions in space." (W *Star & News*, 6/21/73, A20)

- It was "nigh impossible" for the U.S.S.R. to cancel its Tu-144 supersonic transport program, Dr. Sarah S. White, consultant on Soviet science, said in a *New Scientist* article. Despite "faint rumblings on the environmental side," the "disquietening financial burden," and the prestige set-back of the June 3 crash of the Tu-144, the program would continue. "Because of its crucial military significance, the government in Moscow has always given top priority to its aviation industry. It is not surprising, then, to hear that work on the Tu-144 is progressing only marginally behind schedule—and that designers have already started evaluating a second generation sst capable of flying at speeds said to be considerably greater than Mach 2–3." (*New Scientist*, 6/21/73, 763-4; *Who's Who in Science in Europe*)
- Rep. Paul G. Rogers (D–Fla.) introduced H.J. Res. 632 to redesignate Cape Kennedy Cape Canaveral. (*CR*, 6/20/73, H5158)

June 21–23: NASA considered but decided against plans to launch a special spacecraft to intercept and study at close range Comet Kohoutek, discovered in March. Kohoutek, 10 times brighter than any comet on record, would be visible from the earth in late 1973.

The NASA Science Advisory Committee on Comets and Asteroids recommended a program "to maximize the scientific return from the unique opportunity" by using new and modified instruments and modified schedules on the Skylab Workshop orbited May 14; other spacecraft already in orbit, such as *Oao 3* (*Copernicus*, launched Aug. 21, 1972) and *Oso 7* (launched Sept. 29, 1971); and sounding rockets, balloons, aircraft, and ground-based observations. NASA would also use a Radio Astronomy Explorer, an Atmospheric Explorer, and *Isis 1* (launched Jan. 30, 1969).

The agency had considered launching a scientific payload on a Thor-Delta or Atlas-Centaur launch vehicle to examine a comet from the outer reaches of the solar system at short range for the first time, but decided that time was insufficient to prepare the specialized instrumentation required. (Adv Com minutes; NASA OSS; Sullivan, *NYT*, 6/16/73; *CSM*, 6/18/73; *NASA Activities*, 9/15/73, 161)

June 22: President Nixon sent a telegram to the *Skylab 2* astronauts following their successful splashdown at the end of a 28-day earth orbital mission [see May 14–June 22]: "The successful completion of the first mission of Skylab is a source of intense pride for the American people. You have demonstrated that just as man can conquer the elements of earth, he can cope with the exigencies of space. You have given conclusive evidence that, even with the most advanced scientific and technological support in the world, the courage and resourcefulness of good men are still central to the success of the human adventure. On behalf of the American people, I welcome you home from the Skylab spaceship to spaceship earth. I also look forward to seeing you at San Clemente on Sunday." (*PD*, 6/25/73, 828)

- NASA held two *Skylab 1–2* postmission briefings at Johnson Space Center. Dr. James C. Fletcher, NASA Administrator, told a NASA press conference that, "as near as we can tell, essentially all of the objectives that were anticipated for this mission have been completed. And needless to say none of us really dreamed that this could be done at the time that the meteoroid shield failed to deploy. So it exceeded our wildest expectations at that time." *Skylab 1-2* could be symbolized by a hammer, and the *Skylab 2* astronauts could be characterized as "the master tinkerers of space." The mission's anomalies had made no major impact on NASA's budget.

Dr. Willard R. Hawkins, JSC Deputy Director of Life Sciences for Medical Operations, said at a crew medical status briefing that the astronauts had "looked rather wobbly and unsteady when they got out of the spacecraft." The crew—aware of the increased pull of gravity—had said they felt as if they were "very heavy . . . like about a two-G load." All three men had experienced dizziness, lightheadedness, and some nausea. Astronaut Dr. Joseph P. Kerwin felt more of these effects than Charles Conrad, Jr., and Paul J. Weitz. His blood pressure had dropped to a critical level 25 min after splashdown and he had to inflate his spacesuit to force the blood back up from his legs to his brain. During medical tests aboard the recovery ship *Ticonderoga*, Conrad had performed at his premission level. Weitz had experienced a sudden drop in blood pressure and pulse rate while on the bicycle ergometer but had recovered within a five-minute rest period. Kerwin had demonstrated more of the vestibular disturbance and was unable to complete any of the tests. (Transcripts)

- President Nixon and Soviet Communist Party General Secretary Leonid I. Brezhnev signed a U.S.–U.S.S.R. Agreement on the Prevention of Nuclear War, at the White House. Both sides would act in such a manner as to prevent development of situations that could cause "dangerous exacerbation of their relations, as to avoid military confrontations, and as to exclude the outbreak of nuclear war between them and between either of the Parties and other countries." (*PD*, 6/25/73, 822–3)

- The House, by a vote of 315 to 21, approved H.R. 8825, FY 1974 Dept. of Housing and Urban Development-Space-Science-Veterans appropriation [see June 19]. A point of order was overruled, to block an amendment by Rep. Bella S. Abzug (D–N.Y.) that would have cut $475 million in NASA research and development funds and forbid any funding for the space shuttle. (*CR*, 6/22/73, H5191–5238)

- Winners of an Army competition to develop the new advanced attack helicopter were announced by Secretary of the Army Howard H. Callaway. Bell Helicopter Co. and Hughes Helicopters & Hughes Aircraft Co. would compete in a flyoff with two prototypes per contractor. The flyoff winner would install and test the required subsystems for night vision, fire control, navigation, and communications. Values of the initial development contracts were $44.7 million to Bell and $70.3 to Hughes. (DOD Release 316–73)

June 22–July 28: During unmanned operations of the Orbital Workshop, launched May 14, the Apollo Telescope Mount observed the sun's chromospheric network, prominences, coronal transients, solar limbs, lunar libration clouds, and solar eclipses. Active ATM operations were discontinued July 16 because the experiment pointing and control system's primary pitch-rate gyroscope failed.

An astrophysics particle collection experiment exposed outside the

Workshop would be retrieved by the Skylab 3 crew, scheduled for July 28 launch. Sample panels of thermal coatings exposed on the airlock module also would be returned. Neutron analysis detectors, a student experiment, were monitored. (NASA prog off)

June 23: NASA held a *Skylab 2* crew medical status briefing at Johnson Space Center. Dr. William R. Hawkins, JSC Deputy Director of Life Sciences for Medical Operations, said that the condition of Astronauts Charles Conrad, Jr., Dr. Joseph P. Kerwin, and Paul J. Weitz had improved over their June 22 splashdown condition. Conrad was in excellent shape, Weitz had no residual effects from the vestibular disturbances of the previous day, and Kerwin was "a thousand percent better" from the cardiovascular standpoint but still had some residual vestibular effects. (Transcript)

- Newspaper editorials commented on the successful conclusion of *Skylab 2* the day after splashdown.

  Baltimore *Sun:* "The Skylab experience, it is said, shows once again the value of man over computer, or rather of man-plus-computer over computer alone, in the exploration of space; and that may be one of its lessons. The larger lesson is that human curiosity and human intelligence are still operating in regions of concern beyond the day-by-day, and will keep on so operating." (B *Sun*, 6/23/73)

  New York Daily News: "The venture proved . . . that man is needed in space to repair and regulate equipment when something goes wrong. But for the skill, coolness and courage of astronauts Conrad, Kerwin and Weitz, the entire Skylab project would have wound up as an expensive washout." (*NY News*, 6/23/73)

- Secretary of Transportation Claude S. Brinegar and Soviet Minister of Civil Aviation Boris P. Bugayev signed a protocol in Washington, D.C., to expand air service between the U.S. and the U.S.S.R. Pan American World Airways, Inc., and the Soviet airline Aeroflot would have accelerated operating rights from and to New York, Washington, D.C., Moscow, and Leningrad, with optional intermediate points in Europe. (*PD*, 7/2/73, 831–2)

- The National Academy of Sciences–National Research Council's Committee on Atmospheric Sciences published *Weather and Climate Modification: Problems and Progress*. The report called for a civilian program to promote understanding of the techniques and effects of man-induced changes in weather and climate, to bring to fruition technology to mitigate severe weather hazards to man and agriculture, and to stimulate earnest consideration of public-policy issues inherent in weather modification. A 1980 target date was set for development of determinative information on trends of climate changes inadvertently induced by man. (NAS–NRC–NAE *News Report*, 8–9/73, 4–5)

June 24: *Skylab 2* Astronauts Charles Conrad, Jr., Paul J. Weitz, and Dr. Joseph P. Kerwin were greeted by President Nixon and Soviet Communist Party General Secretary Leonid I. Brezhnev at the San Clemente, Calif., White House. The astronauts were en route to Ellington Air Force Base, Tex., after their successful June 22 splashdown off San Diego. (*PD*, 7/2/73, 876)

- Newspaper editorials commented on the success of the May 14–June 22 *Skylab 1–2* mission.

  New York Times: "This first group of astronauts did much more . . . than demonstrate man's biological capability to live and work in space for four weeks. They proved that men in space can do what machines cannot do, they can repair their vehicle." It was conceivable that "when all the Skylab data are available, the field of solar astronomy may be advanced as radically as lunar astronomy has been in recent years." At Skylab's completion "a strong economic and scientific case will doubtless have been made for creating one or more permanent manned stations in space. Any such orbiting laboratories should surely be United Nations ventures in international cooperation, not new instruments for senseless and expensive national rivalries." (*NYT*, 6/24/73)

The *Philadelphia Inquirer:* It had been shown that "enterprising spacemen can do more than routine work for long periods of time. They can deal with the unexpected, correct serious malfunctions and adjust routine to changed circumstances. In space, as on earth, things do not always go well. There, as here, adversity is the true test of men. The Skylab pioneers have met that test with distinction." (*P Inq*, 6/24/73)

- The U.S.S.R. announced the start of operations at the world's first nuclear station within the Arctic Circle. The plant, in the Kola Peninsula, had been designed to bolster the electrical supply of a rich mining region. (*NYT*, 6/25/73)

June 24–25: The status of the joint U.S.–U.S.S.R. Apollo-Soyuz Test Project for a July 1975 launch was described in a joint communique dated June 24 and released June 25 from San Clemente, Calif., at the close of summit meetings between President Nixon and Soviet Communist Party General Secretary Leonid I. Brezhnev: "Preparations for the joint space flight . . . are proceeding according to an agreed timetable. The joint flight of these spaceships for a rendezvous and docking mission and mutual visits of American and Soviet astronauts in each other's spacecraft are scheduled for July 1975." (*PD*, 7/2/73, 840-48)

June 25: Salvage of the *Skylab 1-2* mission [see May 14–June 22] had illustrated the need for both men and automated space activities in the same vehicle, NASA Associate Administrator for Manned Space Flight Dale D. Myers said in a speech before the National Space Club in Washington, D.C. "To argue that we don't need man in space is like saying we don't need a fork at the table because we have a knife." He praised aerospace industry cooperation in salvaging Skylab: "It has not been very well publicized, but many of the components on the sun shield that saved Skylab were fabricated by contractors at their own plants. The response from everyone was tremendous."

Skylab 1-2 had shown that the astronauts went through a rapid learning curve in adjusting to zero g and that "often times one man is fully capable of accomplishing tasks that we had originally planned for two." NASA also had learned "that for a man to perform efficiently in zero gravity he must have adequate, well-designed body restraints, and that this was one area that we probably underdesigned in the Skylab system." The space shuttle would merge the capabilities of manned and automated space activities in one vehicle "where now they usually complement one another in separate vehicles." The issue of manned versus unmanned space flight had been dubious from the start. "What is really the question is whether we want to develop the *full* potentials of space flight or only a *part* of them. Because if you eliminate man from the loop, you significantly reduce our options to explore and use space." (NSC Notice, 6/6/73; *NASA Activities*, 7/15/73, 114-6)

- The dizziness and nausea of *Skylab 2* Astronauts Charles Conrad, Jr., Dr. Joseph P. Kerwin, and Paul J. Weitz on their return to the earth June 22 were probably due to problems of the inner ear, the *Wall Street Journal* quoted Dr. Willard R. Hawkins, Johnson Space Center Deputy Director of Life Sciences for Medical Operations, as saying. The symptoms had occurred particularly when the men moved their heads from side to side. Dr. Hawkins had theorized that the fluid in the semicircular canal, whose movement transmitted impulses via tiny hairs in the ear to the brain, was out of action because of the long period of weightlessness. When the astronauts had come under the sudden influence of gravity, the inner-ear mechanism became unusually sensitive to almost any head movement. Ordinary movements could produce the symptoms of motion sickness. The inner-ear balancing mechanism could adapt to gravity quickly. Dr. Kerwin, the most affected, had returned to normal after 12 hrs. (Bishop, *WSJ*, 6/25/73)
- Dr. George M. Low, NASA Deputy Administrator, accepted for NASA a Silver Plaque presented by the British Interplanetary Society at the opening of the 13th European Space Symposium in London. The plaque was awarded in recognition of the Apollo program's "triumphant achievements" and to acknowledge "the dedicated service of many thousands of scientists and engineers which culminated in this great result." (*SF*, 10/73, 396)
- *Apollo 17* Astronauts Eugene A. Cernan, Ronald E. Evans, and Dr. Harrison H. Schmitt had arrived in Indonesia to present a moon rock to Indonesian President T. N. J. Suharto and lecture at the Indonesian Institute for Outer Space, the Associated Press reported. They were on an 11-country tour of Asia, the Middle East, Africa, and the Pacific. (*Birmingham News*, 6/25/73)
- New York Federal Court Judge Edward Wienfeld dismissed the $2-million libel suit brought by Dr. W. Ross Adey, Univ. of California at Los Angeles brain research specialist, against New York-based animal welfare organization United Action for Animals [see Jan. 22]. The organization's bulletin had criticized Dr. Adey's treatment of Bonny, the instrumented monkey used to measure life functions under weightlessness during the NASA-Air Force *Biosatellite 3* experiment (launched June 28, 1969). Judge Wienfeld said that, as a public official, Dr. Adey "must accept harsh criticism, ofttimes unfair and unjustified." (UPI, *W Post*, 6/27/73, A10)

June 26: Johnson Space Center's selection of a United Aircraft Corp. Pratt & Whitney Div. TF33-P-7 engine for use on the space shuttle orbiter in atmospheric flight was announced by NASA. The engine was similar to those used on the Air Force C-141 Starlifter transport aircraft. The USAF would secure 25 engines for NASA use during horizontal flight-testing of the orbiter and for ferry flight when the shuttle was operational. (NASA Release 73-121)
- The House Committee on Science and Astronautics' Subcommittee on Manned Space Flight held a hearing on the General Accounting Office's June 1 report on space shuttle program costs. NASA's Deputy Associate Administrator Willis H. Shapley testified: "NASA believes that the results of our cost benefit analyses are important and valid elements supporting the decision to develop the Space Shuttle, even though the principal justification for the Space Shuttle is in the new capabilities it will provide. In our view, the GAO review, which we welcome as an independent

review and check of what we have attempted to do, has not found or demonstrated any substantial reasons for questioning the correctness of the decision which has been made to proceed with the development of the Space Shuttle."

The Subcommittee concluded, following their testimony by NASA and GAO representatives: "a. The five noneconomic issues cited by GAO as major considerations in the decision to develop a space shuttle are valid and proper elements in the original decision to proceed and in the future evaluation of the progress and pace of space shuttle development. b. Cost, performance and schedule goals and estimates are both a valid and essential element in the decision making process of the Congress and the GAO studies in these areas . . . must have sufficient depth to encompass all major cost considerations. c. The fragmentary arguments with respect to costs advanced by GAO reports are so incomplete as to fail to validate the GAO conclusions drawn from them. d. That GAO should continue review of the space shuttle program development in conjunction with the legislative and oversight activities of the responsible committees of Congress with emphasis on evaluation of cost, performance and schedule as the space shuttle development program progresses." (Transcript)

- Maj. Gen. Vladimir A. Shatalov, U.S.S.R. director of cosmonaut training, denied Western press reports of recent Soviet space failures and said there would be manned Soviet launches before the 1975 U.S.–U.S.S.R. Apollo-Soyuz mission, during a Moscow interview with U.S. correspondents. Of *Soyuz 2* (launched April 3 and reported to have disintegrated in space) he said, "This craft this time was not intended for manned flight. The experiment was successfully completed and we are satisfied with the results." *Salyut 2* had had a "much more narrow purpose" than *Salyut 1* (launched April 19, 1971, as a space station). *Salyut 2* was "intended to finalize some design peculiarities. As soon as it fulfilled its task, it finished its existence." Several manned space flights were planned as part of the Soviet space program and to test design changes in the Soyuz spacecraft for the Apollo-Soyuz Test Project (ASTP). The U.S.S.R. was considering admitting Western journalists to their space facilities for the first time and the establishment of a press center at Baykonur Cosmodrome, but newsmen would not be permitted beyond 7 km (4.3 mi) of the launch site. (UPI, *Intl Herald Tribune*, 6/27/73; B *Sun*, 6/27/73, A2)

- Flight Research Center announced the award of a cost-reimbursable $1 600 000 contract to the Charles Stark Draper Laboratory at Massachusetts Institute of Technology to provide software for an improved control system for NASA's digital fly-by-wire experimental aircraft. The system would advance digital fly-by-wire technology by replacing the single-channel digital system undergoing flight tests with a dual system that used hardware developed originally for the Apollo program. (FRC Release 15–73)

- The Univ. of California at Davis successfully completed a 15-wk, NASA-sponsored study to determine how astronauts on extended space missions would cope with isolation [see May 18]. Six male students, in groups of three, lived in 3- by 5-m (11- by 17-ft) rooms with bathrooms and closets. Interior lighting was controlled to interchange day-night cycles and provide long periods of uninterrupted light barely bright enough to read by. Each room was monitored by closed-circuit TV. The students, selected from 100 volunteers, were paid $1600 each to participate. Dr. Don A. Rockwell, codirector of the study, said it had proved that, while space explorers should be able to live without sex for extended periods, they needed someone outside their spacecraft to whom "they could blow their tops" from time to time. (AP, *W Post*, 6/27/73)

- A *Los Angeles Times* editorial commented on the success of the May 14–June 22 *Skylab 1-2* mission: "By overcoming the initial difficulties which plagued the mission, and going on to complete most of the experiments assigned to them, the men of the first Skylab mission have made an impressive contribution toward the goal of putting space technology to work here on earth." (*LA Times*, 6/26/73)
- The Air Force announced award of a $1 700 000 definitive contract to United Aircraft Corp. Pratt & Whitney Div. for fabrication of an advanced gas dynamic laser for the airborne laser laboratory of the Air Force Special Weapons Center at Kirkland Air Force Base, N. Mex. (DOD Release 320-73)

June 27: *Cosmos 576* was launched by the U.S.S.R. from Plesetsk into orbit with a 331-km (205.7-mi) apogee, 203-km (126.1-mi) perigee, 89.8-min period, and 72.9° inclination. The satellite reentered July 9. (GSFC *SSR*, 6/30/73; 7/31/73; *SBD*, 6/29/73, 333)

- The U.K. Trades Union Congress announced that it was asking its 9.5 million members to boycott French products for a week in protest against French nuclear testing in the South Pacific. (*W Post*, 6/28/73, B19)

June 27–28: *Pioneer 11*, launched by NASA April 5 toward a December 1974 rendezvous with the planet Jupiter, became the second spacecraft to cross the orbit of Mars. By June 28 the probe was 76 781 000 km (47 711 000 mi) from the earth, traveling at 108 000 km (67 110 mi) per hr. The earth had passed between the spacecraft and the sun, permitting scientists to look for the earth's magnetic tail, the extension of the magnetic field blown away from the earth by the solar wind. Data had suggested a malfunction in the spacecraft radio transmitter but were insufficient to identify the problem. The backup traveling-wave tube was being used.

Flight controllers had fired the thrusters on *Pioneer 10*—twin probe launched March 2, 1972, toward December 1973 meeting with Jupiter—on June 21 to trim its course slightly and ensure passage behind Jupiter's moon Io. *Pioneer 10*, traveling at 43 470 km (27 000 mi) per hr, was 533 540 000 km (331 542 000 mi) from the earth. It had crossed the orbit of Mars in 1972 and finished crossing the Asteroid Belt in February. (NASA Release 72-122)

June 28: NASA held three *Skylab 1-2* postflight reviews at Johnson Space Center.

At a summary review of the May 14–June 22 mission, Skylab Program Office Manager Kenneth S. Kleinknecht said that *Skylab 1-2* had accomplished the intent of its 48 experiments—6 earth resources, 15 medical, 5 astronomical, 17 scientific and technological, and 5 student-proposed. Space station operation experience had been gained and there were no constraints on continuing the Skylab program as originally planned. More data had been handled in real time than ever before and gave "no reason to believe that all of the things that we had previously said we accomplished in flight are not going to give us creditable scientific data for analysis by scientists."

Dr. Verl R. Wilmarth, earth resources experiment package project scientist, said during an EREP review that the EREP experiments had gathered good data for 145 principal investigators on 13 km (8 mi) of magnetic tape. Data were collected over six foreign countries—Mexico, Brazil, Colombia, Bolivia, Nicaragua, and Venezuela. There were 186 individual task sites covered, including the Puerto Rican trench, the Gulf of Mexico, the Wabash River area, and the Amazon River basin.

During a review of the *Skylab 2* crew health, Dr. Willard R. Hawkins, Deputy Director of Life Sciences for Medical Operations, said medical experiments on the Workshop had yielded very interesting results. All three crew members had stayed within their preflight baseline levels on the bicycle ergometer study. In the motion-sensitivity study—with two crew members participating—the preflight threshold level (the point at which motion sickness symptoms would begin) of science pilot Dr. Joseph P. Kerwin was $12\frac{1}{2}$ rpm in the rotating chair, at which speed he could perform 50 head movements. Pilot Paul J. Weitz had a preflight threshold of 15 rpm and 40 to 50 head movements. Both men could do 150 head movements in flight at preflight speeds without symptoms. During tests immediately following splashdown both men had experienced vestibular disturbances with even the slightest head movement. The symptoms had disappeared within two days and both crewmen had performed at preflight speeds on the third day following recovery, completing 150 head movements without symptoms. (Transcripts)

- The House-Senate Conference Committee reported out its compromise Conference Report on H.R. 7528, recommending a $3.065-billion FY 1974 NASA authorization. The Senate agreed to the report, clearing it for House action.

    The bill provided $2.246 billion for research and development, up $14.5 million from the June 19 Senate-approved authorization but $9 million less than the May 23 House-approved bill. It provided $112 million for construction of facilities, the same as the House authorization but $2 million more than the Senate's, and $707 million for research and program management, unchanged from the House and Senate versions.

    The R&D allocation included $555.5 million for space flight operations, $475 million for the space shuttle, $1.5 million for advanced missions, $63.6 million for physics and astronomy, $311 million for lunar and planetary exploration, $177.4 million for launch vehicle procurement, $161 million for space applications, $180 million for aeronautical research and technology (including $14 million for the JT3D refan retrofit research program), $72 million for space and nuclear research and technology, $244 million for tracking and data acquisition, and $4.5 million for technology utilization.

    Construction of facilities funds included $67.2 million for space shuttle facilities at various locations. (*CR*, 6/28/73, D773; Rpt 93-272)

- The Senate Committee on Appropriations reported H.R. 8825, FY 1974 Dept. of Housing and Urban Development-Space-Science-Veterans appropriation bill that included a $3.002-billion NASA appropriation, $13 million above the $2.989-billion version passed by the House June 22. The bill increased the appropriation for construction of facilities from $89 million to $101 million, including $56.3 million for space shuttle facilities at various locations. (*CR*, 6/28/73, D771; S Rpt 93-272)

- NASA and the Environmental Protection Agency signed an agreement to participate in a three-year Automotive Gas Turbine Technology Program to develop and demonstrate a car powered by a gas turbine engine that met or bettered the 1976 Federal emission standards, was economical to run, and performed well. The agreement was signed by Roy P. Jackson, NASA Associate Administrator for Aeronautics and Space Technology, and Robert L. Sansom, EPA's Assistant Administrator for Air and Water Programs. Lewis Research Center would manage the program for NASA. (NASA Release 73-123; NASA NMI 1052.179)
- The House unanimously passed H. Con. Res. 223, requesting the President to proclaim the week of July 16–22, 1973, as United States Space Week. (*CR*, 6/28/73, H5574)
- A helmet designed by Ames Research Center and Aerotherm Corp. to keep the heads of helicopter pilots cool had been tested successfully by stock car racing driver Richard Petty, NASA announced. The helmet, lined with polyurethane through which water circulated, kept the pilot's head cool, lowering heat stress on his entire body. (NASA Release 73-120)
- A *Japan Times* editorial praised the successful completion of the May 14–June 22 *Skylab 1-2* mission: "The accomplishments of the first Skylab mission were many and significant not only for the future of space travel but in increasing man's knowledge of the sun and his own earth. We look forward to the even longer missions of Skylab and to the knowledge which will be obtained for the benefit of all of us on earth." (*Japan Times*, 6/28/73)

June 29: The importance of exercise in space was stressed by *Skylab 2* Astronaut Charles Conrad, Jr., at a Johnson Space Center Skylab crew press conference: "I do think that the bicycle ergometer ... contributed to our well being." Subsequent Skylab crews "probably should exercise longer periods of time, and ... add to the exercise some muscular exercises rather than strictly cardiovascular."

Dr. Joseph P. Kerwin said that "it was a continuous and pleasant surprise" to find out "how easy it was to live in zero-g, and how good you felt. It's easy to move, it's easy to work." He said that future missions could go on open-ended as long as the individual crew member could "pool a reasonable amount of blood in his legs ... and provided his response to exercise, his appetite, his body weight, general feeling of well being, and ability to do work" continued without significant change.

Paul J. Weitz said that the easy adjustment in space was due partially to the "extremely high fidelity equipment" on which they trained. "There were many times that you could look around and you really didn't know ... if you were in the trainer or you were in the vehicle." There was no difference in the extravehicular activity, it could just as well have been in the water tank because of the good training gear. (Transcript)

- *Explorer 48* Small Astronomy Satellite, launched for NASA by an Italian crew from the San Marco Facility Nov. 16, 1972, was adjudged a success. It had satisfied its primary objective of measuring the spatial and energy distribution of primary galactic and extragalactic gamma radiation for six months. *Explorer 48* had observed a weak but finite component of high-energy gamma ray photons at galactic latitudes greater

June 29:

than 20° and a positive flux of gamma rays from the Crab Nebula, established as a source of 60 to 200 mev radiation.

Spacecraft-control-section performance had excelled except for a degraded star sensor. An anomaly in the experiment readout system in late May had caused some degradation in the scientific data and a June failure in the experiment low-voltage power-supply circuitry had halted experiment data transmission. The first anomaly had corrected itself; attempts were being made to correct the second. (NASA prog off)

- President Nixon issued Executive Order 11726 from San Clemente, Calif., establishing a Federal Office of Energy Policy. He announced the appointment of Colorado Gov. John A. Love as its Director and released a statement outlining a $10-billion, five-year program for energy research and development. The President established a goal for the next 12 mos for energy consumption reduction of 5% nationwide, and 7% within the Federal Government. (*PD*, 6/7/73, 867–75)

June 30: A seven-minute, four-second, total solar eclipse, one of three seven-minute-plus total solar eclipses in the 20th century, began at sunrise in Brazil and crossed Guyana, Surinam, the Atlantic Ocean, the Cape Verde Islands, Mauretania, Mali, Algeria, Niger, Chad, the Central African Republic, Sudan, Uganda, Kenya, and the Somali Republic. It ended at sunset in the Indian Ocean between India and Madagascar. A partial eclipse was visible for 3200 km (2000 mi) on both sides of the totality track. Previous total solar eclipses had been June 8, 1937, for seven minutes four seconds, and June 20, 1955, for seven minutes eight seconds. The next solar eclipse in the current "saros," or 18-yr period, would be July 11, 1991, for 6 minutes 54 seconds, but the June 30 eclipse's duration would not be exceeded for 177 years, British astronomer Peter MacDonald said later in a *New Scientist* article.

The African eclipse was witnessed by an estimated 5000 amateur astronomers and tourists and by 2500 international scientists on land, on sea, and aboard research aircraft. Concorde 001, the specially instrumented Anglo-French supersonic transport aircraft, carried seven U.S., U.K., and French scientists across North Africa in lunar shadow at altitudes to 17 000 m (56 000 ft) during a two-hour six-minute flight. They observed total eclipse conditions for 74 min.

The *Skylab 1* Workshop, unmanned since *Skylab 2*'s June 22 splashdown, saw only about 10% of the eclipse because its orbital position and telescope mechanisms were not suited for maximum effectiveness. Its cameras operated from 7 am EDT until midnight, and eclipse films taken would be retrieved by Skylab 3 astronauts scheduled for July 27 launch. Other scientific observations of the eclipse by at least 28 nations included an instrumented U.S. Air Force aircraft and rockets fired from the Soviet research ship *Professor Vize* near the Cape Verde Islands [see May 29]. Univ. of Texas observations by telescopes in the Sahara desert tested Einstein's theory of relativity by measuring the bend of the sun's gravity. (Sullivan, *NYT*, 5/15/73, 6; 6/30/73, 32; 7/1/73, 1; *New Scientist*, 7/12/73, 90–91; *SF*, 8/15/73, 321)

- The Senate, by a vote of 73 to 1, passed H.R. 8825, the FY 1974 Dept. of Housing and Urban Development-Space-Science-Veterans appropriation bill that included a $3.002-billion NASA appropriation [see June 28]. (*CR*, S12611–47)

- NASA launched three sounding rockets in a series from Wallops Station carrying Goddard Space Flight Center payloads to determine the ion composition and electron density of the ionospheric D-region.

  A Nike-Cajun was launched to an 86.2-km (53.6-mi) altitude. The rocket and instrumentation performed satisfactorily.

  An Astrobee D was launched 18 min later to a 92.5-km (57.5-mi) altitude. The rocket performance was less than planned. The instrumentation was satisfactory except for the Gerdien trap, which ceased providing data at 22 sec.

  A Nike-Apache was launched three minutes after the Astrobee D to a 77.4-km (48.1-mi) altitude. The rocket performed satisfactorily but no useful scientific data were received from the ion mass spectrometer and the scientific objectives of this flight were not achieved. (NASA Rpts SRL)

- Aircraft operations at the 364 Federal Aviation Administration control towers increased 1% in FY 1973, from 53 620 706 in FY 1972 to 53 992 674. Chicago's O'Hare International Airport again recorded the largest number of takeoffs and landings—682 984. Santa Ana, Calif., followed with 608 361. The most air carrier operations were at O'Hare, Atlanta International, Los Angeles International, New York's Kennedy International, and San Francisco International. Busiest air route traffic control centers were Cleveland with 1 664 634 aircraft handled, Chicago with 1 603 157, and New York with 1 564 570. (FAA Release 73–218)

- Lockheed Aircraft Corp. retained its ranking as the Nation's largest defense contractor during FY 1973. A Dept. of Defense annual list of 100 top contractors showed Lockheed with $1.66 billion in DOD procurement awards, 5.3% of all prime contracts over $10 000. General Electric was second with $1.42 billion and Boeing Co. was third with $1.23 billion. McDonnell Douglas Corp. was fourth with $1.14 billion and Grumman Corp. fifth with $909 million. (DOD Release, 11/2/73)

- Kenyan Director of National Parks Perez Olindo asked the East African Directorate of Civil Aviation to ban supersonic transport aircraft from flying over East Africa because of "untold hazards" to human health and to wildlife. (Agence France Presse, *W Post*, 6/30/73, A5)

During June: NASA retirements during the month totaled 778 agency-wide, under the law signed by President Nixon June 12 to permit early retirements and a 6.1% cost-of-living bonus for employees retiring by June 30. Of 76 in NASA Hq.—including reduction-in-force and voluntary retirements—59 were in grades GS–13 and above.

Retirees at Marshall Space Flight Center peaked at 172, an all-time monthly high. An additional 18 employees applied for disability retirement. Previous monthly high for MSFC retirements was 106 in June 1972. June 1973 MSFC retirements, added to 24 retired Jan. 1–May 30, brought the 1973 total to 170.

Retirements at other Centers during June numbered 142 for Lewis Research Center, 120 for Langley Research Center, 76 for Goddard Research Center, 56 for Kennedy Space Center, 53 for Johnson Space Center, 50 for Ames Research Center, 15 each at Flight Research Center and Wallops Station, and 3 at Space Nuclear Systems Office.

Hq. retirees included Jacob E. Smart, Assistant Administrator for DOD and Interagency Affairs; Dr. Fred Schulman, Special Assistant to the Manager and Asst. Manager, Space Nuclear Systems Office; Leonard Carulli, Management Analyst, Office of Management and Development;

Jocelyn R. Gill, Space Scientist for Physics and Astronomy Programs; Ernest W. Brackett, Special Assistant to the Associate Administrator for Industry Affairs; Ralph E. Ulmer, Program Analyst, Advanced Concepts and Mission Div.; Raymond Einhorn, Special Assistant to the Associate Administrator for Organization and Management; Dr. Harvey Hall, Principal Scientist, Advanced Missions; and Charles M. Hochberg, Director, Budget Operations, Budget Operations Div. (NASA Personnel Off; NASA Admin. Off; NASA Ann, 6/13/73; MSFC Release 73-89; Causey, *W Post*, 6/15/73, D17)

- The National Academy of Sciences released *International Magnetospheric Study: Guidelines for United States Participation.* A study panel of the Committee on Solar-Terrestrial Research and the Space Science Board of the National Research Council had identified unanswered questions on the magnetosphere and appraised opportunities for U.S. participation in the IMS, scheduled for 1976–1978. The report, initiated at NASA and NSF request, recommended that the U.S. endorse the IMS "and participate with a coordinated research program of ground-based, balloon, rocket, and satellite observations that includes the NASA/ESRO [European Space Research Organization] Mother-Daughter-Heliocentric missions (International Magnetospheric Explorers)." To facilitate the U.S. program, the report recommended establishing a program office within NSF with a representative designated in each participating governmental agency to coordinate IMS-related projects. (Text)
- Successful flights of Air Force Airborne Warning and Control System (AWACS) testbed aircraft had demonstrated AWAC's potential to increase effectiveness of the North Atlantic Treaty Organization command and control system, the Air Force System Command's *Newsreview* reported. The Central European and Mediterranean flights also had ensured AWAC's suitability for the European environment. (AFSC *Newsreview*, 6/73, 1)

July 1973

July 1: The National Aeronautics and Space Council and the Executive Office of Science and Technology were abolished under terms of Reorganization Plan No. 1 of 1973 [see Jan. 26]. The plan went into effect after Congress had taken no veto action before the prescribed date. OST functions were transferred to the National Science Foundation and its Director, Dr. H. Guyford Stever [see July 10]. (White House PIO; White House Ann, 7/10/73)

- Dr. Thomas O. Paine, NASA's third Administrator, became a senior vice president and member of the corporate executive staff of General Electric Co. He had resigned from NASA Sept. 15, 1970, to join GE as Vice President and Group Executive, Power Generation Group. (*Av Wk*, 7/2/73; GE PIO)
- NASA Public Information Director Richard T. Mittauer died of Hodgkin's disease in Alexandria, Va., at age 46. He had joined NASA in 1959 and had received a NASA Exceptional Service Medal for his work on the 1971 Mariner Mars mission. (W *Star & News*, 7/2/73, B5)

July 2: The Skylab 3 mission to carry the second crew to work and live aboard the *Skylab 1* Workshop (launched May 14) would be launched no earlier than 7:08 am EDT July 28, NASA announced. Skylab 3 Astronauts Alan L. Bean (commander), Dr. Owen K. Garriott (science pilot), and Jack R. Lousma (pilot) would conduct scientific and technical experiments for 56 days. Splashdown was scheduled for 8:38 pm EDT Sept. 22 in the Pacific. (NASA Release 73-125)

- Dr. James R. Schlesinger, former Atomic Energy Commission Chairman, was sworn in by President Nixon as Secretary of Defense. (*PD*, 7/9/73, 885)
- Plans for a space-shuttle-like "astroplane" or "transport ship," a project of the U.S.S.R. and several other countries, were discussed during a Moscow radio broadcast to Germany. The spacecraft would take off vertically and consist of two winged stages. The 1st stage would separate from the 2nd at a 120-km (75-mi) altitude and start to descend on a ballistic course. At 30 to 40 km (19 to 25 mi) from the earth the engine and steering mechanism would switch on to control an aircraftlike descent to an airfield. The 2nd stage would accelerate after separation and move into orbit at 200 km (124 mi) to dock with an orbital station, unload, and reload. After separation from the station, the 2nd stage would leave orbit and also return to the earth. (FBIS-Sov, 7/11/73, L1)
- NASA announced the appointment of Dr. Alois W. Schardt as Director of Physics and Astronomy Programs in the Office of Space Science. Schardt, who had been Deputy Director since 1970, succeeded Jesse L. Mitchell, who was retiring after 26 yrs with NASA and the predecessor National Advisory Committee for Aeronautics (NACA). (NASA Release 72-124)

July 2
- NASA launched an Aerobee 170 sounding rocket from White Sands Missile Range carrying a Columbia Univ. x-ray astronomy experiment to a 175.9-km (109.3-mi) altitude. The rocket performed satisfactorily but the telemetry data were severely compromised by radio frequency interference. (GSFC proj off)
- The Air Force was redesigning an F-15 jet fighter engine part for the third time, possibly creating a cost-delay in the $7.5-billion program, the Washington *Evening Star and News* reported. The Chief F-15 Project Engineer, L/C Frederick A. Rall, had said in an interview that the aircraft could perform satisfactorily at speeds and altitudes required for air-to-air combat and that the engine part redesign was to improve performance even further. Other sources had said the redesign had to succeed before the aircraft could perform as a fighter. (Kelly, W *Star & News*, 7/2/73)
- A Federal District Court jury in Fort Worth, Tex., found General Dynamics Corp. innocent of Government charges that it had conspired to "cheat" the Air Force out of $114 000. The charges had led to a May 30, 1972, indictment. The Government had said General Dynamics had destroyed $114 000 worth of flawed F-111 aircraft components produced by the Selb Manufacturing Co. and then filed a claim with the Air Force for repayment. The jury accepted General Dynamics' claim that it would not have tried to cheat the Air Force out of $114 000, since it had been working under a $4.5-billion contract to produce the aircraft. (AP, *W Post*, 7/3/73, A1)

July 2-14: European scientists participated in a National Academy of Sciences Summer Study on the use of the space shuttle and Spacelab, at Woods Hole, Mass. (NASA Release 73-191)

July 3: NASA announced it had negotiated cost-sharing, no-fee contracts for the ground-and-flight-test phase of its program to reduce jet engine noise of DC-9 and Boeing 727 aircraft in U.S. commercial service. United Aircraft Corp. Pratt & Whitney Div., McDonnell Douglas Corp. Douglas Aircraft Co., and Boeing Co. had completed design studies showing the engines could be quieted using existing technology without degrading engine or aircraft performance. In Phase 2 of the program, they would develop and test a modified JT8D engine with a larger single-stage fan to reduce exhaust velocity, incorporating acoustic treatment to muffle fan noise in new nacelles for the refan engines, and adding two booster stages to the low-pressure compressor to maintain proper airflow to the engine core.

Pratt & Whitney would receive $14.6 million to modify six engines, make analyses and design studies of the JT8D engine and components, fabricate engine-modification hardware, and support tests by the airframe manufacturers. Boeing Co., under a contract to be signed later, would study design of the sound-absorbing nacelle, identify modifications to the 727 aircraft to accommodate the nacelle, ground-test refanned 727 aircraft, and evaluate nacelle configurations with varying acoustic treatments. Douglas Aircraft would receive $6.9 million to develop experimental installation of the refanned JT8D engine on DC-9 aircraft and to flight-test the refanned DC-9. The contracts were for two years. First flight tests of refanned engines on DC-9s were scheduled for early 1975. (NASA Release 73-126)

- President Nixon submitted to the Senate the nomination of Under Secretary and Acting Secretary of the Air Force John L. McLucas to be

Secretary of the Air Force succeeding Dr. Robert C. Seamans, Jr., who had resigned in May to become President of the National Academy of Engineering. McLucas had been Deputy Director of Defense Research and Engineering 1962–1964 and Executive Secretary General for Scientific Affairs of the North Atlantic Treaty Organization (NATO) in Paris 1964–1966.

The President also submitted to the Senate the nomination of Gen. George S. Brown to be Chief of Staff, U.S. Air Force, for four years beginning Aug. 1. Gen. Brown—to succeed Gen. John D. Ryan, who would retire July 31—had been Commander of the Air Force Systems Command since September 1970.

Both nominations were confirmed July 14. (*PD*, 7/9/73, 884; 7/16/73, 903; *CR*, 7/14/73, D831)

- Dr. Philip Handler, President of the National Academy of Sciences, told newsmen in Moscow that he had informed President Mstislav V. Keldysh of the Soviet Academy of Sciences that there would be a time when the U.S. would need "more definitive information" about the Soviet space program. Dr. Handler had referred to the 1975 joint U.S.–U.S.S.R. Apollo-Soyuz Test Project. Concluding a two-week tour of Soviet scientific establishments as head of an NAS delegation, Dr. Handler said U.S. science was "a cut ahead" of Soviet science but the distance between the countries was "not very great." (Reuters, *W Post*, 7/4/73, A24)
- Twelve sensors mounted on an Air Force NC–135 aircraft flying 11 km (7 mi) above the Sahara desert had found no evidence of the shadow bands which had swept over the landscape just before the June 30 total solar eclipse, the *New York Times* reported. A Canadian aircraft flying in the troposphere, the lowest region of the atmosphere, had seen the bands—one of the most unusual manifestations of the eclipse—clearly and had photographed them on the aircraft's wing. Astronomers had theorized that the bands resulted from refraction of the narrow slit of sunlight that shone through atmospheric layers when the sun was almost totally eclipsed. Another theory was that the sunlight had been defracted by the edge of the moon. (Sullivan, *NYT*, 7/3/73, 7)

July 3–6: The United Nations Educational, Scientific and Cultural Organization (UNESCO) and the French government sponsored a conference, "The Sun in the Service of Mankind," in Paris. More than 600 international scientists discussed solar power as an answer to man's energy needs. A working group of 13 scientists from developed and developing countries suggested that "a massive internationally funded and directed research and development programme for the attainment of specific objectives, should receive wide discussion among scientists in [UNESCO] member states." The group asked whether the development and exploitation of satellites to collect solar energy above the atmosphere should be left to one country "or should there be a cooperative worldwide attack on the problem with a provision for beaming the collected energy to receiving stations in many places in the world?"

U.S. energy experts reiterated recommendations made by NASA and the National Science Foundation in their December 1972 report *Solar Energy Research: A Multidisciplinary Approach*. NASA had recommended that the U.S. undertake a 10- to 15-yr program to develop the use of solar energy to generate electricity. NASA and the NSF had called for Federal investment of $3 billion in solar energy research. NSF scientist William H. Wetmore told the conference that the U.S. estimated its deficit in oil supplies alone would reach from 10 to 16 million barrels a day by 1990 and would cause a trade imbalance of $15 billion annually.

U.S. industry representatives reported their progress with solar climate control research to use solar energy for heating houses and factories. *Washington Post* columnist Claire Sterling commented later on the impact of the industry reports: "Perhaps more than any single factor at this conference, the willingness of these big American companies to actually put money into solar research and development has brought about a stunning change in thinking. From an essentially 'do-good' movement of scientists hoping to help the poor countries . . . with cheap solar cookers there has suddenly emerged a rich-state research and development movement on a supremely sophisticated technological level." (Lewis, *NYT*, 7/2/73, 12; *New Scientist*, 7/12/73, 71–2; *W Post*, 7/16/73, A22; *A&A 1972*)

July 4: L/C Charles H. Manning (USAF, R.) flew his HU–16B Albatross aircraft from Homestead Air Force Base, Fla., to a new record altitude for an amphibian aircraft, 9600 m (31 500 ft). The previous record had been held by Boris Sergievsky, who flew a Sikorsky S–43 amphibian from Stratford, Conn., to 7600 m (24 950 ft) April 14, 1936. (NAA *News*, 8/73, 2)

- A Kennedy Space Center ceremony celebrated the Center's selection by the Bicentennial Commission of Florida as a historic site on the Bicentennial Trail. Florida Lt. Gov. Tom Adams presented a Bicentennial flag to NASA. (KSC Release 147–73; KSC PIO)

July 5: The Air Force Space and Missiles Organization (SAMSO) announced it was issuing a $1 334 319 cost-plus-fixed-fee contract to TRW Systems Group, TRW Inc., to maintain a history of all operational testing of Minuteman missiles. (DOD Release 341–73)

- The U.S.S.R. had launched meteorological sounding rockets from the Soviet research ship *Academician Korolev* at the 180° meridian, Tass reported. The rockets had obtained data for a study of the interaction of the high-altitude atmosphere and the ocean on the frontier of the Eastern and Western Hemispheres. (FBIS–Sov, 7/20/73, L1)

July 5–6: NASA test pilot John A. Manke made ground taxi tests of the X–24B lifting body at Flight Research Center to check for nosegear shimmy and evaluate nosewheel steering and ground handling characteristics. The successor to lifting bodies flown in a joint NASA and Air Force program between 1966 and 1972 would undergo ground functional tests of its main and subsystems before scheduling of its first flight. A modification of the X–24A, the new vehicle had an extended nose, flattened underside, and small blended wings. The NASA–USAF program was developing technology for future aircraft that could cruise at hypersonic speeds at the edge of space. (FRC *X–Press*, 7/6/73, 2; NASA photos 72–H–1395, 73–H–722)

July 6: An automated water-sampling station developed at Langley Research Center for use in remote marshlands had been made available to the Virginia Institute of Marine Science, the *Langley Researcher* reported. The prototype sampler could be deployed and retrieved by helicopter. (*Langley Researcher*, 7/6/73, 1)

July 8: Loading of the Apollo command and service module's hypergolic propellants was completed at Kennedy Space Center in preparation for the July 28 launch of Skylab 3. (KSC Release 163–73)

- Miami chemist Harry Bennett received U.S. Patent No. 3 738 374 for a tobaccoless cigarette filled with graphitized carbons developed by NASA for space use. The carbon, which would emit normal combustion gases and produce an ash without nicotine or tar, had been used by NASA to resist heat and strengthen lightweight structural components in spacecraft. (Pat Off PIO; B Sun, 7/17/73)

July 9: NASA announced it had invited U.S. and foreign scientists to propose experiments to be carried on the Apollo spacecraft during the July 1975 Apollo-Soyuz mission to rendezvous and dock U.S. and Soviet spacecraft in space. The U.S. would carry 180 kg (400 lbs) of scientific and applications experiments and NASA also would conduct experiments from the earth orbit in space science, applications, medicine, and technology. Proposals for experiments were due July 23. (NASA Release 73-127)

- Boeing Co. announced it was preparing preliminary designs for a multipurpose vehicle that could skim across Arctic ice and snow at high speeds under a $771 000 Navy contract. The arctic surface effects vehicle (ASEV) would operate like an air cushion vehicle but would be larger, faster, and able to cope with extremely harsh terrains and environments. It could be used as a mobile scientific station, cargo carrier, and a vehicle for military missions. (Boeing Release A-0448)

- Several U.S. correspondents had been permitted by the U.S.S.R. to visit the Soviet cosmonaut training center at Zvezdny Gorodok (Star City) near Moscow, *Time* magazine reported. *Time* reporters included in the tour had quoted Maj. Gen. Vladimir A. Shatalov, chief of cosmonaut training, as saying that similarities rather than differences predominated in the U.S. and U.S.S.R. approach to manned space flight. The similarities included a preference for jet pilots as spacecraft commanders; "training is shaped by requirements, just as the shape of the aircraft is decided by its speed." Shatalov had recalled watching Skylab astronauts practicing in a simulator. "It was the same as we simulate here, not more, not less. We are treading the same paths." (*Time*, 7/9/73)

- An *Aviation Week & Space Technology* editorial saluted Skylab as "a triumph of man's ingenuity, endurance and indomitable spirit." The May 14-June 22 *Skylab 1-2* mission had been "a tremendous demonstration of why man is vital to the broad expansion of space missions that looms for the next decade, and it demonstrated on live television for every taxpayer who cared to watch exactly what the space shuttle is all about." The first Skylab crew had salvaged 80% of a successful mission and a $2.5-billion investment from the prospects of total disaster, opened new vistas on the scientific work that could be accomplished from a space platform, and proved that man could live and work usefully for extended periods in space. (Hotz, *Av Wk*, 7/9/73, 7)

July 9-20: Apollo-Soyuz Test Project specialists from NASA and the Soviet Academy of Sciences met at Johnson Space Center. They continued development of trajectories and flight plans, agreed tentatively to a scientific experiment plan, and familiarized the Soviet crew of the July 1975 mission with the Apollo spacecraft. Participants included ASTP technical directors Glynn S. Lunney for the U.S. and Academician Konstantin D. Bushuyev for the U.S.S.R., Apollo and Soyuz flight crews, the Working Group on Mission Model and Operational Plans, and members of the Working Group on Control and Guidance. The Soyuz cosmonauts [see May 24] attended lectures on and demonstrations of the Apollo spacecraft, docking model briefings, and crew activity discussions. Lunney and Bushuyev reviewed ASTP milestones and reaffirmed that they were on schedule.

On July 14 the Soviet delegation heads and cosmonauts toured assembly and checkout facilities at the Rockwell International Corp. plant in Downey, Calif., where ASTP work was being carried out.

Meeting results were announced by NASA in a July 20 communique. Progress had been reported on crew activities, control center operations, trajectories, and other operational aspects. Agreements were reached on November familiarization of the U.S. flight crews with Soyuz equipment in the U.S.S.R., a preliminary cosmonaut training plan in the U.S. in April 1974 and February 1975, and astronaut training in the U.S.S.R. in July 1974 and March 1975. The training plan was expected to be completed in November and flight procedures finalized by the end of 1974. Final selection of joint experiments would be made in October 1973. Flight safety reports would be exchanged. The U.S.S.R. had specified the Moscow Center of Control of Manned Space Flight near Moscow as ASTP control center and had announced that Cosmonaut Aleksey S. Yeliseyev would be U.S.S.R. flight director. The U.S. technical director and several U.S. working groups would attend U.S.S.R. ASTP meetings in October. (NASA Release 73-93; NASA Communique)

July 10: Dr. James C. Fletcher, NASA Administrator, announced that future spacecraft technical requirements would permit a phasedown beginning in mid-1974 of a Deep Space Network (DSN) facility and a Spaceflight Tracking and Data Network (STDN) facility—NASA's two major tracking and data-acquisition facilities near Johannesburg, South Africa. The 64-m (210-ft) antenna facilities at Goldstone, Calif., and the new large-dish facilities in Canberra, Australia, and Madrid, Spain, with smaller 26-m (85-ft) antennas at these stations, would be used for the new planetary spacecraft probing further into the solar system and beyond after June 1974. (NASA Release 73-128)

• President Nixon announced the designation of Dr. H. Guyford Stever, National Science Foundation Director, as Chairman of the Federal Council for Science and Technology. Dr. Stever, who had been Acting Chairman since May 14, would also serve as Presidential Science Adviser, pursuant to the President's Jan. 26 Reorganization Plan No. 1 of 1973, which abolished the Executive Office of Science and Technology effective July 1. A former President of Carnegie-Mellon Univ., Dr. Stever had been NSF Director since 1972. The Federal Council for Science and Technology had been established by Executive Order 10807 of March 13, 1959, to recommend measures for effective implementation of Federal policies on the administration and conduct of Federal programs and science and technology. (Off of White House Press Secy Press Notice)

• Sen. Frank E. Moss (D-Utah), Chairman of the Senate Committee on Aeronautical and Space Sciences, addressed a joint symposium on space program planning of the American Institute of Aeronautics and Astronautics, the American Society of Mechanical Engineers, and the Society of Automotive Engineers in Denver, Colo. Points paramount for the U.S. space program, in his mind, were: "*First,* despite a year with all the trappings of potential disaster, NASA is coming through in pretty good shape. *Second,* now is the time for space planners—inside and outside the NASA family—to get serious about how best to use the space shuttle. *Third,* the single most important need for a healthy future space program is a substantially higher NASA budget request for Fiscal Year 1975. *Fourth,* Congress is willing to leave overall planning for future space options to the Executive Branch, but cavalier disregard for specific congressional decisions—such as we witnessed last January—is not likely to be so lightly accepted in the future." (*CR,* 7/14/73, S13479)

- A NASA investigation board had found an electrical malfunction to be the major factor in the May 10, 1972, accident over Bergstrom Air Force Base, Tex., in which Astronaut Charles Conrad, Jr., had ejected safely from a T-38 jet trainer aircraft, Johnson Space Center announced. The malfunction had caused loss of instruments in severe weather and the aircraft had been destroyed. (JSC Release 73-91)
- U.S. Patents Nos. 3 744 480 and 3 744 794 were granted to Marshall Space Flight Center physicist Raymond L. Gause and engineers Raymond A. Spier and Bobby G. Bynum. They had invented the ergometer installed in *Skylab 1* (launched May 14) to evaluate the astronauts' muscular output and heart rate and also the harness that held the astronaut to the ergometer in zero-g or earth environment. (Pat Off PIO; Jones, *NYT,* 7/14/73, 31)

July 10-11: The NASA Committee on Remote Manipulator Systems and Extravehicular Activity met at Marshall Space Flight Center to review technology and discuss using remote manipulator systems and extravehicular activity in the space shuttle, Large Space Telescope, and space tug projects. Representatives of the Army, Navy, and Air Force attended. (MSFC Release 73-90; *Marshall Star,* 7/11/73, 4)

July 11: The U.S.S.R. launched *Molniya II-6* communications satellite from Plesetsk into orbit with a 39 284-km (24 410-mi) apogee, 441-km (274.0-mi) perigee, 705-min period, and 65.4° inclination. The satellite would help provide a system of long-range telephone and telegraph radio communications in the U.S.S.R. and would transmit Soviet central TV programs to the Orbita network. Western observers said later that *Molniya II-6* also might transmit communications during a manned mission. Reports that Soviet tracking ships used for manned space flights had been sighted heading for their stations in the Atlantic possibly indicated an upcoming manned launch. (GSFC *SSR,* 7/31/73; FBIS-Sov, 7/12/73, L1; *SBD,* 7/12/73, 58; *NYT,* 7/12/73, 4)

- The House agreed to the conference report on H.R. 7528 recommending a $3.056-billion FY 1974 NASA authorization [see June 28]. The bill was cleared for President Nixon's signature. (*CR,* 7/11/73, H5934)
- Publication of *Federal Plan for Meteorological Services and Supporting Research, Fiscal Year 1974* was announced by the National Oceanic and Atmospheric Administration. FY 1974 programs would cost $504 659 000, an increase of $22 032 000 over FY 1973. The increase would be used to employ computers, satellites, ocean buoys, automatic weather stations, remote sensing, and automated data-handling techniques to improve weather services with reduced manpower. NOAA would increase its expenditures for weather radars, satellite operations, new computers, and expanded weather-dissemination systems. The Dept. of Defense would add more weather radars, expand its satellite data-processing programs, and install advanced instrumentation aboard storm reconnaissance aircraft. The DOD surface and upper-air weather observing program would be decreased because of curtailed military activities.

July 11   The Federal Aviation Administration planned 100 more unmanned Flight Service Stations to provide weather information for pilots. The FAA's En Route Flight Advisory Service would be expanded from 4 to 25 stations. NASA launch of two Synchronous Meteorological Satellites, prototypes for the Geostationary Operational Environmental Satellites (GOES), would begin a new phase in the environmental satellite program. Subsequent GOES spacecraft would be funded by NOAA. Both NOAA and DOD were readying ground facilities for the GOES system. NOAA's series of polar-orbiting operational satellites would be maintained with launches scheduled as needed. ITOS-E was to be launched by NASA in July to become *Noaa 3*. (NOAA Release 73–150)

- The Senate agreed by a vote of 82 to 10 to S. Res. 71 to seek agreement with other governments to prohibit the use of environmental or geophysical modification activity as a weapon of war. (*CR*, 7/11/73, S13101–02)

July 12: Dr. James C. Fletcher, NASA Administrator, outlined potential uses of space age technology before the Western Assn. of State Game and Fish Commissioners in Salt Lake City, Utah. He said earth surveying could find hidden resources, such as formations likely to have petroleum beneath them. (Evans, *Desert News*, 7/12/73)

- A European Space Conference in Brussels failed to reach agreement on future programs, including the establishment of a European space agency. *Spaceflight* magazine later attributed the failure to divergent interests. The U.K. was concentrating on its maritime satellite, West Germany on the Spacelab, and France on the L–3S launcher. Sweden and the Netherlands required more time, and France was disappointed that the U.K. would not join West Germany in supporting the L–3S project. The meeting would reconvene in Brussels July 31 to meet NASA's Spacelab decision deadline of Aug. 10. (*SF*, 9/15/73, 321)

- President Nixon transmitted *The World Weather Program Plan for Fiscal Year 1974* to Congress. U.S. participation in the world program, coordinated by the National Oceanic and Atmospheric Administration, included the Depts. of Defense, State, and Transportation; Atomic Energy Commission; Environmental Protection Agency; National Science Foundation; and NASA. In addition to programs presented in NOAA's Federal Plan [see July 11], the Global Atmospheric Research Program (GARP) would conduct the GARP Atlantic Tropical Experiment (GATE) June 15–September 1974 to study tropical atmosphere meteorology and its effects on circulation of the earth's atmosphere. And the first GARP Global Experiment, scheduled for 1977, would include two polar-orbiting satellites, one provided by the U.S. and one by the U.S.S.R., and five earth-synchronous spacecraft—two operated by NOAA for the U.S., one over the Indian Ocean by the U.S.S.R., one over the western Pacific by Japan, and one over Africa and the eastern Atlantic by the European Space Research Organization (ESRO). (NOAA Release 73–168; *PD*, 7/16/73, 900)

- The B–1 development program director, M/G Douglas T. Nelson (USAF), told the press in Washington, D.C., that construction of the swing-wing strategic bomber had fallen behind schedule, forcing a nine-month postponement of a decision on whether to produce B–1s at an estimated $54 million each. The decision date had been rescheduled for May 1976. First flight of a B–1 prototype had been rescheduled from April to June 1974. B–1 prime contractor Rockwell International Corp. had been "falling noticeably behind schedule" in installing subsystems because it had had to spend more time than anticipated on assembling the airframe. By slowing the development program the Air Force hoped to avoid a program cost increase during the current fiscal year, but development problems would result in a $78-million total increase, to bring the total development cost to $2.79 billion. (*WSJ*, 7/13/73)

July 13: The Air Force launched an unidentified satellite from Vandenberg Air Force Base by a Titan IIID booster into orbit with a 273.6-km (170-mi) apogee, 154.5-km (96-mi) perigee, 88.7-min period, and 96.2° inclination. Press reports termed the spacecraft a "Big Bird" photo-reconnaissance satellite. The satellite reentered Oct. 12. (Pres Rpt 74; *Av Wk*, 8/27/73, 30)

- The *Skylab 1* Investigation Board appointed to determine the cause of the anomaly that lost the Orbital Workshop's meteoroid shield and one solar array wing during launch May 14 reported its findings and recommendations to Dr. James C. Fletcher, NASA Administrator. The Board, chaired by Lewis Research Center Director Bruce T. Lundin, said the most probable cause was inadequate venting of the pressure in a tunnel beneath the shield. Differential pressure buildup in the tunnel had forced the shield away from the Workshop shell and into the supersonic air stream. The shield then broke the tiedowns holding the solar array wings onto the spacecraft. About 10 min into the flight one solar array wing was torn away completely when it was struck by the exhaust plume of the 2nd-stage retrorockets. The remaining solar array wing would not deploy until the astronauts cut the strap which had curled around the wing and penetrated the metal fairing housing the array.

  The Board had found that "the significance of the aerodynamic loads on the meteoroid shield during launch was not revealed by the extensive review process" and recommended that complex, multidisciplinary systems like the shield "should have a designated project engineer who is responsible for all aspects of analysis, design, fabrication, test and assembly." There was no evidence that the shield's design deficiencies had resulted from or been masked by the Skylab management system. "On the contrary, the rigor, detail, and thoroughness of the system are doubtless necessary for a program of this magnitude." (NASA Release 73-135; Lundin testimony before Sen Com on Aero and Space Sci, 7/30/73)

July 14: Sweden had handed the 25-nation Geneva disarmament conference a list of 925 nuclear weapon tests carried out from World War II through April 25, 425 of them after the August 1963 signing of the treaty banning all experimental nuclear blasts above ground, Reuters reported. The majority of the 425 had been underground tests, but China and France, which did not sign the treaty, had continued atmospheric nuclear testing. (*NYT*, 7/15/73, 7)

July 15: The countdown for the July 28 launch of Skylab 3 began at Kennedy Space Center. A Saturn IB launch vehicle would boost the Apollo spacecraft and its three-man crew into earth orbit to rendezvous and dock with the Skylab Orbital Workshop (launched May 14), where astronauts would live and work 56 days in space. The countdown was the first for a manned Saturn launch in which the flight crew would not participate with the KSC launch team in a dress rehearsal. The rehearsal had been eliminated because of the performance record of the Saturn IB and V. The countdown would include a simulated T-zero, ignition, and liftoff with a fully fueled launch vehicle on Complex 39, Pad B. (KSC Release 163-73)

- A 1-million-cu-m (36-million-cu-ft) balloon—launched for NASA from Fort Churchill, Canada, by the U.S. Office of Naval Research—reached its planned 46 000-m (150 000-ft) altitude, but failed to land in western Manitoba as planned when its descent system did not react to radio commands from the ground. The balloon, carrying a 408-kg (900-lb) scientific payload that included a cosmic ray detector to measure distribution of electrons and positrons in primary cosmic rays, continued to drift over the Pacific Ocean. It entered Soviet airspace July 18 and tracking data later indicated that it had landed in eastern Siberia. (NASA Release 73-195)

- Janet Lee, 17-yr-old representative of the Republic of China (Taiwan) on the December 1972 NASA-conducted International Youth Science Tour of America, had written to tour coordinator Mrs. Lillian Levy saying she could sometimes see Skylab "flying over my head like a little sparkling star in the evening," *NASA Activities* reported. Mrs. Levy also had heard from tour participant Karl Muller of Mbabane, Swaziland: "It is very pleasing to see that Swaziland is enlisting the aid of satellites in agriculture. The idea of using satellites for detecting forest fires is particularly interesting, since Swaziland has the world's largest man-made forest." (*NASA Activities*, 7/15/73, 116)

July 16: ITOS-E Improved Tiros Operational Satellite, launched by NASA for the National Oceanic and Atmospheric Administration, failed to reach orbit after a 10:10 am PDT liftoff from Western Test Range on a two-stage, long-tank, thrust-augmented Thor-Delta booster. Early flight data indicated that abrupt cessation of the hydraulic pump output 270 sec after 2nd-stage ignition had caused loss of hydraulic pressure and thrust vector control. The spacecraft tumbled and failed to achieve orbital velocity. Investigation was begun immediately at Goddard Space Flight Center.

The meteorological spacecraft had been intended for sun-synchronous orbit to make day and night cloud-cover observations, to make continuous observations of weather features from its synchronous altitude, and to obtain global-scale quantitative measurements of the earth's atmospheric structure.

ITOS-E, which was to have been christened *Noaa 3*, was the fourth spacecraft in a series of second-generation meteorological satellites. *Noaa 1* (ITOS-A) was launched Dec. 11, 1970, and was operational for nine months. ITOS-B was launched Oct. 21, 1971, but did not achieve orbit because of a Delta 2nd-stage failure. *Noaa 2* (ITOS-D) was launched successfully Oct. 15, 1972, and had been NOAA's primary operational spacecraft. (NASA prog off)

- Marshall Space Flight Center issued a request for proposals for space shuttle solid-fueled rocket motor (SRM) development to Aerojet Solid Propulsion Co., Lockheed Propulsion Co., Thiokol Chemical Corp., and United Technology Center. Technical proposals were due Aug. 27 and cost proposals Aug. 30. The program would include increments for design, development, and test (including production of sufficient hardware for six development flights); for production of new and refurbished SRMs for 54 flights; and for delivery of new and refurbished units for 385 flights. (NASA Release 73-133)

- NASA announced the appointment of Harold E. Pryor to succeed Dr. Dudley G. McConnell as Director, Scientific and Technical Information Office, effective July 22. Dr. McConnell had been named Assistant Administrator for NASA's Equal Opportunity Programs. Pryor, Management Systems Office Director, had joined NASA in 1964. (NASA Hq *WB*)
- A lunar sample brought back from the moon by *Apollo 17* (Dec. 7–19, 1972, lunar mission) was presented to President Henryk Jablonski of the Polish Council of State by U.S. Ambassador to Poland Richard T. Davies in Warsaw. Ambassador Davies invited Poland to participate in the U.S. Bicentennial celebration in 1976. (Warsaw PAP, FBIS–Poland, 7/17/73, G1)
- The U.S.S.R. announced that its atomic breeder reactor at Shevchenko on the Caspian Sea had started commercial operation. It had been undergoing tests since its startup in December 1972. The power-generating and water-desalinization facility—first power source of its kind—would yield 150 000 kw of electrical power and convert 113 400 cu m (30 million gal) of sea water into fresh water daily on reaching its full design capacity of 350 000 kw. (Shabad, *NYT*, 7/17/73, 9)
- Prospects were immense for solar energy practical for area heating and cooling, "but not yet for central power stations," a *New York Times* editorial said. "The R&D work in this field is more one of spreading the word than inventing the devices, for solar equipment is technologically at hand and in use in other countries." The greatest long-term promise would be in generating power through nuclear fusion. "The agreement with the Soviet Union to coordinate both superpowers' fusion research, concluded during Mr. Brezhnev's visit to Washington, could give long-needed impetus to development of this energy source even before the end of the century." (*NYT*, 7/16/73, 28)
- Capt. John T. Geary (USN) became Director of the Naval Research Laboratory, succeeding Capt. Earle W. Sapp who was reassigned. Capt. Geary had been Deputy Commander for Planning, Programming and Resources with the Naval Electronics System Command. (*Naval Research Reviews*, 8/73, 27; NRL PIO)

July 17: Skylab Astronauts Charles Conrad, Jr., Dr. Joseph P. Kerwin, and Paul J. Weitz testified on the May 14–June 22 *Skylab 1–2* mission in a joint hearing before the Senate Committee on Aeronautical and Space Sciences and the House Committee on Science and Astronautics. Conrad said the mission had proved the feasibility of maintaining a large space station for manufacturing in space. "I think the data will show that you can do all the things up there that you can do down here and many of them perhaps better." *Skylab 1–2* earth resources and solar telescope data would show "that eventually space has a very definite role in helping mankind and in solving some of these problems that are in front of us. After all, the Sun is probably the most single efficient nuclear device that we know about. We do have an energy crisis. We need to better understand that. We just scratched the surface with our 30,000 photographs we brought back. I think you will find that the Earth resources data, when coupled with . . . ERTS [Earth Resources Technology Satellites] and those types of sensors will . . . as an overall systems design, tie all of these things we talk about with different satellites into a big picture." At the mission's completion the *Skylab 1–2* crew could hand over to the Skylab 3 crew a Workshop that could harbor "a 100-percent successful 56-day mission."

Dr. Kerwin said *Skylab 1–2* medical experience enabled him to extrapolate a 56-day mission "without difficulty." He felt the inflight portion would be "a piece of cake." Postflight, the Skylab 3 astronauts might "feel a bit loggier than we did." But *Skylab 1–2* had demonstrated that this was a "minor illness . . . that you get from being 28 days in space and that you can return very rapidly to normal."

Weitz said Skylab's earth resources survey equipment permitted a more selective data acquisition than that of unmanned ERTS. "I think we have shown in the past with unmanned satellites that part of the problem is that you have a continuous data flow, not all of which is usable, but all of which must be processed and reduced and analyzed in some form." With Skylab equipment "we can track point sites" and "be selective . . . pick out specific sites, fields, or a bend in the river, if you want to study silting." (Transcript)

- NASA launched a Javelin sounding rocket from Wallops Station carrying a Univ. of Pittsburgh aeronomy experiment to a 724.2-km (450-mi) altitude. The rocket and experiment performed satisfactorily. (GSFC proj off)
- The U.S. defense outlay continued to increase despite improving relations with the U.S.S.R. and the People's Republic of China, the U.S.–U.S.S.R. agreement to limit strategic arms, the Vietnam cease-fire agreements, and sharp cutbacks in U.S. military forces, the *Wall Street Journal* noted. FY 1974 spending was projected at $79 billion, up $4.2 billion from FY 1973, "a shade above the Vietnam-war peak and a shade below the World War II pinnacle. While the White House talks detente, the Pentagon speaks of growing Soviet military power and the likelihood, given further price and pay inflation, of $100 billion-plus military budgets by the end of the decade." (Levine, *WSJ*, 7/17/73, 1)

July 17–18: NASA launched a series of 26 meteorological experiments on rockets and balloons from Wallops Station to provide an in-depth comparison between remote sensors on satellites and sensors on rockets and balloons from a mid-latitude site. The project—divided into a day and night series to provide data both with and without solar energy input—was conducted in conjunction with *Nimbus 5* (launched Dec. 10, 1972) and *Noaa 2* (launched Oct. 15, 1972) satellites.

Four Nike-Cajun sounding rockets were launched carrying Goddard Space Flight Center acoustic grenades to measure temperature and winds. The grenades were ejected and detonated in flight, one at a time, from 30- to 95-km (19- to 59-mi) altitude. The first Nike-Cajun was launched July 17 and reached an altitude of 117.4 km (72.9 mi). The rocket performed satisfactorily. Thirty of the thirty-one grenades were successfully detonated and the sound returns were received by the ground microphone array.

The second Nike-Cajun launched 79 minutes later reached an altitude of 116.2 km (72.2 mi) and detonated all of its 12 grenades successfully.

The third Nike-Cajun was launched July 18 to a 114.9-km (71.4-mi) altitude. Rocket performance was satisfactory and all 31 grenades were successfully detonated.

The final Nike-Cajun was launched two hours later to an altitude of 120.7 km (75 mi). The rocket performance was satisfactory and all 12 grenades were detonated successfully.

In addition to the Nike-Cajuns, the launches included 4 balloonsondes, 3 Viper Darts, and 15 Super Lokis instrumented to obtain density and temperature data. Participating in the program with Wallops Station were White Sands Missile Range, the Air Force, and GSFC. (WS Release 73-7; NASA Rpts SRL)

July 17, 19, 23, 24: The House Committee on Science and Astronautics held hearings on Federal policy, plans, and organization for science and technology. Dr. H. Guyford Stever on July 17 explained his dual role as National Science Foundation Director and, as of July 1, Science Adviser to the President: "I am aware of the need to maintain a strong, independent position as Science Adviser. I do not believe there will be insurmountable difficulties in dealing with this challenge. The newly established Science and Technology Policy Office will be the strong staff arm to help me carry out these new duties. . . . . I have structured this office to maintain a maximum degree of objectivity and impartiality on science policy matters. . . . I intend to call upon the significant and broad capabilities existent within the Foundation. . . . I will also call upon other Government agencies for assistance whenever appropriate."

Dr. Edward E. David, Jr., Executive Vice President for Research and Development and Planning with Gould, Inc., and former Presidential Science Adviser, on July 24 commented on NSF's new role as "the Government's highest level policy and advisory body with technical competence." He saw the new arrangement at NSF as unstable, with "formidable and demanding" tasks. He saw two possible resolutions: NSF could restrict itself to science and academic research or it could transcend its historic past and achieve the national stature necessary to influence agencies and departments on technical programs that cut across operating boundaries and to serve as the technological beacon for other agencies.

Science, Dr. David said, had not been downgraded. "Science and technology will continue to be the warp and woof of our society," but the national scene had changed. The former White House apparatus, 1957-1965, had been "more weighted toward space and military matters. Importantly, the apparatus was also responsible for development of programs to strengthen the infrastructure of science and engineering. . . . In recent years we have seen the emphasis shifting toward consumer and public-oriented technologies in energy, transportation, health, education, natural resources, ecology and environment, and social systems. This shift brings with it a powerful new set of policy issues which illustrate the necessity to include not only science and technology, but also economic, social, legal, and political factors"—consumer markets, public preferences, and political beliefs, "matters with which the Federal establishment has little experience or expertise. It is in such matters that the new NSF office must exert leadership and influence." (Transcript)

July 18: RCA Corp. announced that its subsidiaries RCA Global Communications, Inc., and RCA Alaska Communications, Inc., had signed a $3 750 000 contract with McDonnell Douglas Corp. to develop an uprated Thor-Delta launch vehicle to launch domestic satellites. It was the first time private industry had set design requirements and provided funds to modify a launch vehicle. Under contract terms, the Thor-Delta payload capacity would be augmented by 30% to enable RCA to place high-capacity domestic communications satellites into geostationary orbit above the equator economically. (RCA Release)

- Langley Research Center engineers and technicians were cooperating with investigators from West Virginia Univ. in a summer study to improve rural West Virginia housing with space age technology. The preliminary study was to lead to eventual construction and demonstration of a self-contained house or small community, "Space Station West Virginia," incorporating modern technology to supply energy, water and sewage, fire control, and systems analysis. (NASA Release 73-132)

July 18–27: The American Institute of Aeronautics and Astronautics sponsored an aeronautical technology display at the Soviet Exhibition of Economic Achievement in Moscow. AIAA also cosponsored, with the U.S.S.R. Committee for Science and Technology, a joint Symposium on Aeronautical Technology at the exhibition July 23 to 27. Advanced equipment to modernize civil air traffic control systems was displayed or depicted by 27 U.S. companies and reviewed by an estimated 25 000 aviation specialists from the U.S.S.R. and its East European allies. The joint symposium—opened by Federal Aviation Administrator Alexander P. Butterfield and the Soviet Minister of Civil Aviation, Gen. Boris P. Bugaev—covered air traffic control, airport and air system facilities, and new aircraft technology. Among participants were members of the House Committee on Science and Astronautics' Subcommittee on International Cooperation in Science and Space. (AIAA Release 7/20/73; Program; Shabad, *NYT*, 7/22/73, 43)

July 19: NASA's X-24B lifting body [see July 5–6], flown by NASA test pilot John A. Manke, successfully completed its first captive flight from Flight Research Center, attached to a B-52 aircraft. Erratic B-52 power resulted in unintelligible telemetry traces and forced a switch to X-24B internal power. The telemetry cleared and excellent traces were returned. X-24B performance was adjudged satisfactory but the scheduled glider flight was postponed until July 24 to allow for battery recharge.

The X-24B had been built around the existing basic structure and subsystems of the X-24A, which, like other lifting bodies, had been designed for reentry from space flight. But the X-24B had a new double-delta shape with small blended wings and three vertical tails—parts of a configuration representing a forerunner of future aircraft capable of hypersonic (above mach 5) flight. Like its lifting-body predecessors, the X-24B would be air-launched by a B-52 aircraft from 12 000-m (40 000-ft). Subsequent flights would use the XLR-11 rocket engine to increase speed and altitude performance in the joint NASA–USAF program at Flight Research Center. (NASA prog off; NASA Release 73-130)

- Samples from three materials-processing experiments performed during the *Skylab 2* mission (manning the Orbital Workshop May 25 to June 22) had been turned over to the principal investigators at Marshall Space Flight Center, MSFC announced. The small metallic spheres and welding specimens represented the first effort to obtain basic engineering and scientific data for materials processing in space. The samples would be analyzed and the results compared with similar data from earth-processed samples to learn how to fabricate and repair structures in space, develop unique or improved materials for use on the earth, and provide new knowledge of material properties and performance. (MSFC Release 73-96)

- President Nixon submitted to the Senate the nomination of *Apollo 8* Astronaut William A. Anders, Executive Secretary of the National Aeronautics and Space Council 1969–1973, to be a Commissioner of the

Atomic Energy Commission for the term expiring June 30, 1978. The nomination was confirmed Aug. 2. Anders would succeed James T. Ramey, whose term had expired. (PD, 7/23/73, 914, 920; CR, 8/2/73, D960)

- U.S. heart specialist Dr. Paul Dudley White sent the first telegram via satellite between the U.S. and the People's Republic of China, to inaugurate direct communications between Peking's General Administration of Telecommunications and the New York headquarters of Western Union International Inc. Dr. White, one of the first U.S. physicians to visit the PRC in 1972, sent greetings from U.S. physicians to their colleagues in China, via *Intelsat-IV F-4* (launched Jan. 22, 1972). (*NYT*, 7/20/73, 27; ComSatCorp PIO)
- A Tass photo published by the *New York Times* showed Soviet Apollo-Soyuz Test Project Cosmonauts Valery N. Kubasov and Aleksey A. Leonov practicing moving from a Soyuz spacecraft in the water to a rescue raft. The site of the training was not identified. (*NYT*, 7/19/73, 1)
- Aérospatiale President Henri Ziegler told a Paris news conference that the Concorde Anglo-French supersonic transport aircraft would visit the U.S. for the first time in September for the opening of the Dallas-Fort Worth Airport in Texas. The Concorde was also scheduled to stop at Dulles International Airport in Chantilly, Va., for a demonstration on its return from Caracas, Venezuela, in September. (B *Sun*, 7/20/73, A5)
- The award of a $7 125 497 firm-fixed-price contract to Litton Systems, Inc., for avionics for the F-4 jet fighter aircraft was announced by the Air Force. (DOD Release 360-73)

July 20: Skylab 3, scheduled for launch July 28 as the second manned mission to crew the Skylab Workshop, would be extended from 56 days to 59 days to provide a better recovery posture, NASA announced. Splashdown would be 6:26 pm EDT Sept. 25, in the Pacific Ocean 547 km (340 mi) southwest of San Diego.

A comprehensive review of the medical findings of *Skylab 2* (launched May 25) had indicated that a 59-day mission was medically feasible. The medical data obtained on Skylab 3 would be reviewed at midmission point and weekly reviews would be scheduled for the remainder of the flight. Following each review, approval for continuation of the flight would be made for the next seven days by Skylab Program Director William C. Schneider.

Simulated T-zero ignition and liftoff with the fully fueled Saturn IB and Apollo spacecraft on its launch pad was successfully completed in the Skylab 3 countdown, begun July 15. (NASA Release 73-136; KSC Release 163-73)

- Fourth anniversary of *Apollo 11*'s first manned lunar landing July 20, 1969. Reps. Don Fuqua (D-Fla.) and Lou Frey, Jr. (R-Fla.), entered tributes in the *Congressional Record*. Fuqua said: "Since July of 1969 when Neil Armstrong first set foot on the lunar surface our national space program has had an ever-increasing wealth of technology and new knowledge and practical information of direct benefit to the people of the Nation and of the world." In remembering the first lunar landing, "we need to be reminded of the need for strong support for our national space program today and in the years ahead." Frey said: "The people of our space program are not resting on the laurels of this great achievement but are pressing on with the Skylab and the space shuttle to make space an ever-increasing contributor to the well-being of our Nation and the world." (CR, 7/20/73, E4957, 4961)

- Dr. Charles A. Lundquist, Assistant Director of the Smithsonian Institution's Astrophysical Observatory, had been named Director of Marshall Space Flight Center's Space Sciences Laboratory, MSFC announced. Before joining the Smithsonian, Dr. Lundquist had been Chief of the Physics and Astrophysics Branch in MSFC's former Research Projects Div. He had been transferred to MSFC in 1960 from the Army Ballistic Missile Agency. (MSFC Release 93–99)

- The Senate confirmed *Skylab 2* astronaut Cdr. Paul J. Weitz, Jr. (USN), for permanent promotion to the grade of captain. (CR, 7/20/73, S14273)

- The Air Force announced the award of a $2 352 060 cost-plus-fixed-fee contract to General Dynamics Corp. for advanced development of conceptual hardware for an undesignated lightweight fighter aircraft. (DOD Release 364–73)

- The National Science Foundation published *Selected Characteristics of Five Engineering and Scientific Occupational Groups, 1972* (NSF 73–306). The bulk of the 1.735 million persons in the five major occupational groups were engineers (72%). Physical scientists accounted for 11% and social scientists, life scientists, and mathematicians, about 9% each. Engineering continued male-dominated with men making up nearly 99%; women had made their greatest penetration in mathematics, more than 27%. (Text)

July 21: France exploded the first nuclear device of her latest series at Mururoa Atoll in the South Pacific. The announcement of the explosion—of a bomb hung from a balloon—came from the New Zealand frigate *Otago*, which had been dispatched to protest the nuclear testing. Reuters said later that the explosion was the 27th in France's nuclear program in the Pacific. (*NYT*, 7/22/73, 1)

July 22: The U.S.S.R. launched its *Mars 4* probe from Baykonur Cosmodrome at 12:31 am local time (3:31 pm EDT, July 21) into a parking orbit from which it was inserted into a 470-million-km (290-million-mi) trajectory toward Mars. All systems performed satisfactorily, Tass reported. The spacecraft would reach Mars in February 1974 to continue scientific exploration of the planet and the space near it begun by *Mars 2* (launched May 19, 1971) and *Mars 3* (launched May 28, 1971). United Press International reported Western sources as saying the apparent aim of *Mars 4* was to softland research instruments and TV cameras on the Martian surface.

Mars 4 would be the fourth Soviet spacecraft to reach Mars. *Mars 1* (launched Nov. 1, 1962, to fly by Mars) had transmitted data to the earth for more than four months during its journey, before communications went dead. *Mars 2* had been put into orbit around the planet and transmitted data to the earth into August 1972. An instrumented capsule from the spacecraft crashlanded on the Martian surface Nov. 27, 1971, becoming the first man-made object to reach the surface. *Mars 3* had also attempted to softland an instrumented capsule, during a dust storm. TV pictures were transmitted to the earth for 20 sec before transmission ceased abruptly. The spacecraft ended its mission in August 1972. (GSFC *SSR*, 7/31/73; FBIS–Sov, 7/23/73, L1; UPI, *NYT*, 7/23/73, 8; *A&A 1962, 1963, 1971, 1972*)

- Jet Propulsion Laboratory was fashioning three large photomosiac spheres of Mars from thousands of photos taken by *Mariner 9* (launched by NASA May 30, 1971, to become the first spacecraft to orbit another planet Nov. 13, 1971). Two 1.2-m (4-ft) globes would be completed, one for NASA Hq. and one for Lewis Research Center. A 1.8-m (6-ft) globe would be exhibited in JPL's museum. (JPL PIO; Miles, *LA Times*, 7/22/73)
- The U.S. airline industry would have to spend an estimated $1 billion to $3 billion over four to five years for modifications necessary to meet new Environmental Protection Agency requirements of lowered aircraft noise and emission levels, aviation consultant Selig Altschul said in a *New York Times* article. (*NYT*, 7/22/73, 6:3)
- Japan, Australia, New Zealand, Canada, and Sweden protested France's July 21 detonation of a nuclear device over Mururoa Atoll in the Pacific. (AP, *NYT*, 7/23/73, 1)

July 23: Skylab 3 Astronauts Alan L. Bean, Jack R. Lousma, and Dr. Owen K. Garriott received their final major medical examinations at Johnson Space Center before their July 28 launch to join the Skylab Orbital Workshop in near-earth orbit. Dr. Paul Buchanan, Chief of the JSC Health Maintenance Branch, said the astronauts "look good and they are ready to fly." They would rendezvous and dock with the orbiting laboratory to live and work in it for 59 days before returning to earth. (Wilford, *NYT*, 7/24/73, 10)

- President Nixon signed H.R. 7528 into Public Law 93–74, NASA's $3.065-billion FY 1974 authorization. Approved by the House-Senate Conference Committee June 28, by the House May 23, and by the Senate June 19, the bill provided $2.246 billion for research and development, $112 million for construction of facilities, and $707 million for research and program management. (*PD*, 7/30/72, 936; P.L. 93–74)
- The House Committee on Science and Astronautics' Subcommittee on Energy held a hearing on H.R. 8348 [see June 4] and H.R. 9133, duplicate bills to authorize the National Science Foundation to designate certain institutions of higher learning national energy research centers. Dr. Raymond L. Bisplinghoff, NSF Deputy Director, described the NSF program of Research Applied to National Needs (RANN) devoted to energy: "It includes research on analytic techniques and potential conservation measures directed at meeting U.S. energy requirements while satisfying environmental quality restraints; energy conversion research directed at developing technologies that allow more efficient use of our energy resources; research on superconducting transmission lines and other aspects of energy and fuel transportation; and research on advanced coal technology, geothermal energy, and perhaps most importantly, on solar energy." (Transcript)
- First anniversary of the launch of *Erts 1* Earth Resources Technology Satellite from Western Test Range July 23, 1972. Dr. James C. Fletcher, NASA Administrator, at an anniversary celebration at the Franklin Institute in Philadelphia, said the use of space technology to solve the earth's problems was "a second giant stride for mankind." *Erts 1* had had "a profound effect on the thinking of the world, particularly on our approach to the emerging problems of protecting our environment and maintaining the quality of life for all of Earth's people. I think it means that we do have to be sensitive to how we use these precious resources; not just clean air and water, but clean land." In the program to follow,

"the ocean is also going to be a precious resource that we must conserve." Dr. Fletcher said 38 countries were participating in the ERTS program, with 300 principal investigators—200 in the U.S. and 100 in foreign countries—examining ERTS data.

Dr. William Nordberg, Chief of Goddard Space Flight Center's Laboratory for Meteorology and Earth Sciences, reviewed *Erts 1*'s year-end status: its multispectral scanner was fully operational, the return-beam-vidicon TV cameras had been turned off but could be reactivated, one of the two tape recorders was still operating, although at reduced capacity. More than 60 000 ERTS scenes—equivalent to an area more than four times the globe—had been imaged, 333 000 unique messages had been relayed from some 140 platforms at remote sites, and nearly 3 million photos had been reproduced at the NASA Data Processing Facility. (*NASA Activities,* 8/15/73, 134–6)

- Capt. Edward V. Rickenbacker (U.S. Army Air Corps, Ret.), the most decorated U.S. pilot of World War I, died in Zurich, Switzerland, after a heart attack at age 82. One of the top automobile racing drivers in the U.S. by the age of 21, Rickenbacker enlisted in the Air Service at the start of World War I. He received the Croix de Guerre with four palms, Legion of Honor, Distinguished Service Cross with nine Oak Leaf Clusters, Medal of Merit, and the Congressional Medal of Honor for downing 26 German aircraft and at least 4 enemy balloons. He became General Manager of Eastern Airlines, Inc., in 1934, and the same year piloted a Douglas airliner from Los Angeles to Newark, N.J., in 12 hrs 3 min 50 sec for a coast-to-coast passenger aircraft record. He was President and General Manager of Eastern Airlines, from 1938 to 1953, when he became Chairman of the Board. He retired in 1963 and was elected to the Aviation Hall of Fame in 1965.

  Throughout his life Rickenbacker had advocated greater use of military aircraft. In 1922 he supported his friend Gen. Billy Mitchell in urging a separate U.S. Government department for aeronautics. Following Rickenbacker's death, President Nixon issued a statement: " 'Captain Eddie' Rickenbacker was an American original—a celebrated racing car driver in the early years of the 20th Century, the leading American fighter pilot in World War I, a pioneer of commercial aviation, and a generous, patriotic citizen in both war and peace." (Hailey, *W Post,* 7/24/73, C4; *PD,* 7/30/73, 924)

July 23–28: The first U.S. women's helicopter team competed in the II World Helicopter Championships at Middle Wallop, England. The championships were organized by the Helicopter Club of Great Britain for the Royal Aero Club of the United Kingdom on behalf of the Federation Aéronautique Internationale (FAI). (NAA *News,* 8/73, 1)

July 24: NASA might delay the Skylab 4 launch from October to November so that its crew could observe Comet Kohoutek swing around the sun, Thomas O'Toole in the *Washington Post* quoted NASA officials as indicating. No firm decision had been made, but most officials favored a delay. NASA Associate Administrator for Manned Space Flight Dale D. Myers had said, "Comets this size come this close to us once in a century. It really looks like the kind of thing you can't pass up." (*W Post,* 7/24/73, A12)

- The maiden glide flight of NASA's X–24B lifting body was postponed for the second time [see July 19] because of a malfunction in the control gyros. (NASA prog off)

- The 1972–73 edition of *Jane's Fighting Ships* was published in London. In the foreword, editor Capt. John Moore said the Soviet navy was "a pre-eminent force with its own air power and with a greater range of submarines than any other. Above all, it has a far greater proportion of ships less than 10 years old than any other major fleet." (AP, *P Inq*, 7/26/73, 11)

July 24–Aug. 8: A summer school on manned space flight—sponsored at Nottingham Univ., England, by the European Space Research Organization (ESRO)—was attended by 63 graduate scientists and engineers from 13 countries and by NASA representatives. Its theme, "The Implications for European Space Programmes of the Possibilities of Manned Mission," was based on use of the space shuttle Spacelab. (*ESRO Newsletter*, 10/73, 1)

July 25–26: The U.S.S.R. launched its *Mars 5* probe from Baykonur Cosmodrome at 11:56 pm local time (2:56 pm EDT) into an earth parking orbit. At 79 min after launch the spacecraft was inserted into a 470-million-km (290-million-mi) trajectory for Mars. All systems functioned normally, Tass reported. The spacecraft was similar in design and purpose to *Mars 4* launched toward Mars July 22. Simultaneous investigations by the two spacecraft—scheduled to arrive at Mars in February 1974—would make it possible to obtain more data about the planet and the dynamics of physical processes occurring in cosmic space. On July 26, at 4:00 am Baykonur time (7:00 pm EDT, July 25), *Mars 5* was 66 000 km (41 000 mi) from the earth. *Mars 4* was 1.46 million km (0.9 million mi) from the earth. (FBIS–Sov, 7/26/73, L1; Reuters, *W Post*, 7/25/73, B19; *SBD*, 7/30/73, 147)

July 25: The U.S.S.R. launched *Cosmos 577* from Plesetsk into orbit with 308-km (191.4-mi) apogee, 170-km (105.6-mi) perigee, 89.2-min period, and 65.4° inclination. The satellite reentered Aug. 7. (GSFC *SSR*, 7/31/73; 8/31/73; *SBD*, 7/30/73, 147)

- Skylab 3 Astronauts Alan L. Bean, Jack R. Lousma, and Dr. Owen K. Garriott concluded training for the second manned mission to the Skylab Orbital Workshop, launched May 14. They practiced key flight maneuvers in a spacecraft simulator at Johnson Space Center and the complex countdown procedures to be used during the final two and one half hours before launch July 28. In an interview with the *New York Times*, Skylab Program Director William C. Schneider said, "Everything seems to be going great." (Wilford, *NYT*, 7/26/73; AP, *B Sun*, 7/26/73, A10)

- The Senate received the nomination of L/G Samuel C. Phillips, Director of the National Security Agency and former NASA Apollo Program Director, to be a general, USAF. (*CR*, 7/25/73, S14742)

- NASA launched two Aerobee 200 sounding rockets from White Sands Missile Range. The first carried a Goddard Space Flight Center payload to a 30.6-km (19-mi) altitude to test rocket performance. The rocket performed unsatisfactorily. The second carried a GSFC heat-pipe experiment to a 275.7-km (171.3-mi) altitude. The rocket and instrumentation performed satisfactorily. (GSFC proj off)

- A *New York Times* editorial criticized continued nuclear testing by France and the People's Republic of China: "Even underground nuclear tests no longer have real justification. Both the United States and the Soviet Union have more than enough nuclear weapons to devastate the world, and they have been testing so long that at most they can be gaining only marginal increases of knowledge. At this late date the old arguments about the need for on-site inspection seem irrelevant as barriers to a universal halt. If Moscow and Washington want really to demonstrate that the proclaimed Soviet-American detente is real, they have no simpler or more effective means than to announce that they are amending the 1963 Moscow Treaty to ban all nuclear tests." (*NYT*, 7/25/73, 38)

July 26: The Skylab 3 astronauts arrived at Kennedy Space Center to prepare for their July 28 launch to join the Skylab Orbital Workshop (launched into earth orbit May 14) for a 59-day mission. During a press briefing at Patrick Air Force Base, Astronaut Bean said, "We're very anxious to get up there and put out 100 percent of what we have to do." (O'Toole, *W Post*, 7/27/73, A1)

• Skylab science and medical briefings were held by NASA at Johnson Space Center. Results from the first Orbital Workshop mission, *Skylab 1-2* (launched May 14 and May 25), were discussed and science plans for the Skylab 3 mission, scheduled for launch July 28, were announced.

At an Apollo Telescope Mount briefing Dr. Robert A. MacQueen, a Skylab principal investigator and High Altitude Observatory scientist, said that a first look at available data indicated that the objectives—to examine the solar corona on a regular basis, deduce the three-dimensional structure and geometry of the corona, understand the evolution of the corona and its relationship to the solar disk, and understand the outflow of the solar wind from the corona—would be met. The most important conclusion drawn was that "in appearance the corona is now a less dynamic subject than other solar phenomena." Dr. Edward M. Reeves, Harvard College Observatory scientist, said that an ATM scanner had observed "velocity changes in whole large areas of the Sun in lines of the transition region and the corona which have never before been seen in this part of the spectrum."

At the medical science briefing, Dr. Michael W. Whittle, Principal Coordinating Scientist, said that the weight losses of the three *Skylab 2* astronauts—1.8 kg (4 lbs) for Charles Conrad, Jr., 2.7 kg (6 lbs) for Dr. Joseph P. Kerwin, and 3.6 kg (8 lbs) for Paul J. Weitz—were "pure and simple adaptation to weightlessness. . . . Increased nitrogen, calcium, and phosphorus, the major constituents of muscle tissue, had appeared in the urine and indicated the muscle loss. Dr. Stephen L. Kimsey, a Skylab principal investigator, said that experiments had shown "no abnormal changes in red cell metabolism" in amount of energy stored, change in the integrity of the cell membrane, or change in the red cells' ability to transfer oxygen from the lungs to the tissue.

Dr. Verl R. Wilmarth, earth resources experiment package project scientist, said during an EREP briefing that the objective of mapping the depression of the sea surface of the Puerto Rican trench using altimeter data had been met.

At a briefing on the corollary experiments Dr. Ernst Stuhlinger, Associate Director for Science at Marshall Space Flight Center, said that the materials processing and welding experiments would help verify methods of construction and assembly in weightless conditions. Among the first things built in space might be a reflector for microwave energy, to transmit large amounts of energy from one point of the earth to another. A reflector of the size needed could not be launched in one piece but would have to be assembled in space.

Dr. Robert A. Parker, astronaut and Skylab program scientist, said new experiments that would be flown on Skylab 3 included a vehicle disturbance experiment; a survey of celestial sources of x-rays; two circadian rhythm experiments, one with vinegar gnats and the other with desert mice; movement tests of astronauts with two different extravehicular activity backpacks; a gallium arsenide crystal growth experiment; observation of cellular mechanisms and metabolism in zero g; observations of the earth's ozone and airglow in the upper atmosphere; the collection of micrometeoroids; and an evaluation of the spectral capability of the human eye while observing earth resources from space. (Transcripts)

- Tenth anniversary of NASA's July 26, 1963, launch of *Syncom 2*, the first communications satellite to relay digital data, telephone conversations, picture facsimile, and TV programs from synchronous earth orbit. The first telephone conversations were between President John F. Kennedy and Nigerian Prime Minister Sir Abubaker Tafawa Balewa and other messages between U.S., Nigerian, and United Nations officials, Aug. 23, 1963. Transmission on *Syncom 1* (launched Feb. 14, 1963) had failed on insertion into synchronous orbit. *Syncom 3* was launched successfully Aug. 19, 1964. (NASA Release 73-140; *A&A 1963*)

- A House-Senate Conference Committee favorably reported a compromise version of H.R. 8825, the FY 1974 Dept. of Housing and Urban Development-Space-Science-Veterans appropriation bill that included a $3.002-billion NASA appropriation. NASA funds were $13.9 million below the budget request and $62.4 million below the $3.0645-billion FY 1974 authorization [see July 23]. The compromise appropriation was similar to the bill approved by the Senate June 30 and $13 million above the $2.989 billion approved by the House June 22. (*CR*, 7/26/73, H6754; H Rpt 93-411)

- The American Museum-Hayden Planetarium in New York commemorated the fourth anniversary of the July 20, 1969, first manned lunar landing, by *Apollo 11* (launched July 16, 1969). The Planetarium exhibited the Goddard Memorial Collection of autographed pencil portraits of outstanding contributors to the U.S. space effort by artist William J. Numeroff and showed "The Salvage of Skylab I," films and narration of the *Skylab 2* astronauts' repair in space of *Skylab 1* spacecraft anomalies [see May 14–June 23]. Participants in the program included *Skylab 2* backup commander Astronaut Russell L. Schweickart; Mrs. Robert H. Goddard, widow of the U.S. rocket pioneer; and author Arthur C. Clarke. (American Museum Release; PIO)

- President Nixon submitted to the Senate the nomination of William D. Ruckelshaus to be Deputy Attorney General and the nomination of Russell E. Train, Chairman of the Council on Environmental Quality, to succeed Ruckelshaus as Administrator of the Environmental Protection Agency. (*PD*, 7/30/73, 931, 936)

July 27: In a Skylab 3 prelaunch press briefing at Kennedy Space Center, NASA Skylab Program Director William C. Schneider said that the launch vehicle was "in good shape" with no open anomalies during the countdown. The crew "is eager, waiting" for their 7:10 am July 28 liftoff. The condition of the unmanned Orbital Workshop (launched May 14) was good. The temperatures were under control and the electrical system had improved since the *Skylab 2* mission (May 25–June 22). Col. Alan R. Vette of the U.S. Air Force recovery forces

July 27

- NASA efforts to study Comet Kohoutek [see June 21–23] were described in a *Science* article. NASA had established a special task force to help prepare for infrared, optical, and ultraviolet observations from spacecraft, aircraft, and ground. *Oao 3* Orbiting Astronomical Observatory (launched Aug. 21, 1972) would observe Kohoutek in the months before and after its closest approach to the sun about Dec. 29, when Kohoutek was expected to be as bright as a full moon. During the passage near the sun, *Oso 7* Orbiting Solar Observatory (launched Sept. 29, 1971), various sounding rockets, and possibly the Skylab Orbital Workshop (launched May 14) would be trained on the comet, which astronomers believed might be insulated from the rest of the solar system. (Metz, *Science*, 7/27/73, 333–4)
- An advertisement in *Science* for Ehrenreich Photo-Optical Industries' Nikon Instrument Div. said its Model H microscope aboard the Skylab Orbital Workshop (launched May 14) was "the first microscope in space." It would be used by scientist-astronauts in several of the 19 life science experimental studies planned to evaluate the physical reaction of the human body during extended weightlessness. (*Science*, 7/27/73, 298–9)
- An era of global scarcity had begun, a *Science* editorial noted. "The advanced countries are vulnerable in one or more respects." Japan had been enjoying a boom but needed to import raw materials, fossil fuels, and food. In Mexico people lived under conditions like those of 100 yrs ago, "with a disregard for sanitation," but with "Coca Cola signs and transistor radios everywhere." Much of the world was aware of U.S. living standard and gadgets. "They yearn for both, while the population explosion continues. We are soon going to witness dramatic and miserable confrontations of aspirations, expectations, and limitations." (Abelson, *Science*, 7/27/73, 303)

July 28–September 25: The *Skylab 3* Apollo spacecraft carrying a three-man crew was launched into near-earth orbit, rendezvoused with the *Skylab 1* Orbital Workshop (launched May 14), and docked with the space laboratory for the second manned mission in the Skylab program. A series of problems threatened to disrupt the mission, but corrective actions and workarounds permitted it to proceed successfully, setting new manned flight records. The astronauts collected significant data in medical experiments, solar astronomy, and earth resources surveys for 59 days before undocking for a safe return to earth Sept. 25.

July 28–31: *Skylab 3* (SL-3) lifted off on time at 7:11 am EDT from Kennedy Space Center Launch Complex 39, Pad B, on a two-stage Saturn IB launch vehicle, watched by some 100 000 persons on Florida beaches and at KSC, as well as by U.S. and foreign TV viewers. The 14 168-kg (31 234-lb) Apollo command and service module (CSM–117) and its crew—Allan L. Bean (commander), Jack R. Lousma (pilot), and Dr. Owen K. Garriott (science pilot)—soared through ground fog into broken clouds.

The CSM/S–IVB combination was placed in a phasing orbit with 224-km (139.2-mi) apogee and 150-km (93.2-mi) perigee at 10 min 2 sec after launch. Eight minutes later the CSM separated from the stage as planned. The galactic x-ray mapping experiment mounted in the instrument unit (IU) was deployed. At 5 hrs 19 min 53 sec after launch the S-IVB/IU deorbit maneuver was commanded, and the stage impacted the predicted Pacific Ocean area after planned maneuvers.

During CSM maneuvers to rendezvous with the Workshop, a leak was detected in the CSM reaction control system (RCS) quad B thruster, one of four assemblies of rockets used to stabilize the spacecraft or to change its velocity and also used as a backup for bringing the CSM back from orbit. Analysis indicated the oxidizer valve was stuck open or partially open. Quad B was isolated; analysis showed that all deorbit mission rules could be satisfied.

The CSM rendezvoused with the Workshop at 3:38 pm EDT on the fifth orbit. Rendezvous, fly-around, and docking were covered by 23 min of TV. The crew entered the multiple docking adapter (MDA) at 4:00 pm EDT and began Workshop activation, but activation was curtailed because the crew quickly began to feel motion sickness. Lousma was experiencing nausea and Bean and Garriott were uncomfortable. Bean said he and Garriott tended "to be fairly careful of how we move. And since we are moving rather slowly, then it's taking a little bit longer than we'd planned."

Motion sickness continued the second day, July 29, and the astronauts were taking scopolamine-dexedrine tablets. Bean communicated, "We feel okay, except in the stomach. . . . And our thinking is that we're not going to be able to work much faster than we are right now. . . . There's a desire to maybe take a break for an hour or two and get in the bunk and stay still for a while." Flight Director Neil B. Hutchinson told the press at JSC that extravehicular activity (EVA) to install a new sun shield and retrieve and install Apollo Telescope Mount (ATM) film probably would be delayed from July 31 to Aug. 1 at the earliest. Four to five hours of work were carried over from July 28 and 29.

Still suffering from motion sickness, the astronauts continued Workshop activation July 30 (Mission Day 3). The first EVA was delayed to Aug. 2 and later rescheduled to Aug. 4 to ensure the astronauts' complete recovery. Activation was completed July 31 (MD-4) and research began as the crew improved slightly. Bean told ground control: "We're getting better; it's just taking a little longer than we thought. . . . Everybody feels pretty good between meals. Then mealtime comes, and we're caught . . . in the squeeze. We're trying to decide whether we ought to eat all we have to . . . to keep our strength up, or eat about what we want to, which isn't very much."

A TV broadcast from the space station July 31 showed minnows for a biomedical experiment swimming in their plastic aquarium in tight spirals and sometimes apparently trying to swim toward the surface, but with their heads pointed straight down. Garriott said, "Apparently they can't make up their minds which way is up and down, either." He also turned a spider named Arabella loose in a plastic cage to observe her web-spinning abilities in zero-g conditions. Later in the mission Garriott reported baby minnows hatched in space performed in water "as if they'd already adapted while they were still in the egg." They "felt right accustomed to zero-g."

August 1–September 24: Bean told ground controllers after breakfast Aug. 1 (MD-5): "This is the first morning we've been up here . . . that after a meal everybody felt good. Everybody feels real good right now. Best since we've been here." During live TV the astronauts ate a hearty lunch to music. Bean demonstrated his proficiency in eating while hanging upside down.

A medical briefing below at Johnson Space Center suggested that moving quickly from the confined CM to the large Workshop might have contributed to the motion sickness. The milder nausea experienced on the May 25–June 22 *Skylab 2* mission could have been caused by a more gradual transition into the Workshop. "I think that it is merely a matter of room in which you can exercise your vestibular mechanisms, and that the workshop simply presents more of that space. . . . perhaps it was too much too soon."

On Aug. 2 a master alarm buzzer awakened the astronauts at 6:45 am EDT. Pressure and temperature were falling rapidly in the CSM reaction control system quad D because of an oxidizer leak. Because of quad B's July 28 leak, only A and C of the four quads remained fully operational. The quad D leak lasted two hours, losing an estimated 5 kg (12 lbs) of nitrogen tetroxide oxidizer. Apollo Flight Director Glynn S. Lunney said at a morning press briefing that the spacecraft was flyable with the loss of the two quads but "it would be flown in a mode which is not entirely standard." The main concern was "that we have had two problems which are related, at least they seem to be related." A flaw common to the whole RCS could necessitate the crew's immediate return to the earth or accelerated preparation of the Skylab 4 spacecraft to make it available as a rescue vehicle should the CSM become unusable.

Full activation of Skylab-rescue-vehicle preparation was authorized. Launch and checkout crews began working a 24-hr-day, 7-day-week schedule, preparing two additional couches and other equipment for installation. Earliest possible date for a rescue mission was established as Sept. 5. The EVA rescheduled for Aug. 4 was again postponed, for 24 to 48 hours to free the crew for system checks, to free ground controllers to troubleshoot the problem, and to give time to work out EVA procedures for the new spacecraft configuration. While reentry procedures were being rewritten for the new configuration, the astronauts continued scientific experiments.

By the afternoon of Aug. 2 Dr. Christopher C. Kraft, Jr., JSC Director, was able to tell the astronauts that troubleshooting the RCS had indicated no apparent generic problem, but two unique quad failures, and there was now confidence that the two remaining quads would ensure attitude control should immediate reentry become necessary. "We're proceeding . . . as if we're going to have a normal mission." Astronaut Bean replied: "Fine. You just said the right words. . . . we're pretty happy with the way things are going, at the moment . . . and we agree 100 percent with what you just announced. . . . everybody's happy up here."

On Aug. 3 Skylab Program Director William C. Schneider said at a JSC press briefing that continuing investigation indicated the two quad failures were unrelated and reexamination of reentry capabilities showed three ways of returning to the earth: using the primary service propulsion system, using the RCS with the two operational quads, and using the hybrid RCS in the CM. "We're feeling considerably better than . . . yesterday morning." A slowdown was ordered in the emergency rescue preparations. The Apollo CM would be outfitted in a Skylab 4 configuration, with three astronaut couches rather than the five needed for a rescue mission. The slowdown would delay rescue readiness until Sept. 10.

Astronauts Bean, Garriott, and Lousma conducted their first full day of experiments and prepared for an earth resources experiment package pass. Garriott held the TV camera at the window to telecast a good view of the U.S. from North Dakota to North Carolina. Later the first of 26 EREP passes was made, to gather data for oil exploration, ocean temperatures, mapping, land use surveys, crop inventories, and pollution studies.

A master alarm sounded in the CSM Aug. 4, later attributed to a short-circuit in Skylab's solar telescope system. The short knocked out one of the two TV systems in the ATM but did not affect Skylab's main power system. Ground crews investigated problems in the primary and secondary coolant systems Aug. 5. One had been slowly leaking since the *Skylab 1* launch May 14 and was expected to become inoperative in 16 to 20 days. The secondary loop was thought to have a small leak, begun during the *Skylab 2* mission and giving the loop a predicted 60-day lifetime. Possibly additional coolant would have to be carried up by the next manned mission, Skylab 4, but Program Director Schneider said there was "no immediate danger to the crew." By Aug. 6 evaluations showed that only the primary loop was leaking; the second was sound and could operate for the rest of *Skylab 3* as well as for Skylab 4.

Flight surgeon Paul Buchanan's nightly bulletin said Aug. 5 that the crew "continues to report good health and optimistic outlook" and had asked about an increase in their menus "to match their growing appetites." Fluid intake and output was "optimal."

The first EVA was performed Aug. 6. Garriott and Lousma stepped outside the orbiting laboratory at 1:30 pm EDT to install the 3.6- by 7.3-m (12- by 24-ft), paper-thin, aluminized Mylar awning over the Workshop to reflect the sun's rays away from the spacecraft skin, damaged when the meteoroid shield was torn off during the May launch. Garriott, standing in the Skylab's open hatch, pieced together the two 16.8-m (55-ft) poles from metal sections and handed them to Lousma. Lousma attached them to a bracket to form a "V" over the Workshop. The sunshade was slowly pulled up by a system of pulleys. One rope kept twisting around the poles and the sunshade folds did not straighten immediately, but after four hours—twice the time expected—the sunshade was successfully deployed. Workshop temperatures were lowered before the EVA ended. Astronaut Bean said Garriott and Lousma worked "at a slow and steady pace" and were not tired. Lousma retrieved and replaced film in four ATM experiments, installed two new cameras, removed a door latch from an ATM experiment, and fixed the particle-collection experiment to the rim of the ATM.

Lousma also inspected the two leaking quads in the CSM RCS and looked for coolant leaks. He noted some evidence of micrometeorite pitting but no clues to causes of the malfunctions. The astronauts completed the EVA at 8:01 pm EDT, after 6 hrs 31 min, surpassing the 3-hr 30-min record set June 7 by *Skylab 2* Astronauts Charles Conrad, Jr., and Dr. Joseph P. Kerwin.

Garriott, first solar physicist in space, focused the ATM cameras on the sun Aug. 7 for the first time during the mission. TV audiences watched him work at the ATM panel.

By Aug. 8 Dr. Lawrence F. Dietlein, JSC Deputy Director for Life Sciences, reported the astronauts were "in excellent spirits and have no health problems." They had completely recovered from motion sickness. Their weights had stabilized after a loss of 1.8 to 2.3 kg (4 to 5 lbs) each. Data obtained on the rotating chair indicated their sensitivity to motion had lifted. Since deployment of the sunshade, temperatures in the Workshop had dropped to 296 K (74°F).

The crew photographed a coronal transient with the ATM camera two hours Aug. 10. It was the first time a transient—protons and free electrons expelled violently from the sun's limb at a speed of about 312 km (194 mi) per sec, a phenomenon that occurred two or three times a year—had been observed at that range and time resolution.

Program Director Schneider announced Aug. 14 that *Skylab 3* would continue its full 59-day mission. From analyses of the CSM and Workshop, he had "concluded that there is no imminent need for rescue, and . . . we have made the decision to proceed with a normal test flow on an accelerated basis at Cape Kennedy." Skylab 4 was being prepared for a normal mission but would be available for a rescue mission if necessary. The earliest launch date could be Sept. 25. Schneider said, "We're reasonably confident that [CSM] 117 is a good vehicle." "Looks like they got a renewed burst of confidence down there," Bean said when told of the decision. "We feel that way up here." JSC announced Aug. 21 that the astronauts would install a new "six-pack" of gyroscopes during an Aug. 24 EVA, to replace faulty ones. The six-pack had been carried up on the *Skylab 3* CM.

Meanwhile the crew continued with experiments, requesting more work Aug. 14 because all had spare time and "you just can't sleep 8 hours up here." They photographed tropical storm Brenda Aug. 19 as it moved across the Yucatan Peninsula toward the Gulf of Mexico.

On Aug. 21 Astronaut Bean discovered and photographed through ATM telescopes what Skylab principal investigator Dr. Robert MacQueen called "certainly one of the biggest solar events" seen on the Skylab missions. Bean said the complex series of events on the solar limb looked "like a big bubble" sitting on the edge of the disc. He observed it with the white light coronagraph and x-ray instruments and reported it to be three fourths the size of the sun. NASA scientists alerted other facilities to make observations from ground instruments and other satellites. Dr. MacQueen said the event might have been caused by an explosion on the back side of the sun. "It was a beautiful event." The crew "did a fine job."

Schneider announced Aug. 23 (MD-27) that the *Skylab 3* crew had been given permission to continue the mission for another seven days. Dr. Willard R. Hawkins, JSC Deputy Director of Life Sciences for Medical Operations, told the press that he saw nothing in the medical data thus far "that would indicate that we could not press on with the mission, and . . . we feel we can go the full 59 days as planned."

EVA-2 began at 12:24 pm EDT Aug. 24 with AM depressurization by Lousma and Garriott. Lousma completed installation of the six redesigned gyros mounted inside the spacecraft by connecting cables to a computer system outside, and the new gyro system was put into operation. The astronauts installed a sail sample on a hand rail to test durability, replaced ATM film, retrieved and stored cassettes from one experiment, and removed the aperture door ramp latch from two ATM experiments, concluding EVA-2 at 4:31 pm EDT, after 4 hrs 30 min.

At 8:01 am EDT Aug. 25 (MD-29) the *Skylab 3* astronauts surpassed the single-mission endurance mark of 28 days 50 min set in June by *Skylab 2*. They had completed 404 earth orbits and had traveled 18.5 million km (11.5 million mi). And at 6:11 am EDT Sept. 5 (MD-40) Astronaut Bean surpassed the space flight record for one man of 49 days 3 hrs 38 min 36 sec, set in June by Charles Conrad, Jr., on *Skylab 2*.

EVA-3 began at 7:18 am EDT Sept. 22. Bean and Garriott moved through space to retrieve and replace ATM film and retrieve experiments and parasol samples. EVA operations ended at 9:59 am EDT, after 2 hrs 41 min.

During the mission, spiders Arabella and Anita first spun webs in corners of their cage, without the normal circular pattern. Each day, however, each improved web formation, adding more radials, adjusting to weightlessness. Garriott said of Arabella Aug. 9 that "she learned very rapidly in zero-g without the benefit of any previous experience." She "figured out a very nice solution to the problems of zero gravity." Later ground controllers told Garriott to share his filet mignon with the spiders to keep them alive for the entire mission. Anita, however, was found dead Sept. 16, possibly from inadequate diet.

September 24–25: The Workshop was deactivated and, on Sept. 25, the CSM powered up. The CSM undocked from the Workshop at 3:34 pm EDT, separating at 3:50. The service propulsion system engine made the deorbit maneuver at 5:38 pm EDT and the CM splashed down upside down at 6:20 pm EDT 400 km (250 mi) southwest of San Diego, Calif.

The astronauts righted the CM and remained inside until the recovery ship U.S.S. *New Orleans* maneuvered alongside and a crane lifted the CM aboard. Immediate medical examinations found the crew in excellent spirits and surprisingly good condition. Blood pressures and heart rates were within normal range. The astronauts would remain aboard ship in San Diego for physical checks and would fly to Houston Sept. 26. They would be quarantined for one week.

The *Skylab 3* mission achieved its primary objectives of performing unmanned Saturn Workshop operations, reactivating the Skylab orbital assembly in earth orbit, obtaining medical data for use in extending manned space flights, and performing inflight experiments. Despite mission problems, 150% of scheduled ATM observing time was accomplished and EREP sensors were operated on 39 Z-axis local vertical data runs and 5 solar inertial data runs, completing 218 of the 386 earth resources task sites. All medical experiments were successfully carried out. All but one of the corollary experiments achieved or exceeded objectives. Of 13 student experiments, 9 were completed. Added unscheduled tests and experiments had significantly increased the data yield above premission planning and Skylab was left in an excellent position to complete the final SL-4 mission. *Skylab 3* was officially adjudged a success Oct. 11.

Major records set by the mission included the longest-duration manned space flight to date, 59 days 11 hrs 9 min; the longest cumulative time in space for one man, 69 days for Bean (who had spent 10 days on the Nov. 14–24, 1969, *Apollo 12* mission); the longest distance in orbit for a manned mission, 41 000 000 km (25 500 000 mi); the longest EVA, 6 hrs 31 min on Aug. 6; and the longest cumulative orbital EVA time, for one astronaut, 13 hrs 40 min for Garriott.

In the first mission to man the Orbital Workshop May 15–June 22, *Skylab 2* astronauts had conducted medical, solar, and earth resources experiments for 28 days, as well as repairing the Workshop to salvage the mission. The Workshop would be operated unmanned until the November Skylab 4 launch for a third and final mission. (NASA prog off; JSC PAO press briefing transcripts, 7/28/73–9/25/73; *NYT*, 7/28/73–9/26/73; *W Post*, 7/28/73–9/26/73; *W Star-News*, 7/28/73–9/27/73; B *Sun*, 8/16/73, A5; 8/20/73, A3)

July 28: NASA launched a Nike-Tomahawk sounding rocket from White Sands Missile Range carrying a Univ. of Colorado aeronomy experiment to a 120.2-km (74.7-mi) altitude. The rocket performed successfully but the experiment data were saturated and not useful. (GSFC proj off)

July 29: Fifteenth anniversary of the signing of the National Aeronautics and Space Act of 1958 by President Dwight D. Eisenhower. The act provided for the establishment of NASA, which officially came into being Oct. 1, 1958. (NASA Release 73–134)

July 30: Dr. James C. Fletcher, NASA Administrator, praised NASA's successful efforts to save the *Skylab 1–2* mission during testimony before the Senate Committee on Aeronautical and Space Science hearing to review the Skylab Investigation Report [see July 13]: "We were able to save Skylab because of the tremendous depth and breadth of the NASA team and the dedication and skills of thousands of people on that team. There certainly isn't another country that could have reacted as we did. I doubt whether there is another team within this country that could have reacted in this manner. Our team could do it because of years of training, working hard on highly exacting projects, and working as a team to drive toward a technical objective. The performance of the team on the ground, and of course the magnificent performance, skill and courage of the astronauts, saved Skylab. I would hope that this is remembered as the real story of Skylab, not the failure which caused it to happen." NASA was "taking all the steps we know how to prevent a future failure due to similar causes in other programs."

Director Bruce T. Lundin of Lewis Research Center testified that, throughout six years of reviews and certifications, "the principal attention devoted to the meteoroid shield was that of achieving a satisfactory deployment. Never did the matter of aerodynamic loads on the shield or aeroelastic interactions between the shield and its external pressure environment during launch receive the attention and understanding during the design and review process which in retrospect it deserved." The omission had not been surprising. "From the beginning, a basic design concept and requirement was that the shield be tight to the tank." The question of whether the shield would remain tight against the tank under the dynamics of flight through the atmosphere "was simply not considered in any coordinated manner." Possibly contributing to the oversight "was the basic view of the meteoroid shield as a piece of structure." No full-time subsystem engineer had been assigned to the shield. "While it is recognized that one cannot have a full-time engineer on every piece of equipment, it is . . . possible that the complex interactions and integration of aerodynamics, structure, rigging procedures, ordnance, deployment mechanisms, and thermal requirements of the meteoroid shield would have been enhanced by such an arrangement." (Testimony)

- Navy development of hydrofoils and surface effect ships (SESs) with speeds to 177 km (110 mi) per hr was seeking to make the Soviet navy obsolete, the *Washington Star-News* reported. R/A George G. Halvorson, Director of the USN's Surface Ship Acquisition Div., had said the USN was already ahead of the U.S.S.R. in "fast navy" technology. "We are now on the verge of the biggest advance in ships since the change from sail to steam." Four hydrofoils and two SESs of patrol size were already in Navy service and the first of two destroyer-size SESs, with speeds to 177 km per hr, was to be added by 1978. The USN was buying 2 of 30 hydrofoil patrol boats planned for Italian, German, and U.S. fleets in Europe. From the success of test ships, Adm. Halvorson could visualize a nuclear-powered, 177-km-per-hr aircraft carrier. Existing destroyers and carriers had a top speed of 64 km (40 mi) per hr.

  Hydrofoils rode above the water on stilts supported by underwater wings. A jet aircraft engine powered a pump to force water from jets below the stern to drive the ship. SESs, or hovercraft, glided on a cushion of air. Their sides extended underwater where propellers or water jets drove the ships. (*W Star-News*, 7/30/73, 31)

- President Nixon presented the Distinguished Service Medal to Gen. John D. Ryan, Air Force Chief of Staff, during a White House ceremony. Gen. Ryan was retiring after 35 yrs of service. (*PD*, 8/6/73, 955)

- Rep. Harold L. Runnels (D–N. Mex.) introduced H.R. 9606 to establish an Office of Solar Energy Research in the Dept. of the Interior.

  Rep. Richard G. Shoup (R–Mont.) introduced H.R. 9691 to authorize additional funds for the National Science Foundation for research in geothermal and solar energy. (*CR*, 7/30/73, H6897, E5208)

- An *Aviation Week & Space Technology* editorial commented on the July 23 anniversary of the launch of NASA's *Erts 1* Earth Resources Technology Satellite. Its first year of operation "has proved that space technology has produced yet another useful tool for man to improve his life on earth. The ERTS experiment has shown—at a relatively modest cost of $160 million—that it is feasible to measure, catalogue and monitor the contents of the earth's crust." It had also yielded "more pertinent information faster and cheaper than all previous methods of tackling this job in the limited areas where it was possible." *Erts 1*'s most important contribution probably would be "the tremendous capability it can provide for man to detect, monitor and control the sources of pollution that are poisoning his environment." (Hotz, *Av Wk*, 7/30/73, 7)

July 30–August 3: Wallops Station used a Boeing 737 jet aircraft from Langley Research Center to measure aircraft noise characteristics for comparison with existing noise prediction techniques. The primary objective of the study—part of a Federal effort to understand and reduce aircraft noise pollution—was to obtain data on noise from aircraft landing and takeoff procedures close in and at long distances from airports. The data would be used to update existing analytical techniques. (Wallops Release 73–8)

July 31: At a European Space Conference meeting in Brussels, aerospace ministers from 11 countries decided to proceed with Spacelab development, linking with NASA's shuttle program, as part of a European space "package deal." The agreement also included development of the proposed French L–3S launcher, development of the MAROTS experimental maritime communications satellite, and creation of a new single European Space Agency (ESA) by April 1, 1974. The new agency would merge the European Space Research Organization (ESRO) and the European Launcher Development Organization (ELDO). (NASA Release 73–191; Gatland, *CSM*, 8/28/73; *W Post*, 8/73/A3)

- Sen. Pete V. Domenici (R–N. Mex.) announced to the Senate the intention of the City of Alamogordo, N. Mex., to establish an International Space Hall of Fame. Dr. Charles Stark Draper, President of the National Academy of Aeronautics and Astronautics, had offered the Academy's service in selecting nominees. (*CR*, 7/31/73, S15277)

- The National Science Foundation released *Graduate Student Support and Manpower Resources in Graduate Science Education* (NSF 73–304). Graduate science enrollment in doctorate departments applying for NSF traineeships declined 3% from 1970 to 1971 in all scientific areas except psychology and the social sciences. The number of students enrolling for the first time dropped 4% in 1970 and 8% in 1971. Graduate enrollment of foreign students declined 2% in 1971, after having increased 10% from 1967 to 1968 and another 11% by 1969. The enrollment dropoff of nearly 2% in full-time graduate science students was attributed to an acute financial squeeze, during which several leading universities began restricting enrollment, and to student disenchantment with science and technology because of rising unemployment among scientists and engineers. (Text)

July 31–August 1: A five-member Universities Scientific Research Assn. committee to advise on the production of glass in space held a meeting at Marshall Space Flight Center. The committee discussed an MSFC study dealing with the basic phenomena of glass production and toured MSFC facilities. (*Marshall Star*, 8/1/73, 2)

During July: Ames Research Center and Hersh Acoustical Engineering Co. began wind-tunnel experiments to test the effect of serrated rotor edges on the noise level of a simulated jet engine compressor. The concept had been inspired by observations of the owl. A British naval officer had theorized that the owl's unique serrated wing might help explain the bird's quietness in flight. ARC engineer Paul Soderman had discovered through five years' study and testing that comb-like edges did reduce noise at low tip speeds, but the tests had been limited because they were made only statically to simulate a helicopter's hover. (NASA Release 73–144)

- Preliminary results of the Naval Research Laboratory's S–009 experiment aboard the *Skylab 1* Workshop (launched by NASA May 14) were published by *Naval Research Reviews*. NRL scientist Dr. Maurice M. Shapiro, principal investigator for S–009, had said the experiment, to study the heavy atomic nuclei among the cosmic ray particles that constantly bombarded the earth from outer space, had obtained more accurate measurement of the relative abundances of nuclei in the primary cosmic ray influx than had previously been obtained from the exposure of sensitive detectors flown on high-altitude balloon flights. Comparisons had already been made between the source composition of cosmic rays and the element abundances in the solar system and in nearby stars. These suggested that certain nuclear processes occurring on a very short time-scale during the explosion of a supernova could be important to the production of cosmic ray nuclei. (*Naval Research Reviews*, 7/73, 32)

- Neilsen Engineering and Research, Inc., began testing the Ames Research Center-developed Randomdec computer-and-analysis method of application to detecting structural weakness in bridges. Under a NASA contract, monthly readings would be taken of bridges at the Highway 85 and 101 interchange in the San Francisco Bay area for one year. The experiment was jointly funded by NASA and the Federal Highway Administration. (NASA Release 73–161)
- *Compendium of Meteorological Satellites and Instrumentation* (NSSDC 73-02) was published by the National Space Science Data Center, Goddard Space Flight Center. Prepared by the U.S. Air Force Environmental and Technical Applications Center's Air Weather Service, the book included an overview of programs and brief descriptions, orbital information, and operating status of 98 satellites launched or planned since 1959 by the U.S., U.S.S.R., France, and U.K., with some 200 experiments. (Text)
- NASA published *The Quiet Sun* (NASA SP–303) by Skylab 4 Astronaut Edward G. Gibson. In the foreword Dr. Leo Goldberg, Kitt Peak National Observatory Director, said the up-to-date textbook of solar physics was "written from the point of view of a physicist seeking to understand and interpret solar observations in the framework of theoretical physics. This orientation as well as the primary purpose of the book combine to make it a unique and valuable contribution to the literature of solar physics, and a superb textbook for college-level and graduate students." (Text)
- A survey of knowledge about technology and research at the Univ. of Michigan by Chairman David Fradin of the Federation of Americans Suporting Science and Technology (FASST) was reported in *Air Force Magazine*. Of student respondents, 50% to 75% knew the Federal Government was the major source of research and development funds, that it took 6 to 10 yrs to research and develop a new product, that the U.S. had a positive balance of trade in high-technology and agriculture products, and that the trend in U.S. trade had been negative. More than 75% of respondents did not know the amount of Federal expenditure for defense or for human and physical resources, or that aerospace was the Nation's largest manufacturing employer. Scientific effort was deemed unnecessary for national defense by 25% to 50%. From 50% to 75% felt the space program had helped U.S. technical growth, but that money spent in going to the moon had been wasted; they considered tangible space program benefits like new knowledge of the environment to be more important than aerospace leadership, national prestige, or keeping ahead of the U.S.S.R. (*AF Mag*, 7/73, 90–91)

August 1973

August 1: *Cosmos 578* was launched by the U.S.S.R. from Plesetsk into orbit with a 251-km (156-mi) apogee, 202-km (125.5-mi) perigee, 89.0-min period, and 65.4° inclination. The satellite reentered Aug. 13. (GSFC *SSR*, 8/3/73; *SBD*, 8/3/73, 175)

- NASA test pilot John A. Manke successfully piloted the X-24B lifting body on its first glide flight, launched at 12 000-m (40 000-ft) altitude from a B-52 aircraft from Flight Research Center. Manke maneuvered the X-24B to evaluate its flight characteristics and made a practice approach at 9000 m (30 000 ft) that ended successfully in a 320-km-per-hr (200 mph) unpowered landing on the dry lake bed near FRC, after the four-minute flight. The first captive flight of the double-delta-shaped vehicle, in the joint NASA and Air Force research program studying possible hypersonic aircraft configurations, had been made July 19. (NASA prog off; NASA Release 73-147; LA *Her-Exam*, 8/2/73)

- The House Committee on Science and Astronautics' Subcommittee on Manned Space Flight held a hearing on the *Skylab 1* Investigation Report [see July 13]. A Senate hearing had been held July 30. Lewis Research Center Director Bruce T. Lundin, *Skylab 1* Investigation Board Chairman, testified that the failure to recognize design deficiencies in the Orbital Workshop's meteoroid shield that led to its loss during the May 14–June 22 *Skylab 1-2* mission and the failure to communicate the importance of proper venting of the shield must be attributed to "an absence of sound engineering judgment and alert engineering leadership . . . over a considerable period of time." There had been no evidence to indicate that "the design, development and testing of the meteoroid shield were compromised by limitations of funds or time. The quality of workmanship applied to the shield was adequate for its intended purpose." Testing emphasis on ordnance performance and shield deployment had been appropriate. "Engineering and management personnel on Skylab, on the part of both contractor and Government, were available from the prior Saturn development and were thus highly experienced and adequate in number."

The Investigation Board had suggested corrective action: omit the meteoroid shield on any future Workshops, coat the Workshop for thermal control, and rely on meteoroid protection from the Workshop tank walls. If further protection was required, the Board had favored a fixed, nondeployable shield. To reduce the probability of separation failures as had occurred at Saturn S-II interstage second separation plane, linear-shaped charges should be detonated simultaneously from both ends, and all ordnance applications should be reviewed for a similar failure mode. Structural systems that had to move or deploy, or that required other mechanisms or components for operation, should not be considered solely as structure; and complex, multidisciplinary systems should have a designated project engineer.

NASA Associate Deputy Administrator Willis H. Shapley described NASA's actions to prevent recurrence of the Skylab Orbital Workshop anomalies: "We established an independent Investigation Board, have accepted the Board's report, and have directed all our program and project people to examine their projects in light of the findings of the Board. We will follow up to see that changes in formal management systems are made when required." (Transcript)

- The House passed by a vote of 401 to 9 the conference report on H.R. 8825, FY 1974 Dept. of Housing and Urban Development-Space-Science-Veterans appropriations bill that included a $3.002-billion NASA appropriation [see July 26]. (CR, 8/1/73, H7109–16)

- U.S. Embassy officials in Moscow presented to U.S.S.R. President Nikolay V. Podgorny, on behalf of President Nixon, a lunar sample brought from the moon by the Dec. 7–19, 1972, *Apollo 7* mission. Podgorny said in accepting the sample that preparations were "successfully continuing" for the July 1975 U.S.–U.S.S.R. Apollo-Soyuz Test Project mission. He wished "all success" to the *Skylab 3* astronauts (launched July 28 to work in the Orbital Workshop launched May 14). (*W Star-News*, 8/2/73, A2)

- The Senate confirmed the nomination of L/G Samuel C. Phillips, Director of the National Security Agency and former NASA Apollo Program Director, to be an Air Force general. Gen. Phillips became Commander of the Air Force Systems Command. (CR, 8/1/73, S15382; DOD Release 368–73)

- The British Interplanetary Society issued a statement to the press following the July 31 decision by the European Space Conference to create a European Space Agency by April 1, 1974: "Acceptance of a European mini-NASA now promises great rewards for European industry. Long-standing neglect of central management in European space affairs has led to waste of limited resources and frustrated Europe's entry into potentially profitable markets. With ESA, Europe now has a chance to work at the space frontier alongside the United States and to develop other space capabilities of enormous social and economic importance." (Gatland, *SF*, 10/73, 384–5)

- NASA launched two Nike-Apache sounding rockets from Kiruna, Sweden, carrying Dudley Observatory and Univ. of Stockholm experiments to collect and identify particulate matter from noctilucent clouds in the upper atmosphere. The first rocket reached a 109-km (67.7-mi) altitude. The second rocket, launched 30 min later, reached 111 km (69.0 mi). Rockets and instrumentation performed satisfactorily and all experimental objectives were achieved. (NASA Rpts SRL)

- The Senate Committee on Armed Services voted 8 to 7 to cut $885 million from the $1.7 billion requested by the Nixon Administration for the Trident missile-launching submarine program in FY 1974. The Committee was considering S. 1263, an FY 1974 military procurement authorization bill. (CR, 8/1/73, D950; Finney, *NYT*, 8/3/73, 8)

- Rep. Bill Gunter (D–Fla.) and Rep. Charles A. Vanik (D–Ohio) introduced H.R. 9785 to provide for "a coherent rational program of energy research and development." Under the proposed program a five-member Energy Research and Development Commission would develop a national energy R&D policy and the National Science Foundation would collect resources information in a research data base. (CR, 8/1/73, E5279–81)

- Federal support to universities and colleges rose $643 million, or 18%, to $4.1 billion from FY 1971 to FY 1972, the National Science Foundation reported. Federal academic science funding increased by $256 million, or 11%, to $2.6 billion. Federal support of academic research and development activities rose by $302 million, or 19%, to $1.9 billion. The life sciences received $896 million in 1972—21%, more than in 1971; environmental sciences received $187 million, up 38%; and engineering received $193 million, up 22%. (NSF *Highlights*, 8/1/73)

August 1-10: Two Soviet research ships and one U.S. and one Mexican vessel participated in the GARP International Sea Trial (GIST) project, part of the Global Atmospheric Research Program to improve weather forecasting. The National Oceanic and Atmospheric Administration's research vessel *Researcher* met Soviet ships *Academician Korolev* and *Ernst Krenkel* in the Atlantic Ocean east of the Virgin Islands to make a series of simultaneous instrumented observations of the atmosphere and the ocean. The 10-day project would be followed in June 1974 by a 100-day effort in which research ships from 11 countries, 11 aircraft, and 5000 scientists would participate. (NSF Release 73–174; NSF PIO)

August 2: A *Today* editorial commented on energy conservation efforts at Kennedy Space Center, while the *Skylab 3* astronauts, in orbit since July 28, coped with a reaction-control-system oxidizer leak in the command module. Newspaper readers—noting that lights at Launch Complex 39 had remained on throughout July 28, *Skylab 3* launch day—had questioned NASA's compliance with the Nixon Administration's energy conservation edict. One caller had queried all-night lights at the Manned Spacecraft Operations Building parking lot. "A call to Kennedy Space Center confirmed our suspicion: they are human and at times forgetful." But KSC had reduced July energy consumption by 7%, "and it was done very simply: air conditioning equipment was turned off at the end of the working day and one half of the electric light bulbs in halls, corridors and lobbies were removed." KSC had said they would check launch pads and parking lots "to determine if a few hundredths of a percentage point can be added to the seven already realized." (*Today*, 8/2/73)

- NASA launched a Nike-Cajun sounding rocket from Kiruna, Sweden, carrying a Goddard Space Flight Center payload to an 88-km (54.7-mi) altitude. The objective was to study ion composition and density of the ionospheric D- and E-regions at polar latitudes. The data would be compared with data from two Nike-Apaches launched Aug. 1. The rocket and instrumentation performed satisfactorily and all experimental objectives appeared to have been achieved. (NASA Rpt SRL)

- The anti-environmental backlash was in full swing, the *Wall Street Journal* said. Environmentalists were being blamed for the energy crisis, lingering unemployment, high prices, high taxes, food shortages, and the falling birth rate. Several state legislatures had killed ecology-oriented bills they would have passed a session or two before, "and a few others have voted actually to relax environmental-protection laws enacted in the heyday of concern." Environmental groups were encountering serious fund-raising problems. "Inevitably, businessmen, union officials and others who have felt the lash of environmental power in recent years will now be tempted to hit back hard." This could be a risky, "almost fatal" calculation. While the environmentalists might no longer win all the battles, "chances are, too, that sooner or later a fresh disaster—a new oil spill, a heavy smog—will rebuild environmentalists' strength." (Otten, *WSJ*, 8/2/73)

August 3: A *New York Times* editorial commented on the Aug. 2 discovery of an oxidizer leak that threatened the safety of the *Skylab 3* astronauts (launched July 28 to man the Orbital Workshop launched May 14): "Even though it finally proved unnecessary to order three men back to earth immediately, two points did emerge with special clarity from this short-lived scare situation. First, not all difficulties that can arise in manned space missions can be repaired by men on a troubled vehicle." Second, "and more fundamental, this latest incident again brings into sharp focus the need for standby rescue rockets for manned space ventures. The lack of rescue vehicles in the past has come in for justified and severe criticism. This time NASA has demonstrated that it has learned the lessons of history." An Apollo spacecraft was being readied as a rescue vehicle for the *Skylab 3* crew if a new emergency arose. "Men in space are no longer forced to be so self-sufficient and independent of help from earth as was the case not long ago, especially during the Apollo voyages to the moon." (*NYT*, 8/3/73, 30)

- Sen. Frank E. Moss (D–Utah), Chairman of the Senate Committee on Aeronautical and Space Sciences, introduced S. 2350 to amend the National Aeronautics and Space Act of 1958 "to provide for the coordinated application of technology to civilian needs in . . . earth resources survey systems," to establish an Office of Earth Resources Systems within NASA, and for other purposes. The bill was referred to the Senate Committee on Aeronautical and Space Sciences. (*CR*, 8/3/73, S15593)

- Results of *Apollo 14* and *16* (launched Jan. 31, 1971, and April 16, 1972) heavy-particle dosimetry experiments were reported in *Science* by scientists from Johnson Space Center and the General Electric Research and Development Center. Earlier measurements using Apollo helmets as dosimeters had indicated that a significant 1% of certain nonregenerative motor-control cells in the bodies of astronauts in space would be killed by radiation from densely ionizing, heavy atomic nuclei with the current Apollo spacecraft shielding. Doses of heavy particles penetrating the CMs of *Apollo 8, 12, 14,* and *16* correlated well with calculated effects of solar modulation of the primary cosmic radiation. Differences in doses at different positions inside the CMs had indicated that the redistribution of mass within the spacecraft could enhance safety from the biological damage that would otherwise be expected on manned, deep-space missions. (Fleischer *et al.*, *Science*, 8/3/73, 436–8)

- Appointment of David R. Scott, former astronaut, as Deputy Director of Flight Research Center was announced by FRC. Scott had flown on *Gemini 8* (March 16–17, 1966) and *Apollo 9* (March 3–13, 1969) missions and was commander of *Apollo 15* (July 26–Aug. 7, 1971). After leaving the astronaut corps in 1972, he had served as Technical Assistant to the Apollo Program Manager and as Special Assistant for Mission Operations and Government Furnished Equipment in the Apollo Spacecraft Office at Manned Spacecraft Center (which became Johnson Space Center Feb. 17). He held two NASA Distinguished Service Medals, the NASA Exceptional Service Medal, two Air Force Distinguished Service Medals, the Air Force Distinguished Flying Cross, the David C. Schilling Trophy of the Air Force, and the Robert J. Collier Trophy for 1971. (FRC Release 21–73)

- Samples of the material used by NASA for Skylab's makeshift sunshade during the May 14–June 22 *Skylab 1–2* mission had been subjected during the mission to high-intensity ultraviolet radiation at the Air Force Systems Command's Arnold Engineering Development Center, AFSC announced. The tests, to determine the degree of degradation the material experienced from solar UV radiation, had been requested by Marshall Space Flight Center. This accelerated "aging" of the samples enabled NASA engineers to decide that it was not necessary to replace the shield during the *Skylab 1–2* mission. (AFSC Release 01P 104.73)
- Marshall Space Flight Center announced receipt of proposals from five firms for a contract for design, procurement, fabrication, installation, and checkout of a data-acquisition system to support space shuttle structural testing. The firms were Avco Corp., Computer Science Corp., Grumman Aerospace Corp., Systems Engineering Laboratories, and Wyle Laboratories. The contract would be awarded in October. (MSFC Release 73–107)
- Rep. Mike McCormack (D–Wash.), with 23 cosponsors, introduced H.R. 9974 to establish a Department of Energy. (*CR*, 8/3/73, H7404)
- NASA launched a Nike-Apache sounding rocket from Wallops Island carrying a Univ. of Illinois payload to a 184.3-km (114.5-mi) altitude. The objective was to measure electron temperature and electron density profile of intense blanketing of the sporadic E-region at 100- to 120-km (60- to 75-mi) altitude. The rocket and instrumentation performed satisfactorily. All experimental objectives appeared to have been achieved. (NASA Rpt SRL)
- President Nixon submitted to the Senate the nomination of John Eger to be Deputy Director of the Office of Telecommunications Policy. Eger, legal assistant to the Federal Communications Commission Chairman, would succeed George F. Mansur, Jr., who had resigned effective April 1. (*PD*, 8/6/73, 951, 957)

August 5: The U.S.S.R. launched its *Mars 6* probe from Baykonur Cosmodrome at 10:46 pm local time (1:46 pm EDT) into a parking orbit from which it was inserted into a trajectory toward the planet Mars. Tass reported *Mars 6* would reach Mars in March 1974 to join *Mars 4* and *5* (launched July 22 and 25) in the exploration of the planet, its surrounding space, and the characteristics of interplanetary space. *Mars 6* differed somewhat in design from *Mars 4* and *5*. In addition to Soviet equipment it carried instruments designed and manufactured by French specialists for the joint Soviet-French experiment Stereo, to study the sun's radioemission in one-meter wavelength and characteristics of the solar plasma and cosmic rays. (GSFC *SSR*, 8/31/73; FBIS–Sov, 8/6/73, L1; *SBD*, 8/7/73; Shabad, *NYT*, 8/7/73, 19)

- Discovery of huge, shallow craters at the near-equatorial zone of Venus was announced by Jet Propulsion Laboratory radar astronomers headed by Dr. Richard A. Goldstein. The team had produced a map showing a 1500-km (930-mi) section of Venus, about the size of Alaska, that showed 12 craters up to 160 km (100 mi) across. The discovery had been made with high-intensity radar beams from Goldstone Tracking Station in the Mohave desert. The resolution, about 10 km (6 mi), was five times better than previously obtained. (JPL Release 665; NASA Release 73–145)
- A Baltimore *Sun* editorial commented on the plight of *Skylab 3* as the astronauts (launched July 28 to man the Orbital Workshop launched

May 14) prepared for Aug. 6 extravehicular activity and ground crews studied coolant system problems in the ows: "The public has grown increasingly sanguine about NASA's manned missions, almost considering them routine, because of the agency's perfect safety record on manned missions." The Aug. 2 oxidizer leak had shown "that the missions are anything but free of hazard, and the mental stress imposed upon the three astronauts must be intense. We hope, of course, that the mission will continue for the full 59 days and that the astronauts will then make it home safely. But if there is any serious question about the condition of the Apollo ship, then NASA certainly should exercise its option of sending a rescue ship up to Skylab and ending the mission in advance of schedule." (B *Sun*, 8/5/73)

August 6: A Chicago *Sun-Times* editorial commented on space rescue capability as the *Skylab 3* astronauts, manning the Orbital Workshop since July 28, conducted their first extravehicular activity: "There is a certain amount of calculated risk involved at this period of space experimentation. In a real emergency Russia might be able to help out with one of its spacecraft. When the shuttle program becomes effective, a reusable winged spaceship could be made ready for launching in two hours." Meantime, "all we can do is trust that the good luck, astronauts' skill and inventiveness of the space program people that have provided magnificent results to date continue during the current series of missions." (C *Sun-Times*, 8/6/73)

- Langley Research Center and the Virginia Institute of Marine Science (VIMS) were using aerial photography to identify Tidewater Virginia vegetation in a project to aid Virginia in using its marshlands. High-resolution photos from helicopters would provide accurate knowledge of vegetation species to enable VIMS to classify plant life in specific marshes. VIMS also was working with Wallops Station to use multispectral aerial photography to identify features of specific vegetation by low-resolution photos. (NASA Release 73-146)

- The House attitude toward solar energy as a factor in relieving the energy crisis was discussed in a *Washington Post* guest column by Rep. James W. Symington (D–Mo.), Chairman of the Subcommittee on Space Science and Applications, and Rep. Mike McCormack (D–Wash.), Chairman of the Subcommittee on Energy, both of the House Committee on Science and Astronautics. Members of the subcommittees had expressed the urgent need for an accelerated program in solar energy research and development and had encouraged NASA and the National Science Foundation to use funds provided to develop "an expanded, mission-oriented program." But they warned against expecting that solar energy would provide any significant relief from the energy crisis during the immediate future: "We cannot emphasize strongly enough that this nation must depend primarily upon the combustion of coal, gas, and oil, and on nuclear fission, for energy for the balance of this century. No amount of enthusiasm for solar energy, and no expansion of solar research and development, should keep us from pursuing and expanding the more immediate and urgent research and development programs in coal gasification and liquification and in research and development associated with nuclear reactors and nuclear breeders." (*W Post*, 8/6/73, A23)

- NASA launched two Nike-Apache sounding rockets from Kiruna, Sweden. The first carried a Goddard Space Flight Center payload to a 123-km

(76.4-mi) altitude to study ion composition and density in the ionospheric D- and E-regions at polar latitudes. A second Nike-Apache carried a Dudley Observatory and Univ. of Stockholm payload to a 103-km (64.0-mi) altitude to collect background data during the absence of noctilucent clouds for comparison with data obtained by two Nike-Apaches launched from Kiruna Aug. 1 and a Nike-Cajun launched from Kiruna Aug. 2. Both rockets and instrumentation performed satisfactorily and experimental objectives appeared to have been achieved. (NASA Rpts SRL)

- The National Science Foundation released *Immigrant Scientists and Engineers in the United States* (NSF 73–302). A mid-1970 survey of characteristics and attitudes of foreign scientists and engineers who had immigrated to the U.S. between 1964 and 1969 showed most had come seeking a higher standard of living. Nearly 6 of every 10 held a Ph.D. or Master's degree. One of every five had continued graduate training. More than 90% were employed in professional positions, and more than one half were in research and development. Although the majority were under 40 yrs of age, 35% had authored scientific papers in the U.S. and 8% held U.S. patents. (Text)

August 7: The earth had passed between NASA's *Pioneer 10* (launched toward Jupiter March 2, 1972) and the sun for the second time during its 1-billion-km (620-million-mi) journey, Ames Research Center announced. *Pioneer 10* was 850 million km (530 million mi) from the earth and was scheduled to make history's first visit to Jupiter in December. When the spacecraft, the earth, and the sun were aligned in inferior conjunction, *Pioneer 10*'s sun sensor would look almost directly at the sun, interfering with the sensor's count of the spacecraft rotation rate. To avoid the problem, ARC flight directors were keeping the pointing direction of the spacecraft's radio beam and sun sensor 1.5° from the sun and 1° from the earth.

Pioneer 11 (launched toward Jupiter April 5, 1973) was 150 million km (93 million mi) from the earth. Both spacecraft were functioning well. (ARC Release 73–84)

- Johns Hopkins Univ. Applied Physics Laboratory held the first public demonstration of its new heart pacemaker that used electric and electronic components designed by NASA for spacecraft use. The new pacemaker, smaller and longer lasting than conventional models, could be implanted more quickly and easily. It had no life-limiting components or radioactive emissions, was immune from outside interference sources, and could be recharged by patients at home. It incorporated modified space satellite power cells. (NASA Release 73–150; Cohn, *W Post*, 8/8/73)

August 8: Spare-time diversions of the astronauts aboard the Orbital Workshop in the May 25–June 22 *Skylab 2* mission and in *Skylab 3* since July 28 launch were described in a *Christian Science Monitor* article. *Skylab 3* Astronaut Dr. Owen K. Garriott, the first physicist in space, had been fascinated by the zero-g antics of his spiders Arabella and Anita and his mummichog minnows. *Skylab 2* Astronaut Dr. Joseph P. Kerwin had experimented with a big blob of water which floated before him instead of splashing on the spacecraft deck. The blob had taken on the form of the sun and planets as he spun it in space. He had blown it into a bubble with a straw and then had blown bubbles within the bubble. His *Skylab 2* colleague, Astronaut Charles Conrad, Jr., had

August 8

discovered that centrifugal force could impel him around the weightless spacecraft upright if he got himself off to a fast enough start racing around the Workshop. Other astronaut pastimes included "marooning" (doing acrobatics in remote sections of the spacecraft), "hamming it up" for TV and "home movies," reading novels and poetry, ricocheting weightless rubber balls around the laboratory, and flying paper airplanes. (Salisbury, *CSM*, 8/8/73)

- Components of a navigation system similar to that of the lunar roving vehicles launched aboard *Apollo 15, 16,* and *17* (July 26, 1971, and April 16 and Dec. 7, 1972) to explore the lunar surface had been turned over by Marshall Space Flight Center to the Univ. of Kentucky for possible incorporation into a remotely controlled mine-surveillance vehicle, the *Marshall Star* announced. (*Marshall Star*, 8/8/73, 2)
- A *New York Times* editorial noted that "works aimed at exploring the planets by instruments of the most diverse sort" continued activity after the close of manned exploration of the moon. It commented on the discovery by Jet Propulsion Laboratory scientists that the Venus equatorial area was dominated by huge craters [see Aug. 5]. (*NYT*, 8/8/73, 32)

August 9–16: The U.S.S.R. launched its *Mars 7* probe from Baykonur Cosmodrome at 10:00 pm local time (1:00 pm EDT) into an earth parking orbit from which it was inserted into a trajectory toward Mars. The parameters of the trajectory were as planned, Tass reported. *Mars 7* was the fourth in a series of spacecraft launched by the U.S.S.R. toward Mars during the month-long window when the earth and Mars were in alignment for a space launch. *Mars 4* and *5* were launched July 22 and 25. *Mars 6* was launched Aug. 5.

Mars 7, which would reach its destination in mid-March 1974, was similar in design and mission to *Mars 6*. Both spacecraft—intended for exploration of Mars, near-Mars space, and the interplanetary medium—carried Soviet- and French-built scientific equipment. All four Mars probes would carry out research in combination. Tass said *Mars 4* would perform experiments using instruments on *Mars 6* while *Mars 5* would work in conjunction with *Mars 7*.

On Aug. 10, at 4:00 am Baykonur time (7:00 pm EDT, Aug. 9), *Mars 4* was 6 432 000 km (3 997 000 mi) from the earth; *Mars 5*, 5 067 000 km (3 148 000 mi) from the earth; *Mars 6*, 1 535 000 km (954 000 mi) from the earth; and *Mars 7*, 102 000 km (63 000 mi) from the earth. Tass reported that all spacecraft were proceeding according to their programs.

A midcourse correction of the *Mars 6* trajectory Aug. 13 and of the *Mars 7* trajectory Aug. 16 put the flight paths "within the pre-set limits." Tass said onboard systems of all four probes were functioning normally and measurements were being made of physical characteristics of interplanetary space. (GSFC *SSR*, 8/31/72; FBIS–Sov, 8/10/73, L1; 8/16/73, L1; 8/17/73, L1; *SBD*, 8/13/73, 217)

August 9: Marshall Space Flight Center announced the award of identical, parallel, $800 000, 17-mo contracts to Itek Corp. Optical Systems Div. and Perkin-Elmer Corp. for competing preliminary designs and program definition of the Large Space Telescope (LST). The large, multipurpose, orbiting, optical astronomical facility would be launched by the space shuttle, which also would return it to the earth for major maintenance and refurbishment. The LST would use a 3-m (10-ft) primary mirror and—because of its size, quality, and location above the earth's atmosphere—would be able to resolve objects 10 times smaller and 100 times fainter than those observed by ground telescope. (MSFC Release 73-110)

- Ames Research Center earth resources survey aircraft were monitoring tussock moth damage to Douglas fir trees in Washington and Oregon at the request of the Washington State Dept. of Natural Resources. Two flights over the Blue Mountain area, covering parts of each state, would help determine damage done by the moths, which also attacked Ponderosa trees. (NASA Release 73-149)
- NASA launched two sounding rockets. A Black Brant IVB, launched from Wallops Island, carried an Air Force Cambridge Research Laboratory and Goddard Space Flight Center payload to a 603.5-km (375-mi) altitude to flight-test a modified version of the Black Brant IVB. The rocket and instrumentation performed satisfactorily.

  A Black Brant VC, launched from White Sands Missile Range, carried a Harvard College Observatory experiment to a 268.8-km (167-mi) altitude to collect data on solar activity to collaborate with data collected by the Apollo Telescope Mount instruments aboard the Skylab Orbital Workshop (launched May 14), now manned by the *Skylab 3* crew (launched July 28). A capsule containing film was parachuted to the earth and recovered 160 km (100 mi) from the launch site. One detector failed but the experiment was 80% successful. The rocket performed satisfactorily. (NASA proj off; Reuters, B *Sun*, 8/10/73, A6)

August 10: NASA announced it had granted a five-year patent license to K–H Enterprises to manufacture and sell a burglar-intrusion detection system based on an Ames Research Center-developed capacitance pressure transducer. The transducer had been developed by ARC scientist Grant Koon to measure pressures on wind-tunnel models. In its application as an aircraft intrusion detection system, the transducer sensed the physical mass of an approaching person in the aircraft's vicinity as a change in capacitance and activated an alarm. (NASA Release 73-148)

- NASA launched a Nike-Apache sounding rocket from Wallops Island carrying a Univ. of Illinois payload to a 187.8-km (116.7-mi) altitude. The objective was to measure electron temperature and electron density profile of intense blanketing of the sporadic E-region at 100 to 120 km (60 to 75 mi). Data would be compared with data from a Nike-Apache launched Aug. 3. The rocket and instrumentation performed satisfactorily. All experimental objectives appeared to have been achieved. (NASA Rpt SRL)
- Remarks by French oceanologist Dr. Jacques-Yves Cousteau during a lecture series sponsored by Ames Research Center, San Francisco Univ., San Jose State Univ., and the Astronomical Society of the Pacific were quoted by the *San Francisco Examiner:* A "new awareness, a space-age consciousness especially among younger people" could "cause mankind to abandon violence, money and false material pleasures based on uncontrolled production and consumption, and to start preparing planet Earth for a cosmic awareness and a personal immortality." (*SF Exam*, 8/10/73)

August 13: Academic research and development expenditures had reached $3.3 billion in FY 1972, the National Science Foundation reported. Since 1964, R&D development at universities and colleges had doubled.

The increase, from $1.6 billion to $3.3 billion, represented an average annual growth of 9.7%. The Federal share declined from a high of 60% in 1968 to 55% in 1972. Institution funds increased from $615 million to $924 million, or 10.7% annually 1968–1972. The $3.3 billion total for academic R&D in 1972 was only 11% of the national R&D total, but the $2.5 billion allocated to basic research amounted to 59% of U.S. basic research. (NSF *Highlights*, 8/13/73; NSF Sci Resources Studies Div)

- Gen. Bruce K. Holloway (USAF, Ret.), former Commander of the Air Force Strategic Air Command, joined NASA as a consultant to the Administrator. (NASA Ann, 9/10/73)

August 14: An Apollo command module atop its Saturn IB launch vehicle was rolled out from the Vehicle Assembly Building to Launch Complex 39, Pad B, at Kennedy Space Center. Vehicle readiness had been ordered by Skylab Program Director William C. Schneider for a possible rescue mission of *Skylab 3* astronauts (launched July 28) when two of the four quadrants of the service module's reaction control system, used for maneuvering the command and service module, failed and raised doubts about the ability of the CM to return safely to earth. If a rescue system were not ordered, the vehicle would carry Skylab 4 astronauts—Gerald P. Carr, Dr. Edward G. Gibson, and William R. Pogue—to a Nov. 9 rendezvous with the Skylab Orbital Workshop. (*Marshall Star*, 8/15/73, 1; Thomas, *Today*, 8/14/73)

- An Intergovernmental Agreement on U.S. and European cooperation in space shuttle development was signed in Paris by the U.S., Belgium, France, West Germany, Switzerland, and the United Kingdom. The agreement included European funding and development of the Spacelab for the space shuttle. A NASA-European Space Research Organization (ESRO) Memorandum of Understanding that supplemented the agreement was initialed by NASA and ESRO representatives in Europe. The memorandum would be signed in Washington, D.C., by Dr. James C. Fletcher, NASA Administrator, and Dr. Alexander Hocker, ESRO Director General, Sept. 24. (NASA Release 73-191; NASA Int Aff)

- Government investigations of the *Skylab 1–2* mission (May 14–June 22) had indicated "we have put too much reliance on computers, on management systems, on brilliant paper designs and on theory," columnist Ray Crowley said in the *Washington Star-News*. Neglected had been "human intuition, communications between the men working on different aspects of major programs and practical experience." The system had submerged the individual. Missing had been "the old-fashioned 'chief engineer' with few managerial duties who brought his total experience and spent most of his time integrating all elements of the system." A major cause of *Skylab 1–2* anomalies had been "lack of discussion between the groups of men working on the project." The investigators had cautioned that management must counteract the natural tendency of engineers to believe a drawing is the real world. One team had found that extensive use of computers for complex analyses could serve to remove the analyst from reality. Attention to rigor, detail, and thoroughness, "if carried too far, can inject an undue emphasis on formalism and documentation." This caused too little thought to be given to "how does it work." (*W Star-News*, 8/14/73, A11)

- NASA and the Soviet Academy of Sciences had approved the report of the June 7–15 Moscow meeting of the U.S.–U.S.S.R. Joint Editorial Board

preparing the publication "Foundations of Space Biology and Medicine," NASA announced. Publication of the first volume was planned for May 1, 1974, and the second and third by July 1, 1974. (NASA Release 73-151)
- Ames Research Center selected California Airmotive Corp. for negotiations leading to a replacement for the Convair 990 instrumented aircraft *Galileo* lost in an April 12 mid-air collision over Moffett Field, Calif. Negotiations would be based on NASA's acquisition of a Convair 990 aircraft with spare parts and modifications to prepare the aircraft for airborne scientific missions. If negotiations were successful, NASA would continue its airborne science program with an identical aircraft. (NASA Release 73-156)
- NASA launched a Nike-Apache sounding rocket from White Sands Missile Range carrying a Univ. of Pittsburgh payload to a 117.1-km (72.8-mi) altitude. The experiment studied mesopheric composition with an optical mass spectrometer measuring both the minor atomic and the molecular species in the presence of major atmosphere constituents. The rocket and instrumentation performed satisfactorily. (NASA Rpt SRL)
- A Cleveland *Plain Dealer* editorial commented on the first anniversary of the *Erts 1* Earth Resources Technology Satellite: "For Ohio alone, ERTS is telling truer tales about strip mining and pollution. For others it is doing such diverse things as charting safe routes for ocean shipping and forecasting water requirements of inland areas. The list of what already has been done and is yet to be done is almost endless." (Cl *PD*, 8/14/73)

August 15: NASA held an earth resources experiment package briefing at Johnson Space Center. Dr. Verl R. Wilmarth, EREP project scientist at JSC, said that *Skylab 3* Astronauts Alan L. Bean, Dr. Owen K. Garriott, and Jack R. Lousma, launched July 28 to crew the Orbital Workshop, had completed 10 EREP passes and had exceeded four hours of data collection. They had photographed 125 test sites over the U.S., South America, Thailand, East Malaysia, and Australia under the conditions required by the individual primary investigators. Wilmarth said the equipment was "working great" with "essentially no problem at all with any of the camera systems." (Transcript)
- Ames Research Center light aircraft and various ground stations were being used to document air pollution in the San Francisco Bay area under a two-year National Science Foundation project. ARC, the Bay Area Pollution Control District, and the Atomic Energy Commission's Lawrence Livermore Laboratory were collaborating in the project. ARC would make atmospheric measurements and computer studies in photochemisty to determine how sunlight affected pollution. The Livermore Laboratory, operated by the Univ. of California, would use weather and pollution data to verify a model under development. Radiosonde balloons launched by students at San Jose State Univ. also would gather data under an ARC grant. Another grant, with the Statewide Air Pollution Research Center at the Univ. of California at Riverside, would analyze hydrocarbons. (ARC Release 73-90)
- California Lt. Gov. Edwin Reinecke and Dr. Hans M. Mark, Ames Research Center Director, had concluded an agreement under which ARC would assist the California Office of Emergency Services in developing a system for rapid evaluation of emergency situations, NASA announced. The agreement was part of an ARC pilot program to develop disaster-assessment systems for regional or Federal agencies. ARC would provide

high-altitude multispectral photography, thermal infrared imagery, and radiometric surveys using earth resources research techniques to survey fire, flood, earthquake, landslide, oil spill, air pollution, or other disasters. (NASA Release 73-157)
- Findings that solar power held vast potential to supplement energy needs were presented at Marshall Space Flight Center by university faculty fellows in an MSFC Systems Engineering Design Program. The joint MSFC-Auburn Univ. program was to provide information and experience for developing multidisciplinary systems engineering design courses and programs. The fellows had made a systems approach study of the application of solar technology and research to meet the energy crisis. (MSFC Release 73-109; MSFC PAO)
- Grumman Aerospace Corp. was surveying Langley and Lewis Research Centers under a $60 000 NASA contract to identify any clearly inefficient energy-use practices, NASA announced. Areas of major potential energy savings would be identified and corrective action taken. (NASA Release 73-159)
- President Nixon announced his intention to nominate John R. Quarles, Jr., as Deputy Administrator of the Environmental Protection Agency succeeding Robert W. Fri, whose resignation he accepted. The President also designated Quarles, Acting EPA Deputy Administrator, as Acting Administrator pending the confirmation and appointment of Russell E. Train as Administrator. Quarles' nomination was submitted to the Senate Sept. 5 and confirmed Sept. 14. (PD, 8/20/73, 982; 9/10/73, 1070; EPA PIO)

August 16: NASA announced tentative plans to observe the Comet Kohoutek during the Skylab 4 mission, planned for Nov. 9 launch to join the Skylab Orbital Workshop. The Apollo Telescope Mount instruments provided a broad capability for comet observations over a range of spectral bands. Observations would begin in mid-November and continue until shortly before mission completion, scheduled for Jan. 4, 1974. The tentative plans would be reviewed after the completion of the *Skylab 3* mission (launched July 28). If technical problems necessitated an earlier launch to minimize the length of unmanned operations, it would not be possible to observe the comet. A possible 10-day delay of the launch beyond Nov. 10 would also be considered, to permit observation of the comet until Jan. 14.

NASA also deleted the requirement to maintain the capability to launch the backup Workshop, which originally was to have been maintained to Sept. 30. (NASA Release 73-162)
- NASA announced selection of Martin Marietta Corp. for negotiation of a $107-million cost-plus-award-fee contract for the design, development, test, and evaluation of the space shuttle external tank. The development phase would include fabrication of three ground-test tanks and six developmental flight tanks. The contract would run through 1978. (NASA Release 73-163)
- College student participants in a 10-week Summer Institute for Biomedical Engineering at Goddard Space Flight Center presented results of the study cosponsored by GSFC and Howard Univ. Ten students had worked to solve problems in health care by applying NASA space age technology. Experiments had included obtaining data from 40 electrocardiogram electrodes simultaneously without using bulky and time-consuming equipment, disposing of anesthetic fumes exhaled by patients during

surgery, using space tools to weigh premature infants without removing them from a controlled environment, measuring stresses in human bones, and researching deafness. (NASA Release 72-152)
- NASA launched two Arcas sounding rockets from Antigua, West Indies, carrying Goddard Space Flight Center experiments to measure ozone distribution in the upper atmosphere, monitor anomalous ultraviolet absorption, and extend the data base for a climatology of stratospheric ozone in the tropics. Both launches were in conjunction with a *Nimbus 4* satellite (launched April 8, 1970) overpass. The first Arcas reached a 52.7-km (32.7-mi) altitude. Ejection of the payload-parachute unit was accomplished, but the nosecone remained attached to the instrumentation section shielding the sensors from the atmosphere. No data were acquired. The second Arcas, launched 80 min later, reached a 51.2-km (31.8-mi) altitude but the payload-parachute unit did not eject and no data were received. (NASA Rpts SRL)

August 17: The Air Force launched an unidentified satellite on a Thor-Burner II booster from Vandenberg Air Force Base into orbit with an 853-km (530.0-mi) apogee, 808-km (502.1-mi) perigee, 101.5-min period, and 98.8° inclination. (GSFC *SSR*, 8/30/73; Pres Rpt 74)
- NASA test pilot John A. Manke successfully completed the second glide flight of the X-24B lifting body from Flight Research Center. The X-24B, launched from a B-52 aircraft at an altitude of 14 000 m (45 000 ft), completed all objectives: flying in transonic configuration, verifying longitudinal trim curve, obtaining stability and control data, making a pressure survey of the fin and rudder end, and studying landing gear dynamics. Part of the flight was flown with dampers off. All planned maneuvers were completed. (NASA prog off)
- NASA announced selection of United Aircraft Corp. Sikorsky Aircraft Div. for negotiation of a $25-million cost-reimbursement contract to design, fabricate, flight-test, and deliver two rotor systems research aircraft (RSRA). The aircraft, helicopters with wings and auxiliary engines, would be used as flying laboratories to develop advanced rotor system technology and flight-testing of new rotorblade concepts in a jointly funded NASA and Army research program. (NASA Release 73-165)
- The U.S.S.R. had successfully flight-tested missiles with multiple warheads that could be directed to separate targets, Dr. James R. Schlesinger, Secretary of Defense, told a Washington, D.C., press conference. One of the intercontinental missiles, the SS-18, had carried at least six hydrogen warheads in the one-megaton range. (Binder, *NYT*, 8/18/73, 1)
- Arizona State officials would use photos taken by NASA, from earth resources survey aircraft in the Arizona land use experiment, part of a three-way agreement by NASA, the Dept. of the Interior, and Arizona, NASA announced. The photos would be used to compile 1800 detailed maps of Arizona, update county maps, analyze high-accident areas for the Arizona Highway Dept., develop a land-use and vegetation-cover map for the Arizona Game and Fish Dept., study land areas requiring lease application decisions for the Arizona Land Dept., and develop a state airport systems plan. (NASA Release 73-158)
- Astro Met Associates would market a fluoride-metal composite material that self-lubricated at high temperatures under a 10-yr exclusive license covering two NASA patents, one for material composition, the other for its manufacture, NASA announced. The license was granted to create the incentive to develop and market the material—fluorides impreg-

nated into porous nickel, cobalt, or iron alloys—in the shortest possible time. The composite, designed for self-lubrication of moving parts in machines and equipment at high operating temperatures, might be used in rotary engines. (NASA Release 73-155)

- President Nixon announced his intention to nominate Dr. Walter B. LaBerge Technical Director of the Naval Weapons Center at China Lake, Calif., to be Assistant Secretary of the Air Force for Research and Development. The President also announced his intention to nominate Norman R. Augustine, Vice President for Advanced Development with LTV Aerospace Corp.'s Vought Missiles and Space Co., to be Assistant Secretary of the Army for Research and Development and David S. Potter, Director of Research for the Detroit Diesel Allison Div. of General Motors Corp., to be Assistant Secretary of the Navy for Research and Development. The nominations were sent to the Senate Sept. 5 and confirmed Sept. 14. (PD, 8/20/73, 998; 9/10/73, 1070; DOD PIO)

August 18: Pioneer 11, launched April 5 on a 1-billion-km (620-million-mi) journey to Jupiter, entered the Asteroid Belt 300 million km (190 million mi) from the earth. The spacecraft would travel seven months through the 251-million-km (156-million-mi) belt of rocky celestial bodies orbiting the sun and was scheduled to exit March 12, 1974. Its sister spacecraft, *Pioneer 10* (launched toward Jupiter March 2, 1972), had safely completed crossing the Asteroid Belt in February and was 544 million km (338 million mi) from the earth. *Pioneer 10* would reach Jupiter in December 1973; *Pioneer 11* would reach the planet in December 1974. (NASA Release 73-164)

August 19: Former Presidential Staff Assistant Guy W. Simpler assumed the newly established NASA position of Special Assistant to the Assistant Administrator for Institutional Management (for Executive Recruiting). The position had been established to provide for key personnel recruitment at top level. Simpler had been responsible for search assignments for subcabinet-level positions at the White House. (NASA Hq *WB*, 8/31/73)

August 20: Twentieth anniversary of the first successful Redstone missile launch Aug. 20, 1953, from an improvised pad at Cape Canaveral (now Cape Kennedy), Fla. The Redstone, developed for the Army by Dr. Wernher von Braun and his rocket team, was the first truly ballistic missile in the U.S. The Mercury-Redstone vehicle launched NASA's *Freedom 7* May 5, 1961, on the mission carrying Astronaut Alan B. Shepard, Jr., as the first American in space. (MSFC Release 73-112)

- The National Science Foundation released *Immigrant Scientists and Engineers Decline in FY 1972; Physicians Increase Sharply* (NSF 73-311). The 11 300 immigrant scientists admitted to the U.S. in FY 1972 were 14% below the 13 100 in FY 1971. A record 13 300 had entered the U.S. in 1970. Immigrant physicians and surgeons admitted in FY 1972 totaled 7100, the largest influx over the past 20 yrs. The major factor in the decrease of immigrant scientists and engineers appeared to be revisions by the Dept. of Labor in regulations for certifying scientists and engineers applying for immigration. Since Feb. 4, 1971, these immigrants were certified only if they had a job offer for which U.S. scientists and engineers were not available and their employment did not adversely affect wages and working conditions of U.S. scientists and engineers. (NSF *Highlights*, 8/20/73)

- Newspaper editorials and columnists commented on the Aug. 17 Dept. of Defense announcement that the U.S.S.R. had flight-tested multiple independently targeted reentry vehicles (MIRVs).

    New York Times: "Since the Soviet MIRV tests have been anticipated for so long, there is no justification for a panic reaction from Washington, or a further fattening of the strategic weapons budget. The latest Soviet tests increase the urgency of concluding a MIRV limitation accord, but they also add to the already overwhelming mutuality of interest in achieving it." (*NYT*, 8/20/73, 20)

    Washington Post: "The *potential* of the MIRV technology coupled with the Soviets' giant missiles—especially given their present numerical advantage in certain key categories—could be menacing indeed." It was "precisely this Soviet potential which argues for a serious, effective and non-wasteful approach to the circumstances in which we now find ourselves. This . . . is no time to proceed with turkeys and lemons, or to box ourselves into investments in weapons systems whose cost is exceeded only by their irrelevance to such danger as we would be confronting a few years hence if the 'worst case' regarding Soviet armament came true." Before the issue of phasing out the present U.S. arsenal in favor of less vulnerable sea- and air-based forces, "one reaches the more immediate question of how this Soviet development can or cannot be tamed within the context of the current phase of the arms negotiations." (*W Post*, 8/20/73, 24)

August 21: The Air Force launched an unidentified satellite from Vandenberg Air Force Base by a Titan IIIB-Agena booster into orbit with a 39 132-km (24 315.5-mi) apogee, 392-km (242.6-mi) perigee, 11-hr 41-min period, and 63.3° inclination. The press later reported that the spacecraft would test techniques to be used in the USAF satellite communications system that provided polar coverage for the Strategic Air Command bombers. (UN Reg; *Av Wk*, 9/10/73, 17)

- The U.S.S.R. launched *Cosmos 579* from Plesetsk into orbit with a 307-km (190.8-mi) apogee, 174-km (108.1-mi) perigee, 89.3-min period, and 65.4° inclination. The satellite reentered Sept. 3. (GSFC *SSR*, 8/31/73; 9/30/73; *SBD*, 9/6/73, 10)

- Discovery by use of NASA's *Oao 3* (*Copernicus*) Orbiting Astronomical Observatory that the density of the universe was much lower than expected was announced on the first anniversary of the satellite's launch. Princeton Univ. scientists Dr. John B. Rogerson and Dr. Donald G. York reported they had used *Oao 3*'s telescope to determine the relative abundance of deuterium (a heavy form of hydrogen) in interstellar gas to be one deuterium atom for every 100 000 atoms of gas. Knowing what fraction of hydrogen became deuterium and accepting the theory that deuterium had been created moments after the "big bang" believed to have started the expansion of the universe, the scientists had calculated the amount of matter created and the average density of the current universe. The extremely low density indicated that there was not a relatively large amount of invisible mass in the universe. (*NYT*, 8/21/73, 13; *A&A 1972*)

- The Federal Communications Commission's Common Carrier Bureau petitioned the FCC to force Communications Satellite Corp. into an immediate 25% rate cut. The Bureau, representing consumers in the eight-year-old FCC investigation of ComSatCorp, said that increasing satellite traffic and the use of more efficient plant had increased ComSatCorp's annual earning rate from 10% in 1971 to about 25%. (Aug, *W Star-News*, 8/25/73, A1; FCC PIO)

August 22: The U.S.S.R. launched *Cosmos 580* from Plesetsk into orbit with a 490-km (304.5-mi) apogee, 273-km (169.6-mi) perigee, 92.1-min period, and 70.9° inclination. The satellite reentered April 1, 1974. (GSFC *SSR*, 8/31/73; 6/30/74; *SBD*, 9/6/73, 10)

- *Apollo 13* Astronaut Fred W. Haise, Jr., was critically burned when a BT-13 aircraft he was landing incurred engine failure near Galveston, Tex. Doctors reported later that Haise, who had walked away from the accident, had suffered second degree burns over 50% of his body and his condition was "critical but stable." (*W Star-News*, 8/23/73, A3)

- The Institute for Advanced Computations at Ames Research Center held a press seminar on the ILLIAC IV computer. ILLIAC IV, developed and funded by the Advanced Research Projects Agency (ARPA) under a contract with Burroughs Corp., had been installed at ARC under a 1970 joint NASA–ARPA agreement and the Institute for Advanced Computations had been established to operate the system. ILLIAC IV was being integrated into a larger remote-access computer system being developed at ARC as a process and data-storage resource on the nationwide ARPA computing network, accessible to universities and Government agencies. ILLIAC IV would make applications in global climate dynamics, distant seismic-event detection, multisensor processing, fluid dynamics, and simulation and optimization problems in logistics and economics. (ARC Releases 73-88, 73-89; ARC PAO)

- NASA launched an Aerobee 200A sounding rocket from White Sands Missile Range carrying a Goddard Space Flight Center astrophysics experiment to a 213.2-km (132.5-mi) altitude. The flight was nominal until 250 sec into the flight, when high-voltage problems developed. Final results would be determined after recovery of the payload. (NASA proj off)

August 23–September 13: *Intelsat-IV F-7* communications satellite was launched by NASA for Communications Satellite Corp. on behalf of the International Telecommunications Satellite Organization (INTELSAT). The 1379-kg (3041-lb) satellite was launched from Eastern Test Range at 6:57 pm EDT by an Atlas-Centaur launch vehicle into a highly accurate transfer orbit with a 35 819.5-km (22 257.2-mi) apogee, 548.1-km (340.6-mi) perigee, and 27.3° inclination.

NASA's primary objectives were to conduct design, performance, and flight-readiness reviews for the Federal Communications Commission; to ensure compatibility of INTELSAT spacecraft with NASA launch vehicles and launch environmental conditions; and to launch *Intelsat-IV F-7* into a transfer orbit that permitted the apogee-kick motor to insert the spacecraft into a synchronous orbit on the equator over the Atlantic Ocean. During a design life of seven years the spacecraft would add 12 TV channels to those available between the U.S. and other countries and would increase the telephone capacity by 3500 two-way circuits.

The apogee-kick motor was fired at 8:30 am EDT Aug. 25 and *Intelsat-IV F-7* entered a near-circular orbit with a 35 778-km (22 231.4-mi) apogee, 35 773-km (22 228.3-mi) perigee, 1435.6-min period, and 0.4° inclination. The spacecraft was at 327° east longitude, drifting at 0.15° per day, and was expected to reach its planned station of 330° east longitude over the Atlantic Ocean within 10 days. NASA adjudged the mission a success Sept. 13.

Intelsat-IV F–7 was the fifth communications satellite in the Intelsat IV series. *Intelsat-IV F–4* (launched Jan. 22, 1972) was operating satisfactorily over the Pacific. *Intelsat-IV F–5* (launched June 13, 1972) was operating over the Indian Ocean. The launch of Intelsat-IV F–6, postponed in May, was tentatively planned for December. (NASA prog off; GSFC *SSR*, 8/31/73; ComSatCorp, Tech Ops Off & Launch Rep)

August 24: The U.S.S.R. launched *Cosmos 581* from Baykonur Cosmodrome into orbit with a 282-km (175.2-mi) apogee, 171-km (106.3-mi) perigee, 89-min period, and 51.6° inclination. The spacecraft reentered Sept. 6. (GSFC *SSR*, 8/31/73; 12/31/73; *SBD*, 9/6/73, 10)

- Direct satellite communications via *Intelsat-IV F–4* began between Thailand and the People's Republic of China. (UPI, *W Star-News*, 8/24/73, A2; ComSatCorp PIO)

- Ames Research Center scientists had developed a miniaturized radio transmitter in pill form to monitor temperatures of the human body, NASA announced. Used in the space program to monitor persons in a simulated spacecraft environment, the transmitter could detect minute variations in temperature which often indicated to doctors the presence of infection. Produced in volume, the transmitter pill could be sold for 50 cents. ARC was working to decrease the pill's size still farther and to increase its transmitting power without increasing its cost. (NASA Release 73–167)

- Demonstration electromechanical devices developed for emergency aid to astronauts in space had been adapted, tested, and installed in a Huntsville (Ala.) Hospital room by Marshall Space Flight Center, NASA announced. The devices permitted remote control by severely handicapped or paralyzed patients of communications systems and appliances for comfort and recreation. (NASA Release 73–169)

August 25: A *New York Times* editorial praised *Skylab 3* (launched July 28 to man the Orbital Workshop launched May 14) as the mission passed the 28-day mark set by *Skylab 1–2* (May 14–June 22), to become the longest manned space mission to date: The *Skylab 3* astronauts "already have much more than mere longevity in space to their credit. Their two long walks in space have permitted them to do more useful work in orbit outside their vehicle than human beings ever have done before. Their scientific observations and experiments—particularly the study of the sun—have added much to human knowledge." With its midpoint near, *Skylab 3* had already "proved itself one of the most productive manned space missions since the first sputnik went into orbit almost a generation ago." (*NYT*, 8/25/73, 22)

August 26: Albion, Mich., police had reported the recovery of a spacesuit worn by Astronaut Alan L. Bean during the *Apollo 12* mission (Nov. 14–24, 1969), the Associated Press reported. The suit, valued at $6000, had been on loan from the Smithsonian Air and Space Museum when it was taken from the Kingman Museum in Battle Creek, Mich., a week earlier. (*W Post*, 8/27/73, A22)

August 27: NASA's Manned Spacecraft Center was officially rededicated as Lyndon B. Johnson Space Center in ceremonies at JSC, attended by Dr. James C. Fletcher, NASA Administrator; Texas Gov. Dolph Briscoe; and Mrs. Lyndon B. Johnson. A bust of the late President Johnson by sculptress Jimilu Mason, donated by the Houston Chamber of Commerce and the Albert Thomas Space Hall of Fame, was unveiled. A Johnson Room opened in the JSC Visitor Center displayed the original

U.S. copy of the Treaty for the Peaceful Uses of Outer Space, signed in 1967 during President Johnson's Administration.

At the dedication Dr. Fletcher said that President Johnson had needed "great political courage" 1965–1967 to fight for funds for NASA programs of the 1970s such as Skylab and Viking. "The decision to continue with a strong space program after Apollo may . . . prove to be the most important of all the decisive actions Lyndon Johnson undertook on behalf of space progress."

Earlier in the day Astronaut Alan L. Bean, commander of the *Skylab 3* crew orbiting the earth after July 28 launch to join the Orbital Workshop for 59 days, had radioed: "We think the work in which we are engaged right now would not have been possible except for [Johnson's] strong support and leadership in the Senate and the Presidency."

President Johnson himself had said—on a March 1, 1968, visit to MSC—referring to the National Aeronautics and Space Act of 1958: "The one legislative enactment that I suppose I am proudest of is the bill that I wrote and introduced that made possible NASA, that brought into existence this great facility and others in the program throughout the nation."

President Nixon had signed the law making the name change Feb. 17. Mr. Johnson had died Jan. 22. (Transcript; JSC *Roundup*, 8/17/73; 8/31/73; McElheny, *NYT*, 8/28/73, 24)

- NASA announced publication of *NASA Patent Abstracts Bibliography* (NASA SP-7039). Second in a semiannual series to implement the liberalized Federal patent policy, the bibliography was accompanied by an index that cross-referenced patent abstracts published in the first two issues. (NASA Release 73-168)

August 28: The U.S.S.R. launched *Cosmos 582* from Plesetsk into orbit with a 542-km (336.8-mi) apogee, 519-km (322.5-mi) perigee, 95.2-min period, and 74.0° inclination. (GSFC *SSR*, 8/31/73; *SBD*, 9/6/73, 10)

- The 128-nation International Civil Aviation Organization (ICAO) convened in Rome to consider proposals for new agreements to curb aircraft hijacking. Neither The Hague Convention of 1970 for the suppression of unlawful seizure of aircraft nor the Montreal Convention of 1971 for suppression of unlawful acts against the safety of civil aviation had yet been ratified by the necessary two-thirds majority of ICAO membership, the *New York Times* said later. "This means that there is still no international legal machinery requiring governments to impose sanctions against air piracy." (Hoffman, *NYT*, 8/29/73, 58)

August 29: The Federal Communications Commission granted a waiver to RCA Global Communications, Inc., ITT World Communications, Inc., and Western Union International, Inc., permitting participation in construction and operation of a maritime communications satellite system proposed by Communications Satellite Corp. FCC had granted permission April 11 to ComSatCorp to contract for three maritime comsats, but had stipulated that ComSatCorp allow other carriers to invest and participate in the system. (AP, *W Star-News*, 8/30/73, D6; FCC PIO)

- A *Huntsville Times* editorial commented on the progress of *Skylab 3* (launched July 28 to man the Orbital Workshop launched May 14): "While the Skylab flights may seem almost commonplace by now, we have a strong hunch that time will show that all the scientific research being performed and all the new pieces of information and know-how being gained up there in earth orbit were anything but routine. With each circuit completed, we can almost hear the added bits of data being dropped into man's ever-expanding bank of knowledge about his earth and its surroundings." (*Huntsville Times*, 8/29/73)

August 30: The U.S.S.R. launched its *Molniya 1-24* communications satellite into orbit from Plesetsk. Tass announced a 37 970-km (23 593.5-mi) apogee, 480-km (298.3-mi) perigee, 679-min period, and 65.3° inclination. The satellite would ensure the operation of a system of long-range telephone and telegraph radio communications in the U.S.S.R. and transmit Soviet central TV programs to the Orbita network. (GSFC *SSR*, 8/31/73; FBIS–Sov, 8/30/73, U1; *SBD*, 9/6/73, 10)

- The U.S.S.R. launched *Cosmos 583* from Baykonur Cosmodrome into orbit with a 298-km (185.2-mi) apogee, 204-km (126.8-min) perigee, 89.5-min period, and 64.9° inclination. The satellite reentered Sept. 12. (GSFC *SSR*, 8/31/73; 9/30/73; *SBD*, 9/6/73, 10)
- NASA launched a Black Brant VC sounding rocket from White Sands Missile Range carrying a Univ. of Colorado solar physics experiment. Rocket and instrumentation performed satisfactorily. (GSFC proj off)
- The 24-nation Geneva disarmament conference ended its 1973 session with no progress to report. Some members had charged that the conference had reached a state of complete paralysis, the *New York Times* later reported. (*NYT*, 8/31/73, 4)

August 31: The third glide flight of the X-24B lifting body was completed successfully near Flight Research Center by NASA test pilot John A. Manke. Objectives were to evaluate stability and control, roll and damper-off handling qualities, effects of water and alcohol jettison, landing gear dynamics, and control system operations with only two hydraulic pumps. Manke reported an excellent flight and good response with damper off. Touchdown was five minutes after air-launch. (NASA prog off)

- Proposals for developing space shuttle solid-fueled rocket motors (SRMs) had been received from Aerojet Solid Propulsion Co., Lockheed Propulsion Co., Thiokol Chemical Corp., and United Technology Center, Marshall Space Flight Center announced. The space shuttle motor program would be conducted in three increments: design, development, and test, ending in September 1979; production of new and refurbished SRMs for 54 flights (108 units) beginning early in 1978; and delivery of new and refurbished units for 385 flights (770 units) beginning in July 1980. (MSFC Release 73-125)
- The Navy's Fleet Weather Facility (FWF) had said that satellite imagery from NASA-developed weather satellites was becoming "indispensable" to Arctic and Antarctic shipping operations, NASA reported. The location of ice masses from satellite photos might ultimately lead to extension of shipping operations through the entire six-month polar night. Weather satellite imagery was provided the Navy by Goddard Space Flight Center. (NASA Release 73-174)
- NASA announced that it had issued three one-year grants averaging $20 000 each to predominantly black Central State Univ. in Ohio for studies of air and water pollution and the metallurgical properties of pure tungsten. (NASA Release 73-171)

During August: NASA's SPHINX Space Plasma High Voltage Interaction Experiment spacecraft, to be launched early in 1974, underwent vibration tests and began simulation tests. Tests would continue until November, when the spacecraft was scheduled to be shipped to Kennedy

Space Center. During its mission, SPHINX would travel in a highly elliptical earth orbit and would be exposed to space plasma (charged particles) ranging from 10 to 10 000 particles per cc. (*Lewis News*, 8/24/73, 3)

- The Naval Research Laboratory was converting a number of Minuteman I missile 2nd stages into large-payload-capacity, high-performance sounding rockets for x-ray astronomy research, *Naval Research Reviews* reported. The surplus 2nd-stage motors were being adapted with specially designed nosecones and scientific payloads for forthcoming experiments. Instruments aboard the new sounding rockets, named the Aries, would be used by NRL scientists to determine ultraviolet background levels of the stars from above the earth's atmosphere. (*Naval Research Reviews*, 8/73, 27–8)

- *Office of Technology Assessment: Background and Status*, a report to the House Committee on Science and Astronautics, was published by the Library of Congress Congressional Research Service, Science Policy Div. No appropriations had been approved as yet for the OTA, which had been established "as an aid in the identification and consideration of existing and probable impacts of technological application" under Public Law 92–484 Oct. 13, 1972. The Senate Committee on Appropriations had requested $289 000 for OTA salaries and expenses under H.R. 7447, an FY 1973 supplemental appropriations bill. No funds had been requested by the House, and the requested appropriation had been deleted in conference. The Senate Committee on Appropriations had recommended a $3 980 000 FY 1974 OTA appropriation. The FY 1974 legislative branch appropriations bill (H.R. 6691) had been passed July 19, but the conference committee had not yet met to resolve differences between the House and Senate versions. OTA funding would not be allotted until the appropriation was available. (Text)

- Appointment of Roald Sagdeyev as Director of the Soviet Space Research Institute was announced in *Vestnik*, monthly publication of the U.S.S.R. Academy of Sciences. Sagdeyev replaced Georgy I. Petrov, who went into voluntary retirement. Sagdeyev, a nuclear physicist, had been made a corresponding member of the Academy of Sciences in 1964 and had become a full member in 1968. Before his appointment as Director, Sagdeyev had specialized in plasma physics at the Nuclear Physics Institute in Novosibirsk. (AP, *W Post*, 9/23/73, A1; 9/24/73, A21)

September 1973

September 2–7: The 66th Annual General Conference of the Fédération Aéronautique Internationale (FAI) was held in Dublin, Ireland. U.S. participants included *Apollo 17* commander Eugene A. Cernan and *Apollo 16* commander John W. Young. The 1972 FAI Gold Space Medal was presented to Cernan in recognition of the Dec. 7–19, 1972, final Apollo mission which had "accomplished all of its goals and magnificently concluded the extraordinary and highly successful Apollo program." Cernan also received the 1972 De la Vaulx Medal for his absolute world record of 21 hrs 32 min 44 sec for "Duration of Stay Outside his Lunar Module 'Challenger' Spacecraft," set during *Apollo 17*. Young received the Yuri Gagarin Gold Medal for "remarkable leadership, skill, daring and dedication" as commander of the April 16–27, 1972, *Apollo 16* mission and the V. M. Komarov Diploma, on behalf of himself and *Apollo 16* Astronauts Thomas K. Mattingly II and Charles M. Duke, Jr., for successful completion of all but one of the mission's 33 major experiments.

The Groupe Diplome d'Honneur for Aeronautics was awarded to the U.S. Federal Aviation Administration "in recognition of its outstanding contribution to air traffic efficiency and safety by the development of the now fully operational Automated Radar Terminal System" (ARTS). The Groupe Diplome d'Honneur for Astronautics went to Boeing Co. Aerospace Group for design and production of the lunar roving vehicle (LRV) used during the *Apollo 15, 16,* and *17* missions. The Paul Tissandier Diploma was presented to Kay A. Brick for "her outstanding contributions to women's aviation over the past 23 years" as a member and Chairman of the Board of the All-Women's Transcontinental Air Race "Powder Puff Derby." (NAA *News*, 9/73, 1–2; FAA Release 73–155; *A&A 1972*)

September 3: The U.S. had "discreetly blocked" Soviet attempts to purchase 10 ILC Industries' Apollo lunar-configured spacesuits, *Aviation Week & Space Technology* said. The U.S.S.R. had been told that U.S. export of such technology could not be approved. (*Av Wk*, 9/3/73, 11)

September 4: World atlases were inaccurate in delineating outer boundaries of polar icecaps according to imagery returned by *Nimbus 5* (weather satellite launched by NASA Dec. 10, 1972), NASA announced. A Goddard Space Flight Center study had shown that synoptic photos from *Nimbus 5* clearly indicated polar regions underwent large-scale changes in a short time. Boundaries between the multiyear icepack around the North Pole and the large areas of first-year ice varied significantly within one freezing season. Even greater changes occurred around Antarctica.

Nimbus 5 also was measuring daily rainfall distribution over the earth's oceans to indicate to meteorologists how much energy was being released into the global atmosphere. An electrically scanning microwave radiometer aboard the spacecraft was measuring rainfall rate and heat thus released to enable forecasters to understand a tropical storm or hurricane and predict its 24- to 48-hr intensity. (NASA Releases 73-176, 73-175)

- Assignment of L/C William E. Barry (USAF), a medical doctor, as Special Assistant to the NASA Administrator was announced. Dr. Barry had been selected from the White House Fellows Program, which provided highly gifted young persons experience in U.S. government and leadership. Dr. Barry had been an aeronautical engineer, a test pilot, and a flight surgeon in the Air Force. He had been named Air Force Systems Command Flight Surgeon of the Year and had won the Aerospace Medical Assn.'s Julian E. Ward Award for outstanding contributions to the art and science of aerospace medicine in 1972. (NASA Ann)

- NASA launched a Black Brant VC sounding rocket from White Sands Missile Range carrying a Naval Research Laboratory and Harvard College Observatory solar physics experiment to a 235.1-km (146.1-mi) altitude. The rocket and instrumentation performed satisfactorily. (GSFC proj off)

- A *Chicago Sun-Times* editorial praised the "mapmaking" of the *Skylab 3* astronauts (launched July 28 to man the *Skylab 1* Orbital Workshop launched May 14): "We were particularly impressed to read about their assigned job of photographing a swamp in southern Sudan as they zipped over North Africa." Such tasks "point up the way in which a space shuttle can be used to facilitate jobs on Earth and, no less important, encourage co-operation between nations." (*C Sun-Times*, 9/4/73)

September 5: President Nixon submitted to Congress *Science Indicators, 1972*, the fifth annual report of the National Science Board of the National Science Foundation. The proportion of the gross national product spent for research and development 1963–1971 declined in the U.S., France, and the United Kingdom, but increased in the U.S.S.R., Japan, and West Germany. By 1971, U.S. R&D expenditures were 2.6% of GNP, Soviet 3.0%, U.K. and West Germany 2.0%, and Japan and France 1.8% each. The number of scientists and engineers engaged in R&D per 10 000 persons declined in the U.S. after 1969, but continued to increase in the U.S.S.R., Japan, West Germany, and France. All countries reduced the national defense proportion of their R&D expenditures significantly between 1961 and 1969. U.S. defense expenditures dropped from 65% to 49% of total Government R&D spending. Increases in the U.S. and most other countries occurred in space, community services, and economic development.

The U.S. had a favorable but declining "patent balance" between 1966 and 1970, with a reduced number of patents of U.S. origin in France, West Germany, and the U.K. and increased U.S. patents of Japanese origin. The U.S. had an increasingly favorable position in the sale of "technical know-how"—including patents, techniques, formulas, franchises, and manufacturing rights—during 1960–1971; Japan was the major purchaser, surpassing all of Western Europe after 1967. The favorable U.S. balance of trade in technology-intensive products grew 1960–1971, but was increasingly negative in nontechnology-intensive areas. Within technology-intensive areas, products with the fastest rising trade surplus were aircraft, computers, and plastics. (*PD*, 9/10/73, 1069; Text)

September 5, 7, 8, 10, 26: The House Committee on Science and Astronautics' Subcommittee on Manned Space Flight held hearings on NASA's proposed space tug. The hearings sought to determine the tug's role in the space shuttle program, the tug's fiscal impact on the program and the overall manned space flight budget, the operational impact on NASA's latest proposed mission model and on projected cost-per-flight of the shuttle, the Dept. of Defense role in development and use of the tug, and NASA's and industry's progress in tug conceptual design. Charles V. Donlan, NASA Deputy Associate Administrator for Manned Space Flight, and Capt. Robert F. Freitag (USN, Ret.), Deputy Director of Advanced Programs, presented an overview of the tug program Sept. 5 but made no recommendations, pending completion of a study. NASA would later recommend program approach and supporting rationale.

Representatives of General Dynamics Corp. Convair Div., McDonnell Douglas Astronautics Co., Lockheed Missiles & Space Co., Martin Marietta Co. Denver Div., and Grumman Aerospace Corp. presented results of efforts to date, with supporting data to substantiate the options afforded NASA in proceeding with tug development. Options under study included liquid-storable and cryogenic systems for a growth-stage tug, the Centaur as an expendable or recoverable launch vehicle, an expendable Thor-Delta vehicle, an expendable or recoverable Agena vehicle, and an expendable or growth-stage Transtage vehicle. (Transcript)

September 6: The U.S.S.R. launched *Cosmos 584* from Plesetsk into an orbit with a 331-km (205.7-mi) apogee, 210-km (130.5-mi) perigee, 89.9-min period, and 72.9° inclination. The satellite reentered Sept. 20. (GSFC *SSR,* 9/30/73; *Sov Aero,* 9/10/73, 110)

- Selection of 18 experiments for the July 1975 U.S.–U.S.S.R. Apollo-Soyuz Test Project mission was announced by NASA. The experiments had been selected from 145 proposals by U.S. and foreign scientists, with those which could include Soviet participation contingent upon Soviet Academy of Sciences agreement.

  Four astronomy and space physics experiments would search for sources of extreme ultraviolet radiation in the night sky, measure intensity and distribution of helium-fluorescent radiation in night-sky regions, measure concentration of atmospheric constituents, and map celestial x-ray emissions in the 0.1- to 1.0-kev range.

  Six space applications processing experiments would use a small multipurpose electric furnace to study properties and processes in zero gravity. Two other applications experiments would study electrophoresis and the structure of the earth's gravity field. Five life sciences experiments would investigate effects of weightlessness on human cells important in protecting against infection and would collect data on microbial exchange, radiation effects, and light flashes. (NASA Release 73–178)

- President Nixon transmitted to Congress *U.S. Participation in the UN: Report by the President to the Congress for the Year 1972.* In his transmittal message, the President said there had been "a growing cooperation" in outer space. "A United Nations working group cooperated in making available to other nations data from our first experimental satellite designed to survey earth resources [*Erts 1,* launched by NASA July 23, 1972] and the Convention on International Liability for Damage Caused by Space Objects, which had been negotiated by a United Nations committee, entered into force on September 1." (*PD,* 9/10/73, 1058–60)

- The Atomic Energy Commission announced the adoption of regulation amendments to permit U.S. citizens and companies to transmit unclassified, published information to Communist bloc destinations. Previously, specific AEC authorization had been required. The amendments would allow U.S. businessmen to conduct promotional nuclear sales activities using published material in bloc countries. They also would permit U.S. citizens to file nuclear activity patent applications in bloc countries. (AEC Release R-374)

- President Nixon announced his intention to nominate Richard J. O'Melia to be a member of the Civil Aeronautics Board succeeding Secor D. Browne, who had resigned effective March 1. The nomination was submitted to the Senate Sept. 7 and confirmed Nov. 20. (*PD*, 9/10/73, 1056, 1070; CAB PIO)

September 7: NASA held a Skylab solar activity briefing at Johnson Space Center, while *Skylab 3* Astronauts Alan L. Bean, Dr. Owen K. Garriott, and Jack R. Lousma completed their 42nd day working aboard the *Skylab 1* Orbital Workshop. National Oceanic and Atmospheric Administration scientist Joseph Hirman said that a flare photographed by the Skylab 3 crew was about 17 times the cross-sectional area of the earth and "optically the biggest flare we've had during ATM and perhaps the biggest for this entire year." Within 30 min after the event "high energy protons made it from the Sun to the earth as a consequence of this flare." Confirmation had also been given by x-ray satellites and by short-wave fade-out and other ionospheric disturbances on the ground of a very intense x-ray event. Within two days, with the arrival of subsequent particles, an aurora would be visible at the latitude of Chicago and enough geomagnetic activity would occur to disrupt power lines in high latitudes. Hirman said that the flare had created enough energy to "run everything in the United States and the rest of the world combined for 500 years." (Transcript)

- The Dept. of Defense and Grumman Aerospace Corp. had agreed on a new price for 50 F-14 jet fighter aircraft, nearly $10 million below that estimated earlier, the *Washington Star-News* quoted informed sources as saying. Deputy Secretary of Defense William P. Clements, Jr., had written Sen. Howard W. Cannon (D-Nev.), Chairman of the Subcommittee on Tactical Air Power of the Senate Committee on Armed Services, that DOD would need $693 million rather than the $703 million it had requested in its FY 1974 budget. (Kelly, *W Star-News*, 9/7/73, A12)

September 8: *Cosmos 585* was launched by the U.S.S.R. from Plesetsk into an orbit with a 1407-km (874.3-mi) apogee, 1375-km (854.4-mi) perigee, 113.5-min period, and 74.0° inclination. (GSFC *SSR*, 9/30/73; *SBD*, 9/13/73, 55)

September 9: Publication of *Nondestructive Testing, a Survey* (NASA SP-5113) was announced by NASA. The survey summarized NASA-developed methods and techniques to determine the integrity, quality, durability, and effectiveness of a material or product and to maintain quality control over products mass-produced by automation. (NASA Release 73-172)

September 10: President Nixon sent a message to Congress on national legislative goals. Of defense spending he said: "I continue to be adamantly opposed to attempts at balancing the overall budget by slashing the defense budget. . . . In constant dollars, our defense spending in this fiscal year will be $10 billion less than was spent in 1964, before the Vietnam war began. . . . Further cuts would be dangerously irresponsible and I will veto any bill that includes cuts which would imperil our national security. . . . The arms limitation agreement signed with the Soviet Union last year has at last halted the rapid growth in the numbers of strategic weapons. Despite this concrete achievement, much needs to be done to ensure continued stability and to support our negotiation of a permanent strategic arms agreement. A vigorous research and development program is essential to provide vital insurance that no adversary will ever gain a decisive advantage through technological breakthrough and that massive deployment expenditures will therefore not become necessary. Yet the Congress is in the process of slashing research and development funding below minimum prudent levels, including elimination of our cruise missile and air defense programs. The Trident and B–1 programs, which are critical to maintaining a reliable deterrent into the next decade, are also facing proposals to cut them to the bone."

The President reiterated his request that Congress authorize creation of Dept. of Energy and Natural Resources and "a new, independent Energy Research and Development Administration so that we can make the very best use of our research and development funds in the future." R&D efforts could produce "the most helpful solutions to the energy problem." He called for a separate and independent Nuclear Energy Commission. (*PD*, 9/17/73, 1074–9)

- Gen. Bruce K. Holloway (USAF, Ret.) became a Special Assistant to the NASA Administrator. He would also serve as Acting Assistant Administrator for DOD and Interagency Affairs. Gen. Holloway, former Commander of the Air Force Strategic Air Command, had joined NASA as a consultant to the Administrator Aug. 13. (NASA Ann)
- The Senate confirmed Russell R. Train as Environmental Protection Agency Administrator. (*CR*, 9/10/73, D994)
- A Baltimore *Sun* editorial commented on NASA's problems with the Office of Management and Budget: OMB had promised NASA about $3.4 billion annually, but OMB impoundment of funds had cut NASA's FY 1974 budget to about $3 billion and "it appears OMB will hold the space agency to about $3.235 billion in fiscal 1975." Inflation had aggravated NASA's problems and $85 million had been cut from the FY 1974 shuttle development budget. "But now NASA is suggesting what some aerospace contractors are stating explicitly: That if the OMB squeeze stays on, then NASA will have to cut back on new starts in scientific programs. This is an odd irony, indeed, since the very purpose of the shuttle is to serve these new programs." The *Sun* felt "that if NASA accountants are told to sharpen their pencils they can keep the shuttle alive as well as provide for initiation of the new scientific programs. The Nixon OMB consistently has given science short shrift, however, and it would be sad indeed if the space program were hobbled at the very time it is beginning to show immense scientific promise." (*B Sun*, 9/10/73, A11)
- An unidentified flying object was reported falling from the skies, burning a hole in the ground, and vanishing in a cloud of steam near Griffin, Ga. Dr. O. E. Anderson, head of the Georgia State Experiment Station's Agronomy Dept., later said the object was probably "something in the nature of a small meteorite or a piece of space hardware" impacting at very high temperature. (UPI, *W Post*, 9/15/73, A3)

September 11: Checkout operations for Skylab 4, scheduled for Nov. 11 launch to crew the *Skylab 1* Orbital Workshop launched May 14, went into a hold period at Kennedy Space Center. Operations had been accelerated so that Skylab 4 could act as a rescue vehicle if needed for

Skylab 3, docked with the Workshop since July 28. The acceleration made necessary a hold until Sept. 25 so checkout operations could resume their normal flow [see July 28–Sept. 25]. (KSC Release 207-73)

September 11–12: The third colloquium of users of NASA's Structural Analysis (NASTRAN) computer system was held at Langley Research Center. Some 300 industry and Government representatives heard 40 papers and talks on the uses of NASTRAN to integrate project disciplines, to manage design work data and make it available during any stage of a design cycle, and as a source of computer analyses of structural stress, buckling, vibration, and transient-response of large and complex structures. (NASA Release 73-179; LaRC PAO)

September 11, 13, 18: The House Committee on Science and Astronautics' Subcommittee on Energy held hearings on geothermal energy. Dr. Alfred J. Eggers, Jr., Assistant Director for Research Applications with the National Science Foundation, testified Sept. 11 that geothermal energy was recognized as a "currently viable and relatively unpolluting energy resource." Its importance depended upon development of new technologies and "resolution of institutional constraints." NSF was recommending a program to develop necessary technologies and "to synthesize these into pilot plants capable of generating power from a variety of geothermal sources." An advantage of geothermal energy was "that construction of power plants can be carried out in a relatively short time." With solution of the technical problems in extracting the earth's heat and converting it into electricity, "the commercial impact of geothermal energy could begin to be felt within this decade. Geothermal energy is, therefore, counted on to contribute at least regionally to a resolution of the energy crisis in the short term as well as to provide sizable blocks of energy in the remaining years of this century." (Transcript)

September 12: The Federal Communications Commission approved five applications for permission to build domestic satellite networks—from American Satellite Corp., jointly owned by Fairchild Industries, Inc., and Western Union International, Inc.; RCA Global Communications, Inc., and RCA Alaska Communications, Inc.; GTE Satellite Corp. and National Satellite Services, Inc., a Hughes Aircraft Co. subsidiary; American Telephone & Telegraph Co.; and Communications Satellite Corp. GTE also was authorized to begin interstate toll telephone service in competition with AT&T. ComSatCorp was given interim authorization to proceed with procurement of high-capacity satellites for lease to AT&T "at its own risk." FCC said ComSatCorp appeared not to have complied with an FCC order to form a separate corporation to engage in domestic satellite activities. (FCC PIO; ComSatCorp Release 73-48; *W Post*, 9/13/73, D12)

- The Navy F-14B swing-wing supersonic fighter aircraft flew its first test flight from the Grumman Aircraft Corp. facility at Calverton, N.Y. The F-14B reached mach 0.8 during a 67-min flight. (*Av Wk*, 9/17/73, 23)
- Wallops Station held a conference on the use of lasers for hydrographic studies. U.S. and Canadian scientists, engineers, and technicians reviewed the use of the laser light-detection-and-ranging (lidar) system for over-water observations from aircraft. The conference was sponsored jointly by NASA, the National Oceanic and Atmospheric Administration, the Environmental Protection Agency, and the Navy. (Wallops Release 73-9)

September 13: Ten Soviet scientists and engineers arrived at Johnson Space Center to help evaluate the docking system for the July 15, 1975, U.S.-U.S.S.R. Apollo-Soyuz Test Project mission. Vladimir S. Syromyatikov of the Soviet State Research Institute of Machine Building led the team, which would assist in evaluating the U.S. and Soviet sections of the docking module for proper mating and in checking pressure integrity and structural strength of the combined system. The U.S.-U.S.S.R. team also would make dynamic tests of the docking. The system would be studied under thermal extremes. JSC was testing full-scale development hardware built by both countries for compatibility and operation. (JSC Release 73-119)

- NASA 427, an instrumented C-54 aircraft from Wallops Station, flew to the German Air Force Base at Leck, West Germany, to participate with European scientists in the Joint North Sea Wave Project Sept. 17-30. Daily research flights from Leck would study surface waves and prevailing wind relationships in the second of a series conducted by the West German Hydrographic Institute, Hamburg. The C-54 carried Langley Research Center's S-band radiometer and laser profilometer; the Naval Research Laboratory's K- and Ka-band radiometers, narrow-pulse radar, and inertial navigation system; and Wallops' Hasselblad and flight research cameras. Participants included the National Center for Atmospheric Research, National Oceanic and Atmospheric Administration, NRL, and universities.

  The NASA 427 itinerary had been planned to fly beneath Skylab passes over Newfoundland and the Azores Sept. 13. (Wallops Release 73-10)

- President Nixon submitted to the Senate the nomination of Dr. Dixy Lee Ray, Atomic Energy Commission Chairman, to be U.S. Representative to the Sept. 18-24 International Atomic Energy Agency General Conference in Vienna. Among five alternate representatives nominated was *Apollo 8* Astronaut and former National Space Council Chairman William A. Anders. (*PD*, 9/17/73, 1128, 1132)

- Military records damaged in a July fire at the Military Personnel Records Center in Overland, Mo., were being salvaged through space-age technology, McDonnell Douglas Corp. officials told the press in St. Louis. A space chamber used to simulate temperatures and pressures encountered in Mercury and Gemini manned space missions was being used to reclaim records. The scorched and water-soaked records had been sealed in the chamber and put through a freeze-drying process. "The records came through the process legible," an official had said. (AP, *NYT*, 9/14/73, 11)

- NASA launched an Aerobee 170 sounding rocket from White Sands Missile Range carrying a Goddard Space Flight Center solar physics experiment. Rocket and instrumentation performed satisfactorily. (GSFC proj off)

- President Nixon accepted the resignation of Dr. Eberhardt F. Rechtin as Assistant Secretary of Defense for Telecommunications. (*PD*, 9/17/73, 1131)

- Military and aerospace expenditures in the laser field were expected to average about $317 million annually in the next five years, according to a study by technological market research organization Frost & Sullivan, Inc., the *New York Times* reported. The FY 1973 laser market would be about $290 million, more than double earlier estimates. Operational thermal (laser) weapons were probably 10 yrs in the future, but prototype experiments were expected to be made by 1975. Technology had progressed to high-energy, gas-dynamic lasers that produced light beams when their internal gases were heated, expanded, and forced through small nozzles at supersonic speeds. The laser's most notable military contribution thus far had been bombs directed through the air toward targets illuminated by laser lights. A new system, target-recognition attack (TRAM), was being developed to spot targets day and night with an infrared detector bore-sighted with a laser range finder to illuminate the target. (Middleton, *NYT*, 9/13/73, E22)

September 13–October 17: Ames Research Center conducted a five-week experiment to determine the effects of weightlessness and reentry into earth gravity on female space shuttle passengers. The experiment, with 12 Air Force nurse volunteers participating, was a follow-on to similar studies on men made in 1972. Following two weeks of orientation and medical studies, eight of the nurses would simulate weightlessness with absolute bedrest while four acted as ambulatory controls. After two weeks of immobility, the eight would be subjected to g forces expected when the shuttle entered the atmosphere at a mission's end. The final week would be for recovery and testing. (ARC Release 73–107; ARC Biomed Research Div)

September 14: Cosmos 586 was launched by the U.S.S.R. from Plesetsk into an orbit with a 1007-km (625.7-mi) apogee, 969-km (602.1-mi) perigee, 104.8-min period, and 82.9° inclination. (GSFC *SSR*, 9/30/73; *SBD*, 9/17/73, 65)

• Climatic changes on Mars indicated by *Mariner 9* (launched by NASA May 30, 1971) were described in *Science* by Cornell Univ. astronomer Dr. Carl E. Sagan and team: "The equatorial sinuous channels on Mars detected by Mariner 9 point to a past epoch of higher pressures and abundant liquid water. Advective instability of the martian atmosphere permits two stable climates—one close to present conditions, the other at a pressure of the order of 1 bar [100 kilonewtons per sq mi] depending on the quantity of buried volatiles. Variations in the obliquity of Mars, the luminosity of the sun, and the albedo of the polar caps each appear capable of driving the instability between a current ice age and more clement conditions. Obliquity driving alone implies that epochs of much higher and of much lower pressure must have characterized martian history. Climatic change on Mars may have important meteorological, geological, and biological implications." (*Science*, 9/14/73, 1045–8)

• The mechanical and structural qualifications program for the ATS–F Applications Technology Satellite (scheduled for April 1974 launch) was completed at Johnson Space Center with a successful deployment test of the parabolic reflector. The reflector on the thermal-structure-model spacecraft had been deployed under simulated space conditions of vacuum and zero g. (*Goddard News*, 10/73, 5)

• Kennedy Space Center announced the award of two study contracts. Florida Institute of Technology would study hypergolic propellant liquid- and vapor-disposal methods for use in space shuttle operations under a $75 000 contract. The Univ. of Florida would receive $25 420 to study the application of remote-sensing devices for temperature evaluation to assist Florida fruit and vegetable growers. (KSC Releases 209–73, 210–73)

• A $110-million "fail safe" satellite system to monitor the 1290-km (800-mi) Alaskan oil pipeline was being planned by RCA Alaska Communications, Inc., United Press International quoted RCA Corp. Board Chairman Robert M. Sarnoff as saying in Anchorage, Alaska. (*NYT*, 9/15/73, 11)

- Univ. of Texas physicists A. A. Jackson IV and Michael P. Ryan, Jr., proposed in a *Nature* article that the event that leveled forests in a wide region of Siberia on June 30, 1908, was the passage of a black hole that had plunged completely through the earth. They suggested that ships' logs for that date be examined for evidence of an unusual occurrence. The object that fell on Siberia had been referred to as the Tunguska meteorite, although it had left no crater or other clearly identifiable meteorite residue. "We suggest that a black hole of substellar mass . . . could explain many of the mysteries associated with the event." (*Nature*, 9/14/73, 88–89)

September 14–23: The first major international aerospace show in Latin America was held in Sao Paulo, Brazil. More than 250 firms from 11 nations displayed products from satellite equipment to metal forgings, aircraft, helicopters, and small components. Aircraft exhibited included the British Hawker Siddeley Harrier VSTOL [vertical or short takeoff and landing] fighter, Fokker F-28 twin-jet transport, Israel Aircraft Industries' Arava cargo transport, Aeronautica Macchi MB-326 jet trainer-fighter, Aerospatiale A-300B Airbus, Canadair CL-215 amphibian, Northrop F-5E light jet fighter aircraft, and Lockheed C-5A heavy logistic transport. Of the nations invited to participate by the sponsoring Brazilian Ministry of Aeronautics, only Japan and the U.S.S.R. had declined. (*Av Wk*, 9/17/73, 22; *Interavia*, 9/73, 954)

September 17: The liquid-fueled "core" vehicle of the Titan-Centaur launch vehicle was moved from the Vertical Integration Building to the Solid Motor Assembly Building in the Titan complex at Kennedy Space Center. Twin 26-m (85-ft) solid-fueled-rocket strap-on boosters would be attached to the core. The Titan-Centaur, scheduled to be launched on its proof flight in January 1974, mated the Titan III core vehicle and its twin solid-fueled rockets with the Centaur high-energy liquid-fueled final stage. The 49-m (160-ft), 3610-kg (7959-lb) configuration was designed to carry heavy payloads on orbital and planetary missions, such as the Viking Mars mission and the joint U.S. and West German Helios mission to the sun, during the mid- and late 1970s. The liftoff thrust of the twin solid-fueled boosters was 10.8 million newtons (2.4 million lbs).

On its proof flight the Titan-Centaur would carry a mass model of the Viking spacecraft and a 66-kg (145-lb) Space Plasma High Voltage Interaction Experiment (SPHINX) spacecraft, to measure in orbit for one year the interaction of space plasmas with high-voltage surfaces. The main purpose of the flight would be to demonstrate that the launch vehicle and launch facilities could support operational missions and that the Centaur could perform an operational two-burn mission and an operational three-burn mission carrying a payload to a synchronous altitude of 35 888 km (22 300 mi).

The effort to integrate the Titan with the Centaur had begun in the mid-1960s when NASA recognized the need to fill a performance and cost gap between the Atlas-Centaur and Saturn launch vehicles. (KSC Release 215-73)

- NASA's hydrogen injection program—to decrease automobile and aircraft engine pollution by injecting hydrogen gas into the gasoline and air mixture before combustion—was described by Dr. James C. Fletcher,

NASA Administrator, in a speech before the Economic Club of Detroit. The program, being carried out by Jet Propulsion Laboratory with assistance from Lewis Research Center, was "an exciting experimental effort to bring old and new technology to bear on one of the major problems of modern society."

In Phase 1, begun in April for conclusion in December, small amounts of hydrogen gas were being injected into a lean gasoline and air mixture to be burned in a standard automobile engine, first in a laboratory and later in an operating automobile with bottled hydrogen carried in its trunk. Preliminary laboratory tests had shown "a significant reduction in pollution without an increase in fuel consumption. In fact, our preliminary figures indicate that hydrogen injection may even produce a fuel *saving*. This comes both from the direct effect of the hydrogen on the combustion process and from the better atomization of the fuel mixture." Work was progressing at LeRC on a laboratory model of a generator in which the hydrogen would be produced from gasoline and water. Phase 2 would attempt to solve engineering problems in integrating the hydrogen injection system, generators, and fuel controls into a smooth-running automobile. (*NASA Activities*, 10/15/73, 183–4)

- Representatives of major U.S. automobile manufacturers began visiting Jet Propulsion Laboratory for demonstrations of a NASA-developed system to enable the automotive industry to meet legal limitations on automobile engine emissions [see above]. JPL engineers had had promising results in reducing pollution while increasing engine efficiency in laboratory tests of the system, but the work was in its early stages. Dr. William H. Pickering, JPL Director, had said automotive company representatives had been invited "to assess the utility of the system with a view to the possibility that they might wish to work cooperatively with us." (NASA Release 73-184; Witkin, *NYT*, 9/17/73, 1)
- Continued Soviet harassment of intellectual U.S.S.R. dissidents was causing a reaction that could retard development of scientific exchanges and trade between the U.S. and the U.S.S.R., *Aviation Week & Space Technology* reported. Little effect was expected on the joint Apollo-Soyuz Test Project mission scheduled for July 1975 launch, however. (*Av Wk*, 9/17/73, 11)

September 18: Sen. Frank E. Moss (D-Utah), Chairman of the Senate Committee on Aeronautical and Space Sciences, introduced S. Con. Res. 45 to observe NASA's 15th anniversary by requesting President Nixon to proclaim the week of Oct. 1 as National Space Week. (*CR*, 9/18/73, S16777)

- French President Georges Pompidou had offered the U.S. the possibility of joining France and the United Kingdom in developing a replacement for the Anglo-French Concorde supersonic transport aircraft, French Transport Minister Yves Guena said in Paris. Guena told a symposium of the seven-nation Western European Union that the offer had been made when Nixon and Pompidou met at Reykjavik, Iceland, in June 1973. (Agence France-Presse, *P Inq*, 9/19/73)

September 18–21: Dr. Charles A. Berry, NASA Director of Life Sciences, was elected president of the International Academy of Aviation and Space Medicine during its 21st Congress at Munich, West Germany. (AP, *B Sun*, 9/22/73, A7)

September 19: Communications Satellite Corp. announced the award of a $65 900 000 contract to Hughes Aircraft Co. for construction of high-

capacity satellites for lease to American Telephone & Telegraph Co. to provide domestic satellite communications services. Delivery of the first of four satellites was scheduled for late 1975. (ComSatCorp Release 73-48)
- Rep. George E. Brown (D–Calif.) introduced H.R. 10392 to authorize NASA to develop fuel-efficient, low-polluting engines for light automobiles to "capitalize on NASA's recognized preeminence in both high technology itself and the management of advanced technology development programs." (CR, 9/19/73, H8137-8; Off Rep Brown)
- The U.S. Civil Service Commission ordered the Air Force to rehire A. Ernest Fitzgerald, a financial analyst who had been fired in 1970 after he had disclosed a $2-billion cost overrun on development of the Air Force C-5A jet transport. The Air Force also was ordered to pay Fitzgerald more than $100 000 in back pay. (Ripley, *NYT*, 9/19/73, 1)
- NASA launched an Arcas sounding rocket from Antigua, West Indies, carrying a Goddard Space Flight Center experiment to a 49.4-km (30.7-mi) altitude. The objectives were to measure ozone distribution in the upper atmosphere, monitor anomalous ultraviolet absorption, and extend the data base for a climatology of stratospheric ozone in the tropics. The sounding rocket was launched in conjunction with a *Nimbus 4* weather satellite (launched April 8, 1970) overpass. The rocket and instrumentation performed satisfactorily. (NASA Rpt SRL)

September 19–21: The exhibition "Technology in the Service of Man" was opened at Lewis Research Center by Dr. James C. Fletcher, NASA Administrator. Some 1400 representatives of business, industry, labor, professions, and Government viewed exhibits that included a Skylab slide show, Saturn launch vehicle models, a Skylab Student Project exhibit, mockups of future space shuttle payloads, a lunar roving vehicle, and material on the High Energy Astronomy Observatory (HEAO). (MSFC Release 73-129; LeRC Release 73-42; *Lewis News*, 9/21/73, 1)

September 19–28: Five Soviet radiation specialists toured U.S. industrial process radiation facilities under the Memorandum on Cooperation in the Peaceful Uses of Atomic Energy between the Atomic Energy Commission and the U.S.S.R. State Committee on the Utilization of Atomic Energy. The visit reciprocated a Soviet tour by a delegation of U.S. radiation specialists in July 1972. (AEC Release R-389)

September 20: President Nixon presented Harmon International Aviation Trophies to chief test pilots Brian Trubshaw and André Turcat of the Anglo-French Concorde supersonic transport aircraft during White House ceremonies that coincided with the Concorde's first U.S. visit. Trubshaw of Great Britain and Turcat of France received the 1971 Aviator's Trophy for their Concorde flights. The 1972 Aviatrix Trophy went to Geraldine Cobb for humanitarian flights in the Amazon River Basin area of South America in 1971. Air Force pilots L/C Thomas B. Estes and L/C Dewain Bick received the 1972 Aviator's Trophy for their April 26, 1972, SR-71 reconnaissance jet aircraft flight that set a record for sustained speed after a 24 100-km (15 000-mi) flight from Beale Air Force Base, Calif., at mach 3-plus speeds.

In the first visit of a supersonic transport aircraft to the U.S., the Concorde arrived at Dallas-Fort Worth Regional Airport at Grapevine, Tex., to participate in Sept. 21 dedication ceremonies for the world's largest jet airport. The Concorde—carrying 32 aviation officials, press representatives, and a 10-man crew—was flown from Caracas, Venezuela, to Texas by Concorde test pilot Jean Franchi in two and one half hours. The passengers had flown from Texas to Caracas the previous day in a conventional jet aircraft in four and one half hours. (*PD*, 9/24/73, 1170; *NYT*, 9/21/73, 1, 43; *A&A 1972*)

- Flight Research Center was evaluating a side-stick flight controller for the Air Force YF-16 lightweight fighter aircraft on the FRC digital fly-by-wire aircraft, a modified F-8 fighter with an electronic control system. The new side-stick controller was stationary and used pilot-applied force to maneuver the aircraft. The inflight studies were to aid fighter pilots to control highly maneuverable aircraft, provide information for the Air Force lightweight fighter program, and provide a base for future research. The side-stick control replaced conventional sticks between the pilot's legs, impractical in future high-performance aircraft where the pilot would fly in supine position. (FRC Release 27-73)

- Lewis Research Center received two *Industrial Research Magazine* IR-100 Awards for 1973 at ceremonies in Chicago's Museum of Science and Industry. The Center also was presented a special award acknowledging that NASA had accumulated 10 or more IR-100 awards—prizes given by the magazine for what it deemed the 100 most significant new products developed during the year. LeRC had won 9 out of the 12 awards presented to NASA and had earned at least one IR-100 Award every year except one since first entering the competition in 1966. LeRC 1973 awards were for an x-ray photographic system with potential application for industry and basic research and for a shaft seal for very-high-speed turbines and compressors. (LeRC Release 73-39)

- Rep. Don Fuqua (D-Fla.) introduced for himself and cosponsors H. Con. Res. 305 and 306 designating the week Oct. 1-7 as National Space Week. (*CR*, 9/20/73)

September 21: The U.S.S.R. launched *Cosmos 587* from Plesetsk into an orbit with a 289-km (179.6-mi) apogee, 177-km (110-mi) perigee, 89.1-min period, and 65.4° inclination. The satellite reentered Oct. 4. (GSFC *SSR*, 9/30/73; 10/31/73; *SBD*, 10/1/73, 152)

- NASA held two Skylab press briefings at Johnson Space Center, while the *Skylab 3* crew (launched into orbit July 28) completed its 56th day of work in the Orbital Workshop.

At an Apollo Telescope Mount briefing, Skylab principal investigator James E. Milligan said that the *Skylab 3* crew had operated the ATM console photographing the sun for 300 hrs. It was "remarkable how few mistakes they have made." The sun also had been cooperative. "During this mission we've had everything from one of the quietest Suns I have ever seen to a Sun that was as active as it ever is, at least during a moderately active day during Sunspot maximum." The crew had taken "thousands and thousands and thousands" of photographs of "on the order of one hundred" flares, coronal transients, and prominences.

Jack H. Waite, Marshall Space Flight Center's manager of corollary experiments, said at a corollary briefing that all the planned objectives of the corollary experiments had been completed. MSFC's Material Processing Program Manager Robert L. Adams said the materials experiments had indicated that both electron-beam welding and the exothermic-tracing techniques could be used successfully in space for the assembly and repair of large structures. Experiments in the formation of metallic spheres had generally indicated more spherical and less porous products than their one-g counterparts could be produced. (Transcript)

- *Pioneer 10*—launched March 2, 1972, toward Jupiter—had traveled farther into space than any man-made object, NASA announced. The spacecraft, when it swung by Jupiter on Dec. 3 to give man his first closeup look at that planet, would set a speed record of 132 000 km (82 000 mi) per hr, breaking its own earlier mark of 51 683 km (32 116 mi) per hr set at launch. After the Jupiter encounter, *Pioneer 10* would be hurled by the planet's gravity beyond the solar system toward the red star Aldebaran in the constellation Taurus. It was expected to cross the orbit of Pluto in 1987 still transmitting scientific data to the earth. The onboard experiments had been designed to yield useful data as far as 30 to 40 astronomical units, 4.5 billion–6.0 billion km (2.8 billion–3.7 billion mi), away. The spacecraft had been tracked and controlled so precisely that it would arrive at Jupiter within less than a minute of the time predicted at launch.

  NASA had submitted three applications to the National Aeronautic Association and the Fédération Aéronautique Internationale for *Pioneer 10*'s unmanned interplanetary space flight records: maximum distance traveled from the sun, maximum distance of communications from the earth, and duration of the mission during which the spacecraft had been functioning. (NASA Release 73–183)

- Sen. Alan Bible (D–Nev.), with cosponsors, introduced S. 2456, the Geothermal Energy Act of 1973, to "authorize the Secretary of the Interior to guarantee loans for the financing of commercial ventures in geothermal energy and to coordinate Federal activities in geothermal energy exploration, research, and development." Under Title II of the bill, NASA would be directed to prepare a proposal for the use of space technologies and NASA services to explore and map geothermal resources. (*CR*, 9/31/73, S1726–9)

- NASA launched an Aerobee 200 sounding rocket from White Sands Missile Range carrying a Goddard Space Flight Center solar physics experiment to a 290.8-km (180.7-mi) altitude. The rocket performed satisfactorily. Preliminary estimates indicated the experiment was 90% successful. (GSFC proj off)

September 22: One of four U.S.S.R. spacecraft—*Mars 4, 5, 6,* or *7,* launched toward Mars July 22 and 25 and Aug. 5 and 9—would soft-land near the Mars south polar cap to test the physical properties of Martian soil and surface rocks and check the possibilities of transmitting TV pictures of the surrounding terrain to earth, Roald Sagdeyev, Director of the Space Research Institute of the Soviet Academy of Sciences, said in a Tass interview. *Mars 4* and *5* were 19 515 000 km (12 126 000 mi) and 18 584 000 km (11 548 000 mi) from the earth and would reach Mars in February 1974. *Mars 6* and *7* were 16 664 000 km (10 354 000 mi) and 15 623 000 km (9 708 000 mi) from the earth and would reach Mars in March 1974. Radiation and magnetic measurements were being made and cosmic rays were being studied, Sagdeyev said. The stations also would investigate solar noise in the meter wave band as well as register the flux of solar plasma.

United Press International reported later that Western space experts had speculated that a landing craft might carry a Marsokhod roving vehicle to explore the Martian surface. UPI also reported that it was believed that the U.S.S.R. departure from the usual practice of not announcing mission goals was to demonstrate Soviet willingness to cooperate with the U.S. in view of the plans for a 1975 joint U.S.–U.S.S.R. mission. (Tass, FBIS–Sov, 9/24/73, U1; AP, *W Post*, 9/24/73, A21)

- The U.S.S.R. was beginning a new major program to strengthen its underground missile silos against nuclear attack, according to senior U.S. officials quoted by the *Washington Post*. The unidentified officials thought the U.S.S.R. might be planning to use a new "popup" missile-launching technique as part of this effort. (Getler, *W Post*, 9/22/73)

September 23: Japanese amateur astronomer Tsuomi Seki reacquired and photographed Comet Kohoutek at his specially designed comet observatory in Tokyo. The comet—last observed by its discoverer, Dr. Lubos Kohoutek, May 5—was falling toward the sun almost exactly as predicted and was approaching the earth and the sun in a long, sweeping orbit. (Am Mus–Hayden Planetarium Release, 10/1/73)

- A *New York Times* editorial praised *Skylab 3*'s solar observations: "The special value of the observations on the sun arises from that body's extraordinary behavior this month." An unexpected burst of tremendous solar activity had occurred. "An unprecedented wealth of data on this startling phenomenon [had] been gathered . . . , which should contribute significantly to helping scientists understand the physics and chemistry of the vast celestial body that is the source of almost all the energy available on earth." (*NYT,* 9/23/73)

September 24: A Memorandum of Understanding on international cooperation in NASA's space shuttle program was signed in Washington, D.C., by Dr. James C. Fletcher, NASA Administrator, and Dr. Alexander Hocker, Director General of the European Space Research Organization (ESRO). Nine ESRO-member countries would design, develop, manufacture, and deliver a Spacelab flight unit to be carried in the space shuttle orbiter. The Spacelab would comprise a pressurized manned laboratory module and an instrument platform, or pallet, to support telescopes, antennas, and other equipment requiring direct space exposure. The Spacelab would be transported to and from orbit in the orbiter payload bay and would be attached to and supported by the shuttle orbiter throughout 7- to 30-day missions. At the mission's end, the Spacelab would be removed and prepared for its next mission.

The estimated $300-million to $400-million Spacelab cost would be borne by Belgium, Denmark, France, West Germany, Italy, The Netherlands, Spain, and Switzerland—the European nations that signed, with the U.S., the Intergovernmental Agreement for space shuttle cooperation [see Aug. 14]. The agreement was open for signatures by other nations as well. Under Memorandum of Understanding terms NASA would manage all operational activities, including crew training and flight operations, following delivery of the Spacelab unit in late 1978, about one year before the space shuttle's scheduled first operational flight. NASA planned to carry a European flight crew member on the first shuttle flight and subsequent European flight crew opportunities would be provided in conjunction with ESRO or ESRO-member-government projects. NASA would make the shuttle available for Spacelab missions on either cooperative or cost-reimbursable basis. NASA would procure from ESRO any additional Spacelab units needed for U.S.

programs and would not develop any further units duplicating the Spacelab's design and capabilities.

Final signature of the Memorandum of Understanding having established the cooperative Spacelab program, NASA had decided to use Spacelab terminology. Steps were instituted to change designations of the Sortie Lab Task Force and Task Team at Marshall Space Flight Center to Spacelab Task Force and Task Team. (NASA Release 73-191; NASA Shuttle Off, OMSF; *Spacelab Newsletter*, 9/26/73)

- Soviet authorities had returned data tape and altitude recorder film salvaged from the scientific payload aboard a NASA balloon that fell in the Siberian wilderness after its July 15 launch from Fort Churchill, Canada, NASA announced. Recovery had followed a request for Soviet assistance from Dr. John E. Naugle, NASA Associate Administrator for Space Science, to Chairman Boris N. Petrov of the Soviet Council for International Cooperation in the Exploration and Use of Outer Space. Dr. Naugle and Academician Petrov were cochairmen of a NASA and Soviet Academy of Sciences Joint Working Group on Near-Earth Space, the Moon, and the Planets. The cosmic ray detector and other instruments aboard the balloon had been damaged in the crash, but the data tape and film flown from Moscow had been pronounced usable by U.S. investigators. The experiment had been designed to measure distribution of electrons and positrons in primary cosmic rays. (NASA Release 73-195)

- NASA announced the award of a $31-million contract to General Dynamics Corp. Convair Aerospace Div. for nine high-energy Centaur rockets. Four Centaurs would be used by NASA to launch satellites built under the direction of Communications Satellite Corp. for an American Telephone & Telegraph Co. domestic communications network. ComSatCorp would reimburse NASA for all costs of the rockets and launch services. Three Centaurs would be used to launch improved Intelsat IV satellites. NASA would use the remaining two Centaurs to launch twin Mariner spacecraft in 1977 for Jupiter and Saturn flybys. (NASA Release 73-189)

- NASA announced negotiation of a $67-million cost-plus-incentive contract with McDonnell Douglas Corp. for procurement of 20 Thor-Delta launch vehicles. (NASA Release 73-188)

- Johnson Space Center had issued a request for proposals for a cost-plus-fixed-fee contract to build two modified aircraft for flight crew training simulating space shuttle operations, NASA announced. Proposals were due Oct. 9, with work to be completed in two and one half years. The shuttle training aircraft (STA) were to simulate space shuttle orbiter flying characteristics, performance, and trajectory during atmospheric flight. (NASA Release 73-194)

- The election of Dr. John V. Harrington, Director of the Center for Space Research at Massachusetts Institute of Technology, as Vice President, Research and Engineering, with Communications Satellite Corp. was announced. He would join the corporation in October. (ComSatCorp Release 73-49)

- A new technique to measure the burn rate of solid rocket propellant on launched vehicles had been developed by the Air Force Systems Command Air Force Rocket Propulsion Laboratory, AFSC announced. An acoustic emission system picked up the sounds of burning propellant on a very sensitive microphone, and acoustic emissions were fed to a device that automatically measured the burn rate. Previously an oscillograph trace had been measured with a ruler, a time-consuming job that sometimes produced errors. (AFSC Release OIP 129.73)

- A *Huntsville News* editorial commented on the planned Oct. 1 celebration of NASA's 15th anniversary and the start of Space Week: "The spinoff from space exploration continues to help Americans and mankind from all kinds of new equipment to watching happenings in Europe or Asia right in your own living room because of the launch of satellites. Years will see new developments come from data already collected." (*Huntsville News*, 9/24/73)

September 25: President Nixon sent a message to *Skylab 3* Astronauts Alan L. Bean, Dr. Owen K. Garriott, and Jack R. Lousma following their splashdown ending the record July 28–Sept. 25 mission: "The record of your Skylab mission combines the traditions of those great explorers of history who have faced the uncharted reaches of the physical unknown with the traditions of those men of science who have unlocked the secrets of the universe and have thus opened the doors to man's future progress. By your scientific endeavor and your physical endurance, you have converted a space vehicle into a repository of more scientific knowledge than mankind can immediately consume. In doing so, you have provided the basis for a quantum jump in human knowledge." (*PD*, 10/1/73, 1197)

- The *Skylab 3* astronauts rested aboard recovery ship U.S.S. *New Orleans* following their splashdown [see July 28–Sept. 25]. Dr. James C. Fletcher, NASA Administrator, told the press at Johnson Space Center that their 59-day mission had proved man's usefulness in earth orbit as the Apollo missions had proved his use of the moon. "In the 15 years since its founding, NASA has successfully placed more than 250 payloads in earth orbit. Of all these, the mission that we completed today will perhaps prove to be the most fruitful of all." *Skylab 3* had proved "how great an advantage it is to have the human mind on the scene to make judgments, to observe, to respond to unexpected developments, and to effect corrective measures. All these were demonstrated on Skylab."

  Skylab Program Director William C. Schneider said the *Skylab 3* astronauts had surpassed their scientific objectives by 150%. "The mission can only be described in superlatives." (O'Toole, *W Post*, 9/26/73, 1A; AP, *B Sun*, 9/26/73)

- NASA announced the signing at Johnson Space Center of a $6 618 500 cost-plus-fixed-fee contract with IBM Federal Systems Div. to design, develop, and maintain avionics software for the space shuttle orbiter data system. The contract would be effective through April 10, 1975. IBM also was selected by Rockwell International Corp. to develop and produce two other key electronic units for the orbiter—an orbiter general-purpose avionics computer and an input-output unit to work with the computer. The two contracts had raised IBM's new business in August to nearly $22 million. (NASA Release 73–192)

- Johnson Space Center was evaluating a Lockheed Aircraft Corp. proposal to test the feasibility of using a C–5A aircraft to ferry the space shuttle orbiter. Lockheed had proposed to mount a scale model of the orbiter in latest configuration on a C–5A scale model for tests in its low-speed tunnel. (NASA Release 73–193)

- A technique of predicting probabilities that had been "absolutely mandatory" to the success of NASA's Apollo program was being used by the Atomic Energy Commission to help determine the risk to the public of nuclear power plants, AEC member and *Apollo 8* Astronaut William A. Anders said in testimony before the Congressional Joint Committee on Atomic Energy. The complicated technique of forecasting the reliability of mechanical systems and human beings using available data from similar activity components and subsystems was described by Massachusetts Institute of Technology nuclear engineer Dr. Norman C. Rasmussen during the nuclear reactor safety hearing. (Willard, *W Post*, 9/26/73)

September 26: Newspapers commented on the successful Sept. 25 splashdown of *Skylab 3* [see July 28–Sept. 25].

New York Times: "The word 'triumphant' is entirely appropriate because even at this early point it is evident that their 59-day stay in orbit was not only the longest manned space flight in history but also the most productive." Skylab results to date had made two conclusions incontrovertible: "First, man can live and work effectively for long periods in space, a point particularly emphasized by the astronauts' pleas in the last days of their journey to be allowed to remain longer in orbit. Second, stations in space are potentially as productive of scientific and technological benefits as the most optimistic predictions had suggested. There is every reason now to suppose that these benefits will range from the discovery of new mineral deposits on earth and early warnings about the formation and movement of storms to new and more effective methods of predicting the turbulence on the sun that interferes seriously with electronic communications on earth." It could be taken for granted that "orbiting laboratories will become permanent fixtures in the heavens before too many years have passed. (*NYT*, 9/26/73, 38)

Chicago Daily News: The *Skylab 3* astronauts had "earned their keep—and a place in the history of man's conquest of the universe." (*C Daily News*, 9/26/73)

- Alloys and crystals formed in weightlessness during *Skylab 3* (launched July 28 to crew the *Skylab 1* Orbital Workshop launched May 14) might temporarily be the world's most precious metals, the *Wall Street Journal* reported. They were mixtures of elemental metals that could not be combined on the earth because gravity would make them separate as they cooled or melted. The metals included three alloys and several semiconductor crystals used to make transistors and other electronic devices. Twelve industrial and university scientists were waiting to determine from them whether manufacturing in space could produce stronger, purer, and more reliable, or even new materials. (*WSJ*, 9/26/73)

- Award of a $7.3-million contract to Rockwell International Corp. Space Div. to provide a site for the final assembly and checkout of the space shuttle orbiter was announced by NASA. The contract included modifications and additions to facilities at Air Force Plant 42, Palmdale, Calif. (NASA Release 73–196)

- Concorde 02, reproduction model of the Anglo-French supersonic transport, flew from Washington, D.C., Dulles Airport to Paris' Orly Field in 3 hrs 33 min. The aircraft had participated in dedication of the Dallas-Fort Worth airport Sept. 21. (*Concorde Bull* 12, 1/1/74)

September 26–27: Sen. Warren G. Magnuson (D–Wash.) explained S. 2495, a bill to apply NASA's expertise to solving domestic problems, during the first day of two-day hearings on the state of the aerospace industry by the Senate Committee on Aeronautical and Space Sciences. Sen. Magnuson, with Sen. Frank E. Moss (D–Utah) and Sen. John V. Tunney (D–Calif.), introduced the bill Sept. 27.

The bill, the Technology Resources Survey and Applications Act, would amend the National Aeronautics and Space Act of 1958 to provide "an organization within NASA to carry out programs . . . to resolve critical national problems. This would expand an activity already underway within NASA and enable us to better utilize a capability which has already exhibited an impressive record of success." The programs would be determined by a National Technology Resource Council of Cabinet members and heads of technological agencies, which would make its recommendations to the President.

Sen. Magnuson suggested changing NASA's name to the National Applications of Science Administration. "I do believe the name . . . would accurately reflect the new role which my bill attempts to create for NASA."

Dr. Karl G. Harr, Jr., President of the Aerospace Industries Assn. of America, Inc., testified that foreign competition for U.S. and international aerospace markets was currently "of a scope, determination, and quality that we have not had to face since World War II. The underpinning of strong research and development programs—vital to our technological advance—is being downgraded to a serious degree. There is an alarming scarcity of private sources of financing for the development and production of new commercial aircraft. As a low-profit industry, traditionally averaging a profit about half that of other manufacturing industries, we lack the resources for reinvestment, cushioning against risk, and attracting investors. Above all, there is uncertainty in this transitional period about future national goals which might utilize our capabilities. Will we continue to contribute primarily to space exploration, defense, and long-range air travel, or will we next find ourselves emphasizing such efforts as environmental control, law enforcement, health delivery services, housing, and urban mass transit?"

Dr. Wernher von Braun—Vice President, Engineering and Development, Fairchild Industries, Inc., and former NASA Deputy Associate Administrator for Planning—testified on the future of satellite communications: "There is no field in which the goals of space applications are expanding at a faster rate." Ever since NASA demonstrated "that radio signals bounced off passive reflection satellites such as the Echo balloon or rebroadcast by active repeater spacecraft such as Early Bird, worldwide communications by satellite have been growing. . . . Satellites, by providing international telephone and television service, are an important force in overcoming regional and national barriers. It is difficult to maintain hostility and isolation in the presence of free communications." Satellite communications would "soon be commonplace in our country." With the advent of the reusable space shuttle to service and repair satellites in the 1980s, "even nations with limited fiscal resources can put in a network of smaller, far less expensive ground stations." Thanks to NASA programs initiated a few years ago, "the United States is still quite active in the area of pioneering new technology for communications satellites." (Transcript)

September 27–29: The U.S.S.R. launched *Soyuz 12*, carrying cosmonauts L/C Vasily G. Lazarev and flight engineer Oleg G. Makarov, from Baykonur Cosmodrome at 5:18 pm local time (3:18 pm Moscow time; 8:18 am EDT) into earth orbit with 229-km (142.3-mi) apogee, 180-km (111.8-mi) perigee, 88.6-min period, and 51.7° inclination. Soviet news agency Tass announced the objectives of the two-day mission were to test improved flight systems, test manual and automatic control in various flight conditions, and obtain spectrographic data of separate sections of the earth "for the solution of economic problems." The flight would further improve development of manned spacecraft.

After six hours in flight the cosmonauts reported they had adjusted to weightlessness and were carrying out all scheduled activities. All spacecraft systems were functioning normally. During the first day, Lazarov—45-yr-old air force test pilot and physician—and Makarov—40-yr-old engineer who had taken part in developing the control board and flight program for Yuri A. Gagarin's *Vostok 1* mission of April 12, 1961—maneuvered their spacecraft in orbit to test manual and automatic control and piloting techniques. The spacecraft was moved into a higher and nearly circular orbit, with parameters reported by Tass as 345-km (214.4-mi) apogee, 326-km (202.6-mi) perigee, 91-min period, and 51.6° inclination.

The cosmonauts also made spectrographs of natural formations of the earth's surface. They were reported Sept. 28 to be in high spirits and good physical condition. Tass reported a new spacecraft design and a new spacesuit, "very simple and comfortable," were being tested. Spacesuits were worn for launch and descent but not during the flight itself.

On Sept. 29 the crew oriented their spacecraft and fired the braking engine. The landing module separated from the spacecraft and entered descent trajectory. The parachute system was deployed at 7.6 km (4.7 mi) above the earth and engines softlanded the craft 400 km (250 mi) southwest of Karaganda in the Kazakhstan Steppe at 2:34 pm Moscow time (7:34 am EDT). A medical check at the landing site showed "the state of [the cosmonauts'] health is good."

Soyuz 12 was the first Soviet manned space flight since the 1971 *Salyut 1–Soyuz 11* mission. *Soyuz 10* (launched April 23, 1971, carrying three cosmonauts) had docked with the orbiting *Salyut 1* space laboratory (launched April 19) but the cosmonauts had not boarded. American observers believed the Soyuz spacecraft had spun out of control, causing the cosmonauts to become ill and necessitating an early return to earth. Three cosmonauts launched June 6, 1971, on *Soyuz 11* had docked with and boarded *Salyut 1*. They worked successfully in orbit for 24 days but were found dead on landing, from pressure loss in the landing craft. (GSFC *SSR*, 9/30/73; FBIS–Sov, 9/28/73, U1-4; 10/1/73, U1-3; *SBD*, 9/28/73, 138; Kaiser, *W Post*, 9/28/73, A1; Shabad, *NYT*, 9/30/73, 1; Seeger, *W Post*, 9/30/73, 145)

September 27: The Air Force launched an unidentified satellite on a Titan IIIB-Agena launch vehicle from Vandenberg Air Force Base into an orbit with a 384.6-km (239-mi) apogee, 128.8-km (80-mi) perigee, 89.8-min period, and 110.5° inclination. The satellite reentered Oct. 29. (Pres Rpt 74; Wilson, *W Post*, 10/16/73, A7)

September 27

- *Skylab 3* Astronauts Alan L. Bean, Jack R. Lousma, and Dr. Owen K. Garriott arrived in San Diego, Calif., aboard the recovery ship U.S.S. *New Orleans* and were flown to Houston, Tex., on an Air Force C-141 troop carrier aircraft. Blood samples from all three astronauts were flown to Johnson Space Center for analysis. The astronauts were under limited quarantine for one week to protect them from infection. (*W Post*, 9/28/73; AP, *B Sun*, 9/28/73)
- In a *Skylab 3* postflight medical briefing at Johnson Space Center, Dr. Willard R. Hawkins, JSC Deputy Director of Life Sciences for Medical Operations, said that the condition of the *Skylab 3* astronauts, who splashed down Sept. 25 after a 59-day mission aboard the *Skylab 1* Orbital Workshop, "did slightly exceed our expectations." They were in better shape than the *Skylab 2* crew (launched May 25 on a 28-day mission). Alan L. Bean had lost just over 3.6 kg (8 lbs); Dr. Owen K. Garriott and Jack R. Lousma had both lost between 2.7 and 3.6 kg (6 and 8 lbs). All three crew members had lost between 2.5 and 3.8 cm (1 and 1.5 in) in calf circumference. They had been able to complete the full routine through the maximum stress level of -50 mm of mercury in the lower-body pressure experiment. Heart rates and blood rates were well above their preflight measurements. The crew was experiencing slight vestibular disturbances, particularly associated with any head movement, but these were gradually disappearing. (Transcript)
- Newspaper editorials commented on the safe Sept. 25 return of *Skylab 3*.

  Baltimore *Sun:* The mission had been "an immense scientific success which produced unprecedented and extremely valuable data especially on solar activity." It would be "mere superstition to believe nature is co-operating with man by producing solar flares and comets just when we would most like to have them. But the Skylab missions so far have had a lot of good luck." (*B Sun*, 9/27/73, A14)

  Miami Herald: "All told, the Skylab missions appear to promise rewards even greater than the enlargement of knowledge created by the Apollo moon voyages. The outreach of the human mind is inexorable." (*M Her*, 9/27/73)

  Chicago Sun-Times: "With casual aplomb, the astronauts performed their sophisticated experiments, floated weightlessly through, outside and around the Skylab, and fixed their spacecraft like mundane mechanics when something went haywire. It all seemed so routine. In the austere domain of space, as on Earth, how quickly does the miraculous become absorbed into the commonplace." (*C Sun-Times*, 9/27/73)

  Christian Science Monitor: "Even those most intimately involved with the space program, slated now for more international participation as well as applause, would admit we may now but dimly grasp the full significance of our space sorties." The names of the *Skylab 3* astronauts would, nonetheless, be added "to the list of welcomed home heroes of mankind's young space chapter." (*CSM*, 9/27/73)

  Philadelphia Inquirer: "Space exploration has matured to space work. During 59 days in earth orbit—more than doubling the space flight record set earlier this year by three astronauts in the same Skylab—assignments were carried out with diligence and precision. Scientific experiments and observations could "produce long-range dividends, not immediately apparent." Despite NASA's extraordinary safety record space was "a hostile environment posing hazardous challenges to courageous men. But the richest rewards in space will come not from derring do but from workmanlike execution of assignments with down-to-earth relevance to needs of mankind." (*P Inq*, 9/27/73)

- General Dynamics Corp. Chairman and Chief Executive Officer David S. Lewis told the press in St. Louis, Mo., that he had signed an agreement with Dzherman M. Givishiany, Vice-chairman of the Soviet Committee for Science and Technology, for scientific and technical cooperation. The agreement covered commercial shipping, telecommunications equipment, asbestos mining, aircraft, computer-operated microfilm equipment, and navigation and weather buoys. (AP, *Today*, 9/28/73)

- L/C Harold E. Turner (USAF, Ret.), the first commanding officer of White Sands Proving Ground (which became White Sands Missile Range in 1958) died in El Paso, Tex., at age 75. He had been transferred from White Sands to Patrick Air Force Base, Fla., where a proving ground was to be built on Cape Canaveral. There Col. Turner supervised the firing of Cape Canaveral's first rocket, the two-stage Bumper 8 (a V-2 first stage with a Wac Corporal 2nd stage) July 24, 1950. He was later medically retired after suffering cerebral damage during an accident at the Cape. (AP, *W Post*, 9/29/73, D1; *A&A 1915-60*)

September 28: Processed samples from *Skylab 3* experiments carried on the 59-day mission (launched July 28 to man the Orbital Workshop, launched May 14) were returned to Marshall Space Flight Center. They were opened, x-rayed, and photographed before being delivered to principal investigators who would report within six months. Among the samples were 11 of solidification (crystal growth performed in the M-518 multipurpose electrical furnace system) and the remains of spiders Anita and Arabella, who died after spinning webs in zero g for the first time. The spiders' bodies would be given to the Smithsonian Institution for display later in the week, following examination by a student experimenter. Preliminary examination had shown that Arabella died of malnutrition. Anita had died in space of probably the same cause Sept. 15. (MSFC Release 73-146)

- Newspaper editorials commented on the Sept. 25 successful splashdown of *Skylab 3* [see July 28–Sept. 25].

    The *Washington Star-News:* The astronauts couldn't have selected a worse evening to return to the earth "in the whole 20th Century." Their reentry had been "upstaged, unfortunately," by Vice President Spiro T. Agnew's vain attempts to clear himself of allegations of bribery and income tax evasion, "and, of course, the Watergate hearings persisted as a distraction. But these latest Skylab voyagers still must receive the recognition they deserve for the longest and most voluminously rewarding space flight in history." Skylab was "the proper program in space for the years just ahead, because it promises many more revelations about the earth and seas—the condition and potentials of our planet." (*W Star-News*, 9/28/73, A18)

    The *Wall Street Journal* noted that the *Skylab 3* crew could anticipate a small pleasure that eluded most returning travelers. "Unlike the rest of us, they have numbers of people who are absolutely hungering to see the photographs they brought back, despite the fact that they've returned with no fewer than 77,600 shots of the sun and 16,800 of the earth." (*WSJ*, 9/28/73)

September 28

- Johnson Space Center announced a third-year extension had been awarded to its contract with Lockheed Electronics Co., Inc., Houston Aerospace Systems Div. to provide general electronic, scientific, and computing center support services. The cost-plus-award-fee contract extension would total $32 733 000. (NASA Release 73-198; JSC Release 73-129)

September 29: A *Washington Post* editorial praised the Skylab program: "The prospect of permanently orbiting manned space laboratories used to be the stuff of dreams. The Skylab program underscores how rapidly such science fiction is becoming scientific fact. It shows how quickly man is becoming more at home in space and more accustomed to the reality of extended space flight—to the extent that the most impressive human and technological feats have come to be regarded by laymen as almost routine. Since the first 'giant leap for mankind,' NASA's achievements have produced a quantum jump in man's comprehension of the universe. The task now is to build on that new foundation, and to secure an equal measure of political progress to insure that this new capability will be used for universal benefit." (*W Post*, 9/29/73, A14)

- Tass special correspondent Aleksander Romanov quoted a spokesman for the U.S.S.R. flight control team of the Sept. 27–29 *Soyuz 12* manned mission as saying that in the future space flights "will become longer and will be made farther and farther from the Earth. Therefore, it is important to improve constantly methods of autonomous navigation. In the opinion of the crew members the instruments designed for these purposes have given a good account of themselves. The night vision apparatus has their preliminary approval." (FBIS–Sov, 10/1/73, U2–3)

September 30: *Soyuz 12* cosmonauts L/C Vasily G. Lazarev and Oleg G. Makarov arrived in Moscow from Karaganda, following their successful Sept. 27–29 orbital mission. The crew was met at the airport by members of a state commission, specialists from flight control, scientists, engineers, relatives, and friends. Before leaving Karaganda, the cosmonauts had held a press conference, in which they said they had implemented their flight program "in full."

Gen. Georgy T. Beregovoy—cosmonaut and chief of the Soviet cosmonaut training center—speaking in Zvezdny Gorodok, said Lazarev and Makarov had displayed "high professional skills," carrying out a comprehensive checkup of improved onboard systems, further development work on manual and automatic control of the ship in different flight modes, and spectrography of areas of the earth's surface. (Moscow Domestic Service, FBIS–Sov, 10/1/73, U3, Tass, U4)

- Astronaut Thomas P. Stafford—speaking in Beverly Hills, Calif.—congratulated Soviet Cosmonauts Vasily G. Lazarev and Oleg G. Makarov on the successful completion of their Sept. 27–29 *Soyuz 12* mission. Stafford, U.S. crew commander for the U.S.–U.S.S.R. Apollo-Soyuz Test Project's planned 1975 space flight, spoke highly of the Soyuz flight and said he was happy that it had ended with success. He noted that many of the new systems that would be used on the 1975 joint flight were tested by *Soyuz 12*. (Tass, FBIS–Sov, 10/1/73, B8)

- A *New York Times* editorial commented on the successful Sept. 27–29 *Soyuz 12* manned mission: "The purpose . . . was not primarily to set new records. Its main goal was to test the latest improved version of the Soyuz spacecraft. . . . Americans as well as Soviet citizens have reason to rejoice. . . . In 1975 present plans call for the historic docking of a Soviet Soyuz craft with a United States Apollo vehicle. Thus Soviet and

American astronauts must want each other's space ships to be as safe as possible. And beyond the docking experiment in 1975 lies the prospect of a future fully integrated Soviet-American joint space program." (*NYT*, 9/30/73, 12)

During September: Soviet plans for a reusable space shuttle had been revealed by Cosmonaut Gherman S. Titov and Soviet Academician Andrey Shikarin, according to a Vienna correspondent quoted by *Interavia*. The Soviet shuttle would consist of two vehicles coupled pick-a-back, each in delta configuration and designed for hypersonic operating speeds. One would be the booster, the other the orbiter. The spacecraft was designed for horizontal takeoff from a runway. The manned booster would carry the orbiter until it attained a 22-km-per-sec (4900 mph) speed corresponding to a 2- to 3-g acceleration. The stages would separate at about a 30-km (19-mi) altitude and the booster would descend in a controlled glide to land on the runway. The orbiter would accelerate to a more than 100-km (60-mi) altitude, complete its mission, and glide back to the earth. Construction was expected to begin during the 1970s. (*Interavia*, 9/1973, 945)

- The Subcommittee on International Cooperation in Science and Space of the House Committee on Science and Astronautics published *Research and Development Collaboration with the U.S.S.R. and Japan*. The report summarized discussions with Soviet and Japanese officials during the Subcommittee's July visit to the U.S.S.R.–U.S. Aeronautical Technology Symposium and Display in Moscow and a subsequent visit to Tokyo. The Subcommittee had found that "although military and space parity had been achieved with the United States, the U.S.S.R. is far behind in virtually all consumer products. There are also glaring weaknesses in certain areas of high technology." Soviet leaders had accepted the fact that, "if their plans for the massive expansion and improvement of their commercial air transportation system are to succeed within the proposed time frame, assistance will be required from abroad where superior technology exists. We believe the U.S. aerospace industry is the best, though not the only, source for such high technology products, and evidently the responsible Soviet officials think so, too." As one of the world's most highly developed technological countries, Japan had "much to offer to any cooperative venture in science or space." The Subcommittee urged the National Science Foundation to explore possible collaboration with the Japanese government in its Sunshine Project, a proposed large-scale effort to develop new energy sources. (Com Print)

- U.S. scheduled airline passenger traffic had increased 7.3%, to 76 400 million revenue-passenger-miles, during the first half of 1973, *Interavia* reported. Traffic for local service carriers alone rose 13.2%, to 4600 million revenue-passenger-miles, and for scheduled domestic service, 7.7%, to 60 400 million. Available seat-miles increased 7.5%, to 120 200 million. Noting the financial and economic improvement of the local service carriers, the Civil Aeronautics Board had instituted an investigation of their class subsidy rate and planned a minor reduction in the subsidy. (*Interavia*, 9/73, 919)

- Tracking of earth-orbiting satellites by the long-range perimeter acquisition radar (PAR) of the Safeguard Ballistic Missile Defense facility at Grand Forks, N.D., had been achieved ahead of schedule, *Armed Forces Journal* reported. PAR construction had begun in 1970; power testing, in June. The site was scheduled to be handed over to the Armed Air

Defense Command in October 1974 and to become operational in early 1975. (*AFJ*, 9/73, 24)
- The "much vaunted European challenge" to U.S. aerospace industry would "remain a myth," an *Interavia* editorial said. Recent agreements for U.S. companies to provide European manufacturers access to their high technology might seem against U.S. interests, but "such agreements are in fact usually followed by orders for U.S.-designed aircraft." It was "not simply that the currency crisis has made the US product much cheaper than the competitive European design, but also that the Americans are in a position to contract out production work on a proven sales success." European commercial aircraft industry products would "never achieve the success expected of them." Europe's total share of the Western market for commercial aircraft—just over 9%—was "steadily dropping as the early model BAC 111s and Caravelles are withdrawn from service and there are no advanced versions to replace them." However tempting technologically the advanced commercial aircraft offered by European industry might be, "they are not what many airlines want in the present economic climate." There "seems little or no inclination in Europe to accept inter-dependence in high technology areas. There is apparently more of a readiness to cooperate across the Atlantic than across the European frontiers. If the European aerospace industry wishes to remain a viable, independent force on the world scene, then the time has surely come when something more than lip service must be paid to the European ideal." (*Interavia*, 9/73, 939)
- The Hawker Siddeley 146, the first major new British aircraft launched in a decade, was to be developed under a program funded jointly by Hawker Siddeley and the United Kingdom government, *Armed Forces Journal* reported. The four-engine aircraft would carry 70 to 100 passengers for 1900 km (1200 mi). (*AFJ*, 9/73, 24)
- U.S.S.R.'s Aeroflot had transported 82.5 million passengers and 21 000 million kg (46 300 million lbs) of freight and mail, and its crop-spraying aircraft had treated 833 389 sq km (321 773 sq mi) of land in 1972, *Interavia* reported. Forecasts for 1973 included the transportation of 87 million passengers and 2100 million kg (46 300 million lbs) of mail and freight and the treatment of 869 800 sq km (335 830 sq mi) of land. The Aeroflot network continued to be extended and the number of scheduled services had risen from 2974 in 1972 to 3068 in 1973. Moscow alone was linked with 200 cities in the U.S.S.R. (*Interavia*, 9/73, 919)
- The National Science Foundation released *Federal Scientific, Technical, and Health Personnel, 1971* (NSF 73–309). In 1971, as in most years since 1966, Federal scientific, technical, and health professional personnel increased little compared with the early 1960s. The Dept. of Defense remained, as in prior years, the largest employer of scientific and technical personnel. About 29% of Federal scientists and engineers performed research and development. The greatest concentration of R&D personnel was at NASA. More than one half of NASA's scientists and engineers performed R&D activities. Salaries of Federal scientists and engineers increased between 1966 and 1971, largely to meet statutory requirements that they be comparable with non-Federal salaries. Women, accounting for 4% of all Federal scientists and engineers, appeared to be advancing in grade faster than men. (Text)

October 1973

October 1: NASA's 15th Anniversary. A National Space Club dinner honored leaders who had implemented the National Aeronautics and Space Act of 1958, which established NASA Oct. 1, 1958. Honored guests included Dr. T. Keith Glennan, NASA's first Administrator, *Apollo 11* Astronaut Neil A. Armstrong, the first man on the moon (July 20, 1969), and Astronaut Alan B. Shepard, Jr., the first American in space (aboard *Freedom 7*, May 5, 1961).

NASA's future role was described by Dr. James C. Fletcher, NASA Administrator: "Unless and until Congress sees fit to change our role, I think NASA should continue to have this one primary mission: to do the necessary research and development in space and aeronautics to identify promising opportunities and meet national needs. In performing this primary mission, we will continue to work closely with user agencies in government and with the aerospace industry." NASA also was transferring its technology and experience to nonaerospace sectors of the economy and helping other Government agencies solve societal problems. "It so happens that we have the reservoir of technology, and also some available manpower, at a time when the Environmental Protection Agency and other government departments need our help." NASA had "unique capabilities to help bring into being a pollution-free, resource-saving hydrogen economy, and to facilitate extensive use of the ultimate energy source, solar power." NASA's traditional function as the Nation's civilian space and aeronautics R&D manager "assures us a challenging and rewarding future during NASA's next 15 years." (Program; Prepared text)

Anniversary ceremonies also included open house at NASA Centers and annual awards presentations. (*NASA Activities*, 11/73)

During its first 15 yrs NASA had completed 277 major U.S. and international launches, with 210 successful and one still under evaluation. With the Mercury, Gemini, and Apollo programs successfully completed and Skylab progressing, NASA had established 15 768 man-hours in space. Astronauts had spent about 40 hrs in inflight extravehicular activity and more than 90 hrs on the moon. NASA manned spacecraft had spent nearly 400 hrs on the moon and had returned more than 385 kg (850 lbs) of lunar material. Of 38 Americans who had flown in space, 3 had made four space flights, 4 had made three, 10 had made two, and 22 had made one flight each. A dozen astronauts had walked on the moon, and 24 had flown around it. Man had learned he could live and work in space.

In unmanned space exploration NASA had launched thousands of sounding rockets carrying U.S. and foreign experiments. *Mariner 9* probe (launched May 30, 1971, and still orbiting Mars) had produced 7300 pictures of the planet, showing it to be dynamic and geologically active. Satellites had studied stars beyond the solar system through x-rays, gamma rays, and cosmic rays. *Oao 1* (launched Dec. 7, 1968) and *Oao 3* (launched Aug. 21, 1972) had returned a wealth of information on the Milky Way and more distant galaxies. *Pioneer 10* (launched March 2, 1972, toward Jupiter) had revolutionized man's study of the sun. The entire electromagnetic spectrum had been opened to man's examination by study of stars in ultraviolet and x-ray. Observations from NASA-launched weather satellites had enabled forecasters to observe weather systems from space as they developed and to forecast their effects as they moved into inhabited parts of the world. Man had learned what instruments in space could provide in information about the earth's resources and their efficient use to facilitate decision making, improve the lot of developing countries, and preserve and improve conditions in countries like the U.S.

NASA had provided technological advancements to preserve the U.S. position as world leader in civil and military aeronautics. The joint NASA, Air Force, and Navy X–15 rocket-powered aircraft program had, during 10 yrs of flight, studied effects of hypersonic flight on aircraft skin friction and thermal expansion, pioneered the use of ablative coatings, aided efficient structural design, and served as a workhorse test-bed for nearly 40 experiments. NASA contributions also included the single-pivot variable-sweep wing, which allowed efficient flight at both high and low speeds; supercritical wing and fuselage shapes that permitted an aircraft to cruise at 15% higher speed without increased fuel consumption; and the world's first digital fly-by-wire control aircraft with a computer-controlled electronic system. The clean combustion program was providing new concepts to reduce pollution from engine exhaust. The aircraft noise-abatement program focused on development of an avionics system, refinement of operational procedures, and the demonstration of two-segment flight paths for quieter landing approaches. The refan program was demonstrating JT8D engine and nacelle modifications to reduce noise of existing jet aircraft.

In applications the knowledge gained in 15 yrs was being applied in science, medicine, navigation, communications, agriculture, crime prevention, and commerce. NASA-developed technologies were producing safer buildings, bridges, and aircraft and improved tires, roads, and railbeds. In international cooperation, NASA had conducted 18 cooperative satellite and space probe projects and flown 25 international experiments on its spacecraft. The global system of Intelsat communication satellites comprised 12 satellites launched by NASA.

Ahead were completion of Skylab 4, the third and final manned Skylab mission; the July 1975 joint U.S.–U.S.S.R. Apollo-Soyuz mission to dock a U.S. and a Soviet spacecraft in space; and the introduction of the revolutionary space shuttle with its European-built Spacelab in the 1980s. In unmanned exploration *Mariner 10*, scheduled for launch toward Mercury and Venus in early November, was expected to provide the first photos of Mercury. *Pioneer 10* was headed for a year-end rendezvous with Jupiter, and *Pioneer 11* (launched April 5, 1973) was scheduled for a second reconnaissance of Jupiter in December 1974. Work was proceeding on two Mariner spacecraft for 1977 launch toward Jupiter and Saturn.

Major thrusts expected for the 1980s were outlined in a statement released to the press by Dr. Homer E. Newell, NASA Associate Administrator. NASA would be vigorously developing and demonstrating systems and technologies that would make space activities useful to national and world communities, monitoring the environment and resources, developing electric power from the sun, and using space as a laboratory for new technology and processes. Continued scientific exploration of the earth, space, and the solar system seemed inevitable. And the 1980s might well be a period of preparation for continued manned exploration and exploitation of the solar system, the two-way direct transfer of information between satellites, a long-duration manned space laboratory, automated outer planet orbiters and atmosphere probes, and probably, before the end of the century, a program to establish a manned lunar base for exploration and research. (NASA Release 73-190)

- Sen. Frank E. Moss (D–Utah), Chairman of the Senate Committee on Aeronautical and Space Sciences, said on the Senate floor that NASA had been "second to none in its record of accomplishments during its first decade and a half." The past 15 yrs had seen the U.S. go "from the depths of postsputnik depression to undreamed of heights in both the exploration and exploitation of the benefits of space." The *Skylab 3* crew (launched July 28 to man the Orbital Workshop launched May 14) had "just completed 59 highly successful days in space, in a mission already being hailed as perhaps the most productive in our history. Eighty-three nations have joined in a global communications satellite system [Intelsat] now carrying three-quarters of the world's intercontinental telecommunications traffic." (*CR*, 10/1/73, S18200)

- Dr. Alexander Hocker, European Space Research Organization (ESRO) Director General, sent a message to Dr. James C. Fletcher, NASA Administrator, on NASA's 15th anniversary: "The resourcefulness, courage and remarkable achievements of NASA personnel have been an inspiration to us in Europe and ESRO is proud to be associated with some of NASA's most stimulating endeavours." (*ESRO Newsletter*, 10/73, 3)

- NASA, on its 15th birthday, found itself "in the throes of an identity crisis," John N. Wilford said in a *New York Times* article. After a "fast and glorious youth, full of promise and widely heralded achievement," could NASA, "knowing that it could be flying to the moon and beyond, adjust to a more mundane existence?" Dr. James C. Fletcher, NASA Administrator, had said in a recent interview: "Our charter is clear: We have to make the country strong in space. But NASA has to move back from the spectacular. We will become more like one of the service agencies of government." (*NYT*, 10/1/73, 62)

- The opening of a direct telephone link via satellite between Canada and the People's Republic of China was announced by the Canadian Overseas Telecommunications Corp. Calls would be transmitted via *Intelsat-IV F–4*, launched Jan. 22, 1972, by NASA for the Communications Satellite Corp. on behalf of the International Telecommunications Satellite Organization. Previously, calls originating in Canada had been routed through submarine cables to Tokyo and relayed to China from there. (AP, *W Post*, 10/2/73, C2)

- Military sources quoted by Reuters news agency said Navy plans to purchase 50 Grumman F-14 jet fighter aircraft each fiscal year through 1977 had been approved by Dr. James R. Schlesinger, Secretary of Defense. The number of F-14s to be purchased was increased from 313 to 334, but Dr. Schlesinger had said after 1977 the Navy must begin buying the VFX fighter, which would be less expensive than the $15-million-each F-14. (*NYT*, 10/2/73, 12)

- M/G John B. Hudson (USAF), Deputy Chief of Staff for Systems of Air Force Systems Command, became a lieutenant general and AFSC Vice Commander. He succeeded L/G Edmund F. O'Connor, who became Vice Commander, AF Logistics Command. (AFSC *Newsreview* 10/73, 1)

October 1–6: National Space Week was observed by 32 states to honor NASA's 15th anniversary [see Oct. 1]. The governors of the states had responded to a request by members of the Federation of Americans Supporting Science and Technology (FASST). Illinois Gov. Daniel Walker said: "The declared policy of the United States, that 'activities of space should be devoted to peaceful purposes for the benefit of all mankind,' has established our country as the scientific and technological leader in space. NASA has coordinated the efforts of scientists and engineers in every field, the aerospace industry and educational institutions." (FASST Release, 10/30/73)

October 1–18: A 47-member NASA delegation led by Apollo-Soyuz Test Project Director Glynn S. Lunney met with Soviet counterparts in Moscow for two weeks of discussions of technical problems of the joint U.S.–U.S.S.R. mission scheduled for July 1975. Discussions culminated in an Oct. 18 midterm review of the project by Dr. George M. Low, NASA Deputy Administrator, and Academician Boris N. Petrov, Chairman of the Intercosmos Council of the Soviet Academy of Sciences. Dr. Low had arrived in Moscow Oct. 14; his stay included visits to the Institute of Geochemistry, the Institute of Space Research, the Cosmonaut Training Center, and the Space Flight Control Center.

In an Oct. 18 communique Dr. Low and Academician Petrov confirmed their satisfaction with the ASTP project status. All major milestones were being met on schedule and, Dr. Low said, progress made indicated that the scheduled launch date could be met. Soyuz orbital and Apollo docking module drawings had been exchanged and agreement had been reached that U.S. specialists would participate in preflight final checkout of flight equipment at the Soviet launch site. Specialists would be exchanged to fit-check compatible equipment at U.S. and U.S.S.R. launch sites. Academician Petrov told the press that U.S. specialists would be permitted at the Soviet launch site "up to the very minute of launching."

During the visit U.S. officials learned the deaths of three cosmonauts on *Soyuz 11* June 30, 1971, were caused by accidental triggering open of an exhaust valve rather than by a hatch leak. The valve tripped open just after the reentry capsule separated from the larger orbiting module in a maneuver including firing of 12 explosive bolts. At least two crew members tried to close the valve but were unable to move fast enough. Unconsciousness from rapid decompression came in less than 10 sec; ground tests showed it would take 27 sec to close the valve by hand. U.S. officials said they were satisfied that subsequent modifications to the valve and crank mechanisms—tested on two unmanned flights, *Cosmos 496* and *573*, and on the manned Sept. 27–29 *Soyuz 12* mission —were sufficient to prevent reoccurrence of the accident. (NASA Releases 73-199, 73-224; Smith, *NYT*, 10/19/73; JSC, ASTP Briefing transcript, 11/19/73; O'Toole, *W Post*, 10/29/73, A1; *Av Wk*, 11/5/73, 20–21)

October 2: NASA held three Skylab press briefings at Johnson Space Center.

At a postflight crew press briefing, *Skylab 3* Astronaut Alan L. Bean —who had splashed down Sept. 25, after a 59-day mission working in the *Skylab 1* Orbital Workshop (launched May 14) with crew members

Dr. Owen K. Garriott and Jack R. Lousma—said that the crew's health had been "good the whole time." Sufficient exercise, meals on time, and plenty of sleep had accounted for their happiness and well-being. Having a scientist-astronaut aboard had created a "greater variety of interchange around the dinner table," making "a better mission for everybody." The mission would have been "50% less productive if Owen had not been there. He added a great amount to it . . . because his point of view is just different." Garriott said that a longer mission would necessitate "a somewhat less strenuous program," but he could have spent eight hours a day at the wardroom window with a camera in each hand and a good supply of film "and never have any tendency to become bored." Lousma said he still felt "a little less energetic" than before the mission, but within two weeks "I'll be back to the same condition."

Dr. Willard R. Hawkins, JSC Deputy Director of Life Sciences for Medical Operations, said that the crew was "in very good shape" but still had dizziness with specific head movements. Strength of the flexor and extensor muscles had decreased 20%. The astronauts' weights were steadily climbing toward their preflight weights. Early blood tests had shown the red cell mass was down an average of 12% for all three crew members. The blood plasma was also down 15% to 20%.

In an overview of Skylab 4, the third and final manned Skylab mission, Skylab Program Office Manager Kenneth S. Kleinknecht said that the spacecraft (Command and Service Module 118) and crew members Gerald P. Carr, Dr. Edward G. Gibson, and William R. Pogue were being readied for Nov. 11 launch. Final decision on the launch date would be made after further evaluation of the mission. In preparation, 140 kg (300 lbs) of operational equipment had been deleted from stowage. It had been replaced with another 150 kg (300 lbs) of operational equipment, 60 kg (130 lbs) for Skylab experiments, and 70 kg (150 lbs) of special equipment for Comet Kohoutek observations. After Oct. 8–10 flight readiness tests, only servicing, stowage, pyrotechnic installations, countdown demonstrations, and launch would remain. CSM-119, the rescue vehicle, was being prepared and would be available for a Jan. 6 launch if needed.

Alfred A. Bishop, Manager of the JSC Missions Office, said that, after the first two weeks of acclimation, the Skylab 4 crew would perform 28 man-hours of experiments each day, an increase from the 19 to 22½ man-hours on *Skylab 3*. The increase would provide an additional 200 experiment man-hours. The crew would observe Comet Kohoutek, perform 12 experiments using the Apollo Telescope Mount, observe Mercury crossing the sun Nov. 10, observe a solar eclipse, and perform 10 to 14 new earth resources experiment package passes. (Transcripts)

- The U.S.S.R.'s *Luna 19* automatic station (launched Sept. 28, 1971, toward lunar orbit) had discovered cosmic plasma on the near side of the moon, Tass reported. Mikhail Kolesov of the Institute of Radio Engineering and Electronics of the U.S.S.R. Academy of Sciences had said the maximum concentration of matter was at 10 km (6 mi) from the lunar surface. Origin of the plasma was unknown but the plasma was believed to be greatly affected by solar wind. (FBIS-Sov, 10/3/73, U1)

October 3: The U.S.S.R. launched eight Cosmos satellites on a single booster from Plesetsk. Orbital parameters were:

    Cosmos 588—1495-km (929-mi) apogee, 1450-km (901-mi) perigee, 115.3-min period, and 74.0° inclination.

October 3

Cosmos 589—1489-km (925.2-mi) apogee, 1417-km (880.5-mi) perigee, 114.9-min period, and 73.9° inclination.
Cosmos 590—1489-km (925.2-mi) apogee, 1435-km (891.7-mi) perigee, 115.1-min period, and 74.0° inclination.
Cosmos 591—1487-km (924-mi) apogee, 1349-km (838.2-mi) perigee, 114.1-min period, and 73.9° inclination.
Cosmos 592—1486-km (923.4-mi) apogee, 1333-km (828.3-mi) perigee, 113.9-min period, and 74.0° inclination.
Cosmos 593—1487-km (924-mi) apogee, 1366-km (848.8-mi) perigee, 114.3-min period, and 73.9° inclination.
Cosmos 594—1487-km (924-mi) apogee, 1383-km (859.4-mi) perigee, 114.5-min period, and 74.0° inclination.
Cosmos 595—1487-km (924-mi) apogee, 1400-km (869.9-mi) perigee, 114.7-min period, and 73.9° inclination.

Kenneth W. Gatland of the British Interplanetary Society later said in a *Christian Science Monitor* article that the satellites possibly were being used as a maritime communications system "to keep shore bases and command centers in contact with far-ranging elements of the Soviet fleet." Since 1971 two eight-satellite systems had been orbited each year four to five months apart. Gaps of seven to nine months occurred between launches during the winter, when the Soviet northern fleet was hampered by sea conditions. The last Soviet eight-satellite launch was June 8, into similar orbits. (GSFC *SSR*, 10/31/73; *CSM*, 11/13/73; *SBD*, 10/4/73, 170; Tass, FBIS–Sov, 10/3/73, U1)

The U.S.S.R. also launched *Cosmos 596* from Plesetsk, into orbit with a 268-km (166.5-mi) apogee, 205-km (127.4-mi) perigee, 89.2-min period, and 65.4° inclination. The satellite reentered Oct. 9. *Cosmos 596* was the first of seven Soviet satellites launched and returned in quick succession, leading to Western speculation they were reconnaissance craft monitoring the Arab-Israeli war [see Oct. 15]. (GSFC *SSR*, 10/31/73; *SBD*, 10/9/73, 192)

- Maj. Michael Love (USAF) successfully completed a captive flight of the X–24B lifting body attached to a B–52 aircraft from Flight Research Center. The flight was the first pilot checkout flight on the X–24B for Maj. Love, principal Air Force pilot in the NASA-USAF lifting body program. (NASA prog off; AFSC *Newsreview*, 11/73, 3)

- President Nixon announced 1973 recipients of the National Medal of Science, including Dr. Richard T. Whitcomb, Jr., Head of the 8-Foot Tunnels Branch at NASA's Langley Research Center. Dr. Whitcomb was cited for "his discoveries and inventions in aerodynamics which have provided and will continue to provide substantial improvements in the speed, range, and payload of a major portion of high performance aircraft produced throughout the country." The other recipients were: Dr. Daniel I. Arnon, Univ. of California at Berkeley physiologist; Dr. Carl Djerassi, Stanford Univ. chemist; Dr. Harold E. Edgerton, Massachusetts Institute of Technology Professor Emeritus of electrical engineering; Dr. William Maurice Ewing, Univ. of Texas at Galveston scientist; Dr. Arie Jan Haagen-Smit, Cal Tech biochemist; Dr. Vladimir Haensel, Vice President for Science and Technology, Universal Oil Products Corp.; Dr. Frederick Seitz, President of Rockefeller Univ.; Dr. Earl W. Sutherland, Jr., Univ. of Miami biochemist; Dr. John W. Tukey, Executive Director of Bell Telephone Laboratories, Inc., Research Communication Principles Div. and Princeton Univ. statistician;

and Dr. Robert Rathbun Wilson, Director of the National Accelerator Laboratory at Batavia, Ill. (*PD*, 10/8/73, 1224)

- Langley Research Center scientists had redesigned a Cessna Cardinal aircraft with a smaller, experimental wing in a research program to apply advanced aerodynamic technology to small general-aviation aircraft, NASA announced. Renamed the Redhawk, the experimental aircraft had been flown across the U.S. and operated by aviation writers and pilots from general-aviation companies. LaRC research pilots had praised the aircraft's low response to air turbulence and the control made possible by spoilers and flap systems on the wings. In the program's second phase, a Piper twin-engine Seneca aircraft was being converted to study a new airfoil section developed by Dr. Richard T. Whitcomb, Jr., Head of the 8-Foot Tunnels Branch at LaRC. (NASA Release 73–197)

- NASA launched an Aerobee 170A–E sounding rocket from White Sands Missile Range carrying a Goddard Space Flight Center galactic astronomy experiment to a 152.4-km (94.7-mi) altitude. The rocket and instrumentation performed satisfactorily. (GSFC proj off)

- Marshall Space Flight Center announced the selection of the Avco Electronics Operation of Avco Precision Products Div., Avco Corp., for negotiation of a $3 252 845, 23-mo contract for modifications to the MSFC Structures and Mechanics Laboratory. The work would provide a data-acquisition system for space-shuttle-hardware structural and fatigue tests. (MSFC Release 73–141)

October 4: President Nixon issued a statement about NASA's 15th anniversary [see Oct. 1]: "At mid-twentieth century, the United States found itself drawn by international necessity, and driven by national pride and pioneering instinct to view the heavens no longer as the stuff of romance, but as a practical environment for the extension of man's dominion over the earth. Since then, four administrations have committed our industrial resources, our technological and scientific capacities and the national will to the proposition that America can and must play a role in outer space equal to its responsibilities as a great nation." People of all nations stood in awe of NASA's accomplishments. "The men of NASA have undertaken one of the greatest human adventures with such consummate skill, cool exactness in the face of untold dangers, and professional disdain for the sensational that we have come already to take for granted the legitimacy of man's place in an environment entirely alien to him. But neither the courage, skill and devotion of the men of NASA, nor the benefits of their efforts, should be taken for granted. In every phase of our national life, new products, new techniques and new understanding have resulted from our efforts in space. And in relations between nations, the prospects for a lasting peace have been enhanced by the now absolute awareness that the primary interests of mankind are identical and shared in common. On the fifteenth anniversary of NASA, we look ahead to the same ordered progress into our space future which has characterized our first tentative steps into space." (Text)

- Maj. Michael Love (USAF) successfully completed the fifth captive flight of the X–24B lifting body attached to a B–52 aircraft from Flight Research Center. The flight checked out the pilot and evaluated handling qualities, forward and aft fuselage pressure, and acoustic noise and vibration. (NASA prog off)

- Aerospace Industries Assn. of America, Inc., issued its semiannual analysis of aerospace employment. Continued restraints in Government spending and an increase in foreign competition for existing markets were cited as causes of an expected continued decline in aerospace industry employment through June 1974. Employment had dropped by 33% since 1968. Nearly 950 000 persons had been employed in June 1973, but by June 1974 payrolls were expected to have been cut by about 32 000. The most significant decline would be among scientists and engineers—a reduction of 6000, 3.6% below December 1972. Employment of production workers was expected to decline by 13 000 jobs, leaving 460 000 employed by the end of FY 1974. Missiles and space employment was expected to decline by 28 000 workers between December 1972 and June 1974. Aircraft group employment faced a similar period of decline, from 556 000 to 533 000. Employment in the helicopter and general-aviation field was expected to increase 500 from June 1973, resulting in a work force of 22 500 by June 1974. Employment in the general-aviation aircraft field alone was expected to rise more than 20% during FY 1974. (Text)

October 5: The *Skylab 2* crew of Navy men—Capt. Charles Conrad, Jr., Capt. Joseph P. Kerwin, and Capt. Paul J. Weitz—was honored with Distinguished Service Medals by Secretary of the Navy John W. Warner in a Dept. of Defense ceremony. Citations were for "exceptionally meritorious service to the Government of the United States in a duty of great responsibility (as crewmen) for the Skylab mission from May 25 to June 22, 1973." Through their professional competence and dedications, they had "contributed immeasurably to the success of this historic mission, the results of which will furnish information of great significance to man's continuing quest for further knowledge and understanding of himself and the universe." (DOD Release 480–73)

- Federal Aviation Administrator Alexander P. Butterfield presented honorary FAA Airframe and Powerplant Mechanic's Certificates with an especially established "spacecraft rating" to *Skylab 2* Astronauts Charles Conrad, Jr., Dr. Joseph P. Kerwin, and Paul J. Weitz during Washington, D.C., ceremonies. Butterfield said: "The in-flight repair job you performed on the Skylab orbital workshop was one of the most spectacular achievements of the entire space program. It transformed a potentially disastrous mission into an outstanding success and demonstrated once again man's ingenuity in the face of adversity. I'm sure every aviation mechanic in the country is proud to have you as honorary members of their select and highly skilled fraternity." The awards were sponsored by the Aviation Maintenance Foundation in cooperation with FAA and the aviation industry. (FAA Release 73–173)

- Johnson Space Center had invited Boeing Co. to submit a proposal for wind-tunnel tests using a 747 transport aircraft to ferry the space shuttle orbiter, NASA announced. The orbiter would be mounted pickaback on the 747 in tests to determine feasibility of the plan. Lockheed Aircraft Corp. had been invited earlier to submit a proposal for use of a C–5A cargo aircraft as a ferry during wind-tunnel tests [see Sept. 25]. (NASA Release 73–204)

- A *Science* editorial commented on the altered structure of Federal science advisory apparatus arising from the demise of the Office of Science and Technology and the July 10 appointment of the National Science Foundation Director, Dr. H. Guyford Stever, as Presidential Science Adviser and Chairman of the Federal Council for Science and Technology. Dr. Stever had met with presidents or principal officers of most of the major scientific and engineering organizations and had indicated "that he would welcome policy advice and recommendations concerning personnel. As specific problems arise, the appropriate organizations will be tapped. Thus it seems that a major difference between the new and the old apparatus will be the replacement of a small, formal, elite group by more broadly based ad hoc groups. The concept is worth a try. Whether it will be viable will depend on how effectively it is implemented by both sides." (Abelson, *Science*, 10/5/73, 13)

October 6: The U.S.S.R. launched *Cosmos 597* from Plesetsk into an orbit with a 214-km (133-mi) apogee, 163-km (101.3-mi) perigee, 88.3-min period, and 65.4° inclination. The satellite reentered Oct. 12. *Cosmos 597* was the second of seven Soviet satellites launched and returned in quick succession, leading to Western speculation that they were reconnaissance craft monitoring the Arab-Israeli war [see Oct. 15]. (GSFC *SSR*, 10/31/73; *SBD*, 10/10/73, 193)

• The Nippon Electric Co. of Tokyo had won a contract to design and install the antenna and communications equipment for the first Swiss communications satellite station, the *New York Times* reported. The station was being built above the Rhone Valley in the Valais canton. (Lusinchi, *NYT*, 10/6/73)

October 7–13: The International Astronautical Federation (IAF) held its 24th Congress in Baku, U.S.S.R. In a message of greeting, the Soviet Council of Ministers said: "Nowadays, when important positive changes take place in international situations, ever more favourable opportunities for development of scientific communication, exchange of experience and uses of scientific achievements for practical needs of mankind open to scientists." Academician Sergey Vernov of the Soviet Academy of Sciences said Soviet artificial satellites, placed simultaneously in different orbits, had facilitated studies of many regions of earth's radiation belts, including those where no space station had been before. He hypothesized the existence of an "unstable radiation zone" beyond the outer radiation belt where Soviet Electron satellites had recorded a flux of electrons with energies over 150 000 ev. He believed the zone to be a necessary link in the formation of the earth's radiation belts.

Honorary guests Astronaut Thomas P. Stafford, U.S. Apollo-Soyuz Test Project commander, and Soviet Cosmonauts Georgy T. Beregovoy, Vitaly I. Sevastyanov, and Vladimir A. Shatalov participated in an open discussion. Shatalov, head of cosmonaut training, said the U.S. and the U.S.S.R. could be good and equal partners in conquering space. Stafford said cooperation in ASTP had already solved colossal technical problems. "And if we can solve problems in space we can, consequently, solve considerable problems on earth."

Soviet scientists reported on a man-plant interaction experiment in which four researchers spent six months in a closed biological environment simulating the earth's biosphere. The Siberian test project, sponsored by the Krasnoyarsk Institute of Physics, had demonstrated that man could live in an artificial atmosphere created by green algae, fast-growing wheat, and a variety of vegetables. The plants had been irrigated with sewage waste and with condensed moisture from the atmosphere. The researchers had breathed oxygen released by the plants, eaten the vegetables, and baked bread from the wheat during a simulated space trip in an underground chamber.

A paper by Soviet specialists reported that an electric field might divert a flux of charged particles from a spacecraft. The spacecraft's exterior surface might act as one of the field's electrodes while the vacuum of outer space acted as an insulating medium, providing an effective radiation shield. Dr. Ernst Stuhlinger, Marshall Space Flight Center Associate Director for Science, cochaired two sessions on scientific spacecraft systems.

The 7th International History of Astronautics Symposium included the papers "From Back Fire to Explorer I" by Dr. Kurt H. Debus, Director, Kennedy Space Center; "R. H. Goddard: Accomplishments of the Roswell Years (1930–1941)" by Frederick C. Durant III, Assistant Director for Astronautics, National Air and Space Museum, Smithsonian Institution; "High Energy Propulsion at NACA Lewis Engine Research Laboratory" by John L. Sloop, former NASA Assistant Associate Administrator for Aeronautics and Space Technology; and "Early Photography from Rockets" by Frank H. Winter of the Air and Space Museum.

Dr. Maxime A. Faget, Johnson Space Center Director of Engineering and Development, received the 1973 Daniel and Florence Guggenheim International Astronautics Award of the International Academy of Astronautics for "playing a major role in developing the basic ideas and original design concepts that have been incorporated into all the manned spacecraft flown by the United States. . . . An expert on vehicles suitable for reentering the Earth's atmosphere, he is particularly noted for his contributions to the basic configuration of the command module and to the development of the pressure-fed hypergolic engines used on the Apollo modules."

The meeting, attended by 1500 scientists from 29 countries, elected Dr. Charles Stark Draper, Director of the Charles Stark Draper Laboratory at Massachusetts Institute of Technology, IAF President. (Tass, FBIS–Sov, 10/8/73; MSFC Release 73–137; NASA Release 73–200; NASA Hist Off; Agence France-Presse, *W Post*, 10/12/73, A30)

October 8: Jet Propulsion Laboratory scientists began to measure stratospheric constituents up to 24 km (80 000 ft) using a Fourier interferometer aboard the British prototype of the Anglo-French Concorde supersonic transport aircraft. Eight scheduled flights from Fairford, England, sponsored by the U.S. Dept. of Transportation, were to obtain information on the potential pollution effect of supersonic transports on the atmosphere. In earlier flights the interferometer had detected nitric oxide between 12 and 24 km (39 000 and 80 000 ft) at less than one part per billion. Nitric oxide was believed to be vital in maintaining ozone balance in the stratosphere. (JPL Release 670; JPL PIO)

• Navy plans for collecting temperature and cloud-cover data directly from Air Force weather satellites were reported by *Aviation Week & Space Technology*. A receiver and computer processing center at Monterey, Calif., would prepare data for worldwide dissemination to the fleet and would process weather photos and atmospheric vertical-temperature profiles. (*Av Wk*, 10/8/73, 9)

October 9: The original name Cape Canaveral was restored to Cape Kennedy by the Dept. of the Interior, acting on a unanimous recommendation by an interagency committee of the Board on Geographic Names.

Historians believed the Cape had been named Canaveral by Spanish explorer Ponce de Leon more than 400 yrs before, the Orlando *Sentinel Star* later reported. The action made no change in the name of NASA's Kennedy Space Center, named for the late President John F. Kennedy. (Dept Int Release, 10/9/73; CR, 10/10/73, H8880; O Sen Star, 10/10/73, 1)

- Johnson Space Center announced it would issue a $1.2-million cost-plus-fixed-fee supplemental agreement to an existing Skylab Payload Integration contract with Martin Marietta Corp. Denver Div. The agreement was for earth resources experiment package hardware evaluation, including evaluation of EREP hardware performance during Skylab missions. (JSC Release 73-133)

- U.S. technological progress had proceeded "almost without interruption" during 70 yrs of flight, Marquis Childs said in his *Washington Post* column. The doubters had "always been proven wrong." In astronautics, Skylab was continuing successfully and work was progressing on the space shuttle. "As for the scientific benefits, perhaps only a few specialists can give a proper evaluation. But the cynics who spoke derisively of spending billions to put some clown in the sky are wrong on several scores." The aircraft production record was equally impressive. In 1972 the U.S. had made 79% of the aircraft being operated by the world's airlines, exclusive of the U.S.S.R. and the People's Republic of China. The achievements in U.S. production, discovery, and invention were "in painful contrast to the failures in self-government illustrated by the grim mess in Washington, It may not be too much to say that if we find a way out of the morass, the swamp of intrigue, deceit and doubt, it will be thanks to American productivity." The Smithsonian Institution's new Air and Space Museum was to be dedicated July 4, 1976. "That no one can predict what manner of man will preside over the occasion is a melancholy commentary on the gap between technology and politics." (*W Post*, 10/9/73)

October 9-12: A radical new growth hormone theory evolved from Ames Research Center studies was introduced by ARC scientist Dr. Richard E. Grindeland during a Baltimore, Md., symposium on growth hormones. Dr. Grindeland's theory that the growth hormone in blood was chemically different from that in the pituitary gland had originated from ARC research to determine effects of physiological stresses like acceleration, simulated weightlessness, chemical agents, thermal extremes, and nutritional status on the human body. The theory, which the symposium suggested would have profound effect on the understanding of growth hormone physiology, showed that radio immunological methods detected only a minute fraction of the concentration of growth hormone in the blood. (ARC *Astrogram*, 10/25/73, 1)

October 9-19: A 37-member delegation from the European Space Research Organization (ESRO) visited the U.S. to discuss plans for the space shuttle Spacelab, which ESRO would build under the Memorandum of Understanding between the U.S. and nine European nations signed Sept. 24. The delegation attended an Oct. 9-10 Spacelab Preliminary Design and Definition Study at Marshall Space Flight Center; visited the Downey, Calif., facility of Rockwell International Corp., prime shuttle contractor; toured Johnson and Kennedy Space Centers; and attended an Oct. 18-19 summary discussion meeting at MSFC. (MSFC

Release 73-143; KSC Release 238-73; *Marshall Star*, 10/3/73, 1; MSFC PAO)

October 10: The U.S.S.R. launched *Cosmos 598* from Plesetsk into orbit with a 331-km (205.7-mi) apogee, 204-km (126.8-mi) perigee, 89.9-min period, and 72.8° inclination. The satellite reentered Oct. 16. *Cosmos 598* was the third of seven Soviet satellites launched and returned in quick succession, leading to Western speculation that they were monitoring the Arab-Israeli war [see Oct. 15].

- Skylab 4 Astronauts Gerald P. Carr, Dr. Edward G. Gibson, and William R. Pogue, scheduled for Nov. 11 launch to crew the *Skylab 1* Orbital Workshop (launched May 14), completed training in the Neutral Buoyancy Simulator at Marshall Space Flight Center. The astronauts had spent 80 hrs since April 1972 performing 24 different exercises in the simulated weightless environment to prepare them for the three extravehicular activities scheduled for their mission. The backup crew —Vance D. Brand, Dr. William B. Lenoir, and Dr. Don L. Lind—would complete their training at MSFC Oct. 12. (MSFC Release 73-145)

- President Nixon presented National Medals of Science to recipients announced Oct. 6, during a White House ceremony. He said: "We all know that because the United States needed a concentration on defense at a critical time, and then later a concentration on space, that this opened broad, new vistas in the area of science, and this also resulted in a much greater Federal contribution and the justification for it from a budgetary standpoint, but now as we turn from war to the works of peace, we must not cut back on that research. What we must do is to channel the efforts in the field of research to peaceful uses." (*PD*, 10/15/73, 1236-7)

October 11: An estimated $2.8 million in laboratory equipment used to process lunar samples had been shipped from Johnson Space Center's Lunar Receiving Laboratory to the Atomic Energy Commission's Los Alamos Scientific Laboratory in New Mexico, JSC announced. AEC would use the equipment—vacuum chambers and pumps, stainless steel pipes, air-tight processing cabinets, and electronic monitoring devices— to investigate the use of lasers to produce energy through controlled thermonuclear reactions. The Lunar Receiving Laboratory had closed in June after processing the last lunar samples from the Apollo program. All sample material had been transferred to the Lunar Sample Curatorial Facility at JSC. The laboratory equipment had been declared excess property. (JSC Release 73-135)

- President Nixon announced he would add $115 million to the FY 1974 budget for energy research and development. The increase would raise total energy R&D funding for FY 1974 to about $1 billion, a 37% increase over FY 1973. He said, "Our hopes for advancing research and development also rest upon my proposed legislation to create a Department of Energy and Natural Resources and an independent Energy Research and Development Administration." (*PD*, 10/15/73, 1239-40)

- The 1973 Nobel Prize in physiology or medicine was awarded in Stockholm to three pioneers in ethology, the comparative study of behavior. Austrian-born Dr. Karl von Frisch, retired, and Dr. Konrad Lorenz of the Max Planck Institute for Behavioral Physiology, shared the $121 000 cash prize with Netherlands-born Dr. Nikolaas Tinbergen, professor of animal behavior at Oxford Univ. (*NYT*, 10/12/73, 1)

October 12: NASA research pilot Einar K. Enevoldson conducted the first flight of a new remotely piloted research vehicle (RPRV). A three-eighths scale model of an Air Force high-performance F–15 fighter—air-launched from a B–52 aircraft from Flight Research Center at 14 000 m (45 000 ft)—was controlled by Enevoldson from a ground cockpit with flight instruments and a TV screen. His commands were fed through a computer programmed with F–15 flight characteristics to the RPRV; flight information from the RPRV and video signals from its TV camera were transmitted to the ground control station. After a nine-minute flight, the RPRV descended 5000 m (15 000 ft) and a series of parachutes were deployed to slow the descent rate. An Air Force helicopter recovered the RPRV in mid-air and lowered it to the ground for reuse. Eventually the 7-m (23-ft) model would be flown through stall and spin maneuvers using the new FRC-developed flight research technique to reduce the cost of flight-testing experimental aircraft and spacecraft. (FRC Release 28–73; AP, *Bakersfield Californian*, 10/13/73)

- The Pratt & Whitney engine designed for the Air Force F–15 Eagle fighter aircraft passed a 150-hr endurance test on the third try at West Palm Beach, Fla. The success cleared the way for approval of a production contract. (UPI, *NYT*, 10/14/73, 7)
- Pascagoula, Miss., Sheriff Fred Diamond told the press that two local shipyard workers had reported being taken aboard a glowing blue spacecraft by buzzing creatures with silvery, wrinkled skin. Workers Charles Hickson and Calvin Parker had said they had seen the spacecraft approaching as they fished off a pier Oct. 11. (UPI, *W Star-News*, 10/13/73, A3)

October 13: The work of Abe Karen, NASA physicist and consultant to New York City's Budget Dept., was described in a *New York Times* article. Karen worked under a "one-man program called the NASA/New York City Applications Project," administered by NASA's Office of Technology Utilization. Karen, paid by NASA, had introduced Jet Propulsion Laboratory's Silent Communication Alarm Network (SCAN) security system to protect life and property in two crime-ridden New York high schools and an Ames Research Center device to detect heroin in urine samples. He was eying other NASA-funded innovations for road-patching, bridge structural testing, fire warning, and removing graffiti from public buildings. (Darnton, *NYT*, 10/13/73, 37; NASA Off TU)

- After a 35-yr local dispute, the "Great White Sands Missile Range Lost Gold Treasure Affair" had "blossomed into a national mystery," the *New York Times* reported. Attorney F. Lee Bailey recently had been engaged by 50 unidentified claimants who sought prospecting rights to gold bars and treasure said to be buried on the missile range and to be worth up to $1.5 billion. In 1963 the Army, insisting there was no treasure, had banned prospectors from the range because of danger from exploding target-practice shells. Rep. Harold L. Runnels (D–N. Mex.) had proposed a joint Federal-state search for the treasure, said to have originated in various ways. (Sterba, *NYT*, 10/13/73, 37)

October 14: Doctors in Galveston, Tex., had pronounced the condition of *Apollo 13* Astronaut Fred W. Haise, Jr., "good and stable" following skin grafts for burns he received in an Aug. 22 aircraft accident, United Press International reported. Haise had been burned over 55% of his body and had suffered a lung ailment following the crash of a

World War II training aircraft he had been piloting for an air show group. (UPI, *M Her*, 10/14/73)

October 15: The U.S.S.R. launched *Cosmos 599* from Baykonur Cosmodrome into orbit with a 276-km (171.5-mi) apogee, 208-km (129.3-mi) perigee, 89.3-min period, and 65.0° inclination. The satellite reentered Oct. 28. Western observers speculated that *Cosmos 599*, with three other Soviet satellites—*Cosmos 596, 597,* and *598,* launched Oct. 3, 6, and 10—had monitored the Arab-Israeli war that had broken out Oct. 6. The quick succession of launches and recoveries were seen as evidence of almost continuous Soviet observation of the fighting by orbiting cameras. A Washington, D.C., source had said that the Cosmos orbits had permitted them to pass over the Middle East several times at altitudes of 196 to 322 km (122 to 200 mi) before reentry. The *New York Times* said the U.S. also could have been receiving information on the war from an Air Force satellite launched Sept. 27.

Cosmos 600, 602, and *603* closely followed the first four Oct. 16, 21, and 27 and were also shortly returned. They were believed to be continuing the observation of tank battles, troop movements, aircraft strike damage, and even of supply centers in the U.S. (GSFC *SSR*, 10/31/73; *SBD*, 10/16/73, 228; 10/18/73, 241; 10/24/73, 266; 11/12/73, 60; Wilford, *NYT*, 10/18/73, 21; Gatland, *CSM*, 11/13/73)

• Three aerospace firms had received eight-month National Science Foundation study contracts to explore use of solar energy to heat, cool, and provide hot water for buildings, *Aviation Week & Space Technology* reported. General Electric Co. would receive $547 000, TRW Inc. $486 000, and Westinghouse Electric Corp. $503 000. Also cooperating would be the Univ. of Pennsylvania, Arizona State Univ., Colorado State Univ., and Carnegie-Mellon Univ. (*Av Wk*, 10/15/73, 24)

• The U.S.S.R. had launched its second nuclear-powered icebreaker, *Arktika,* in a Leningrad shipyard, Tass reported. The 14.5-million-kg (32-million-lb) ship was part of a growing fleet of modern icebreakers to keep the Arctic seaway open all year for Soviet trade. (FBIS-Sov, 10/30/73, U3; Shabad, *NYT*, 10/25/73, 4)

• A *New York Times* editorial criticized "boom-and-bust science": Scholarship and research were "caught between the penny-pinching of economy-minded conservatives and the hostility of those, at the other end of the spectrum, who equate science expenditures with support of the military. A more realistic view of national needs should make it evident that none of the nation's and the world's critical problems—from environmental issues to the urgent need for food, energy and transportation—will be solved without the innovative contributions of trained minds. The continued boom-and-bust cycles in support of education and research constitute a costly waste of the nation's human resources. Perhaps the President's recent presentation of the National Medal of Science to eleven top scientists—the first such awards since 1970—represents a signal of Administration desire for a sounder approach." (*NYT*, 10/15/73, 36)

October 16: The U.S.S.R. launched two Cosmos satellites from Plesetsk. *Cosmos 600* entered orbit with a 355-km (220.6-mi) apogee, 192-km (119.3-mi) perigee, 90.0-min period, and 72.8° inclination. The satellite reentered Oct. 23, believed by Western observers to be fifth in a series of reconnaissance spacecraft monitoring the Arab-Israeli war [see Oct. 15].

Cosmos 601 entered orbit with a 1494-km (928.3-mi) apogee, 199-km

(123.7-mi) perigee, 101.8-min period, and 81.9° inclination. It reentered Aug. 15, 1974. (GSFC SSR, 10/31/73; 8/31/74; SBD, 10/18/73, 241–242)

- Rep. Mike McCormack (D–Wash.) introduced H.R. 10952 for himself and Rep. Olin E. Teague (D–Tex.), Rep. Charles A. Mosher (R–Ohio), and Rep. Barry M. Goldwater, Jr. (R–Calif.). The bill would provide for "the early commercial demonstration of the technology of solar heating by NASA in cooperation with the National Bureau of Standards, National Science Foundation, Secretary of Housing and Urban Development, and other Federal agencies and for early development and demonstration of technology for combined solar heating and cooling." The bill was referred to the Committee on Science and Astronautics. (CR, 10/16/73, H9192)

- Soviet news agency Tass reported that Soviet scientists, first in Gorky and later in other parts of the U.S.S.R., recently had picked up signals from outer space that might have been "sent by a technically developed extraterrestrial civilization." Observation of such signals was supervised in the U.S.S.R. by Soviet astrophysicist Vsevolod Troitsky and more than 30 scientists. Signals were monitored at four stations throughout the U.S.S.R., Tass said. (Reuters, B Sun, 10/17/73)

- The National Science Foundation published *Continued Growth Planned for Federal Civilian R&D Programs* (NSF 73–314). Federal support for civilian research and development had risen 9.1% per year for the past five years. The chief areas of growth were health, transportation and communication, environment, and crime prevention and control. In 1969 civilian-oriented programs had constituted nearly 25%, or $3.6 billion, of a $15.6 billion total; they were expected to form 33.3% of the 1974 $17.4 billion total. National security took up more than 50% of the R&D total each year. In 1974, its growth in R&D obligations was expected to be $478 million, making the national security share of the Federal R&D total 54.2%, higher than any other year since 1969. Space remained the second function in funding size 1969–1974, but reflected the greatest loss in priority. Space R&D obligations showed a steady annual decrease, and the share in the R&D total had declined correspondingly, from 23.9% in 1969 to 14.1% in 1974. From 1973 to 1974, total R&D obligations for space were expected to drop by $234 million, the largest decrease for any function. (Text)

October 17: NASA and National Oceanic and Atmospheric Administration researchers had devised a system to anticipate and counter the effects of lightning at Kennedy Space Center, NOAA announced. A network of instruments measured the area's electric field and a weather radar monitored rain-cloud precipitation, and a computer displayed electric-field contour maps of the launch area. KSC personnel would be able to forecast lightning while a launch vehicle was on the pad and during the low-altitude portion of its ascent, so that persons and equipment could be protected. (NOAA Release 73–210)

- Maryland State Police officials ordered troopers to search the sky for any abnormal light phenomena after receiving hundreds of reports of lights in the sky. One report came from a State Police helicopter pilot, who said he had seen a "very large" airborne vehicle with bright lights. Flight controllers at Andrews Air Force Base, Md., later said the object might have been a "special experimental aircraft testing an innovative lighting system for NASA." (B Sun, 10/18/73, A3)

October 17

- Telecare, an 18-kg (40-lb) medical unit equipped to diagnose and treat victims at a scene of emergency, had been developed by SCI Systems, Inc., from space technology, NASA announced. The ambulance-stored unit with two-way voice and telemetry communications permitted medical technicians to administer prompt professional care with special equipment under radio supervision of a doctor miles away. (NASA Release 73-209)

October 18: NASA announced plans to study Comet Kohoutek with an extensive array of electronic eyes. The comet, now some 400 million km (250 million mi) from the sun, was expected to be visible in the Northern Hemisphere around Christmas time. It was expected to be as bright as a full moon and to be larger than Halley's Comet, which had appeared last in 1910. NASA scientists would use optical telescopes, radiotelescopes, and radar to study Kohoutek in visible, ultraviolet, and infrared light from the ground and a high-altitude C-141 aircraft, as well as satellites, sounding rockets, and Skylab 4 telescopes and cameras—to learn more about whether planets were remnants of the formation of the solar system or interstellar matter captured by the sun. Scientists would try to determine possible existence of a solid comet nucleus. Kohoutek was believed to have a 20- to 30-km (12- to 19-mi) nucleus with a probable head diameter of 96 000 km (60 000 mi). (NASA Release 73-207)

- Dr. Fred L. Whipple, Harvard Univ. astronomer and former Director of the Smithsonian Astrophysical Observatory, received the Smithsonian Institution's Henry Medal during Cambridge, Mass., ceremonies honoring his July retirement from the Observatory. Dr. Whipple was cited for his leadership of the Observatory from the early days of the national space program to the present and for his contributions to man's understanding of the solar system. (Smithsonian *Torch*, 11/73, 4)

- Col. Bernt Balchen (USAF, Ret.), aviator and explorer who was chief pilot on Adm. Richard E. Byrd's first South Pole flight in 1929, died in Mt. Kisco, N.Y., at age 74. From 1933 to 1935 Balchen had been chief pilot of the Ellsworth Antarctic Expedition and in 1935 to 1940 chief of inspection for Norwegian Airlines. He was President of Norwegian Airlines in 1946 when it became the parent company of Scandinavian Airlines System. In 1948 Col. Balchen was recalled to active duty in the Air Force and named commanding officer of the 10th Rescue Squadron at Fort Richardson, Alaska. (*NYT*, 10/19/73, 42)

- Dr. Frank T. McClure, Deputy Director of the Johns Hopkins Univ. Applied Physics Laboratory, died in Baltimore, Md., following a heart attack. He was internationally known as an authority on combustion, rockets, and guided missile technology and was inventor of the Navy satellite doppler navigation system. For this invention he had received NASA's first inventions award in 1961. (*W Post*, 10/19/73, C11)

October 19: The U.S.S.R. launched *Molniya II-7* communications satellite from Baykonur into orbit with 39 860-km (24 767.9-mi) apogee, 503-km (312.6-mi) perigee, 717.9-min period, and 62.8° inclination. The satellite would help provide a system of long-range telephone and telegraph radio communications in the U.S.S.R. and would transmit Soviet central TV programs to the Orbita network. (GSFC *SSR*, 10/31/73; Tass, FBIS-Sov, 10/24/73, U1; *Sov Aero*, 1/7/74, 4)

- Concorde 001, the French prototype of the Anglo-French supersonic transport aircraft, was flown by Aérospatiale test pilot André Turcat on its last flight, from Toulouse to Le Bourget Airport near Paris. Turcat had piloted 001's maiden flight March 2, 1969. The aircraft, which had flown 396 flights and 810 hrs (254 hrs at supersonic speeds), would be given to the French Air Museum in Paris. (*NYT*, 10/21/73, 21)

- Election of Howard K. Nason, President of Monsanto Research Corp., as Chairman of NASA's Aerospace Safety Advisory Panel was announced by NASA. He would succeed L/G Carroll H. Dunn (USA), who was retiring after a year of service. (NASA Release 73–213)
- Lewis Research Center scientist Stanley G. Young had applied cavitation—erosion of metal by tiny cavities formed by uneven pressure gradients in liquids—to restoring serial numbers obliterated by filing or grinding, NASA reported. In the process, of use to police in identifying stolen property, an ultrasonic vibrator generated very-high-frequency vibrations in water, creating millions of microscopic bubbles. The bubbles struck the metal surface on which serial numbers had been erased, weakening particles that filled the serial number grooves. Conventional methods of restoring serial numbers used chemicals or mechanical processes that might obliterate remaining traces of the numbers. (NASA Release 73–216)
- Johnson Space Center announced it had authorized Rockwell International Corp. to modify a building at Air Force Plant 42 in Palmdale, Calif., for final assembly, system installations, and checkout stations for two space shuttle orbiters. The modifications would be made under an existing cost-reimbursement-no-fee contract and would bring total contract value to $6 000 000. (JSC Release 73–137)
- NASA launched an Arcas sounding rocket from Antigua, West Indies, carrying a Goddard Space Flight Center meteorology experiment to a 42.7-km (26.5-mi) altitude to measure ozone distribution in the upper atmosphere, monitor anomalous ultraviolet absorption, and extend the data base for a climatology of stratospheric ozone in the tropics. The launch was in conjunction with a *Nimbus 4* satellite (launched April 8, 1970) overpass. Because only one of four optical channels transmitted data and the thermistor was reading open, no significant data were received. (NASA Rpt SRL)

October 20: The U.S.S.R. launched *Cosmos 602* from Plesetsk into orbit with a 332-km (206.3-mi) apogee, 207-km (128.6-mi) perigee, 90.0-min period, and 72.9° inclination. The satellite reentered Oct. 29, believed by Western observers to be sixth in a series of reconnaissance spacecraft monitoring the Arab-Israeli war [see Oct. 15]. (GSFC *SSR*, 10/31/73; *SBD*, 10/24/73, 266)

- Skylab 4 crewmen Gerald P. Carr, Dr. Edward G. Gibson, and William R. Pogue began a 21-day prelaunch isolation period geared to a possible Nov. 10 launch. The astronauts would crew the *Skylab 1* Orbital Workshop (launched May 14) in the third and final manned Skylab mission. The isolation would minimize the exposure of the crew to infectious disease. (NASA Release 73–220)

October 21: Not since the 1947 first report of what came to be known as "flying saucers" had there been "such widespread reports of unidentified flying objects, or UFO's, as in recent days," the *New York Times* said. They had ranged from a report of a flying V formation of lights in Rochester, N.Y., to press accounts of "strange creatures with weirdly shaped heads" stopping cars on Route 90 in Gulfport, Miss. Two men in Pascagoula, Miss., had said they had been taken aboard a UFO by creatures with crab-claw hands [see Oct. 12]. The Federal Aviation Administration had said that no reports of special significance had been submitted by airline pilots, and an Air Defense Command spokesman had denied that any UFO had been detected in the last three weeks. (Sullivan, *NYT*, 10/21/73, 65)

October 23: Two fuel tanks on the Saturn IB launch vehicle scheduled to launch the Skylab 4 astronauts toward a Nov. 10 rendezvous with the Orbital Workshop (launched May 14) buckled slightly after fueling at Kennedy Space Center. The dome-shaped top sections of the tanks sank after ground crews pumped 163 cu m (43 000 gal) of kerosene fuel into the Saturn IB's four giant tanks and then drained some off to prevent expansion or overpressurization. A plastic cover protecting the tanks from a rainstorm prevented air from rushing into the aluminum tanks as the fuel was drained. The resulting vacuum buckled the tanks inward. Metallurgy experts were flown to KSC from Marshall Space Flight Center to assess the problem. (*W Star-News*, 10/24/73, A23; AP, *B Sun*, 10/24/73)

- A five-week Ames Research Center experiment to determine women's qualifications for space flight had shown women to be as physically fit for travel in weightlessness as men, NASA announced. The experiment—in which 8 of 12 Air Force flight nurses had two weeks of total bed rest in weightlessness simulation, while 4 acted as ambulatory controls—was one of a series to investigate responses to space flight conditions, looking forward to use of the space shuttle by other than pilot-trained astronauts. NASA emphasized that the test was to determine women's qualifications for space flight and did not represent a commitment to add women to the astronaut corps. Dr. Harold Sandler, head of the experiment team, said the bedridden nurses were weaker after the experiment and showed a 50% reduction in their tolerance to various stresses. "But the same is true of men astronauts." (NASA Release 73-218; AP, *P Inq*, 10/25/73)

- Resignation of NASA Associate Administrator for Astronautics and Space Technology Roy P. Jackson, effective Oct. 26, was announced. Jackson would rejoin Northrop Corp. as Corporate Vice President, Program Management. He had joined Northrop in 1953 and left to become NASA Associate Administrator for Aeronautics and Space Technology Nov. 2, 1970. (NASA Release 73-221)

- The 1973 Nobel Prizes in physics and in chemistry were awarded in Stockholm. The $122 000 physics award went to two Americans and one Briton for step-by-step proof of how electrons tunneled through conductors to form superconductors of electricity. The winners were Japanese-born Leo Esaki, International Business Machines Co. scientist; Norwegian-born Ivar Giaever, General Electric Co. scientist; and Brian Josephson of Cambridge Univ. in England.

    The $120 000 chemistry award went to Ernst O. Fischer, Munich Technical Univ. scientist, and Geoffrey Wilkinson of London's Imperial College of Science and Technology for "pioneering work, performed independently, on the chemistry of the so-called sandwich compounds" which attached rings of carbon compounds to transitional metals like titanium and vanadium. (O'Toole, *W Post*, 10/24/73, A19; *FonF*, 10/21-27/73, 882)

- Rockwell International Corp. and Admiral Corp. announced plans to merge. RIC would absorb Admiral for approximately $100 million in a stock exchange. The proposal required approval by directors of both companies and by Admiral shareholders. (*W Post*, 10/24/73)

- Dr. Margaret Burbidge, Director of the Royal Greenwich Observatory in England, would resign in November to return to the U.S. as a Univ. of California astronomy professor, the Chicago Tribune Press Service reported. Dr. Burbidge would be succeeded by Dr. A. Hunter, Royal Greenwich Observatory Deputy Director. Announcing Dr. Burbidge's resignation, the British Science Research Council had said she preferred to return to her own research rather than devote a major share of her time to administrative matters. Dr. Burbidge had criticized the Royal Observatory telescope at Herstmonceux, England, because cloudiness made it almost useless. (*C Trib*, 10/23/73)

- A Los Angeles *Herald-Examiner* editoral commented on NASA's 15th anniversary [see Oct. 1]: "A decade and a half ago, the Congress charged NASA to explore outer space and to adapt space-related advances for immediate general use. Both mandates have been fulfilled, and NASA officers predict hundredfold advances over the next 10 years. We salute NASA on its 15th anniversary. When the question is asked 'If we can go to the moon, why can't we . . . ?'—chances are, the answer is 'We can!'" (*LA Her-Exam*, 10/23/73)

- A *Chicago Tribune* editorial commented on the increase in reports of unidentified flying objects [see Oct. 21]: "We read about an Indiana farmer who heard of the UFO upsurge and who joked with his family about little green men from some distant nebula hanging around. The next thing he knew, one of the UFOs was hovering . . . above the family conveyance and following it. Chased the family all the way home. No, you won't catch *us* joking about these little guys with the crablike hands and the pointed ears. We're going on record: We *like* crablike hands and pointed ears." (*C Trib*, 10/23/73)

- B/G William J. Kennard (USAF, Ret.), former air surgeon of the Military Air Transport Service, died at Andrews Air Force Base, Md., at age 67. He had served from 1959 to 1967 as Executive Vice President of the Aerospace Medical Assn. and managing editor of *Aerospace Medicine*. (*W Post*, 10/26/73, C8)

October 24: Establishment of the Spacelab Program Office in the Office of Manned Space Flight was announced by NASA. The new office, to be headed by Douglas R. Lord, would replace the Sortie Lab Task Force. The reorganization followed the Sept. 24 signing of a cooperative agreement by NASA and the European Space Research Organization which called for European design, development, and manufacture of a Spacelab to be flown in the space shuttle orbiter. (NASA Release 73-228)

- Gerald M. Truszynski, NASA Associate Administrator for Tracking and Data Acquisition, said in hearings on NASA's tracking program before the House Committee on Science and Astronautics' Subcommittee on Aeronautics and Space Technology that the percentage of spacecraft requiring real-time control would increase from 70% to 90% in 1975. With increased number of satellites and volume of data, as well, new tracking capability would be required. NASA proposed a new Tracking and Data Relay Satellite System (TDRSS) for the late 1970s—possibly leased from the private sector—to support low-earth-orbit spacecraft. Two synchronous satellites would relay telemetry, voice, and video data from other spacecraft to a single U.S. earth station, or from the station to the spacecraft. TDRSS would support all spacecraft below 5000 km

(3100 mi), including shuttle, Spacelab, and automated spacecraft. The new system would give 88% continuous contact with spacecraft, real-time interactive experiments, and reduced dependence on foreign-based tracking stations. Half the Spaceflight Tracking and Data Network stations could be closed, saving "tens of millions of dollars." (Transcript)
- NASA announced it had granted a five-year exclusive patent license for the automated visual sensitivity tester (AVST), a device to detect eye abnormalities, to Consultants Unlimited. The device—an 8-mm motion picture projector with an electronic control unit and an automatic response plotter—had been developed by Ames Research Center scientists for astronaut eye examinations during long-duration space missions. (NASA Release 73–226)
- NASA launched an Aerobee 200 sounding rocket from White Sands Missile Range carrying a Johns Hopkins Univ. planetary atmosphere experiment. The rocket and instrumentation performed satisfactorily. (GSFC proj off)

October 25–31: NASA launched *Explorer 50* (IMP–J) Interplanetary Monitoring Platform by a long-tank, thrust-augmented Thor-Delta launch vehicle from Eastern Test Range at 10:26 pm EDT to study the cislunar environment during a period of decreasing solar activity. The spacecraft entered orbit halfway to the moon, with a 228 808-km (142 174.7-mi) apogee, 197.5-km (122.7-mi) perigee, 117.1-hr period, and 28.7° inclination.

The apogee kick motor was fired at 1:00 am EST Oct. 29, placing *Explorer 50* in its mission orbit with 288 857-km (179 487.4-mi) apogee, 141 184-km (87 727.7-mi) perigee, 12-day period, and 28.7° inclination. The spacecraft was oriented to the ecliptic and spun up to 23 rpm.

Explorer 50 would make detailed and near-continuous studies of the interplanetary environment for orbital periods comparable to several rotations of active solar regions and study particle and field interactions in the distant magnetotail, including cross-sectional mapping of the tail and neutral sheet.

Over a two-day period beginning Oct. 30, 11 of the 12 experiments were turned on. On Nov. 12 they would be turned off while four 60-m (200-ft) antennas would be deployed. The experiments would be turned back on Nov. 13 and 14 along with the 12th experiment. The experiments were supplied by Goddard Space Flight Center, National Oceanic and Atmospheric Administration, Atomic Energy Commission, and universities.

Explorer 50 was the 10th and final spacecraft in the IMP series, which had monitored solar radiation during an entire solar cycle. The series had provided the first accurate measurements of the interplanetary magnetic field, the magnetosphere boundary, and the collisionless magnetohydrodynamic shock wave associated with the interaction of solar wind with the geomagnetic field. The first spacecraft in the IMP series, *Explorer 18*, was launched Nov. 26, 1963; the most recent, *Explorer 47*, was launched Sept. 22, 1972.

The IMP program was directed by NASA's Office of Space Science and managed by GSFC. Tracking and data acquisition was managed by GSFC under the direction of the Office of Tracking and Data Acquisition. GSFC was also responsible for the Thor-Delta launch vehicle. (NASA prog off; NASA Release 73–211; KSC Release 231–73)

October 25: Two buckled fuel tanks on the Skylab 4 Saturn IB launch vehicle [see Oct. 23], scheduled for a Nov. 10 launch, were snapped back into shape by using the pressure of kerosene fuel and helium. Skylab engineers were testing the aluminum tanks to ensure the buckling had caused no structural damage. (UPI, *LA Times*, 10/26/73)

- NASA's 15th Annual Awards Ceremony was held in Washington, D.C. Dr. James C. Fletcher, NASA Administrator, presented NASA's highest award, the Distinguished Service Medal, to Arnold W. Frutkin, Assistant Administrator for International Affairs; Robert H. Gray, Kennedy Space Center Director of Launch Operations; S. Neil Hosenball, Deputy General Counsel; Roy P. Jackson, Associate Administrator for Aeronautics and Space Technology; and William E. Lilly, Comptroller.

  The NASA Distinguished Public Service Medal, the agency's highest award to non-Government personnel, was presented to Dr. Leo Goldberg, Director of Kitt Peak National Observatory, and Dr. Frank Press, Massachusetts Institute of Technology lunar geophysicist.

  The Group Achievement Award went to the Apollo Lunar Sounder Investigator Team, Jet Propulsion Laboratory; the Communications Cost Reduction Team, NASA Hq.; the Mariner Venus/Mercury 1973 Program Team, JPL; the Scout Project Team, Langley Research Center; and the YF–12 Thermal Loads Calibration Team, Flight Research Center. Other awards included the Exceptional Service Medal to 35 persons and the Exceptional Scientific Achievement Medal to 32. (Program; NASA Release 73–219)

October 26: Skylab 4, the third manned Skylab mission—crewed by Gerald P. Carr, commander; Dr. Edward G. Gibson, science pilot; and William R. Pogue, pilot—was scheduled for launch Nov. 10 at 11:40 am EST, Skylab Program Director William C. Schneider announced. Plans called for a 60-day or more, open-ended mission carrying enough consumables for 85 days. Mission extensions would be considered on the 56th, 63rd, 70th, and 77th days of the flight.

  Up to five extravehicular activities would be scheduled, to install and retrieve Apollo Telescope Mount film and experiment samples, repair an antenna on an earth resources experiment, and photograph the Comet Kohoutek on Dec. 25, just before it swung around the sun. An extended 85-day mission would splash down Feb. 3, 1974, at 12:15 pm EST off the Pacific Coast near San Diego. (NASA Release 73–233)

- President Nixon signed H.R. 8825, FY 1974 Dept. of Housing and Urban Development-Space-Science-Veterans appropriations bill that included a $3.002-billion NASA appropriation and $569.6 million for the National Science Foundation. The funds appropriated were those agreed upon by the House-Senate Conference Committee July 26, but submission of the bill for presidential signature had been delayed by subsequent conference reports of Aug. 1, Sept. 7, and Oct. 11, all dealing with extraneous matters. With the President's signature, the bill became Public Law 93–137. (P.L. 93–137; *PD*, 11/5/73, 1307; *CR*, 10/13/73, S19181–2)

- Dr. James C. Fletcher, NASA Administrator, dismissed Mrs. Ruth Bates Harris, NASA Deputy Assistant Administrator for Equal Opportunity. In a Nov. 2 memorandum to NASA employees he said: "Mrs. Harris has made public a paper she wrote me criticizing NASA's Employment Opportunity Program and recommending the reassignment of the Assistant Administrator." Dr. Fletcher was "deeply and personally committed to the goal of equal employment opportunity for members of minority groups and women" and "was not satisfied with NASA's performance in this field." He said NASA would consider Mrs. Harris' recommendations and suggestions and "will adopt those we feel will be helpful." He appreciated her effort and respected her "for having expressed her opinions." But, "because Mrs. Harris has not demonstrated the degree of administrative and management skill required of her position, because she has been unwilling to share the broader problems of management with her peers, and because she became a seriously disruptive force within her own office, I concluded that she should not continue to serve. . . . In so concluding, I also concluded that Dr. [Dudley G.] McConnell [Assistant Administrator for Equal Opportunity] can successfully do the job he has been assigned provided he has the good will and working support of all NASA employees."

Dr. McConnell, a former civil rights activist, had said most of Mrs. Harris' report had been on public record already "but the time they spent preparing it should have been devoted to more positive kinds of things," the Baltimore *Sun* reported. (O'Brien, *W Post*, 10/29/73, A5; Text, B *Sun*, 10/29/73, A3)

- Robert E. Smylie—Chief, Crew Systems Div., Johnson Space Center—was appointed Deputy Associate Administrator (Technology) for NASA's Office of Aeronautics and Space Technology (OAST). He would be responsible for research and technology offices dealing with space propulsion and power; guidance, control, and information systems; materials and structures; aerodynamics and vehicle systems; aeronautical man-vehicle technology, aeronautical propulsion; and research. (NASA Release 73-234)

- NASA announced selection of General Electric Co. Aircraft Engine Group for negotiations leading to a 56-mo, cost-plus-award-fee, $31-million contract to design, fabricate, test, and deliver two quiet, clean short-haul experimental engines (QCSEE). The contract was part of a program to demonstrate technology to reduce aircraft engine noise and exhaust emissions and relieve airway and airport congestion. (NASA Release 73-232)

- Signing of a $17.9-million, three-year, fixed-price-incentive contract with award fees with LTV Aerospace Corp. for complete system management of the Scout launch vehicle was announced by NASA. Work would be performed by LTV Vought Systems Div. with Langley Research Center monitoring. Processing of 6-12 Scout launches of U.S. and foreign payloads was to be included in the system management. LTV, Scout contractor since 1958, had built some 80 Scout vehicles for NASA. (NASA Release 73-231)

- Dr. James C. Fletcher, NASA Administrator, presented 104 Skylab awards to employees from NASA Hq., Johnson Space Center, Kennedy Space Center, and Marshall Space Flight Center during an MSFC ceremony attended by *Skylab 2* crew Charles Conrad, Jr., Dr. Joseph P. Kerwin, and Paul J. Weitz. Also participating were Dr. George M. Low, Deputy Administrator, Associate Administrator for Manned Space Flight Dale D. Myers, and Skylab Program Director William C. Schneider. (NASA Release 73-156; *Marshall Star*, 10/31/73, 1)

- Wallops Station announced it had awarded the largest contract to date under NASA's Minority Business Enterprise Program. Optimal Data Corp. would receive a first-year increment of $535 300 on a potential $1.7-million contract to operate data-programming and data-reduction services at Wallops' information and data-processing laboratory. (WS Release 73–12)

- A *Science* article commented on President Nixon's Oct. 3 speech announcing National Medal of Science winners [see Oct. 3]: "How serious Nixon is about achieving the objectives cited in his remarks will be indicated by future budgetary activities. In recent months, however, there has been plenty of evidence that the Administration wants to make changes not only in the substance of science policy, but also in the way that policy is made. The most obvious indicator was the shift of the science advisory apparatus from the White House to NSF. Underlying the change seems to have been not just a dissatisfaction with the science advisory machinery, but also with the basic relationship between the scientific community and its federal patrons which has prevailed since World War II." It would be wise for the scientific community to examine carefully the new terms which President Nixon seems to be offering. "However, the recent invitation from the White House did, figuratively, seem to be directed to scientists and engineers in general." (Walsh, *Science*, 10/26/73, 365–8)

October 27: The U.S.S.R. launched *Cosmos 603* from Plesetsk into orbit with a 356-km (221.2-mi) apogee, 204-km (126.8-mi) perigee, 90.0-min period, and 72.8° inclination. The satellite, which reentered Nov. 9, was believed by Western observers to be the seventh reconnaissance Cosmos launched during October to monitor the Arab-Israeli war [see Oct. 15]. (GSFC *SSR*, 10/31/73; 11/30/73; *SBD*, 10/31/73, 307)

- The Skylab 4 spacecraft escaped damage from a five-second earth tremor that rattled windows in a central Florida area including Kennedy Space Center. (*W Star-News*, 10/28/73, A8)

October 28: Countdown toward Nov. 10 launch of Skylab 4, the third and final manned Skylab mission, was begun on schedule at 10:00 pm EDT at Kennedy Space Center. (AP, *W Star-News*, 10/29/73, A10)

October 29: NASA in cooperation with the Air Force launched *Nnss O-20* Transit satellite for the Navy from Vandenberg Air Force Base on a four-stage Scout booster. The satellite, launched as part of the Navy Navigation Satellite System, entered orbit with a 1139.4-km (708-mi) apogee, 902.8-km (561-mi) perigee, 105.6-min period, and 90.2° inclination. NASA would be reimbursed by the Air Force for the cost of the launch vehicle and services. (Pres Rpt 74; NASA OSS; *SBD*, 11/1/73, 6; AP, *NYT*, 10/31/73, 13)

- The U.S.S.R. launched *Cosmos 604* from Plesetsk into orbit with a 636-km (395.2-mi) apogee, 614-km (381.5-mi) perigee, 97.2-min period, and 81.2° inclination. (GSFC *SSR*, 10/31/73; *SBD*, 10/31/73, 307)

- A Soviet delegation of aviation and aerospace industry officials, headed by First Deputy Minister of the Ministry of Aviation Industry Stephan I. Kadishev, had completed a three-week tour of major U.S. air transport and engine manufacturers, *Aviation Week & Space Technology* reported. The visit had been "smooth" despite strained relations between the U.S. and the U.S.S.R. occasioned by the Middle East war. (*Av Wk*, 10/29/73, 24)

- The European Space Research Organization had settled on "Ariane" as the name for the European L-3S launcher, *Aviation Week & Space Technology* reported. ESRO previously had selected "Vega," but some member countries had objected that Vega was a French beer. (*Av Wk*, 10/29/73, 9)

October 29–November 2: NASA and the Federal Aviation Administration flight-tested a United Air Lines 727 transport aircraft at West Coast airports to evaluate the effects of wing tip vortex trailing during two-segment instrument landing approaches. The tests were part of a continuing NASA and FAA study of wing wake vortices, invisible flows of turbulent air streaming from aircraft wing tips in circular or funnel motion. The UAL 727 was flown under contract to Ames Research Center. (NASA Release 73–237)

October 30: The U.S.S.R., with the cooperation of the German Democratic Republic and Czechoslovakia, launched *Intercosmos 10* from Plesetsk into orbit with a 1454-km (903.5-mi) apogee, 259-km (160.9-mi) perigee, 102.1-min period, and 74.0° inclination. The satellite would make geophysical investigations in high latitudes and study the electromagnetic connection between the magnetosphere and ionosphere and effects on the neutral atmosphere. Solar panels were of Soviet design and much of the antenna complex was built by Czechoslovak researchers. Meteorological rockets would be launched and simultaneous ground measurements made in support of the mission. (GSFC *SSR*, 10/31/73; Tass, FBIS–Sov, 10/31/73, U1; *SBD*, 11/1/73, 3; *Av Wk*, 11/26/73, 65)

- NASA test pilot John A. Manke flew the X–24B lifting body on a captive flight from Flight Research Center, attached under a B–52 aircraft wing. Flight objectives, to determine the pylon adapter's dynamic response and pylon loads and to check out the propulsion system, were only partially obtained because failure of the cabin-pressure regulator valve limited the flight altitude to 8000 m (25 000 ft). Pylon damping tests and a modified engine test were completed. (NASA prog off)

- Experiments by 50 European scientists at the European Nuclear Research Center (CERN), near Geneva, had indicated that electromagnetism and the "weak" force of radioactive decay—two of nature's four basic forces—might be expressions of the same phenomenon, the *New York Times* reported. The findings had been supported by experiments at the National Accelerator Laboratory near Batavia, Ill. If accurate, the findings would support the long-sought goal of a "unified field theory" relating the four seemingly diverse forces, which also included gravity and the "strong" force that bound together particles of the atomic nucleus. (Sullivan, *NYT*, 10/30/73, 29)

October 30–31: Dr. James C. Fletcher, NASA Administrator, testified Oct. 30 on the space shuttle status in hearings before the Senate Committee on Aeronautical and Space Sciences: "We have selected the prime contractor for the external hydrogen/oxygen tank and are now in the process of selecting the Solid Rocket Motor contractor. When this choice is made later this fall, all of the prime contractors will have been selected, with the exception of the Solid Rocket Booster integration contractor." On the Spacelab, the European Space Research Organization planned to "select the design concept and cost proposal of one of the two competing prime contractors in time to begin detailed design and fabrication by June of 1974."

NASA and the Dept. of Defense were in the final phase of a joint study to determine the most efficient development method for the shuttle space tug, or upper stage, for high-energy orbit missions. "One consideration is the adaptation of existing stages such as the Agena, Centaur, or Tran-stage, which could be modified for use with the shuttle. Since the modification . . . would not satisfy all of the science, application and DOD requirements that are now projected for the 1980s, we are also examining costs and technical characteristics of new configurations which would have the performance capability required to accommodate the entire range of prospective payloads. It may be possible to phase these two Tug approaches in a way that will provide adequate transportation capability when needed while reducing near-term funding requirements. We . . . expect to reach a final decision by the end of this calendar year."

Dr. Fletcher presented *1973 Payload Model: Space Opportunities 1973–1991*, a study of the number and kinds of payloads that could be launched through 1991 with the space shuttle and Spacelab using budgets at the FY 1973 level. The study was not a program plan but an estimate of kinds of payloads that could be flown. Dr. Fletcher felt "the full range of the space capability offered" by the shuttle was even greater than the model indicated.

Anticipated possible annual space payloads for all NASA programs and the user community, excluding DOD, began with 12 in 1973 and moved through 18 in 1974, 38 in 1980 (when the shuttle was to begin its operational phase), 61 in 1985, to 70 in 1990 and 59 in 1991. The 70 payload opportunities in 1990 included 21 automated and 27 sortie (Spacelab) payloads for NASA and 17 automated and 5 sortie for non-NASA, non-DOD users. Anticipated requirements through 1991 totaled 810 payloads, 572 for NASA and 238 for non-NASA, non-DOD users. NASA and other non-DOD agencies were expected to use 69% of the shuttle traffic and DOD 31%.

The model anticipated NASA automated payloads through 1991 might include 77 for astronomy, 43 for physics, 49 for planetary exploration, 8 lunar, 26 life science, 53 earth observations, 22 earth and ocean physics, 2 communications and navigation, and 6 space technology. Anticipated user totals were 120 automated communications and navigation payloads, 59 earth observations, and 9 earth and ocean physics.

Possible NASA sortie payloads were forecast at 58 in astronomy, 52 in physics, 24 in earth observations, 43 in space processing, 24 in earth and ocean physics, 11 for communications and navigation, 28 in life science, and 46 in space technology. User sortie payloads might be 10 for space manufacturing and 40 foreign payloads. (Transcript)

October 31: The first powered flight of the X–24B lifting body was aborted when the igniters failed to function properly during the prelaunch sequence check. The B–52 aircraft and the captive X–24B returned to Edwards Air Force Base, Calif., without mishap. (NASA prog off)

October 31–November 22: The U.S.S.R. launched biological satellite *Cosmos 605* from Baykonur Cosmodrome into orbit with a 400-km (248.6-mi) apogee, 214-km (133-mi) perigee, 90.7-min period, and 62.8° inclination. Tass reported that *Cosmos 605* carried white rats, steppe turtles, insects, fungi, and instruments to monitor their condition. Objectives of the mission were to study the effects of space on living organisms and to test life-sustaining systems for biological subjects. The spacecraft was recovered Nov. 22.

Dr. Avetik I. Burnazyan, U.S.S.R. Deputy Minister of Public Health, said in a *Pravda* article on Nov. 9 that the satellite was a continuation of earlier Soviet experiments designed to gain statistically reliable data on the effects of weightlessness on individual cells of organisms and the effects of radiation on the genetic properties of organisms. Previous Soviet biological satellites included *Cosmos 110* (launched Feb. 22, 1966, carrying two dogs for a 22-day mission) and *Cosmos 368* (launched Oct. 8, 1970, carrying yeast cells, seeds, and onion bulbs all of which were exposed to radiation during a 6-day mission). (GSFC *SSR*, 10/31/73, 11/30/73; FBIS–Sov, 11/1/73, U1; 11/9/73, U1; 11/13/73, U1–U3; *NYT*, 11/25/73, 23; *SBD*, 11/2/73, 10)

During October: Selection of Sen. Barry M. Goldwater (R–Ariz.) to receive the Wright Brothers Memorial Trophy for 1973 was announced by the National Aeronautics Assn. The trophy would be presented Dec. 17 in Washington, D.C., for "leadership and inspiration to all elements of aviation in the United States, both military and civilian, and for serving as an articulate spokesman for American aviation and space in the Congress and throughout the world." (NAA *News*, 10/73, 1)

- The Air Force System Command's Space and Missiles Systems Organization (SAMSO) announced the establishment of a system program office to develop the Defense Navigation Satellite System (DNSS). The space-based radio navigational system would permit suitably equipped users to determine precisely three-dimensional position, velocity, and time information globally. On successful completion of testing, the Dept. of Defense might approve deployment of DNSS to provide a universal positioning and navigational capability for military and civilian users on land, sea, and air. (AFSC *Newsreview*, 10/73, 11)

- The European Space Research Organization had awarded the British Aircraft Corp. Electronic and Space Systems—on behalf of the STAR consortium of European electronic and aerospace companies—a $24-million contract for design, development, and manufacture of Europe's first geostationary scientific satellite, GEOS, *Spaceflight* reported. The three-year development program would include 15 companies in 10 European countries. GEOS, scheduled for a summer 1976 launch on a Thor-Delta booster, would carry nine scientific experiments to measure electric and magnetic fields and particle densities and distributions for two years. (*SF*, 10/73)

- A need "to shift gears and return to the main line of development in space flight" was underlined by Michael A. G. Michaud in a *Spaceflight* article: The purposes of space flight "go far beyond our lifetime. Ultimately, our journey away from Earth will determine the future of the human race; this millennial, evolutionary effort must not founder now for lack of a satisfying rationale." The basic purpose of space flight was "to expand the realm of Man." Reasons were to challenge human abilities, learn more about the universe we live in, search for extraterrestrial intelligence and prepare for contact with it, develop and test technologies for earth use, stimulate and employ advanced sectors of the economy, improve orbital applications of space technology with immediate benefits, develop and test military technology, and encourage international cooperation. Michaud proposed a medium-term program to the year 2000 to assemble a manned space station in earth orbit, build a base on the moon, land human beings on Mars by 2000, prepare for interstellar flight, search for planets in other systems, and begin the search for extraterrestrial intelligence. A long-term program beyond 2000 would plan to send unmanned probes to the most promising nearby stars, create a permanent colony on Mars, experiment in planetary engineering—possibly on Venus, "where orbiting space vehicles could introduce materials into the atmosphere that might change its composition"

—launch the first manned interstellar flights, and communicate with extraterrestrial intelligence. (*SF*, 10/73, 362-5)

- The Navy was investigating shipboard launch and recovery of remotely piloted vehicles (RPVs), the Office of Naval Research *Reviews* reported. The RPVs could save in staffing and pilot training and could perform reconnaissance, search, radio relay, and antisubmarine duties if fail-safe recovery on vessels with limited deck space could be developed. A command-and-control data link between the vehicle and a ship command room would track the RPV, monitor its performance, and guide it to touchdown. Candidate systems included vertical takeoff and landing (VTOL), including a stoppable rotor which worked as a helicopter rotor during takeoff and landing, a retractable rotor used with auxiliary wings, and a fixed-wing vehicle with rotating jet exhausts to provide takeoff thrust. (ONR *Rev*, 10/73, 32-3)

- The National Science Foundation released *Graduate Science Education: Student Support and Postdoctorals, Fall 1972* (NSF 73-315). Graduate enrollment in science doctorate-granting institutions declined 2% from 1971 to 1972; first year enrollment declined 3%. Engineering enrollment dropped at the greatest rate, 5%, followed by physical sciences—4%. There were 10% fewer students receiving Federal support in 1972 than in 1971. Full-time students receiving fellowships for trainee support declined by 18% from 1971 to 1972. The loss was reflected in every area of science. (Text)

November 1973

November 1: A variable-sweep-wing testbed F-111 aircraft with NASA's supercritical wing made its first flight, in a joint Air Force and NASA program at Flight Research Center. Prime pilots were USAF Maj. Stu R. Boyd and NASA research pilot Einer K. Enevoldson. The transonic aircraft technology (TACT) program would investigate application of supercritical wing technology to high maneuverable aircraft at transonic speeds. Wind-tunnel studies of the new airfoil shape—based on research by Dr. Richard T. Whitcomb, Head of Langley Research Center's 8-Foot Tunnels Branch, and almost directly opposite to conventional shapes—had shown it could delay rise in aerodynamic drag and allow an aircraft to perform more efficiently at higher speeds. The new wing, with flat top and upward-curving rear portion of the bottom side, was expected to improve the tactical advantage of military aircraft in air-to-air combat. (NASA Release 73-215; NASA Photo 73-H-1073)

- Former Congressman Emilio Q. Daddario was named Executive Director of the new congressional Office of Technology Assessment for a six-year term by the 12-member joint Congressional Technology Assessment Board chaired by Sen. Edward M. Kennedy (D–Mass.) (*A&A*, 12/73, 12; Off Tech Ass)

November 2: The U.S.S.R. launched *Cosmos 606* from Baykonur Cosmodrome into orbit with a 39 708-km (24 673.4-mi) apogee, 635-km (394.6-mi) perigee, 717.5-min period, and 62.7° inclination. (GSFC SSR, 11/30/73; Sov Aero, 1/7/74, 4)

- NASA launched two Aerobee 170 sounding rockets from Woomera, Australia. The first carried a Univ. of Wisconsin galactic astronomy experiment. The second, launched two hours later, carried a California Institute of Technology astronomy experiment. Both rockets and instrumentation performed satisfactorily. (GSFC proj off)

November 3–13: NASA's *Mariner 10* (Mariner-J) Venus-Mercury probe was launched into a parking orbit from Eastern Test Range at 12:45 am EST by an Atlas-Centaur D1-A launch vehicle. After a 25-min coast, a 130-sec burn of the Centaur engines injected the spacecraft into a heliocentric orbit that would allow *Mariner 10* to pass within 5300 km (3300 mi) of Venus Feb. 5, 1974. Using the gravitational field of Venus to change the speed and flight path of the spacecraft, *Mariner 10* would travel toward a March 29 encounter with Mercury, approaching the sun more closely than any other planetary mission. *Mariner 10* would be the first spacecraft to use gravity assist in a dual-planet mission and the first to explore Mercury.

The primary objective was to measure Mercury's environment, atmosphere, surface, and body characteristics and to make similar investigations of Venus. Secondary objectives were to perform experiments in the interplanetary medium and to obtain experience with a dual-planet gravity-assist mission.

The seven experiments aboard the octagonal 503-kg (1108-lb) spacecraft included a TV camera and an infrared radiometer to investigate the planet's surface; plasma science experiment, charged particle telescope, and magnetometer to investigate the environment around the planet and the interplanetary medium; dual-frequency-radio science experiment and ultraviolet spectrometer to measure characteristics of the Mercury neutral atmosphere and ionosphere if they existed; and celestial mechanics experiments to measure planetary mass characteristics and test the theory of relativity. All experiments had been designed and selected for their Mercury scientific objectives but also would obtain important data at Venus and during the cruise phases.

The Tidbinbilla, Australia, tracking station was unable to maintain command lock with the spacecraft following spacecraft injection. Acquisition of the star Vega by the spacecraft Canopus tracker also was delayed, but two-way communications were established within an hour and, after repeated commands, Vega was acquired. Solar panels, high-gain and low-gain antennas, and plasma science and magnetometer booms were deployed. On Nov. 4–5 four moon and three earth TV picture mosaics were completed.

By Nov. 6 the spacecraft was 1 394 000 km (866 000 mi) from the earth, traveling at a speed of 4.32 km (2.68 mi) per sec. All science experiments except the infrared radiometer and electrostatic analyzer had been turned on. The scanning electron spectrometer portion of the plasma science experiment was not operating properly and the problem was being analyzed.

Four trajectory correction maneuvers were scheduled during the mission. The first, Nov. 13, successfully adjusted the flight path for closest approach to Venus on Feb. 5, 1974, as planned. A second would be made Jan. 18, 1974. On Feb. 9 and March 1, 1974, after the Venus encounter, maneuvers would target the *Mariner 10* to pass Mercury at 1000-km (620-mi) altitude.

The Mariner program, including mission operations and tracking and data systems, was managed by Jet Propulsion Laboratory under the direction of NASA's Office of Space Science. Lewis Research Center, with contractor General Dynamics Corp. Convair Div., was responsible for the launch vehicle. Previous missions had been made to Mars, including *Mariner 4, 6, 7,* and *9* (launched Nov. 28, 1964; Feb. 24, 1969; March 27, 1969; and May 30, 1971). *Mariner 9* had been the first spacecraft to orbit a planet other than the earth. All others had been flyby missions. (NASA prog off; JPL Release 673)

November 3–January 9, 1974: NASA's *Pioneer 10* interplanetary probe (launched March 2, 1972) became the first spacecraft to reach Jupiter during a two-month encounter, returning new findings about the giant planet and continuing toward Uranus and eventual exit from the solar system.

November 3–December 3: Encounter operations were begun Nov. 3 by Ames Research Center controllers and scientists and on Nov. 4 *Pioneer 10* began sending photopolarimeter images of Jupiter to the earth three to eight hours a day. By Nov. 6 the spacecraft, 25 000 000 km (15 500 000 mi) from Jupiter, had begun measuring the planet's atmosphere above Jupiter's orange-and-blue-striped cloud tops and the characteristics of particles making up the dense clouds.

On Nov. 8 the spacecraft crossed the orbits of Hades and Poseidon, the first two of Jupiter's outer moons, and on Nov. 9 and 11 it crossed the orbits of Pan and Andrastea.

At a Nov. 19 press briefing at ARC Dr. John E. Naugle, NASA Associate Administrator for Space Science, said that low temperature in that area of the solar system slowed evolution and scientists hoped to find objects not very different from what they were at their formation. "The passage of *Pioneer 10* past Jupiter also marks the gateway to the exploration of the outer solar system. This is an exploration which is critical if we truly want to understand the formation and evolution of our solar system." Dr. William H. Kinard, Langley Research Center scientist, said that *Pioneer 10* had encountered 100 times more space dust than expected during the journey between the Asteroid Belt and Jupiter. He expected that 30 times more particles would be detected as the spacecraft neared Jupiter.

Pioneer 10 crossed the orbits of Jupiter's moons Demeter, Hera, and Hestia Nov. 22 and was 11 400 600 km (7 084 000 mi) from Jupiter.

By Nov. 25 a charged particle detector aboard the spacecraft had begun to show traces of particle radiation. Preliminary indications were that the probe was beginning to cross into Jupiter's magnetosphere, estimated to be 20 times stronger than the earth's. The planet's radiation belts contained particle concentrations up to 1 million times higher than the earth's Van Allen Belt and were considered by some scientists to be extremely dangerous to the spacecraft's electronics. At a distance of 8 336 400 km (5 180 000 mi) from Jupiter, spacecraft instruments began returning pictures of Jupiter's great red spot.

The final spacecraft-attitude change before encounter was made Nov. 26 and a 24-hour day of imaging and polarimetry began. At 3:30 pm EST Nov. 26 *Pioneer 10* successfully crossed Jupiter's bow shock wave at a distance of 7 709 000 km (4 790 000 mi)—with great turbulence when the solar wind hit Jupiter's magnetic field. The spacecraft was unaffected by the turbulence but data from 6 of its 11 instruments changed dramatically. Solar wind speed abruptly decreased to half the normal velocity and solar wind temperatures rose sharply. On Nov. 27, at a distance of 6 760 000 km (4 200 000 mi) from Jupiter, the spacecraft crossed into the planet's true magnetic envelope, the region of interplanetary space into which the solar wind could not penetrate. *Pioneer 10* crossed the magnetopause, entering the magnetic field for the first time at 3:30 pm EST.

At a Nov. 29 press briefing ARC scientists said new data had shown the reach of Jupiter's magnetic field to be greater than predicted. It stretched more than 13 000 000 km (8 000 000 mi) in diameter and its strength was 40 times that of the earth's magnetic field. Its direction was south, rather than north as predicted. The strength of the magnetism at 6 400 000 km (4 000 000 mi) from the planet seemed to rise and fall in a regular 10-hour phase, possibly because of the planet's rotation. Jupiter's mass was also found to be greater than estimated and the spacecraft was being drawn toward Jupiter faster than planned; it would arrive at its closest approach two minutes earlier than predicted.

On Nov. 30 *Pioneer 10* recorded a sharp drop in magnetic and radiation measurements, indicating disappearance of the magnetic field. Eleven hours later the spacecraft again crossed the magnetopause, which had apparently been driven toward the planet by a disturbance in the solar wind. At 1 427 500 km (887 000 mi) from the planet, the spacecraft entered Jupiter's dipole magnetic field. Preliminary data indicated that the outer portions of Jupiter's magnetic field did not rotate with the planet and were more flattened than the earth's field. Also, the magnetic dipole center was appreciably offset, both radially and toward the north from the center of the planet. The magnetic axis was tilted from the spin axis.

Pioneer 10 crossed the orbits of Jupiter's innermost moons—Callisto, Ganymede, Europa, and Io—Dec. 2–3. As it came within 446 000 km (277 000 mi) of Ganymede, the infrared radiometer indicated the moon's surface temperature was 125 K ($-235°F$). Six hours before its closest approach to Jupiter the probe entered the dangerous region of intense radiation and, soon after, the imaging system ceased returning picture data. Radio commands switched the system back to normal operation within 30 minutes, however. The press later reported that NASA attributed the mishap to an unidentified flight control room guest who had inadvertently pressed against a cathode ray display, pushing a button that cut out eight instruments.

December 3–11: As the gold and silver spacecraft approached Jupiter and radiation intensity increased 10-fold every two hours, reaching 65 million ev 90 min before periapsis, tense scientists paced the floor of ARC's Mission Control with fingers crossed. At 9:25 pm EST Dec. 3, after a flight of 641 days and 826 000 000 km (513 000 000 mi), *Pioneer 10* sped past Jupiter at an altitude of 130 000 km (81 000 mi), traveling 155 140 km per hour (96 400 mph), the fastest speed ever achieved by a man-made object. The spacecraft's 11 instruments sent back readings and pictures that came in loud and clear, taking 46 minutes to reach a jubilant team on earth.

"Needless to say, we're elated," Dr. John Wolfe, Pioneer Project Scientist, told cheering newsmen.

The *Washington Post* later described the scene at ARC: "So deluged were Pioneer scientists with transmissions of the spacecraft's instruments, they were literally changing their minds about Jupiter's physics and chemistry every hour." Dr. James C. Fletcher, NASA Administrator, said, "Some of us have been looking through telescopes at Jupiter since our early teens. This is more than we ever dreamed of."

The spacecraft suffered no apparent failures and continued to function as planned. However, preliminary estimates of radiation belt parameters and observed effects on the spacecraft and instrumentation indicated that the margin of survival had been small.

As *Pioneer 10* flew by Jupiter, the spacecraft was accelerated and slung by the planet's gravity into a new trajectory that would take it past the orbits of Saturn, Uranus, Neptune, and Pluto and, traveling at a constant 40 000 km per hr (25 000 mph), out of the solar system toward the constellation Taurus. *Pioneer 10* would be the first spacecraft to leave the solar system. By 7:00 pm EST Dec. 4, it had traveled nearly 1 600 000 km (1 000 000 mi) beyond Jupiter.

December 11–January 9: ARC Mission Control ended Jupiter picture-taking Dec. 11 and returned to routine interplanetary operations. The spacecraft recrossed the bow shock wave Dec. 12 and continued into interplanetary space. By Dec. 14 the spacecraft was 10 881 000 km (6 761 000 mi) from Jupiter, traveling at 83 525 km per hr (51 900 mph) toward the edge of the solar system.

Preliminary findings from the encounter included the discovery of an ionosphere (implying a tenuous atmosphere) surrounding Jupiter's moon Io, detection of helium in Jupiter's atmosphere, new density measurements of the entire Jovian system that would affect current theories on the formation of the universe, and new temperature measurements of Jupiter that indicated little difference between average daytime and nighttime temperatures.

Pioneer 10 was expected to continue collecting and returning data until it reached the orbit of Uranus in 1980, almost 3 200 000 000 km (2 000 000 000 mi) from the earth. Data returned would determine whether the second Jupiter probe, *Pioneer 11* (launched April 5 and crossing the Asteroid Belt since August), could safely be retargeted closer to the planet. The *Pioneer 10* mission was officially adjudged a success Jan. 9, 1974.

The Pioneer program, begun in 1958, was directed by ARC. *Pioneer 6* and *7* (launched Dec. 16, 1965, and Aug. 17, 1966, to study the heliocentric space environment) and *Pioneer 8* and *9* (launched Dec. 13, 1967, and Nov. 8, 1969, to study interplanetary phenomena) continued to supply data from heliocentric orbit on solar plasma, magnetic and electric fields, and cosmic rays. (NASA prog off; ARC Release, 11/6/73; NASA Releases 73–243K, 73–256, 73–279; Wilford, *NYT*, 11/25/73; 11/30/73, 1; 12/4/73, 1; 12/5/73, 13; O'Toole, *W Post*, 12/4/73, A1; AP, *W Star-News*, 11/28–29/73; AP, *B Sun*, 12/3–4/73; Miles, *LA Times*, 12/4/73; 12/6/73)

November 3: The Skylab 4 launch crew successfully completed a simulated launch of the Saturn IB booster at Kennedy Space Center. The booster was scheduled to lift the Apollo command and service module and crew members Gerald P. Carr, Dr. Edward G. Gibson, and William R. Pogue into orbit Nov. 10 to dock with the Orbital Workshop (launched May 14) and begin NASA's third manned Skylab mission. Following the perfectly timed simulation, the fuel was drained from the launch vehicle and preparations for the final countdown were begun. (AP, B *Sun*, 11/3/73, A7)

- France had begun work to hold future nuclear tests underground at two uninhabited atolls in the Pacific, French Adm. Christian Claverie said in Papeete, Tahiti. French nuclear tests previously had been conducted above the French-governed Tahitian atoll Mururoa, against international protests. (UPI, *NYT*, 11/4/73, 77)

November 5: Skylab 4 Astronauts Gerald P. Carr, Dr. Edward G. Gibson, and William R. Pogue completed ground practice of docking maneuvers at Johnson Space Center. The astronauts were scheduled for Nov. 10 launch on the third and final mission to man the Skylab Orbital Workshop (launched May 14). (UPI, *NYT*, 11/6/73, 25)

- The return payload limit on the space shuttle orbiter had been increased 28%, up from 11 300 kg to 14 500 kg (from 25 000 lbs to 32 000 lbs), following system analysis by Rockwell International Corp., *Aviation Week & Space Technology* reported. The increase could be made without modifying the orbiter structure, by permitting the orbiter to land at a faster speed, and without exceeding the 68 000-kg (150 000-lb) liftoff weight limit. The orbital payload limit remained 29 500 kg (65 000 lbs) in easterly orbits and 13 600 (30 000 lbs) in polar orbits. (*Av Wk*, 11/5/73, 9; NASA OMSF)

November 5

- Sen. Lowell P. Weicker, Jr. (R–Conn.), and Sen. Frank E. Moss (D–Utah) introduced S. 2658, Solar Heating and Cooling Demonstration Act of 1973, as a companion bill to H.R. 10952 introduced in the House Oct. 16. (Sen Rpt 93–734; Text)
- Unusual pulsed-noise signals detected by Soviet scientists at Gorky Univ. [see Oct. 16] might be from experimental U.S. Air Force communications satellites in highly elliptical Molniya-like orbits, *Aviation Week & Space Technology* said. Possible sources were satellites orbited March 21, 1971, and Aug. 21, 1973—thought to be test vehicles for the Air Force's planned Satellite Data System. (*Av Wk*, 11/5/73, 47)

November 6–29: NASA launched *Noaa 3* (ITOS–F) National Oceanic and Atmospheric Administration meteorological satellite from Western Test Range at 6:02 am PST on a two-stage, long-tank, thrust-augmented Thor-Delta 0300 launch vehicle. The satellite entered near-nominal orbit with a 1509.2-km (937.8-mi) apogee, 1500-km (932.1-mi) perigee, 116.09-min period, and 102.08° inclination.

The primary objectives were to place the spacecraft in sun-synchronous orbit with a local equator crossing time of 8:30 am to permit the spacecraft to accomplish its operational mission requirements, make an in-orbit evaluation, and check out the spacecraft before turning operational control over to NOAA's National Environmental Satellite Service (NESS).

Operational responsibility for the spacecraft was transferred to NOAA Nov. 29. Excellent data were being returned. The orbit would permit regular, dependable daytime and nighttime meteorological observations in both direct-readout and stored modes of operation in support of the National Operational Meteorological Satellite System (NOMSS). *Noaa 3* was the second operational spacecraft to provide temperature soundings of the earth's atmosphere as well as direct-readout and globally recorded cloud-cover data. The three-axis-stabilized satellite carried redundant very-high-resolution radiometers, vertical-temperature-profile radiometers, and scanning radiometers.

Noaa 3 was the fourth in the Improved TIROS Operational Satellite series of second-generation meteorological satellites. The prototype had been *Itos 1* (launched Jan. 23, 1970), and the first NOAA-funded spacecraft, *Noaa 1*, had been launched Dec. 11, 1970. *Noaa 2* (launched Oct. 15, 1972) continued to provide meteorological data. The first series of spacecraft had included nine satellites—*Essa 1* through *Essa 9* (launched 1966–1969).

The TOS program was a joint NASA and Dept. of Commerce effort under a Jan. 30, 1964, agreement. Goddard Space Flight Center was responsible for the design and development of the spacecraft and ground systems, launch checkout and evaluation, tracking and data acquisition, and the Thor-Delta launch vehicle. NESS was responsible for operating the spacecraft and processing the data. (NASA prog off; GSFC proj off)

November 6: Fourteen hairline cracks were found in the eight stabilizing fins of Skylab 4's Saturn IB launch vehicle during a routine inspection at Kennedy Space Center. The cracks were discovered just after the Saturn rocket had been loaded with 157 cu m (41 500 gal) of fuel. Six of the fins had two cracks each, the two others one crack each, all at the juncture of the 5-m (16-ft) aluminum fins and the body of the launch vehicle. During rocket inspection, Skylab 4 Astronauts Gerald

P. Carr, Dr. Edward G. Gibson, and William R. Pogue practiced launch abort procedures in a simulator at Johnson Space Center. (O'Toole, *W Post*, 11/8/73, A3; AP, *W Star-News*, 11/7/73, B8)

- Skylab's contribution to better understanding of solar processes, leading to possible generation and control of solar energy for earth use, was described by NASA Associate Administrator for Manned Space Flight Dale D. Myers in a speech before the Poor Richard Club in Philadelphia: The sun was the "original model of nuclear energy," understanding of which could lead to fusion control and its efficient use for power generation. "The Skylab investigators are concerned with this and also with explaining how solar events affect the Earth, particularly solar flares that trigger auroras and disrupt radio transmission. Because sunspot activity correlates with temperature and density variations in the Earth's upper atmosphere, it is conceivable that the injection of energy into the atmosphere by solar particles may trigger world-wide weather phenomena. Our astronauts saw flares shooting up 100 000 miles [160 000 kilometers] above the surface and they watched giant snake-like clouds erupting in the corona that disturbed radio transmissions here on Earth. Two massive flares exploded on the Sun, producing auroral lights on Earth; . . . each one could supply the world's energy needs for the next 500 years." (Text)

- The James H. Wyld Propulsion Award for 1973 was presented by the American Institute of Aeronautics and Astronautics to Gerard W. Elverum, Jr., Manager of TRW Inc. Energy Systems Operations Div., and Norman C. Reuel, Vice President for Liquid Rocket Programs and Production Operations of Rockwell International Corp. Rocketdyne Div., during Las Vegas, Nev., ceremonies. Elverum was cited for "outstanding leadership and technical contribution to the deep throttling rocket engine for the Apollo lunar mission." Reuel was cited for "outstanding leadership and technical contribution to the liquid hydrogen rocket engine for the Apollo lunar mission." (*Av Wk*, 11/12/73, 9; AIAA PIO)

November 6, 9, 13: U.S.-U.S.S.R. Strategic Arms Limitation Talks (SALT) were held in Geneva. (FBIS-Sov, 11/19/73, AA1)

November 7: A five-day delay in the launch of Skylab 4, third and final manned Skylab mission, was announced by Skylab Program Director William C. Schneider at a Kennedy Space Center press briefing. The delay was caused by the discovery of cracks in the Saturn IB launch vehicle fins [see Nov. 6]. The earliest rescheduled launch time would be 9:27 am EST Nov. 15. Director of Launch Operations at KSC Walter J. Kapryan said the 3.8-cm-long (1.5-in-long) cracks probably were caused by stress erosion—exposure of the material to the elements combined with stress from the weight of the fueled vehicle—and would have caused the breakup of the vehicle at 12 000 m (40 000 ft), where the rocket underwent maximum aerodynamic stress. All eight fins would be replaced. New fins were being shipped from Michoud Assembly Facility for replacement on the launch pad. (Transcript; O'Toole, *W Post*, 11/8/73, A3)

- NASA Associate Administrator for Applications Charles W. Mathews testified on the scheduled spring 1974 launch of the Synchronous Meteorological Satellite (SMS) during House Committee on Science and Astronautics' Subcommittee on Space Science and Applications hearings on research into short-term weather phenomena. Based on utility demonstrated by Applications Technology Satellites for storm monitoring, the first operational prototype geostationary meteorological satellite was a forerunner of the National Oceanic and Atmospheric Administration's Geostationary Operational Environmental Satellite (GOES) system. Once the satellites were placed in orbit and checked out by NASA, they would be turned over to NOAA as a part of the operational system, which could observe the earth's surface day and night, using visible-light and infrared sensors, and acquire *in situ* data from ocean buoys, aircraft, ships, river stage stations, and other ground-based remote stations. The GOES system—of two satellites in geostationary orbit viewing the western U.S. and Pacific Ocean area and the eastern U.S. and Atlantic area—would provide near-continuous surveillance of all tropical Atlantic storms, all tropical storms of the central and eastern Pacific Ocean, and all storms of observable size within the limits of the U.S., its territories, and its surrounding coastal waters, except Alaska. Improved sensors were expected to provide new and improved data on severe storm causes and development. (Transcript)

- Sen. Frank E. Moss (D–Utah), Chairman of the Senate Committee on Aeronautical and Space Sciences, told Senate colleagues and the press that he had learned of consideration in the Executive branch of a proposal to transfer authority over all U.S. weather satellites from civilian to military control. He believed plans also were under consideration for placing all earth observation satellites under military control. He called for "serious reflection" over any such plans, noting that civilian control of peaceful space activities had been a standing U.S. policy for 15 yrs. (Sen Moss Release 73–304)

- President Nixon addressed himself to the energy crisis in a speech over nationwide radio and TV: "We are heading toward the most acute shortages of energy since World War II. Our supply of petroleum this winter will be at least 10 percent short of our anticipated demands, and it could fall short by as much as 17 percent. Now, even before war broke out in the Middle East, these prospective shortages were the subject of intensive discussions among members of my Administration, leaders of the Congress, Governors, mayors, and other groups. From these discussions has emerged a broad agreement that we, as a Nation, must now set upon a new course. In the short run, this course means that we must use less energy—that means less heat, less electricity, less gasoline. In the long run, it means that we must develop new sources of energy which will give us the capacity to meet our needs without relying on any foreign nation." The President announced steps to lower energy consumption. He asked the Atomic Energy Commission to speed licensing and construction of nuclear plants and would ask development of an emergency energy act. (*PD*, 11/12/73, 1312–8)

- The first direct TV transmission from the U.S.S.R. to Cuba was made. The Soviet Red Square Parade, telecast from Moscow, was transmitted via a Molniya satellite and a ground station in eastern Cuba. Regular TV and communications traffic was expected to begin in 1974. (*Av Wk*, 11/12/73, 9)

- Rockwell International Corp. Rocketdyne Div. had received an $8.7-million Air Force contract for production and support services for Atlas propulsion systems, the *Wall Street Journal* reported. The systems would launch communications satellites for Communications Satellite Corp. (*WSJ*, 11/7/73, 10)

- McDonnell Douglas Corp. had confirmed that it was offering three additional versions of its DC-10 wide-bodied jet airliner, increasing available versions to eight, the *Wall Street Journal* reported. Four versions already were in production; orders had not been placed for the others. (*WSJ*, 11/7/73, 14)
- NASA launched an Aerobee 170 sounding rocket from Woomera, Australia, carrying a Massachusetts Institute of Technology astronomy experiment to a 167.4-km (104-mi) altitude. The rocket and instrumentation performed satisfactorily. (GSFC proj off)

November 7–8: Scientists from the European Space Research Organization (ESRO), NASA Hq., Jet Propulsion Laboratory, and four U.S. universities met at Marshall Space Flight Center on project SOREL, a sun-orbiting relativity experiment. SOREL, a satellite to orbit the sun and measure relativistic gravity phenomena, had been proposed by ESRO in 1970. Final results of the mission definition study and SOREL's scientific merits and technical feasibility were discussed. (MSFC Release 73–161; MSFC PAO)

- A symposium at Goddard Space Flight Center sought to define critical measurements needed to determine interrelationships of solar and meteorological activity, the mechanisms needed to explain them, and the measurements that could best be made from space. The symposium was sponsored by GSFC with the University Corp. for Atmospheric Research and the American Meteorological Society. (NASA Release 73–240; GSFC PAO)

November 8: Ground crews at Kennedy Space Center began replacing the damaged stabilizing fins of the Skylab Saturn IB launch vehicle [see Nov. 6]. After the fuel was drained from the 1st stage, the rocket was jacked up. One fin was removed and additional corrosion was discovered where the fin had been attached. Corrosion also was found on the replacement fins shipped from NASA's Michoud Assembly Facility. Workmen cleaned the affected areas with nitric acid, buffed them, and applied a zinc chromate primer. Shims, or strips of very thin aluminum, were added to the rocket surface to ensure proper fit. Crews worked around the clock to replace the eight 215-kg (473-lb) fins to meet a Nov. 15 launch date. (JSC press briefing transcript, 11/9/73; Wilford, *NYT*, 11/9/73, 2)

- A contract for basic design of a geostationary meteorological satellite to be launched by NASA for Japan by the end of 1976 had been awarded by Japan's National Space Development Agency to Nippon Electric Co., *Electronics* reported. Nippon was expected to receive follow-on contracts to complete a detailed design and fabricate and test two satellites —one for mission and one for backup. The mission would include gathering cloud pictures, collecting and distributing meteorological data, and making space measurements. Nippon would use Hughes Aircraft Co. technology, through its technical tie-in with Hughes Aerospace Div., to design the Japanese satellite. (*Electronics*, 11/8/73)

November 9: Ground crews at Kennedy Space Center replaced one of the eight damaged fins discovered Nov. 6 on the Skylab 4 Saturn IB launch vehicle. At a KSC press briefing Director of Launch Operations Walter J. Kapryan said that work on fin replacement was "running almost a day behind" but he thought a Nov. 15 launch was still possible. Drew F. Evans, aerospace technologist in KSC's Materials Laboratories Div., said tests confirmed that the cracks were caused by stress corrosion. (Transcript; Wilford, *NYT*, 11/10/73, 48)

- *Noaa 2* (ITOS-D) meteorological satellite, launched Oct. 15, 1972, for the National Oceanic and Atmospheric Administration by NASA, was adjudged a success. NASA had achieved prelaunch objectives of launching the spacecraft into sun-synchronous orbit with a local equator crossing time between 9:00 and 9:20 am and of conducting in-orbit engineering evaluation before turning the operational control of the spacecraft over to NOAA's National Environmental Satellite Service *Noaa 2* had completed one year of flawless operation, making regular daytime and nighttime meteorological observations in both readout and stored modes of operation in support of the National Operational Meteorological Satellite System and for incorporation into worldwide daily weather charts and reports. Day and night global scanning radiometer automatic picture transmissions were being supplied on direct readout to more than 700 worldwide users. (NASA prog off)

- A NASA Hq. memorandum to employees requested compliance with energy conservation measures set forth for Government buildings by the General Services Administration in accordance with Presidential guidelines. Office temperatures would fall to 292 to 293 K (65° to 68°F). Kilowatt consumption was to be cut by extinguishing office lights during vacancy. (NASA Hq Memo, 11/9/73)

November 10: The Air Force launched three unidentified satellites on a single Titan IIID booster from Vandenberg Air Force Base. The first entered orbit with a 265.5-km (165-mi) apogee, 162.5-km (101-mi) perigee, 88.7-min period, and 96.9° inclination and reentered March 13, 1974. The second entered orbit with a 505.3-km (314-mi) apogee, 487.6-km (303-mi) perigee, 94.5-min period, and 96.3° inclination. Orbital parameters for the third satellite were 1458.1-km (906-mi) apogee, 1417.8-km (881-mi) perigee, 114.6-min period, and 96.9° inclination. (Pres Rpt 74; *SBD*, 11/20/73, 112; GSFC *SSR*, 3/31/74)

- The U.S.S.R. launched *Cosmos 607* from Plesetsk into orbit with a 292-km (181.4-mi) apogee, 217-km (134.8-mi) perigee, 89.6-min period, and 72.2° inclination. The satellite reentered Nov. 22. (GSFC *SSR*, 11/30/73; *SBD*, 11/13/73, 66)

- NASA launched two Aerobee 170 sounding rockets from Woomera, Australia. The first carried a California Institute of Technology astronomy experiment to a 169.0-km (105-mi) altitude. The second, launched 45 min later, carried a Massachusetts Institute of Technology astronomy experiment to a 174.6-km (108.5-mi) altitude. Both rockets and instrumentation performed satisfactorily. (GSFC proj off)

November 11: Four of eight damaged stabilizing fins discovered on the Skylab 4 Saturn IB launch vehicle Nov. 6 had been replaced by Kennedy Space Center ground crews, Skylab Program Director William C. Schneider said at a KSC press briefing. Despite rain and high wind the work was accelerating to meet the Nov. 15 launch date. (Transcript)

November 12: Kennedy Space Center ground crews working to replace eight damaged fins discovered on the Skylab 4 Saturn IB launch vehicle Nov. 6 discovered additional cracks in seven of eight structural beams in the aft intersection of the rocket. Because of the new discovery and the subsequent necessary examinations and evaluations of other susceptible sections of the rocket, Skylab Program Director William C. Schneider announced at a KSC press briefing a further postponement of the Nov. 15 launch. A new launch date would be set after the situation was assessed. Director of Launch Operations Walter J. Kapryan said that the newly discovered cracks might have been the "result of residual stresses left in the forgings from the fabrication process." (Transcript)

- Richard C. McCurdy, Associate Administrator for Organization and Management since Oct. 1, 1970, had resigned effective Dec. 12, NASA announced. In accepting the resignation, Dr. James C. Fletcher, NASA Administrator, said: "He brought much to NASA in managerial know-how from his former capacity as President and Chief Executive Officer of the Shell Oil Company. . . . He made NASA a better organization—in our dealing with industry and in our handling of internal affairs." (NASA Ann)

- Ames Research Center biologists Paul H. Deal and Kenneth A. Souza had discovered an earth organism that could survive and grow in an environment resembling that of the outer planets of the solar system, NASA announced. The rod-shaped bacteria were swimming, growing, and reproducing in a highly alkaline solution of sodium hydroxide. The atmospheres of Jupiter, Saturn, and Uranus were believed by some scientists to be highly alkaline. The discovery renewed questions of possible life on Jupiter, since the planet's extreme alkalinity had been cited as a major deterrent to life. (NASA Release 73-238)

- Johnson Space Center announced a $4 374 813, one-year, cost-plus-fixed-fee extension to its contract with Boeing Co. for reliability, quality assurance, and safety engineering support services at JSC. The extension increased funding of the multiyear contract to $12 452 722. (JSC Release 73-146)

- The National Science Foundation released *Research and Development in Industry 1971: Funds, 1971; Scientists & Engineers, January 1972* (NSF 73-305). Industry spent $18.3 billion on R&D activities in 1971—a gain of 1% over the 1970 level. Company R&D funds increased by 4% 1970-1971, with the companies financing 58% of total U.S. R&D effort; R&D decreased only in the aircraft and missiles industry. Aircraft firms spent 6% less on R&D in 1971 than in 1970. The ratio of R&D spending to net sales had declined steadily from its 1964 peak. In 1971 the ratio was 3.5%—the lowest since 1957. The number of R&D scientists and engineers in industry also continued to decline. Between January 1969 and January 1972 the number decreased by 9%, to 325 200. During this period R&D professionals in the aircraft industry dropped by 27%. (Text)

- NASA launched two Aerobee 170 sounding rockets from Woomera, Australia. The first carried a Univ. of Wisconsin astronomy experiment to a 171.9-km (106.8-mi) altitude. The second, launched 50 min later, carried a Naval Research Laboratory astronomy experiment to a 154.5-km (96.mi) altitude. Both rockets and instrumentation performed satisfactorily. (GSFC proj off)

November 13: The decision to launch Skylab 4, the third and final mission to man the Orbital Workshop (launched May 14), at 9:01 am EST Nov. 16 was announced at a Kennedy Space Center press briefing by Skylab Program Director William C. Schneider. The replacement of eight damaged fins discovered on the Saturn IB launch vehicle Nov. 6 was complete and additional precaution had been taken by retrofitting an aluminum block over two lower brackets on each of the tail fins. Cracks found on the aft interstage structural beams Nov. 12 were left

November 13

untouched after stress tests showed that the cracks would not worsen even under the most extreme aerodynamic stress. "We know of nothing today that would keep us from going," Schneider said.

Skylab 4 Astronauts Gerald P. Carr, Dr. Edward G. Gibson, and William R. Pogue arrived at Patrick Air Force Base, Fla., from Houston to participate in the countdown, which was ordered resumed. (Transcript; O'Toole, *W Post*, 11/14/73, A24)

- NASA activities in solar heating and cooling were summarized by Dr. James C. Fletcher, NASA Administrator, in testimony before the House Committee on Science and Aeronautics' Subcommittee on Energy in hearings on H.R. 10952, the Solar Heating and Demonstration Act. Thermal control of spacecraft had been a NASA challenge since the space program's inception. NASA and its contractors had developed competency in materials, thermal control coatings, heating and cooling technology, and thermal analysis technology which bore directly on the collection and use of solar energy. In the Skylab program Marshall Space Flight Center had developed a thermal coating with unique properties that increased solar collection efficiencies, resulting in higher operating temperatures. The higher temperatures allowed more efficient operation of absorption refrigeration cycles. NASA's work in solar heating and cooling included advanced research and technology, a residential systems engineering demonstration program, and the planned use of a large new office building as a test bed to obtain realistic engineering and operating experience. (Transcript)
- The first rocket-powered flight of the X-24B lifting body, to be piloted by NASA test pilot John A. Manke, was aborted because of overcast conditions at the launch and recovery site. At B-52 takeoff the weather was clear, but it worsened steadily during the flight-to-launch period. An igniter test was performed with satisfactory results. The flight was rescheduled tentatively for Nov. 14. (NASA prog off)
- Sen. Marlow W. Cook (R-Ky.) introduced for himself and others S. 2694, a bill to establish an Energy Research, Development, and Demonstration Administration. The agency would conduct a national program to resolve critical energy shortages. (*CR*, 11/13/73, S20248)

November 14: The U.S.S.R. launched *Molniya 1-25* communications satellite from Baykonur Cosmodrome into orbit with a 39 798-km (24 729.3-mi) apogee, 566-km (351.7-mi) perigee, 717.9-min period, and 64.8° inclination. The satellite would help provide a system of long-range telephone and telegraph radio communications in the U.S.S.R. and would transmit Soviet central TV programs to the Orbita network. (GSFC *SSR*, 11/30/73; Tass, FBIS–Sov, 11/15/73; *SBD*, 11/15/73, 96)

- The countdown for the twice-delayed launch of the Skylab 4 mission [see Nov. 6, 12, and 13] proceeded smoothly toward Nov. 16 launch. The Apollo spacecraft's electrical system was activated and the 1st stage of the Saturn IB launch vehicle was loaded with 157 cu m (41 500 gal) of fuel. Skylab 4 Astronauts Gerald P. Carr, Dr. Edward G. Gibson, and William R. Pogue inspected the spacecraft and received final briefings on the Apollo systems and the solar telescopes in the Orbital Workshop (launched May 14). (KSC press briefing transcript, 11/14/73; Wilford, *NYT*, 11/15/73, 6)
- Dr. Dixy Lee Ray, Chairman of the Atomic Energy Commission, read a message from President Nixon during a speech before the annual meet-

ing of the Atomic Industrial Forum and American Nuclear Society in San Francisco: It was "important that the AEC aggressively pursue research and development of the harnessing of thermonuclear fusion to produce electric power. While the fusion program is a longer-range effort, we have a responsibility to future generations to do our very best to perfect this process, since it promises to provide mankind with a virtually limitless supply of clean energy. Nuclear power in all its forms —both present fission and breeder reactors and fusion—must accept a greater share of the world's energy load in the years ahead." (Text)

- The Federal Aviation Administration released *Census of U.S. Civil Aircraft* for 1970 and 1971. Civil aircraft at the end of 1971 totaled 166 785, of which 131 870 were listed as "active." End-of-1969 figures had shown 190 749 aircraft, of which 133 814 were listed as "eligible to fly." (FAA Release 73–197)
- A planned layoff of 2500 workers—including scientists, engineers, and blue collar personnel—had been announced by Lockheed-California Co., the Associated Press reported. The work force reduction of 9% of the 29 000 employees would follow earlier signing of an agreement with Eastern Airlines, Inc., in which the airline asked for delayed delivery of wide-bodied Lockheed aircraft. Eastern had lost $10 million during the first eight months of 1973. (*W Post*, 11/15/73, C13)

November 15: Final reviews of all launch preparations had been completed and Skylab 4 was "now GO for launch at 9:01 tomorrow morning," Skylab Program Director William C. Schneider announced at a Kennedy Space Center news briefing. Leland F. Belew, Skylab Program Manager at Marshall Space Flight Center, said the *Skylab 1* Orbital Workshop (launched May 14) had experienced no major problems during the unmanned period and was in "excellent shape to support a 60-day open-ended up to 84-day mission." Col. Alan R. Vette, U.S. Air Force recovery forces, said that all Dept. of Defense recovery forces were in position and ready for launch. The prime recovery ship U.S.S. *New Orleans* was in port in San Diego and would sail in early January to Hawaii to support target points in the Pacific beginning Jan. 13. Any splashdown before that date would be covered by Air Force helicopters and fixed-wing aircraft out of Hickham Air Force Base in Hawaii. (Transcript)

- NASA test pilot John A. Manke successfully completed the first powered flight of the X-24B lifting body. The experimental craft, being flown in a USAF-NASA hypersonic-aerospace-craft research program, was air-launched from a B-52 aircraft from Flight Research Center at 14 000 m (45 000 ft). Three chambers of the XLR-11 engine propelled the lifting body to mach 0.85 and 16 000-m (52 000-ft) altitude. A series of maneuvers obtained stability control data. A preliminary data check indicated excellent maneuvers had been performed. (NASA prog off)
- *Skylab 3* scientist-astronaut Dr. Owen K. Garriott, Jr., received an honorary Doctor of Science degree from Phillips Univ. during Homecoming Day ceremonies in his home town of Enid, Okla. (*Oklahoma Journal*, 11/15/73)

November 16–February 8, 1974: The *Skylab 4* Apollo spacecraft carrying a three-man crew was launched into near-earth orbit and rendezvoused and docked with the *Skylab 1* Orbital Workshop (launched May 14) for the third and final manned mission in the Skylab program. Despite persistent problems with the spacecraft's onboard control moment gyroscopes, the crew proceeded successfully, making four space walks and setting new manned flight records. The astronauts collected significant data in medical experiments, astronomy (including 121 observations of Comet Kohoutek), and earth resources surveys for 84 days before undocking for a safe return to earth Feb. 8, 1974.

November 16–21: Skylab 4 (SL–4) lifted off on time at 9:01 am EST from Kennedy Space Center's Launch Complex 39, Pad B, on the two-stage Saturn IB–208 launch vehicle. The Apollo command and service module (CSM–118) and its crew—Gerald P. Carr, Dr. Edward G. Gibson, and William R. Pogue—roared into a clear blue sky. Traveling at 1600 km per hr (1000 mph) the spacecraft successfully endured maximum dynamic pressure at 72 sec after liftoff, proving the reliability of the Saturn IB launch vehicle and relieving the tense atmosphere in Mission Control at KSC. During the prelaunch period cracks discovered in the fins and body of the launch vehicle had twice postponed the launch date for replacement of the Saturn IB's eight stabilizing fins [see Nov. 6, 12, and 13]. Director of Launch Operations Walter J. Kapryan said he "personally sweated out those first 72 seconds through max q."

The CSM/S–IVB combination was placed in a phasing orbit with a 222.3-km (138.1-mi) apogee and 148.2-km (92.1-mi) perigee 9 min 45 sec after launch. Eight minutes later the CSM separated from the stage. The S–IVB/IU deorbit maneuver was commanded on the third orbit and the stage impacted in the Pacific Ocean at 3:00 pm EST.

After five rocket maneuvers the CSM rendezvoused with the Orbital Workshop during the fifth orbit. Docking was achieved at 5:02 pm EST after two unsuccessful attempts. Carr radioed down to Mission Control, "Great to be home."

During rendezvous and docking Pogue suffered from motion sickness and vomited. Later, in a conversation unintentionally taped by the astronauts and transmitted to the ground the astronauts decided not to report the incident to ground controllers. Capsule communicator Alan B. Shepard later told the crew, "We think you made a fairly serious error in judgment" in the attempt to cover up Pogue's condition. Carr agreed, "It was a dumb decision." Pogue recovered by late evening, Carr suffered only minor motion sickness problems, and Gibson remained in good health.

Carr, Gibson, and Pogue entered the multiple docking adapter at 9:32 am EST Nov. 17 (Mission Day 2) and began Workshop activation. They found a welcoming note on the teleprinter: "Jerry, Ed and Bill, welcome aboard the space station Skylab. Hope you enjoy your stay. We're looking forward to several months of interesting and productive work. Signed, Flight Control." In addition they found three dummies dressed in tan flight suits left by *Skylab 3* Astronauts Alan L. Bean, Dr. Owen K. Garriott, and Jack R. Lousma.

Pogue replenished the fluid in the Workshop's primary cooling system, which had developed a leak during the *Skylab 3* mission [see July 28–Sept 25]. The 5½-hr task required Pogue to punch a hole into the plumbing lines to inject the fluid. Spacecraft activation was completed Nov. 21.

November 22–February 4: The first extravehicular activity (EVA–1) began at 12:44 pm EST Nov. 22. Pogue and Gibson moved through space to install film cassettes in the Apollo Telescope Mount (ATM) cameras, deploy experimental hardware, and repair the control system for an earth-scanning antenna that had jammed during the *Skylab 3* mission. Gibson, anchored in a portable foot restraint on the bottom of the spacecraft, held onto Pogue while Pogue rewired the instrument, bypassing a short circuit and restoring the antenna to 80% of its original capacity. Gibson and Pogue reentered the Workshop and began pressurization at 6:19 pm EST. The 6-hr 35-min EVA surpassed the 6-hr 31-min record set Aug. 6 by *Skylab 3* Astronauts Garriott and Lousma.

At 3:45 am EST Nov. 23 one of three moment gyros which controlled the spacecraft overheated and was ordered shut down by Flight Controller Philip C. Shaffer. Shaffer told the press at Johnson Space Center that a near-normal mission could still be flown but it would take more time and fuel to make maneuvers. Pogue began extensive photographic study of Comet Kohoutek. By the end of the mission Comet Kohoutek had been observed and recorded 121 times as it approached, looped around, and retreated from the sun.

The first full day of scientific research was scheduled for Nov. 27. While photographing a barium cloud—created by the launch of a Black Brant IVA sounding rocket from Poker Flats, Alaska—the spacecraft's thruster attitude-control system used three times the amount of nitrogen gas fuel anticipated. The extra fuel use, in addition to poor weather conditions, forced postponement of the first earth resources experiment package (EREP) pass while ground teams designed improved techniques for managing attitude control and fuel consumption. For the first time Carr pointed the space station's six large solar telescopes at an unusually active region of the sun.

The first EREP pass was made Nov. 29 during a 23-min 8850-km (5500-mi) sweep from the southwestern U.S. to South America. In a Johnson Space Center press conference Skylab Program Director William C. Schneider said that the ground team had concluded that the Nov. 28 problem had been caused by an incorrect input of initial conditions into the fuel management program. Restrictions on maneuvers were eliminated and, with proper fuel management, the mission could "continue essentially as planned."

On Dec. 1 the astronauts photographed a laser beam pointed at them by Goddard Space Flight Center scientists who were testing the lasers as a reference point for possible use in spacecraft navigation and communications. Earlier attempts to spot the flashing green laser had failed.

In a Dec. 13 Mission Day 28 Review of the first third of the mission Program Director Schneider said the *Skylab 4* crew had accomplished 84 hrs of ATM solar viewing, 45 separate observations of Comet Kohoutek, 80 earth observation experiments, and 12 EREP passes, although a significant portion of the data from 9 passes had been lost because of wrong camera filters. Schneider said that the crew was in good physical and mental health and that their performance had been outstanding, but flight planners had needed to adjust work schedules to the pace of the crew.

A spectacular limb prominence on the sun, estimated to be the largest observed in 25 yrs, was recorded by the crew Dec. 18-19.

Astronauts Carr and Pogue opened the hatch for EVA-2 at 11:55 am EST Dec. 25, Christmas Day. Pogue mounted three special cameras to photograph the sun and Comet Kohoutek before it passed behind the sun. Photographs were taken of the comet's tail and coma. Carr replaced film cassettes in the ATM and repaired a jammed filter on an ATM camera. The constant movement of Carr and Pogue outside the spacecraft disrupted Gibson's efforts to aim the laboratory cameras at Kohoutek and the thruster attitude control system used more fuel than planned. Carr and Pogue reentered the Workshop and closed the hatch at 6:56 pm EST after a 7-hr 1-min spacewalk that surpassed Gibson and Pogue's Nov. 22 record.

The Dec. 28 perihelion passage of Comet Kohoutek was marked with a live ground-to-air televised conference between Dr. Lubos Kohoutek, discoverer of the comet (1973F), and the *Skylab 4* crew. Gibson said that the comet, as seen on the white-light coronagraph, was growing dramatically. The tail fanned out 20° from the central axis and could be seen back to a distance three times the diameter of the coma. Dr. Kohoutek congratulated the "first human beings studying a comet from outer space" and wished them a flawless splashdown in February.

The crew began EVA-3 at 12:29 pm EST Dec. 29. Carr and Gibson moved outside to photograph the sun and Comet Kohoutek as it began its trajectory away from the sun. Gibson said the comet had a spike and a tail. "That spike is very evident. It is not 180 degrees out from the tail, but more like 160 degrees. It is yellow and orange . . . just like a flame. It seems to be the same distance out as the tail, and there is a diffuse amount of material which goes out and joins up with the tail." The astronauts recorded other astronomical phenomena, retrieved samples from the magnetospheric particle collector, and retrieved the comet observation camera. At 3:58 pm EST Carr and Gibson moved back inside the Workshop and closed the hatch after 3 hrs 29 min.

On Jan. 10 the mission was given a go-ahead for a seven-day extension, with weekly reviews through the end of a tentative 84-day total mission. In the 56-day mission review at JSC, Skylab Program Director Schneider said the crew was "in good spirits and excellent physical condition and the spacecraft is in good shape to continue."

On Jan. 13 and 14 the *Skylab 4* crew photographed a solar flare from beginning to end—the first opportunity for scientists to study the complete life cycle of a solar flare.

At 10:10 pm EDT Jan. 14 Carr, Gibson, and Pogue had been in orbit 59 days 11 hrs 9 min, surpassing the record manned flight time in space on a single flight set by *Skylab 3* Astronauts Bean, Garriott, and Lousma during their July 28–Sept. 25 mission.

Mission controllers struggled with gyroscope problems with increasing frequency throughout the mission. On Jan. 22 the second control moment gyro was in the distress position for six hours, causing Skylab officials to order the recovery force into port at San Diego three days earlier than planned. EREP passes were limited and other maneuvers reduced to conserve fuel and reduce stress. Despite nearly three dozen gyro 2 stress periods, gyros 2 and 3 held up for the rest of the mission.

At 1:46 pm EDT Jan. 25 (MD-71) the crew passed the record cumulative time in space of 69 days 15 hrs 45 min set by Bean during the *Skylab 3* mission.

The hatch was opened for EVA-4 at 11:19 am EDT Feb. 3. Gibson and Carr measured the atmosphere surrounding the Skylab solar instruments to evaluate the light-scattering produced when photos of the sun were taken. The astronauts also photographed the sun with an x-ray sensitive camera, retrieved film from ATM cameras, and removed a plate and piece of rubberized material from the side of the spacecraft for a study of effects of nine-month exposure to the space environment. EVA operations ended at 4:38 pm EDT after 5 hrs 19 min.

February 5–8: Deactivation of the Workshop was begun Feb. 5 (MD–82). A three-minute firing of the steering rockets at 4:48 pm EDT Feb. 6 raised the Workshop's altitude to about 450 km (280 mi) to reduce atmospheric drag and extend the orbital lifetime to an estimated nine years, increasing chances for a future revisit to the laboratory, although it could not be inhabited again. The crew left behind samples of food, film, teleprinter paper, and electrical cables for possible inspection during the 1975 U.S.–U.S.S.R. Apollo-Soyuz Test Project mission or during U.S. space shuttle flights planned to begin in 1979.

The CSM undocked from the Workshop at 6:34 am EDT Feb. 8 (MD–85) and made a fly-around inspection of the Workshop. One of two redundant CSM rocket thruster systems developed a leak during reentry preparation, forcing the astronauts' to reenter without a backup thruster system. Just before SM jettison at 10:45 am EDT, a procedural error caused circuit breakers to open, blocking the stabilization and control thrusters from firing. Carr quickly took manual control for reentry. The CM splashed down upside down at 11:17 am EDT Feb. 8, 282 km (175 mi) southwest of San Diego and 5 km (3 mi) from the recovery ship U.S.S. *New Orleans*—ending 1214 revolutions of the earth. The capsule was righted by flotation bags while the carrier maneuvered alongside, and the CM was raised by a crane to the hangar deck about 40 min after landing. Immediate medical examinations found the crew had no apparent severe effects from their long exposure to weightlessness. They had grown one to two inches taller from stretched spinal columns in space and had lost some muscle mass, with body fluids readjusted for zero gravity; but they lost most of the slim-trim look on return to the earth. They were back to normal in a few days.

The astronauts would remain aboard ship in San Diego for medical examinations and, after a brief dockside ceremony Feb. 10, would fly to Houston. The men would be allowed to rejoin their families but contact with the general public would be restricted for one week.

The *Skylab 4* mission achieved its primary objectives of performing Saturn Workshop operations, reactivating the Skylab orbital assembly in earth orbit, obtaining medical data for use in extending the duration of manned space flights, and performing inflight experiments. All 58 planned scientific and technical experiments in biomedicine, solar astronomy, solar physics, and engineering were accomplished. EREP instruments were operated during 39 passes over the earth and the crew spent 338 hrs photographing the sun and Comet Kohoutek with ATM instruments. *Skylab 4* was officially adjudged a success Mar. 29, 1974.

Major records set by the mission included the longest-duration manned space flight to date, 84 days 1 hr 16 min; the longest cumulative time in space for a man, 84 days each for Carr, Gibson, and Pogue; the longest orbital EVA, 7 hrs 1 min Dec. 25; the longest cumulative inflight EVA time for one mission, 22 hrs 24 min; and the longest distance in orbit for a manned mission, 55.5 million km (34.5 million mi).

Skylab 4 was the third and final mission to man the Skylab Orbital Workshop. *Skylab 2* astronauts had worked aboard the laboratory 28 days, May 15–June 22, conducting experiments as well as making repairs to the orbital assembly to salvage the Skylab program. The

Skylab 3 crew collected data in medical experiments, solar astronomy, and earth resources surveys for 59 days, July 28–Sept. 25. There were no plans to send another crew to inhabit the Workshop, which would be g-dormant—lacking in power, air conditioning, and atmosphere—by Feb. 9. *Skylab 1* was expected to remain in orbit for nine years before reentering the earth's atmosphere and burning up. (NASA prog off; JSC press briefing & communications transcripts, 11/16/73–2/2/74; JSC Skylab Off; JSC, *NASA Facts;* NASA Release 74–8; Wilford, *NYT*, 11/17/73, 1; UPI, *W Star-News*, 12/16/73, F1; AP, *B Sun*, 12/27/73, A6; McElheny, *NYT*, 2/7–9/73)

November 16: President Nixon issued a statement following the launch of *Skylab 4:* "The Skylab flight that begins today marks the conclusion of the single most productive program in the history of man's quest for knowledge about himself and his world. The crew of Skylab will mark in space the 16th anniversary of America's first step toward the heavens. From the launch of a 30-pound [13.6-kg] space satellite, we progressed to the exploration of the Moon, which is helping to determine Earth's place in the universe. Now we are moving to determine through space technology man's own place in the universe, and to greater understanding of our own planet. As we are grateful for the success of our previous ventures into space, and as we look to the success of that which is imminent, let us never take for granted the skill and courage and devotion of those who labor on behalf of the United States space program. Let us rather pause to reflect with pride upon what we have done, asking God's blessings upon our efforts, and God's grace upon those who bravely place their lives at the service of peace for all mankind." (*PD*, 11/19/73, 1338)

- A *Today* editorial commented on *Skylab 4* on the day of launch: "We now talk about the strong probability of an 85-day mission. It will be the extreme test of man's ability to function in a world of weightlessness, and it should settle once and for all the question of whether man can begin to seriously think of visiting other planets in his solar system." (*Today*, 11/16/73)

- *Science* articles presented geological data from the Dec. 7–19, 1972, *Apollo 17* mission. The *Apollo 17* Preliminary Examination Team described characteristics of lunar samples in one article. Petrographically, the samples were "the most variable returned by any mission. Some have the cataclastic, highly crushed textures common in *Apollo 16* [April 16–27, 1972] return. Many are crystalline breccias whose petrographic characteristics indicate varying degrees of recrystallization or partial melting. Others are friable and dark gray like the many regolith breccias of previous missions. Others display features typical of the lavas returned from the *Apollo 11* [July 16–24, 1969], *12* [Nov. 14–24, 1969] and *15* [July 26–Aug. 7, 1971] mare sites and a few have the coarse-grained igneous textures typically developed during the slow crystallization from basaltic melts." Chemically, nearly all the characteristics of the *Apollo 17* rocks could also be found in samples from previous Apollo missions. "However, unusually high zinc concentrations in the orange soil and the exceptionally low Ni content of the basalts suggest different source materials than for previously returned igneous rocks. The trace element contents of the anorthositic rocks are significantly different from nearly all those previously returned, again suggesting variations in the source regions."

The Apollo Field Geology Investigation Team reported on the *Apollo 17* Taurus-Littrow landing site. Impact-generated breccias underlay the massifs adjacent to the valley "and basalt has flooded and leveled the valley floor. The dark mantle inferred from orbital photographs was not recognized as a discrete unit; the unusually thick regolith of the valley floor contains a unique high concentration of dark glass beads that may cause the low albedo of much of the surface."

Apollo 17 scientist-astronaut Dr. Harrison H. Schmitt described the geological investigation of the Taurus-Littrow valley as reported in "the last major report of observations by the crews during the Apollo explorations of the earth's moon." He was confident "that the future holds many other such reports as man continues his exploration of the earth's frontier and his use of the space environment." He believed that "man's abilities and spirit will continue to be the foundation of his evolution into the universe. Full satisfaction from this evolution only comes with being there." (*Science*, 10/16/73, 659-90)

- Soviet Academician Boris N. Petrov said in Moscow that India's first scientific satellite—to be launched in late 1974—would study solar neutrons, gamma rays, and x-ray emissions from outer space and the earth's ionosphere. It would be launched by a Soviet booster of the kind used for Intercosmos satellites. Subsequent Indian satellites would include those for meteorology and the study of natural resources. (*SF*, 2/74, 41)

- An experiment by Greek scientists in Athens had shown that Greek inventor Archimedes could have used solar energy to burn the Roman fleet in about 212 B.C., the *New York Times* reported. In a recent test, the scientists had lined up 50 to 60 sailors on a pier at the Skaramanga naval base, near Athens. Each sailor held a 1.5-m (5-ft) oblong mirror. On command the sailors reflected the sun's rays onto a wooden boat 49 m (160 ft) away. Within seconds the boat had caught fire. (Modiano, *NYT*, 10/16/73, 16)

- A *Science* editorial commented on the energy crisis: "A combination of environmental concerns and energy shortages is fostering wide-spread beggar-your-neighbor attitudes. Everyone wants cheap, unlimited energy, but all are prepared to fight tooth and nail to protect the environment and to prevent the location of energy facilities in their vicinity. It so happens that we are going to have neither cheap energy nor a perfect environment. Moreover, if we do not learn how to think nationally rather than parochially and to balance energy needs against environmental concerns, we are headed for trouble." (Abelson, *Science*, 10/16/73, 657)

November 17: Soviet Apollo-Soyuz Test Project Cosmonaut Aleksey A. Leonov discussed progress and problems of the ASTP to date in an interview published in *Sotsialisticheskaya Industriya*. Generally, everything was progressing according to plan, but "as yet there is no unified search and docking system either on our or the American spacecraft. Only now are designers creating such a system on Apollo and Soyuz." Spacecraft design and technical difficulties were being overcome, but there had been many problems. Personal relations between U.S. and Soviet crews and experts were very good: "The American astronauts are knowledgeable and highly skilled specialists. Suffice it to say that my opposite number, Colonel [Thomas P.] Stafford, has flown in space three times. Among ourselves we call the program . . . 'the meeting on the Elbe' in space.

It was in 1945 that Soviet and American soldiers marked their joint victory over fascism by meeting on the Elbe. Thirty years later we dream of marking a victory over the accursed 'cold war' with a unique space 'meeting on the Elbe.' " The language barrier had not yet been surmounted. "We are learning English, the Americans are learning Russian. It is said that Russian is the more difficult. But Stafford is courageously surmounting the barrier. Recently we were visited by Dr. [George M.] Low, deputy director of NASA, Eugene [N.] Cernan and Tom Stafford. Stafford made an entire speech on the space meeting [see Oct. 1–18] in relatively fluent Russian. And I assumed the role of translator." (FBIS–Sov, 12/6/73, B3)

- Thomas G. Pownall, President of Marietta Aerospace Div. of Martin Marietta Corp., was elected Chairman of the Board of Governors of the Aerospace Industries Assn. of America, Inc. He succeeded W. Paul Thayer, Chairman and Chief Executive Officer of LTV Corp. Dr. Karl G. Harr, Jr., was reelected President. (AIAA Release)

November 18: A *Miami Herald* editorial commented on the effect of the end of Skylab launches on Brevard County, "Florida's broad window on the space age." Of *Skylab 4* (launched Nov. 16), there was little more to say than that "as usual . . . the celebrated Cape [Canaveral] has another predictable success; that the space program will slow up to catch its breath, and finally that the United States is inescapably committed to space and thus to the place that all of it began. Florida has reason to be proud of its oldest point in geographical time and its youngest venture into the future." (*M Her*, 11/18/73)

- NASA launched a Javelin sounding rocket from Natal, Brazil, carrying a Univ. of California barium cloud experiment. The rocket and instruments performed satisfactorily. (GSFC proj off)

November 18–Dec. 1: U.S. flight crews for the July 1975 U.S.–U.S.S.R. Apollo-Soyuz Test Project mission visited the U.S.S.R. for familiarization with the Soyuz spacecraft and discussions with Soviet counterparts on the flight plan. The discussions, held mostly at the Yuri A. Gagarin Center for Cosmonaut Training near Moscow, included joint crew activities and onboard documentation. Astronauts participating were Thomas P. Stafford, U.S. ASTP commander; Vance D. Brand, command module pilot; and Donald K. Slayton, docking module pilot; backup crew Alan L. Bean, Ronald E. Evans, and Jack R. Lousma; support crew Robert F. Overmyer and Karol J. Bobko; and Eugene A. Cernan, Special Assistant to the U.S. ASTP Technical Director. At a press conference following the Nov. 29 concluding meeting, Maj. Gen. Vladimir A. Shatalov, chief of cosmonaut training, said the U.S.S.R. planned further manned space flights—individual Soyuz missions and linkups of Soyuz with the Salyut orbital platform—before the joint U.S.–U.S.S.R. mission. Stafford told newsmen he was completely satisfied that the Soyuz would be "a great ship to fly." He said there would be two joint training sessions in 1974, in Moscow in July and at Johnson Space Center in September. (NASA Release 73–250; Smith, *NYT*, 11/30/73, 9)

November 19: The Soviet Mission Control Center at Star City, near Moscow, was described by U.S. Apollo-Soyuz Test Project Director Glynn S. Lunney during a Johnson Space Center briefing on the Oct. 1–18 visit of a U.S. delegation to work with Soviet counterparts on the July 1975 joint docking mission. Soviet Mission Control was "not unlike the control centers ... in this country." It was "characterized by consoles with some ... large display screens down front and television display systems on the individual console." The building housing Mission Control was in two sections. "One is sort of a wing where the main operations Control Room is and some auxiliary rooms and equipment." Another wing held equipment for teletype, intercom, and TV.

Lunney said a U.S.-built docking system and a U.S.S.R. docking system, "actually almost two docking systems, that were brought over by members of the Soviet test team," had been undergoing tests at JSC since September. "The tests are almost complete with the mandatory test points that have to be taken. There's a period ... where some additional test points may be taken ... and that will probably go on for the rest ... of November." To date, the tests had gone smoothly. "I think what they are telling us is that the assumptions that went into the design appear to be verified and that indeed we can expect that same kind of results when we use the production hardware in the qualification test next year."

Lunney said U.S. delegates had learned that the Sept. 27–29 *Soyuz 12* flight had tested fixes to the spacecraft to prevent reoccurrence of the fatal June 30, 1971, *Soyuz 11* accident in which three cosmonauts died from decompression. He said that "all the fixes worked properly." Modifications included use of spacesuits for launch and reentry, a gas supply and hose system, and a strong cap on the valve that had been triggered open accidentally; the cap now had to be fired pyrotechnically to release. Firings to separate the reentry module were modified and a manual override on the valve was made more accessible to the crew. Discussions had given "a fairly full treatment of the problem, to the point where we are satisfied that we know what happened and we know what the fixes are and they appear indeed to be very substantial." (Transcript)

- George J. Vecchietti, NASA Director of Procurement since May 1964, became Assistant Administrator for Procurement. He would report directly to the Associate Administrator for Organization and Management. Vecchietti had received the NASA Medal for Outstanding Leadership in 1966 and the NASA Exceptional Service Medal in 1969. He had joined NASA in 1960 as Assistant Director of Procurement. (NASA Ann)
- A device to identify paper money by its sound "signature" had been developed for the blind from early NASA technology for semiautomatic inspection of microfilm records, NASA announced. A bill was passed under a light source on the small, inexpensive device. A phototransistor measured changes in the bill's light patterns and the changes were converted into "beeping" signals that differed with each denomination. (NASA Release 73-247)

November 20: The U.S.S.R. launched *Cosmos 608* from Plesetsk into orbit with a 498-km (309.4-mi) apogee, 269-km (167.2-mi) perigee, 92.2-min period, and 70.9° inclination. The satellite reentered July 10, 1974. (GSFC *SSR*, 11/30/73, 8/31/74; *SBD*, 11/21/73, 117)

- NASA announced the selection of Thiokol Corp. for negotiation of a $106-million, cost-plus-award-fee contract for design, development, test, and evaluation of the solid-fueled rocket motors for the space shuttle. The six-year contract would run through 1979. (NASA Release 73-258)
- Marshall Space Flight Center was drop-testing the space shuttle solid-fueled rocket booster (SRB) scale model and a three-parachute re-

November 20

covery system to determine the feasibility of keeping parachutes attached to the SRB rather than releasing them on impact with the water. (MSFC Release 73-172)

- Establishment of the Laser Geodynamic Satellite (LAGEOS) Task Team in the Marshall Space Flight Center Program Development Directorate was announced by Dr. Rocco A. Petrone, MSFC Director. LAGEOS, first in a series of satellites for application to earth and ocean dynamics investigations, would investigate physical motions and distortions of the solid earth that caused earthquakes, tidal waves, volcanic eruptions, mineral differentiation, and mountain building. D. R. Bowden—former chief of the Engineering Branch, Saturn Workshop Project, in the Skylab Program Office—would be LAGEOS Task Team Manager. (MSFC Release 73-171)
- The Senate confirmed *Skylab 3* Astronaut Jack R. Lousma to be a lieutenant colonel in the Marine Corps. He had held the rank of major. (*CR*, 10/20/73, S20947)
- Thiokol Chemical Corp. changed its corporate name to Thiokol Corp. following approval by a shareholders' meeting. (*Aerospace Facts*, 10-12/73, 13)

November 21: The U.S.S.R. launched *Cosmos 609* from Plesetsk into orbit with a 330-km (205.1-mi) apogee, 173-km (107.5-mi) perigee, 89.5-min period, and 69.9° inclination. The satellite reentered Dec. 4. (GSFC *SSR*, 11/30/73; 12/31/73; *SBD*, 11/26/73, 123)

- The Navy launched six Phoenix air-to-air missiles nearly simultaneously from a Grumman F-14A air-superiority fighter off Pt. Mugu, Calif. Four of six target drones were intercepted successfully at a range exceeding 80 km (50 mi). Two targets were destroyed and two damaged; one missile experienced hardware failure during flight and one drone veered off course and was not intercepted. (*Av Wk*, 12/3/73, 30)
- Appointment of Joseph M. Jones, Marshall Space Flight Center Chief of Public Information, as MSFC Public Affairs Director was announced by Dr. Rocco A. Petrone, MSFC Director. (NASA Release 73-259)

November 22: British Aircraft Corp. announced that Britain's *Ariel 4* (launched by NASA Dec. 11, 1971) had been reactivated to support five Skylark sounding rocket launches from Andoeya, Norway, to study magnetic substorms associated with the Aurora Borealis. (*SF*, 2/74, 41)

- NASA launched a Javelin sounding rocket from Natal, Brazil, carrying a Univ. of California and Max Planck Institute ionospheric physics experiment. The rocket and instruments performed satisfactorily. (GSFC proj off)

November 23: Recent discoveries had brought the number of kinds of molecules detected in the Milky Way galaxy to 23, the *New York Times* reported. Many were precursors to the chemistry of life and almost all were termed "organic" compounds because it had long been thought that they could be produced only by living organisms. While it had later been shown that they could be synthesized, scientists had been surprised that the simpler molecules, at least, had been formed in deep-space vacuum. Univ. of Minnesota scientist Dr. Philip M. Solomon had said, "The question of how far chemical evolution in the interstellar medium has proceeded toward biochemistry is not yet answered, but it has clearly gone much further than anyone would have estimated five years ago." (Sullivan, *NYT*, 11/23/73, 1)

- NASA launched two sounding rockets. An Arcas from Antigua, West Indies, carried a Goddard Space Flight Center meteorological experiment to measure ozone distribution in the upper atmosphere, monitor anomalous ultraviolet absorption, and extend the data base for a climatology of stratospheric ozone in the tropics. The launch was in conjunction with an overpass of the *Nimbus 4* satellite (launched April 8, 1970), but the mission was unsuccessful because of the rapid fall rate, caused by abnormal parachute performance, and because of the lack of radar track. (NASA Rpt SRL)

  An Aerobee 200A from White Sands Missile Range carried a Univ. of Hawaii solar physics experiment to a 228.4-km (141.9-mi) altitude. The rocket performed satisfactorily but, because of failure of the solar-pointing-attitude reaction control system, the experiment was unsuccessful. (GSFC proj off)

November 24: A *Washington Post* editorial criticized NASA's dismissal of Mrs. Ruth Bates Harris [see Oct. 26]: "NASA has compiled a dismal record with respect to female and minority employment. Dr. Fletcher himself admits as much. The question is, what to do about it? Dr. Fletcher says that every effort is being made to correct the problem. By now NASA should have learned that institutionalized sexism and racism give way to neither simple pieties nor eloquent declarations of principle. Achieving equitable employment opportunities for women and minorities in large American institutions requires skill, determination and sustained effort, just as a successful space program does. That is a lesson for the 1970s that all major American institutions must learn if the tragedies of the 1960s are to be avoided in this country's future." (*W Post*, 11/24/73, A14)

- Soviet helicopter designer Nikolay I. Kamov died at age 71. He had received the title of Hero of Socialist Labor for his services in the development of Soviet helicopter design and construction. (Tass, FBIS–Sov, 10/28/73, U1)

November 25: Compelling evidence of the existence of a previously predicted black hole in space had been discovered by a Univ. College team in London, using NASA's *Oao 3* Orbiting Astronomical Observatory (launched Aug. 21, 1972), NASA announced. The scientists, headed by Dr. R. L. F. Boyd with Dr. Peter Sanford, had tied the binary supergiant star system HDE 226868 to the x-ray source Cygnus X–1 and had detected evidence of the structure of the star system's gas clouds swirling around and into the x-ray source, or black hole. Using *Oao 3*, they had also observed extended x-ray sources identified with clusters of galaxies like those in the Perseus, Coma, Virgo, and Centaurus constellations. They believed enormous upheavals in the nucleus of the Seyfert galaxy were affecting the entire Perseus cluster of several thousand galaxies. They also believed the x-ray emission detected in the Seyfert galaxy might be associated with the inverse Compton effect, in which high-velocity electrons traveling near the speed of light collided with radio and optical photons. Their conclusions suggested that x-rays originated in the most active galaxies of the cluster. (NASA Release 73–251)

- President Nixon, in a nationwide address, announced new measures to conserve energy, including a 15% reduction in jet aircraft fuel consumption for passenger flights, and restated the Government's overall objective in the energy crisis: "What I have called Project Independence–1980 is

November 25

a series of plans and goals set to insure that by the end of this decade Americans will not have to rely on any source of energy beyond our own." (*PD*, 12/3/73, 1363–6)

- A *Los Angeles Times* article by James Oberg urged that a fifth Skylab mission be undertaken: "There are some unique opportunities this proposed mission could take advantage of." A Skylab in February or March "would fill the gap in the cycle of seasonal Skylab coverage. A complete year could be surveyed at the selected Skylab observation sites. This would be particularly valuable for agricultural and ecological studies." (*CR*, 12/11/73, E7920)

November 26: Rep. Don H. Clausen (R–Calif.) introduced H.R. 11554, a bill to provide for early commercial demonstration of solar heating technology by NASA with the National Bureau of Standards, National Science Foundation, Secretary of Housing and Urban Development, and other Federal agencies and for the early development and commercial demonstration of combined solar heating and cooling technology. The bill was referred to the House Commitittee on Science and Astronautics. (*CR*, 11/26/73, H10153)

- An *Aviation Week & Space Technology* editorial commented on the energy crisis: "There is an urgent and permanent need to develop energy sources that are not dependent on fossil fuels. The aerospace industry is going to be hit hard by the current oil shortage, but it also has a great deal of new technology to offer in developing permanent new energy sources." The technology of harnessing solar energy had advanced enormously in the space program. "It offers immediate promise for light-cloud-cover areas of the earth and eventual promise for enduring production of solar energy in space for continuous transmission to earth under any conditions. NASA has long had plans to move in this direction that have been annually chopped by the President's Office of Management and Budget." (*Av Wk*, 11/26/73, 11)

- President Nixon submitted the nomination of G. Joseph Minetti to be reappointed as a member of the Civil Aeronautics Board for six years ending Dec. 31, 1979. (*PD*, 12/3/73, 1370, 1381)

November 27: The U.S.S.R. launched *Cosmos 610* from Plesetsk into orbit with a 546-km (339.3-mi) apogee, 515-km (320-mi) perigee, 95.2-min period, and 74.0° inclination. (GSFC *SSR*, 11/30/73; *SBD*, 11/29/73, 146)

- Dr. James C. Fletcher, NASA Administrator, met in Cleveland with Ohio newspaper editors and businessmen to discuss the space program's future and applications of NASA technology. He denied rumors that Lewis Research Center would be delegated responsibility as the national center for energy research, but said LeRC would continue its important role in Government energy research. Dr. Fletcher predicted the next steps in space would be establishment of a permanently inhabited space station, probably by the U.S.S.R.; establishment of a permanent base on the moon; and a manned mission to Mars. "But I don't see a manned Mars mission in this century, unless it is a cooperative venture with the Russians." He said the U.S.S.R. had "let up on their secrecy enormously in recent months" in preparation for the July 1975 joint U.S.–U.S.S.R. Apollo-Soyuz docking mission. (Cl *PD*, 11/28/73)

- NASA launched a Black Brant IVA sounding rocket from Poker Flats, Alaska, carrying a Univ. of Alaska aeronomy and barium-cloud-release experiment to a 552.0-km (343-mi) altitude. The cloud release was observed and tracked by ground camera sites and the *Skylab 4* crew, launched Nov. 16 to man the Orbital Workshop. The rocket and instruments performed satisfactorily. (GSFC proj off)

November 28: The U.S.S.R. launched two Cosmos satellites from Plesetsk.

Cosmos 611 entered orbit with a 481-km (298.9-mi) apogee, 269-km (167.2-mi) perigee, 92.0-min period, and 71.0° inclination and reentered June 19, 1974.

Cosmos 612 entered orbit with a 346-km (215.0-mi) apogee, 205-km (127.4-mi) perigee, 90.0-min period, and 72.8° inclination and reentered Dec. 11. (GSFC *SSR*, 11/30/73; 12/31/73; 6/30/74; *SBD*, 11/30/73, 160)

- President Nixon submitted to the Senate the nomination of James W. Plummer, Vice President of Lockheed Aircraft Corp. and Lockheed Missiles & Space Co. and General Manager of Lockheed Missiles & Space Co.'s Space Systems Div., to be Under Secretary of the Air Force. He would succeed John L. McLucas, who became Secretary of the Air Force July 19. (*PD*, 3/12/73, 1378, 1381)

- The next 10 yrs in transportation could be the decade of the hovercraft, *Jane's Surface Skimmers* said in its 1973–74 edition. Editor Roy McLeavy noted that the Soviet Ekranoplan already carried 900 fully armed troops at more than 350 km per hr (220 mph) over almost any surface. Hovercraft production would almost certainly develop into "one of the world's major industries." U.S. Navy surface ships might cost as much as $50 million; France was studying "hoverfrigates"; and the U.S.S.R. was joining in "this more sophisticated approach to naval strategy." (AP, *W Post*, 11/29/73, A32)

November 28–30: The U.S.S.R. Joint Commission on Scientific and Technical Cooperation held its second session in Moscow. The first session had been held in Washington, D.C., March 19–21. At the opening meeting, Chairman Vladimir A. Kirillin of the Soviet State Committee for Science and Technology said useful work had been done by the Joint Commission, particularly in power engineering and chemical catalysis. Dr. H. Guyford Stever, National Science Foundation Director and Presidential Science Adviser, said he was certain that U.S.-Soviet cooperation in science and technology represented a major contribution to the progress of all mankind. An accord signed Nov. 30 called for more specific scientific cooperation. Dr. Stever said hundreds of scientists would be exchanged before the Commission's next meeting, scheduled for Washington, D.C., in October 1974. (Tass, FBIS–Sov, 11/29/73, B7; Wren, *NYT*, 12/1/73, 6)

November 28–December 1: An International Colloquium on Mars was held at Cal Tech and Jet Propulsion Laboratory by NASA, the American Astronautical Society, and the American Geophysical Union to evaluate new ideas stimulated by the successful *Mariner 9* mission to Mars (launched by NASA May 30, 1971). Four hundred scientists from 10 countries participated in the program, which discussed Martian surface history and evolution, atmospheric history as related to interactions with the surface, and future exploration of Mars. (NASA Release 73-255; JPL PIO)

November 29: Dr. James C. Fletcher, NASA Administrator, gave personal views on the existence of extraterrestrial life in a speech at the Univ. of Maryland: "There is no hope of finding intelligent life elsewhere in the solar system. As far as this particular corner of the universe is concerned, we are it." Chances were that primitive life forms would be found on other planets or in their atmospheres, "and this will still be one of the great discoveries of all time. This discovery will shed light on how life arose on Earth. It will strengthen our conviction that intelligent life must exist on the planets of millions, or even billions, of other star systems in the universe." Dr. Fletcher doubted that any other civilizations would exchange visits with the earth because of the great distances. "We could not begin to build the kind of spaceship that would be needed. But who can say what new technologies, what new understandings of the physical laws of the universe, future generations will develop —or hear about on the inter-galactic radio network?" He was "very optimistic" about long-range chances of communicating with very advanced civilizations in the Milky Way or other galaxies: "When you think how many stars there are which could have planets like Earth, and how long they have been in existence, it is quite easy to believe that there must be many advanced civilizations broadcasting in our direction. And I am sure we have the technology needed to intercept such signals and eventually answer them." (Text)

- A team of Ames Research Center scientists, headed by Dr. James B. Pollack, had discovered that the upper levels of Venus's brilliant clouds consisted of droplets of sulfuric acid more concentrated than the acid in an automobile battery, ARC announced. The team had measured the infrared "color" of Venus from a Learjet aircraft and compared results with a computer simulation of the color properties of various substances. (ARC Release 73-128)

- Naval Air Test Center development of a rescue system to pinpoint the position of any downed aircraft anywhere on earth was reported by R/A Roy M. Isaman, Test Center Commander, in a telephone interview with the Associated Press. A distressed aircraft would use a transmitter of less than five watts to bounce a signal off a communications satellite in stationary orbit back to one of eight Navy navigation stations under construction. (B Sun, 11/30/73, A3)

- Sen. Pete V. Domenici (R–N. Mex.) introduced S. 2755, a bill to require the NASA Administrator to study the feasibility of entering into international cooperative programs using space technology and applications. The NASA Administrator would work with the Secretary of State, Secretary of Defense, and the National Science Foundation Director to study the desirability of international cooperation and cost sharing to develop a solar energy collection and conversion system. (CR, 11/29/73, S21370)

November 30: The U.S.S.R. launched *Cosmos 613* from Baykonur Cosmodrome into orbit with a 274-km (170.3-mi) apogee, 188-km (116.8-mi) perigee, 89.7-min period, and 51.6° inclination. Western observers identified the spacecraft as an unmanned Soyuz. It was brought down by parachute near Karaganda Jan. 30, 1974, after orbiting the earth 60 days, a record length of flight for any Soviet spacecraft recovered intact. *Cosmos 613* flew the same orbital track as other manned Soviet spacecraft but at a higher altitude, leading to speculation that the purpose of the mission was to test the ability of Soyuz spacecraft to fly higher and the durability of its systems as a prerequisite to a possible two-month manned mission. (GSFC SSR, 11/30/73; 1/31/74; SBD, 12/3/73, 164; O'Toole, *W Post*, 12/16/73, A25; 2/10/74, A27; *Av Wk*, 2/11/74, 32)

- The U.S.S.R. launched *Molniya 1-26* communications satellite into orbit with a 40 830-km (25 370.6-mi) apogee, 618-km (384-mi) perigee, 740.0-min period, and 62.9° inclination. The satellite would help provide a system of long-range telephone and telegraph radio communications in the U.S.S.R. and would transmit Soviet central TV programs to the Orbita network. (GSFC *SSR*, 11/30/73; Tass, FBIS–Sov, 12/5/73, U1)
- *Skylab 2* Astronaut Charles Conrad, Jr., was retiring from NASA and the Navy effective Feb. 1, Johnson Space Center announced. He would become Vice President, Operations, and Chief Operating Officer, of the American Television and Communications Corp., a Denver-based cable TV firm. (JSC Release 73–158)
- NASA announced award of a $41.8-million, cost-plus-award-fee contract to General Dynamics Corp. Convair Div. to continue management and engineering services in the Atlas-Centaur and Titan-Centaur launch vehicle program. The procurement extended an earlier contract two years. (NASA Release 73–271)
- The U.S.S.R. was using onboard computers on multiple independently targetable reentry vehicles for the first time, providing "the approaching technological option over both MIRVing and having greater accuracy in . . . MIRVed warheads," Dr. James R. Schlesinger, Secretary of Defense, said at a Dept. of Defense briefing. The technological advantages held by the U.S. would "tend to wane" as the Soviets acquired improved warheads, improved guidance, and improved MIRVs. There was no immediate change in strategic balance but "with these technologies available . . . , one can look out to the period beyond 1976, say, and with marrying of these technologies to the throw-weight available to the Soviet Union, it could create an unbalance in the strategic area." (Transcript)
- The Rev. Walter J. Miller, Fordham Univ. astronomer and former assistant papal astronomer at the Vatican Astronomical Observatory, died in New York following a heart attack, at age 69. He had specialized in the study of faint variables, pulsating stars that varied their apparent magnitude periodically. He was known particularly for his observations of Cygnus in the Milky Way galaxy. (*NYT*, 12/2/73, 85)

November 30, December 3, 4: The Federal Aviation Administration sponsored an International Microwave Landing System Symposium in Washington, D.C., to exchange information between the International Civil Aviation Organization (ICAO) member nations sponsoring MLS programs and other ICAO states interested in MLS efforts. Presentations by the U.S., United Kingdom, France, Australia, and West Germany showed progress in the use of MLS. Canada reported on its short takeoff and landing (STOL) aircraft program. MLS provided pilots precise course and glide-slope guidance to runways during landing approach. Broad areas covered permitted operational procedures to increase airport acceptance rate and reduce noise over surrounding communities. (FAA Release 73–199; FAA PAO)

December 1973

December 1: Dr. Dixy Lee Ray, Atomic Energy Commission Chairman, submitted the report *The Nation's Energy Future* to President Nixon in response to the President's June 29 request for recommended energy research and development programs for inclusion in the FY 1975 budget. Recommendations were to establish an operational Energy Research and Development Administration by July 1, 1974, to plan and coordinate the total program and direct the major share of the Federal program; conduct an annual program review to reallocate funds among programs as required; ensure full consideration of energy consequences of all Federal actions to achieve nonenergy goals; promote private-sector participation in the conduct, review, and evaluation of the Federal ER&D program; initiate in FY 1975 a Synthetic Fuels Pioneer program of privately funded construction of commercial plants for producing synthetic fuels from coal, using existing technologies; and accelerate ongoing work in environmental effects research, basic research, and manpower development. The congressional Joint Committee on Atomic Energy held hearings on the report Dec. 11. (Transcript)

- A *Science & Government Report* editorial commented on Federal science policy response to the energy crisis: "Following on the pattern of the Sputnik trauma, the energy debacle is inevitably evoking prescriptions for still another reorganization of federal scientific and technical advisory services." The essential fact about science in Washington was "that knowledge is power only when the political element accepts the knowledge as being politically palatable." Rearranging the Government was one of Washington's favorite indoor pastimes, "but the utility of it is open to serious doubt." (*Sci & Gov Rpt*, 12/1/73)

December 2: The drive-through tour of Kennedy Space Center and Cape Kennedy Air Force Station was suspended because of the energy conservation program. NASA's Visitor Information Center would continue its escorted bus tours. (KSC Release 266–73)

- A real estate boom apparently would replace the space boom with the end of the Skylab program at Cape Canaveral, the *Atlantic Journal Constitution* said. Layoffs following Skylab's end would reduce aerospace employment on the Cape to its lowest point in a decade; a total work force of less than 10 000 had been projected for December 1974. But unlike the cutbacks in 1969 and 1970, there was no panic. An influx of retirees from the North and from southern Florida had caused a shortage of homes. Unemployment in Brevard County was down to 4.2%; the area had once led the nation with more than 8% unemployed. (Hesser, *Atlanta JC*, 12/2/73)

December 2–10: The U.S.–U.S.S.R. Joint Working Group on Space Biology and Medicine met at Johnson Space Center to continue exchanging medical data and results from U.S. and Soviet manned flight experience. The Group—cochaired by Dr. Charles A. Berry, NASA Director of Life Sciences, and Dr. N. N. Gurovsky of the Soviet Academy of Sciences—discussed vestibular disturbances and the characteristics of human blood circulation under space flight conditions and developed recommendations for using common biomedical procedures. The Soviet delegation reported on pre- and postflight cardiovascular evaluations with lower-body negative pressure during the *Soyuz 12* mission (launched Sept. 27). The report would permit the first direct comparison of Apollo, Skylab, and Soyuz data. The Soviet delegation visited JSC Mission Operations Control to observe facilities for monitoring the health of Skylab crews, Skylab one-g trainers, and the Skylab mobile laboratories. The Group's recommendations would be forwarded to the NASA Administrator and the President of the Soviet Academy of Sciences for confirmation. (NASA Release 73–273)

December 3: NASA's *Pioneer 10* (launched March 2, 1972) became man's first spacecraft to reach Jupiter as it passed 130 000 km (81 000 mi) from the planet at 9:25 pm EST, after traveling 641 days and 826 000 000 km (513 000 000 mi) [see Nov. 3–Jan. 9]. (NASA prog off)

- President Nixon wired congratulations to scientists at Ames Research Center on *Pioneer 10*'s encounter with the planet Jupiter: "Our nation takes profound pride in so impressive a scientific achievement. On behalf of the American people, I want to express my admiration and gratitude to you." The President said *Pioneer 10* had "demonstrated that man's ability to explore the heavens is on the threshold of the infinite." (O'Toole, *W Post,* 12/4/73, A1; *W Star-News,* 12/4/73, A3)

- The Skylab Saturn IB/Apollo rescue vehicle, readied for rescue of *Skylab 4* astronauts (launched Nov. 16), was rolled out from Kennedy Space Center's Vehicle Assembly Building to Launch Complex 39, Pad B. By Dec. 19 the vehicle would complete flight readiness tests, the two-day final overall test of spacecraft systems and ground support equipment, and a simulated launch. On Dec. 21 the rescue vehicle would be within nine days of launch readiness and would hold at that point until needed in an emergency or returned to the VAB at the end of the mission. The rescue command module was equipped with two additional crew couches and additional life support and communications connections to permit it to return the three *Skylab 4* crewmen and the two rescue crew members, Vance D. Brand and Dr. Don L. Lind. (KSC Release 283–73)

- President Nixon conferred the four-star rank on Adm. Hyman G. Rickover (USN, Ret.) in a White House ceremony. The President said: "He, of course, will be remembered . . . because he leaves as a monument not only the fact that there are now more ships and a stronger Navy, but that superb breakthrough in technology—*Polaris, Poseidon, Trident.* No one can ever speak of these breakthroughs without thinking of Admiral Rickover." (*PD,* 12/10/73, 1385)

- The House passed H.R. 9430, a bill naming a Federal building and carrying an amendment that repealed the 1972 naming of Jet Propulsion Laboratory after former Rep. H. Allen Smith (R–Calif.). (*CR,* 12/3/73, H10468)

- President Nixon accepted the resignation of John A. Love as Assistant to the President and Director of the Energy Policy Office. (*PD,* 12/10/73, 1448)

December 4: The U.S.S.R. launched *Cosmos 614* from Plesetsk into orbit with an 807-km (501.5-mi) apogee, 769-km (477.8-mi) perigee, 100.6-

min period, and 74.0° inclination. (GSFC *SSR*, 12/31/73; *SBD*, 12/6/73, 186)
- A *New York Times* editorial commented on the flight of *Pioneer 10* to within 130 000 km (81 000 mi) of Jupiter: Never before had "the mind of man projected its sensors so far out in space toward such a distant heavenly body." Already, preliminary results had "surprised scientists and challenged existing views." Climaxing 350 years of scientific study of Jupiter, the clear and detailed data from *Pioneer 10* inaugurated "a qualitatively new stage in man's knowledge of this giant planet and his understanding of how its physical and other properties relate to the larger problem of the origin of the solar system and of the universe." (*NYT*, 12/4/73, 42)
- Latin America opposed direct and uncontrolled satellite transmission of TV programs during the Second World Broadcasting Conference in Buenos Aires, Argentina. The U.S. delegation had maintained that censorship of direct transmission would be restriction of the freedom of information. Argentine jurist Augusto Ferrer replied that "without governmental control . . . individual countries would find themselves subject to various forms of intervention in their cultural sovereignty, in their national security, and in their psychological development." Conference President Jose A. Castro said the rest of the Latin American countries had come out against direct broadcasting. (FBIS–Inter-American Affairs, 12/21/73, A1)
- President Nixon signed Executive Order 11748 establishing the Federal Energy Office in the Executive Office of the President. He announced the appointment of Deputy Secretary of the Treasury William E. Simon as Executive Director. The President said: "I have decided to bring together in one agency the major energy resource management functions of the Federal Government to provide the centralized authority we must have for dealing with the energy crisis." He was assuming chairmanship of the Energy Emergency Action Group, "which will continue to oversee all major policy issues relating to energy." He would ask Congress to create a Federal Energy Administration "and, in the Executive Office of the President, a Federal Energy Office to carry out all energy-related functions." (*PD*, 12/10/73, 1388–92)
- Sen. Henry M. Jackson (D–Wash.), at the Nixon Administration's request, introduced S. 2776 to establish a Federal Energy Administration, an independent executive agency to deal with the energy emergency for two years. (*CR*, 12/4/73, S21776–81)
- Rep. Benjamin A. Gilman (R–N.Y.) introduced H.R. 11762 to direct the National Science Foundation to start and support basic geothermal energy research and to amend the National Aeronautics and Space Act of 1958 to direct NASA to carry out a demonstration program for the development of the commercial use of geothermal resources. (*CR*, 12/4/73, E7752–3)
- The Air Force Systems Command announced it had signed a $3 450 000 agreement to initiate a study for the Dept. of the Interior Office of Coal Research on a more efficient method of generating electrical power from coal with less pollution. Electrical energy would be extracted from a high-temperature ionized gas—similar to the flaming exhaust of a rocket or jet engine—by directing it through a magnetic field. Work on the process would be conducted at AFSC's Arnold Engineering Development Center, using a magnetohydrodynamic generator built in the 1960s

as an electrical power source for a wind tunnel. (AFSC Release 01P 157.73)
- NASA launched a Black Brant IVA sounding rocket from Poker Flats, Alaska, carrying a Univ. of Alaska aeronomy and barium cloud-release experiment to a 569.7-km (354-mi) altitude. The rocket and instrumentation performed satisfactorily. The barium cloud was viewed by Alaskan ground sites but the planned support by *Skylab 4* mission (launched Nov. 16) was not available because of scheduling problems. (GSFC proj off)

December 5: Newspaper editorials commented on the Dec. 3 rendezvous of *Pioneer 10* (launched March 2, 1972) with Jupiter.

Chicago Tribune: "American technology has done it again. The spacecraft Pioneer 10 has soared past the planet Jupiter, weathering its barrage of radiation and extracting from it secrets of its nature that have tantalized earth people since their earliest days. The flood of data obtained by television photography and transmitted back to earth is so voluminous that no one can yet assay its meaning." (*C Trib,* 12/5/73)

Philadelphia Inquirer: "We have taken space spectaculars so much for granted in recent years that it may be hard to grasp the amazing feat of Pioneer 10." The spacecraft had been launched "more than three months before Watergate—and has been whizzing through the void ever since." It had provided "a wealth of new knowledge about the solar system's largest planet." (*P Inq,* 12/5/73)

- NASA announced plans for a 1976 launch of a Laser Geodynamic Satellite (LAGEOS). The satellite would provide information for mathematical models on the earth's movements and strains which caused natural disasters such as earthquakes. The measurements made by the spacecraft would be accurate within 2 cm (0.75 in) of actual earth movement. LAGEOS would be a solid sphere weighing about 380 kg (835 lbs) and measuring 0.6 m (2 ft) in diameter. It would be fitted with 600 laser retroreflectors designed to permit accurate laser ranging from ground stations. Marshall Space Flight Center, under the overall direction of NASA's Office of Applications, had been assigned definition and development responsibility for the satellite. (NASA Release 73-261)
- Marshall Space Flight Center announced two contract awards. A $1 176 872 extension was made in a $5 188 357 contract with Bendix Corp. to provide management, sustaining engineering, and logistics support for the ST-124M stabilized platform for the July 1975 mission of the U.S.-U.S.S.R. Apollo-Soyuz Test Project. The contract extension, bringing the total contract value to $6 365 229, was from May 1, 1974, through July 31, 1975. A $160 000 contract was awarded Norman Engineering Co. for architect-engineering services to modify a test stand for space shuttle testing. (MSFC Releases 73-190, 73-191)

December 6: Concorde 201—Air France's first Concorde supersonic airliner, the first production Concorde and the fifth Concorde to fly—made its maiden flight, from Toulouse, France, including 45 min supersonic flight. The aircraft reached mach 1.57, more than 1600 km (100 mi) per hr, at 1280 m (42 000 ft) during its 2-hr 50-min flight. Command pilot was Aerospatiale's André Turcat. (*Concorde Bull* 12, 1/1/74)

- Sen. Gaylord Nelson (D-Wis.), on behalf of himself and Sen. Henry M. Jackson (D-Wash.), introduced S. 2782 to establish a National Energy Information System, to authorize the Dept. of the Interior to undertake an inventory of U.S. energy resources on public lands and elsewhere, and for other purposes. (*CR*, 12/6/73, S22002)

December 7: A major reorganization plan that included two key directorates at Johnson Space Center had been announced by Dr. Christopher C. Kraft, Jr., JSC Director, the JSC *Roundup* reported. Effective immediately, the reorganization would not be implemented fully until February 1974. Flight Crew Operations and Flight Operations would be combined into a new Flight Operations Directorate devoted entirely to space flight and aircraft activities. A new Data Systems and Analysis Directorate would provide institutional and programmatic data systems and analysis and the onboard software for the space shuttle. Pilot astronauts in the former Flight Crew Operations Directorate would be assigned to the Astronaut Office in the new Flight Operations Directorate. Scientist-astronauts would be assigned to Astronaut Offices in the Science and Applications and Life Sciences Directorates.

Skylab Program Manager Kenneth S. Kleinknecht and Flight Control Div. Chief Eugene K. Kranz would be Director and Deputy Director of the new Flight Operations Directorate. They also would continue in their present assignments through completion of the Skylab program. Astronaut Alan B. Shepard, Jr., would be Chief of the Flight Control Astronaut Office. Howard W. Tindall, Jr., and Lynwood C. Dunseith had been named Director and Deputy Director of the Data Systems and Analysis Directorate. Astronaut Owen K. Garriott was Deputy to Director Anthony J. Calio of the Science and Applications Directorate; Astronaut Harrison H. Schmitt would be Chief of the Science and Applications Astronaut Office. Astronaut Joseph P. Kerwin would be Chief of the Life Sciences Astronaut Office, which also would include astronaut physicians F. Story Musgrave and William E. Thornton.

In addition to their duties as space shuttle crew candidates, Astronauts Fred W. Haise, Jr., Charles M. Duke, Jr., and Eugene A. Cernan had been assigned duties in project management. Haise would be Technical Assistant to the Space Shuttle Orbiter Project Manager, Duke would be Technical Assistant to the Acting Manager for Space Shuttle Systems Integration, and Cernan would be Special Assistant to the Apollo-Soyuz Test Project Manager. Of 37 astronauts on active flight status, 26 were pilots. Of these, 16 would participate in space shuttle activities; 10 had been assigned to the 1975 U.S.–U.S.S.R. Apollo-Soyuz Test Project mission. Eight scientist-astronauts would work in the Science and Applications Office, and three in Life Sciences. (*Roundup*, 12/7/73, 1)

- The Senate passed S. 1283 by unanimous vote of 82. The Energy Research and Development Policy Act would provide support for programs in R&D of fuels and energy. (*CR*, 12/7/73, S22190)
- Sir Robert Watson-Watt, Britain's "father of radar," had died in Scotland at age 81, the *Washington Post* reported. Watson-Watt was credited with directing and accelerating development of radar immediately preceding World War II. He was best known for his work in the use of radio waves to detect aircraft, but also had developed an underwater detection system for use against German U-boats during the Battle of the North Atlantic. (Hailey, *W Post*, 12/7/73, C15)

December 8: The risk to future development of man's life outside the earth was "that the public will come to denigrate what Skylab has done and rate it below the achievements" of *Pioneer 10*, an *Economist* article said. *Pioneer 10*'s pictures, "passable photographs" of Jupiter, were "better than no pictures at all, but if the evidence of the first space shots past Mars are anything to go by, useful results require the cameras to go in much closer than Pioneer was able to do, and to stay in orbit round the planet for long enough to take . . . a set of controlled mapping pictures. Pioneer 10 has shown that it is going to be technically possible to do this; the next stage is to build the space probe that will do it. When it comes to this sort of detailed work, the instruments on board tend to need the fine adjustments that only men can give them." The U.S. should be planning "another Skylab with instruments calibrated to look further into space than the present ones. The results could be even better than Pioneer 10's." (*Economist*, 12/8/73, 20)

- A *Detroit News* editorial commented on *Pioneer 10*'s Dec. 3 rendezvous with Jupiter: "Jupiter is lucky. It appears to radiate three times as much energy as it receives from the sun. When all the knowledge Pioneer has transmitted has been absorbed and categorized by NASA scientists, it should provide clues to new energy sources for earth, particularly solar energy and the sun's rays. All these possibilities are beyond the comprehension of an energy-conscious nation that in four years has grown blasé about man walking on the moon. But they open the door for the scientists to apply on earth the potential resources of the solar system." (*D News*, 12/5/73)

December 9: A partial lunar eclipse, one of three visible from North America during 1973, shaded 10% of the moon. Cape Canaveral residents could view the obscurity at its peak at 8:45 pm EST. (*Today*, 12/10/73)

December 10: Establishment of a NASA Task Force on Energy Conservation and designation of Gen. Bruce K. Holloway (USAF, Ret.), Special Assistant to the Administrator and Assistant Administrator for DOD and Interagency Affairs, as Director of Energy Conservation were announced by Dr. George M. Low, Deputy Administrator. In a memorandum to Center Directors and program and staff office heads Dr. Low said the Task Force would direct NASA activities to reduce energy use, determine where further energy reductions could be made in NASA and its programs, and apply technical and management ingenuity to finding new and imaginative ways in which NASA and other agencies could conserve energy. (Text)

- Kennedy Space Center requested bids from 50 construction firms on the 4600-m (15 000-ft) runway to be built for the space shuttle northwest of the Vehicle Assembly Building. The runway would be the first facility built in reshaping Launch Complex 39 for its new role in the shuttle program. Bids were to be opened Feb. 22, 1974. (KSC Release 7-74)

- NASA launched a Black Brant VC sounding rocket from White Sands Missile Range carrying a Harvard College Observatory solar physics experiment to a 268.7-km (167-mi) altitude. The launch was also one of the series of calibration rockets (CALROC) launched in support of Skylab missions [see Jan. 22]. The rocket and instrumentation performed satisfactorily. (GSFC proj off)

- Skylab backup hardware was being stored in plastic tents with controlled environments at Marshall Space Flight Center for later use or disposal. (MSFC Release 73-193)

- Former astronaut John H. Glenn, Jr., announced in Cleveland he would make a third bid for the Democratic nomination to the Senate from Ohio. He had lost the 1970 Democratic senatorial primary; a head injury had forced him to end his 1964 Senate campaign. Glenn, as pilot of *Friendship 7* (launched Feb. 20, 1962), was the first American to make an orbital space flight. (*CSM*, 12/11/73; *A&A 1962*)
- Dr. Wolf V. Vishniac, Univ. of Rochester microbiologist and biological scientist in NASA's Viking Mars lander program, died at age 51 after falling down a steep slope during a scientific expedition to Antarctica for NASA. Dr. Vishniac was studying how microbes multiplied in the arid Antarctic soil. He had been a member of NASA's Lunar and Planetary Missions Board 1967–1970. He had served on the Office of Manned Space Flight's Joint Editorial Board on Space Biology and Medicine since 1965 and was coediting Volume One of "Space Biology and Medicine." He had been a member of the Viking Biology Team since 1969, had served on the Lunar Sampling Analysis Team in 1968 and 1969, and had been active on the Space Sciences Board of the National Academy of Sciences. (NASA Release 73-277; AP, *NYT*, 12/13/73, 48)

December 10–14: More than 120 papers by scientists worldwide were presented at the Third Earth Resources Technology Satellite Symposium sponsored by Goddard Space Flight Center. The Symposium highlighted data from *Erts 1* Earth Resources Technology Satellite (launched by NASA July 23, 1972), which had provided more than 100 000 images of the earth's surface. Also emphasized was the extension of scientific techniques and findings discussed in a March symposium, including applications directly benefiting the public in dealing with local and regional resource problems. (NASA Release 73-271; GSFC PIO)

December 11: *Mariner 10*, NASA's two-planet probe launched Nov. 3 toward Venus and Mercury, had discovered that the gas remnants of the exploding star Gamma Velorium in the Gum Nebula were twice as hot as scientists had expected, the *Washington Post* reported. Instruments aboard the spacecraft had detected ultraviolet radiation from the nebula so intense that it corresponded to temperatures as high as 56 000 K (100 000°F), suggesting that some stars were undergoing even more violent nuclear reactions than expected. Dr. Bruce Murray, California Institute of Technology scientist, was surprised that so old a nebula was still radiating any detectable heat. The discovery might be "the start of a whole new field of astronomy. . . . Never before have we been able to detect interstellar radiation associated with this temperature range." (O'Toole, *W Post*, 12/11/73, A10)
- Radio signals from Comet Kohoutek, racing toward a Dec. 28 pass behind the sun were picked up by scientists using the Kitt Peak, Ariz., radio telescope, NASA reported. It was the first strong evidence of radio emissions—produced by electrically excited molecules of methyl cyanide, a gas previously observed in the region of the Milky Way where new stars were formed—from any comet and the first evidence that methyl cyanide molecules existed in comets. The press quoted a Goddard Space Flight Center spokesman as saying, "Comet Kohoutek appears to be bringing a sample of the distant interstellar matter into the solar system for the first time." (UPI, *W Post*, 12/12/73, A23)
- Appointment of Capt. Clyde T. Lusk, Jr. (USCG), as Acting Director of the newly created Office of Transportation Energy Policy in the Dept. of

Transportation was announced by Secretary of Transportation Claude S. Brinegar. Myron Miller would be Acting Deputy Director. (DOT Release 82-73)

- A *Wall Street Journal* editorial commented on Comet Kohoutek: "The mind can only boggle at how little we still really know about the universe where comets like Kohoutek may number in the billions. Reflecting on this may not lead to any change in our conduct here on earth anymore than the appearance of Halley's Comet in 1910 seems to have significantly improved earthly conduct. But the implications of Kohoutek, and of the scientific void into which it will soon return, are reminders of the cosmic scope of man's environment." (*WSJ*, 12/11/73)

December 12: NASA test pilot John A. Manke made the seventh flight and second powered flight of the X-24B lifting body, launched from a B-52 aircraft from Flight Research Center. Primary objectives of flight-envelope expansion to mach 0.95, stability and control at mach 0.9, and performance survey with aileron bias at 11° and secondary objectives of fin- and rudder-pressure survey and an acoustic noise and vibration experiment were met. (NASA prog off)

- Dr. Henry A. Kissinger, U.S. Secretary of State, said in a speech before The Pilgrims in London that the 1973 energy crisis should become "the economical equivalent of the Sputnik challenge of 1957. Only this time, the giant step for mankind will be one that America and its closest partners take together." He proposed "that the nations of Europe, North America and Japan establish an energy action group of senior and prestigious individuals, with a mandate to develop within three months an initial action program for collaboration in all areas of the energy problem." (Text)

- Czechoslovakian astronomer Dr. Lubos Kohoutek, disembarking from a three-day comet-watch cruise aboard the liner *Queen Elizabeth 2*, told the press in New York that his viewing of the comet he discovered, Kohoutek, had been hampered by clouds and seasickness. He said he believed he had seen the comet with his naked eye "just above the horizon before 5:30 a.m." (Reuters, *C Trib*, 12/13/73)

- Unidentified astronomers were quoted as saying Comet Kohoutek was not likely to become visible to the naked eye for at least another week. After it rounded the sun Dec. 28, it would appear in the evening sky after sunset. It might be fragmented by solar gravity and solar radiation with the comet and its tail becoming more brilliant. A Naval Observatory astronomer predicted the comet would be no brighter than Comet Ikeya-Seki in 1965 or Comet Bennett in 1970. Smithsonian Astrophysical Observatory astronomers did not expect Kohoutek to be visible without binoculars during the current week. (AP, *NYT*, 12/13/73)

- Office of Management and Budget Director Roy L. Ash said in a letter to Rep. Ralph S. Regula (D-Ohio) that immediately upon the establishment of the proposed Energy Research and Development Administration, OMB would urge the ERDA Administrator to undertake "a thorough review of all NASA personnel and facilities that might be closed because of NASA's post-Apollo 'scaling down.'" Ash said he had been "concerned for some time about the possible loss from government service of these valuable skills and resources." (*CR*, 12/14/73, E8084)

- The Cost of Living Council, following a court order, directed five aerospace firms to pay a 17-cent hourly wage increase that had not been paid in 1971 and 1972. The increase, to some 108 000 workers, had been delayed by the Administration's Pay Board in 1971 because it exceeded the Government's anti-inflation guidelines. The Council said payments could total $75 million to $85 million and could ultimately affect 250 000 workers, since many other firms followed pay practices of the five large companies. The companies were Boeing Co., Lockheed Aircraft Corp., McDonnell Douglas Corp., Rockwell International Corp., and LTV Aerospace Corp. (*W Post*, 12/13/73; *WSJ*, 12/13/73)

- The Air Force announced contract awards: a $5 992 290 cost-plus-fixed-fee award to General Dynamics Corp. for automatic-flight-control technology to enhance the maneuvering performance of fighter aircraft and a $5 406 500 fixed-price-incentive contract for A-7D aircraft mission simulator, aerospace ground equipment, spare parts, and data. (DOD Release 597-73)

December 13: The Air Force launched *Dscs F-3* and *Dscs F-4* communications satellites from Eastern Test Range on one Titan IIIC booster, after two postponements to correct faulty instrumentation in the transtage of the launch vehicle. *Dscs F-3* entered orbit with a 36 124.9-km (22 447-mi) apogee, 35 799.9-km (22 245-mi) perigee, 24-hr 5-min period, and 2.0° inclination. *Dscs F-4* entered orbit with a 36 137.8-km (22 455-mi) apogee, 35 944.7-km (22 335-mi) perigee, 24-hr 9-min period, and 2.0° inclination. On Dec. 19 the press reported Air Force officials as saying that the satellites—part of the Defense Satellite Communications System Phase 2—were operating well in preliminary tests and would be positioned on operational stations at 13° W and 175° E in synchronous orbit approximately 80 days after launch. They would replace two DSCS 2 satellites launched Nov. 2, 1971, which had ceased operating. (Pres Rpt 74; *SBD*, 12/13/73, 226; 12/19/73, 250)

- The U.S.S.R. launched *Cosmos 615* into orbit with a 829-km (515.1-mi) apogee, 269-km (167.2-mi) perigee, 95.6-min period, and 71.0° inclination. (GSFC *SSR*, 12/31/73)

- NASA announced the selection of Grumman Aerospace Corp. for negotiation of a contract to modify two Gulfstream II aircraft for crew trainers for the space shuttle orbiter. The contractor's proposed cost was $19.5 million. (NASA Release 73-278)

- Air Force Secretary John L. McLucas outlined the Air Force role in space in a speech before the Military Order of the World Wars in Memphis, Tenn. Twenty-six military communications satellites orbited as a research effort had proved so successful they were converted to a basic operational system in 1967. Half of these were still operational. An experimental tactical communications program was essentially complete; satellites and some receivers for aircraft, ships, trucks, jeeps, and a man-pack had been tested. The first operational Fleet Satellite Communications System would improve control for Navy ships and Air Force strategic aircraft.

    A second military use of space provided data for weather analysis and forecasting. McLucas had announced in March that information from an advanced meteorological satellite program was being made available to the public through the National Oceanic and Atmospheric Administration. Satellites also provided early warning of missile attack, detection of nuclear explosions in the atmosphere, and navigation data. A single satellite system to meet all requirements was under study. (*Av Wk*, 2/18/74)

- Sen. Mike Gravel (D–Alaska) introduced S. 2806, the Energy, Revenue and Development Act of 1973, to establish an energy trust fund funded by a tax on energy sources, to establish a Federal Energy Administration, to provide for the development of domestic sources of energy, and for more efficient use of energy. (*CR*, 12/13/73, S22725–52)
- Reference to the U.S. space program was made in the editorial "In Praise of America" by Canadian commentator Gordon Sinclair, reprinted by the *Washington Post*. Sinclair had said over Toronto radio: "You talk about American technocracy, and you find men on the moon—not once but several times—and safely home again." The editorial defended the U.S. against foreign critics. (*W Post*, 12/13/73, A18)
- A *Philadelphia Inquirer* editorial commented on the modesty of Dr. Lubos Kohoutek, Czechslovakian astronomer who had described his discovery as "just a comet": The Kohoutek comet "will be far more eventful than 'something normal in the night.'" Its closeness and brightness were incidental "to the more important fact that astronomers such as Lubos Kohoutek patiently search with telescopes night after night, year upon year—observing the known and looking for the unknown. Their work does not match spaceship launchings in attracting public attention but even in the space age, as for centuries past, it is the patience of astronomers, seeking they're not sure what, that continues to unlock the secrets of the universe." (*P Inq*, 12/13/73)
- Wendell S. Smith, Head of the Experiment Management Office of Goddard Space Flight Center's Laboratory for Meteorology and Earth Sciences, died in Baltimore, Md., of cancer at age 38. He had joined NASA in 1959 and had worked on early investigations of the meteorology of the upper atmosphere. In early 1973 he had headed a NASA group in a cooperative experiment with Soviet scientists to verify corresponding satellite measurements. (*W Star-News*, 12/16/73, C7)

December 14: A NASA C–45 aircraft equipped as a remote-sensing platform flew from Redstone Arsenal Airstrip, Ala., on the first flight in a series to study geological phenomena and water quality. The aircraft, obtained free from the Army and equipped by Marshall Space Flight Center with multispectral cameras and other fittings, covered three sites in Alabama with high- and low-altitude multispectral photography. The aircraft later would be used to investigate water resources, agriculture, land resources, ecology, and archaeology. It would also support the Earth Resources Technology Satellite and the Skylab earth resources experiment package. A thermal scanner would be added to the aircraft's equipment early in 1974. (MSFC Release 73-201)

- First results of Skylab earth resources experiments were reported by Skylab Program Deputy Director John H. Disher at an American Institute of Astronautics and Aeronautics Capital Section luncheon meeting in Washington, D.C.: "Perhaps the most significant of these so far in terms of economic value is the identification of a potential mineral deposit some 12 miles [19 kilometers] north of Ely, Nevada, which was not previously discernible from ground or aircraft surveys. Other early results have been the identification of the path of citrus fruit fly infestations in Mexico, successful use of remote sensing for snow mapping, and demonstration of the use of imagery from orbit for inventory of vegetation patterns in California. The high spatial resolution provided by the Skylab film systems is particularly applicable to studies of regional planning and land use in urbanized areas, and photographs of the Baltimore-Washington area and 12 other cities are being used by the Department of the Interior to test their use for updating the 1970 census. Other examples of high spatial resolution such as the identification of fields of an acre or smaller and the detection of impounded water have further indicated the need for a selective, high resolution capability for future earth survey systems." (Text)

- Sen. Barry M. Goldwater (R–Ariz.) received the Wright Brothers Memorial Trophy [see During October] at a Washington, D.C., dinner commemorating the 70th anniversary of the Wright brothers' first aircraft flight, made Dec. 17, 1903. Honored guests included Prof. Willy Messerschmidt, German aircraft designer, and 85-year-old Grover Loening who had flown with Orville Wright in 1913 and 1914. (McCardle, *W Post*, 12/15/73, D1)

- The theory that comets might be growing from material dispersed through the solar system as quickly as they were being destroyed by the sun was advanced in *Nature* by astrophysicists Dr. Hannes O. Alfvén and Dr. Asoka Mendis of the Univ. of California at San Diego. The predominant view, that comets like Kohoutek that approached the sun for the first time had probably remained in the outer fringes of the solar system since formation of the sun and the planets, was offered by Dr. L. Biermann of the Max Planck Institute for Physics and Astrophysics in Munich. Many astronomers hoped the Dec. 28 closest approach of Kohoutek to the sun would indicate which theory was correct, Walter Sullivan said later in the *New York Times*. (*Nature*, 12/14/73, 400–402, 410–11; *NYT*, 12/24/73)

- Lockheed Aircraft Corp. announced it was in danger of running out of cash in 1974 and might have to seek new short-term credits against already built trijet aircraft on which airline customers had obtained agreements to delay delivery. Lockheed said it had retained an investment-banking firm, Lazard Freres & Co., to help find a solution to its financial difficulties, including a possible merger with a healthier company. Lockheed disclosed that it had already drawn $20 million more than the $250 million in Government-guaranteed loans granted to it in 1971 to save the firm from bankruptcy. (Witkin, *NYT*, 12/15/73, 1)

December 15–20: NASA launched *Explorer 51* (AE–C) Atmosphere Explorer on a two-stage, thrust-augmented Thor-Delta launch vehicle from Western Test Range at 10:18 pm PST (1:18 am Dec. 16 EST). The satellite entered an almost perfect elliptical orbit with a 4304-km (2674.4-mi) apogee, 157-km (97.6-mi) perigee, 132.5-min period, and 68.1° inclination. *Explorer 51*, third of five AE missions and first of second-generation AE spacecraft designed to extend the knowledge of energy absorption and photochemical processes of the atmosphere, would obtain data relating solar ultraviolet activity to atmospheric composition in the lower atmosphere, extending measurements down to altitudes where most solar extreme UV radiation was absorbed by the upper atmosphere. Onboard propulsion would vary perigee and apogee altitudes on command, permitting excursions into the largely unexplored region of the lower thermosphere.

By Dec. 19 command and telemetry systems were checked out, the memory of the onboard computer was loaded and read out, and the tape recorder, atmosphere density accelerometer, and cylindrical electrostatic probe were switched on. All spacecraft systems were checked out by Dec. 20 and were functioning normally. Experiment instrumentation checkout would be completed by Dec. 31.

The *Explorer 51* mission would consist of two phases. During Phase 1, excursions would be made at two- to four-week intervals, each lasting several days. The spacecraft perigee would be lowered in steps through a range of 150 to 125 km (95 to 80 mi). At each perigee step, telemetry data would be acquired. Phase 2 would begin after eight months and last four months; the spacecraft would be placed in a series of circular orbits ranging between 250 and 800 km (155 and 500 mi) in altitude.

Explorer 51 carried 14 scientific instruments provided by a research team of scientists from nine institutions. In addition to the investigation of solar UV activity, the instrumentation would examine particle fluxes, airglow intensities, plasma densities, and temperatures and magnetic fields at low altitudes.

To use the full spacecraft and instrument capabilities, a new rapid-response data system would provide data reduced to physical parameters, allowing the spacecraft to operate as a laboratory-in-space periodically reprogrammed to make optimum measurements in the light of interim results rather than gathering data in predetermined modes.

Explorer 17 (launched April 2, 1963), the first AE satellite, had confirmed that the earth was surrounded by a belt of neutral helium. The second, *Explorer 32* (launched May 25, 1966), was placed in a higher orbit than planned but the higher altitude enhanced the resolution of the measured atmospheric parameters. The spacecraft continued to return useful data for 10 mos.

The AE program was managed by Goddard Space Flight Center under the direction of the NASA Office of Space Science. (NASA prog off; NASA Release 73-269)

December 15: President Nixon issued Proclamation 4257 calling for nationwide observance of Wright Brothers Day Dec. 17: "It is both fitting and proper . . . , on this 70th anniversary of powered flight, that we should commemorate the achievements of two resourceful and farsighted men who have come to symbolize America's inventive genius." (*PD*, 12/17/73, 1462)

- More than 100 Congressmen had endorsed legislation to start a $50-million, five-year program to find practical ways of harnessing solar energy to heat and cool U.S. buildings, according to an Associated Press report quoted in *NASA Activities*. Several senators planned to introduce identical and similar versions of the bill. House hearings would start in two weeks, and sponsors hoped Congress would pass the bill by early 1974. (*NASA Activities*, 12/15/73, 237)

- President Nixon signed H.R. 11324 into law, placing the U.S. on daylight saving time for approximately two years beginning Jan. 6, 1974. The action was taken to alleviate the energy crisis. (*PD*, 12/17/73, 1462)

December 17: The U.S.S.R. launched *Cosmos 616* from Plesetsk into orbit with a 326-km (202.6-mi) apogee, 206-km (128-mi) perigee, 89.8-min period, and 72.9° inclination. The satellite reentered Dec. 28. (GSFC SSR, 12/31/73; SBD, 1/2/74, 2)

- The first scientific report on the findings of the Soviet remotely controlled lunar explorer *Lunokhod 2* (launched Jan. 8 aboard *Luna 21*) during its 37-km (23-mi) traverse of the moon's surface had been published in *Pravda*, the *New York Times* reported. Chemical analysis of the soil had shown a decrease in iron content and an increase in aluminum as Lunokhod had traveled from the plains into the foothills of the Taurus Mountains. Samplings taken near the landing site in the Lemonnier Crater at the eastern end of the Sea of Serenity had shown 6% iron and 9% aluminum. The ratio changed to 4% iron and 12% aluminum near the uplands. The change in content of aluminum was a distinctive feature of lunar geology; on the earth the amount of aluminum remained constant in continents and the ocean bottoms. Soviet scientists hoped to compare the lunar soil data with samples returned by the *Apollo 17* astronauts (launched Dec. 7, 1972), who explored the Taurus-Littrow Valley 225 km (140 mi) away. (Shabad, *NYT*, 12/17/73, 66)

- Sen. Hubert H. Humphrey (D-Minn.) introduced S. 2819 to establish an Office of Solar Energy Research within the Atomic Energy Commission to conduct research and development to ensure the use of solar energy for national energy needs. (*CR*, 12/17/73, S23017–24)

- Dr. Charles Greeley Abbot, scientist, inventor, and fifth Secretary of the Smithsonian Institution, died in Washington, D.C., at the age of 101. He had been honored on his 100th birthday in 1972 by having a site on the moon named after him by the International Geophysical Union. Dr. Abbot, an astrophysicist, had specialized in solar energy. In 1938 he had patented a solar engine that focused sunlight, using mirrors and sapphire lenses, into a steel cylinder where superheated air performed the same function as expanding steam in a locomotive or turbine. He refined the engine in a later patent, but never found anyone to build a prototype. At his 100th birthday celebration he said, "The sun will be there long after we need it. I wish I had gotten to work on this sooner." (Hailey, *W Post*, 12/18/73, B11; *A&A 1972*)

December 18–26: The U.S.S.R. launched *Soyuz 13* carrying cosmonauts Maj. Pyotr Klimuk (commander) and Valentin Lebedev (flight engineer) from Baykonūr Cosmodrome at 4:55 pm local time (2:55 pm Moscow time; 6:55 am EST) into earth orbit with a 246-km (152.9-mi) apogee, 188-km (116.8-mi) perigee, 88.9-min period, and 51.6° inclination. With *Skylab 4* astronauts in earth orbit since Nov. 16, the launch of *Soyuz 13* made it the first time in the history of space travel that the U.S. and U.S.S.R. had men orbiting the earth simultaneously.

The Soviet news agency Tass announced the objectives of the mission were to observe stars in the ultraviolet range using a special system of telescopes, survey separate sections of the earth's surface and obtain data "for the accomplishment of economic tasks," continue comprehensive verification of onboard systems, and test manual and automatic controls and methods of autonomous navigation in various flight conditions. Western observers also expected the cosmonauts to study Comet Kohoutek as it approached the sun. The Western press reported that *Soyuz 13* was the same kind of modified Soyuz capsule as would be used for the U.S.–U.S.S.R. Apollo-Soyuz Test Project scheduled for 1975. It would test a new navigational system that would increase the cosmonauts' ability to fly the spacecraft themselves rather than relying on ground control, ensuring that the quality of the spacecraft support system would match that of the Apollo.

During the fifth orbit the spacecraft moved into a higher orbit with parameters reported by Tass as 272-km (169-mi) apogee, 225-km (139.8-mi) perigee, 89.2-min period, and 51.6° inclination. At 5:30 pm Moscow time (9:30 am EST) *Soyuz 13* had made 18 orbits of earth.

The cosmonauts reported they had adjusted to weightlessness and were carrying out their scheduled scientific program.

Tass reported Dec. 20 that the crew had harvested the first crop of nutritive protein mass in an onboard experimental biological system, Oasis 2. In a closed system of two kinds of microorganisms, the waste products from the synthesis of the first were used by the second to accumulate protein mass. During the eight-day mission, the cosmonauts also aimed the telescope at preset areas of the sky, photographing constellations; studied cerebral circulation under weightless conditions; studied natural resources on the earth; took x-ray photos of the sun; and photographed light effects before sunset and sunrise along the line of the terrestrial horizon to further development of autonomous navigation for future interplanetary journeys.

Soyuz 13 landed successfully in a snowstorm and cyclonic winds at 11:50 am Moscow time (3:50 am EST) Dec. 26, 200 km (125 mi) southwest of Karaganda in the Kazakhstan Steppe. A medical examination indicated the cosmonauts were in good health.

Soyuz 13 was the second Soviet manned launch in 1973. *Soyuz 12*, during a successful two-day mission Sept. 27–29, had tested improved flight systems, tested manual and automatic control in various flight conditions, and obtained spectrographic data of separate sections of the earth. *Soyuz 12* had been the first Soviet manned flight since the death of three cosmonauts on *Soyuz 11* in June 1971. (GSFC *SSR*, 12/31/73; Tass, FBIS–Sov, 12/21/73, U1; 12/26/73, U1; 12/27/73, U1; Kaiser, *W Post*, 12/19/73, A26; Parks, B *Sun*, 12/19/73, A1; Wren, *NYT*, 12/21/73, 56; *SBD*, 12/21/73, 267)

December 18: Dr. Kurt H. Debus, Kennedy Space Center Director, announced that KSC would go on a single daylight shift from 8 am to 4:30 pm, effective Jan. 6, 1974, when the U.S. returned to daylight saving time by Presidential order. Other KSC measures to conserve energy were establishing a computer system to promote car pooling, limiting automobile speeds to 50 mph, curtailing drive-through tours, reducing lighting, and rescheduling visitor tours. (KSC Release 284–73)

December 19: The U.S.S.R. launched eight Cosmos satellites on a single booster from Plesetsk. Orbital parameters were:

Cosmos 617—1485-km (922.7-mi) apogee, 1336-km (830.2-mi) perigee, 114.0-min period, and 74.0° inclination.

Cosmos 618—1488-km (924.6-mi) apogee, 1445-km (897.9-mi) perigee, 115.2-min period, and 74.0° inclination.

Cosmos 619—1490-km (925.8-mi) apogee, 1424-km (884.8-mi) perigee, 115.0-min period, and 74.0° inclination.

Cosmos 620—1496-km (929.6-mi) apogee, 1460-km (907.2-mi) perigee, 115.4-min period, and 74.0° inclination.

Cosmos 621—1487-km (924-mi) apogee, 1407-km (874.3-mi) perigee, 114.8-min period, and 74.0° inclination.

Cosmos 622—1487-km (924-mi) apogee, 1371-km (851.9-mi) perigee, 114.3-min period, and 74.0° inclination.

Cosmos 623—1487-km (924-mi) apogee, 1389-km (863.1-mi) perigee, 114.5-min period, and 74.0° inclination.

Cosmos 624—1487-km (924-mi) apogee, 1353-km (840.7-mi) perigee, 114.2-min period, and 74.0° inclination.

The Western press reported that the satellite system would operate in support of Soviet naval forces, providing maximum communications between ship and fleet commanders and naval operations within the U.S.S.R. (GSFC SSR, 12/31/73; Sov Aero, 12/24/73, 96)

- NASA announced Kennedy Space Center had awarded a $14 670 525 extension to a contract with Rockwell International Corp. in support of Apollo-Soyuz Test Project operations. The extension brought total contract value to $62 885 682. It provided support in processing the command and service modules, spacecraft lunar module adapter, launch escape system, and docking module for the July 1975 U.S.-U.S.S.R. rendezvous and docking mission. (NASA Release 73-283)

- United Aircraft Corp. Pratt & Whitney Aircraft Div. announced a $42-million cooperative program with nine electric utility companies to develop a 26 000-kw fuel cell as a new way to generate electricity. The utility companies had provisionally ordered 56 cells for possible delivery by 1978 and had options on 112 more. Orders totaled more than $250 million. The program did not include Government support, but Pratt & Whitney was discussing the program with Federal energy officials and "would welcome government participation," Pratt & Whitney President Bruce N. Torell told a New York press conference. A fuel cell produced electrical power directly from fuel through a chemical reaction. It could be an efficient system for generating electrical power with little pollution. Pratt & Whitney had produced small cells for NASA's Apollo and Skylab spacecraft and for the space shuttle. It had also tested larger units for home use. (WSJ, 12/20/73, 24; P&W PIO)

- The House Committee on Science and Astronautics unanimously approved H.R. 11864, Solar Heating and Cooling Demonstration Act of 1973, introduced Dec. 10 as a clean bill incorporating amendments to H.R. 10952 [see Oct. 16] and 17 similar bills. Language changes more explicitly delineated agency responsibilities and two distinct program phases. NASA would have major responsibility for developing solar heating and cooling equipment in the first phase. In the second, the Dept. of Housing and Urban Development would be responsible for installation, testing, and evaluation. A major amendment would require HUD to establish a Solar Heating and Cooling Information Data Bank. If a law created a permanent Federal energy agency, responsibility assigned to NASA and the National Science Foundation would be transferred to that agency. (House Rpt 93-769, reprint in Com Print, 2/74)

- The Senate, by a vote of 86 to 2, and the House, by 355 to 25, passed bills to establish the Federal Energy Administration. The Senate bill was S. 2776, and the House, H.R. 11510. (CR, 12/19/73, D1461, D1464)

- NASA launched an Aerobee 200A sounding rocket from White Sands Missile Range carrying a Lockheed Missiles & Space Co. solar physics experiment to a 220.3-km (136.9-mi) altitude. The experiment was performed in collaboration with a *Skylab 4* (launched Nov. 16) Apollo Telescope Mount experiment. The rocket performed satisfactorily. The experiment was 75% successful. (GSFC proj off)

December 20: The Air Force announced it had awarded Catalytic, Inc., a $2 600 000 firm-fixed-price contract for technical studies and investigations and preliminary design of a high-mach-number wind tunnel. The contract contained an option for Phase 2 completion of the total design. (DOD Release 610-73)

December 21: The U.S.S.R. launched *Cosmos 625* from Plesetsk into orbit with a 340-km (211.3-mi) apogee, 187-km (116.2-mi) perigee, 89.6-min period, and 72.8° inclination. The satellite reentered Jan. 3, 1974. (GSFC SSR, 12/31/73; 1/31/74; SBD, 1/2/74, 2)

- Establishment of a Spacelab Program Office at Marshall Space Flight Center to manage NASA's activities in the international project was announced by Dr. Rocco A. Petrone, MSFC Director. Thomas J. Lee had been appointed Spacelab Program Manager. (MSFC Release 73-202)
- Dr. Charles A. Berry, NASA Director of Life Sciences, would retire from NASA to become President of the Univ. of Texas Health Science Center April 1, 1974, NASA announced. The new position had been created to bring all the health-oriented schools and services of the University under one administrator. Dr. Berry, an Air Force flight surgeon since 1951, had been Chief of the Dept. of Aviation Medicine at the School of Aerospace Medicine and Chief of Flight Medicine in the Surgeon General's Office before coming to NASA on loan in July 1962. He accepted appointment as Chief of the Manned Spacecraft Center's Medical Operations Office in August 1963. Dr. Berry had participated in the early medical evaluations leading to selection of the original seven Mercury astronauts in 1958 and had continued to work with NASA throughout the project. He was appointed Director for Life Sciences at NASA Hq. in September 1971. His honors included the NASA Exceptional Service Medal and the NASA Distinguished Service Medal. Dr. James C. Fletcher, NASA Administrator, said Dr. Berry's "work in determining the ability of man to function and work in space for long periods . . . is an important contribution to the future of the space program. The University of Texas is to be congratulated in obtaining the services of an outstanding medical researcher and administrator." (NASA Release 73-284; NASA PAO)
- The Federal Aviation Administration announced it had adopted fuel-saving regulations permitting airlines to use flight simulators more in pilot training and flight checking. The new rules could save 200 000 cu m (50 million gal) of fuel annually and reduce flights by 9300. (FAA Release 73-215)

December 23: Dr. Gerald P. Kuiper, astronomer and space pioneer, died of a heart attack at the age of 68, while visiting Mexico City. During his early career, Dutch-born Kuiper had discovered groups of dwarf stars, satellites of Uranus and Neptune, and an atmosphere surrounding Saturn's largest moon, Titan. In the 1950s he became the first astronomer to measure Pluto's diameter accurately and charted the equator and position of the poles of Venus. In the 1960s Dr. Kuiper became one of the foremost authorities on the moon, working in NASA's Ranger and Surveyor projects and, as head of the Ranger scientific team, directed the photographic analysis to select landing sites for the Apollo astronauts. (McFadden, *NYT*, 12/25/73, 24)

December 24: Fifth anniversary of the Christmas Eve reading of Genesis by the moon-orbiting crew of *Apollo 8* (launched Dec. 21, 1968) to TV and radio audiences that included one of every four persons on earth. The crew of *Skylab 4* (launched Nov. 16 to man the Orbital Workshop launched May 14), second crew to spend Christmas in space, devoted Christmas Eve to searching for and opening gifts hidden for them in the Apollo command module before launch by colleagues in Mission Control. The packages included a 91-cm (36-in) green cloth Christmas tree beneath the gear they would need for their scheduled Christmas Day space walk and a surprise package stored in the CM locker. Mission Control also planned to transmit over a teleprint circuit a Christmas card depicting the sun, Skylab, gifts, and Santa's sled. (*A&A 1968;* UPI, *W Star-News,* 12/24/73, A1)

- *Skylab 4* Astronauts Gerald P. Carr, Dr. Edward G. Gibson, and William R. Pogue—in the 39th day of their mission aboard the Orbital Workshop [see Nov. 16–Feb. 8, 1974]—were participating in new tests developed by NASA scientists and heart specialists at the National Heart and Lung Institute to determine the effects of prolonged weightlessness on the heart, the *New York Times* reported. Space program doctors wanted data to help explain the smaller heart size, sudden drops in blood pressure when shifting quickly from a prone to upright position, and the smaller output of blood during exercise experienced by astronauts immediately after space flight. The new technique, called echo-cardiography, used sonarlike sound waves of extremely high frequency to produce internal pictures of the heart. The *Skylab 4* crew had been tested using echo-cardiography for a week preceding launch. Repeat studies would be made aboard the recovery ship within hours after splashdown. The astronauts would be studied over the course of weeks and possible months to record the complete readaptation of the heart to gravity. (Schmeck, *NYT,* 12/24/73, 1)

December 25: The U.S.S.R. launched *Molniya II–8* communications satellite into orbit with a 40 809-km (25 357.5-mi) apogee, 488-km (303.2-mi) perigee, 12-hr 17-min period, and 62.8° inclination. The satellite would help provide a system of long-range telephone and telegraph radio communications in the U.S.S.R. and would transmit Soviet central TV programs to the Orbita network. (GSFC *SSR,* 12/31/73; Tass, FBIS–Sov, 12/26/73, U1)

- U.S. Patent No. 3 781 647 was awarded to Peter E. Glaser, Vice President for Engineering Sciences of Arthur D. Little Co., for a system to produce electricity with satellite solar power stations. The proposal was to orbit satellites at 37 500-km (23 300-mi) altitude in geostationary position to convert radiation received by solar cells to direct current and then to microwaves for transmission to an earth station. There the microwave beam would be converted to electric power. The patent said a belt of solar cells 5 km [3 mi] wide around the earth could provide more than 200 times the projected world electrical energy requirements for 1980. (Jones, *NYT,* 12/29/73; Pat Off PIO)

December 26: The U.S.S.R. launched *Oreol 2* from Plesetsk into orbit with a 1974-km (1226.6-mi) apogee, 399-km (247.9-mi) perigee, 109.1-min period, and 73.9° inclination. The satellite, a Soviet-French cooperative mission, would continue the program begun by *Oreol 1* (launched Dec. 27, 1971) to study the upper atmosphere at high latitudes and the nature of the polar lights. (GSFC *SSR,* 12/31/73; Tass, FBIS–Sov, 12/27/73, U1; UN Registry)

- The retirement, effective Dec. 31, of Dr. Homer E. Newell, NASA Associate Administrator, and Vincent L. Johnson, Deputy Associate Administrator for Space Science, was announced by Dr. James C. Fletcher, NASA Administrator.

  Dr. Newell had transferred to NASA in October 1958 from the Naval Research Laboratory, where he had been Science Program Coordinator for Project Vanguard, the Nation's first satellite program, and Acting Superintendent of the Atmosphere and Astrophysics Div. He served four years as NASA Associate Administrator for Space Science and Applications before becoming Associate Administrator in 1967. Dr. Newell held the American Institute of Aeronautics' Pendray Award, the American Astronautical Society's Space Flight Award, the President's Award for Distinguished Federal Civilian Service, and the NASA Distinguished Service Medal, among other honors.

Dr. Fletcher said: "Dr. Newell's efforts in the space program antedate the birth of NASA. He has served this agency with distinction and dedication and he will be sorely missed."

Johnson, who had joined NASA from the Navy in 1960, had served as Deputy Associate Administrator of the Office of Space Science and Applications from 1970. He had assumed his present post in 1971. "Mr. Johnson has been an able and conscientious leader in advancing our unmanned space programs and we are sorry to see him go," Dr. Fletcher said. (NASA Ann; NASA Releases 74-4, 74-5)

- Mobile launcher 3, used for five manned Apollo/Saturn V launches, was being deactivated at Kennedy Space Center. A task team of NASA and contractor employees, organized by the KSC Support Operations Directorate, was removing and storing reusable equipment from the platform that was two stories high, 49 m (160 ft) long, 41 m (135 ft) wide. Nine swing arms had already been removed; other items to go included an acceptance checkout equipment room; a computer station; racks of electronic, timing and communications gear; and instrument cables. Cable salvage alone was expected to cover half the deactivation cost. With completion of deactivation in mid-1974, equipment worth millions would have been salvaged for future use in space shuttle and other programs. The basic launcher structure would be modified for use in space shuttle launches. (KSC Release 288-73)

- A Univ. of Alaska Geophysical Institute team headed by Dr. Gerd Windler was examining potential of high-resolution satellite images in providing information on Alaska's environmental conditions and hazards in a one-year pilot project of the National Oceanic and Atmospheric Administration, NOAA announced. The project was using radiometer images received twice daily at NOAA's Gilmore Creek Command Data and Acquisition Center near Fairbanks from *Noaa 2* and *3* satellites launched for NOAA by NASA Oct. 15, 1972, and Nov. 6, 1973. (NOAA Release 73-243)

December 27: The U.S.S.R. launched *Cosmos 626* from Baykonur Cosmodrome into orbit with a 264-km (164-mi) apogee, 250-km (155.3-mi) perigee, 89.6-min period, and 65.0° inclination. (GSFC SSR, 12/31/73; SBD, 1/3/74, 10)

- Comet Kohoutek would not be as spectacular as expected, Dr. Stephan P. Maran, director of Goddard Space Flight Center's Operation Kohoutek said at a Johnson Space Center news briefing. Dr. Sdeneka Sekanina, Smithsonian Observatory scientist, said that although the comet might be a disappointment to the general public, it would be the first comet to be studied during its first close pass of the sun and therefore, scientifically, "the most important comet we've had since Halley's." The comet would be difficult to see by naked eye from ground because it would be low on the horizon and because of light pollution from the earth and lack of light-reflecting dust. (Transcript)

- The approach of Comet Kohoutek had taken on special significance in conservative religious circles, the New York Times News Service reported. Its Christmas season timing had revived speculation about the nature of the star reported to have led the Wise Men to Bethlehem. One theory had been that it was a comet similar to Kohoutek. Some Protestant fundamentalists had ascribed religious significance to Kohoutek and saw it as a possible sign of the second coming of Christ. Others interpreted it as a warning of impending doom. (B *Sun*, 12/27/73)

December 28: Comet Kohoutek passed perihelion, 21 million km (13 million mi) from the sun, at 5:24 am EST, completing the sunward swing of its estimated 75 000-yr voyage, and began its journey back into distant space. (AP, B *Sun*, 12/28/73, A7)

- NASA released details of Langley Research Center's planned program to heat and cool a 5000-sq-m (53 000-sq-ft) Systems Engineering Building by solar energy. The program had been announced by Dr. James C. Fletcher, NASA Administrator, in Nov. 13 testimony before the House Committee on Science and Astronautics' Subcommittee on Energy. The program was to obtain realistic engineering and operating experience with a source of clean power as an alternate to fossil fuels and other conventional energy sources. Planned for mid-1975 completion, the building—believed to be first of its size for which solar energy would provide a significant part of its heating and cooling system—would use a 1400-sq-m (15 000-sq-ft) solar collector to provide most of the heating and some of the cooling requirements. It also would test energy-storage capacity. NASA believed the solar collector to be the pacing component in the system and had established heating and cooling cost goals of $1 to $2 per 0.09 sq m (1 sq ft) over a 15-yr life span. (NASA Release 73–282)

- Dr. Robert R. Gilruth, NASA Director of Key Personnel and former Director of Manned Spacecraft Center, retired. With NASA since its beginning, Dr. Gilruth had been Director of MSC (now Johnson Space Center) from its establishment in November 1961—and through five moon landings of the Apollo program—until he came to Headquarters in January 1972. Earlier he had been at the Langley Memorial Aeronautical Laboratory, predecessor of Langley Research Center, and was appointed Assistant Director in 1952. In 1958 he was assigned to manage the Nation's first manned space program, NASA's Project Mercury, as Director of the Space Task Group at Langley. (NASA Personnel Off; NASA Biog; NASA Release 72–11)

- Kennedy Space Center announced a $10 517 967 contract extension, through July 31, 1975, to McDonnell Douglas Astronautics Corp. to process the Saturn IB 2nd stage for launching the Apollo spacecraft to rendezvous and dock with the Soviet Soyuz spacecraft during the July 1975 Apollo-Soyuz Test Project mission. The award brought total contract value to $101 995 628.

  A six-month, $7 909 148 contract extension to Bendix Corp. Launch Support Div. brought to $240 857 165 the total value of a contract for launch support services, including operation and maintenance of Launch Complex 39. An $11 831 263 extension to an International Business Machines Corp. contract would provide launch support services to Saturn IB booster instrument units, for possible Skylab rescue vehicles and the ASTP launch. The extension, through July 1975, brought total contract value to $58 772 009. (KSC Releases 290–73, 291–73, 292–73)

- Dr. James R. Schlesinger, Secretary of Defense, had directed the Air Force to plan safe demonstration launches of four Minuteman II missiles without warheads from operational silos in Montana during the winter of 1974–75, the Dept. of Defense announced. In Project Giant Patriot, the missiles would enter space shortly after launch and their trajectory would carry them up to 560 km (350 mi) over portions of Montana, Idaho, Washington, Oregon, and California en route to an ocean target in the Phoenix Islands, southwest of the Hawaiian Islands. No nuclear components would be carried; a safety destruction package would be carried in place of the normal warhead. The launches were to provide important test data and demonstrate the effectiveness and reliability of the Minuteman strategic deterrent force. (DOD Release 615–73)

December 29: The U.S.S.R. launched *Cosmos 627* from Plesetsk into orbit with a 1019-km (633.2-mi) apogee, 973-km (604.6-mi) perigee, 105.0-min period, and 83.0° inclination. (GSFC *SSR*, 12/31/73; *SBD*, 1/3/74, 10)

December 30: U.S.S.R. Chief Marshal of Aviation Konstantin A. Vershinin died at the age of 74 after a long illness. Vershinin had directed Soviet military aviation during World War II. After the war he had commanded the antiaircraft defense forces and had been Chief of the Air Force and U.S.S.R. Deputy Minister of Defense. (Tass, FBIS–Sov, 1/4/74, V1; *NYT*, 1/3/74, 36)

During December: NASA employee retirements totaled 215 during the month, including 37 at Headquarters, 58 at Marshall Space Flight Center, 50 at Langley Research Center, 22 each at Goddard Space Flight Center and Johnson Space Center, 18 at Kennedy Space Center, and 8 at Lewis Research Center.

Among retirees were Dr. Homer E. Newell [see Dec. 26], NASA Associate Administrator; Dr. Robert R. Gilruth [see Dec. 28], NASA Director of Key Personnel Development; Vincent L. Johnson [see Dec. 26], Deputy Associate Administrator for Space Sciences; Dr. Charles A. Berry [see Dec. 21], NASA Director of Life Sciences; General Counsel Arthur D. Holzmann; William T. O'Bryant, Director of Lunar Programs in the Office of Space Science; C. Dixon Ashworth, Manager of the Astronomy and Solar Observatory Programs; Director Adelbert O. Tischler of the Low Cost Systems Office; and Dr. Hermann H. Kurzweg, Research Council Chief in the Office of Science and Technology. (NASA Off Personnel; NASA Off Comptroller)

- Space program objectives would be served by broadening NASA's charter, Daniel J. Fink—Vice President of General Electric Co., General Manager of GE Space Div., and 1974 President of the American Institute of Aeronautics and Astronautics—said in an *Astronautics & Aeronautics* article. By 1973 the U.S. "had attained or surpassed all the objectives set forth" in the Space Act of 1958; now NASA's emphasis should shift from space techniques to "mission management and the science and applications technologies, allowing industry to assume fuller responsibility for subsystem design and development." Congress also should charter NASA "to design and manage a broad-based, *goal*-oriented science program" to maximize the return from the achieved capability in lunar, planetary, and astronomical exploration.

  NASA would best fill its role if given "clearly defined responsibilities in nurturing space applications until they reach the most efficient handover point; if it is directed toward system integration and management rather than subsystem engineering and design; if it becomes the executive agency for an integrated space-science program; if it maintains its role as the advanced space-systems application-technology house for all users; and if it remains the active working partner of the aerospace industry in all programs, domestic or international, cooperative or competitive." (*A&A*, 12/73, 16–19)

- The present total electric power demands of the U.S. could be supplied by solar energy plants with a 2000-km (1240-mi) total area, assuming 30% efficiency, Associate Director Walter E. Morrow, Jr., of the Massachusetts Institute of Technology's Lincoln Laboratory said in a *Technology Review* article. "This is about 0.03 per cent of the U.S. land area devoted to farming and about 2 per cent of the land area devoted to roads; and it is about equal to the roof area of all the buildings in the U.S." (*Tech Rev*, 12/73, 31–43)

Summary

During 1973: The U.S. put 26 payloads into orbit in 23 launches and the U.S.S.R. orbited 107 payloads in 86 launches. The U.S. total included 13 spacecraft orbited by the Dept. of Defense in 10 launches (2 of them multiple launches) and 13 by NASA. NASA's 13 successful launches out of 14 attempts nearly matched the agency's 1972 record of 17 flawless launches out of 17 attempts.

NASA's Skylab program successfully launched the first U.S. experimental space station—the Orbital Workshop—and three successive crews to man it. *Skylab 4*, the final mission and the longest-duration manned space flight to date (84 days 1 hr 16 min), also set records for man's cumulative time in space (84 days), duration of extravehicular activity (7 hrs 1 min), cumulative orbital EVA time for one mission (22 hrs 21 min), and distance in orbit for a manned mission (55.5 million km; 34.5 million mi). By the year's end the Workshop had circled the earth more than 3350 times and housed nine astronauts. The *Skylab 2* crew's repair in space of the *Skylab 1* Workshop, after aerodynamic forces damaged it during liftoff, salvaged the endangered program and proved man could do difficult construction work in space. The *Skylab 2* crew spent 28 days in space and the *Skylab 3* crew 59 days, each mission a record at the time. The missions proved man could live and work in space for extended periods; expanded solar astronomy beyond earth-based observations, collecting much valuable new data that might revise understanding of the sun and its effects on the earth; and improved techniques for surveying earth resources from space, returning more information than planned.

In NASA's unmanned program, *Pioneer 10* (launched in 1972) swept past Jupiter in December 1973 at a distance of 130 000 km (81 000 mi). The first spacecraft to fly beyond the orbit of Mars and to penetrate the Asteroid Belt capped its two-year, 826-million-km (514-million-mi) journey by returning more than 300 closeup photos of Jupiter and its inner moons and providing new information on the planet's atmosphere and magnetic field. By the year's end *Pioneer 10* was beginning its five-year extended mission to reach Saturn's orbit in 1976 and the orbit of Uranus, the limit of spacecraft communications with the earth, in 1979. Thereafter it was to become the first man-made object to escape the solar system. *Pioneer 11*, launched in April, had passed safely through three fourths of the Asteroid Belt on its way to its December 1974 encounter with Jupiter. *Mariner 10*, headed toward Mercury via Venus, was more than 14.5 million km (9 million mi) from the earth; it would pass by Venus in February 1974 and would provide the first closeup view of Mercury in March.

NASA also selected landing sites for the Viking project's landing of two spacecraft on Mars in 1976. Orbiter and lander critical design reviews, spacecraft thermal and structural tests, and other hardware tests were completed during the year.

Scientific satellites launched in 1973 included *Explorer 49* in lunar orbit to measure galactic and solar radio noise using occultation of the moon to reduce background interference, *Explorer 50* to complete a study of the interplanetary environment over an 11-yr solar cycle, and *Explorer 51* to study the earth's outer atmosphere. As December ended, scientists were studying Comet Kohoutek with optical telescopes, radiotelescopes, and radar from the ground; high-flying aircraft; and instruments on unmanned satellites, sounding rockets, and *Skylab 4*.

Noaa 3, launched by NASA for the National Oceanic and Atmospheric Administration, would provide atmospheric soundings and very-high-resolution, day-and-night cloud-cover imaging. Two communications satellites, *Intelsat-IV F-7* for the International Telecommunications Satellite Organization and *Anik 2* Canadian domestic comsat, were successfully launched, as well as *Nnss O-20* Transit navigational satellite for the Navy.

Erts 1, launched in mid-1972 as the first Earth Resources Technology Satellite, completed its 7000th photo-taking orbit of the earth, demonstrating the great potential of remote-sensing of the earth's resources by providing data for use throughout the world.

All but one prime contractor for the space shuttle had been selected and design was proceeding on schedule. Most of the shuttle orbiter subcontractors were lined up, with fabrication to begin in FY 1975. Nine nations in the European Space Research Organization agreed to build the Spacelab for a wide variety of applications as a shuttle payload. NASA and DOD made joint system studies for the space tug and tentative agreement was reached for DOD to develop an initial tug stage for the shuttle.

The first two powered flights of the X-24B lifting body were successfully made in the NASA and Air Force hypersonic aerospacecraft research program. NASA also vigorously pursued solutions to problems of aircraft noise, pollution, safety, and operational efficiency and technology for future-generation aircraft. Flight tests of NASA's digital fly-by-wire advanced flight control system indicated the system improved handling qualities of aircraft in all flight conditions, with the reduced aircraft weight increasing payload potential and reducing fuel consumption. The JT8D engine refan program progressed to scheduling ground tests for February 1974. The engine modifications—in 727, 737, and DC-9 aircraft, models that made up 60% of the domestic commercial fleet—were expected to reduce areas affected by aircraft noise by 75%. A 67% reduction in 727 aircraft landing noise was anticipated from a two-segment landing procedure being tested in the NASA and Federal Aviation Administration noise abatement program. Automobile engines were used to demonstrate an internal combustion concept for reducing pollution emissions while increasing engine efficiency for general-aviation aircraft. New flight research programs included the remotely piloted research vehicle project in which the pilot flew scale models from a ground-based cockpit, eliminating risk to the test pilot and reducing test-flight costs more than 50%.

In NASA's sounding rocket program, more than 80 flights studied atmosphere, ionosphere, auroras and airglow, geomagnetic storms, meteor streams, and trapped radiation fluctuations; male astronomical observations in x-ray, ultraviolet, and radio regions of the electromagnetic spectrum; and supported other programs. Rocket launches provided a reference for calibration of equipment on Skylab's Apollo Telescope Mount. Forty-nine balloons were flown for scientific research and for development of new technology for long-duration orbital missions.

DOD's year included further development and implementation of satellite communications and navigation systems, development of new military aircraft, and construction of advanced medium STOL prototypes.

The U.S.S.R.'s record 86 launches topped 83 in 1971 and 74 in 1972. The 107 Soviet payloads included 85 Cosmos satellites, 2 Intercosmos, 1 Luna, 2 Meteor, 1 Prognoz, 1 cooperative Soviet-French Oreol, 2 manned Soyuz, 1 Salyut, 4 Mars probes, and 4 Molniya I and 4 Molniya II comsats. The *Luna 21* probe landed the *Lunokhod 2* self-propelled vehicle on the moon for lunar studies. *Soyuz 12* carried two cosmonauts into orbit on a two-day mission in September, the first Soviet manned flight since 1971, and *Soyuz 13* carried two cosmonauts on an eight-day mission in December. Spacecraft system tests and experiments were reported successful on both flights. (Pres Rpt 1974; NASA Release 73–281; *NASA Activities*, 1/15/74; *A&A 1973*; GSFC *SSR*, 12/31/73; GSFC Sounding Rocket Div)

- NASA's 15th anniversary year—memorable for advances in earth resources technology, aeronautics progress, expanded international cooperation in space, and increased promise for future manned and unmanned planetary exploration—boosted the U.S. space program from the spectacular orbit of Apollo lunar exploration to a steady program of extended service to science and mankind. Skylab's triumph heralded the approach of the shuttle era that would weld with the earth the vast regions of near-earth space, extending man's transportation system into regions beyond land, sea, and air. The newest Pioneer probes began penetration into the mysteries of outer planetary space.

    Budget restrictions in line with President Nixon's January $250-billion target for total Government spending slowed space shuttle manpower buildup, suspended work on the High Energy Astronomy Observatory, and required phase-out of communications satellite work and curtailment of work on nuclear propulsion. But Skylab, the space shuttle, the scheduled U.S.-U.S.S.R. 1975 Apollo-Soyuz Test Project mission, the Viking, the Mariner Jupiter-Saturn mission, and many applications and aeronautics projects were retained in a balanced and productive program.

    Dr. Homer E. Newell, NASA Associate Administrator for the past seven years and member of the NASA staff since October 1958, retired at the end of the year. Dr. Newell had been Science Program Coordinator in the Nation's first satellite program, the Naval Research Laboratory's Project Vanguard, before establishment of NASA.

    Tight funding and the near completion of the Skylab missions brought a net reduction in NASA Civil Service employees of 1646, including retirements, during the calendar year, with a total 25 598 on board Jan. 5, 1974. Headquarters employees dropped by a net change of 80, Kennedy Space Center by 117, Johnson Space Center by 110, Marshall Space Flight Center by 374, Ames Research Center by 49, Flight Research Center by 13, Langley Research Center by 62, Lewis Research Center by 494 (with the phasing down of Plum Brook Station), Goddard Space Flight Center by 289, and Wallops Station by 14. AEC–NASA Space Nuclear Systems Office was closed with the end of the NERVA program. Further reductions of 628—including 405 at MSFC and

200 at LeRC—were scheduled in 1974, to bring the FY 1974 end-of-year total to the required 24 970 ceiling.

Aerospace industry employment in the Nation increased to 946 000 in June but declined to 935 000 by the year's end.

Techniques developed for aerospace activities were applied to earthquake prediction, development of residential solar heating, and fuel-saving integrated systems for housing. NASA added 550 technical briefs to its more than 5000 technical innovations already available for U.S. industry use. More than 68 000 U.S. industrial and commercial firms requested additional data on these innovations. Some 3000 firms spent $625 000 for special assistance from NASA's Regional Dissemination Centers; 1500 firms purchased 1891 computer programs and related documentation for $221 530. NASA liberalized its patent-licensing procedures to speed technology transfer and during 1973 five exclusive patent licenses, the largest number since the program's inception, were granted. A NASA liaison office was opened in New York City to work directly with city agencies on public safety, drug detection, and fire prevention.

A rechargeable heart pacemaker was developed under NASA auspices by the Johns Hopkins Univ. Applied Physics Laboratory. Electronic components designed for space use were incorporated in the device, which could be recharged at home by the patient, eliminating the need for surgery every two years to replace conventional devices. A new radio transmitter temperature pill allowed doctors to check localized internal body temperature changes. A portable visual testing device permitted early detection of eye disorders. A test hospital room for paralyzed patients used space-developed devices for remote operation of communications systems and appliances.

The NASA FY 1974 appropriation of $3.002 billion was $13.9 million below the budget request, $62.4 million below the $3.0645-billion FY 1973 authorization, and $406 million below the FY 1973 appropriation of $3.408 billion. (*A&A 1972; A&A 1973; Aerospace*, 2/74, 6–11; NASA Off Comptroller; NASA Off of Admin; NASA SP–5119; NASA Release 73–281)

- In its program of international cooperation, NASA launched *Anik 2* domestic communications satellite for Telesat Canada and *Intelsat-IV F–7* for Communications Satellite Corp. on behalf of the International Telecommunications Satellite Organization.

    A significant milestone was the Intergovernmental Agreement on U.S. and European cooperation on Spacelab, signed by the U.S., Belgium, France, West Germany, Switzerland, and the United Kingdom. To be funded and developed by nine members of the European Space Research Organization, the Spacelab would be carried into orbit by NASA's space shuttle orbiter and would accommodate experiments and personnel on missions of 7 to 30 days.

    Notable progress was made in the Apollo-Soyuz Test Project, with the joint U.S.–U.S.S.R. rendezvous and docking mission on schedule for July 1975 launch. Working groups met in both countries and a November Moscow meeting familiarized eight U.S. astronauts with the Soviet Soyuz spacecraft. Docking system designs were essentially completed and all hardware construction was on schedule. Joint dynamic testing of test units began at Johnson Space Center. Flight crews were selected in both countries.

Eighty-three nations entered into Definitive Agreements for INTELSAT, permanently establishing the global communications satellite system envisioned in the U.S. Communications Satellite Act of 1962.

Scientists from other nations cooperated in NASA studies ranging from investigations of the solar system to practical applications of space technology. *Erts 1* and Skylab environment and resources data were used throughout the world. Scientists from 37 countries and the United Nations participated in *Erts 1* experiments and Brazil and Canada established their own ERTS data-acquisition facilities. Four foreign experiments were carried on Skylab missions and foreign scientists participated in the Apollo solar telescope program and in correlated astronomy sounding rocket programs. Scientists in 19 countries and the U.N. participated in 40 Skylab data-analysis investigations.

German and Australian scientists were coinvestigators for *Pioneer 11* experiments and French and U.K. scientists for *Mariner 10*. Investigators from 40 nations and ESRO were selected to participate in the *Apollo 17* lunar sample program. A joint working group of the U.S.S.R. Academy of Sciences and NASA held the third and fourth meetings on space biology and medicine, exchanging information on manned space flight.

Parallel communications channels using Intelsat and Molniya comsat systems were planned for 1974 to support the U.S.–U.S.S.R. Direct Communications Link, with construction of earth stations in both countries for access to each other's satellites.

Work went forward for future cooperative satellite projects. Agreement was reached with Italy to fly San Marco C–2 in 1974. Negotiations were underway with West Germany for 1974 launch of the AEROS–B aeronomy satellite and work continued on the experimental CTS comsat to be launched with Canada, the HELIOS solar probe with Germany, the INTASAT ionospheric beacon with Spain, the UK–5 stellar x-ray satellite, Canada's Anik C, and U.K.'s X4 technology spacecraft.

Negotiations were also underway on NASA launch services for the French-German Symphonie comsat in 1974 and two ESRO test comsats in 1976 and 1977. The U.S. and ESRO were about to enter agreement for an air traffic control and navigation satellite system.

During the year NASA launched 30 sounding rockets from ranges in other countries, including rockets carrying foreign payloads or cooperative experiments. New sounding rocket agreements were signed with Australia, Brazil, Germany, The Netherlands, Norway, and Sweden. (Pres Rpt 1974; NASA Release 73-281; *NASA Activities*, 1/15/74, 2-3)

- A burgeoning worldwide energy shortage led President Nixon to propose in June the establishment of a Cabinet-level Dept. of Natural Resources, an Energy Research and Development Administration, and a Nuclear Energy Commission. The shortage reached crisis proportion with the embargo by Arab nations on oil shipments to the U.S. in retaliation for U.S. support to Israel during the October Arab-Israeli war.

In a November speech President Nixon noted the U.S. was on the verge of "the most acute shortage of energy since World War II." He called for reduced use of heat, electricity, and gasoline; announced Federal Government steps to lower energy consumption; asked the Atomic Energy Commission to accelerate licensing and constructing nuclear power plants; and asked for an emergency energy act.

Nine energy bills, passed in rapid succession by the first session of the 93rd Congress, included authorizations for mandatory fuel allocation programs, construction of the trans-Alaska pipeline, a return to year-round daylight saving time, a speed limit on interstate highways, and the National Emergency Act. Soaring prices of all kinds of fuel, long lines of cars and banned Sunday sales at gas stations, and airline fare increases and schedule cutbacks marked closing months of the year.

Economists forecast a possible depression as energy shortages aggravated an already recessed economy. Scientists sought additional sources of fossil fuels and, for the future, new sources of energy to make the U.S. energy-autonomous. The focus turned to NASA and its pioneering expertise in the technologies of solar energy and development of hydrogen as a substitute for gasoline in automobiles and aircraft. At the year's end pending legislation endorsed NASA cooperation with other appropriate agencies in solving the worldwide energy shortage and in providing the U.S. with future sources of non-fossil-fuel energy. (*A&A 1973*; Library of Congress, Congr Research Service, *Major Legislation of the 93rd Congress; Newsweek*, 12/3/73)

- Aerospace industry sales increased to an estimated $24.9 billion, more than 10% above 1972's $23.5 billion. Commercial aerospace sales increased 43.1%, to $5.8 billion, reflecting deliveries of wide-bodied transports, helicopters, and general-aviation aircraft. Major aerospace sales included $13.8 billion to the Dept. of Defense, up from $13.2 billion; missile sales at $5.6 billion, up from $5.2 billion; and aircraft sales at $13.4 billion, up from $11.6 billion.

  Sales in the space area alone declined to $2.9 billion, from $3.0 billion in 1972. Executive and utility aircraft sales increased from $558 million in 1972 to $826 million—up 48%. Units delivered increased 39.1%. Civilian helicopter sales increased from $90 million in 1972 to $121 million in 1973, a gain of 48%.

  Aerospace export dollar value rose 27.6% because of two U.S. dollar devaluations, to a total of $4.9 billion; aerospace imports increased 37.5% to $777 million. Aerospace industry profits were expected to increase by 2.7% of sales after taxes. The 1973 profit rate for all manufacturing industries was expected to be 4.6%. (*Aerospace*, 2/74, 3)

- U.S. scheduled airlines flew 161 billion passenger-miles carrying more than 200 million passengers, nearly 4 million tons of freight and express, and some 16 billion letters for the U.S. Postal Service. The U.S. fleet of 2250 aircraft was composed primarily of jet aircraft, valued at over $12 billion. Air freight accounted for a record billion dollars of operating revenue. Total operating revenue was just under $12 billion, but earnings, hard hit by rising fuel costs, were less than $200 million. (*Aerospace*, 2/74, 6–7)

Appendix A

SATELLITES, SPACE PROBES, AND MANNED SPACE FLIGHTS

A CHRONICLE FOR 1973

The following tabulation was compiled from open sources by Leonard C. Bruno of the Science and Technology Division of the Library of Congress. Sources included the United Nations Public Registry; the *Satellite Situation Report* compiled by the Operations Control Center at Goddard Space Flight Center; and public information releases of the Department of Defense, NASA, NOAA, and other agencies, as well as those of the Communications Satellite Corporation. Russian data are from the U.N. Public Registry, the *Satellite Situation Report*, translations from the Tass News Agency, statements in the Soviet press, and international news services reports. Data on satellites of other foreign nations are from the U.N. Public Registry, the *Satellite Situation Report*, governmental announcements, and international news services reports.

This tabulation lists payloads that have (a) orbited; (b) as probes, ascended to at least the 6500-kilometer (4000-mile) altitude that traditionally has distinguished probes from sounding rockets, etc.; or (c) conveyed one or more human beings into space, whether orbit was attained or not. Furthermore, only flights that have succeeded—or at least can be shown by tracking data to have fulfilled our definition of satellite or probe or manned flight—are listed. Date of launch is referenced to local time at the launch site. An asterisk by the date marks dates that are one day earlier in this tabulation than in listings which are referenced to Greenwich Mean Time. A double asterisk by the date marks dates of Soviet launches which are a day later in this compilation than in listings which are referenced to Greenwich Mean Time.

World space activity increased slightly over last year. Total successful launches increased—109 against 106 in 1972—as did total payloads orbited—133 against 126 in 1972. The difference between launches and payloads is of course accounted for by the multiple-payload launches (DOD, the principal user of this system in the past, made 2 multiple launches in 1973, orbiting 5 payloads; NASA made none, for the first time since 1966; the U.S.S.R. made 3 multiple launches, orbiting 8 satellites in each).

Of the 1973 world total, the United States launched 23 boosters carrying 26 payloads. In 1972 30 U.S. boosters launched 35 payloads. The U.S. booster total for 1973 also declined about 25 percent from the previous year, and both booster and payload totals were the lowest since 1960. Of these 1973 totals, DOD was responsible for 10 launches and 13 payloads. Of NASA's 13 launches, 4 were non-NASA missions—*Anik 2* for Canada, *Intelsat*-

IV F–7 for ComSatCorp, *Noaa 3* for the National Oceanic and Atmospheric Administration, and a navigation satellite, *Nnss O–20,* for the Navy. The Soviet Union once again far exceeded United States space totals, launching 107 payloads with 86 launches—both world records. It had launched 89 payloads with 74 launches in 1972.

The year was dominated by the marathon flights of the U.S. Skylab program. Each three-man crew exceeded the records and accomplishments of the former, and the three separate flights that began in May 1973 and ended in February 1974 proved that man could live and work in space for extended periods. The final Skylab flight lasted a record 84 days and, together, nine U.S. astronauts lived in space for nearly six months.

Pioneer 10, launched in March 1972, became man's first spacecraft to reach Jupiter, passing 130 000 km (81 000 mi) from the planet on Dec. 3, 1973. A second Jupiter probe, *Pioneer 11,* was launched early in 1973 and emerged undamaged from the Asteroid Belt March 20, 1974. It was scheduled to pass Jupiter in December 1974 and to fly by Saturn in 1979. *Mariner 10,* launched late in 1973, photographed Venus during a February 1974 flyby and used that planet's gravity for a first-ever encounter with Mercury during March 1974. NASA launched three different scientific satellites: *Explorer 49* was placed in lunar orbit to investigate galactic and solar radio noise; *Explorer 50* monitored interplanetary radiation from an orbit halfway between the earth and the moon; and *Explorer 51*'s highly elliptical orbit allowed it to dip into the earth's lower atmosphere for aeronomy investigations. The Department of Defense also placed two improved communications satellites in synchronous orbit.

The Soviet Union, in its most prolific year to date, resumed its manned efforts and continued a program of scientific, interplanetary, and applications satellites—not all with success. Early in the year, its unmanned orbital laboratory *Salyut 2* broke up after 10 days in earth orbit. A similar orbital laboratory attempt a month later met with failure in orbit and was called *Cosmos 557.* The two-day *Soyuz 12* mission, carrying two cosmonauts, was the first Soviet manned flight in more than two years and demonstrated Soviet confidence in the redesigned Soyuz capsule. The December two-man flight of *Soyuz 13* lasted nearly eight days in an apparent test of the modified Soyuz craft to be used for the 1975 Apollo-Soyuz docking mission. The Soviets also sent four spacecraft toward Mars, two orbiters and two landers, but only one orbiter performed properly upon arrival in February 1974. *Luna 21* landed successfully on the moon and deposited the unmanned *Lunokhod 2,* which traveled 37 km (23 mi) over its surface. During 1973 the Soviets also launched eight communications satellites, two meteorological satellites, and *Oreol 2,* a cooperative Soviet-French scientific satellite.

This year was the first since 1969 that no third nation launched a satellite of its own.

As we have cautioned in previous years, the "Remarks" column of these appendixes is never complete, because of printing time and the inescapable lag behind each flight of the analysis and interpretation of results.

Launch Date	Name, Country, International Designation, Vehicle	Payload Data	Apogee in Kilometers (and st mi)	Perigee in Kilometers (and st mi)	Period in Minutes	Inclination in Degrees	Remarks
Jan. 8	Luna 21 (U.S.S.R.) 1973-1A Not available	Total weight: Not available. Objective: Continue scientific investigations of the moon and of circumlunar space. Payload: Spacecraft carried Lunokhod 2, 840-kg (1850-lb) self-propelled lunar vehicle, atop descent stage. Improved vehicle resembled large pot-bellied tub with 8 spoked wheels; powered by solar energy and batteries, contained 3 TV cameras, magnetometer, and French laser reflector array.	Lunar orbit, 110 (68.4)	90 (55.9)	118	60	Luna 21 entered lunar orbit 1/12/73. Softlanded inside Le Monnier Crater on eastern fringe of Sea of Serenity 5:35 pm EST 1/15 (1:35 am 1/16 Moscow time) and released Lunokhod 2. Self-propelled vehicle operated during lunar days and hibernated during lunar nights. In cooperative experiment, U.S. successfully located and bounced signal off French retroreflector array on Lunokhod 2. Vehicle ended lunar operations after 4 mos. It had traveled 37 km (23 mi) investigating lunar surface and transmitted 86 lunar panoramas and more than 80 000 TV pictures of surface.
Jan. 11	Cosmos 543 (U.S.S.R.) 1973-2A Not available	Total weight: Not available. Objective: "Continuation of scientific satellite series." Payload: Not available.	309 (192)	202 (125.5)	89.6	65	Reentered 1/24/73.
Jan. 20	Cosmos 544 (U.S.S.R.) 1973-3A Not available	Total weight: Not available. Objective: "Continuation of scientific satellite series." Payload: Not available.	547 (339.9)	510 (316.9)	95.1	74	Still in orbit.
Jan. 24	Cosmos 545 (U.S.S.R.) 1973-4A Not available	Total weight: Not available. Objective: "Continuation of scientific satellite series." Payload: Not available.	492 (305.7)	268 (166.5)	92.1	71	Reentered 7/31/73.
Jan. 26	Cosmos 546 (U.S.S.R.) 1973-5A Not available	Total weight: Not available. Objective: "Continuation of scientific satellite series." Payload: Not available.	613 (380.9)	574 (356.7)	96.5	50.6	Still in orbit.
Feb. 1	Cosmos 547 (U.S.S.R.) 1973-6A Not available	Total weight: Not available. Objective: "Continuation of scientific satellite series." Payload: Not available.	311 (193.3)	201 (124.9)	89.6	65	Reentered 2/13/73.

Launch Date	Name, Country, International Designation, Vehicle	Payload Data	Apogee in Kilometers (and st mi)	Perigee in Kilometers (and st mi)	Period in Minutes	Inclination in Degrees	Remarks
Feb. 3	Molniya I-23 (U.S.S.R.) 1973-7A Not available	Total weight: Not available. Objective: Continue operation of long-range telephone, telegraph, and radio communications system and transmission of U.S.S.R. central television programs to stations in Orbita network. Payload: Spacecraft carries apparatus for transmitting television programs and multichannel radio communications, instruments of control-and-measurement complex, and orientation system, orbit correction system, and power supplies.	39 772 (24 713.2)	578 (359.2)	717.6	65	Molniya I-23 communications satellite still in orbit.
Feb. 8	Cosmos 548 (U.S.S.R.) 1973-8A Not available	Total weight: Not available. Objective: "Continuation of Cosmos scientific satellite series." Payload: Not available.	284 (176.5)	209 (129.9)	89.4	65.4	Reentered 2/21/73.
Feb. 15	Prognoz 3 (U.S.S.R.) 1973-9A Not available	Total weight: 845 kg (1863 lbs). Objective: Explore corpuscular, gamma, and x-ray solar radiation; solar plasma flow; and magnetic fields in near-earth space to determine effects of solar activity on interplanetary medium and magnetosphere of earth. Payload: Carried 15 scientific instruments and radio transmitter with frequency of 928.4 mhz.	200 000 (124 274)	590 (367)	5783	65	Prognoz 3 (Forecast), third Soviet interplanetary monitoring platform, was launched into highly eccentric orbit. Functions as solar observatory. Still in orbit.
Feb. 28	Cosmos 549 (U.S.S.R.) 1973-10A Not available	Total weight: Not available. Objective: "Continuation of Cosmos scientific satellite series." Payload: Not available.	723 (449.3)	516 (320.6)	92.2	73.9	Still in orbit.
Mar. 1	Cosmos 550 (U.S.S.R.) 1973-11A Not available	Total weight: Not available. Objective: "Continuation of Cosmos scientific satellite series." Payload: Not available.	313 (194.5)	204 (126.8)	89.7	65.4	Reentered 3/11/73.

Date	Name	Description	Weight kg (lb)	Orbit info		Remarks
Mar. 6	Cosmos 551 (U.S.S.R.) 1973-12A Not available	Total weight: Not available. Objective: "Continuation of Cosmos scientific satellite series." Payload: Not available.	292 (181.4)	202 (125.5)	89.4	65	Reentered 3/20/73.
Mar. 6	DOD Spacecraft (United States) 1973-13A Atlas-Agena	Total weight: Not available. Objective: Develop space flight techniques and technology. Payload: Not available.	36 679 (22 791.3)	35 855 (22 279.3)	1435	0.2	Still in orbit.
Mar. 9	DOD Spacecraft (United States) 1973-14A Titan IIID	Total weight: Not available. Objective: Develop space flight techniques and technology. Payload: Not available.	263 (163.4)	152 (94.5)	88.6	95.7	Reentered 5/19/73.
Mar. 20	Meteor 14 (U.S.S.R.) 1973-15A Not available	Total weight: Not available. Objective: Acquisition of meteorological information for use by weather service. Payload: Cylindrical body with 2 large solar paddles attached; 3 data-collection systems; TV cameras, infrared sensors, and actinometric scanner; 3-axis attitude control system.	892 (554.3)	872 (541.8)	102.5	81.2	Meteor 14 meteorological satellite still in orbit.
Mar. 22	Cosmos 552 (U.S.S.R.) 1973-16A Not available	Total weight: Not available. Objective: "Continuation of Cosmos scientific satellite series." Payload: Not available.	308 (191.4)	202 (125.5)	89.5	72.8	Reentered 4/3/73.
Apr. 3	Salyut 2 (U.S.S.R.) 1973-17A Not available	Total weight: 18 427 kg (40 625 lbs). Objective: Test improved design, onboard systems, and equipment; conduct scientific and technical experiments. Payload: 20-m (65.6-ft) long, cylindrical module with 4-m (13.1-ft) maximum dia. Consists of several compartments, 3 pressurized: first is transfer compartment about 2-m (6.6-ft) dia, expanding to 3 m (9.8 ft); main habitable compartment is 4-m (13.1-ft) dia; 3rd pressurized compartment contains control and communications equipment, power supply, and life support system; 4th compartment not pressurized, about 2-m (6.6-ft) in dia, contains engine installations and asso-	260 (161.6)	215 (133.6)	89	51.6	Salyut 2, U.S.S.R.'s second orbital workshop, was launched successfully and maneuvered to higher orbit. On 4/14 malfunction separated craft's solar cell panels and other appendages. Tracking data listed 27 orbiting fragments and radio contact was lost. Spacecraft breakup apparently was caused by some form of explosion or severe tumbling produced by wildly firing thruster. Reentered 5/28/73.

ASTRONAUTICS AND AERONAUTICS, 1973

Launch Date	Name, Country, International Designation, Vehicle	Payload Data	Apogee in Kilometers (and st mi)	Perigee in Kilometers (and st mi)	Period in Minutes	Inclination in Degrees	Remarks
		ciated control equipment. Internally, about 100 cu m (3530 cu ft) of space in pressurized compartments. Also contains chemical batteries, oxygen and water reserves, and scientific instruments. Externally, 2 double sets of solar cell panels, placed at opposite ends, extend winglike from small compartments.					
Apr. 5	Molniya II-5 (U.S.S.R.) 1973–18A Not available	Total weight: Not available. Objective: Continue operation of long-range telephone and telegraph radio communications system in U.S.S.R. and transmission of central television programs of U.S.S.R. to stations in Orbita and participating international networks. Payload: Not available.	39 828 (24 748)	525 (326.2)	717.7	65.2	Molniya II-5 communications satellite still in orbit.
Apr. 5*	Pioneer 11 (United States) 1973–19A Atlas-Centaur– TE-M-364-4	Total weight: 259 kg (570 lbs). Objective: Obtain, during 1973 Jovian opportunity, precursory scientific information beyond orbit of Mars with emphasis on: (a) investigation of interplanetary medium, (b) investigation of nature of Asteroid Belt; (c) exploration of Jupiter and its environment. Payload: Hexagonal spacecraft with auxiliary offset hexagonal compartment for scientific instruments and 2.7-m (9-ft)-dia parabolic antenna reflector; radioisotope thermoelectric generators (RTG) on trusses 120° apart extend 1.7 m (5.6 ft) radially beyond periphery of antenna reflector; 4-segment folding magnetometer boom extends radially 4.7 m (15.4 ft) beyond reflector and 120° from RTGS; from top of high-gain antenna feed to bottom of low-gain antenna measures	Heliocentric orbit, later to become solar escape trajectory				Pioneer 11, 2nd in new-generation Pioneer series, was launched successfully by 3-stage Atlas-Centaur. Spacecraft was scheduled to pass Jupiter within 42 000 km (26 000 mi)—three times closer than Pioneer 10—12/3/73 flyby. By 3/20/74, Pioneer 11 had emerged undamaged from 7-mo journey through 80-million-km (50-million-mi) wide Asteroid Belt. Two thruster burns on 4/19–4/20/74 shifted spacecraft aim to left instead of right of Jupiter and will permit Jovian gravity to throw craft toward Saturn for Sept. 1979 first-ever encounter (1740 days after Jovian encounter). Altered trajectory also may provide new information on shape and outer limits of heliosphere. Spacecraft operating normally and encounter with Jupiter's never-explored south polar region scheduled 12/3/74 EST (12/2/74 PST).

ASTRONAUTICS AND AERONAUTICS, 1973

Date	Spacecraft	Description	Weight kg (lbs)	Orbit km (mi)	Period min.	Inclination deg.	Remarks
Apr. 12	Cosmos 553 (U.S.S.R.) 1973-20A Not available	2.9 m (9.5 ft). Twelve scientific instruments measure magnetic fields; plasma; cosmic rays and charged particles; electromagnetic radiation in ultraviolet, visible, and invisible ranges; and asteroid and meteoroid population. Six thrusters. Spin-stabilized. Total weight: Not available. Objective: "Continuation of Cosmos scientific satellite series." Payload: Not available.	486 (302)	270 (167.8)	92	70.9	Reentered 11/11/73.
Apr. 19	Cosmos 554 (U.S.S.R.) 1973-21A Not available	Total weight: Not available. Objective: "Continuation of Cosmos scientific satellite series." Payload: Not available.	329 (204.4)	171 (106.3)	89.4	72.8	Maneuverable satellite exploded into approximately 180 pieces after 1½ week mission. Speculated to have been destroyed by ground control when satellite could not be recovered.
Apr. 19	Intercosmos Copernicus 500 (U.S.S.R.) 1973-22A Not available	Total weight: Not available. Objective: Investigation of solar radio-frequency radiation and characteristics of earth's ionosphere. Payload: Not available.	1 518 (943.3)	199 (123.7)	102.1	48.4	Intercosmos Copernicus 500, built by Soviet and Polish scientists, was launched to mark 500th birthday of Polish scientist Nicolaus Copernicus. Also carried Czechoslovakian instruments. Reentered 10/15/73.
Apr. 20	Anik 2 (Canada-U.S.) 1973-23A Thorad-Delta	Total weight: 540 kg (1200 lbs) at launch; 270 kg (600 lbs) in orbit. Objective: Place satellite in orbit of sufficient accuracy to allow spacecraft propulsion systems to place it in stationary synchronous orbit while retaining sufficient station-keeping propulsion to meet mission lifetime requirements. Payload: Cylindrical spacecraft 1.8 m in dia and 3.4 m high (6 by 11 ft). Spin-stabilized. Optically transparent, 1.5-m (5-ft) antenna weighing 4.1 kg (9 lbs) affixed to top of spacecraft remains stationary, pointed toward Canada, as spacecraft revolves. Spacecraft provides 10 color TV channels or up to 9600 telephone circuits; 23 000 solar cells.	36 480 (22 667.6) After apogee motor firing, 35 709 (22 188.5)	212 (131.7) 35 604 (22 123.3)	1430.7	26.7 0.1	Anik 2 (Telesat-B) was launched by NASA on 3-stage Thor-Delta for Canadian Domestic Communications Satellite System into elliptical transfer orbit. Apogee kick motor fired by Telesat Canada 4/23; spacecraft was placed in stationary equatorial orbit off west coast of South America. American Satellite Corp. (U.S.) signed agreement with Telesat to lease 3 full time and 1 or more part time transponders on Canadian satellite, and on 6/18 conducted first relay of a telecast across U.S. by domestic communications satellite. Still in orbit.
Apr. 25	Cosmos 555 (U.S.S.R.)	Total weight: Not available. Objective: "Continuation of Cosmos sci-	230 (142.9)	216 (134.2)	88.9	81	Reentered 5/7/73.

Launch Date	Name, Country, International Designation, Vehicle	Payload Data	Apogee in Kilometers (and st mi)	Perigee in Kilometers (and st mi)	Period in Minutes	Inclination in Degrees	Remarks
	1973–24A Not available	entific satellite series." Payload: Not available.					
May 5	Cosmos 556 (U.S.S.R.) 1973–25A Not available	Total weight: Not available. Objective: "Continuation of Cosmos scientific satellite series." Payload: Not available.	233 (144.8)	207 (128.6)	88.9	81.4	Reentered 5/14/73.
May 11	Cosmos 557 (U.S.S.R.) 1973–26A Not available	Total weight: Not available. Objective: "Continuation of Cosmos scientific satellite series." Payload: Not available.	249 (154.7)	213 (132.4)	89.1	51.6	Cosmos 557 launched into orbit nearly identical to crippled Salyut 2. Speculation was that mission was to test Salyut-class station. Early decay indicated mission failure. Reentered 5/22/73.
May 14	Skylab 1 (United States) 1973–27A Saturn V	Total weight: 74 783.4 kg (164 869 lbs) at initial earth orbit insertion, including OWS, IU, AM, MDA, and ATM. Objective: Establish Skylab Orbital Assembly in earth orbit; obtain medical data on crew for use in extending duration of manned space flight; perform inflight experiments. Payload: Modified S-IVB stage measures 26 m (85.3 ft) when deployed in orbit and has habitable volume of about 345 cu m (12 200 cu ft). Consists of 4 major modules: Orbital Workshop (OWS), airlock module (AM), multiple docking adapter (MDA), and Apollo Telescope Mount (ATM). OWS has cylindrical body 14.6 m (48 ft) long and 6.6 m (21.7 ft) in dia; 2 solar array wings extend 9.1 m (30 ft) on opposite sides. Has 2 compartments: forward is work and activity area; aft is for crew accommodation. Habitable volume 295 cu m (10 426 cu ft). AM consists of 2 concentric cylinders 5.3 m (17.4 ft) long and contains passageway for crew to OWS from MDA; also has port for	438 (272)	438 (272)	93	50	Skylab 1, 1st U.S. Orbital Workshop, employed for 3 manned visits (Skylab 2, 3, 4). During launch, vehicle vibration tore off meteoroid shield, which caused tearing off of solar array Wing 2 and prevented deployment of solar array Wing 1. Electrical power was reduced and temperature inside Workshop reached 325 K (125°F). Skylab 2's 3-man crew docked with Skylab Workshop 5/25/73, entered next day, and deployed makeshift solar parasol through solar airlock. Temperatures dropped. On 6/7/73, astronauts released jammed solar array wing in 3-hr 30-min EVA, restoring power. Crew conducted solar and earth resources data. CM returned 6/22/73. Total flight time 28 days 50 min. Workshop was again manned by Skylab 3 crew July 28–Sept. 25 and by Skylab 4 crew Nov. 16–Feb. 8, 1974. Each mission set new flight records and returned a great amount of new data. Workshop was successful medical and scientific laboratory and provided habitable environment for nearly 6 mos for 3 separate 3-man crews. Still in orbit. (See also Skylab 2, 3, and 4)

Date	Spacecraft	Description	Weight kg (lb)	Payload kg (lb)	Orbital inclination	Period (min)	Remarks
May 16	DOD Spacecraft (United States) 1973-28A Titan IIIB-Agena	EVA; habitable volume 17.6 cu m (622 cu ft). MDA 5.2-m (17.1-ft) module provides for docking CSM to Workshop while in orbit; has 2 docking ports for redundancy; habitable volume 32.3 cu m (1140 cu ft). ATM consists of 2 concentric elements; outer element is octagonal structure 3.4 m (11 ft) wide and 4.4 m (14.5 ft) high, which supports solar experiment canister, 4 solar arrays, and communications system; inner element, canister, about 3 m (10 ft) long and 2 m (7 ft) in dia, mounted on gimbals for pointing, houses sophisticated solar lab with 8 telescopes. ATM's 4 solar array wings, each 13.7 m (45 ft) long, are deployed in X-shape after launch. Total weight: Not available. Objective: Develop space flight techniques and technology. Payload: Not available.	397 (246.7)	134 (83.3)	89.8	110.4	Reentered 6/13/73.
May 17	Cosmos 558 (U.S.S.R.) 1973-29A Not available	Total weight: Not available. Objective: "Continuation of Cosmos scientific satellite series." Payload: Not available.	491 (305.1)	267 (165.9)	92.1	70.9	Reentered 12/22/73.
May 18	Cosmos 559 (U.S.S.R.) 1973-30A Not available	Total weight: Not available. Objective: "Continuation of Cosmos scientific satellite series." Payload: Not available.	324 (201.3)	204 (126.8)	89.8	65.4	Reentered 5/23/73.
May 23	Cosmos 560 (U.S.S.R.) 1973-31A Not available	Total weight: Not available. Objective: "Continuation of Cosmos scientific satellite series." Payload: Not available.	325 (201.9)	178 (110.6)	89.5	72.8	Reentered 6/5/73.
May 25	Skylab 2 (United States) 1973-32A Saturn IB	Total weight: 30 464.7 kg (67 163 lbs) at initial earth orbit insertion, includes CSM/S-IVB; 13 978.4 kg (30 817 lbs) CSM after S-IVB separation; 87 810.6 kg (193 589 lbs) includes total of docked assembly (CSM/Workshop). Objective: Establish Skylab Orbital Assembly in earth orbit; obtain medical	224 (139.2) After rendezvous and docking, 440 (273.4)	150 (93.2) 424 (263.5)	93.2	50	Skylab 2 carried Astronauts Charles Conrad, Jr., Paul J. Weitz, and Joseph P. Kerwin to Skylab Workshop, launched into earth orbit May 14. After Skylab 2 was launched into phasing orbit, CSM separated from S-IVB stage and rendezvoused with damaged Workshop at 4:30 pm EDT during 5th orbit. After 4 unsuccessful docking attempts, hard docking with Workshop was

Launch Date	Name, Country, International Designation, Vehicle	Payload Data	Apogee in Kilometers (and st mi)	Perigee in Kilometers (and st mi)	Period in Minutes	Inclination in Degrees	Remarks
		data on crew for use in extending duration of manned space flight; perform inflight experiments. Payload: Apollo Block II command and service modules (CSM) modified for Skylab use. Command module conical, 3.9 m (12 ft 10 in) in dia, 3.5 m (11 ft 5 in) high, with habitable volume of 17.9 cu m (630 cu ft). Service module cylindrical 3.9 m (12 ft 10 in) in dia and 7.6 m (24 ft 10 in) long, with 88 960-newton (20 000-lb)-thrust main propulsion system, reaction control system, fuel cells, oxygen, and radiators. Skylab cluster (with docked CSM) 36 m (118 ft) long, with habitable volume of 363 cu m (12 800 cu ft).					achieved at 11:50 pm EDT. Next day crew entered multiple docking adapter (MDA) at 12:45 pm EDT and entered Workshop at 4:30 pm EDT. Makeshift sun shield (parasol) was deployed through solar airlock at 8:30 pm EDT and external temperature began to decrease. Workshop was activated and internal temperature decreased. On June 7 Conrad and Kerwin made 3-hr 30-min EVA; used 7.6-m (25-ft) pole fitted with cutters to sever small bolt jamming remaining solar wing. Wing was released and OWS recovered 3000 w electrical power. On June 19 Conrad and Weitz made 1-hr 31-min EVA to retrieve and replace cassettes. On June 22 CSM undocked from Workshop at 4:55 am EDT and inspected Workshop during fly-around. CSM separated from SM EDT June 22. Despite mission anomalies, 80% of solar data planned was obtained; 12 of 15 earth resources data runs were accomplished; all 16 medical experiments were conducted. Live and recorded TV was transmitted throughout mission. Total EVA time, 5 hrs 6 min (not including one standup-EVA). Total flight time, 28 days 50 min.
May 25	Cosmos 561 (U.S.S.R.) 1973-33A Not available	Total weight: Not available. Objective: "Continuation of Cosmos scientific satellite series." Payload: Not available.	290 (180.2)	205 (127.4)	89.4	65.4	Reentered 6/6/73.
May 29	Meteor 15 (U.S.S.R.) 1973-34A Not available	Total weight: Not available. Objective: Acquisition of meteorological information for use by weather service. Payload: Cylindrical body with 2 large solar paddles attached; 3 data-collection systems; TV cameras, infrared sensors, and actinometric scanner; 3-axis attitude control system.	896 (556.7)	852 (529.4)	102.4	81.2	Meteor 15 meteorological satellite still in orbit.
June 5	Cosmos 562 (U.S.S.R.)	Total weight: Not available. Objective: "Continuation of Cosmos sci-	473 (293.9)	270 (167.8)	91.9	70.9	Reentered 1/7/74.

ASTRONAUTICS AND AERONAUTICS, 1973

		1973-35A Not available	entific satellite series." Payload: Not available.				
June 6	Cosmos 563 (U.S.S.R.) 1973-36A Not available	Total weight: Not available. Objective: "Continuation of Cosmos scientific satellite series." Payload: Not available.	306 (190.1)	190 (118.1)	89.5	65.2	Reentered 6/18/73.
June 8	Cosmos 564 (U.S.S.R.) 1973-37A Not available	Total weight: Not available. Objective: "Continuation of Cosmos scientific satellite series." Payload: Not available.	1 482 (920.9)	1 397 (868.1)	114.6	74	Eight satellites launched with single booster. Still in orbit.
and							
	Cosmos 565 1973-37B	Total weight: Not available. Objective: "Continuation of Cosmos scientific satellite series." Payload: Not available.	1 493 (927.7)	1 448 (899.7)	115.3	74	Still in orbit.
and							
	Cosmos 566 1973-37C	Total weight: Not available. Objective: "Continuation of Cosmos scientific satellite series." Payload: Not available.	1 484 (922.1)	1 435 (891.7)	115	74	Still in orbit.
and							
	Cosmos 567 1973-37D	Total weight: Not available. Objective: "Continuation of Cosmos scientific satellite series." Payload: Not available.	1 486 (923.4)	1 413 (878)	114.8	74	Still in orbit.
and							
	Cosmos 568 1973-37E	Total weight: Not available. Objective: "Continuation of Cosmos scientific satellite series." Payload: Not available.	1 483 (921.5)	1 377 (855.6)	114.4	74	Still in orbit.
and							

Launch Date	Name, Country, International Designation, Vehicle	Payload Data	Apogee in Kilometers (and st mi)	Perigee in Kilometers (and st mi)	Period in Minutes	Inclination in Degrees	Remarks
	Cosmos 569 1973-37F	Total weight: Not available. Objective: "Continuation of Cosmos scientific satellite series." Payload: Not available.	1 480 (919.6)	1 361 (845.7)	114.2	74	Still in orbit.
	and						
	Cosmos 570 1973-37G	Total weight: Not available. Objective: "Continuation of Cosmos scientific satellite series." Payload: Not available.	1 482 (920.9)	1 340 (832.6)	114	74	Still in orbit.
	and						
	Cosmos 571 1973-37H	Total weight: Not available. Objective: "Continuation of Cosmos scientific satellite series." Payload: Not available.	1 480 (919.6)	1 322 (821.5)	113.7	74	Still in orbit.
June 10	Cosmos 572 (U.S.S.R.) 1973-38A Not available	Total weight: Not available. Objective: "Continuation of Cosmos scientific satellite series." Payload: Not available.	286 (177.7)	199 (123.7)	89.4	51.8	Reentered 6/23/73.
June 10	Explorer 49 (United States) 1973-39A Thorad-Delta	Total weight: 330 kg (730 lbs) at liftoff, including lunar insertion motor and vcrs; 200 kg (442 lbs) in lunar orbit. Objective: Make measurements of galactic and solar radio noise at frequencies below ionospheric cutoffs and beyond terrestrial background interference by using moon for occultation, focusing, or aperture blocking for increased resolution and discrimination. Payload: Truncated cylinder 92 cm (36 in) in dia and 79 cm (31 in) high with 4 fixed solar paddles mounted on	Lunar orbit, 1 063 (660.5)	1 052 (653.7)	221.2	38.7	Explorer 49 (Radio Astronomy Explorer 2) was launched successfully into lunar transfer trajectory. One midcourse correction made. Lunar orbit insertion made at 3:21 am EDT 6/15. Lunar orbit circularized 6/18. All experiments turned on. Upper 2 antennas fully extended to 228.6 m (750 ft); lower 2 antennas extended only 182.9 m (600 ft) but mission not jeopardized. Using moon as shield from radio noise from earth, spacecraft will make most extensive study ever undertaken of low-frequency signals from galactic and extragalactic radio sources. Last scheduled U.S. space mission to moon. Still in lunar orbit.

Date	Spacecraft	Description	Weight/Objective/Payload	Perigee km (mi)	Apogee km (mi)	Period min	Inclination deg	Remarks
June 12	DOD Spacecraft (United States) 1973-40A Titan IIIC	main body; outside skin of aluminum honeycomb material; lunar insertion motor attached to upper end of body is jettisoned after lunar orbit insertion. Overall measurements with 2 cameras mounted on solar arrays are 183 cm (72 in) wide, 147 cm (58 in) high, and 160 cm (63 in) long. Spacecraft contains one libration damper boom extendable to 192 m (630 ft) and one retractable 36.6-m (120-ft) dipole antenna for calibration tests. Experiment antennas consist of 2 back-to-back V antennas formed by deploying 4, 1.5-cm (0.6-in) elements up to maximum length of 228.6 m (750 ft) each, measuring 457.2 m (1500 ft) tip-to-opposite-tip. Velocity Control Propulsion Subsystem (VCPS) uses hydrazine fuel thruster containing spontaneous catalyst for velocity correction; jettisoned after spacecraft properly oriented in lunar orbit. Spacecraft also carries tape recorders, battery, and telemetry and command subsystem.	Total weight: Not available. Objective: Develop space flight techniques and technology. Payload: Not available.	35 787 (22 237)	35 778.9 (22 232)	1 436	0.3	Spacecraft was launched into synchronous orbit above Indian Ocean. Still in orbit.
June 15	Cosmos 573 (U.S.S.R.) 1973-41A Not available		Total weight: Not available. Objective: "Continuation of Cosmos scientific satellite series." Payload: Not available.	401 (249.2)	195 (121.2)	90.5	51.6	Cosmos 573 speculated to be test of unmanned Soyuz in preparation for upcoming manned launch. Reentered 6/17/73.
June 20	Cosmos 574 (U.S.S.R.) 1973-42A Not available		Total weight: Not available. Objective: "Continuation of Cosmos scientific satellite series." Payload: Not available.	1 014 (630.1)	985 (612.1)	105	82.9	Still in orbit.
June 21	Cosmos 575 (U.S.S.R.) 1973-43A Not available		Total weight: Not available. Objective: "Continuation of Cosmos scientific satellite series." Payload: Not available.	267 (165.9)	204 (126.8)	89.1	65.3	Reentered 7/3/73.

ASTRONAUTICS AND AERONAUTICS, 1973

Launch Date	Name, Country, International Designation, Vehicle	Payload Data	Apogee in Kilometers (and st mi)	Perigee in Kilometers (and st mi)	Period in Minutes	Inclination in Degrees	Remarks
June 27	Cosmos 576 (U.S.S.R.) 1973-44A Not available	Total weight: Not available. Objective: "Continuation of Cosmos scientific satellite series." Payload: Not available.	331 (205.7)	203 (126.1)	89.8	72.9	Reentered 7/9/73.
July 11	Molniya II-6 (U.S.S.R.) 1973-45A Not available	Total weight: Not available. Objective: Continue operation of long-range telephone and telegraph radio communications system in U.S.S.R. and transmission of central television programs of U.S.S.R. to stations in Orbita and participating international networks. Payload: Not available.	39 284 (24 410)	441 (274)	705	65.4	Molniya II-6 communications satellite still in orbit.
July 13	DOD Spacecraft (United States) 1973-46A Titan IIID	Total weight: Not available. Objective: Develop space flight techniques and technology. Payload: Not available.	273.6 (170)	154.5 (96)	88.7	96.2	Spacecraft speculated to be USAF "Big Bird" photo-reconnaissance satellite. Reentered 10/12/73.
July 22**	Mars 4 (U.S.S.R.) 1973-47A Not available	Total weight: 4536 kg (10 000 lbs). Objective: Continue scientific exploration of planet Mars and surrounding space. Payload: Orbital station contains astrophysical instruments for study of Mars atmosphere, surface, and solar wind; gamma spectrometer; photometer; magnetometer; braking engine for orbital insertion; and "photo-television device."	Heliocentric orbit				Mars 4 1st of 4 spacecraft in U.S.S.R. 1973-1974 Mars mission. Mars 4 and 5 were intended to orbit Mars and relay transmissions from Mars 6 and 7 landers. Mission near total failure; only one spacecraft orbited planet and one capsule was landed but failed to transmit from surface. Mars 4 failed to enter Mars orbit after malfunctioning control device did not operate properly and spacecraft's retrorockets were not fired. Mars 4 passed planet at distance of 2200 km (1367 mi) on 2/10/74 and entered heliocentric orbit.
July 25	Cosmos 577 (U.S.S.R.) 1973-48A Not available	Total weight: Not available. Objective: "Continuation of Cosmos scientific satellite series." Payload: Not available.	308 (191.4)	170 (105.6)	89.2	65.4	Reentered 8/7/73.
July 25	Mars 5 (U.S.S.R.)	Total weight: 4536 kg (10 000 lbs). Objective: Continue scientific explora-	Areocentric orbit,				Mars 5 2nd of 4 spacecraft in U.S.S.R. 1973-1974 Mars mission. Mars 4 and 5 were

ASTRONAUTICS AND AERONAUTICS, 1973

Date	Name/Designation	Description	Weight kg (lbs)	Altitude km (mi)	Inclination	Remarks
	1973-49A Not available	tion of planet Mars and surrounding space. Payload: Orbital station contains astrophysical instruments for study of Mars atmosphere, surface, and solar wind; gamma spectrometer; photometer; magnetometer; braking engine for orbital insertion; and "phototelevision device."	32 500 (20 200)	1 760 (1 094)	1 500	35	intended to orbit Mars and relay transmissions sent from *Mars 6* and 7 landers. Mission near total failure; only one spacecraft orbited planet and one capsule was landed but failed to transmit from surface. *Mars 5*, only survivor of 4-spacecraft mission, successfully entered highly elliptical orbit of Mars 2/12/74. Spacecraft returned clear photographs from Mars surface showing former riverbeds and crater 25 km (15.5 mi) in dia on smooth plain.
July 28	*Skylab 3* (United States) 1973-50A Saturn IB	Total weight: 30 718 kg (67 721 lbs) at initial earth orbit insertion, includes CSM/S-IVB, IU, SLA; CSM, 14 169-kg (31 237 lbs) after S-IVB separation; docked assembly (CSM/Workshop), 87 566 kg (193 050 lbs). Objective: Perform unmanned Saturn Workshop operations, reactivate Skylab orbital assembly in earth orbit, obtain medical data on crew for use in extending duration of manned space flights, perform inflight experiments. Payload: Apollo Block II command and service modules (CSM) modified for Skylab use. Command module conical, 3.9 m (12 ft 10 in) in dia and 3.5 m (11 ft 5 in) high, with habitable volume of 17.8 cu m (630 cu ft). Service module cylindrical 3.9 m (12 ft 10 in) in dia and 7.5 m (24 ft 10 in) long, with main propulsion system of 88 960-newton (20 000-lb) thrust, reaction control system, fuel cells, oxygen, and radiators. Skylab cluster (with docked CSM) about 36 m (118 ft) long, with habitable volume of 363 cu m (12 800 cu ft).	224 (139.2) After rendezvous and docking, 440 (273.4)	150 (93.2) 423 (263.5)	93.2	50	*Skylab 3* 2nd U.S. manned Orbital Workshop mission, carried Astronauts Alan L. Bean, Owen K. Garriott, and Jack R. Lousma. After launch into phasing orbit, CSM separated from S-IVB stage and rendezvoused with Workshop at 3:38 pm EDT during 5th orbit. Crew entered multiple docking adapter (MDA) at 4:00 pm EDT and began Workshop activation. Crew's motion sickness caused rescheduling of tasks. Leak in 2 of 4 clusters of CSM steering rockets endangered safe reentry; rescue preparations were initiated on ground but not required. On 8/6, in 6-hr 31-min EVA, Garriott and Lousma installed 3.6- by 7.3-m (12- by 24-ft) Mylar twin-pole sunshade to augment parasol. Also deployed experiments, inspected CSM RCS, and retrieved film. During 4-hr 30-min EVA-2 on 8/24, Lousma and Garriott completed installation of 6 redesigned gyros. On 9/22 Bean and Garriott, in 2-hr 41-min EVA, retrieved and replaced ATM film and retrieved experiments and parasol samples. On 9/25 CSM undocked from Workshop at 3:34 pm EDT and separated at 3:50. CSM splashed down in Pacific at 6:20 pm EDT 9/25. Experiment accomplishments surpassed all plans; 150% of observing time scheduled for ATM accomplished; EREP sensors operated on 39 data runs, exceeding scheduled 28 passes; all crew medical experiments successfully conducted. Total EVA time 13 hrs 42 min. Total flight time 59 days 11 hrs 9 min.
Aug. 1	*Cosmos 578* (U.S.S.R.) 1973-51A Not available	Total weight: Not available. Objective: "Continuation of Cosmos scientific satellite series." Payload: Not available.	251 (156)	202 (125.5)	89	65.4	Reentered 8/13/73.

377

Launch Date	Name, Country, International Designation, Vehicle	Payload Data	Apogee in Kilometers (and st mi)	Perigee in Kilometers (and st mi)	Period in Minutes	Inclination in Degrees	Remarks
Aug. 5	*Mars 6* (U.S.S.R.) 1973-52A Not available	Total weight: 4536 kg (10 000 lbs). Objective: Continue scientific exploration of planet Mars and surrounding space. Payload: Contains separate landing capsule and cameras, descent engine, parachute system, and French-built experiment Stereo to study solar emissions.	Heliocentric orbit				*Mars 6* 3rd of 4 spacecraft in U.S.S.R. 1973–74 Mars mission. *Mars 4* and *5* were intended to orbit Mars and relay transmissions from *Mars 6* and *7* landers. Mission near total failure; only one spacecraft orbited planet and one capsule was landed but failed to transmit from surface. *Mars 6* ejected capsule which parachuted to Mars surface 3/12/74. Capsule transmitted information 148 sec during descent, but radio contact ceased before surface was reached. While no surface readings obtained, descent capsule found planet's magnetic field 7 to 10 times greater than that of outer space, found atmosphere of Mars contained several times more water vapor than previously thought, and recorded presence of inert gas, assumed to be argon, major atmospheric component.
Aug. 9	*Mars 7* (U.S.S.R.) 1973-53A Not available	Total weight: 4536 kg (10 000 lbs). Objective: Continue scientific exploration of planet Mars and surrounding space. Payload: Contains separate landing capsule and cameras, descent engine, parachute system, and Soviet and French experiments.	Heliocentric orbit				*Mars 7* last of 4 spacecraft in U.S.S.R. 1973–74 Mars mission. *Mars 4* and *5* were intended to orbit Mars and relay transmissions from *Mars 6* and *7* landers. Mission near total failure; only one spacecraft was landed but failed to transmit from surface. *Mars 7* failed to enter Mars orbit and to eject landing capsule because of faulty spacecraft control device. *Mars 7* passed planet at distance of 1300 km (800 mi) on 3/9/74 and entered heliocentric orbit.
Aug. 17	DOD Spacecraft (United States) 1973-54A Thor-Burner II	Total weight: Not available. Objective: Develop space flight techniques and technology. Payload: Not available.	853 (530)	808 (502.1)	101.5	98.8	Still in orbit.
Aug. 21	*Cosmos 579* (U.S.S.R.) 1973-55A Not available	Total weight: Not available. Objective: "Continuation of Cosmos scientific satellite series." Payload: Not available.	307 (190.8)	174 (108.1)	89.3	65.4	Reentered 9/3/73.

Date	Spacecraft		Weight kg (lbs)	Perigee & Apogee km (mi)	Period min	Inclination to equator (deg)	Remarks
Aug. 21	DOD Spacecraft (United States) 1973-56A Titan IIIB—Agena	Total weight: Not available. Objective: Develop space flight techniques and technology. Payload: Not available.		39 132 (24 315.5)	392 (243.6)	701	63.3	Test communications satellite, according to press reports, launched into highly elliptical orbit. Still in orbit.
Aug. 22	Cosmos 580 (U.S.S.R.) 1973-57A Not available	Total weight: Not available. Objective: "Continuation of Cosmos scientific satellite series." Payload: Not available.		490 (304.5)	273 (169.6)	92.1	70.9	Reentered 4/1/74.
Aug. 23	Intelsat-IV F-7 (United States) 1973-58A Atlas-Centaur-TE-M-364-4	Total weight: 1379 kg (3041 lbs) at launch; 700 kg (1544 lbs) after apogee motor fire. Objective: Conduct design, performance, and flight readiness reviews for FCC; ensure compatibility of INTELSAT spacecraft with NASA launch vehicles and launch satellite into transfer orbit permitting spacecraft apogee motor to inject spacecraft into synchronous orbit. Payload: 528-cm-high x 238-cm-dia (208- x 93.7-in) cylindrical satellite that can carry 3000-9000 telephone circuits simultaneously or 12 color TV channels or combinations; spin-stabilized; 12 communications repeaters (transponders); 6 antennas (2 transmit horns, 2 receive horns, and 2 steerable 127-cm [50-in] dish spot-beam antennas); 42 240 solar cells.		35 819.5 (22 257.2) After apogee motor firing, 35 778 (22 231.4)	548.2 (340.6) 35 773 (22 228.3)	1 435.6	27.3 0.4	Launched by NASA into highly accurate transfer orbit for ComSatCorp. Apogee kick motor fired 8/25 stationed satellite in synchronous orbit over Atlantic Ocean at 330° E longitude. Commercial operations began 9/9/73. Still in orbit.
Aug. 24	Cosmos 581 (U.S.S.R.) 1973-59A Not available	Total weight: Not available. Objective: "Continuation of Cosmos scientific satellite series." Payload: Not available.		282 (175.2)	171 (106.3)	89	51.6	Reentered 9/6/73.
Aug. 28	Cosmos 582 (U.S.S.R.) 1973-60A Not available	Total weight: Not available. Objective: "Continuation of Cosmos scientific satellite series." Payload: Not available.		542 (336.8)	519 (322.5)	95.2	74	Still in orbit.
Aug. 30	Molniya 1-24 (U.S.S.R.) 1973-61A Not available	Total weight: Not available. Objective: Continue operation of long-range telephone, telegraph, and radio communications system and transmis-		37 970 (23 593.5)	480 (298.3)	679	65.3	Molniya 1-24 communications satellite still in orbit.

Launch Date	Name, Country, International Designation, Vehicle	Payload Data	Apogee in Kilometers (and st mi)	Perigee in Kilometers (and st mi)	Period in Minutes	Inclination in Degrees	Remarks
		sion of U.S.S.R. central TV programs to stations in Orbita network. Payload: Carries apparatus for transmitting television programs and multichannel radio communications, instruments of control-and-measurement complex, orientation system, orbit correction system, and power supplies.					
Aug. 30	Cosmos 583 (U.S.S.R.) 1973-62A Not available	Total weight: Not available. Objective: "Continuation of Cosmos scientific satellite series." Payload: Not available.	298 (185.2)	204 (126.8)	89.5	64.9	Reentered 9/12/73.
Sept. 6	Cosmos 584 (U.S.S.R.) 1973-63A Not available	Total weight: Not available. Objective: "Continuation of Cosmos scientific satellite series." Payload: Not available.	331 (205.7)	210 (130.5)	89.9	72.9	Reentered 9/20/73.
Sept. 8	Cosmos 585 (U.S.S.R.) 1973-64A Not available	Total weight: Not available. Objective: "Continuation of Cosmos scientific satellite series." Payload: Not available.	1 407 (874.3)	1 375 (854.4)	113.5	74	Still in orbit.
Sept. 14	Cosmos 586 (U.S.S.R.) 1973-65A Not available	Total weight: Not available. Objective: "Continuation of Cosmos scientific satellite series." Payload: Not available.	1 007 (625.8)	969 (602.1)	104.8	82.9	Still in orbit.
Sept. 21	Cosmos 587 (U.S.S.R.) 1973-66A Not available	Total weight: Not available. Objective: "Continuation of Cosmos scientific satellite series." Payload: Not available.	289 (179.6)	177 (110)	89.1	65.4	Reentered 10/4/73.
Sept. 27	Soyuz 12 (U.S.S.R.) 1973-67A Not available	Total weight: 6577 kg (14 500 lbs). Objective: 2-day mission to test improved flight systems; test manual and automatic control in various flight con-	229 (142.3)	180 (111.8)	88.6	51.7	Soyuz 12, 1st U.S.S.R. manned flight since fatal Soyuz 11 mission in June 1971. Cosmonauts Vasily G. Lazarev and Oleg G. Makarov wore new spacesuits, tested manual

		ditions; obtain spectrographic data of separate sections of earth's surface. Payload: 3-unit spacecraft 9.5 m (31 ft) long with 2 crew cabins totaling 8.9 cu m (315 cu ft) of interior space.					and automatic control and piloting techniques, made spectrographs of natural formations of earth's surface. Orbit was changed during flight; set new Soviet altitude record for manned flight. Redesigned Soyuz craft allowed more extensive maneuvering, more distant orbits, and more manual control. Flight believed related to future 2-day ferry missions to orbital stations. *Soyuz 12* reentered successfully and landed southwest of Karaganda in Kazakhstan Steppe at 2:34 pm Moscow time (7:34 pm EDT) 9/29. Total flight time 47 hrs 16 min.
Sept. 27	DOD Spacecraft (United States) 1973–68A Titan IIIB-Agena	Total weight: Not available. Objective: Develop space flight techniques and technology. Payload: Not available.	384.6 (239)	128.8 (80)	89.8	110.5	Reentered 10/29/73.
Oct. 3**	*Cosmos 588* (U.S.S.R.) 1973–69A Not available	Total weight: Not available. Objective: "Continuation of Cosmos scientific satellite series." Payload: Not available.	1 495 (929)	1 450 (901)	115.3	74	Eight satellites launched with single booster. Possibly part of maritime comsat system. Still in orbit.
	and						
	Cosmos 589 1973–69B	Total weight: Not available. Objective: "Continuation of Cosmos scientific satellite series." Payload: Not available.	1 489 (925.2)	1 417 (880.5)	114.9	73.9	Still in orbit.
	and						
	Cosmos 590 1973–69C	Total weight: Not available. Objective: "Continuation of Cosmos scientific satellite series." Payload: Not available.	1 489 (925.2)	1 435 (891.7)	115.1	74	Still in orbit.
	and						
	Cosmos 591 1973–69D	Total weight: Not available. Objective: "Continuation of Cosmos scientific satellite series." Payload: Not available.	1 487 (924)	1 349 (838.2)	114.1	73.9	Still in orbit.

Launch Date	Name, Country, International Designation, Vehicle	Payload Data	Apogee in Kilometers (and st mi)	Perigee in Kilometers (and st mi)	Period in Minutes	Inclination in Degrees	Remarks
and	Cosmos 592 1973–69E	Total weight: Not available. Objective: "Continuation of Cosmos scientific satellite series." Payload: Not available.	1 486 (923.3)	1 333 (828.3)	113.9	74	Still in orbit.
and	Cosmos 593 1973–69F	Total weight: Not available. Objective: "Continuation of Cosmos scientific satellite series." Payload: Not available.	1 487 (924)	1 366 (848.8)	114.3	73.9	Still in orbit.
and	Cosmos 594 1973–69G	Total weight: Not available. Objective: "Continuation of Cosmos scientific satellite series." Payload: Not available.	1 487 (924)	1 383 (859.4)	114.5	74	Still in orbit.
Oct. 3	Cosmos 595 1973–69H	Total weight: Not available. Objective: "Continuation of Cosmos scientific satellite series." Payload: Not available.	1 487 (924)	1 400 (869.9)	114.7	73.9	Still in orbit.
Oct. 3	Cosmos 596 (U.S.S.R.) 1973–70A Not available	Total weight: Not available. Objective: "Continuation of Cosmos scientific satellite series." Payload: Not available.	268 (166.5)	205 (127.4)	89.2	65.4	Believed to be 1st of series of 7 reconnaissance spacecraft to monitor Arab-Israeli war. Frequency of launch and recovery indicate U.S.S.R. maintained continuous space surveillance of Mideast conflict. See Cosmos 597, Cosmos 598, Cosmos 599, Cosmos 600, Cosmos 602, and Cosmos 603. Cosmos 596 reentered 10/9/73.

Date	Name	Details	Weight kg (lb)	Orbit	Period (min)	Inclination	Remarks
Oct. 6	Cosmos 597 (U.S.S.R.) 1973-71A Not available	Total weight: Not available. Objective: "Continuation of Cosmos scientific satellite series." Payload: Not available.	214 (133)	163 (101.3)	88.3	65.4	Thought to be reconnaissance spacecraft to monitor Arab-Israeli war (see Cosmos 596). Reentered 10/12/73.
Oct. 10	Cosmos 598 (U.S.S.R.) 1973-72A Not available	Total weight: Not available. Objective: "Continuation of Cosmos scientific satellite series." Payload: Not available.	331 (205.7)	204 (126.8)	89.9	72.8	Thought to be reconnaissance spacecraft to monitor Arab-Israeli war (see Cosmos 596). Reentered 10/16/73.
Oct. 15	Cosmos 599 (U.S.S.R.) 1973-73A Not available	Total weight: Not available. Objective: "Continuation of Cosmos scientific satellite series." Payload: Not available.	276 (171.5)	208 (129.3)	89.3	65	Thought to be reconnaissance spacecraft to monitor Arab-Israeli war (see Cosmos 596). Reentered 10/28/73.
Oct. 16	Cosmos 600 (U.S.S.R.) 1973-74A Not available	Total weight: Not available. Objective: "Continuation of Cosmos scientific satellite series." Payload: Not available.	355 (220.6)	192 (119.3)	90	72.8	Thought to be reconnaissance spacecraft to monitor Arab-Israeli war (see Cosmos 596). Reentered 10/23/73.
Oct. 16	Cosmos 601 (U.S.S.R.) 1973-75A Not available	Total weight: Not available. Objective: "Continuation of Cosmos scientific satellite series." Payload: Not available.	1 494 (928.3)	199 (123.7)	101.8	81.9	Reentered 8/15/74.
Oct. 19	Molniya II-7 (U.S.S.R.) 1973-76A Not available	Total weight: Not available. Objective: Continue operation of long-range telephone and telegraph radio communications in U.S.S.R. and transmission of central television programs of U.S.S.R. to stations in Orbita and participating international networks. Payload: Not available.	39 860 (24 767.9)	503 (312.6)	717.9	62.8	Molniya II-7 communications satellite still in orbit.
Oct. 20	Cosmos 602 (U.S.S.R.) 1973-77A Not available	Total weight: Not available. Objective: "Continuation of Cosmos scientific satellite series." Payload: Not available.	332 (206.3)	207 (128.6)	90	72.9	Thought to be reconnaissance spacecraft to monitor Arab-Israeli war (see Cosmos 596). Reentered 10/29/73.
Oct. 25*	Explorer 50 (United States) 1973-78A Thorad-Delta	Total weight: 397 kg (875 lbs). Objective: Make detailed, nearly continuous studies of interplanetary environment for orbital periods compara-	228 808 (142 174.7) After apogee	197.5 (122.7)	117.1	28.7	Explorer 50 was 10th and last in IMP (Interplanetary Monitoring Platform) series to study interplanetary radiation throughout 11-yr solar cycle. Spacecraft was launched

Launch Date	Name, Country, International Designation, Vehicle	Payload Data	Apogee in Kilometers (and st mi)	Perigee in Kilometers (and st mi)	Period in Minutes	Inclination in Degrees	Remarks
		ble to several rotations of active solar regions; study particle and field interactions in distant magnetotail. Payload: 16-sided drum-shaped structure, 158 cm high, 135 cm in dia (62 x 53 in). Upper portion of spacecraft contains aluminum honeycomb shelf which supports experiments and spacecraft electronics. Lower portion has 46 cm (18 in) thrust tube for solid-propellant kick motor. Spacecraft consists of aluminum honeycomb RF shield panels and three bands of solar panels (16 panels per band) mounted on aluminum honeycomb substrate. Two diametrically opposed 3.1-m (10-ft) experiment booms and two 1.2-m (4-ft) attitude control system booms spaced 90° from experiment booms, appended to spacecraft exterior and deployed after launch. Two weeks later, four 61-m (200-ft) antennas extended in support of electric and magnetic fields experiments. Spacecraft fitted with 8 equally spaced, RF antennas (4 active, 4 passive turnstile) extending radially. Spin-stabilized. Contains 2 engineering tests and 12 scientific experiments.	motor firing, 288 857 (179 487.4)	141 184 (87 727.7)	17 279.1	28.7	into elliptical transfer orbit; kick motor, fired 10/29, placed spacecraft in orbit halfway between earth and moon. All 12 experiments turned on and returning good data. Still in orbit.
Oct. 27	Cosmos 603 (U.S.S.R.) 1973–79A Not available	Total weight: Not available. Objective: "Continuation of Cosmos scientific satellite series." Payload: Not available.	356 (221.2)	204 (126.8)	90	72.8	Thought to be reconnaissance spacecraft to monitor Arab-Israeli war (see Cosmos 596). Reentered 11/9/73.
Oct. 29	Cosmos 604 (U.S.S.R.) 1973–80A Not available	Total weight: Not available. Objective: "Continuation of Cosmos scientific satellite series." Payload: Not available.	636 (395.2)	614 (381.5)	97.2	81.2	Still in orbit.

ASTRONAUTICS AND AERONAUTICS, 1973

Date	Name	Objective/Payload	Weight kg (lb)	Dimensions/Orbit	Period (min)	Remarks
Oct. 29*	Nnss O-20 (United States) 1973-81A Scout	Total weight: Not available. Objective: Develop space flight techniques and technology. Payload: Not available.	1 139.4 (708)	902.8 (561)	105.6	90.2	Launched by NASA for U.S. Navy as part of Navy Navigation Satellite System. Still in orbit.
Oct. 30	Intercosmos 10 (U.S.S.R.) 1973-82A Not available	Total weight: Not available. Objective: Make geophysical investigations in high latitudes and study electromagnetic connections between magnetosphere and ionosphere. Payload: Cylinder girded with folded solar wings; spherical lattice antennas extend perpendicular to satellite's axis; long spindle needle extends from lower portion with Langmuir sonde electrode at end; magnetometer extended on long rod from satellite.	1 454 (903.5)	259 (160.9)	102.1	74	Intercosmos 10 carried scientific instrumentation built by East Germany, Czechoslovakia, and U.S.S.R. Still in orbit.
Oct. 31	Cosmos 605 (U.S.S.R.) 1973-83A Not available	Total weight: Not available. Objective: Study effects of space on living organisms and test life-sustaining systems for biological subjects. Payload: Not available.	400 (248.5)	214 (133)	90.7	62.8	Biological satellite carried white rats, steppe turtles, fungi, and insects. Reentered 11/22/73 and was recovered.
Nov. 2	Cosmos 606 (U.S.S.R.) 1973-84A Not available	Total weight: Not available. Objective: "Continuation of Cosmos scientific satellite series." Payload: Not available.	39 708 (24 673.4)	635 (394.6)	717.5	62.7	Satellite speculated to be failure in Molniya II communications satellite series. Still in orbit.
Nov. 3	Mariner 10 (United States) 1973-85A Atlas-Centaur— TE-M-364-4	Total weight: 503 kg (1108 lbs) at launch. Objective: Investigate planet Mercury during 1973 opportunity by measuring environment, atmosphere, surface, and body characteristics, and investigate Venus. First priority assigned to Mercury investigations. Payload: Basic structure octagonal magnesium framework with 8 electronic compartments; octagonal structure measures 138 cm (54.5 in) wide diagonally, 46 cm (18 in) high. Two solar panels, each 269 cm (106 in) long and 97.5 cm (38.4 in) wide, attached by outriggers to octagon top; panels have combined photovoltaic cell area of 4.8 sq m (51.7 sq ft). When fully deployed, spacecraft measures 3.7 m	Heliocentric orbit				Mariner 10, 1st dual-planet mission, first to use gravitational attraction of one planet (Venus) to reach another planet (Mercury). Venus encounter at 1:01 pm EST 2/5/74, at 5800 km (3600 mi) from surface. Some 4000 photos of Venus revealed nearly round planet enveloped in smooth cloud layers, with slow rotational period (243 days) and only 0.05% of earth's magnetic field; atmosphere mostly hydrogen resulting from solar wind bombardment. After Venus flyby, spacecraft trajectory bent in toward sun for first exploration of Mercury. Mercury encounter at 4:47 pm EST 3/29/74 at 704 km (437 mi) from planet's surface. Photos revealed intensely cratered moonlike surface and thin atmosphere, mostly helium. Huge iron-rich core makes Mercury most dense planet; iron

Launch Date	Name, Country, International Designation, Vehicle	Payload Data	Apogee in Kilometers (and st mi)	Perigee in Kilometers (and st mi)	Period in Minutes	Inclination in Degrees	Remarks
		(12.2 ft) from top of low-gain antenna to bottom of heat shield for thrust vector control assembly of propulsion system; span is 8 m (26.2 ft) with solar panels extended; 6-m (20-ft) magnetometer boom adds additional 3.3 m (10.8 ft). Low-gain antenna consists of two conical plates slotted into outer end of pair of concentric tubes, each 2.9 m (9.35 ft) long and 5.72 cm (2.25 in) and 2.54 cm (1.0 in) in dia. Motor-driven high-gain antenna consists of aluminum honeycombed parabolic dish reflector 137 cm (54 in) in dia. Two sets of reaction control jets mounted at tips of solar panels and also on outrigger structures supporting high-gain antenna and magnetometer boom. Canopus star tracker assembly on upper ring structure of octagon. Seven experiments include two TV cameras, infrared radiometer, ultraviolet airglow spectrometer, uv occultation spectrometer, two magnetometers, plasma science experiment detector, charged particle telescope, and dual-frequency radio system. Propulsion subsystem is multistart monopropellant hydrazine system.					core also accounts for existence of magnetic field despite planet's extremely slow spin rate. After Mercury flyby, spacecraft entered solar orbit. *Mariner 10* return past Mercury scheduled for 9/21/74 to photograph planet's southern hemisphere.
Nov. 6	*Noaa 3* (United States) 1973-86A Thor-Delta 0300	Total weight: 345 kg (760 lbs). Objective: Launch spacecraft into sun-synchronous orbit of sufficient accuracy to permit spacecraft to accomplish operational mission requirements, make in-orbit evaluation and checkout of spacecraft and, upon completion, turn operational control over to NOAA/NESS. Orbit will have local equator crossing time of 8:30 am to permit regular and dependable daytime and night-	1 509.2 (937.8)	1 500 (932.1)	116.1	102.1	Launched by NASA on 2-stage Thor-Delta into circular polar orbit. Becomes primary satellite in NOAA global weather watch program. First to transmit local-area atmospheric temperature measurements to ground stations around world. Spacecraft functioning normally and turned over to NOAA 11/29/73 for operational use. Still in orbit.

time meteorological observations in both direct readout and stored modes of operation in support of National Operational Meteorological Satellite System.
Payload: Rectangular, box-shaped spacecraft with deployable 3-panel solar array. Base of main body 102 x 102 cm (40 x 40 in), overall height 122 cm (48 in). Total area of array 4.5 sq m (48 sq ft) with each of 3 panels measuring 93 x 162 cm (36.4 x 63.8 in). Three-axis stabilized earth-oriented satellite carries primary sensor complement of 2 very-high-resolution radiometer (VHRR) instruments and two scanning radiometer (SR) sensors for daytime and nighttime coverage, 2 vertical-temperature-profile radiometers (VTPR) for profile over every part of earth's surface at least twice daily, and 1 secondary sensor, solar proton monitor (SPM) for continuous measurements of proton and electron flux activity near earth. Thermal control system; 4 antennas.

Date	Spacecraft	Description					Remarks
Nov. 10	*Cosmos 607* (U.S.S.R.) 1973–87A Not available	Total weight: Not available. Objective: "Continuation of Cosmos scientific satellite series." Payload: Not available.	292 (181.4)	217 (134.8)	89.6	72.2	Reentered 11/22/73.
Nov. 10	DOD Spacecraft (United States) 1973–88A Titan IIID and	Total weight: Not available. Objective: Develop space flight techniques and technology. Payload: Not available.	265.5 (165)	162.5 (101)	88.7	96.9	Three satellites launched with single booster. 88A reentered 3/13/74.
	DOD Spacecraft 1973–88B and	Total weight: Not available. Objective: Develop space flight techniques and technology. Payload: Not available.	505.3 (314)	487.6 (303)	94.5	96.3	Still in orbit.

387

Launch Date	Name, Country, International Designation, Vehicle	Payload Data	Apogee in Kilometers (and st mi)	Perigee in Kilometers (and st mi)	Period in Minutes	Inclination in Degrees	Remarks
	DOD Spacecraft 1973-88D	Total weight: Not available. Objective: Develop space flight techniques and technology. Payload: Not available.	1 458.1 (906)	1 417.8 (881)	114.6	96.9	Still in orbit.
Nov. 14	*Molniya I-25* (U.S.S.R.) 1973-89A Not available	Total weight: Not available. Objective: Continue operation of long-range telephone, telegraph, and radio communications system and transmission of U.S.S.R. central television programs to stations in Orbita network. Payload: Carries apparatus for transmitting TV programs and multichannel radio communications, instruments of control-and-measurement complex, orientation system, orbit correction system, and power supplies.	39 798 (24 729.4)	566 (351.7)	717.9	64.8	*Molniya I-25* communications satellite still in orbit.
Nov. 16	*Skylab 4* (United States) 1973-90A Saturn IB	Total weight: 31 232.6 kg (68 856 lbs) at initial earth orbit insertion, including CSM/S-IVB, IU, SLA; 14 900.1 kg (32 869 lbs), CSM after S-IVB separation; 87 117.5 kg (192 061 lbs), total docked assembly (CSM/Workshop). Objective: Perform unmanned Saturn Workshop operations; reactivate Skylab orbital assembly in earth orbit; obtain medical data on crew for use in extending duration of manned space flights; perform inflight experiments. Payload: Apollo Block II command and service module (CSM) modified for Skylab use. Command module conical, 3.9 m (12 ft 10 in) in dia, 3.5 m (11 ft 5 in) high, with habitable volume of 17.8 cu m (630 cu ft). Service module cylindrical, 3.9 m (12 ft 10 in) in dia and 7.5 m (24 ft 10 in) long, with 88 960-newton (20 000-lb)-thrust main propulsion system, reac-	222.3 (138.1) After rendezvous and docking (Workshop) 442 (274.6)	148.2 (92.1) 422 (262.2)	93.2	50	*Skylab 4*, 3rd and final U.S. manned Orbital Workshop mission, carried Astronauts Gerald P. Carr, Edward G. Gibson, and William R. Pogue. After launch into phasing orbit, CSM separated from S-IVB stage and rendezvoused with Workshop. After 2 unsuccessful docking attempts, hard dock with Workshop achieved at 5:02 pm EST 11/16/73. Next day crew entered Workshop and began activation. During EVA-1 of 6 hrs 35 min, Gibson and Pogue repaired 2 antennas, installed film in ATM, and deployed impact detectors. Mission threatened by failure of gyroscope No. 1 on 11/23/73, but 2nd and 3rd gyros continued operation and flight was able to go full mission duration. On 12/25/73 Carr and Pogue set new duration record with 7-hr 1-min EVA-2. Astronauts made repairs, replaced ATM film, retrieved cassettes, and photographed sun and Comet Kohoutek. EVA-3, by Carr and Gibson 12/29/73, lasted 3 hrs 29 min, photographed sun and

ASTRONAUTICS AND AERONAUTICS, 1973

Date	Designation	Description	Weight kg (lb)	Perigee/Apogee km (mi)	Period (min)	Inclination (deg)	Remarks
		tion control system, fuel cells, oxygen, and radiators. Skylab cluster (with docked CSM) 36 m (118 ft) long with habitable volume of 363 cu m (12 800 cu ft).					Comet Kohoutek. EVA-4, made 2/3/74 by Carr and Gibson, lasted 5 hrs 19 min. On 2/8/74 CSM undocked from Workshop at 6:34 am EDT and inspected Workshop during fly-around. CM separated from SM and splashed down in Pacific at 11:17 am EDT 2/8/74. Crew in excellent physical condition after longest manned flight in space. Flight also set records for total inflight EVA, revolutions of earth (1214), data taken of earth (20 500 pictures and 31 km [19 mi] of recording tape). Total EVA time 22 hrs 24 min. Total flight time 84 days 1 hr 16 min.
Nov. 20	Cosmos 608 (U.S.S.R.) 1973-91A Not available	Total weight: Not available. Objective: "Continuation of Cosmos scientific satellite series." Payload: Not available.	498 (309.4)	269 (167.2)	92.2	70.9	Reentered 7/10/74.
Nov. 21	Cosmos 609 (U.S.S.R.) 1973-92A Not available	Total weight: Not available. Objective: "Continuation of Cosmos scientific satellite series." Payload: Not available.	330 (205.1)	173 (107.5)	89.5	69.9	Reentered 12/4/73.
Nov. 27	Cosmos 610 (U.S.S.R.) 1973-93A Not available	Total weight: Not available. Objective: "Continuation of Cosmos scientific satellite series." Payload: Not available.	546 (339.3)	515 (320)	95.2	74	Still in orbit.
Nov. 28	Cosmos 611 (U.S.S.R.) 1973-94A Not available	Total weight: Not available. Objective: "Continuation of Cosmos scientific satellite series." Payload: Not available.	481 (298.9)	269 (167.1)	92	71	Reentered 6/19/74.
Nov. 28	Cosmos 612 (U.S.S.R.) 1973-95A Not available	Total weight: Not available. Objective: "Continuation of Cosmos scientific satellite series." Payload: Not available.	346 (215)	205 (127.4)	90	72.8	Reentered 12/11/73.
Nov. 30	Cosmos 613 (U.S.S.R.) 1973-96A Not available	Total weight: Not available. Objective: "Continuation of Cosmos scientific satellite series." Payload: Not available.	274 (170.3)	188 (116.8)	89.7	51.6	Cosmos 613, reportedly an unmanned Soyuz spacecraft, maneuvered to orbit 100 km (60 mi) farther out in test of onboard propulsion system. Flight test indicates Soviets possibly moving toward higher operational Soyuz-Salyut regimes to carry out longer missions. Softlanded near Karaganda in Soviet Central Asia 1/30/74.

Launch Date	Name, Country, International Designation, Vehicle	Payload Data	Apogee in Kilometers (and st mi)	Perigee in Kilometers (and st mi)	Period in Minutes	Inclination in Degrees	Remarks
Nov. 30	Molniya 1–26 (U.S.S.R.) 1973–97A Not available	Total weight: Not available. Objective: Continue operation of long-range telephone, telegraph, and radio communications system and transmission of U.S.S.R. central television programs to stations in Orbita network. Payload: Carries apparatus for transmitting television programs and multichannel radio communications, instruments of control-and-measurement complex, orientation system, orbit correction system, and power supplies.	40 830 (25 370.6)	618 (384)	740	62.9	Molniya 1–26 communications satellite still in orbit.
Dec. 4	Cosmos 614 (U.S.S.R.) 1973–98A Not available	Total weight: Not available. Objective: "Continuation of Cosmos scientific satellite series." Payload: Not available.	807 (501.5)	769 (477.8)	100.6	74	Still in orbit.
Dec. 13	Cosmos 615 (U.S.S.R.) 1973–99A Not available	Total weight: Not available. Objective: "Continuation of Cosmos scientific satellite series." Payload: Not available.	829 (515.1)	269 (167.2)	95.6	71	Still in orbit.
Dec. 13	Dscs F–3 (United States) 1973–100A Titan IIIC and	Total weight: 545 kg (1200 lbs). Objective: Continue operational Defense Satellite Communications System. Payload: Cylindrical 2.7-m-dia x 4-m-high (9- x 13-ft) spacecraft; 13.7-m (4.5-ft) unfolding boom with 13.6-kg (30-lb) mass at end located between narrow beam communications antennas to correct dynamic instabilities.	36 124.9 (22 447)	35 799.9 (22 245)	1445	2	Two improved DSCS (Defense Satellite Communications System) satellites launched by single Titan IIIC booster. Both satellites modified to correct design deficiencies of earlier pair. Both performing well and turned over to Defense Communications Agency. Operational stations in synchronous orbit at 13° W and 175° E. Still in orbit.
Dec. 13	Dscs F–4 1973–100B	Total weight: 545 kg (1200 lbs). Objective: Continue operational Defense Satellite Communications System.	36 137.8 (22 455)	35 944.7 (22 335)	1449	2	Still in orbit.

Date	Name	Payload/Objective	Weight kg (lbs)	Apogee/Perigee km (mi)	Period (min)	Incl. (deg)	Remarks
		Payload: Cylindrical 2.7-m-dia × 4-m-high (9- × 13-ft) spacecraft; 13.7-m (4.5-ft) unfolding boom with 13.6-kg (30-lb) mass at end located between narrow beam communications antennas to correct dynamic instabilities.					
Dec. 15*	*Explorer 51* (United States) 1973-101A Thor-Delta 1900	Total weight: 668 kg (1473 lbs). Objective: Investigate photochemical processes accompanying absorption of solar ultraviolet radiation in earth's atmosphere by making closely coordinated measurements of reacting constituents from spacecraft with onboard propulsion to permit perigee and apogee altitudes to be varied by command. Payload: Cylindrical 16-sided satellite 135 cm (35 in) dia and 114 cm (45 in) high. Consists of two shells; inner shell holds 14 scientific instruments and outer shell covered with n-on-p solar cells. Orbit-adjust propulsion subsystem carries 166.5 kg (367 lbs) of hydrazine fuel and employs three thrusters. Spin-stabilized satellite also contains omnidirectional S-band antenna mounted in spacecraft equatorial plane, between 2 solar array hats, and redundant nickel-cadmium batteries.	4 304 (2674.4)	157 (97.6)	132.5	68.1	First of a series of 2nd-generation spacecraft. Placed into elliptical orbit, spacecraft dips as low as 125 km (77.7 mi) with series of maneuvers into largely unexplored region of lower atmosphere. Onboard propulsion system then raises spacecraft's altitude. Orbit eventually will be circularized. All 14 scientific instruments operational and returning excellent data. Still in orbit.
Dec. 17	*Cosmos 616* (U.S.S.R.) 1973-102A Not available	Total weight: Not available. Objective: "Continuation of Cosmos scientific satellite series." Payload: Not available.		326 (202.6) 206 (128)	89.8	72.9	Reentered 12/28/73.
Dec. 18	*Soyuz 13* (U.S.S.R.) 1973-103A Not available	Total weight: Not available. Objective: Astrophysical observation of stars in uv range with Orion 2 telescope system; multispectral observation of separate sections of earth's surface for earth resources purposes; continued detailed verification of on-board systems; further testing of manual and automatic controls and autonomous methods of navigation in various flight conditions. Payload: 3-unit spacecraft 9.5 m (31		246 (152.9) 188 (116.8)	88.9	51.6	*Soyuz 13*—carrying Cosmonauts Pyotr Klimuk and Valentin Lebedev—orbited earth in apparent test of modified Soyuz capsule to be used for rendezvous with U.S. Apollo spacecraft in 1975. Crew photographed earth in various spectral bands, conducted successful biological experiment with first nutritive protein mass grown in space, and took 16 spectrographies of stellar radiation using Orion 2 telescope. Inflight spacecraft maneuvers changed orbit to more circular 272-km (169-mi) apogee and 225-km (139.8-

Launch Date	Name, Country, International Designation, Vehicle	Payload Data	Apogee in Kilometers (and st mi)	Perigee in Kilometers (and st mi)	Period in Minutes	Inclination in Degrees	Remarks
		ft) long with 2 crew cabins totaling 8.9 cu m (315 cu ft) of interior space.					mi) perigee. During Soyuz flight, *Skylab 4* (SL–4) was also in earth orbit—the first time U.S.S.R. and U.S. had manned missions in space simultaneously. *Soyuz 13* softlanded 12/26/73 near Karaganda, Kazakhstan. Total mission duration, 7 days 20 hrs 55 min.
Dec. 19	*Cosmos 617* (U.S.S.R.) 1973–104A Not available	Total weight: Not available. Objective: "Continuation of Cosmos scientific satellite series." Payload: Not available.	1 485 (922.7)	1 336 (830.1)	114	74	Eight satellites launched on single booster. Still in orbit.
and							
	Cosmos 618 1973–104B	Total weight: Not available. Objective: "Continuation of Cosmos scientific satellite series." Payload: Not available.	1 488 (924.6)	1 445 (897.9)	115.2	74	Still in orbit.
and							
	Cosmos 619 1973–104C	Total weight: Not available. Objective: "Continuation of Cosmos scientific satellite series." Payload: Not available.	1 490 (925.8)	1 424 (884.8)	115	74	Still in orbit.
and							
	Cosmos 620 1973–104D	Total weight: Not available. Objective: "Continuation of Cosmos scientific satellite series." Payload: Not available.	1 496 (929.6)	1 460 (907.2)	115.4	74	Still in orbit.
and							

	Cosmos 621 1973-104E	Total weight: Not available. Objective: "Continuation of Cosmos scientific satellite series." Payload: Not available.	1 487 (924)	1 407 (874.3)	114.8	74	Still in orbit.
and							
	Cosmos 622 1973-104F	Total weight: Not available. Objective: "Continuation of Cosmos scientific satellite series." Payload: Not available.	1 487 (924)	1 371 (851.9)	114.3	74	Still in orbit.
and							
	Cosmos 623 1973-104G	Total weight: Not available. Objective: "Continuation of Cosmos scientific satellite series." Payload: Not available.	1 487 (924)	1 389 (863.1)	114.5	74	Still in orbit.
and							
	Cosmos 624 1973-104H	Total weight: Not available. Objective: "Continuation of Cosmos scientific satellite series." Payload: Not available.	1 487 (924)	1 353 (840.7)	114.2	74	Still in orbit.
Dec. 21	Cosmos 625 (U.S.S.R.) 1973-105A Not available	Total weight: Not available. Objective: "Continuation of Cosmos scientific satellite series." Payload: Not available.	340 (211.3)	187 (116.2)	89.7	72.8	Reentered 1/3/74.
Dec. 25	Molniya II-8 (U.S.S.R.) 1973-106A Not available	Total weight: Not available. Objective: Continue operation of long-range telephone and telegraph radio communications in U.S.S.R. and transmission of central television programs of U.S.S.R. to stations in Orbita and participating networks. Payload: Not available.	40 809 (25 357.5)	488 (303.2)	736.9	62.8	Molniya II-8 communications satellite still in orbit.

Launch Date	Name, Country, International Designation, Vehicle	Payload Data	Apogee in Kilometers (and st mi)	Perigee in Kilometers (and st mi)	Period in Minutes	Inclination in Degrees	Remarks
Dec. 26	Oreol 2 (U.S.S.R.-France) 1973-107A Not available	Total weight: Not available. Objective: Continue investigations in earth's upper atmosphere at high latitudes and study nature of polar auroras begun by Oreol satellite in 1971. Payload: Carries equipment for research in spectrum of protons and electrons within wide band of energy, to measure integrated intensity of protons, and to determine ion composition of atmosphere.	1 974 (1 226.6)	399 (247.9)	109.1	73.9	Oreol 2, second satellite in joint Soviet-French program, launched successfully into elliptical orbit. Provided corroborating data (along with Intercosmos 10) for Soviet-French series of 16 high-altitude balloons launched from Sweden. Still in orbit.
Dec. 27	Cosmos 626 (U.S.S.R.) 1973-108A Not available	Total weight: Not available. Objective: "Continuation of Cosmos scientific satellite series." Payload: Not available.	264 (164)	250 (155.3)	89.6	65	Radar-carrying satellite can monitor surface ship traffic around globe. Orbit changed to 990 by 909 km (615.2 by 564.8 mi) on 2/11/74. Still in orbit.
Dec. 29	Cosmos 627 (U.S.S.R.) 1973-109A Not available	Total weight: Not available. Objective: "Continuation of Cosmos scientific satellite series." Payload: Not available.	1 019 (633.2)	973 (604.6)	105	83	Still in orbit.

*Time at launch site; 1 day later by Greenwich Mean Time.
**Time at launch site; 1 day earlier by Greenwich Mean Time.

Appendix B

CHRONOLOGY OF MAJOR NASA LAUNCHES, 1973

This chronology of major NASA launches in 1973 is intended to provide an accurate and ready historical reference, compiling and verifying information previously scattered in several sources. It includes launches of all rocket vehicles larger than sounding rockets launched either by NASA or under "NASA direction" (e.g., in 1973 NASA provided vehicles and launch facilities and launched the Communications Satellite Corporation's *Intelsat-IV F-7* and Canada's *Anik 2*, as well as *Nnss O-20* for the U.S. Navy and *Noaa 3* for the National Oceanic and Atmospheric Administration. During 1973, NASA's only launch failure was NOAA's ITOS-E.

An attempt has been made to classify performance of both the launch vehicle and the payload and to summarize total results in terms of primary mission. Three categories have been used for evaluating vehicle performance and mission results—successful (S), partially successful (P), and unsuccessful (U). A fourth category, unknown (Unk), has been added for payloads when vehicle malfunctions did not give the payload a chance to exercise its main experiments. These divisions are necessarily arbitrary; many of the results cannot be neatly categorized. Also they ignore the fact that a great deal is learned from missions that may have been classified as unsuccessful.

Date of launch is referenced to local time at the launch site. Open sources were used, verified when in doubt with the project offices in NASA Headquarters and with NASA Centers. For further information on each item, see Appendix A of this volume and the entries in the main chronology as referenced in the index. The information was compiled in August 1974 by Leonard C. Bruno of the Science and Technology Division of the Library of Congress.

ASTRONAUTICS AND AERONAUTICS, 1973

Date	Name (NASA Code)	General Mission	Launch Vehicle (Site)	Performance Vehicle	Performance Payload	Performance Mission	Remarks
Apr. 5*	*Pioneer 11* (Pioneer-G)	Scientific interplanetary probe	Atlas-Centaur–TE–M–364–4 (ETR)	S	S	S	Second in new-generation Pioneer series; scheduled to fly by Jupiter within 42 000 km (26 000 mi) 12/3/74 EST (12/2/74 PST). After investigating Jupiter's never-explored south polar region, planet's gravity will throw craft toward Saturn for Sept. 1979 first-ever encounter. By 3/20/74 spacecraft had emerged undamaged from 7-mo journey through 80-million-km (50-million-mi)-thick Asteroid Belt.
Apr. 20	*Anik 2* (Telesat-B)	Operational communications satellite	Thorad-Delta (ETR)	S	S	S	Launched by NASA for Canadian Domestic Communications Satellite System into elliptical transfer orbit. Apogee kick motor fired by Telesat Canada 4/23 and spacecraft placed in synchronous orbit off west coast of South America. On 6/18 relayed first domestic comsat telecast across U.S.
May 14	*Skylab 1* (SL–1)	Orbital Workshop	Saturn V (ETR)	S	S	S	First U.S. Orbital Workshop, employed for three manned visits (*Skylab 2, 3, 4*). Provided environmental and laboratory base for crew's earth resources, solar, and medical experiments. Three separate 3-man crews lived and worked in Orbital Workshop for total of nearly 6 mos.
May 25	*Skylab 2* (SL–2)	Manned Orbital Workshop visit	Saturn IB (ETR)	S	S	S	First manned visit to Orbital Workshop. Astronauts Charles Conrad, Jr., commander; Paul J. Weitz, pilot; Joseph P. Kerwin, science pilot. Made major Workshop repairs; deployed makeshift sun shield and released jammed solar wing. Performed 3 EVAs totaling 5 hrs 6 min. Acquired medical, solar, and earth resources data. CM landed in Pacific near U.S.S. *Ticonderoga* at 9:50 am EDT 6/22. Total mission time 28 days 50 min.
June 10	*Explorer 49* (RAE–B)	Scientific satellite, radio astronomy	Thorad-Delta (ETR)	S	S	S	Last scheduled U.S. space mission to moon. Entered lunar orbit 3:21 am EDT 6/15. Uses moon as shield from extraneous radio noise from earth, to make most extensive study ever of low-frequency signals from galactic and extragalactic radio sources. Still in lunar orbit.
July 16	ITOS–E	Operational meteorological satellite	Thor-Delta 0300 (WTR)	U	Unk	U	Spacecraft was to be 3rd operational satellite of Improved Tiros Operational System (ITOS) series, launched for NOAA. At 270 sec after 2nd-stage ignition, hydraulic pump abruptly ceased output, causing loss of hydraulic pressure and thrust vector control. Vehicle tumbled and did not achieve orbital velocity.

396

ASTRONAUTICS AND AERONAUTICS, 1973

Date	Name	Description	Launch vehicle				Remarks
July 28	Skylab 3 (SL-3)	Manned Orbital Workshop visit	Saturn IB (ETR)	S	S	S	Second manned visit to Orbital Workshop. Astronauts Alan L. Bean, commander; Jack R. Lousma, pilot; Owen K. Garriott, science pilot. Made further Workshop repairs, deployed Mylar twin-pole sunshade to augment makeshift parasol, installed new gyros. Performed 3 EVAs totaling 13 hrs 42 min. Acquired medical, solar, and earth resources data. CM landed in Pacific near U.S.S. New Orleans at 6:20 pm EDT 9/25. Total mission time 59 days 11 hrs 9 min.
Aug. 23	Intelsat-IV F-7	Operational communications satellite	Atlas-Centaur-TE-M-364-4 (ETR)	S	S	S	Launched into elliptical transfer orbit by NASA ComSatCorp. On 8/25 ComSatCorp fired apogee kick motor to circularize synchronous orbit and put satellite over Atlantic Ocean at 330° E longitude. Commercial operations began 9/9/73.
Oct. 25*	Explorer 50 (IMP-J)	Interplanetary monitoring platform	Thorad-Delta (ETR)	S	S	S	Tenth and last in IMP (Interplanetary Monitoring Platform) series to study interplanetary radiation throughout an 11-yr cycle. Placed in orbit halfway between earth and moon. Has returned great amount of data. Still in orbit.
Oct. 29*	Nnss O-20	Transit navigation satellite	Scout (WTR)	S	S	S	Launched by NASA for U.S. Navy as part of Navy Navigation Satellite System. Still in orbit.
Nov. 3	Mariner 10 (Mariner-J)	Scientific interplanetary probe (Venus/Mercury)	Atlas-Centaur-TE-M-364-4 (ETR)	S	S	S	First dual-planet mission; first to use gravitational attraction of one planet (Venus) to reach another planet (Mercury). Took over 4000 photos of Venus on flyby 5800 km (3600 mi) from surface 2/5/74. First-ever Mercury exploration accomplished during flyby at 704 km (437 mi) 3/29/74; photos revealed intensely cratered planet with huge iron-rich core. After flyby, spacecraft entered solar orbit; scheduled for 9/21/74 Mercury return flyby.
Nov. 6	Noaa 3 (TIROS-F)	Operational meteorological satellite	Thor-Delta 0300 (WTR)	S	S	S	Third operational spacecraft in series and first to transmit local-area atmospheric temperature measurements to ground stations around world. Functioning normally and turned over to NOAA 11/29/73 for operational use.
Nov. 16	Skylab 4 (SL-4)	Manned Orbital Work-shop visit	Saturn IB (ETR)	S	S	S	Third manned visit to Orbital Workshop. Astronauts Gerald P. Carr, commander; William R. Pogue, pilot; Edward G. Gibson, science pilot. Collected medical, solar, and earth resources data. Longest manned space flight (84 days 1 hr 16 min) also set record for longest inflight EVA (7 hrs 1 min, Carr and Pogue), total EVA time for one mission (22 hrs 24 min), and most manned revolutions of earth (1214). Obtained first photos of solar flare from beginning to end; observed and photographed Comet Kohoutek. CM landed in Pacific near U.S.S. New Orleans at 11:17 am EDT 2/8/74.

Date	Name (NASA Code)	General Mission	Launch Vehicle (Site)	Performance			Remarks
				Vehicle	Payload	Mission	
Dec. 15*	Explorer 51 (AE-C)	Scientific satellite, aeronomy	Thor-Delta 1900 (WTR)	S	S	S	First of series of 2nd-generation spacecraft. Highly elliptical orbit allows spacecraft to dip as low as 125 km (77.7 mi) into largely unexplored region of lower atmosphere. Onboard propulsion system will eventually circularize satellite's orbit. All 14 scientific instruments operational.

*Time at launch site; 1 day later by Greenwich time.
**Time at launch site; 1 day earlier by Greenwich time.

Appendix C

CHRONOLOGY OF MANNED SPACE FLIGHT, 1973

This chronology contains basic information on all manned space flights during 1973 and—taken with Appendix C to the 1965, 1966, 1968, 1969, 1970, 1971, and 1972 volumes of this publication—provides a summary record of manned exploration of the space environment through 1973. The information was compiled by Leonard C. Bruno of the Science and Technology Division of the Library of Congress.

This was the most active year for manned space flight since 1969. While two U.S. Apollo flights placed 6 men in space during 1972 (the U.S.S.R. had no manned flights last year), 1973 saw three U.S. Skylab flights and two U.S.S.R. Soyuz flights, for a total of 13 men in space.

The year was marked by the record-breaking flights of the U.S. Skylab program and the Soviet resumption, after a hiatus of more than two years, of its manned program. The crew of *Skylab 2* flew nearly a month and proved man can do real work in space, repairing the damaged *Skylab 1* Orbital Workshop. *Skylab 3* flew for almost two months and further demonstrated the flexibility provided by man in space. The *Skylab 4* crew orbited the earth for more than 84 days, performed all their assigned tasks, and returned to the earth in excellent physical condition—ostensibly placing no time limit on man's ability to live and work in space. This final Skylab mission set many records—among them were longest manned flight in space, and longest single EVA in orbit. The U.S. has no more manned flights scheduled until the 1975 Apollo-Soyuz Test Project with the U.S.S.R.

The Soviet Union demonstrated its confidence in the redesigned Soyuz capsule with the successful two-day flight of *Soyuz 12*, its first manned attempt since the fatal flight of *Soyuz 11* during June 1971. Its December two-man *Soyuz 13* mission lasted nearly 8 days in an apparent test of the Soyuz craft to be used for the 1975 Apollo-Soyuz Test Project.

By the end of 1973, the United States had conducted a total of 30 manned space flights—2 suborbital, 19 in earth orbit, 3 in lunar orbit, and 6 manned lunar landings—with a total of 41 different crewmen. Of the 41 American astronauts, 10 had participated in two flights each, 4 had flown three times, and 3 had flown four times. The Soviet Union had made 20 manned flights, all in earth orbit, with 29 cosmonauts. Three had participated in two flights each and two had flown three times. Cumulative totals for manned spacecraft hours in flight had reached 7643 hours 42 minutes for the United States and 2333 hours 27 minutes for the Soviet Union. Cumulative total man-hours in space were 21 851 hours 15 minutes for the United States and 4875 hours 34 minutes for the U.S.S.R.

Data on U.S. flights are the latest available to date within NASA, although minor details are subject to modification as data are refined.

ASTRONAUTICS AND AERONAUTICS, 1973

Date Launched	Date Recovered	Designation (NASA Code)	Crew	Weight in Kilograms (and in lbs)	Revolutions	Maximum Distance from Earth in Kilometers (and st mi)	Duration	Remarks
May 25	June 22	*Skylab 2* (SL-2)	Charles Conrad, Jr. Joseph P. Kerwin Paul J. Weitz	30 464.7 (67 163)	404	440 (273.4)	28 days 50 min	First U.S. manned Orbital Workshop. After launch into phasing orbit, CMS separated from S-IVB and rendezvoused with damaged *Skylab 1* Workshop (launched May 14) at 4:30 pm EDT during 5th orbit. After 4 unsuccessful docking attempts, hard docking with Workshop was achieved at 11:50 pm EDT. Next day crew entered multiple docking adapter (MDA) at 12:45 pm EDT and entered Workshop at 4:30 pm EDT. Makeshift sun shield (parasol) deployed through solar airlock at 8:30 pm EDT May 26 and external temperature began to decrease. Workshop activated and internal temperature decreased. On June 7 Conrad and Kerwin made 3-hr 30-min EVA; used 7.6-m (25-ft) pole fitted with cutters to sever small bolt jamming solar wing. Wing was released and crew recovered 300 w electrical power. On June 19 Conrad and Weitz made 1-hr 31-min EVA to retrieve and replace cassettes. Total astronaut EVA time 5 hrs 6 min. On June 22 CSM undocked from Workshop during fly-around. CM separated from SM and splashed down in Pacific at 9:50 am EDT June 22. Recovery by U.S.S. *Ticonderoga*. Despite mission anomalies, 80% of solar data planned was obtained; 12 of 15 earth resources data runs were accomplished; all 16 medical experiments were conducted. Live and recorded TV were transmitted throughout mission.
July 28	Sept. 25	*Skylab 3* (SL-3)	Alan L. Bean Jack R. Lousma Owen K. Garriott	30 718 (67 721)	858	440 (273.4)	59 days 11 hrs 9 min	Second U.S. manned Orbital Workshop. After launch into phasing orbit, CSM separated from S-IVB stage and rendezvoused with Workshop at 3:38 pm EDT during 5th orbit. Crew entered multiple docking adapter (MDA) at 4:00 pm EDT and began Workshop activation. Crew's motion sickness caused rescheduling of tasks. Leak in 2 of 4 clusters of CSM steering rockets threatened safe reentry, and rescue preparations were initiated on ground but not required.

Date	Mission	Crew	Weight kg (lb)	Orbits	Duration	Remarks
Sept. 27						On 8/6 in 6-hr 31-min EVA, Garriott and Lousma installed 3.6- by 7.3-m (12- by 24-ft) Mylar twin-pole sunshade to augment parasol. Also deployed experiments, inspected CSM RCS, and retrieved film. During 4-hr 30-min EVA-2, Lousma and Garriott completed installation of 6 redesigned gyros. On 9/22 Bean and Garriott made 2-hr 41-min EVA to retrieve experiments and parasol samples. Total astronaut EVA time 13 hr 42 min. On 9/25 CSM undocked from Workshop at 3:34 pm EDT and separated at 3:50 pm. CM splashed down in Pacific at 6:20 pm EDT 9/25. Recovery by U.S.S. *New Orleans*. Experiment accomplishments surpassed all plans; 150% of observing time scheduled for ATM accomplished; EREP sensors operated on 39 data runs, exceeding scheduled 28 passes; all crew medical experiments successfully conducted.
Sept. 29	*Soyuz 12*	Vasily G. Lazarev Oleg G. Makarov	6 577 (14 500)	32	47 hrs 16 min	First U.S.S.R. manned space flight since fatal *Soyuz 11* mission June 1971. *Soyuz 12* launched from Baykonur Cosmodrome at 5:18 pm local time (3:18 pm Moscow time; 8:18 am EDT). With successful launch, Soviets announced mission duration of a manned flight for first time. Cosmonauts wore new spacesuits, tested manual and automatic control and piloting techniques, obtained spectrographic data of natural formations of earth's surface. Orbit was changed during flight and set new Soviet altitude record for manned flight. Redesigned *Soyuz* craft allowed more extensive maneuvering, more distant orbits, and more manual control. Flight believed related to future 2-day ferry missions to orbital stations. Spacecraft reentered successfully and landed 400 km (250 mi) southwest of Karaganda in Kazakhstan Steppe at 2:34 pm Moscow time (7:34 am EDT) 9/29.
Nov. 16	*Skylab 4* (SL-4)	Gerald P. Carr Edward G. Gibson William R. Pogue	31 232.6 (68 856)	1214	84 days 1 hr 16 min	Third and final U.S. manned Orbital Workshop mission. After launch into phasing orbit, CSM separated from S-IVB stage and rendezvoused with Workshop. After 2 unsuccessful docking attempts, hard dock with Workshop was achieved at 5:02 pm EST 11/16/73. Next day crew entered Workshop and began activation. During EVA-1 of 6 hrs 35 min, Gibson and Pogue repaired 2

Date		Designation (NASA Code)	Crew	Weight in Kilograms (and in lbs)	Revolutions	Maximum Distance from Earth in Kilometers (and st mi)	Duration	Remarks
Launched	Recovered							
								antennas, installed film in ATM, and deployed impact detectors. Mission threatened by failure of gyroscope No. 1 on 11/23/73, but 2nd and 3rd gyros continued operation and flight was able to continue to full mission duration. On 12/25/73 Carr and Pogue set new duration record with 7-hr 1-min EVA. Astronauts made repairs, replaced ATM film and retrieved cassettes, and photographed sun and Comet Kohoutek. EVA-3, by Carr and Gibson 12/29/73, lasted 3 hrs 29 min., photographed sun and Comet Kohoutek. Final EVA, on 2/3/74 by Carr and Gibson, lasted 5 hrs 19 min. Total astronaut EVA time 22 hrs 24 min. On 2/8/74 CSM undocked from Workshop at 6:34 am EDT and inspected Workshop during fly-around. CM separated from SM and splashed down in Pacific at 11:17 am EDT 2/8/74. Recovery by U.S.S. *New Orleans*. Crew in excellent physical condition after longest manned flight in space. Set record for longest single inflight EVA, total EVA time for mission, and most manned revolutions of earth.
Dec. 18	Dec. 26	*Soyuz 13*	Pyotr Klimuk Valentin Lebedev	6 577 (14 500)	128	272 (169)	7 days 20 hrs 55 min	Launched from Baykonur Cosmodrome at 4:55 pm local time (2:55 pm Moscow time; 6:55 am EST). With U.S. *Skylab 4* astronauts in earth orbit since 11/16, marked first time U.S. and U.S.S.R. had men orbiting earth simultaneously. Apparently tested modified Soyuz capsule to be used for docking with U.S. Apollo spacecraft scheduled for 1975. Crew photographed earth in various spectral bands and conducted successful biological experiment with first nutritive protein mass grown in space. Took 16 spectrographies of stellar radiation using Orion 2 telescope. Inflight spacecraft maneuvers changed orbit to more circular one. Spacecraft reentered successfully in snowstorm 200 km (125 mi) southwest of Karaganda in Kazakhstan Steppe at 11:50 am Moscow time (3:50 am EST) 12/26.

Appendix D

ABBREVIATIONS OF REFERENCES

Listed here are the abbreviations used for citing sources in the text. Not all sources are listed, only those that are abbreviated.

AAAS Bull	American Association for the Advancement of Science Bulletin
A&A	American Institute of Aeronautics and Astronautics' magazine, *Astronautics & Aeronautics*
A&A 1974	NASA's *Astronautics and Aeronautics, 1974* chronology (this publication)
ABC	American Broadcasting Company
AEC Release	Atomic Energy Commission News Release
Aero Daily	*Aerospace Daily* newsletter
Aero Med	*Aerospace Medicine* magazine
AF Mag	Air Force Association's *Air Force Magazine*
AFFTC Release	Air Force Flight Test Center News Release
AFHF Newsletter	*Air Force Historical Foundation Newsletter*
AFJ	*Armed Forces Journal* magazine
AFNS Release	Air Force News Service Release
AFOSR Release	Air Force Office of Scientific Research News Release
AFRPL Release	Air Force Rocket Propulsion Laboratory News Release
AFSC *Newsreview*	Air Force Systems Command's *Newsreview*
AFSC Release	Air Force Systems Command News Release
AFSSD Release	Air Force Space Systems Division News Release
AIA Release	Aerospace Industries Association of America News Release
AIAA *Facts*	American Institute of Aeronautics and Astronautics' *Facts*
AIAA Release	American Institute of Aeronautics and Astronautics News Release
AIP *Newsletter*	American Institute of Physics *Newsletter*
AP	Associated Press news service
ARC *Astrogram*	NASA Ames Research Center's *Astrogram*
Astro Journ	American Astronomical Society's *Astrophysical Journal*
Atlanta JC	*Atlanta Journal Constitution* newspaper
Av Daily	*Aviation Daily* newsletter
Av Wk	*Aviation Week & Space Technology* magazine
B News	*Birmingham News* newspaper
B Sun	*Baltimore Sun* newspaper
Bull Atom Sci	Education Foundation for Nuclear Science *Bulletin of the Atomic Scientists*
Bus Wk	*Business Week* magazine
C Daily News	*Chicago Daily News* newspaper
C Trib	*Chicago Tribune* newspaper
Can Press	Canadian Press news service
CBS	Columbia Broadcasting System
C&E News	*Chemical & Engineering News* magazine
Cl *PD*	Cleveland *Plain Dealer* newspaper
Cl *Press*	*Cleveland Press* newspaper
Columbia J Rev	*Columbia Journalism Review* magazine

403

ComSatCorp Release	Communications Satellite Corporation News Release
CQ	*Congressional Quarterly*
CR	*Congressional Record*
CSM	*Christian Science Monitor* newspaper
CTNS	Chicago Tribune News Service
D News	*Detroit News* newspaper
D Post	*Denver Post* newspaper
DASA Release	Defense Atomic Support Agency News Release
DJ	Dow Jones news service
DOC PIO	Department of Commerce Public Information Office
DOD Release	Department of Defense News Release
DOT Release	Department of Transportation News Release
EOP Release	Executive Office of the President News Release
FAA Release	Federal Aviation Administration News Release
FBIS–Sov	Foreign Broadcast Information Service, Soviet number
FonF	*Facts on File*
FRC Release	NASA Flight Research Center News Release
FRC X–Press	NASA Flight Research Center's *FRC X–Press*
GE Forum	*General Electric Forum* magazine
Goddard News	NASA Goddard Space Flight Center's *Goddard News*
GSFC Release	NASA Goddard Space Flight Center News Release
GSFC SSR	NASA Goddard Space Flight Center's *Satellite Situation Report*
GT&E Release	General Telephone & Electronics News Release
H Chron	*Houston Chronicle* newspaper
H Post	*Houston Post* newspaper
JA	*Journal of Aircraft* magazine
JPL Lab-Oratory	Jet Propulsion Laboratory's *Lab-Oratory*
JPL Release	Jet Propulsion Laboratory News Release
JPRS	Department of Commerce Joint Publications Research Service
JSC Release	NASA Lyndon B. Johnson Space Center (Manned Spacecraft Center until Feb. 17, 1973) News Release
JSC Roundup	NASA Lyndon B. Johnson Space Center's *Space News Roundup*
JSR	American Institute of Aeronautics and Astronautics' *Journal of Spacecraft and Rockets* magazine
KC Star	*Kansas City Star* newspaper
KC Times	*Kansas City Times* newspaper
KSC Release	NASA John F. Kennedy Space Center News Release
LA Her-Exam	Los Angeles *Herald-Examiner* newspaper
LA Times	*Los Angeles Times* newspaper
Langley Researcher	NASA Langley Research Center's *Langley Researcher*
LaRC Release	NASA Langley Research Center News Release
LATNS	Los Angeles Times News Service
LC Info Bull	Library of Congress *Information Bulletin*
LeRC Release	NASA Lewis Research Center News Release
Lewis News	NASA Lewis Research Center's *Lewis News*
M Her	*Miami Herald* newspaper
M News	*Miami News* newspaper
M Trib	*Minneapolis Tribune* newspaper
Marshall Star	NASA George C. Marshall Space Flight Center's *Marshall Star*
MJ	*Milwaukee Journal* newspaper
MSC Release	NASA Manned Spacecraft Center (became Lyndon B. Johnson Space Center Feb. 17, 1973) News Release
MSC Roundup	NASA Manned Spacecraft Center's *Space News Roundup*
MSFC Release	NASA George C. Marshall Space Flight Center News Release
N Hav Reg	*New Haven Register* newspaper
N News	*Newark News* newspaper
N Va Sun	*Northern Virginia Sun* newspaper
NAA News	National Aeronautic Association *News*
NAA Record Bk	National Aeronautic Association's *World and U.S.A. National Aviation-Space Records*

NAC Release	National Aviation Club News Release
NAE Release	National Academy of Engineering News Release
NANA	North American Newspaper Alliance
NAS Release	National Academy of Sciences News Release
NAS–NRC Release	National Academy of Sciences–National Research Council News Release
NAS–NRC–NAE *News Rpt*	National Academy of Sciences–National Research Council–National Academy of Engineering *News Report*
NASA Ann	NASA Announcement
NASA HHR–39	NASA Historical Report No. 39
NASA Hist Off	NASA Historical Office
NASA Hq *WB*	NASA Headquarters *Weekly Bulletin*
NASA Int Aff	NASA Office of International Affairs
NASA *LAR*, XIII/8	NASA *Legislative Activities Report*, Vol. XIII, No. 8
NASA Leg Off	NASA Office of Legislative Affairs
NASA prog off	NASA program office (for program reported)
NASA proj off	NASA project office (for project reported)
NASA Release	NASA Headquarters News Release
NASA Rpt SRL	NASA report of sounding rocket launching
NASA SP–4018	NASA Special Publication No. 4018
Natl Obs	*National Observer* magazine
Nature	*Nature Physical Science* magazine
NBC	National Broadcasting Company
NGS Release	National Geographic Society News Release
NMI	NASA Management Instruction
NN	NASA Notice
NOAA Release	National Oceanic and Atmospheric Administration News Release
NR Release	North American Rockwell Corporation (became Rockwell International Corporation Feb. 16, 1973) News Release
NR *Skywriter*	North American Rockwell Corporation's *Skywriter* (became Rockwell International Corporation's *Skywriter* Feb. 16, 1973)
NSC Release	National Space Club News Release
NSC *News*	National Space Club *News*
NSC *Letter*	National Space Club *Letter*
NSF *Highlights*	National Science Foundation's *Science Resources Studies Highlights*
NSF Release	National Science Foundation News Release
NSTL Release	NASA National Space Technology Laboratories News Release
NY News	*New York Daily News* newspaper
NYT, 5:4	*New York Times* newspaper, section 5, page 4
NYTNS	New York Times News Service
O Sen Star	*Orlando Sentinel Star* newspaper
Oakland Trib	*Oakland Tribune* newspaper
Omaha W-H	*Omaha World-Herald* newspaper
ONR *Rev*	Navy's Office of Naval Research *Reviews*
OST Release	Office of Science and Technology News Release
P *Bull*	Philadelphia *Evening* and *Sunday Bulletin* newspaper
P *Inq*	*Philadelphia Inquirer* newspaper
PAO	Public Affairs Office
PD	National Archives and Records Service's *Weekly Compilation of Presidential Documents*
PIO	Public Information Office
PMR *Missile*	USN Pacific Missile Range's *Missile*
PMR Release	USN Pacific Missile Range News Release
Pres Rpt 74	*Aeronautics and Space Report of the President: 1973 Activities*
RI Release	Rockwell International Corporation (North American Rockwell Corporation until Feb. 16, 1973) News Release
RI *Skywriter*	Rockwell International Corporation's *Skywriter* (North American Rockwell Corporation's *Skywriter* until Feb. 16, 1973)

SAO Release	Smithsonian Astrophysical Observatory News Release
SBD	*Defense/Space Business Daily* newspaper
Sci Amer	*Scientific American* magazine
Sci & Govt Rpt	*Science & Government Report*, independent bulletin of science policy
SciServ	Science Service news service
SD	*Space Digest* magazine
SD Union	*San Diego Union* newspaper
SET Manpower Comments	Scientific Manpower Commission's *Scientific, Engineering, Technical Manpower Comments*
SF	British Interplanetary Society's *Spaceflight* magazine
SF Chron	*San Francisco Chronicle* newspaper
SF Exam	*San Francisco Examiner* newspaper
Sov Aero	*Soviet Aerospace* newsletter
Sov Rpt	Center for Foreign Technology's *Soviet Report* (translations)
SP	*Space Propulsion* newsletter
Spaceport News	NASA John F. Kennedy Space Center's *Spaceport News*
Spacewarn	IUWDS World Data Center A for Rockets and Satellites' *Spacewarn Bulletin*
SR list	NASA compendium of sounding rocket launches
SSN	*Soviet Sciences in the News*, publication of Electro-Optical Systems, Inc.
St Louis G-D	*St. Louis Globe-Democrat* newspaper
St Louis P-D	*St. Louis Post-Dispatch* newspaper
T-Picayune	New Orleans *Times-Picayune* newspaper
Tech Rev	Massachusetts Institute of Technology's *Technology Review*
Testimony	Prepared congressional testimony
Text	Prepared report of speech text
Transcript	Official transcript of news conference or congressional hearing
UN Reg	United Nations Public Registry of Space Flight
UPI	United Press International news service
US News	*U.S. News & World Report* magazine
USGS Release	U.S. Geological Survey News Release
USPS Release	U.S. Postal Service Release
W Post	*Washington Post* newspaper
W Star-News (*W Star & News*)	*Washington Star-News* newspaper (Washington *Star and News* until July 18, 1973)
WH Release	White House News Release
WJT	*World Journal Tribune* newspaper
WS Release	NASA Wallops Station News Release
WSJ	*Wall Street Journal* newspaper

INDEX AND LIST OF ABBREVIATIONS AND ACRONYMS

A

A-7D (Corsair II tactical fighter), 343
A-9 (close air support aircraft), 18
A-10, 18, 63
A-300B (European airbus), 113, 163, 263
AAS. See American Astronautical Society and American Astronomical Society.
AAAS. See American Assn. for the Advancement of Science.
Abbot, Dr. Charles Greeley, 347
ABM. See Antiballistic missile system.
ABMA. See Army Ballistic Missile Agency.
Abourezk, Sen. James S., 4, 153
Abzug, Rep. Bella S., 161, 192
AC-47 (military aircraft), 137
Academician Korolyev (U.S.S.R. research ship), 206, 237
Academy of Sciences, U.S.S.R. See Soviet Academy of Sciences.
Acapulco, Mexico, 175
Accelerometer, 345–346
Accident (see also Fire protection)
  aircraft
    BT-13, 250
    Convair 990, 115, 116–117, 117–118, 127–128, 131, 181, 245
    simulation, 107–108
    P-3C, 131
    trainer, 209, 291–292
    Tu-144, 163, 173, 175, 176–177, 179, 184, 191
  astronaut, 209, 250, 291–292
  automobile, 168
  balloon, 269
  cosmonaut, 103, 153, 282, 327, 348
  NASA personnel, 275
  scientist expedition, 341
  spacecraft
    Castor and Pollux, 159
    Lunokhod 2, 175
    Salyut 2, 133, 141–142, 182
    Skylab. See *Skylab 1*.
    Soyuz 11, 103, 153, 273, 282, 327, 348
Accomack County, Va., 129
ACDA. See U.S. Arms Control and Disarmament Agency.
Acheson, David C., 53–54
Acoustics, 269–270, 285, 342
Active repeater spacecraft, 273
Active-ride-control system, 139
Actuator, 83–84

ACV. See Air-cushion vehicle.
Adams, Robert L., 266–267
Adams, Lt. Gov. Tom (Fla.), 206
ADC. See Aerospace Defense Command.
Adey, Dr. W. Ross, 22–23, 195
Admiral Corp., 296–297
Advanced Airborne Command Post Program, 56
Advanced medium STOL transport (AMST), 90, 96, 359
Advanced missions, 219
Advanced Research Projects Agency (ARPA), 250
"Advanced Technology and the Space Shuttle" (lecture), 9
Advanced transport technology (ATT) program, 72
AE. See Atmosphere Explorer.
AE-A. See *Explorer 17*.
AE-B. See *Explorer 32*.
AE-C. See *Explorer 51*.
AEC. See Atomic Energy Commission.
AEC-NASA Space Nuclear Systems Office (SNSO), 7–8, 201–202, 359–360
AEDC. See Arnold Engineering Development Center.
Aermacchi MB-326 (jet trainer-fighter), 263
Aerobee (sounding rocket)
  150, 58
  150A, 161
  170, 75, 111–112, 165, 204, 261, 307, 315, 316, 317
  170A, 92, 98, 172, 285
  200, 19, 221, 267, 298
  200A, 181, 250, 329, 349
Aerodynamics, 71, 72, 211, 284–285, 300, 313, 317–318, 357
Aeroflot (U.S.S.R. national airline), 172–173, 184, 192–193, 278
Aerojet Solid Propulsion Co., 212, 253
Aeronautica Macchi S.p.A. See Aermacchi MB-326.
Aeronautical Technology Symposium and Display. See U.S.–U.S.S.R. Aeronautical Technology Symposium and Display.
Aeronautics (see also Federal Aviation Administration)
  aerospace industry. See Aerospace industry.
  air pollution. See Air pollution.
  anniversary, 345, 346

awards and honors, 9–10, 10–11, 80, 102, 113, 128, 135, 183, 255, 265–266, 304, 345, 351–352
cooperation, 266
employment, 143, 189
exhibit, 163, 263
funds
   appropriation, FY 1974, 29–30, 96, 219
   authorization, FY 1974, 56–57, 139–140, 167
   congressional hearings, 56–57, 68, 72, 74, 89–90
   constant level budget, 2, 29–30, 359
   reduction, 5
   request, FY 1974, 27–28, 30
general aviation. See General aviation.
materials patent, 136–137
meeting, 9, 10–11, 11–12, 216, 333
military (see also U.S. Air Force, aircraft), 359
NASA program. See National Aeronautics and Space Administration, programs.
Nixon, President Richard M., report, 86–87
press comment, 289
research (see also individual aircraft; Lifting body; and Wing, aircraft), 68, 86–87, 280, 300
   advanced flight systems, 95
   advanced transport technology, 72
   composite fuselage structure, 92
   control systems, 74–75, 107–108, 139, 160, 266, 280, 343, 358
   hypersonics, 67, 74, 206, 235, 280
   landing methods, 7, 72–73
   monitoring systems, 40, 172, 332
   quiet engine technology, 68–69, 72, 90, 280, 300, 358
   wake vortices, 302
safety. See Safety, aircraft.
Aeronautics and Space Report of the President: 1972 Activities, 86–87
Aeronomy, 42–43, 56, 58, 214, 228, 330–331, 338, 361
Aeros (West German satellite), 22
AEROS–B, 361
Aerospace Defense Command (ADC), 92, 176
Aerospace Facts and Figures, 1973/74, 158
Aerospace Industries Assn. of America, Inc. (AIA), 83, 105–106, 113, 140, 156, 158, 189, 286
   Board of Governors, 326
   International Committee, 113
Aerospace industry
   achievements, 42, 105–106, 166, 354–355
   contracting and procurement, 278, 292
   cooperation, 70, 194, 279, 282, 286, 301
   employment, 42, 189, 233, 286, 319, 335, 342–343, 360
   energy crisis impact, 330

Europe, 278, 304
exhibit, 263
international competition, 272, 277
NASA technology use, 42, 167–168, 261–262, 360
press comment, 1, 289, 330
Regional Dissemination Center use, 360
research and development, 58–59, 317
trade, 72–73, 83, 158, 166, 167–168, 362
U.S. leadership, 70, 72–73, 105–106, 277
Aerospace Medical Assn., 256
Aerospace Medical Research Laboratory, Ohio, 19
Aerospace medicine. See Space biology and medicine.
Aérospatiale (Société Nationale Industrielle Aérospatiale), 102, 114, 217, 263, 294–295, 338
Aerostat (tethered balloon), 9
Aerotherm Corp., 199
AFCRL. See Air Force Cambridge Research Laboratories.
AFETR. See Eastern Test Range.
Affirmative Action Plan, 10
AFL–CIO. See American Federation of Labor–Congress of Industrial Organizations.
Africa, 172–173, 179, 195, 200, 201, 205, 210, 212, 256
AFSATCOM. See Air Force Satellite Communications System.
AFSC. See Air Force Systems Command.
AFSF (Air Force Satellite Facility). See Space and Missile Systems Organization.
AFWTR. See Western Test Range.
Ag-Cat (agricultural aircraft), 2
Agena (booster upper stage), 257, 302–303
Agnew, Vice President Spiro T., 21, 24, 275
Agreement. See International cooperation; International cooperation, space; Memorandum of Understanding.
Agriculture, 66, 212, 280
AIA. See Aerospace Industries Assn. of America, Inc.
AIA Survey of Aerospace Employment, 189
AIAA. See American Institute of Aeronautics and Astronautics.
Ainsworth, R/A Herbert S. (USN), 116–117
Air and Space Museum, 109, 116, 251, 288, 289
Air cargo, 67, 278
Air-cushion vehicle (ACV), 176, 207
Air Defense Command, 277–278, 295–296
Air Force, Department of. See U.S. Air Force.
Air Force Avionics Laboratory, 158
Air Force Bureau of International Scientific and Technological Affairs, 114

ASTRONAUTICS AND AERONAUTICS, 1973

Air Force Cambridge Research Laboratories (AFCRL), 118, 220, 243
Air Force Eastern Test Range. See Eastern Test Range.
Air Force Environmental and Technical Applications Center, Air Weather Service, 233
Air Force Logistics Command, 282
Air Force Museum, 59, 118
Air Force Plant 42, Palmdale, Calif., 271, 295
Air Force Satellite Communications (AFSATCOM) System, 113
Air Force Satellite Facility (AFSF). See Space and Missile Systems Organization.
Air Force School of Aerospace Medicine, 48
  Aviation Medicine Dept., 350
Air Force Space and Missile Systems Organization. See Space and Missile Systems Organization.
Air Force Space and Missile Test Center, 142–143
Air Force Systems Command (AFSC)
  Aeronautical Systems Div., 85
  AFSATCOM development, 113
  aircraft research, 19
  Arnold Engineering Development Center (AEDC), 239, 337–338
  computer communications, 99
  cooperation, 117, 239, 337–338
  Electronic Systems Div., 169
  Flight Dynamics Laboratory, 139
  Instrument landing system installation, 117
  oxygen contamination research, 48
  personnel, 170, 205, 236, 282
  Rocket Propulsion Laboratory, 269–270
  weather forecasting, 220
Air Force Weather Service, 169
Air France, 33, 338
Air Piracy. See Hijacking of aircraft.
Air pollution
  aircraft, 20, 28, 59, 68–69, 72, 74, 86, 101, 219, 264, 280, 288
  automobile, 58, 263–264, 265
  control, 28–29, 62, 87, 178, 219–220
  data analysis, 42, 61–62, 245–246, 253
  modular, integrated utility system (MIUS), 188–189
  monitoring, 20, 27, 29–30, 86, 161, 178
  nuclear, 219
  technology utilization, space, 88–89
Air-route-control-system, 46
Air-route-surveillance radar. See ARSR-3.
Air Show, 291–292
Air traffic control (ATC)
  Airport congestion, 43, 201, 300
  FAA program, 72–73, 129, 137, 153
  radar monitoring systems, 46, 255, 339
  satellite technology, 332, 361
  U.S.–U.S.S.R. Aeronautical Technology Symposium and Display, 216, 277
Air turbulence. See Clear-air turbulence.

Airborne Warning and Control System (AWACS), 202
Aircraft (see also individual aircraft)
  accident. See Accident, aircraft.
  advanced airborne command post, 143
  Advanced medium STOL transport AMST, 90, 359
  air pollution. See Air pollution, aircraft.
  amphibious, 206, 263
  antisubmarine patrol, 115, 131
  automated handling, 67
  biplane, 158
  bomber, 63, 84, 112–113, 176, 210–211, 216, 258–259, 319
  cargo, 33, 263, 265, 270, 274, 278, 286
  carrier, 186
  civil, 108, 184, 319, 362
  close air support, 18, 63
  commercial, 95, 128, 272, 278, 329–330, 358, 362
  communications, 56, 332
  contract, 4, 40, 132, 172, 181, 204, 269, 291, 342
  control-configured vehicle (CCV), 139
  cost, 204, 210–211, 258, 265
  crop-spraying, 278
  digital fly-by-wire aircraft, 74–75, 196, 266, 280, 358
  drone. See Drone.
  earth observation, 63, 70, 94, 243, 247
  engine. See Engine, aircraft.
  executive, 362
  exhibit, 163, 263
  experimental, 59, 285
  ferry, 270, 286
  fighter, 4, 11, 13, 28, 33, 40, 58, 69, 74–75, 86, 96, 111, 127, 139, 155, 157, 163, 168, 176, 183, 189, 204, 218, 258, 260, 263, 266, 281, 291, 307, 328, 343
  flight tests, 107, 135, 139, 232, 291, 345, 358
  foreign, 7, 23, 74, 80, 111, 144, 159, 163, 203, 220, 272, 278, 338, 354
  funds, 27–28, 50, 167, 258
  general aviation, 107–108, 138, 285, 358, 362
  helicopter, 152, 189, 192, 199, 206, 220, 232, 240, 247, 263, 291, 293, 305, 319, 329
  high performance, 284–285, 266
  highly maneuverable, 132, 266
  hijacking. See Hijacking of aircraft.
  hypersonic, 67, 74, 206, 235, 280
  industry, 2, 32–33, 86–87, 189, 204, 210–211, 256, 278, 285, 286, 289, 301, 317, 362
  infrared photography, 86, 92–93
  instrumentation, 119, 120, 183, 191, 205, 224, 237, 243, 260, 261
  landing methods, 117, 302, 358
  lifting body. See Lifting body.
  light aircraft, 245
  meteorological satellite support, 313–314

409

military, 24–25, 59, 74–75, 86–87, 137, 163, 168, 175, 220, 263, 265, 270, 281, 286, 305, 343, 344, 359
missile. See Missile.
modified commuter, 107–108
navigation, 59
noise. See Noise, aircraft.
QUESTOL, 5, 50, 90, 140
reconnaissance, 50, 59, 144, 209–210, 265–266
record, 220, 265–266, 345
remote sensing, 87, 130, 344
remotely piloted research vehicle (RPRV), 291, 358
remotely piloted vehicle (RPV), 160, 305
research. See Aeronautics research.
rocket research, 280
rotor system research (RSRA), 92, 247
rotorcraft, 80
safety. See Safety, aircraft
shuttle training aircraft (STA), 343
STOL, 20, 90, 114, 187, 333, 359
Stratofortress, 216, 235, 247, 284, 285, 291, 302, 319, 342
subsonic/sonic, 72
super-stable, 139
supercritical wing. See Wing, aircraft, supercritical.
supersonic. See Supersonic transport; Concorde; Tu-144;
survey, 110, 130, 358
symposium on, 216, 333
tilt-rotor, 118–119
traffic control. See Air traffic control.
training. 135, 144, 291–292
transport, 7, 20, 22, 31, 49, 56, 59, 72, 80, 89–90, 95, 134, 135–136, 137, 143, 163, 182, 263, 265, 270, 286, 345, 358, 362
trimotor, 122
troop carrier, 274, 294
utility, 362
V/STOL, 117, 263
VTOL, 305
wide body, 108, 319
wing. See Wing, aircraft.
Airframe and Powerplant Mechanic's Certificate (FAA), 286
Airglow, 38, 223
Airlines (see also individual airlines), 359
EPA requirements, 219
fuel cost and conservation, 350, 361–362
hijacking. See Hijacking of aircraft.
industry, 219, 319, 362
passenger service, 108, 277, 278, 362
security, 40
supersonic transport (SST), 97
Airlock module (AM), 90, 145, 146, 192–193
Airports (see also Air pollution; Air traffic control; Noise, aircraft; and individual airports), 216, 247
Aviation Advisory Commission recommendations, 3
congestion, 43, 300

dedication ceremony, 265–266
fog dispersal study, 13–14
instrument landing system (ILS), 117
instrumentation contract, 178–179
microwave landing system (MLS), 86, 333
noise reduction, 50, 182, 333
offshore technology, 129
planning, 28, 247
security, 40
takeoff and landing procedures, 182, 231, 302
Alabama, 86, 103, 344
Alabama Regional Council of Governments, 86
Alamogordo, N. Mex., 232
Alaska, 313–314
environment study, 352
petroleum discoveries, 134–135
satellite data, 66–67, 79, 262–263
sounding rocket launch, 85, 89, 90, 95, 98, 321, 330–331, 338
Alaska pipeline, 66–67, 262–263, 361–362
Alaska, Univ. of, 66, 330–331, 338, 352
Alba (Mars landing site), 138
Albatross. See HU-16B aircraft.
Albert, Rep. Carl B., 186
Albert Thomas Space Hall of Fame, 319
Albertazzi, Col. Ralph D., 41
Albion, Mich., 251
Albuquerque, N. Mex., 22, 23
Aldebaran (star), 267
Aleutian Islands, 176
Alfvén, Dr. Hannes O., 345
Algae, 287
Algeria, 200
Alkalinity, 317
All Union Znaniye Society, 130, 170
All-Women's Transcontinental Air Race, 255
Allen, Sen. James B., 54–55
Allen, M/G Lew, Jr., 170
Alloys, metal, 247–248, 271
Alpha particle, 52
Alsop, Joseph, 17
Altschul, Selig, 219
Aluminum, 136–137, 150, 296, 299, 315, 317–318, 346–347
AM. See Airlock module.
Amazon Basin, 46, 198, 265
Amazon valley, 37–38
Ambulance, 294
American Airlines, Inc., 33
American Assn. for the Advancement of Science (AAAS), 130, 170, 189
American Astronautical Society (AAS), 116, 331, 351–352
American Astronomical Society, 12
American Aviation Corp., 2
American Chemical Society, 14–15
American Federation of Labor–Congress of Industrial Organizations (AFL–CIO), 70
American Geophysical Union, 119, 331

American Institute of Aeronautics and Astronautics (AIAA), 9, 11–12, 167–168, 182, 190, 208–209, 216, 313, 351–352
  Cleveland Chapter, 18
  Houston Section, 10
  New York Section, 129
American Meteorological Society, 315
American Museum. See Hayden Planetarium.
American Nuclear Society, 318–319
American Philosophical Society, 15
American Photographic Interpretation Award, 83
American Physical Society, 126
American Satellite Corp. See Western Union International, Inc.
American Science and Engineering, Inc., 75, 172, 185
American Society of Mechanical Engineers (ASME), 208–209
American Society of Photogrammetry, 83
American Telephone & Telegraph Co. (AT&T), 73, 92, 179, 260, 264–265, 269
Ames Research Center (ARC, NASA)
  air pollution research, 245
  aircraft research, 62, 72, 181, 182, 232, 245, 302
  automated visual sensitivity tester, 298
  CALROC program, 22, 121
  capacitance pressure transducer, 243
  cooperation, 81, 182, 199, 245–246, 302
  disaster-assessment systems, 245–246
  drug detection, 81, 291
  employment, 10, 30, 359–360
  fire index measurement, 161
  helmet design and testing, 199
  Institute for Advanced Computations, 250
  lecture series sponsorship, 243
  Nixon, President Richard M., message and statements, 336
  oceanography, 41
  personnel, 115, 201
  Pioneer program, 33–34, 47, 107, 160, 185, 241, 308
  press conference, 116–117, 250, 309
  Randomdec computer-and-analysis method, 233
  space biology and medicine, 251, 262, 289, 296, 317
  space shuttle, 262
  Venus atmosphere research, 332
  wind-tunnel tests, 232
Amino acid, 115–116
AMST. See Advanced medium STOL transport.
An-30 (U.S.S.R. reconnaissance aircraft), 144
Anaheim, Calif., 186
Analysis of Cost Estimates for the Space Shuttle and Two Alternate Programs (GAO report), 172
An Analysis of Federal R&D Funding: FY 1963- (NSF 72-313) 38

Anders, L/C William A. (USAF, R.), 21, 216–217, 261, 271
Anderson, Sen. Clinton P., 4
Anderson, Dr. O. E., 259
Andoeya, Norway, 15–16, 26, 40, 328
Andrastea (Jovian moon), 308–309
Andrews AFB, Md., 293
Andreyev, Boris, 162
Anglo-French supersonic transport. See Concorde.
Anik 1 (Telesat A; Canadian comsat), 123
Anik 2 (Telesat B), 1, 14, 123, 186, 358, 360
Anik C, 361
Animal experiments, space, 22–23, 232, 303–304, 344–345
Anita (spider), 229, 241–242, 275
Anniversary
  ABMA, 33
  Apollo 8 Christmas celebration, 350–351
  Apollo 11 lunar landing, 54–55, 217–218
  ComSatCorp, 35–36
  Copernicus, birth, 4, 50, 92, 112, 122, 124
  Erts 1 launch, 219–220, 231, 245
  Explorer 1, 33
  first nonstop flight over U.S., 135
  Freedom 7, 137
  NASA, 230, 264, 270, 279, 281, 285, 297, 299, 359
  Oao 3 (Copernicus) launch, 249
  Redstone (missile), first launch, 248
  space program, national, 324
  Syncom 2 launch, 223
  U.S.S.R., 8
  Vostok 1, 114–115
  Wright Brothers flight, 345, 346
Annual Space Congress, 10th, 114
Anorthosite (lunar surface sample), 324–325
Antarctic Ocean, 48
Antarctica, 12, 41, 51, 54, 62, 87, 104, 253, 255, 341
Antenna
  comsat station, 287
  spacecraft, 160, 178, 180, 185–186, 268, 298, 302, 308
    Skylab, 299, 320–321
    tracking, 64–65, 116, 208
Antiballistic missile system, (ABM), 94, 276–277
Antigua, 46, 51, 174, 247, 265, 295, 329
Antihelium, 159
Antimatter, 11–12
Antitritium, 159
Antonov, Oleg, 144
AP. See Associated Press.
APIB. See NASA Applications Program Integration Board.
Apollo (program)
  achievements, 17, 21, 42, 55, 86–87, 96, 255
  astronaut preparations, 52, 350

awards and honors, 66, 104, 135, 136, 195, 313
conclusion, 6, 21, 27, 279
cost, 17
experiments, 81–82, 110
funds, 27
Johnson, Lyndon B., 21, 25
launch, 352
launch vehicle, 36
lunar samples, 15, 65–66, 290
management, 21, 25
medical aspects, 153, 335–336
Nixon, President Richard M., 49–50
personnel, 15, 58, 126, 136, 221, 238, 353
power systems, 42, 271
press comment, 166, 238
results, 91, 110, 359
significance, 21, 105–106, 109–110, 166, 270, 274
surplus hardware use, 74–75
USAF history course, 153
Apollo (spacecraft)
  Apollo-Soyuz Test Project, 35, 84, 207–208, 276–277, 282, 347, 353
  command and service module. See Command and service module.
  command module. See Command module.
  contract, 47
  equipment, 1, 8–9, 10, 144, 288, 349
  exhibit, 93, 163
  launch schedule, 10
  lunar module. See Lunar module.
  press comment, 276–277
  Skylab 3 mission, 206, 211–212, 217, 224–230, 238, 239–240
  Skylab 4 mission, 226–227, 318, 319–320
Apollo 4 mission, 162–163
Apollo 8 mission, 7, 10–11, 121–122, 216–217, 238, 261, 271, 350–351
Apollo 9 mission, 46, 238
Apollo 10 mission, 93
Apollo 11 mission
  anniversary, 54–55, 217–218
  astronaut, 109, 121–122, 152, 279
  commemoration of, 40–41, 162–163, 217–218, 223
  launch preparations, 162–163
  lunar samples, 162, 324
  significance, 20–21, 96
Apollo 12 mission
  astronaut, 152, 251
  experiments, 46, 82, 238
  lunar samples, 162, 324
  personnel, 120
Apollo 13 mission, 77–78, 158–159
  astronaut, 7, 117, 121–122, 153, 250, 291–292
  awards and honors, 120
  press comment, 153
Apollo 14 mission, 96, 238
Apollo 15 mission
  astronaut, 106, 238
  lunar roving vehicle, 242, 255

  lunar samples, 52, 324
  management, 120
  significance, 96
  spacesuit, 93
  spectrometer, 53
Apollo 16 mission
  achievements, 86
  astronaut, 121–122, 255
  experiments, 7, 238, 255
  lunar roving vehicle, 242, 255
  lunar samples, 6–7, 52, 84, 164, 324
  spacesuit, 93
  spectrometer, 53
  tracking, 83
Apollo 16 Preliminary Examination Team, 6–7
Apollo 17 mission
  achievements, 86–87
  astronaut, 5, 18, 19, 20–21, 49, 50–51, 52, 76, 105, 107, 130, 163, 186, 195, 255, 325, 346–347
  awards and honors, 52, 63–64, 104
  Congress, report to, 20–21, 112
  experiments, 49, 55, 110
  funds, 27
  lunar landing site, 325
  lunar module. See Lunar module.
  lunar roving vehicle, 242, 255
  lunar samples, 2–3, 5, 32, 45, 49, 65–66, 84, 162, 213, 236, 324, 346–347, 361
  management, 105, 107, 120
  medical aspect, 55
  photographs, 110
  press comment, 26
  press conference, 5
  significance, 1, 3, 5, 9
  spacesuit, 93
  tracking and data acquisition, 83
Apollo 17 Preliminary Examination Team, 324
Apollo Field Geology Investigation Team, 325
Apollo Lunar Landing Commemorative Trust Fund (proposed), 40–41
Apollo Lunar Sounder Investigator Team (JPL), 299
Apollo-Soyuz Test Project (ASTP), 124
  agreements, 1, 76, 84, 134, 194, 282
  astronaut, 31–32, 35, 190, 194, 276, 287, 325–326, 339, 360
  benefits, 36, 55, 91
  Congress, 71–72
  contract, 338, 349, 353
  cosmonaut, 35, 71–72, 84, 162, 163, 190, 194, 207–208, 217, 276, 325–326, 360
  docking, 8–9, 35, 84, 134, 261, 276–277, 325–326, 326–327, 349, 353, 360
  employment impact, 91–92
  equipment, 1, 8–9, 326–327, 338
  experiments, 102–103, 207, 257
  funds, 5, 27, 359
  language, 99
  launch preparations, 182, 196, 207–208, 236, 282, 330
  management, 15, 64, 105, 124, 339, 360
  Mars mission prototype, 133

meeting, 8-9, 71-72, 134, 207-208, 282
Mission Control Center, U.S.S.R., 326-327
plans, 194, 124, 205, 280, 323
political implications, 36, 264
press comment, 141, 156, 276-277
press conference, 35, 84, 326-327
rendezvous, 8-9, 35, 134
spacecraft, 35, 84, 93, 163, 196, 347
system testing, 84, 276, 326-327, 360
training, 326
working group, 84, 360
Apollo Telescope Mount (ATM)
CALROC reference, 22
equipment calibration, 121, 358-359
press conference, 184, 266
Skylab 1 mission, 145, 169, 173, 182-183, 222
Skylab 2 mission, 146-147, 148, 149, 151, 222
Skylab 3 mission, 152, 225, 227, 228-229, 258, 266
Skylab 4 mission, 246, 283, 299, 320-321, 322, 349
solar observations, 78, 149, 151, 152, 243
Appalachia, 18
Application of Area Navigation in the National Airspace System (FAA report), 59
Applications satellite program (NASA), 62
Applications Technology Satellite. See ATS, Ats 3, ATS-E, ATS-F, and ATS-G.
Arab-Israeli war, 284, 287, 290, 292-293, 295, 301, 314, 361
Arabella (spider), 225, 229, 241-242, 275
Arava (cargo transport aircraft), 263
ARC. See Ames Research Center.
Arcas (sounding rocket), 46, 51, 174, 247, 265, 295, 329
Super Arcas, 15-16, 37, 51
Archaeology, 344
Archimedes, 325
Arctic, 194, 253, 292
Arctic surface effects vehicle (ASEV), 207
Area Navigation Task Force, 59
Argentina, 48, 51, 62, 97, 103, 337
Argon (lunar atmosphere), 48, 51, 62, 97, 103, 337,
Ariane (L-3S French booster), 301
Ariel 4 (U.K. satellite), 328
Aries (sounding rocket), 254
Aristarchus Crater (moon), 52
Arizona, 12, 247
Arizona State Univ., 292
Arizona, Univ. of, 109
Arkansas, 130
Arktika (U.S.S.R. icebreaker ship), 292
Arms Control and Disarmament Agency. See U.S. Arms Control and Disarmament Agency.
Arms limitation. See Disarmament.
Armstrong, Neil A., 9, 20-21, 217-218
award, 128-129, 279
Armstrongite (mineral), 128-129
Army. See U.S. Army.
Army Air Corps, 220
Army Air Mobility Research and Development Laboratory, 92
Army Ballistic Missile Agency (ABMA), 33, 218
Arnold Engineering Development Center (AEDC), 239, 337-338
Arnon, Dr. Daniel I., 284-285
ARPA. See Advanced Research Projects Agency.
ARSR-3 (air-route-surveillance radar), 41
ARTS. See Automated radar terminal systems.
Asbestos, 275
ASEV. See Arctic surface effects vehicle.
Ash, Roy L., 342
Ashley, Dr. Holt, 9
Ashworth, C. Dixon, 354
Asia, 34, 183, 195, 270
Askew, Gov. Reubin (Fla.), 167
ASME. See American Society of Mechanical Engineers.
Aspin, Rep. Leslie, 155
Asteroid, 106
Asteroid Belt, 47, 78, 106, 174, 197, 248, 309, 311, 357
ASTP. See Apollo-Soyuz Test Project.
Astro Met Associates, 247-248
Astrobee D (sounding rocket), 201
Astronaut (see also individual missions and programs; Cosmonaut; Scientist-astronaut; and Space biology and medicine), 18, 190
accident. See Accident.
achievements, 3, 281, 323
appointments and promotions, 109, 117, 218, 238, 328
authorship, 125, 233
awards and honors, 51, 52, 66, 104, 107, 128-129, 221, 223, 238, 255, 279, 286, 287, 319, 322
Comet Kohoutek observations, 319-320, 321
Congress, visit to, 20-21, 56
emergency aid devices, 251
experiments, 312, 319-320, 321, 330-331
extravehicular activity, 112, 160-161, 279, 319-321 322, 323
FAI Annual General Conference, 255
film, 223
former, 238, 341
good will tour, 186, 195
hazards, 168, 238, 239-240
man-hours in space, 7, 82, 152, 279, 319-320, 323
medical aspects, 55, 82, 106, 133, 134, 136, 153, 158, 180, 193, 194, 195, 196-197, 198, 209, 219, 222, 225-226, 227, 228, 229, 238, 274, 282-283, 298, 320, 321
memorial, 51
Mercury (program), 350
mission readiness, 90, 223-224, 283
Nixon, President Richard M., messages and statements, 162, 216-217, 324

performance, 191, 194, 225, 228, 256, 274
photographs, 185, 321, 322–323
pilot, 339, 341
press comment, 153, 166, 169, 179, 180, 193–194, 238, 239–240, 251, 256, 271, 274, 275, 276–277
press conference, 6, 35, 199, 222, 223, 274, 282–283
private communications, 134, 171
public appearances, 13, 76, 130
quarantine, 126, 295, 323
rescue, 91
retirement, 7, 333
safety, 238
scientist-astronaut, 6, 81–82, 174–175, 223, 224, 227, 241–242, 282–283, 319, 325, 339
training, 17, 52, 112, 158–159, 181, 199, 207–208, 221, 290, 311, 312, 326, 360
White House visit, 63–64
Astronautics Corp. of America, 107–108
Astronautics Engineer Award (NSC), 76
Astronauts Memorial Commission (proposed), 51
Astronomical Society of the Pacific, 243
Astronomy (see also individual celestial bodies; observatories; probes; sounding rockets; telescopes)
  anniversary, 112, 122, 124
  Asteroid Belt, 78, 106, 174
  award, 186
  cosmic ray, 22, 106, 174, 212, 221
  exhibit, 4
  galactic, 39, 165, 174, 199–200, 280, 285, 342
  gamma ray, 11–12, 35, 47, 78–79, 128, 199–200
  high-energy, 11–12
  infrared, 115–116, 224, 332
  international cooperation, 124, 138–139, 361
  light pollution, 98
  NASA program
    Apollo, 110, 112
    balloon, 103
    budget, FY 1974, 5, 26–27, 36, 38–39, 56–57, 140, 219
    satellite, 13, 47, 131, 106–107, 119, 124, 131, 138–139, 160, 249, 279–280, 341
    Skylab, 142, 145, 149, 184–185, 192–193, 222, 224, 227, 228, 230, 319–320, 322, 361
    sounding rocket, 42–43, 111–112, 165, 172, 204, 285, 307, 315, 316, 317
    space shuttle, 125, 303
  personnel, 297, 354
  physical laws and requirements, 115–116
  radar, 64, 239
  solar (see also individual Solar headings), 193–194, 224, 268, 323–324, 357
  stellar
    black hole, 125, 263, 329
    radiation, 341

supernova, 11–12, 39, 232, 341
  ultraviolet. See Ultraviolet.
  universe, 47, 128
  U.S.S.R., 47, 65, 139
  x-ray. See X-ray.
Astronomy and Astrophysics for the 1970's, Vol. 2, *Report of the Panels* (NAS–NRC report)115–116
AT&T. See American Telephone & Telegraph Co.
ATC. See Air traffic control.
Athens, Greece, 325
ATI. See Automated Typographics, Inc.
Atlanta, Ga., 184
Atlanta International Airport, Ga., 201
Atlantic Ocean (see also South Atlantic magnetic anomaly), 24, 102, 167, 183, 200, 209, 210, 237, 250, 278, 313–314
Atlas (booster), 314
  Atlas-Agena, 67
  Atlas-Centaur, 192, 250, 263, 333
    D1-A, 307
    3rd stage, 106
ATM. See Apollo Telescope Mount.
Atmosphere
  biosphere, 287
  cloud study, 236
  infrared, 65
  nuclear testing, 211
  parameter measurement, 346
  planetary. See individual planets.
  temperature sounding, 22, 61–62, 71, 288, 312
  upper (earth), 92, 115–116, 166, 189, 90, 237, 261–262, 325, 346
    air glow, 223
    composition, 10, 46, 51, 59, 103, 104, 174, 212, 223, 245, 247, 255–256, 257, 295, 345–346
    ionosphere, 15–16, 122, 201, 237, 239, 240–241, 243, 258, 302, 208, 325, 328, 358–359, 361
  pollution. See Air pollution.
  solar activity effects, 258, 312
  SST impact, 59, 74
  stratosphere, 46, 51, 74, 120–121, 174, 247, 265, 288, 295, 329
  troposphere, 205, 345–346
  U.S.S.R. program, 71
  weather. See Meteorological satellites and Meteorology.
Atmosphere Explorer (see also AE, *Explorer 17*, *Explorer 32*, and *Explorer 51*), 191
Atoll. See Mururoa Atoll.
Atomic bomb, 155
Atomic breeder reactor, 213
Atomic energy, 20–21, 77, 189–190
Atomic Energy Commission (AEC; see also AEC–NASA Space Nuclear Systems Office)
  environmental experiments, 298
  funds, 5, 28, 39, 129–130
  Lawrence Livermore Laboratory, 245
  licensing, 361–362

Los Alamos Scientific Laboratory (LASL), 290
Memorandum on Cooperation in the Peaceful Uses of Atomic Energy, 265
MIUS development, 188–189
The Nation's Energy Future, 335
nuclear research, 14, 117, 140, 271, 314, 318–319
Office of Solar Energy Research (proposed), 347
personnel, 23, 40, 117, 141, 216–217, 261, 335
Plowshare program, 140, 157, 161
regulations, 257–258
World Weather Program for 1974, 210
Atomic Industrial Forum, 14, 318–319
ATS (Applications Technology Satellite), 313–314
Ats 3, 41, 48, 50, 51
ATS-F, 27–28, 71, 262
ATS-G, 5
ATT. See Advanced transport technology program.
AU: astronomical unit
Auburn, Mass., 5
Auburn Univ., 246
Augustine, Norman R., 248
Aurora, 16, 32, 37, 41, 63, 258, 313, 328, 358–359
Aurora, Colo., 106
Aurora 7 mission, 35
Austin, Tex., 20
Australia, 104, 116, 159–160, 169, 185–186, 219, 245, 333, 361
Austria, 139
Automated Radar Terminal System (ARTS), 46, 255
Automated Typographics, Inc. (ATI), 95
Automated visual sensitivity tester (AVST), 298, 360
Automobile, 58, 168, 171–172, 199, 220, 263–264, 265, 343, 357, 362
Automotive Gas Turbine Technology Program, 199
Ava (hurricane), 150, 175, 181–182
Avco Corp., Avco Precision Products Div., Avco Electronics Operation, 285
Aviation. See Aeronautics; Civil aviation; Commercial aviation; General aviation.
Aviation Advisory Commission (President's), 3, 31, 38
Aviation Forecasts, Fiscal Years 1973–1984 (FAA report), 108
Aviation Maintenance Foundation, 286
Aviation/Space Writers Assn., 135
Aviator's Trophy, 189
Aviatrix Trophy, 265–266
AVST. See Automated visual sensitivity tester.
AWACS. See Airborne warning and control system.
Awards
educational grants, 52
foreign, 80, 220
government, 52, 73, 111, 120, 286, 351–352

NASA. See National Aeronautics and Space Administration.
institutions, 9–11, 128, 137, 294, 319
international, 186, 255, 288
military, 122, 135, 220, 231, 238, 256, 286
society
achievements, 83, 136
aeronautics, 102, 113, 128, 135, 183, 255, 265–266, 304, 345, 351–352
astronautics, 52, 255, 288, 313, 351–352
science, 157, 256, 266, 284–285, 290, 292, 296, 301
A-X (close-support aircraft), 18, 40, 63
Azores, 261

B

B-1 (advanced strategic bomber), 85, 176, 210–211, 258–259
B-52 (Stratofortress), 291, 342
lifting body test, 174, 235, 247, 284, 285, 304, 303, 318, 319
B-57 (bomber), 63
BAC. See British Aircraft Corp.
BAC-111 (U.K. aircraft), 278
Backscatter UV experiment (BUV), 51
Bacteria, 317
Bailey, F. Lee, 291
Baker, Sen. Howard H., Jr., 31
Baker, Russell, 166
Baku, U.S.S.R., 287
Balchen, Col. Bernt (USAF, Ret.), 294
Balewa, Prime Minister Sir Abubaker Tafawa (Nigeria), 223
Balloon, 104
accident, 269
applications, 9, 273, 323
foreign, 218, 220
research, 26, 90–91, 191, 212, 232, 245, 269, 358–359
Balloonsonde, 214–215
Baltic Sea, 87
Baltimore, Md., 59, 289, 344–345
Barium cloud experiment, 321, 326, 330–331, 338
Barnum, John W., 168
Barry, L/C William E. (USAF), 256
Bartlett, Sen. Dewey F., 13
Basalt (lunar surface sample), 323–324
Basic Principles of Negotiations on the Further Limitation of Strategic Offensive Arms (U.S.–U.S.S.R. agreement outline), 189–190
Bastido, Col. William G. (USAF), 114
Bathyscaphe, 38
Battin, Dr. Richard H., 10–11
Battle Creek, Mich., 251
Battle of the North Atlantic (World War II), 339
Baumgardner, Dr. Marian, 66
Bay Area Pollution Control District, San Francisco, Calif., 245
Bayh, Sen. Birch, 18

Baykonur Cosmodrome, U.S.S.R.
  launch
    Cosmos. See individual Cosmos satellites.
    Luna 21, 8
    Mars 4, 218
    Mars 5, 221
    Mars 6, 239
    Mars 7, 242
    Molniya I-23, 38
    Molniya I-25, 318
    Molniya II-5, 107
    Molniya II-7, 294
    Prognoz 3, 47
    Salyut 2, 102–103
    Soyuz 12, 273
    Soyuz 13, 347
  press center, 196
Beale AFB, Calif., 265–266
Beall, Sen. J. Glenn, Jr., 31
Bean, Capt. Alan L. (USN) 224–228, 251, 252, 258, 270, 282–283, 320
  ASTP backup crew, 31–32, 326
  experiments, 222, 227, 228, 245
  medical aspects, 219, 225, 274
Beechcraft (NASA aircraft), 92–93
Behr, Robert M., 134
Belew, Leland F., 319
Belgium, 17–18, 175–176, 244, 268–269, 360
Belgium, Univ. of, 37
Belgrade, Yugoslavia, 56
Bell, Rep. Alphonzo, 4, 23
Bell Helicopter Co. See Textron, Inc.
Bell Telephone Laboratories, Inc., 2
Bellonte, Maurice, 158
Bendix Corp.
  contract, 86, 338
  Launch Support Div., 353
Bennett (comet), 342
Bennett, Rep. Charles E., 18
Bennett, Harry, 207
Bentsen, Sen. Lloyd M., 25
Beregovoy, Gen. Georgy T. (U.S.S.R.), 54, 276, 287
Beresford, Spencer M., 156
Bergen, William B., 21
Bergland, Rep. Bob S., 23
Bergstrom AFB, Tex., 209
Bering Sea, 22, 47, 49, 74, 116–117
Berry, Dr. Charles A., 55, 82, 105, 264, 335–336, 350, 354
Bethesda, Md., 129
Bethpage, N.Y., 4, 157
Beverly Hills, Calif., 276
bev: billion electron volts.
Bhutto, Zulfikar Ali, 186
Bible, Sen. Alan, 267
Bicentennial, 213
Bicentennial Trail, 206
Bick, L/C Dewain (USAF), 265–266
Biemiller, Andrew, 70
Biermann, Dr. L., 345
Bierwirth, John C., 75, 157
Big Bang theory, 249
"Big Bird" (reconnaissance satellite), 45

Binoculars, 342
Biochemistry, 55, 328
Biological Impact of Increased Intensities of Solar Ultraviolet Radiation (NAS report), 59
Biosatellite 3, 22–23, 195
Biosphere, 287
Bird Co., 97
Birds, 232
BIS. See British Interplanetary Society.
Bishop, Alfred A., 283
Bisplinghoff, Dr. Raymond L., 219
Black Brant (sounding rocket)
  Black Brant IIIB, 37
  Black Brant IVA, 321, 330–331, 338
  Black Brant IVB, 32, 41, 243
  Black Brant VB, 37
  Black Brant VC, 22, 42–43, 121, 158, 173, 182–183, 243, 256, 340
Black hole (space phenomenon), 125, 263, 329
Black Sea, 87, 102–103, 159
Blagonravov, Dr. Anatoly A., 87
Blahnik, J. E., 12
Blasingame, Dr. B. Paul, 105
Blindness, 327
Blue Mountains, 243
BOAC. See British Overseas Airways Corp.
Board of Geographic Names, 288–289
Bobko, Maj. Karol J. (USAF), 31–32, 326
Boeing 707 (jet passenger transport) 32, 41, 89–90, 95, 138
Boeing 727, 89–90, 95, 182, 204, 302, 358
Boeing 737, 89–90, 95, 358
Boeing 747, 86, 95, 138, 286
Boeing 747-200B (advanced airborne command post aircraft), 143
Boeing 747B, 56
Boeing Co.,
  Boeing Aerospace Group, 105, 255
  Boeing Associated Products, 80
  Boeing Commercial Airplane Co., 95
  contract
    aircraft, 56, 83, 95, 108, 132, 139, 201, 204, 207
    missile, 38, 73, 140, 176, 185
    space shuttle, 101, 169, 286
    support services, 317
  noise abatement, 138
  wage increase order, 342–343
Boelkow, Ludwig, 113
Boilerplate spacecraft (Skylab), 10
Boland, Rep. Edward P., 187
Bolivia, 198
Bonny (space monkey), 22–23, 195
Boomerang, Project, 9, 26, 90–91
Booster. See individual boosters and stages, such as Agena, Ariane, Atlas, Centaur, Diamant, Europa, Juno, Jupiter C, L-3S, Mercury-Redstone, Saturn, Scout, Thor-Delta, Thorad, Titan, Transtage.
Borman, Col. Frank (USAF, Ret.), 133, 165
Bowden, H. R., 328

Boyd, Dr. R. L. F., 329
Boyd, Maj. Stu R. (USAF), 307
bps: bits per second.
Brackett, Ernest W., 201-202
Brand, Vance D., 31-32, 35, 290, 326, 336
Brandt, Dr. John C., 12
Bray, Charles W., III, 70
Brayton dynamic conversion system, 80
Brazil, 24, 37-38, 59, 85, 95, 168, 197, 200, 263
Brazilian Institute for Space Research (Institute de Pesquisas Espaciais; INPE), 95, 108-109
Brazilian Ministry of Aeronautics, 263
Breccia (lunar surface sample), 324
Breguet Co., 158
Brenda (tropical storm), 228
Breslow, Jerome W., 53-54
Brevard County, Fla., 25, 40-41, 56, 183, 326, 335
Brezhnev, Leonid I., 176, 184, 186, 187, 189-191, 193, 194, 213
Bribery, 275
Brice, Neil M., 119
Brick, Kay A., 255
Bridge (structure), 233
Brinckloe, W. D., 129
Brinegar, Claude S., 19, 39, 46, 83, 86, 95, 193, 317
Briscoe, Gov. Dolph (Tex.), 251-252
Bristol Aerojet Ltd., 60
British Aircraft Corp. (BAC), 97, 304, 328
British Gold Medal for Aeronautics, 80
British Interplanetary Society (BIS), 195, 236, 284
British Overseas Airways Corp. (BOAC), 33
British Science Research Council, 297
British Silver Medal for Aeronautics, 80
Brooke, Sen. Edward W., 18
Brown, Rep. George E., Jr., 23, 205, 265
Browne, Secor D., 5, 55, 258
Brussels, Belgium, 76, 175-176, 210, 231-232
BT-13 (aircraft), 250
Buchanan, Dr. Paul, 219, 227
Buenos Aires, Argentina, 337
Bugayev, Gen. Boris P., 193, 216
Bulletin of the Atomic Scientists, 155
Bumper 8 (rocket), 275
Buoy, 24, 183, 275, 313-314
Burbidge, Dr. Margaret, 297
Burch, Dean, 48
Burchard, Lt. Col. Eduard (West Germany), 81
Bureau of Standards. See National Bureau of Standards.
Burglar-intrusion detection system, 243
Burke, Rep. J. Herbert, 51
Burnazyan, Dr. Avetik I., 303
Burroughs Corp., 250
Bush, Dr. Vannevar, 170
Bushuyev, Konstantin D., 84, 207-208
Business and Employment Council of the Governor of Ohio, 70

Butterfield, Alexander P., 55, 79, 83, 123, 216, 286
BUV. See Backscatter UV experiment.
Bynum, Bobby G., 209
Byrd, Sen. Harry F., Jr., 13
Byrd, Adm. Richard. E., 294
Byrd, Sen. Robert C., 25

C

C-5A (military cargo transport aircraft), 33, 134, 263, 265, 270, 286
C-45 (research aircraft), 344
C-47 (World War II transport aircraft), 137
C-54, 261
C-130 (Hercules; military transport), 59, 96, 175
C-141 (troop carrier aircraft), 274, 294
CAB. See Civil Aeronautics Board.
Cacolla, India, 26
Calibration rocket. See CALROC.
California, 48, 53, 220
 accident, 115, 131, 245
 airport, 129
 astronaut visit, 13, 130, 276
 bridges, 233
 disaster-assessment systems, 245-246
 Div. of Forestry, 161
 earth resources photographs, 149, 167-168
 earth station, 73
 missile test route, 353-354
 solar explosion observation, 12
 vegetation patterns, 344-345
California Airmotive Corp., 245
California Institute of Technology (Cal Tech; see also Jet Propulsion Laboratory)
 Environmental Quality Laboratory, 154
 International Colloquium on Mars, 331
 sounding rocket experiment, 92, 307, 316
California, Univ. of, 43, 77, 297, 326, 328
 Berkeley, 137
 Davis, 158
 Livermore, 245
 Riverside, 245
Calio, Anthony J., 2-3, 339
Callaway, Howard, A., 134, 192
Callisto (Jovian moon), 310
CALROC (calibration rocket), 22, 104, 121, 158, 173, 182-183, 340
Cal Tech. See California Institute of Technology.
Calypso (oceanographic research vessel), 48, 51, 62
Cambridge, Mass., 294
Camera (see also Photography and Television)
 aircraft earth survey, 181
 Apollo-Soyuz Test Project, 84
 Apollo Telescope Mount, 227, 228, 320-323

flight research, 261
ground site, 330–331
Hasselblad, 261
high-resolution, 45
infrared, 81
Kohoutek (comet) study, 294, 321–323
multispectral, 344
satellite, 292
Skylab, 143, 155, 245, 283, 294
television, 220, 227, 291, 308
tracking, 155
Cameron, Dr. Roy E., 12
Camp, Rep. John N. Happy, 23
Campbell, James W., 10
Canada
aircraft, 20, 205, 263, 333
balloon launch, 103, 269, 361
earth resources, 37–38, 361
GATE participation, 24
ICAO convention of 1971, 252
international cooperation, 1, 14, 24, 260, 281
protest, 219
satellite, 1, 14, 27, 123, 358, 360, 361
sounding rocket launch, 37, 41
space object count, 185
telephone communications, 281
U.S. space program reaction, 344
Canadair Ltd., 263
Canadian National Research Council, 32, 37, 41
Canadian Overseas Telecommunications Corp., 281
Canary Islands, 114
Canberra, Australia, Tracking Station, 116, 208
Cannon, Sen. Howard W., 4, 31, 258
Canopus (star), 308
Cape Canaveral, Fla. (see also Cape Kennedy, Fla.)
accident, 275
employment, 335
lunar eclipse, 340
name restoration, 18, 51, 92, 120, 132, 141, 167, 191, 288–289
personnel, 48–49
press comment, 326
Skylab News Center, 127
Cape Kennedy Air Force Station, 335
Cape Kennedy, Fla., 3
anniversary, 248
Labor–Management Relations Council, 48–49
Man in Space National Historic Site, 36
name change, 18, 51, 92, 120, 132, 141, 167, 191, 288–289
Pioneer–G shipment, 78
Cape Verde Islands, 200
Capitol. See U.S. Capitol.
Caracas, Venezuela, 265–266
Caravelle (twin-jet transport aircraft), 278
Carbon, 207, 296
Carbon dioxide, 42, 144
Cardiac pacemaker, 42, 360
Cardiology, 82, 88–89, 351

Career Service Award for Sustained Excellence, 136
Caribbean Sea, 3
Carnegie–Mellon Univ., 208–209, 292
Carolina Bays, 129
Carr, L/C Gerald P. (USMC)
achievements, 320
echo-cardiography, 351
mission preparations, 295, 299, 311, 317–318
press conference, 283
rendezvous, 244
training, 290, 311, 312–313
Carson, Rachel, 170
Carulli, Leonard, 201–202
Cas 1 (Cooperative Applications Satellite). See *Eole*.
CAS–C, 27
Case, Sen. Clifford P., 18, 63
Caspian Sea, 213
Castor (French satellite), 159
Castro, Jose A., 337
CAT. See Clear-air turbulence.
Catalytic, Inc., 349
Catholic Univ., 52
Cavitation, 295
CCV. See Control-configured vehicle.
CDDT. See Countdown demonstration test.
Census (1970), 344–345
Census of U.S. Civil Aircraft, 319
Centaur (booster upper stage), 257, 263, 269, 302–303, 307, 333
Centaur (Indian sounding rocket), 38
Centaurus (constellation), 329
Center for Space Research (MIT). See Massachusetts Institute of Technology.
Central African Republic, 200
Central Intelligence Agency (CIA), 23, 40, 117, 141, 170
Central Meteorological Institute (Peking), 159
Central State Univ., Ohio, 253
CERN. See European Nuclear Research Center.
Cernan, Capt. Eugene A. (USN)
Apollo 17 mission, 5, 18, 19, 21, 49, 130, 163, 255
Apollo-Soyuz Test Project, 325–326
assignment, 339
awards and honors, 63–64, 105, 107, 255
Congress, visit, 20–21
FAI 66th Annual General Conference, 255
good will tour, 186, 195
KSC visit, 50–51
press conference, 5
public appearance, 13, 76
Cessna Cardinal (aircraft), 285
Chaco Canyon National Monument, N. Mex., 12
Chad, 114, 200
Challenger (*Apollo 17* LM). See Lunar module.

Chamber of Commerce, Houston, Tex. See Houston Chamber of Commerce.
Charged particle detector, 309
Charyk, Dr. Joseph V., 53-54, 108, 154
Chemistry, 284-285, 296, 328
Cheng, Dr. Sin-i, 175
Cherry, George W., 69, 72-73, 74, 81
Chesapeake Bay, Md., 181-182
Chi Sheng-ying, 159
Chicago, Ill., 184
Chiles, Sen. Lawton M., Jr., 18, 36, 141
Chicago, Univ. of, 137
Chile, 172-173
China Lake, Calif., 248
China, People's Republic of (P.R.C.)
  aircraft, 33, 289
  communications, 44, 217, 281
  disarmament, 51
  General Administration of Telecommunications, 217
  international cooperation, 32-33, 52, 135, 214, 281
  meteorological satellite, 159
  missile program, 17, 61
  nuclear test, 51, 211, 221-222
  press comment, 221-222
  scientists, 34
  space object count, 185-186
  U.S. detente, 26
China, Republic of (Nationalist), 212
Christmas, 350-351, 352-353
Chromagraphic analyzer, 48
Chryse Valley (Mars landing site), 138
Chrysler Corp., 101, 133
  Space Div., 105, 169
Churchill Research Range, Canada, 32, 37, 41, 58, 63
CIA. See Central Intelligence Agency.
CIAP. See Climatic-Impact Assessment Program.
Cigarette, 207
Circadian rhythm, 223
Cislunar space, 298
Citron, Robert A., 127
Civil Aeronautics Board (CAB), 5, 55, 113, 156, 258, 330
Civil aviation
  Census of U.S. Civil Aircraft, 319
  safety, 252
  security, 40
  service carriers, 277
  statistics, 108, 362
  Tu-144 use, 184
  U.S. leadership, 3
Civil Aviation Board, 277
Civil Defense, Dept. of, 56
Civil Service Commission (CSC), 10, 265
CL-215 (amphibian aircraft), 263
Clarke, Arthur C., 223
Clausen, Rep. Don H., 330
Claverie, Adm. Christian (France), 311
Clayton, Joseph F., 105
Clean combustion program, 280
Clear-air turbulence (CAT), 59, 220, 285
Clear Lake City, Tex., 120

Clements, William P., 23, 96, 175, 176, 187, 258
Cleveland, Ohio, 2, 201, 330, 341
Climate. See Meteorology.
Climatic-Impact Assessment Program (CIAP), 74
Clouds
  Pioneer Venus probe, 183-184
  satellite data, 70-71, 88, 166, 358
    geostationary satellite, 315
    global coverage, 62, 212, 312, 316
  solar, 313
  sounding rocket experiment, 236, 240-241
  stellar, 115, 329
  telescope interference, 297
  Venusian, 183, 332
CM. See Command module.
CNEE. See National Commission for Outer Space, Mexico.
CNES. See National Center for Space Studies, France.
Coal, 240, 335
Coast Guard. See U.S. Coast Guard.
Coatings, thermal, 318
Cobalt, 247-248
Cobb, Geraldine, 265
Cobra Dane (phased-array radar station), 176
Cocoa Beach, Fla., 114
Cohen, Aaron, 121-122
Colby, William E., 141
Cold war, 325-326
Colino, Richard R., 54
Collier, Robert J., Trophy, 183, 238
Collins, B/G Michael (USAF, R.) 109
Collins Radio Co., 43
Colombia, 197
Colorado, 92, 130, 140, 161
Colorado Springs, Colo., 92
Colorado State Univ., 177, 292
Colorado, Univ. of, 53, 158, 228
Columbia Univ., 128, 157, 204
Coma (constellation), 329
Comet (see also individual comets, such as Bennett, Halley's, Ikeya-Seki, and Kohoutek), 4, 220, 345
Comision Nacional del Espacio Exterior, (CNEE), Mexico, 87
Command and service module (CSM)
  Apollo 13 (Odyssey), 77-78
  Apollo 17 (America), 20-21
  Apollo-Soyuz Test Project, 349
  Skylab 2, 116, 145, 146, 147, 149-150, 151, 152
  Skylab 3, 206, 224-225, 226, 227, 228, 229, 244
  Skylab 4, 90, 283, 311, 320, 323
Command module (CM), 288
  Apollo program, 238
  Apollo 13 (Odyssey), 77-78
  Apollo 17 (America), 20-21
  Skylab 2, 147, 148, 152
  Skylab 3, 180, 225-229, 237, 244
  Skylab 4, 226-227, 323, 326, 336, 350-351

Commerce, Dept. of (DOC), 28–29, 83
Commercial aviation, 108, 187, 220, 272, 277, 358, 362
Committee for the Global Atmospheric Research Program, 59
Committee on Space Research (COSPAR), 162
Committee to Reelect the President, 154–155
Common Carrier Bureau (FCC), 249–250
Communications
　astronaut, 134, 171
　NASA technology application, 280
　press comment, 271
　research and development, 293
　U.S.S.R., 102–103, 283, 294, 348–349
Communications Cost Reduction Team (NASA), 299
Communications satellite (see also individual satellites, such as *Anik 2, Intelsat IV F-4, Molniya II-5*)
　anniversary, 223
　benefits, 31
　Brazil, 85
　Canada, 358, 360
　contract, 57, 63, 114, 215, 264–265, 314
　failure, 343
　DOD, 359
　domestic, 5, 44, 92, 186, 260
　earth station, 5, 141, 287
　Europa 2 launcher, 37, 131
　European experimental maritime, 231–232
　experimental, 118
　FCC policy, 14, 92
　funds, 27, 48
　high frequency radio link, 71
　international cooperation, 1, 5, 14, 35–36, 37, 67, 217, 231–232, 250, 251, 361
　laser transmission, 132
　launch
　　Anik 2, 122, 358
　　Dscs F-3, 343
　　Dscs F-4, 343
　　Intelsat-IV F-7, 250, 358, 360
　　Molniya I-23, 38
　　Molniya I-24, 253
　　Molniya I-25, 318
　　Molniya I-26, 333
　　Molniya II-6, 209
　　Molniya II-7, 294
　　Molniya II-8, 351
　leasing services, 154
　maritime, 252, 284
　military, 312, 343
　multifrequency, 65
　NASA program, 5, 359
　ocean research, 183
　Presidential report, 86
　public service, 14, 18
　reliability, 91
　shipping service, 115
　technology utilization, 62, 67–68
　telephone, 281
　use, 41, 135, 273, 332

　U.S.S.R., 284, 294, 348–349, 351, 361
Communications Satellite Act of 1962, 35–36, 48, 361
Communications Satellite Corp. (ComSatCorp; see also individual Intelsat satellites)
　anniversary, 35–36
　Annual meeting of shareholders, 10th, 154
　AT&T stock sellout, 92
　COMSAT General Corp., 53–54
　contract, 44, 63, 252, 264–265
　cooperation, 59, 118, 250, 269, 281
　domestic satellite network, 260
　earth station, 5
　FCC regulations, 73, 114, 115, 249–250
　high-capacity satellite procurement, 260
　International Systems Div., 54
　launch support services, 314
　leasing services, 154, 179
　maritime communications satellite system (proposed), 65, 252
　multifrequency satellite system, 65
　personnel, 108, 269
Communist block, 257–258
Compendium of Meteorological Satellites and Instrumentation (NSSDC 73-02), 233
Composite materials, 247–248
Comprehensive Designers, Inc., 131–132
Compressor, 266
Compton effect, 329
Comptroller General's Report to the Congress: Analysis of Cost Estimates for the Space Shuttle and Two Alternate Programs, 172
Computer (see also Data processing and individual computers, such as IBM 9020, ILLIAC IV, Randomdec)
　aircraft navigation, 46, 59, 74–75, 291
　contract, 250, 270, 276, 300–301
　KSC, 293, 348, 352
　meteorology and environment, 54, 64, 209, 245, 250, 288, 293
　missile, 63, 90
　recycling, 63
　simulation, 77–78, 332
　spacecraft, 228–229, 244, 270, 345–346
　technology utilization, 64, 67–68, 233, 250, 260, 360
　U.S. trade surplus, 256
　U.S.S.R. cooperation, 99, 275
Computer Sciences Corp., 178–179, 239
ComSatCorp. See Communications Satellite Corp.
Concorde (Anglo-French supersonic transport)
　acquisition delay, 176
　Congressional report, 85
　design, 7
　donation to French Air Museum, 294–295
　exhibit, 163
　flight test, 73
　future of, 42, 179
　international visitation, 217, 265–266
　press conference, 217

production, 97, 102, 113
replacement, 264
sales options, 33
solar eclipse, 114, 200
stratosphere measurement, 288
Concorde 001, 294–295
Concorde 02, 271
Concorde 201, 338
Congress, 354
  aerospace industry, 1
  aircraft authorization, 32, 40, 58, 75, 86, 176
  astronaut praise, 177–178
  bills passed, 361–362
  defense debate, 112–113, 124
  energy crisis consideration, 314, 361–362
  Executive reaction, 208–209
  Federal Energy Administration (proposed), 337
  Federal Energy Office (President's; proposed), 337
  Joint Committee on Atomic Energy, 203, 271, 335
  Joint Conference Committee, 198, 219, 223, 299
  Joint Economic Committee, 85
  Joint Technology Assessment Board, 307
  NASA programs consideration, 279
  Office of Assistant Secretary of Transportation for Congressional and Intergovernmental Affairs, 39
  Office of Science and Technology, 203, 208–209
  Office of Technology Assessment, 63
  Presidential message
    aeronautics and space report, 86–87
    budget, 26–27
    national legislative goals, 258–259
    NSF achievements, 80
    Reorganization Plan No. 1, 24–25
    Science Indicators, 1972, 256
    U.S. foreign policy, 134
    U.S. participation in the U.N., 257
  role given NASA, 279, 297
  Skylab program, 84
  solar energy legislation, 246
  space program, 21, 36, 42
  supersonic transport, 42
  reports submitted to, 172, 256
Congress, House of Representatives, 155
  amendments blocked, 192
  amendments carried, 336
  astronauts report to, 20–21
  bills introduced, 4, 19, 40–41, 57, 95, 189, 231, 236, 239, 265, 293, 330, 337
    NASA budget, 187
  bills passed, 192, 336, 349
    NASA budget, 161, 192, 209, 236
  Committee on Appropriations
    Subcommittee on Defense Appropriations, 94–95
    Subcommittee on Dept. of Housing and Urban Development–Space–Science–Veterans, 93, 101, 187
  Committee on Armed Services, 94, 189
  Committee on Finance
    Subcommittee on International Trade, 70
  Committee on Foreign Affairs, 109
  Committee on the Judiciary, 73, 161
  Committee on Manned Space Flight, 195–196
  Committee on Science and Astronautics, 4, 189
    astronauts report to, 213
    bill approved, 349
    bills referred to, 293, 330
    committee members, 23, 73, 117
    energy hearings, 135, 177, 219, 240
    The Federal Government and Energy R&D: Historical Background, 122
    NASA budget, 88, 90, 139–140, 161
    reports submitted to, 254
    resolutions referred to, 26, 40
    science and technology, 215
    Subcommittee on Aeronautics and Space Technology, 68, 72, 74, 76–77, 80, 83, 298
    Subcommittee on Energy, 154, 160, 177, 219, 240, 260, 318, 353
    Subcommittee on International Cooperation in Science and Space, 216, 277
    Subcommittee on Manned Space Flight, 55, 57–58, 61, 69–70, 71–72, 75–76, 81–82, 235, 256–257
    Subcommittee on Space Science and Applications, 160
  OMB energy proposal, 342
  resolutions introduced, 18, 23, 31, 26, 40, 51, 161, 266
  resolutions passed, 26, 40, 92, 199
  Select Committee on Committees, 156
  solar energy legislation, 231, 293, 330, 246
Congress, Senate
  announcement to, 232
  astronaut report to, 20–21
  bills defeated, 254
  bills introduced, 5, 13, 36, 87, 120, 133–134, 238, 267, 332, 337, 338–339, 344
    NASA budget, 181
  bills passed, 4, 131, 254, 339, 349
    NASA budget, 187, 198
  Committee on Aeronautical and Space Sciences
    aerospace industry hearing, 272
    astronauts report to, 213
    bills approved, 167
    committee members, 4, 6, 13, 21, 208–209, 314
    history of, 20
    NASA budget, 167
    inquiry, 30
    testimony, 56–57, 67, 71, 78, 81, 89–90, 91, 103, 105–106, 112
  Skylab 1 investigation, 153, 155, 211, 230–231
  Skylab 2, 160–161, 177–178

space shuttle hearings, 302–303
space tug hearings, 302–303
U.S.S.R. space program, 182
Committee on Appropriations, 198, 254
  Subcommittee on Defense Appropriations, 94–95
  Subcommittee on Dept. of Housing and Urban Development-Space-Science-Veterans, 18, 39–40
Committee on Armed Services, 19, 236
  Preparedness Investigating Subcommittee, 20
  Subcommittee on Tactical Air Power, 258
Committee on Commerce
  Subcommittee on Aviation, 31
Committee on Government Operations
  Subcommittee on Executive Reorganization and Government Research, 51–52
Committee on Labor and Public Welfare
  Special Subcommittee on the National Science Foundation, 181
Democratic seat campaign, 341
Johnson, Lyndon B., leadership, 252
military promotion confirmed, 328
nominations approved and confirmed, 19, 23, 31, 79, 109, 117, 174, 204–205, 216–217, 218, 236, 258, 259
nominations submitted to, 134, 141, 168, 221
reports submitted to, 281
resolutions introduced, 25, 54–55, 120, 141, 264
resolutions passed, 4, 13, 36–37, 40, 210
solar energy legislation, 332, 246, 347
Special Committee on Space and Astronautics, 20
Congressional Medal of Honor, 220
Conlan, Rep. John B., 23
Connecticut, 73
Conrad, Capt. Charles, Jr. (USN)
  aircraft accident, 209
  award, 286, 300
  presidential reception for, 193
  retirement, 333
  Skylab 1 mission, 144, 145
  Skylab 2 mission
    antics, 241–242
    communications from space, 177, 179
    earth resources data, 149, 181–182, 213
    EVA, 149, 227
    launch, 147, 164
    manufacturing in space, 213
    medical aspects, 126, 142, 148, 150, 192, 193, 195, 222, 178
    mission preparations, 90, 112, 139, 146
    press conference, 193, 199
    record, 150, 152, 229
    Skylab 1 entry, 148

Skylab 2 repairs, 147, 150
visit to Congress, 56
Consejo Nacional de Ciencia y Technologia of Mexico (inter-American meeting), 189
Constance, West Germany, 162
Constellation (see also individual constellations such as Centaurus, Coma, Cygnus, Perseus, Taurus, and Virgo), 267, 310, 329, 333, 348
Consultants Unlimited, 298
Contamination, 48, 97, 154
Conte, Rep. Silvio O., 114
Continued Growth Planned for Federal Civilian R&D Programs(NSF 73-314), 293
Contract. See individual agencies and industries.
Control-configured vehicle (CCV), 139
Convair 990 (*Galileo*; NASA research aircraft)
  accident, 115, 116–117, 117–118, 127–128, 131, 181, 186, 187, 189, 245
  Bering Sea study, 22, 47, 49, 74
  GATE, 24
  replacement, 167, 181, 187, 245
Convair Div. See General Dynamics Corp.
Convention on International Liability Caused by Space Objects, 257
Conventional takeoff and landing aircraft. See CTOL.
Conyers, Rep. John, Jr., 161
Cook, Sen. Marlow W., 31, 318
Cook, Richard W., 126
Cooper, Henry S. F., Jr., 77–78
Cooperative Applications Satellite. See CAS–C and *Eole.*
Copernican Quincentennial, 125
Copernicus, Nicholaus, 4, 50, 92, 112, 122, 124
Copernicus, Nicholaus, Week, 92, 112
Copernicus. See *Oao 3.*
Copernicus 500. See *Intercosmos Copernicus 500.*
Corliss, William R., 33–34
Cornell Univ., 119, 186, 188, 262
Corona. See Solar corona.
Coronagraph, 228, 322
Corsair II. See A-7D.
Cortright, Edgar M., 9, 108, 137
Cosmic plasma, 283
Cosmic ray
  balloon experiment, 269
  biomedical effects, 238
  composition, 22
  detection, 174, 212
  earth surface effects, 90–91
  HEAO study, 38–39
  Mars exploration, 267
  Pioneer 6, 7, 8, 9 data, 33–34, 311
  Pioneer 11 experiment, 106
  Skylab 1 experiment, 232
  stellar research, 279–280
  stereo experiment, 239
Cosmonaut, 159, 190

accident. See Accident.
Apollo-Soyuz Test Project. See Apollo-Soyuz Test Project.
astronaut, 190
   death, 102–103, 116, 153, 273, 282, 348
   honorary guest, IAF 24th Congress, 287
   medical aspects, 82, 273, 348
   meeting, 163
   memorials to, 56
   navigation instrumentation, 276
   press comment, 276–277
   Salyut 2 docking, 102–103
   Skylab 2 crew congratulations, 186–187
   Soyuz 10 mission, 103, 273
   Soyuz 11 mission, 103, 151, 273, 282, 327, 348
   Soyuz 12 mission, 273, 276, 359
   Soyuz 13 mission, 347, 359
   space shuttle, 277
   training, 54, 55, 84, 108, 196, 207, 287, 326
Cosmonaut Day, 114–115
Cosmonaut Training Center, 282
Cosmos (U.S.S.R. satellite), 45, 65, 359
Cosmos 110, 303–304
Cosmos 368, 303–304
Cosmos 496, 282
Cosmos 543, 12
Cosmos 544, 19
Cosmos 545, 23
Cosmos 546, 24
Cosmos 547, 35
Cosmos 548, 41
Cosmos 549, 56
Cosmos 550, 61
Cosmos 551, 67
Cosmos 552, 90
Cosmos 553, 114–115
Cosmos 554, 122
Cosmos 555, 127
Cosmos 556, 137
Cosmos 557, 141–142, 159
Cosmos 558, 156
Cosmos 559, 158
Cosmos 560, 161
Cosmos 561, 164
Cosmos 562, 173
Cosmos 563, 175
Cosmos 564, 177
Cosmos 565, 177
Cosmos 566, 177
Cosmos 567, 177
Cosmos 568, 177
Cosmos 569, 177
Cosmos 570, 177
Cosmos 571, 177
Cosmos 572, 180
Cosmos 573, 184, 282
Cosmos 574, 188
Cosmos 575, 189
Cosmos 576, 197
Cosmos 577, 221
Cosmos 578, 235
Cosmos 579, 249
Cosmos 580, 250
Cosmos 581, 251
Cosmos 582, 252
Cosmos 583, 253
Cosmos 584, 257
Cosmos 585, 258
Cosmos 586, 262
Cosmos 587, 266
Cosmos 588, 284
Cosmos 590, 284
Cosmos 591, 284
Cosmos 592, 284
Cosmos 593, 284
Cosmos 594, 284
Cosmos 595, 284
Cosmos 596, 284, 292
Cosmos 597, 287, 292
Cosmos 598, 290, 292
Cosmos 599, 292
Cosmos 600, 292–293
Cosmos 601, 292–293
Cosmos 602, 292, 295
Cosmos 603, 292, 301
Cosmos 604, 301
Cosmos 605, 303
Cosmos 606, 307
Cosmos 607, 316
Cosmos 608, 327
Cosmos 609, 328
Cosmos 610, 330
Cosmos 611, 331
Cosmos 612, 331
Cosmos 613, 332
Cosmos 614, 336–337
Cosmos 615, 343
Cosmos 616, 346
Cosmos 617, 348–349
Cosmos 618, 348–349
Cosmos 619, 348–349
Cosmos 620, 348–349
Cosmos 621, 348–349
Cosmos 622, 348–349
Cosmos 623, 348–349
Cosmos 624, 348–349
Cosmos 625, 349–350
Cosmos 626, 352
Cosmos 627, 354
COSPAR. See Committee on Space Research.
Cost Growth in Major Weapons Systems (GAO report), 94
Cost of Living Council, 342–343
Costes, Col. Dieudonne, 158
Cotter, Rep. William R., 23
Cotton, Sen. Norris, 31
Cottonwood, Ariz., 113
Coughlin, Rep. Lawrence R., 23, 73
Council for International Cooperation in the Exploration and Use of Outer Space, U.S.S.R., 269
Council on Energy Policy, 131
Council on Environmental Quality, 223
Countdown demonstration test (CDDT), 136
Cousteau, Jacques-Yves, 41, 48, 51, 62, 243
Crab Nebula, 199–200
Crew escape module, 85

Crime prevention, 280, 293
Crimea Astrophysical Observatory, 19
Criminal Code Reform Act of 1973 (proposed), 119
Crippen, L/C Robert L. (USN), 31–32
Croix de Guerre (France), 220
Cronin, Rep. Paul W., 23
Crop-spraying aircraft, 278
Crystals, 64–65, 271, 275, 324
CSC. See Civil Service Commission.
CSM. See Command and service module.
CTOL (conventional takeoff and landing) aircraft, 72
CTS (Canadian comsat), 361
Cuba, 133, 314
Cuiba, Brazil, 95
Culbertson, Philip E., 69–70, 111
Curaçao, Dutch West Indies, 133
Currency crisis, 278
Currie, Dr. Malcolm R., 174
Curtin, M/G Robert H., (USAF, Ret.), 70, 76–77
Curtis, Sen. Carl T., 13
CVN–70 (aircraft carrier), 28
Cydonia (Mars landing site), 138
Cygnus (constellation), 333
Cygnus X–1 (X-ray source), 329
Czechoslovakia, 98, 122, 172–173, 302, 342, 344

D

Daddario, Rep. Emilio Q., 307
Dallas–Fort Worth Airport, Grapevine, Tex., 217, 265–266, 271
Dallas, Tex., 184
Data processing, 95, 209–210, 250, 300–301
David, Dr. Edward E., Jr., 2, 19, 215
Davies, Richard T., 213
Davis, D. R., 12
Davis, Rep., John W., 23, 63
Davis, Sammy, Jr., 63–64
DC–3 (jet transport), 135–136, 137
DC–8, 89–90
DC–9, 89–90, 204, 358
DC–10, 138, 315
Deal, Paul H., 317
Death Valley National Monument, 13
DeBellevue, Capt., Charles B. (USAF), 135
Debus, Dr. Kurt H., 32, 33, 135, 288, 348
Deep Space Network (DSN), 83, 116, 208
Defense, Dept. of (DOD; see also U.S. Air Force; U.S. Army; U.S. Navy; U.S. Marine Corps)
 aircraft. See Aircraft.
 awards, 135, 286
 budget, 28, 96, 104, 130, 258
 Computer Institute, 116
 contract, 201, 258
 cooperation, 59, 71, 210, 229, 332
 employment, 278
 Giant Patriot Project, 353–354
 launch record, 357
 Manned Space Flight Support Office, 142–143
 missile program, 7
 personnel, 23, 112, 141, 259
 satellite, 5, 113, 209–210, 314, 359
 Shuttle-User Committee, 103
 Skylab 142–143, 155, 319
 space shuttle, 71, 103, 104, 302–303, 358
 space tug development, 256–257
 Spacelab, 303
 weapon system, 7, 112–113, 333
 World Weather Program for Fiscal Year 1974, 210
Defense Navigation Satellite System (DNSS), 304
Defense Research and Engineering, 174, 205
Defense Satellite Communications System (DSCS), 104
 Phase 2 (DSCS 2), 343
Definitive Agreements for INTELSAT, 361
DeFlorez Training Award, 10
De Gaulle, President Charles (France), 136
de Havilland Aircraft Co. Ltd., 20
De la Vaulx Medal (France), 255
Delta launch vehicle. See Thor-Delta.
Dembling, Paul G., 136
de Mendonca, Dr. Fernando, 95
Demeter (Jovian moon), 309
Democratic National Committee Headquarters, 154–155
Democratic Party, 341
Denmark, 268–269
Denver, Colo., 104, 208–209
Deoxyribonucleic acid (DNA), 14–15, 59
Descartes (lunar landing site), 7, 84
Detente, U.S.–U.S.S.R., 221–222
Detroit Diesel Allison Div. See General Motors Corp.
Deuterium, 249
Development Sciences, Inc., 132
DFBW. See Digital fly-by-wire aircraft.
DHC–7 (Canadian turboprop STOL transport), 20, 114
Diamant-B (French booster), 159
Diamond, Fred, 291
Dietlein, Dr. Lawrence F., 228
Digital fly-by-wire aircraft (DFBW), 74–75, 196, 266, 207, 358
Dipole experiment, 178, 180
Disarmament, 51, 97, 139, 189–190, 192, 214, 249, 253, 258–259, 313
Disaster-assessment systems, 245–246
Disher, John H., 171, 344–345
Distinguished Federal Civilian Service Award (President's), 351–352
Distinguished Flying Cross (USAF), 238
Distinguished Public Service Medal (NASA), 105, 299
Distinguished Service Award (NASA), 105
Distinguished Service Cross (USA), 220
Distinguished Service Medal (NASA), 157, 238, 286, 299, 350, 351–352

Distinguished Service Medal (USAF), 231, 238
District of Columbia. See Washington, D.C.
Djerassi, Dr. Carl, 284–285
DNA. See Deoxyribonucleic acid.
DNSS. See Defense Navigation Satellite System.
DOC. See Commerce, Dept. of.
Docking
   ASTP. See Apollo-Soyuz Test Project.
   Cosmos, 159
   Skylab 2, 144–145, 147
   Skylab 3, 211–212, 219, 224, 225
   Skylab 4, 311, 319–320
   Soyuz 10-Salyut 1, 103, 273
   Soyuz 11-Salyut 1, 103, 273
   Soyuz-Salyut, 326
DOD. See Defense, Dept. of.
Dog, 303–304
DOI, See Interior, Dept. of
Dollfus, Adouin, 186
Domenici, Sen. Pete V., 13, 232, 332
Domestic communications satellite network, 260, 264–265, 269
Dominick, Sen. Peter H., 63, 133–134
Donlan, Charles V., 256–257
Donnelly, John P., 171
Doolittle, L/G James H. (USAF, Ret.), 137
Doppler radar, 118, 220, 294
Dosimetry, 238
DOT. See Transportation, Dept. of.
Douglas Aircraft Co. See McDonnell Douglas Corp.
Douglas, Justice William O., 153
"Down to Earth Space Program Application" (program), 9–10
Downey, James A., III, 39
Downing, Rep. Thomas N., 23
Drake Passage, 48, 51, 62
Draper, Dr. Charles Stark, 232, 288
Draper, Charles Stark, Laboratory (MIT), 10–11, 288
Drone, 328
Drug detection, 291, 360
Dryden, Dr. Hugh L., Memorial Fellowship Award, 11–12, 52
Dryden Research Award, 11
DSCS. See Defense Satellite Communications System.
DSCS 2 (communications satellite). See Defense Satellite Communications System, Phase 2.
Dscs F-3, 343
Dscs F-4, 343
DSN. See Deep Space Network.
Dublin, Ireland, 255
Ducander, Charles F., 117
Dudley Observatory, Albany, N.Y., 236, 240–241
Duke, Col. Charles M., Jr. (USAF), 255, 339
Dulles International Airport, Va., 138, 217, 271
Dulski, Rep. Thaddeus J., 57
Dunn, L/G Carroll H. (USA), 295

Dunseith, Lynwood C., 339
Durant, Frederick C., III, 288
Dyal, Dr. Palmer, 52
Dynalectron Corp., 94
Dzhanibekov, Vladimir, 162

E

Early Bird (comsat). See *Intelsat I.*
"Early Photography from Rockets" (symposium paper), 288
Early warning satellite system, 104
Earth, 331–332, 271
   atmosphere. See Atmosphere, upper.
   evolution, 5
   photographs, 24, 36, 157, 275, 313–314
   quake, 301, 338
   radiation belt, 287
   rotation, 184–185, 185–186
   solar effects, 47, 136, 313
   spectrographic data, 273, 276
   storm formation, 271
   structure, 164, 165, 328
   tectonic plate motion, 66–67, 81–82
   thermal mapping, 104
   "When Earth Became a Planet" (exhibit), 4
Earth and ocean physics applications program (EOPAP), 79, 80–81, 162, 303
Earth Observatory Satellite (EOS), 27, 80–81
Earth resources, 359
   contract 139
   international cooperation, 37–38, 42, 45, 61–62, 87, 95, 108–109, 162, 170, 219–220, 245, 257, 280–281, 341, 361
   management, 25, 85, 92–93, 110, 267, 341
   mapping, 130, 227, 247
   monitoring, 161, 210, 238, 243, 257, 280–281, 344–345
   NASA program. See National Aeronautics and Space Administration, programs.
   photography, 46, 61–62, 127, 240, 247
   remote sensing, 30, 70, 80–81, 108–109, 129, 181, 245–246, 344, 352, 358
   Skylab activities, 44, 142, 145, 197, 213, 224, 230, 231, 271, 319–320, 323–324, 344–345, 357
   U.S.S.R. program, 102–103, 348
Earth resources experiment package (EREP), 87, 181–182, 198
   aircraft support, 344
   instrumentation, 143, 190, 289
   press conference, 143, 181–182, 222
   Skylab 2 mission, 149, 181–182, 222
   Skylab 3 mission, 227, 228, 245
   Skylab 4 mission, 283, 321, 322, 323
Earth resources survey aircraft, 243, 247
Earth Resources Technology Satellite (ERTS; see also *Erts 1,* ERTS–B)
   aircraft support, 344
   applications, 61–62, 167–168, 213, 257
   award, 76
   funds, 167

international cooperation, 162, 219–220, 341, 361
Earth Resources and Technology Satellite Symposium, 66, 341
Earth station, 272–273
  air pollution studies, 245
  communications satellite system, 154, 186, 287, 314
  installation, 56, 73, 83, 141, 176
  solar power transmission, 351
  tracking, 122, 184, 298, 313–314
  U.S.-U.S.S.R. Direct Communications Link, 361
  U.S.S.R., 293
Earthquake, 64, 78, 81–82, 154, 189, 245–246, 328, 338, 360
Eastern Airlines, Inc., 33, 319
Eastern Hemisphere, 206
Eastern Test Range (ETR, USAF)
  launch
    Anik 2, 123
    Dscs F-3 (comsat), 343
    Dscs F-4, 343
    Explorer 49, 179
    Explorer 50, 298
    Intelsat IV, 250
    Mariner 10, 307
    Pioneer 11, 106
    schedule, 1
    unidentified satellite, 67, 181
EC–135 (communications aircraft), 56
Echo 1 (communications satellite), 273
Echo-cardiography, 351
Ecology (see also Air pollution; Noise, aircraft), 12, 37–38, 41, 46, 162, 170, 330, 344
Economic Club of Detroit, 263
Economics, 250
Edgerton, Dr. Harold E., 284–285
Education
  graduate studies, 34, 232, 305
  NASA program, 219, 245
  NSF program, 219, 245
  press comment, 292
Edwards, Rep. Don, 117–118
Edwards AFB, Calif., 111, 303
EEO. See Equal Employment Opportunity.
Eger, John, 239
EG &G, Inc., 5
Eggers, Dr. Alfred J., Jr., 154, 260
Ehrenreich Photo-Optical Industries, Nikon Instrument Div., 224
Ehrlichman, John D., 154–155, 168
Einhorn, Raymond, 201–202
Eisenhower, President Dwight D., 2, 230
Ekranoplane (U.S.S.R. hovercraft), 331
Elbe River, 325
ELDO. See European Launcher Development Organization.
Electric connector, 52–53
Electric field, 287, 293, 311
Electric Research Council, 69–70
Electrical energy, 5, 133, 155, 160–161, 177, 188–189, 194, 260, 279, 314, 318–319, 337–338, 349, 351, 355, 361–362

Electrical furnace, 275
Electrically scanning microwave radiometer (ESMR), 104
Electromagnetism, 279–280, 302, 358–359
Electromechanics, 251
Electron (U.S.S.R. satellite), 287
Electron-beam welding, 266–267
Electronic and Space Systems. See British Aircraft Corp.
Electronics, 71, 118, 182, 271, 276
Electrophoresis, 238, 257
Ellington AFB, Tex., 152, 193
Elliott, James R., 19–20
Ellsworth Antarctic Expedition, 294
Elverum, Gerald W., Jr., 313
Ely, Nev. 344–345
Emergency Energy Act (proposed), 314, 361–362
Employment Opportunity Program (NASA), 299–300
Energy action group (proposed), 342
Energy and Natural Resources, Dept. of (proposed), 258–259, 290
Energy crisis
  congressional consideration, 134–135, 141, 239, 240, 318, 361–362
  international cooperation, 342
  KSC measures, 348
  Presidential consideration, 200, 314, 337, 346, 361–362
  press comment, 237–238, 325, 330, 335
  technology utilization, 42
Energy, Dept. of (proposed), 239
Energy Development and Supply Act of 1973, 95
Energy Development and Supply Commission (proposed), 95
Energy Emergency Action Group, 337
Energy Management Project (proposed), 87
Energy Policy Act of 1973, 131
Energy Policy, Federal Office of, 200, 336
Energy research and development (ER &D; see also nuclear energy and solar energy),
  budget, 140, 200, 290
  conservation, 188–189, 246, 265, 340, 350
  contract, 246
  energy source development, 115–116, 154, 159, 260, 267, 337–338, 338–339, 349
  The Federal Government and Energy R.&D.: Historical Background, 99
  international cooperation, 342
  LeRC activities, 330
  The Nation's Energy Future (AEC report), 335
  plans, 200, 258–259
  press comment, 137, 160, 240, 292
Energy Research and Development Administration (ERDA; proposed), 258–259, 290, 335, 342, 361–362
Energy Research and Development Commission (proposed), 236

Energy Research and Development Policy Act, 339
Energy Research, Development, and Demonstration Administration (proposed), 318
Energy, Revenue and Development Act of 1973, 344
Enevoldson, Einar K., 291, 307
Engine (see also individual engines and motors, such as JT3D, JT8D, Pegasus 15, TF33-P-7, XLR-11), 137, 247–248, 266
 aircraft (See also individual aircraft, such as Supersonic transport, X-15 hypersonic aircraft), 13–14, 27–28, 50–41, 72–73, 74, 80–81, 232, 263–264
 automobile, 87, 199, 263–264, 265, 358
 hydrofoil, 185
 rocket, 80, 313
 space shuttle. See Space shuttle.
 spacecraft, 288
Engineer, 165, 178, 282
 aircraft research, 107–108
 Apollo program, 66, 110
 Apollo-Soyuz Test Project, 8–9, 83–84, 163, 261
 automotive research, 264
 conference, 53, 260
 employment, 54, 58–59, 109–110, 126, 137–138, 143, 188, 189, 218, 232, 256, 264, 278, 286, 317, 319
 foreign, 34, 221, 241, 248
 OTA services, 63
 science and technology policy changes, 301
 Skylab, 158–159, 211, 235, 239, 244, 299
 Soyuz 10, 162
 space shuttle, 76–77
 technology utilization, 215–216
Engineering, 4–5, 140, 237, 286–287, 305, 318
 in space, 235, 239, 266–267, 271, 303, 304–305, 323, 353
England. See United Kingdom.
Enid, Okla., 319–320
Enrico Fermi Institute, 164
Environment (see also Air pollution; Noise, aircraft)
 disaster, 170, 245–246, 338
 international cooperation, 134, 164, 187–188, 361
 planetary. See individual planets
 press comment, 170, 237–238, 292, 325
 remote sensing, 85, 95, 209–210, 352
 research, 12, 206
 solar effects. See Solar energy.
 technology utilization, 162, 233, 272–273
Environmental Protection Agency (EPA), 69–70
 budget, 28–29
 conference, 260
 cooperation, 80–81, 110, 188–189, 199
 NASA technology utilization, 58, 279
 personnel, 223, 246, 259
 regulations, 130, 153, 219
 reports submitted to, 61
 World Weather Program for Fiscal Year 1974, 210
Environmental Research Institute of Michigan, 183
Environmental Satellite Service (NOAA), 62
Eole (CAS 1; French cooperative applications satellite), 183
EOPAP. See Earth and ocean physics applications program.
EOS. See Earth Observatory Satellite.
EPA. See Environmental Protection Agency.
epndb: effective perceived noise in decibels.
Epsilon Bootis (star), 130
Equal Employment Opportunity (EEO), 10, 95, 119, 172, 213, 299–300, 300–301, 329
ER&D. See Energy research and development.
ERDA. See Energy Research and Development Administration (proposed).
EREP. See Earth resources experiment package.
Ergometer, 151, 197, 209
Ernst Krenkel (U.S.S.R. research ship), 237
ERNO Raumfahrttechnik GmbH, 119
ERNO-VFW-Fokker, 111
ERTS. See Earth Resources Technology Satellite.
Erts 1, 70, 143
 anniversary, 219–220
 information dissemination, 66, 162
 international cooperation, 95, 257, 361
 malfunction, 187
 photographs, 25, 37–38, 42, 46, 73, 110, 167–168, 170, 358
 Noaa 2 data correlation, 24–25
 press comment, 37–38, 85, 231, 245
 tape recording systems, 98
ERTS-B, 1, 29–30, 70, 140, 187
ESA. See European Space Agency.
Esaki, Leo, 296
ESC. See European Space Conference.
Esch, Rep. Marvin L., 23
ESMR. See Electrically scanning microwave radiometer.
ESRO. See European Space Research Organization.
Essa 1 (Environmental Science Services Administration meteorological satellite, now under NOAA), 312
Essa 2, 312
Essa 3, 312
Essa 4, 312
Essa 5, 312
Essa 6, 312
Essa 7, 312
Essa 8, 62, 312
Essa 9, 312
ESTEC. See European Space Research and Technology Center.

Estes, L/C Thomas B. (USAF), 265–266
ET (external tank). See Space shuttle.
Ethology, 290
ETR. See Eastern Test Range.
Europa (Jovian moon), 310
Europa II (ELDO booster), 37, 131
Europe, 41, 172–173, 202, 270, 280, 281
   communications satellite, 14, 179
   industry, 236, 256, 278
   international cooperation, 210, 244, 268, 278, 304, 342, 360
   scientists, 204, 261, 302
   space program, 77, 192–193, 236
   U.S. military withdrawal, 97
   Western European Union, 264
European Launcher Development Organization (ELDO), 77, 37, 131, 231–232
European Nuclear Research Center (CERN), 77, 302
European Space Agency (proposed), 76–77, 231–232, 236
European Space Conference (ESC), 210, 231–232, 236
European Space Research and Technology Center (ESTEC), 119
European Space Research Organization (ESRO), 281
   contract, 304
   international cooperation, 202, 244, 268–269, 289, 297, 302–303, 315, 358, 360, 361
   international space object count, 185–186
   L-3S launcher, 301
   member nations, 268
   reorganization, 77, 231–232
   personnel, 174, 281
   satellite, 33, 118, 124, 174, 202, 210, 361
   space shuttle participation, 17–18, 55, 61, 76–77, 111, 113, 119, 136, 231–232, 244, 268, 358, 360
European Space Symposium, 13th, 195
EUV. See Ultraviolet, extreme.
ev: electron volt.
EVA. See Extravehicular activity.
Evans, Drew F., 315–316
Evans, Capt. Ronald E. (USN)
   Apollo 17 mission, 5, 18, 19, 130
   ASTP backup crew, 31–32, 326
   awards and honors, 63–64, 105, 107
   Congress visit, 20–21
   good will tour, 186, 195
   meeting, 163
   KSC visit, 50–51
   press conference, 5
   public appearance, 76
Ewing, Dr. William Maurice, 284–285
Exceptional Scientific Achievement Medal (NASA), 105–106, 299
Exceptional Service Medal (NASA), 105–106, 120, 135, 157, 203, 238, 299, 350
Executive Office of the President
   Federal Energy Office (proposed), 337
Exhibit
   Aeronautical Symposium and Display, 277
   Goddard Memorial Collection, 223
   "The Image of the Moon–Galileo to Apollo XI," 120
   International aerospace show, 263
   International Exposition, 93
   lunar sample display, 51
   Paris Air Show, 163
   "The Salvage of Skylab 1," 223
   Soviet Exhibition of Economic Achievement, 216
   "Technology in the Service of Man," 265
   "When Earth Became a Planet," 4
Exothermic-tracing techniques, 266
Explorer 1 (satellite), 33
Explorer 17 (AE-A; Atmosphere Explorer), 346
Explorer 18 (IMP-A; Interplanetary Monitoring Platform), 298
Explorer 27 (Beacon Explorer C), 64
Explorer 29 (Geos 1; Geodetic Earth Orbiting Satellite), 64
Explorer 32 (AE-B), 346
Explorer 36 (Geos 2), 64
Explorer 38 (RAE 1; Radio Astronomy Explorer), 180
Explorer 42 (SAS-A; Small Astronomy Satellite; *Uhuru*), 52
Explorer 47 (IMP-H), 131, 298
Explorer 48 (SAS-B), 35, 199–200
Explorer 49 (RAE-B; RAE 2), 1, 178, 179–180, 358
Explorer 50 (IMP-J), 298, 358
Explorer 51 (AE-C), 1, 345–346, 358
Extraterrestial life, 14–15, 36, 50, 106, 138, 293, 304–305, 317, 331–332
Extravehicular activity (EVA)
   Apollo 17 mission, 255
   ASTP, 35
   LST, 209
   record, 255, 279, 357
   Skylab 2 mission, 112, 145, 146–147, 149–150, 152, 174
   Skylab 3 mission, 223, 225–230, 239–240
   Skylab 4 mission, 290, 319–323, 350–351, 357
   space shuttle, 209
   space tug, 209
Eye test device, 298
Eye of Jupiter (great red spot), 308–309

F

F-4 (Phantom supersonic fighter-bomber aircraft), 11, 183, 217
F-4E, 13, 172
F-4J, 28
F-5E (international fighter aircraft), 163, 168, 263
F-8 (carrier fighter aircraft), 69, 74, 266
F-14 (Tomcat fighter aircraft), 4, 74, 157, 163, 176, 189, 258, 281

F-14A, 328
F-14B, 260
F-15 (Eagle fighter aircraft), 28, 58, 86, 127, 155, 204, 291
F-28 (Fokker twin jet transport), 263
F-105G (Thunderchief jet fighter-bomber aircraft), 172
F-111, 33, 175, 183, 189, 204, 307
FAA. See Federal Aviation Administration.
Facilities. See National Aeronautics and Space Administration, facilities
Faget, Dr. Maxime A., 288
FAI. See Fédération Aéronautique Internationale.
Fairbanks, Alaska, 85, 352
Fairchild Industries, Inc., 18, 21, 40, 63, 97, 273
   American Satellite Corp., 260
Fairford, U.K., 288
Falcon (*Apollo 15* LM). See Lunar module.
FASST. See Federation of Americans Supporting Science and Technology.
Fawn Creek, Colo., 157
FCC. See Federal Communications Commission.
Federal Advisory Committee Act of Oct. 6, 1972, 22
Federal Aviation Act of 1958, 120
Federal Aviation Administration (FAA), 137, 295–296
   air traffic control, 129, 153, 201
   airlines, 320
   airports, 129
   Application of Area Navigation in the National Airspace System, 59
   Automated Radar Terminal Systems (ARTS), 255
   Aviation Forecasts, Fiscal Years 1973–1984, 108
   awards, 255, 286
   budget, 28
   Census of U.S. Civil Aircraft, 319
   contract, 41, 43, 95, 135–136
   cooperation, 110, 182, 286, 358
   cooperation, international, 216
   navigation methods, 59, 209
   noise, aircraft, 95, 182, 358
   personnel, 10, 55, 79, 83, 123
   policy, 3, 40, 350
   research, 13–14, 72–73
   symposium, 216, 333
Federal Communications Commission (FCC)
   AT&T stock sellout, 92
   Common Carrier Bureau, 249–250
   communications satellite, 4, 48, 65, 114, 179
      corporation approval, 33, 115, 135
   cooperation, 250
   policy, 14, 73, 92, 141, 184, 252, 260
Federal Council for Science and Technology, 153, 208, 286–287
Federal Energy Administration, 337, 344, 349
Federal Energy Office (President's; proposed), 337
Federal Funds for Research, Development and Other Scientific Activities, Fiscal Years 1971, 1972, and 1973 (NSF 72–317), 39
The Federal Government and Energy R. & D.: Historical Background, 99
Federal Highway Administration, 233
Federal Mediation and Conciliation Service, 48
Federal Plan for Meteorological Services and Supporting Research, Fiscal Year 1974, 209
Federal Republic of Germany. See Germany, West.
Federal Scientific, Technical, and Health Personnel, 1971 (NSF 73–309), 278
Federal Support to Universities, Colleges, and Selected Nonprofit Institutions, Fiscal Year 1971 (NSF 73-300), 129–130
Federal Transportation Policy: The SST Again, 85
Federal Women's Award, 52, 73
Fédération Aéronautique Internationale (FAI), 220, 255, 267
Federation of Americans Supporting Science and Technology (FASST), 233, 282
Fedotov, Aleksander, 111
Feinstein, Capt. Jeffrey S., (USAF), 135
Ferguson, Harold, 39
Ferranti Ltd., 80
Ferrer, Augusto, 337
Fichtl, George H., 10
Filipchenko, Anatoly V., 162
Fink, Daniel J., 354
Fins, stabilizing, 313, 315–316, 316–317, 320
Fiori, Dr. Franco, 114
Fire, 84, 89, 154, 168, 178, 216, 246, 261, 291, 360
First Across–U.S. Navy Transatlantic Flight of 1919, 10
Fischer, Ernest O., 296
Fischer, Dr. William A., 66
Fischetti, Thomas L., 143
FIT. See Florida Institute of Technology.
Fitzgerald, A. Ernest, 265
Fleet Satellite Communications (FLTSATCOM) System, 104, 113, 343
Fleet Weather Facility (FWF), 48, 51, 253
Fleming, Arthur S., Award, 52
Fletcher, Dr. James C.
   agreement signing, 95, 244
   appointment, 5
   awards and honors, 10, 15, 66, 299, 301
   ceremonies, 21, 107, 116, 179, 265
   congressional testimony, 37, 56–57, 67, 88, 90, 93, 101, 160–161, 231, 302–303, 318
   equal employment opportunity, 329
   letters, 2, 15, 36, 155
   NASA Space Program Advisory Council, 22
   personnel, 57, 299–300, 317, 350, 351

Skylab program, 84, 134, 155, 160–161, 171, 191–192, 211, 230, 270
space program, national, 36, 48, 185, 208, 279, 281, 303, 310, 330, 353
speeches, 17, 29–30, 85, 104, 183, 210, 252, 263–264, 279, 331–332
Flight Research Center (FRC, NASA)
   aircraft, 59, 342
      flight tests, 74–75, 107, 266, 291, 307
   award, 299
   contract, 132, 196
   employment, 30, 359
   lifting body test, 206, 216, 235, 247, 253, 284, 285, 304, 319, 342
   personnel, 111, 127–128, 201, 238
Flight Surgeon of the Year, Award, 256
Florida
   Bicentennial Commission, 206
   Cape Kennedy name change, 18, 51
   Department of Natural Resources, 92–93
   earth resources, 25, 92–93, 262
   earth station, 83
   earth tremor, 301
   employment, 335
   environmental research site, 183
   historic site, 206
   industrial contracts, 15
   land use, 40, 335
   legislature, 132
   lunar eclipse viewing, 340
   meeting, 114
   NASA tracking ship, 97
   press comment, 326
   Senate Natural Resources Committee, 92
   Skylab 1 liftoff, 224
   S.S. *Statendam* cruise, 3
Florida Institute of Technology (FIT), 262
Florida Technological Univ., 178
Florida, Univ. of, 262
Flowers, Rep. Walter, 23
Floyd, Henry B., 127
Floyd, Joseph S., Corp., 94
FLTSATCOM. See Fleet Satellite Communications System.
Fluid dynamics, 250
Fluorides, 247
Fly-by-wire aircraft. See Digital fly-by-wire aircraft.
Fog dispersal, 13–14
Fong, Sen. Hiram L., 18
Fordham Univ., 333
Forestry, 66, 85, 212, 243, 263
Fort Churchill Range, Canada, 212, 269
Fort Lamay, Chad, 114
Fort Myers, Fla., 83
Fort Worth, Tex., 204
Foster, Dr. John S., Jr., 103, 120, 174
"Foundations of Space Biology and Medicine" (U.S.–U.S.S.R. joint report), 245
Fourier interferometer, 288
Foxbat. See MiG-23 aircraft.
Fra Mauro Crater (moon), 7
Fradin, David, 233

France
   aircraft (see also Concorde), 42, 158, 168, 173, 176–177, 184, 217
   award, 186, 265
   employment, 174, 256
   exhibit, 93
   hovercraft, 331
   international cooperation, 24, 51, 56, 114, 159, 256, 264, 333
   international cooperation, space, 24, 65, 71, 77, 79, 244, 268, 351, 359, 360
   meeting, 331
   nuclear tests, 51, 77, 136, 159–160, 197, 211, 218, 219, 221–222, 311
   press conference, 217
   research and development, 256
   scientists, 183, 200, 243, 361
   space program, 8, 159, 183, 185, 233
Franchi, Jean, 266
Franck Report, 155
Franklin Institute, 219
FRC. See Flight Research Center.
Freedom 7 mission, 137, 248, 279
Freitag, Capt. Robert F., (USN, Ret.), 111, 114, 257
French Air Museum, 295
Frey, Rep. Louis, Jr., 23, 217–218
Fri, Robert W., 246
Friedheim, Jerry W., 90, 112
Friedman, Dr. Herbert, 11–12, 126
Friendship 7 mission, 341
Froehlke, Secretary of the Army Robert F., 32, 134
"From Back Fire to Explorer I" (symposium paper), 288
"From Space Station to Orbital Operations in Space Travel Thought, 1895–1951" (essay), 52
Frost & Sullivan, Inc., 261
Fruit fly, 344
Frutkin, Arnold W., 91, 299
Fuel. See Energy crisis and individual fuels, such as Coal, Gasoline, and Hydrogen.
Fuel tank, 296, 299
Fulbright, Sen. J. William, 187
Fullerton, Calif., 274
Fund for Peace, 165
Fuqua, Rep. Don, 23, 161, 217, 266
FWF. See Fleet Weather Facility.

G

Gagarin (U.S.S.R. tracking ship), 102
Gagarin Center for Cosmonaut Training, 55, 326
Gagarin, Col. Yuri A. (U.S.S.R.), 115, 133, 273
Gagarin, Col. Yuri A., Gold Medal, 255
Galaxy (see also Milky Way and Seyfert), 12, 38, 125, 128, 280, 328, 329, 332, 333
Galileo. See Convair 990.
Galilei, Galileo, 4
Gallagher, J. Wes, 124

Gallium, 223
Galveston, Tex., 250, 291
Gamma ray, 12, 35, 39, 47, 53, 78–79, 128, 199–200, 279, 325
Gamma Velorium (exploding star), 341
Ganymede (Jovian moon). See Jupiter.
GAO. See General Accounting Office.
GARP. See Global Atmospheric Research Program.
GARP International Sea Trial (GIST). See Global Atmospheric Research Program.
Garrett Corp., AiResearch Co., 146
Garrick, I. Edward, 10
Garriott, Dr. Owen K., 224–225, 320, 322
  awards and honors, 319
  EVA, 227, 228–229, 320
    experiments, 225, 227, 229
  JSC reorganization plan, 339
  space flight effects, 219, 225, 274, 282–283
  mission preparation, 180, 203, 221
  Presidential message, 270
  press conference, 283
Gas, 249, 337, 341
Gasoline, 135, 263–264, 314, 361–362
GASP. See Global Air Sampling Program.
Gast, Dr. Paul W., 11, 84, 128, 157
GATE. See Global Atmospheric Research Program, Atlantic Tropical Experiment.
Gatland, Kenneth W., 283, 284
Gause, Raymond L., 209
Gavin, Joseph, Jr., 21
GDR (German Democratic Republic). See Germany, East.
GE. See General Electric Co.,
Geary, Capt. John T. (USN), 213
Gemini (program), 25, 261, 279
Gemini 4 mission, 186
Gemini 5 mission, 152
Gemini 6 mission, 186
Gemini 7 mission, 7, 186
Gemini 8 mission, 238
Gemini 9 mission, 186
Gemini 11 mission, 152
Gemini 12 mission, 7, 186
General Accounting Office (GAO), 136, 155, 195–196
  Analysis of Cost Estimates for the Space Shuttle and Two Alternate Programs, 172
  Cost Growth in Major Weapon Systems, 94
General Administration of Telecommunications, Peking, China, 217
General Agreement on Contacts, Exchanges, and Cooperation in the Fields of Science, Technology, Education, and Culture, (U.S.–U.S.S.R.), 187
General aviation, 362
  award, 10, 255, 265, 304
  cooperation, international, 277
  employment 286
  research, 72–73, 107–108, 138, 285
General Dynamics Corp., 33, 92, 132, 183, 204, 218, 275, 343
  Convair Aerospace Div., 46, 97, 269
  Convair Div., 257, 308, 333
General Electric Co., (GE)
  Aircraft Engine Group, 185, 300
  award, 76
  contract, 63, 69, 96, 201, 292
  cooperation, 13, 238
  Gas Turbine Products Div., 137
  personnel, 203, 296
  Power Generation Group, 203
  Research and Development Center, 238
  Space Div., 354
  Valley Forge Space Center, 139
General Motors Corp., (GMC)
  Delco Electronics Div., 105
  Detroit Diesel Allison Div., 248
General Services Administration (GSA), 316
General Telephone & Electronics Corp.
  GTE Satellite Corp., 260
  GTE Sylvania, Inc., 132
Genetics, 303
Geneva Disarmament Conference, 51, 211, 253
Geneva, Switzerland, 302, 313
Geochemical Society, 157
Geodetic Earth Orbiting Satellite. See GEOS, *Explorer 29*, and *Explorer 36*.
Geological Survey. See U.S. Geological Survey.
Geology, 5, 45, 53, 66, 82, 103, 104, 323–324, 344, 346–347
Geomagnetic field, 298
Geomorphology, 45
George Washington Univ.
  School of Engineering and Applied Science, 137
Georgia, 130, 184
Georgia State Experiment Station, 259
GEOS (Geodetic Earth Orbiting Satellite), 162, 304
Geos 1. See *Explorer 29*.
Geos 2. See *Explorer 36*.
Geostationary Operational Environmental Satellite. See GOES.
Geothermal Energy Act of 1973, 267
German Democratic Republic. See Germany, East.
Germantown, Md., 186
Germany, 183–184, 220, 339, 345
Germany, East, 302
Germany, West
  Aeros satellite, 22, 90, 361
  Air Force, 81
  ELDO withdrawal, 37, 131
  employment, 256
  International Academy of Aviation and Space Medicine, 264
  international cooperation, 24, 256, 261, 333, 361
  international cooperation, space, 17–18, 79, 138–139, 162, 210, 244, 263, 268–269, 333, 360, 361

Max Planck Institute for Physics and Astrophysics, 345
 missile purchase, 175–176
 Moscow radio broadcast, 203
 press comment, 156
 scientists, 361
 solar research, 98
 space debris, 185
 Technical Univ., 296
Gesellschaft fur Weltraumforschung, 139
GET: ground elasped time.
ghz: gigahertz (1 billion cycles per second).
G forces, 19, 262
Giaever, Ivar, 296
Giant Patriot Project, 353–354
Gibson, Dr. Edward G.
 authorship, 233
 Skylab 4 mission
  achievements, 244, 319–323
  medical aspects, 351
  mission preparations, 290, 295, 299, 311, 317–318
  press conference, 283
Gill, Jocelyn R., 201–202
Gillam, Manitoba, 37
Gillilland, Whitney N., 5
Gilman, Rep. Benjamin A., 337
Gilmore Creek Command Data and Acquisition Center (NOAA), 352
Gilruth, Dr. Robert R., 353, 354
GISS. See Goddard Institute for Space Studies.
GIST (GARP International Sea Trial). See Global Atmospheric Research Program.
Givishiany, Dzherman M., 275
Glaciology, 45
Glaser, Dr. Peter E., 351
Glasgow Univ., Scotland, 130
Glass (lunar surface sample), 325
Glass Containers Manufacturers Institute, 23
Glassphalt, 23
Glenn, Col. John H., Jr. (USMC, Ret.) 125, 341
Glennan, Dr. T. Keith, 55, 141, 279
Global Air Sampling Program (GASP), 86
Global Atmospheric Research Program (GARP), 27, 59
 Atlantic Tropical Experiment (GATE), 24, 210
 International Sea Trial project (GIST), 237
Glomar Challenger (deep-sea drilling ship), 87
GMC. See General Motors Corp.
Gnat, 223
GNP. See Gross national product.
Gobi Desert, 128–129
Goddard, Mrs. Esther H., 223
Goddard, Dr. Robert H., 5, 288
Goddard Institute for Space Studies (GISS), 136
Goddard Memorial Collections, 223
Goddard Memorial Dinner, 16th Annual, 76
Goddard, Robert H., Award, 11
Goddard, Robert H., Historical Essay Award, 52
Goddard, Robert H., Memorial Trophy, 76
Goddard, Robert H., Scholarship, 52
Goddard Scientific Colloquium, 128
Goddard Space Flight Center (GSFC, NASA)
 AE program management, 346
 Anik 2 management, 123
 Apollo program, 2–3, 53
 Ats 3 photo processing, 51
 cooperation, 246–247, 253, 315
 earthquake research, 64
 employment, 30, 128, 360
 ERTS, 76, 341
 IMP program, 298
 International Projects Office, 139
 ITOS–E investigations, 212
 Laboratory for Meteorology and Earth Sciences, 220
  Experiment Management Office, 344
 meeting, 66, 128
 National Space Science Data Center (NSSDC), 233
 Nimbus 5 monitoring, 255–256
 oceanography, 48
 Operation Kohoutek, 352
 personnel, 52, 53, 143
 RAE project, 180
 Salyut 2 tracking, 102–103
 Skylab tracking, 152, 321
 Small Astronomy Satellite program, 73
 Solar Physics Laboratory, 12
 sounding rocket experiments (see also Sounding rocket)
  astronomy, 285
  astrophysics, 250
  atmospheric data, 201, 237, 240–241, 247, 265, 329
  heat pipe, 221
  magnetospheric physics, 90, 95, 98
  meteorology, 329
  ozone, 46, 51, 174, 295, 329
  performance test, 214–215, 221, 243
  solar physics, 121, 261, 267
 Summer Institute for Biomedical Engineering, 246–247
 symposium, 128, 315, 341
 tracking network, 97
 TOS program responsibilities, 312
 Weather and Climate Center, 62
GOES (Geostationary Operational Environmental Satellite), 210, 313–314
Gold, 291
Gold Space Medal, 255
Gold, Dr. Thomas, 188
Goldberg, Dr. Leo, 233, 299
Goldschmidt, Victor, Award, 157
Goldstein, Dr. Richard A., 239
Goldstein, Dr. Richard M., 64–65
Goldstone Tracking Station, Calif., 64–65, 116, 208, 239

Goldwater, Sen. Barry M., 13, 30, 293, 304, 345
Goldwater, Rep. Barry M., Jr., 23
Gooney. See DC-3 aircraft.
Gorky Univ., U.S.S.R., 312
Gorky, U.S.S.R., 61
Gorman, Harry H., 58
Gould, Inc. 215
Goussainville, France, 163, 173, 176–177
Government Procurement Practices Board (GPPB, proposed), 156
GPPB. See Government Procurement Practices Board.
Grace, Clinton H., 105
Graduate Science Education: Student Support and Postdoctorals, Fall 1972 (NSF 73–315), 305
Graduate Student Support and Manpower Resources in Graduate Science Education (NSF 73–304), 232
Grand Canyon (Mars), 138
Grand Forks, N. Dak., 165, 277–278
Grant, Dr. David A., 130
Grapevine, Tex., 265–266
Gravel, Sen. Mike, 344
Gravity (see also Weightlessness, effects), 19, 162, 195, 261–262, 302, 307, 310, 315, 323–324, 342, 351
Gray, Edward Z., 108
Gray, Rep. Kenneth J., 26, 40, 41
Gray, Robert H., 299
Great Britain. See United Kingdom.
Great Lakes, 108
Great red spot (Eye of Jupiter), 309
Great Salt Lake, 149
"Great White Sands Missile Range Lost Gold Treasure Affair," 291
Greece, 325
Greenish, Thomas S., 44
Greenland, 75
Greer, Robert E., 105
Grenade, acoustic, 214
Gribben, John, 136
Griffin, Ga., 259
Griffin, Gerald D., 120
Griffin, Sen. Robert P., 31
Grindeland, Dr. Richard E., 289
Gromyko, Foreign Minister Andrey A., 187
Gross national product (GNP), 256
Ground station. See Earth station.
Group Achievement Award (NASA), 105, 299
Groupe Diplome d'Honneur for Aeronautics, 255
Grubbs, H. Dale, 120
Grumman Corp., 2, 4, 163, 201
   Grumman Aerospace Corp., 21, 96, 105, 108, 132, 157, 176, 189, 257, 258, 281, 328
      contract, 47, 75, 97, 132, 239, 246, 343
   Grumman Aircraft Corp., 260
   Grumman American Aviation Corp., 2
GSA. See General Services Administration.
GSFC. See Goddard Space Flight Center.
GTE Satellite Corp. See General Telephone & Electronics Satellite Corp.
GTE Sylvania, Inc. See General Telephone & Electronics Corp.
Guam, 147
Gubser, Rep. Charles S., 4
Guena, Transport Minister Yves (France), 264
Guggenheim, Daniel and Florence, International Astronautics Award (1973), 288
Guggenheim, Daniel, Medal, 9–10
Gulf of Guinea, 159
Gulf of Mexico, 149, 197, 228
Gulf Oil Corp.
   Gulf General Atomic Div., 5
Gulfport, Miss., 295
Gulfstream I, (jet aircraft), 2
Gulfstream II, 2, 343
Gum Nebula, 341
Gunter, Rep. Bill, 23, 236
Gurney, Sen. Edward J., 36, 120
Gurovsky, Dr. N. N., 55, 335–336
Guyana, 200
Gyroscope, 319–320, 321, 322

H

Haagen-Smit, Dr. Arie Jan, 284
Hacker, Barton C., 52
Hackes, Peter S., 21
Hades (Jovian moon). See Jupiter.
Haensel, Dr. Vladimir, 284
Hague Convention of 1970 (ICAO), 252
The Hague, Netherlands, 159–160
Haise, Fred W., Jr., 121–122, 250, 291–292, 339
Hale Observatories, 38, 110, 125
Hall, Dr. Harvey, 201–202
Halley's Comet, 294, 342, 352
Halvorson, R/A George G. (USN), 231
Hamburg Observatory, Bergedorf, West Germany, 98
Hamburg, West Germany, 261
Hamilton Standard Div. See United Aircraft Corp.
Hampshire College, 174–175
Handicapped assistance devices, 251, 327, 360
Handler, Dr. Philip, 79, 125, 188, 205
Hanna, Rep. Richard T., 23
Hannah, John W., 25
Hansen, Grant L., 173
Harmon International Aviation Trophies, 265
Harr, Dr. Karl G., Jr., 105–106, 140, 272, 326
Harrier (U.K. V/STOL aircraft), 117, 263
Harrington, Dr. John V., 269
Harris, Mrs. Ruth Bates, 119, 299–300, 329
Hart, Sen. Philip A., 31
Hartke, Sen. Vance, 31

Harvard College Observatory, 22, 121, 222, 243, 256, 340
Harvard Univ., 4, 294
Harvey, Rep. James, 63
Haskell, Sen. Floyd K., 4
Hawaii, 73, 319, 353–354
Hawaii, Univ. of, 329
Hawker Siddeley Group Ltd., 117, 278
  Hawker Siddeley Aviation Ltd., 263
Hawker Siddeley 146 (U.K. aircraft), 278
Hawkins, Dr. Willard R., 148, 150, 192, 193, 195, 197, 228, 274, 283
Hayden Planetarium, 4, 223
Hazeltine Corp., 86
HDE 226868 (binary supergiant star system), 329
Health services, 272, 278, 293
HEAO. See High Energy Astronomy Observatory.
Heath, Gordon R., 167
Hechler, Rep. Ken, 23, 68
Heidman, Marcus F., 11
Heimaey Island, 24–25
Helgafell (volcano), 24
Helicopter, 199, 293
  applications, 206, 240
  design, 192, 232, 247, 305, 329
  exhibit, 263
  industry, 286, 362
  rescue and recovery, 152, 189, 291, 319
  World Helicopter Championships (II), 220
Helicopter Club of Great Britain, 220
Helios (solar probe), 138–139, 263, 361
Helios–A, 138–139
Helios–B, 138–139
Helium, 66, 174, 299, 311, 346
Helms, Sen. Jesse A., 13
HELOS. See Highly Eccentric Lunar Occultation Satellite.
Hematology, 283
Henkin, Daniel J., 112
Henry Award (Smithsonian Inst.), 294
Hera (Jovian moon), 309
Hercules (NASA research transport aircraft). See C–130.
Hero of Socialist Labor (U.S.S.R. honor), 329
Heroin, 291
Hersh Acoustical Engineering Co., 232
Herstmonceux, U.K., 297
Heseltine, Michael R. D. (Minister of Aerospace and Shipping, U.K.), 47
Hestia (Jovian moon), 309
Hickam AFB, Hawaii, 319
Hickson, Charles, 291
High Altitude Observatory, 185, 222
"High Energy Astronomy" (AIAA presentation), 11–12
High Energy Astronomy Observatory (HEAO), 5, 27, 38–39, 58, 265, 359
"High Energy Propulsion at NACA Lewis Engine Research Laboratory" (symposium paper), 288
Highly Eccentric Lunar Occultation Satellite (HELOS), 124

Highway 85, Calif., 233
Highway 101, Calif., 233
Highway safety, 247
Hijacking of aircraft, 9, 17, 252
Hilburn, Earl D., 135
Hill, Louis W., Space Transportation Award, 10–11
Himmel, Dr. Seymour C., 68–69, 72, 74, 80
Hinners, Dr. Noel W., 64
Hirman, Joseph, 258
History Manuscript Contest Award (AIAA), 10
Hislop, Dr. George S., 80
Hoag, David G., 10–11
Hobbs, Leonard S., 10
Hochberg, Charles M., 201–202
Hocker, Dr. Alexander, 244, 268, 281
Hodges, Kenneth E., 101
Hoffman, Dr. John H., 66
Holland. See Netherlands.
Hollings, Sen. Ernest F., 31, 63
Holloman AFB, N. Mex., 85
Holloway, Gen. Bruce K. (USAF, Ret.), 244, 259, 340
Holm, M/G Jeanne M. (USAF), 172
Holzer, Dr. Thomas E., 32
Holzmann, Arthur D., 354
Homecoming Day, Enid, Okla., 319
Homestead AFB, Fla., 206
Hooe, Roy W., 122
Hosenball, S. Neil, 299
Hospitals, 251
Housing and Urban Development, Dept. of (HUD)
  appropriations, 186
  cooperation, 62, 293, 330
  energy research, 349
  MIUS development, 188–189
  Solar Heating and Cooling Information Data Bank (proposed), 349
Houston Chamber of Commerce, Tex., 251–252
Houston, Tex., 62–63, 229, 274, 318, 323
Houston, Univ. of, 37, 120
Hovercraft. See Surface effect ship.
Hovertrain, 47
Howard, Dr. Keith A., 66
Howard, Dr. Robert, 12
Howard Univ., Washington, D.C., 246–247
Hryniewiecki, Edward, 131–132
HU–16B (Albatross; amphibious aircraft), 206
Huber, William G., 46
HUD. See Housing and Urban Development, Dept. of.
Hudson, L/G John B. (USAF), 282
Hughes Aircraft Co., 123, 264–265
  Hughes Aerospace Div., 315
  Hughes Helicopters, 192
  National Satellite Services, Inc., 260
Humphrey, Sen. Hubert H., 21, 63, 347
Hunter, Dr. A., 297
Huntsville, Ala., 124, 251
Hurricane, 85, 150, 175, 182, 189, 255–256

Hutchinson, Neil B., 225
Hydrocarbon, 68, 245
Hydrofoil craft, 140, 185, 231
Hydrogen, 11, 67, 68, 74, 75, 174, 247, 249, 263–264, 302
Hydrogen bomb, 136
Hydrography, 260, 261
Hypergolic propellant, 261–262, 288
Hypersonic aerospace craft research program (NASA–USAF). See X–24B.
HZE-Particle Effects in Manned Spaceflight (NRC report), 133

I

Il-18 (U.S.S.R. transport aircraft), 22, 47, 49, 74
Il-62-M, 163
IAA. See International Academy of Astronautics.
IAEA. See International Atomic Energy Agency.
IAF. See International Astronautical Federation.
IAM. See International Assn. of Machinists.
IBM. See International Business Machines Corp.
IBM 9020 (computer), 61.4
ICAO. See International Civil Aviation Organization.
ICBM. See Intercontinental ballistic missile.
Ice information system, 108, 253, 255
Iceland, 264
Ickle, Dr. Fred C., 109
ICX-International Computer Exchange, Inc., 93
Idaho, 353–354
Ikeya-Seki (comet), 342
ILC Industries, Inc., 93, 255
Iliff, Kenneth W., 111
ILLIAC IV (computer), 250
Illinois, Univ. of, 15, 239, 243
ILS. See Instrument landing system.
"The Image of the Moon–Galileo to Apollo XI" (exhibit), 120
IME. See International Magnetospheric Explorer.
Immigrant Scientists and Engineers Decline in FY 1972; Physicians Increase Sharply (NSF 73–311), 248
Immigrant Scientists and Engineers in the United States (NSF 73–302), 241
Immigration, 241, 248
Immunology, 289
IMP–A (Interplanetary Monitoring Platform). See *Explorer 18*.
IMP-H. See *Explorer 47*.
IMP-J. See *Explorer 50*.
"The Implications for European Space Programmes of the Possibilities of Manned Mission," (summer school theme) 221

Improved Tiros Operational Satellite (ITOS). See *Itos 1*, ITOS–B, *Noaa 1*, *Noaa 2*, *Noaa 3*, and Tiros Operational Satellite system.
IMS. See International Magnetospheric Study.
Independence Project–1980, 329–330
India, 18, 34, 38, 46, 165, 200, 325
Indian Ocean, 56, 200, 210
Indiana, 297
Indianapolis 500 automobile race, 168
Indonesia, 195
Indonesian Institute for Outer Space, 195
Industrial Contractors, Inc., 26, 40
Informatics TISCO, Inc., 95
Infrared (IR) radiation, 48, 65, 86, 115, 181, 224, 245–246, 294
Infrared radiometer, 181, 208, 308, 310, 323
Infrared temperature profile radiometer (ITPR), 104
Inouye, Sen. Daniel K., 18, 31
INPE. See Brazilian Institute for Space Research.
Institute for Medical Biological Problems. See Soviet Academy of Sciences.
Institute of Electrical Engineers, 18
Institute of Geochemistry, U.S.S.R., 282
Institute of High Energy Physics, U.S.S.R. See Soviet Institute of High Energy Physics.
Institute of Nuclear Research, U.S.S.R., 159
Institute of Space Research, U.S.S.R., 36, 282
Instituto Nacional de Técnica Aeroespacial (INTA), Spain, 60
Instrument landing system (ILS), 117
Instrument unit (IU), 145, 224–225, 320
INTA. See Instituto Nacional de Técnica Aeroespacial, Spain
INTA 300 (U.K.–Spain sounding rocket), 60
INTASAT (U.S.–Spain ionospheric beacon), 361
INTELSAT. See International Telecommunications Satellite Organization.
Intelsat (communications satellite series), 154, 280, 281, 361
Intelsat I (*Early Bird* comsat), 37, 273
Intelsat IV, 37, 269
Intelsat IV A, 37
Intelsat-IV F-4, 217, 250–251, 281
Intelsat-IV F-5, 250
Intelsat-IV F-6, 1, 134, 250–251
Intelsat-IV F-7, 1, 250, 358, 360
Intelsat-IV F-8, 1
Intercontinental ballistic missile (ICBM) contract, 73
  MIRV comparison and use, 7, 112–113
  P.R.C., 61
  U.S.S.R., 50, 90, 247
Intercosmos (U.S.S.R. satellite), 65, 122, 325, 359
Intercosmos 10, 302
Intercosmos Copernicus 500, 123

Intercosmos Council. See Soviet Academy of Sciences.
Interferometer, 288
Intergovernmental Agreement on U.S. and European Cooperation in Space Shuttle Development, 76–77, 135, 231–232, 244, 268, 360
Intergovernmental Maritime Consultative Organization
  Communications Subcommittee, 67
Interior, Dept. of the (DOI)
  Cape Canaveral name restoration, 288–289
  census (1970), 344–345
  cooperation, 247, 337–338, 344–345
  geothermal energy, 267
  Man in Space National Historic Site, 36, 162–163
  National Energy Information System (proposed), 338–339
  Office of Coal Research, 337–338
  Office of Solar Energy Research (proposed), 231, 347
International Academy of Astronautics (IAA), 288
International Academy of Aviation and Space Medicine, 264
International Assn. of Machinists (IAM), 48
International Astronautical Federation (IAF), 111, 287
International Atomic Energy Agency (IAEA), 55, 141, 261
International Business Machines Corp. (IBM), 46, 105, 270, 296, 353
International Civil Aviation Organization (ICAO), 17, 252, 333
International Colloquium on Mars, 331
International Conference on Offshore Airport Technology, First, 129
International cooperation (see also Disarmament and Treaty)
  agreement, 89, 130, 170, 187, 189–190, 192, 249, 258–259, 325–326
  aircraft, 32, 114, 117, 135, 172–173, 193, 263, 264, 269, 278, 288
  astronomy, 114, 200
  atomic energy, 77, 302
  congressional hearings, 210
  earth resources, 37–38, 42, 45, 87, 95, 108–109, 143, 170, 219–220, 245, 257, 280–281, 341, 361
  energy research and development, 181, 187–188, 277, 342
  environmental protection, 45, 87, 134, 164, 212
  exhibit, 163, 277
  hijacking of aircraft resolutions, 17, 252
  Lonar Crater origin, 165
  meteorology, 22, 24, 47, 71, 74, 101, 210, 237
  nuclear energy, 165, 221–222, 265
  oceanography, 22, 45, 47, 49, 74, 87, 116–117, 237
  offshore drilling, 87

  press comment, 256
  research and development, 101
  science, 13, 89, 101, 134, 187–188, 244–245, 264, 275, 331
  solar energy, 205, 213
  summit accord, 176, 184, 187, 189–190, 192, 194
  symposium, 216, 264, 333
  Vietnam cease fire, 214
International cooperation, space (see also Apollo-Soyuz Test Project; European Launcher Development Organization; European Space Research Organization; Global Atmospheric Research Program; International Telecommunications Satellite Organization; *Isis 1*; Spacelab), 304–305
  agreement, 16, 26, 36, 44, 77, 134, 186, 231–232, 244, 268, 360
  peaceful uses of space, 186
  astronomy, 361
  Canada-China, 281
  communications, 44, 67, 154, 217, 360
  Convention on International Liability for Damage Caused by Space Objects, 44, 67, 154, 217, 360, 257
  earth resources. See Earth resources.
  Japan–Yugoslavia, 56
  joint working groups. See joint working groups.
  lunar programs, 64, 66, 84, 96
  NASA-
    -Australia, 361
    -Brazil, 108–109, 361
    -Canada, 123, 358, 360, 361
    -ESRO, 61, 77, 202, 221, 279, 289, 297, 302, 315, 358, 360, 361
    -Europe, 114, 119, 268
    -France, 78–79, 96, 183–184
    -Germany, West, 78–79, 81, 138–139, 183–184, 190, 263, 361
    -Italy, 361
    -Japan, 315
    -Mexico, 87
    -Netherlands, 361
    -Norway, 361
    -Soviet Academy of Sciences, 36, 45, 64
    -Sweden, 78–79, 361
    -U.K., 16, 328
    -U.S.S.R. See U.S.-U.S.S.R.
  oceanography, 49, 183
  Pioneer Venus project, 183–184
  press comment, 156, 271, 276–277
  satellite
    AEROS, 190, 361
    communications. See Communications satellite.
    Copernicus 500, 122
    Erts 1, 162, 219–220, 341, 361
    France–Germany, West, 361
    GEOS, 304
    Helios, 138–139, 264
    Intelsat-IV F-7, 250
    launch assistance, 16, 174, 279, 315, 328, 361

MAROTS, 231–232
meteorological. See Meteorological satellite.
Oso 7, 78–79
Oreol, 351, 359
San Marco C-2, 361
SOREL, 315
Skylab program, 81, 142, 271, 361
solar energy research, 205, 332
sounding rocket. See Sounding rocket, international programs.
space biology and medicine, 55, 177, 335–336, 361
space shuttle, 76, 231–232, 244, 268, 303, 360
Switzerland–Japan, 287
U.S. (see also NASA)-
 -Europe, 236, 268
 -France, 24
 -Italy, 114
 -Japan, 70
 -U.K., 17
 -U.S.S.R., 26, 36, 55, 84, 96, 102–103, 134, 163, 182, 184, 186, 189–190, 194, 267, 277, 326–327, 330, 344, 361
U.S.S.R.-
 -Czechoslovakia, 302
 -France, 65, 71, 96, 154, 239, 242, 351, 359
 -Germany, East, 302
 -Poland, 122
International Court of Justice, 159–160
International Exposition, 30th, 93
International Galabert Prize, 186
International Geophysical Union, 347
International History of Astronautics Symposium, 288
International Magnetospheric Explorer (IME), 202
International Magnetospheric Study (IMS), 202
International Microwave Landing System Symposium, 333
International Paper Co., 73
International Space Hall of Fame, 232
International Sun–Earth Physics Satellite (ISEPS) program, 124
International Symposium and Workshop on Gamma-Ray Astrophysics, 128
International Telecommunications Satellite Organization (INTELSAT), 14, 37, 67, 153
 communications satellite launch, 250, 281, 358, 360
 Global Satellite System, 54
 operations, 35–36, 44, 67
International Telephone & Telegraph Corp. (ITT), 86, 182
ITT World Communications, Inc., 252
International Youth Science Tour of America, 212
Interplanetary Monitoring Platform (IMP). See *Explorer 18*; *Explorer 47*; *Explorer 50*.

The Interplanetary Pioneers (NASA SP–278), 33–34
Intrusion detection system, 243
Io (Jovian moon), 197, 310
Ionosphere
 planetary, 208, 311
 satellite monitoring, 122, 302, 361
 solar, 258, 325
 sounding rocket experiments, 15–16, 201, 237, 239, 240–241, 243, 328, 358–359
Iowa Geological Survey, 130
Iowa Science, Engineering and Humanities Symposium, 130
Iowa State Univ., 52
Iowa, Univ. of, 130
IR: infrared.
IR-100 award, 266
Iran, 46
Iran Air (airlines), 33
Ireland, Roger G., 10
Iron, 247–248, 346–347
Irwin, Col. James B. (USAF, Ret.), 106
Isaman, R/A Ray M. (USN), 332
ISEPS. See International Sun–Earth Physics Satellite program.
Isis 1 (International Satellite for Ionospheric Studies), 191
Israel, 284, 287, 290, 292, 295, 301, 314, 361
Israel Aircraft Industries, 263
Italy, 5, 17–18, 35, 77, 176, 185, 252, 268–269, 361
Itek Corp., Optical Systems Div., 242–243
Itos 1 (Improved Tiros Operational Satellite), 312
ITOS–A. See *Noaa 1*.
ITOS–B, 212
ITOS–D. See *Noaa 2*.
ITOS–E. See *Noaa 3*.
ITPR. See Infrared temperature profile radiometer.
ITT. See International Telephone & Telegraph Corp.
ITT Research Institute, 12
IU. See Instrument unit.
Ivanchenko, Aleksander, 162

J

Jablonski, Henryk, 213
Jackass Flats, Nev., 7–8
Jackson, A. A., IV., 263
Jackson, Sen. Henry M., 36, 87, 337, 338–339
Jackson, Nelson P., Aerospace Award, 76
Jackson, Roy P., 68, 72, 80, 89–90, 199, 296, 299
Jaffe, Leonard, 45
Jakarta, Indonesia, 172
JAL. See Japan Airlines.
Jane's Fighting Ships 1972–73, 221
Jane's Surface Skimmers 1973–74, 331
Janus, Tony, Award, 102

Japan, 137, 185, 219, 281
  budget, 31
  industry, 224
  international cooperation, 56, 70, 277, 315, 342
  launch, 26
  research and development, 256
  satellite, 210, 315
Japan Airlines (JAL), 176
Javelin (sounding rocket), 214, 326, 328
Jaycees. See U.S. Junior Chambers of Commerce.
Jeffries, John, Award, 10
Jeffs, George W., 105
Jet Propulsion Laboratory (JPL; Cal Tech)
  Apollo Lunar Sounder Investigator Team, 299
  astronomy, 64–65, 144, 239, 242
  automotive research, 263–264
  awards, 10–11, 15, 299, 300
  cooperation, 97, 154, 288
  environmental research, 12
  Mariner (program), 219, 308
  Mariner Venus-Mercury 1973 Program Team, 299
  meeting, 331
  name change, 41, 336
  personnel, 10–11, 15, 20
  scientists, 53, 108
  security system, 291
  solar energy research, 154
  SOREL Project, 315
  stratosphere study, 288
Jewett, John P., 7–8
Johannesburg, S. Africa, 109, 208
Johns Hopkins Univ., 298
  Applied Physics Laboratory, 241, 360
Johnson, John A., 53
Johnson, Mrs. Lyndon B., 251
Johnson, President Lyndon B., 20, 21, 25, 31, 49–50, 54, 125, 132, 251–252
Johnson, R. Tenney, 156
Johnson, Vincent L., 104, 160, 351–352, 354
Johnson Space Center (JSC, NASA)
  Apollo program, 238
  ASTP. See Apollo-Soyuz Test Project.
  astronauts at, 144
  ATS-E program, 262
  Brezhnev visit, 176, 184
  contract, 39, 49, 54, 83, 94, 133, 139, 188–189, 269, 270, 276, 286, 289, 317
  cooperation, 238
  Data Systems and Analysis Directorate (proposed), 339
  dedication ceremonies, 251–252
  Earth Observations Div., 62–63
  Earth Resources Program Office, 139
  EEO review, 10
  employment, 30, 122, 359
  EREP review, 245
  ESRO delegation, 289
  Flight Control Div., 339
  Flight Medicine Section, 81
  Flight Operations Directorate (proposed), 339
  Health Maintenance Branch, 219
  international cooperation, 87
  Life Science Laboratories, 39
  Life Sciences Directorate, 339
  Lunar Receiving Laboratory, 2–3
  lunar sample analysis, 2–3, 32–33
  Lunar Sample Curatorial Facility, 290
  medical diagnostic system, 116
  meeting, 65–66, 119, 207–208, 335–336
  Mission Control Center, 78, 97, 127, 335–336
  name change, 25, 31, 36–37, 40, 49–50, 54
  Orbiter Projects Office, 121–122
  personnel, 81, 120, 121–122, 201–202, 339, 353, 354
  Planetary and Earth Sciences Div., 157
  press comment, 54
  press conference, 6, 35, 167–168, 169, 171, 181–182, 184–185, 190, 191–192, 193, 199, 222, 266–267, 270, 274, 282–283, 321, 352
  Science and Applications Directorate, 339
  Skylab. See Skylab missions.
  space shuttle, 195, 295
  Space Shuttle Orbiter Project Management, 339
  Space Shuttle Systems Integration Management, 339
  Visitor Center, Johnson Room, 251–252
Joint Chiefs of Staff, 94, 103, 183
Joint Commission on Scientific and Technical Cooperation, U.S.–U.S.S.R., 87, 89, 101, 134, 331
Joint Committee on Cooperation in the Field of Environmental Protection, U.S.–U.S.S.R., 87, 134
Joint Committee on Cooperation in the Peaceful Uses of Atomic Energy, U.S.–U.S.S.R., 190
Joint Editorial Board on Space Biology and Medicine, 177, 244–245, 341
Joint North Sea Wave Project, 261
Joint Oceanographic Institutions for Deep Earth Sampling, 87
Joint Working Group on Interplanetary Exploration, 36
Joint Working Group on the Natural Environment (U.S.–U.S.S.R.), 45, 164
Joint Working Group on Near-Earth Space, the Moon, and the Planets, 64
Joint Working Group on Satellite Meteorology, 22
Joint Working Group on Space Biology and Medicine (U.S.–U.S.S.R.), 55, 122, 335–336, 361
Jones, Joseph M., 328
Jones, Dr. Robert T., 108
Josephson, Brian, 296
JPL. See Jet Propulsion Laboratory.
JSC. See Johnson Space Center.
JT3D (turbofan engine), 89–90, 219

JT8D, 89-90, 204, 280, 358
Juno (booster)
  Juno I, 36
  Juno II, 36
Jupiter (planet)
  atmosphere, 14-15, 125, 308, 311, 357
  composition, 11
  exploration, 78, 83, 269, 357
  great red spot, 309
  Kohoutek comet flyby, 98
  magnetic field, 309, 357
  moons, 197, 309, 310, 357
  Pioneer 10 mission, 36, 47, 54, 174, 197, 241, 248, 267, 280, 308-309, 336, 338, 340, 357
  Pioneer 11 (Pioneer-G) mission, 1, 45, 106, 160, 174, 197, 248, 280
  press comment, 54, 337, 338, 340
  press conference, 309
  radiation, 96, 309-310,
Jupiter C (booster), 36

K

K-band radiometer, 261
Kadishev, Stephan I., 301
Kamov, Nikolay I., 329
Kansas, 130
Kansas, Univ. of, 107-108
Kapryan, Walter J., 145, 313, 315-316, 317, 320
Kapustin Yar, U.S.S.R., 122
Karaganda, Kazakhstan, U.S.S.R., 273, 276, 332, 348
Karen, Abe, 291
Karoonda C4 (meteorite), 137
Katz, Amrom H., 134
Kazakhstan, U.S.S.R., 273, 276, 348
Kearny, N. J., 186
Keldysh, Prof. Mstislav V., 65, 205
Kelley, Lt. Oakley G., 135
Kelly, Thomas J., 11
Kemp, James Furman, Medal, 128, 157
Kennard, B/G William J. (USAF, Ret.), 297
Kennedy, Sen. Edward M., 4-5, 181, 307
Kennedy, President John F., 20, 54, 125, 132, 223, 289
Kennedy International Airport, N.Y., 201
Kennedy Space Center (KSC, NASA), 52-53, 295-296, 301
  ASTP. See Apollo-Soyuz Test Project.
  astronauts at, 50-51, 222
  Bicentennial historic site, 206
  contract, 183, 262, 340, 349, 353
  Earth Resources Office, 92-93
  EEO review, 10
  employment, 30, 32, 91-92, 359-360
  energy conservation measures, 335, 348
  ERTS imagery report, 25
  ESRO delegation to, 289
  Flight Training Building, 106
  government equipment transfer, 56
  Helios Project, 138-139
  Launch Complex 34, 56
  Launch Complex 37, 56
  Launch Complex 39, 58, 72, 118, 123, 145, 147, 162-163, 164, 180, 212, 224-225, 237, 244, 320, 336, 340, 353
  launch services, 180
  launch vehicle development, 263
  lightning prediction system, 293
  memorial, 51
  Mission Control, 320
  mission preparations, 1, 45, 71-72, 76-77, 127, 136, 180, 206, 222
  mobile launcher 3, 352
  name change, 167, 288-289
  personnel,
    awards and honors, 107, 299, 300
    retirement, 135, 201-202, 354
  press comment, 237
  press conference, 142, 143, 223-224, 313, 315-318, 319
  Public Information Office, 127
  Skylab. See Skylab missions.
  Solid Motor Assembly Building, 263
  space shuttle, 69-70
  SPHINX (spacecraft), 253-254
  student seminar, 181
  Support Operations Directorate, 352
  Vehicle Assembly Building (VAB), 10, 106, 180, 244, 263, 336, 340
  Vertical Integration Building, 263
  Visitor Information Center (VIC), 335
Kenton, Thomas H., 190
Kentron Hawaii, Ltd., Continental Operations, 94
Kentucky, Univ. of, 242
Kenya, 200, 201
Kepler, Johannes, 4
Kerguelen Island, 56, 71
Kerosene, 296, 299
Kerwin, Cdr. Joseph P. (USN), 144, 145, 241-242
  awards and honors, 193, 286, 300
  JSC reorganization plan, 339
  mission activities, 147, 148, 149-150, 151, 164, 181-182, 227
  mission preparations, 90
  press comment, 193
  press conference, 199
  space flight effects, 148, 149-150, 151-152, 192, 193, 195, 198, 214, 222
  training, 112, 139, 146
Key Biscayne, Fla., 151
Khodarev, Yu. K., 45
Kimsey, Dr. Stephen, L., 222
Kinard, Dr. William H., 47, 309
Kingman Museum, 251
Kirillin, Vladimir A., 187-188, 331
Kirkland AFB, N. Mex., 197
Kiruna, Sweden, 42-43, 236, 237, 240-241
Kissinger, Dr. Henry A., Secretary of State, 184, 342
Kitt Peak National Observatory, Ariz., 233, 299, 341
Kleinknecht, Kenneth S., 197, 283, 339
Klimuk, Maj. Pyotr, 347
Kline, Richard, 101-102

Knight, Sir Geoffrey E., 102
Kohoutek (comet)
  comet-watch cruise, 342
  composition, 341
  dimension, 294
  discovery, 98, 344
  formation theory, 345
  observation, 220, 224, 246, 268, 283, 294, 299, 319–320, 342, 358
  path, 268, 321–322, 345, 353
  photographs, 321–322, 323
  press comment, 342, 344
  press conference, 352
  significance, 352–353
  solar effects, 342
  tail, 342
  U.S.S.R. study, 347
  visibility, 342, 352
Kohoutek, Dr. Lubos, 98, 268, 322, 342, 344
Kola Peninsula, U.S.S.R., 194
Kolesov, Mikhail, 283
Komarov (U.S.S.R. tracking ship), 102
Komarov, V. M., Diploma, 255
Koon, Grant, 243
Korean war, 137, 153
Kourou Space Center, French Guiana, 159
Kozlov, Mikhail, 163
Kraft, Dr. Christopher C., Jr., 226, 339
Kranz, Eugene K., 78, 339
Krasnoyarsk Institute of Physics, U.S.S.R., 287
KREEP (lunar material), 7
Kremlin Wall, Moscow, 133
Kriegsman, William E., 117
Krier, Gary E., 74–75
Kristian, Dr. Jerome, 38
Krogh, Egil, Jr., 168
KSC. See Kennedy Space Center.
Kubasov, Valery N., 162, 163, 217
Kuiper, Dr. Gerard P., 350
Kurzweg, Hermann H., 354
Kutzer, Ants, 138–139

L

L-3S (French booster), 210, 231–232, 301
LaBerge, Dr. Walter B., 248
Labor, Dept. of, 48–49, 63, 248
LAGEOS (Laser Geodynamic Satellite), 29, 328, 338
LAGEOS Task Team, 328
Lake Champlain, 73
Lake Michigan, 170
Lance (U.S. surface-to-surface missile), 175–176
Land resources management, 45, 183, 219–220, 247, 278, 344–345
Landslide (environment), 246
Langley Research Center (LaRC, NASA)
  aircraft research, 72–73, 83, 92, 107–108, 231, 285
  automated water sampling station, 206
  contract, 95, 300
  cooperation, 216, 240, 300
  EEO review, 10
  employment, 30, 359
  energy research, 246, 353
  Joint North Sea Wave Project, 261
  NASTRAN user's colloquium, 260
  personnel,
    awards and honors, 11, 108, 137, 299
    retirement, 201, 353, 354
  Scout Project Team, 299
  Systems Engineering Building, 353
  wind tunnel, 143
LaRC. See Langley Research Center.
Large Space Telescope (LST), 125, 181, 209, 242–243
Large Space Telescope Task Force, 39
Las Cruces, N. Mex., 12
Las Palmas, Canary Islands, 114
Las Vegas, Nev., 313
Laser
  airborne laser laboratory, 197
  altimeter, 175
  gas-dynamic, 197, 261–262
  reflector, 96, 338
  utilization
    communications, 132, 158, 321
    earthquake detection, 64
    lunar experiments, 8, 24
    meteorology, 181
    nuclear power, 165
    oceanography, 260, 261
    tracking, 94
    weaponry, 169, 261–262
Laser Geodynamic Satellite. See LAGEOS.
Laser light-detection-and-ranging (lidar) system, 59, 260
Laserphoto, 124
LASL. See Los Alamos Scientific Laboratory.
Lassiter Coast, Antarctica, 104
Launch vehicle (see also individual launch vehicles and stages, such as Agena, Atlas, Centaur, Thor–Delta), 17, 27, 219, 244, 269, 293, 296, 299, 325, 349
Lava, 324
Law enforcement, 280, 291, 295
Lawrence Livermore Laboratory, California Univ., Livermore, 245
Lazard Freres & Co., 345
Lazarev, Lt. Col. Vasily G. (U.S.S.R.), 273, 276
Lear, William P., Sr., 143
LearAvia Corp., 143
Learjet (executive transport aircraft), 143, 332
Lebedev, Valentin, 347
Le Bourget Airport, 93, 163, 173, 294
Leck German AFB, West Germany, 261
Lee, Capt. Chester M. (USN, Ret.), 15, 71–72, 105
Lee, Janet, 212

Lee, v/a John M. (USN), 98
Lee, Thomas J., 133, 350
Legion of Honor, 220
Le Monnier Crater (moon), 110–111, 139, 174, 347
Lenin, Vladimir I., 8
Leningrad, U.S.S.R., 130, 167, 193, 292
Lenoir, Dr. William B., 290
Leonov, Lt. Col. Aleksey A. (U.S.S.R.), 162, 163, 217, 325–326
Leovy, Dr. Conway B., 16
LeRC. See Lewis Research Center.
Levy, Lillian, 212
Levy, Maurice M., 76
Lewis, David S., 275
Lewis Research Center (LeRC, NASA; see also Plum Brook Station)
  aircraft research, 69, 89–90
  automotive research, 81, 199, 263–264
  award, 266
  contract, 137
  employment, 30, 143, 359–360
  energy research and development, 154, 181, 330
  exhibit, 219, 265
  ice information system, 108
  Mariner (program), 308
  metallurgy, 295
  personnel, 10, 39, 201, 354
  Propulsion Laboratory, 89
  Skylab 1 Investigation Board, 235
  Special Outplacement Service Office, 54
Library of Congress, Congressional Research Service, Science Policy Research Div., 54, 30, 99, 254
Licensing, 247–248
Licensintorg, Z/O (U.S.S.R. licensing agency), 80
Lidar. See Laser light-detection-and-ranging system.
Life magazine, 76, 125
Lifting body. See individual lifting bodies, such as X-24A and X-24B.
Light pattern identification device, 327
Light pollution, 98
Lightweight fighter program, 266
Lilly, William E., 30, 299
Lincoln (satellite), 5
Lincoln Laboratory (MIT), 355
Lind, Dr. Don L., 290, 336
Linder, Clarence H., 94, 154
Ling-Temco-Vought, Inc. See LTV Aerospace Corp.
Liquid hydrogen, 46, 68
Liquid oxygen, 46
Little, Arthur D., Inc., 117, 351
Little Joe II (booster), 36
Litton Systems, Inc., 217
LM. See Lunar module.
Lockheed Aircraft Corp., 33, 201, 263, 270, 286, 343
  Lockheed-California Co., 50, 83, 132, 319
  Lockheed Electronics Co.
    Houston Aerospace Systems Div., 276
  Lockheed Missiles & Space Co., 19, 181, 257, 349
  Lockheed Propulsion Co., 212, 253
Loening, Grover, 345
Lof, Dr. George O., 177
Logistics, 250
LOI: lunar orbit insertion.
Lonar Crater (India), 165
London, Eng., 67, 135, 195, 221, 329, 342
Long Beach, Calif., 43
Long-duration manned space laboratory (proposed), 281
Long Island Univ., 46
Long-range perimeter acquisition radar. See PAR.
Long-tank, thrust-augmented Thor-Delta (LTTAT). See Thor-Delta
Lord, Douglas R., 61, 297
Lorenz, Dr. Konrad, 290
Los Alamos Scientific Laboratory (LASL), N. Mex., 5, 290
Los Angeles, Calif., 13, 17, 41, 48, 97, 130, 184, 220
Los Angeles International Airport, Calif., 129, 201
Losey, Robert M., Award, 10
Louis, Victor, 172
Lousma, L/C Jack Robert (USMC), 224–225
  ASTP backup crew, 31, 326
  military promotion, 328
  mission activity, 227, 228–229, 245, 258, 322
  mission preparations, 180, 203, 219, 221, 320
  Presidential message, 270
  press conference, 283
  space flight effects, 225, 274
Love, Eugene S., 9, 11
Love, John A., 200, 336
Love, Maj. Michael (USAF), 284, 285
Lovell, Capt. James, A., Jr. (USN, Ret.), 7, 150, 153
Low, Dr. George M.
  Apollo program, 65
  awards and honors, 5, 15, 76, 131, 171, 195, 282, 300
  NASA budget testimony, 81, 88–89
  NASA Task Force on Energy Conservation, 340
  U.S.S.R. visit, 326
Lowrey, H. Douglas, 105
Lox. See Liquid oxygen.
LRL (Lunar Receiving Laboratory). See Johnson Space Center.
LRV. See Lunar roving vehicle.
LSAPT. See Lunar Samples Analysis Planning Team.
LST. See Large Space Telescope.
LTTAT (long-tank, thrust-augmented Thor-Delta). See Thor-Delta.
LTV Aerospace Corp., 132, 326, 342
  Vought Missiles & Space Co., 248

Vought Systems Div., 186, 300
Luna (U.S.S.R. lunar probe), 359
Luna 16, 162
Luna 17, 8
Luna 19, 162, 283
Luna 20, 8, 65, 162
Luna 21, 8, 19, 30, 96, 110, 139, 174, 346, 359
Lunan, Duncan A., 130
Lunar. See Moon.
Lunar module (LM), 11, 52, 255
Lunar and Planetary Missions Board (NASA), 341
Lunar Receiving Laboratory (LRL). See Johnson Space Center.
Lunar roving vehicle (LRV, Rover; see also *Lunokhod 1*), 139, 242, 255, 265
Lunar Samples Analysis Planning Team (LSAPT), 32, 341
Lunar Science Conference, Fourth Annual, 32, 65–66, 89
Lunar Science Institute, 2
Lundin, Bruce T., 10, 39, 160, 211, 230, 235
Lunney, Glynn S., 35, 84, 207–208, 226, 282, 326–327
Lunokhod (U.S.S.R. lunar surface explorer), 124
Lunokhod 1, 8
Lunokhod 2
  accomplishments, 65, 174
  exhibit, 163
  Luna 21 mission, 8, 19, 24, 30, 96, 110–111, 139, 346–347, 359
Lunquist, Dr. Charles A., 218
Lusk, Capt. Clyde T., Jr. (USCG), 341–342

M

M-518 (multipurpose electrical furnace system), 275
McCandless, L/Cdr Bruce, II (USN), 126
McClory, Rep. Robert, 19
McClure, Dr. Frank T., 294
McConnell, Dr. Dudley G., 119, 213, 300
McCormack, Rep. Mike, 23, 239, 240, 293
McCurdy, Richard C., 88–89, 91, 317
McDonald, Peter, 200
MacDonald Observatory, 24
McDonnell Douglas Corp., 50, 70, 117, 138, 146, 261, 315, 343
  contract, 11, 13, 183, 201, 215, 269
  Douglas Aircraft Co., 204
  McDonnell Douglas Astronautics Co., 47, 97, 101, 119, 169, 257, 353
  Skylab, 146
McDonugh, Thomas R., 119
Maceio, Brazil, 59
McElroy, Joseph, 26
McFall, Rep. John J., 23, 41
McIntosh, Patrick S., 132
Mackay Trophy (USAF), 135
McLaughlin, Dr. Edward J., 142
McLean, Va., 114
McLeavy, Roy, 331
McLucas, Secretary of the Air Force, John L., 70, 204–205, 331, 343
McNamara, Joseph P., 105
Macqueen, Dr. Robert A., 185, 222, 228
Macready, Col. John A. (USAF, Ret.), 135
Madagascar, 200
Madeira Islands, 73
Madrid Tracking Station, Spain, 116, 208
MAF. See Michoud Assembly Facility.
Magellanic Gold Medal, 15
Magnetic field, 47, 106, 197, 267, 298, 304, 309–310, 311, 337, 346, 357
Magnetic storm, 328
Magnetohydrodynamics (MHD), 298, 337
Magnetometer, 30, 106
Magnetopause, 309
Magnetosphere (earth), 26, 40, 47, 133, 202, 298, 302, 309, 322
Magnuson, Sen. Warren G., 4, 31, 272
Maharashtra, India, 165
Mailer, Norman, 3
Makarov, Oleg G., 273, 276
Malaga, Joseph F., 88, 91
Malaysia, 245
Malek, Frederick V., 51
Mali, 200
Malkin, Dr. Myron S., 105
Manhattan Project, 155
Man in Space National Historic Site (proposed), 36
Manke, John A., 206, 216, 235, 247, 253, 304, 318, 319, 342
Manned space flight (see also Apollo; Apollo-Soyuz Test Project; Astronaut; Cosmonaut; Gemini; Mercury; Salyut; Skylab; Soyuz; Space biology and medicine; Space shuttle)
  achievements, 17, 55, 66, 166, 169, 213, 216, 228, 255, 271, 274, 279
  anniversary, 115, 137, 248
  awards and honors, 76, 186, 255
  benefits, 112, 144, 157, 164, 168, 270, 304–305
  budget, FY 1974, 27, 44
  criticism, 174–175, 188
  hazards, 153, 158–159, 238, 239–240
  justification, 194
  life support systems, 139
  long duration, 281, 298, 357
  lunar landing. See Moon.
  medical aspects, 82, 196–197, 335–336
  meeting, 221, 361
  policy and plans, 1, 36, 88, 111, 280–281
  press comment, 144, 166, 179, 238, 239–240, 274, 275, 276
  safety aspects, 55
  significance, 87, 275
  training, 112, 116, 207
  U.N. direction, 156
  U.S.S.R., 93, 102–103, 108, 115, 116, 273, 276, 277, 347
  women, 296

Manned Spacecraft Center (MSC, NASA).
  See Johnson Space Center.
Manning, L/C Charles H. (USAF, Ret.),
  206
Mansfield, Sen. Mike, 18
Mansur, George F., Jr., 239
Mapping
  celestial, 257
  earth resources, 93, 104, 130, 182, 227,
    247, 261
  geothermal resources, 267
  lightning prediction, 293
  planetary, 75, 96, 239, 256, 298, 340
  polar, 255
Maran, Dr. Stephen P., 352
March, John O., 112
Mar Del Plata, Argentina, 97
Mare Acidalium (Mars), 138
Mare Imbrium (Sea of Rains; moon), 53
Mare Serenitatis (Sea of Serenity; moon),
  8
Marianetti, Eugene A., 130
Marine Corps. See U.S. Marine Corps.
Mariner (spacecraft), 78, 269
Mariner 4 (Mars probe), 308
Mariner 6 (Mariner-F), 308
Mariner 7 (Mariner-G), 308
Mariner 9 (Mariner I)
  achievements, 16
  climatic data, 262
  mission profile, 308, 331
  ozone variation experiment, 53
  photographs, 75, 119, 219, 279
  press conference, 16
Mariner 10 (Mariner-J; Venus-Mercury
  probe), 1, 64, 77, 280, 307, 341, 357,
  361
Mariner Jupiter-Saturn (spacecraft), 78,
  280
Mariner Jupiter-Saturn 1977 (MJS '77)
  mission, 5, 10, 78, 269, 359
Mariner Mars 1971 mission, 203
Mariner Venus-Mercury 1973 Program
  Team (JPL), 299
Marinin, Yuri, 12-13
Maritime communications satellite system,
  252, 284
Mark II (quiet experimental engine), 72
Mark, Dr. Hans M., 116-117, 245
MAROTS (experimental maritime communications satellite), 231-232
Marov, M. Y., 186
Mars (planet; see also Mariner; Mars
  U.S.S.R. probe; Viking)
  asteroids, 54
  atmosphere, 53, 119, 184, 262
  colony (proposed), 304-305
  exploration
    manned, 51, 133, 304-305
    remotely manned systems, 36, 78, 106,
      132, 138, 174, 197, 262, 263, 267,
      308, 331, 357
    U.S.S.R., 65, 218, 221, 239, 242, 267
  international cooperation, 36, 64, 330
  mapping, 75
  meeting, 331
  photographs, 16, 218, 219, 279
  polar caps, 53, 262, 267
  surface, 16, 262, 267
Mars (U.S.S.R. Mars probe), 359
Mars 1, 218
Mars 2, 64, 65, 218
Mars 3, 64, 65, 218
Mars 4, 218, 221, 239, 242, 267
Mars 5, 221, 239, 242, 267
Mars 6, 239, 242, 267
Mars 7, 242, 267
Marshall Space Flight Center (MSFC,
  NASA)
  aircraft research, 344
  ASTP support, 15
  awards, 10, 300
  CALROC management, 22, 121
  contract, 26, 40, 169, 212, 239, 242-243,
    285, 338
  cooperation, 239, 242, 246
  employment, 10, 30, 124, 143
  Environmental Applications Office, 86
  infrared camera, 81
  LAGEOS, 328, 338
  LST Task Force, 39
  meeting, 22, 71, 119, 209, 232, 315
  Neutral Buoyancy Simulator, 112, 146,
    290
  Operations Support Center, 17
  personnel
    appointment, 133, 174, 218, 328, 350
    award, 300
    retirement, 126, 201, 354
  press comments, 319
  Research and Technology Review, 53
  Skylab
    experiments, 112, 139, 222, 266-267,
      275, 318
    hardware, 50, 146, 152, 340
    repairs, 146, 155, 157, 296
  Skylab Student Project, 127
  sounding rocket experiment, 158
  space shuttle, 71, 212, 239, 253, 327-
    328
  Spacelab, 289
  Spacelab Program Office, 350
  Spacelab Task Force, 269
  Spacelab Task Team, 269
  Structures and Mechanics Laboratory,
    285
  Systems Engineering Design Program,
    246
  technology utilization, 251
Marshlands, 206, 240
Marsokhod (U.S.S.R. roving vehicle),
  267
Martin, Rep. James G., 23
Martin, John L., Jr., 53
Martin Marietta Corp., 17, 46, 146, 246
  Aerospace Div., 101, 169, 326
  Denver Div., 257, 289
Maryland, 182, 289, 293, 294
Maryland, Univ. of, 331
  Laboratory of Chemical Evolution, 14
Mascon (mass concentrations of dense
  material beneath lunar surface), 15

Mason, Jimilu, 251
Mass transit, 272
Massachusetts Institute of Technology (MIT)
  awards, 109, 126, 284, 299
  Draper, Charles Stark, Laboratory, 25, 49, 196
  Laserphoto development, 124
  sounding rocket experiment, 98, 315, 316
Masursky, Harold, 16
Materials technology, 247–248, 266–267
  aircraft, 92
  Skylab, 142, 216, 222, 239, 266–267, 271
Mathews, Charles W., 61–62, 70, 79, 80–81, 114, 143, 313
Mathias, Sen. Charles M., Jr., 18
MATS. See Military Air Transport Service.
Mattingly, Cdr. Thomas K., II (USN), 255
Mauretania, 200
Max Planck Institute, West Germany, 328
Max Planck Institute for Physics and Astrophysics, 345
Maxwell AFB, Ala., 153
MB-326 (jet fighter trainer), 263
MD: Mission day.
MDA. See Multiple docking adapter.
Medal of Freedom Group Achievement Award (President's), 120
Medal of Merit (DOD), 220
Medicine. See Space biology and medicine.
Mediterranean Sea, 202
Meister, Frederick A., Jr., 123
Memorandum of Understanding, 81, 95, 108–109, 136, 244, 268–269, 289
Memorandum on Cooperation in the Peaceful Uses of Atomic Energy, 265
Memphis, Tenn., 46, 343
Mendis, Dr. Asoka, 345
Mentzer, William C., 9–10
Mercury (planet; see also Mariner 10), 1, 78, 180, 280, 307, 341, 357
Mercury (program), 35, 125, 135, 261, 279, 350, 353
Mercury-Redstone (booster), 36, 248
Mercury-Redstone 3, 137
Mesosphere, 245
Messerschmidt, Prof. Willy, 345
Messerschmitt-Boelkow-Blohm GmbH, 111, 113, 119
Metallurgy, 247–248, 253, 271, 295, 296
Meteor, 358–359
Meteor (U.S.S.R. meteorological satellite), 65, 359
Meteor 14, 88
Meteor 15, 166
Meteorite (see also Tunguska meteorite), 137, 259
Meteoroid, 106, 131
Meteoroid shield, 145, 147, 157, 160–161, 211
Meteorological satellite (see also individual satellites, such as *Essa 1*, *Itos 1*, *Nimbus 4*, and *Noaa 1*), 159, 212, 288, 325
  achievements, 86, 279, 315
  cooperation, 313–314, 315, 316
  cost, 29
  military use, 343
Meteorology
  aircraft monitoring, 128, 214
  EOS support, 80–81
  experiments, 46, 51, 206, 265, 329
  foreign programs, 159, 206, 302, 315, 325
  funds, 27, 28–29, 188, 189
  international cooperation, 56, 210, 237, 261
  NASA–NOAA cooperation, 316, 358
  NASA program, 188, 209, 295, 312, 313–314, 315, 316
  radical climate change theory, 75
  solar effects, 313, 315
  USAF satellite system, 70–71
  USAF–USN cooperation, 288
  weather forecasting, 47, 62, 88, 102–103, 188, 220, 237, 253, 255–256, 271, 275, 279, 293, 343
  weather modification, 189, 194
Metric system, 19, 30, 140
Metric Systems Corp., 94
mev: million electron volts.
Mexico, 24, 37, 87, 149, 157, 224, 237, 344
Mexico City, 189, 350
Meyer, R. W., 2
MHD. See Magnetohydrodynamics.
mhz: megahertz (one million cycles per second).
Mice, 223
Michaelis, Anthony, 120
Michigan, 251
Michigan Environmental Research Institute, 183
Michigan, Univ. of, 58, 110
Michoud Assembly Facility (MAF, MSFC, NASA), 69, 313, 315
Microbe, 341
Micrometeorite, 91
Micrometeoroid, 223
Microscope, 224
Microwave landing system (MLS), 86, 333
Microwave radiometer, 181
Middle East war. See Arab–Israeli war.
Middle Wallop, Eng., 220
MiG-23 (Foxbat; U.S.S.R. fighter aircraft), 111
Milford, Rep. Dale, 23
Military Air Transport Service (MATS), 297
Military Order of the World Wars, 343
Military Personnel Records Center, Overland, Mo., 261
Military strength, 94–95, 112–113, 120, 136, 155, 187, 214
Milky Way (galaxy), 23, 109, 279, 328, 332, 333, 341

Miller, Rep. George P., 23
Miller, Dr. John, 66
Miller, Myron, 342
Miller, Rev. Walter J., 333
Milligan, James E., 266
Millionschikov, Mikhail D., 165
Mineralogy, 271, 328, 344
Minetti, G. Joseph, 330
Mining, 242, 245, 275
Ministry of Aeronautics, Brazil. See Brazilian Ministry of Aeronautics.
Minnesota, Univ. of, 85
Minnows, 225, 241
Minority Business Enterprise Program (NASA), 300–301
Minuteman (missile), 7, 38, 73, 90, 94, 175, 206
Minuteman I, 63, 254
Minuteman II, 353–354
Minuteman III, 63, 165
MIRV. See Multiple independently targetable reentry vehicle.
Missile (see also individual missiles, such as Minuteman, Phoenix, Polaris, Poseidon, Trident)
 air-to-air, 189, 328
 antiballistic (ABM), 28, 94
 budget, 28
 contract, 38, 175, 206
 conversion to sounding rockets, 254
 cruise, 259
 foreign
  France, 136
  P.R.C., 17
  U.S.S.R., 90, 181, 187, 247, 249, 268
 guidance systems, 96, 294
 industry, 189, 286, 317, 362
 intercontinental ballistic (ICBM), 7, 50, 73, 90, 113, 247
 multiple independently targetable reentry vehicle (MIRV), 94, 113, 333
 multiple-warhead, 63, 165
 patrol hydrofoil (PHM), 140
 short-range-attack (SRAM), 28
 submarine cargo, 49
 submarine-launched (SLM), 28
 surface-to-surface, 175–176
 undersea long-range missile system (ULMS), 124
 U.S. program, 20
 warning, 28, 343
Missile Sites Labor Committee (President's), 48
Mississippi River, 130
MIT. See Massachusetts Institute of Technology.
Mitchell, Jesse L., 203
Mitchell, B/G William (Billy; USA), 220
Mittauer, Richard T., 203
MIUS. See Modular integrated utility system.
MJS. See Mariner Jupiter-Saturn mission.
MLS. See Microwave landing system.
Mobile launcher 3 (KSC), 352
Model T (automobile), 171
Modular integrated utility system (MIUS), 188–189
Moffett Field, Calif., 115, 117, 131, 181, 245
Mojave, Calif., 239
Molecular chemistry, 328
Molniya (U.S.S.R. comsat series), 65, 312, 314, 361
Molniya I, 359
Molniya I-23, 38
Molniya I-24, 253
Molniya I-25, 318
Molniya I-26, 333
Molniya II, 359
Molniya II-5, 107
Molniya II-6, 209
Molniya II-7, 294
Molniya II-8, 351
Molodezhnaya Observatory (U.S.S.R.), 54
Molton, Dr. Peter M., 14–15
Mondale, Sen. Walter F., 172
Mongolia, 129
Monkey, 195
Monolith, 43
Monopsony: A Fundamental Problem in Government Procurement (Orkand Corp. report), 156
Montana, 353–354
Monterey, Calif., 288
Montreal, Canada, 165
Montreal Convention of 1971 (ICAO), 252
Moon (see also Lunar headings)
 age, 49, 66
 atmosphere, 66
 base, 280, 304, 330
 colonization, 3
 crater (see also specific craters, such as Aristarchus, Le Monnier, and Shorty), 77, 139
 formation, 7
 gas emissions, 52
 landing site, 77, 110–111, 174
 magnetic field, 43, 110–111
 soil sample, 32, 45
 eclipse, 205, 340
 exhibit, 120
 experiments
  corpuscular emissions, 174
  ion detector, 82
  laser reflector, 8, 24, 96
  radio noise measurement, 179
  seismometer, 82
  space shuttle payloads, 303
 exploration
  astronomical contribution, 87, 142, 194
  manned
   Apollo program, 6, 25, 42, 65–66, 76, 81–82, 86, 96, 105, 107, 110, 112, 186, 236, 270, 274
   budget, 27, 173, 219
   hazards, 238
   Kennedy, President John F., 20
   Nixon, President Richard M., statement, 49–50, 324

press comment, 166
significance, 1, 55, 242
statistics, 279
U.S.S.R., 12–13
unmanned
  Explorer 49, 179
  Luna 19, 162, 284
  Luna 20, 65, 162
  Luna 21, 8, 19, 30, 96, 110, 139, 174, 346, 359
  Mars 3, 65
  Ranger and Surveyor probes, 350
  U.S.S.R., 12–13
  Venus 8, 65
history, 89, 93
landing, 8, 233, 350, 353
  anniversary, 54–55, 217–218, 279
  commemoration, 178, 223
landing site
lunar roving vehicle. See Lunar roving vehicle.
maria, 8, 53, 324, 346–347
mascon, 15
massif, 325
meteorite, 82
occultation, 358
origin of, 89, 93, 96
photographs, 2, 8, 45, 110, 112, 174, 325, 350
quake, 52
radioactivity, 53
surface, 8, 49, 139, 174, 242, 283, 324–325
surface sample
  analysis, 11, 93, 157, 213, 290
  Apollo, 2–3, 6–7, 32–33, 49, 65–66, 84, 164, 236, 324–325, 346–347, 361
  Luna, 8, 43, 174, 347
  quantity returned, 279
temperature, 30
volcanism, 6, 49, 66
water, 164
Moore, Capt. John, 221
Moorer, Adm. Thomas H. (USN), 94, 183
Moorhead, John D., 23
Morelli, Frank A., 12
Morgan Guaranty International Council, 42
Morris, George A., Jr., 64
Morrow, Walter E., Jr., 355
Morse, Ralph, 76
Mory, Robert L., 174
Moscow, U.S.S.R., 159, 222, 203, 325
  air service, 172, 193, 278
  Apollo 17 lunar sample presentation, 236
  astronaut visit to, 133
  exhibit, 216
  experimental facility, 9
  meeting, 182
    ASTP, 282
    environment, 45, 87, 164
    science, 331
    space biology and medicine, 55, 122, 177, 244
    space cooperation, 36, 187
  summit agreements, 134, 182, 186, 187
  working groups, 36, 45, 64
Moscow Center of Control of Manned Space Flight, 208, 282, 326
  NAS delegation, 188
  press conference, 276
  TV transmission, 314
  U.S. scientist visit, 130
Mosher, Rep. Charles A., 23, 63, 293
Mosquito, 183
Moss, Sen. Frank E., 4, 31, 155, 182, 281
  Apollo splashdown, 21
  astronauts congratulations, 177
  legislation, 120, 238, 264, 272, 314
  NASA budget inquiry, 30
  space program, 6, 208–209
Moth (tussock), 243
Mother-Daughter-Heliocentric missions. See International Magnetospheric Explorer.
Mt. Wilson Observatory, Calif., 12
Mountainview, Calif., 116
MSC (Manned Spacecraft Center). See Johnson Space Center.
MSFC. See Marshall Space Flight Center.
MSS. See Multispectral scanner subsystem.
Muller, Karl, 212
Muller, Paul M., 15
Multiple docking adapter (MDA), 57, 90, 127, 145, 148, 152, 225, 320
Multiple independently targetable reentry vehicle (MIRV), 7, 90, 94, 113, 249, 333
Multipurpose electrical furnace system. See M–518.
Multispectral scanner subsystem (MSS), 143, 220
Munich, West Germany, 264
Murphy, Attorney General Lionel (Australia), 159–160
Murray, Dr. Bruce C., 119, 341
Mururoa Atoll, 218, 219, 311
Museum. See individual museums.
Museum of Science and Industry, Chicago, Ill., 266
Musgrave, Dr. Franklin Story, 126, 339
Myers, Dale D., 55, 61, 69, 71, 171, 177, 220, 300, 313
Mylar (solar parasol), 148

N

NAA. See National Aeronautic Assn.
NACA. See National Advisory Committee for Aeronautics.
NAE. See National Academy of Engineering.
NAS. See National Academy of Sciences.
NASA. See National Aeronautics and Space Administration.
NASA Aerospace Safety Advisory Panel, 295
NASA Applications Program Integration Board (APIB), 62

NASA Applications Satellite program, 62
NASA Applications Team, 88–89
NASA Committee on Remote Manipulator Systems and Extravehicular Activity, 209
NASA Communications Network (NASCOM), 91
NASA Data Processing Facility, 220
NASA Distinguished Service Medal, 157
NASA Earth Resources Technology Satellite Team, 83
NASA General Counsel for Procurement Matters, 26, 40
NASA Headquarters, 316
   Life Sciences Directorate, 16, 62, 299, 300, 350, 353, 354
NASA Key Personnel Development program, 353, 354
NASA Low Cost Systems Office, 354
NASA Lunar and Planetary Missions Board, 341
NASA Medal for Exceptional Service, 157
NASA Lunar Sampling Analysis Team, 341
NASA Office of Aeronautics and Space Technology (OAST), 68, 72, 74, 80
   General Aviation Technology Office, 138
NASA Office of Applications (OA), 61–62, 70, 80, 338
NASA Office of DOD and Interagency Affairs, 340
NASA Office of Earth Resources Systems (proposed), 238
NASA Office of Equal Employment Opportunity, 119
NASA Office of Industry Affairs and Technology Utilization (OIATU), 108, 119
   Scientific and Technical Information Office (STIO), 213
NASA Office of Institutional Management (proposed), 248
NASA Office of International Affairs, 113
NASA Office of Management Development, 19–20
NASA Office of Management Planning and Review, 19
NASA Office of Management Systems, 213
NASA Office of Manned Space Flight (OMSF), 55, 58, 61, 71, 82, 152
   Apollo/ASTP Program Office
      Lunar Program Office, 64
   Joint Editorial Board on Space Biology and Medicine, 341
   Mission and Payload Integration Office, 69
   personnel, 105, 108, 111, 114, 126
   Sortie Lab Task Force, 61
   Spacelab Program Office, 297
NASA Office of Safety and Reliability and Quality Assurance, 31
NASA Office of Space Science (OSS), 36, 96, 123, 180, 203, 298, 308, 346, 354
   Apollo/ASTP Program Office
      Lunar Program Office, 64
   Lunar Programs Div., 93

NASA Office of Space Science and Applications (OSSA), 352
NASA Office of Technology Utilization, 53, 291
NASA Office of Tracking and Data Acquisition (OTDA), 152, 298
NASA Patent Abstracts Bibliography (NASA SP-7039), 252
NASA Regional Dissemination Centers (RDC), 13, 23, 360
NASA Scientific and Technical Information Facility (STIF), 95
NASA Space Program Advisory Council (SPAC), 22
NASA Space Science and Applications Steering Committee, 181
NASA Structural Analysis Computer System (NASTRAN), 260
NASA Task Force on Energy Conservation, 340
NASA Viking Biology Team, 341
NASC. See National Aeronautics and Space Council.
NASCOM. See NASA Communications Network.
Nason, Howard K., 295
NASTRAN. See NASA Structural Analysis Computer System.
Natal, Brazil, 326, 328
The Nation's Energy Future (AEC report), 335
National Academy of Aeronautics and Astronautics, 232
National Academy of Engineering (NAE), 94, 108, 125, 154, 205
   Commission on Education, 137
National Academy of Sciences (NAS), 79, 94, 125, 188, 202, 204, 205
   Astronomy Survey Committee, 115
   Biological Impact of Increased Intensities of Solar Ultraviolet Radiation, 59
   Committee on Atmospheric Sciences, 194
   Environmental Studies Board, 59
   Panel on Astrophysics and Relativity, 115
   Panel on Optical Astronomy, 115
   Panel on Radio Astronomy, 115
   Panel on Space Astronomy, 115
   Plans for U.S. Clear-Air Turbulence Research in the Global Atmospheric Research Program, 59
   Space Science Board (SSB), 162, 341
      Committee on Space Biology and Medicine
         Radiobiological Advisory Panel, 133
      HZE-Particle Effects in Manned Spaceflight, 133
      United States Space Science Program, 162
National Accelerator Laboratory, Batavia, Ill., 285, 302
National Advisory Committee for Aeronautics (NACA), 203

National Aeronautic Assn. (NAA), 113, 135, 183, 267, 304
National Aeronautics and Space Act of 1958, 20, 24, 40, 136, 230, 238, 252, 272, 279, 337, 354
National Aeronautics and Space Administration (NASA; see also NASA Centers, programs, probes, satellites, and related headings), 9, 24, 169
   accident, 115, 116–117, 117–118, 127–128, 131, 341
   accomplishments, 2, 17, 20–21, 33–34, 86–87, 104, 163, 281
   aircraft (see also individual aircraft, such as Convair 990, and Aircraft), 87, 93, 196, 209, 231, 232, 243, 245, 247, 261, 269, 270, 280, 285, 286, 318, 319, 332, 342, 343, 344, 358
   anniversary, 230, 264, 270, 279, 281, 285, 297, 299, 359
   astronauts. See Astronaut.
   awards and honors, 107, 266, 300
      astronaut, 52, 128–129
      Distinguished Public Service Award, 105, 299
      Distinguished Service Award, 105, 238, 299, 350, 352
      Exceptional Scientific Achievement Medal, 105, 299
      Exceptional Service Medal, 105, 120, 135, 157, 203, 238, 327, 350
      Goddard Memorial Trophy, 76
      Group Achievement Award, 105, 299
      Inventions Award, 294
      Outstanding Handicapped Federal Employee, 111
      Outstanding Leadership Award, 135, 327
      Silver Plaque, 195
   budget, FY 1974, 32, 36, 39, 48, 173, 192, 256–257, 279, 359, 360
      bills introduced, 187
      bills passed, 161, 187, 192, 198, 201, 209, 236
      bills signed, 219, 299
      House consideration
         appropriations, 101, 186, 192, 223
         authorization, 55, 57–58, 61–62, 68, 72, 74, 80, 81–82, 83, 84, 88, 90, 93, 96, 104, 139–140, 161, 198, 209
      impoundment, 50, 259
      press comment, 259
      press conference, 29–30
      reduction, 5, 9, 29–30, 360
      request, 26–27, 29–30, 42, 223, 360
      Senate consideration
         appropriations, 39–40, 198, 223
         authorization, 30, 56–57, 67, 71, 76–77, 91, 105–106, 112, 167, 187, 198, 265
   cash contributions to, 152–153
   contract, 13, 15
      aeronautics, 40, 80, 95, 107, 181, 188, 196, 204, 269
      airborne radar, 183

Apollo spacecraft, 47
ASTP, 338, 349, 353
Boeing 727 aircraft test, 302
communications satellite, 123
construction of facilities, 26, 40, 94, 272, 285, 295
data processing, 94, 233, 300–301
domestic satellite network, 264–265
earth phenomenon reports, 127
earth resources surveys, 139, 183
energy-use practices, 246
gas turbine engine assembly, 137
government production equipment, 4
highly maneuverable aircraft, 132
Large Space Telescope, 242
laser tracking system, 94
launch vehicle, 269, 300, 333, 353
minority business enterprises, 95
MIUS, 188–189
photographic support, 94
quiet engine development, 68–69, 300
remote-control-vehicle, 131–132
remote-sensing devices, 262
RSRA, 92, 247
Skylab Payload Integrator, 289
space shuttle
   avionics software, 270
   data acquisition system, 239
   external tank (ET), 61, 101, 169, 246
   hydraulic actuator, 83
   main engine (SSME), 61
   navigation subsystems, 25–26, 49
   orbiter, 61, 97, 119–120, 358
   propellants, 262
   runway, 340
   shuttle training aircraft (STA), 343
   solid-fueled rocket booster (SRB), 61, 327
   test stand, 338
space tug, 46
SST, 83
support services, 39, 54, 94, 178–179, 276, 353
wind tunnel testing, 133
cooperation, 279, 282, 315, 354–355
AAS, 331
Alaska Univ., 330, 338
American Astronomical Society, 331
American Geophysical Union, 331
Arizona, 247
ARPA, 250
Astro Met Associates, 247–248
AT&T, 269
Bird Co., 97
Cal Tech, 316
Calif. Univ., 156, 328
ComSatCorp, 250, 269, 281
DOC, 312
DOD, 71, 103, 142–143, 302–303, 358
DOT, 62, 74, 97
EPA, 58, 81, 110, 199
FAA, 110, 302, 358
FCC, 250
Federal Highway Administration, 233

Geological Survey, 64
Harvard College Observatory, 340
Hawaii Univ., 329
Hersh Acoustical Engineering Co., 232
HUD, 62, 188–189, 293, 330
Interior, Dept. of, 247, 345
Lockheed Aircraft Corp., 270
MIT, 316
National Heart and Lung Institute, 351
Natural Resources, Dept. of, 92–93
NBS, 293, 330
NOAA, 48, 51, 62, 64, 154, 293, 312, 314, 316
NRL, 317
NSF, 154, 205–206, 240, 293, 330, 332
O'Brien, Hugh, Youth Foundation, 181
Ohio Central State Univ., 253
Princeton Univ., 249
Southwest Research Institute, 81
Stanford Univ., 81
State, Dept. of., 332
USA, 92, 247
USAF, 22–23, 86, 90, 174, 195, 206, 215, 216, 235, 266, 284, 285, 291, 301, 307, 319, 358
USCG, 108, 110
USN, 48, 51, 110, 131
VIMS, 240
Virginia Polytechnic Institute, 129
Wisconsin Univ., 317
cooperation, international. See International cooperation; International cooperation, space; Sounding rocket, international programs.
criticism, 238, 329
employees and employment
  budget considerations, 29–30, 32, 35–36, 359
  EEO program, 39, 213, 329
  increase, 61, 278
  manned space flight, 61, 71, 335, 342
  placement service, 54
  reassignment, 122
  recruitment, 91, 248
  reduction in force
    Civil Service, 5, 29–30, 88, 122, 124, 359
    manned space flight, 30, 58, 91, 143, 335, 342
  training, 88
  transfer, 125
exhibit, 93
facilities
  budget and funding, 76–77, 139–140, 161, 167, 198, 219
  closure, 7–8, 125
  construction of, 187, 260
  maintenance, 5, 54
  modifications, 36, 94, 342
  post-Apollo use, 2–3
  visitation, 265
history, 54
launch
  balloon, 26, 212, 214–215
  failure, ITOS-E, 212
  postponed, 21–22, 134, 144, 319–320
  probe
    Mariner 10 (Mariner–J), 307
    Pioneer 11 (Pioneer–G), 106
  record(1973), 357
  satellite
    Anik 2 (Telesat-B), 123
    Explorer 49 (RAE–B), 179
    Explorer 50 (IMP–J), 298
    Explorer 51 (AE–C), 345–346
    Intelsat-IV F-7, 250
    Noaa 3 (ITOS–F), 312
    Nnss 0-02, 301
  schedule, 1, 10, 71, 209–210
  Skylab 1, 145
  Skylab 2, 147, 164
  Skylab 3, 224
  Skylab 4, 319–320
  sounding rocket
    Aerobee 150, 58
    Aerobee 150A, 161
    Aerobee 170, 75, 111–112, 165, 204, 261, 307, 315, 316, 317
    Aerobee 170A, 92, 98, 172, 285
    Aerobee 200, 19, 221, 267, 298
    Aerobee 200A, 181, 250, 329, 349
    Arcas, 46, 51, 174, 247, 265, 295, 329
    Astrobee D, 201
    Black Brant IVA, 321, 330–331, 338
    Black Brant IVB, 41, 243
    Black Brant VC, 22, 43, 104, 12 158, 173, 182–183, 243, 256, 340
    Javelin, 214, 326, 328
    Nike-Apache, 15–16, 63, 65, 201, 236, 239, 240–241, 243, 245
    Nike-Cajun, 201, 214, 237
    Nike-Tomahawk, 16, 26, 40, 42–43, 85, 89, 90, 95, 98, 228
    Scout, 300
    Super Arcas, 15–16, 37
    Super Loki, 214–215
    Viper Dart, 214–215
management, 235–236, 244, 297
meeting, 18, 22, 23, 53, 65–66, 71, 89, 119, 125, 126, 128, 138–139, 156, 331, 361
National Applications of Science Administration (proposed NASA name change), 272
organization, 19–20, 22, 31, 38–39, 46, 57, 64, 93, 138
patents, 131–132, 243, 247–248, 298, 360
personnel, 236, 244, 281, 295, 309, 310, 312, 313, 315, 328
  appointment, 5, 15, 19–20, 25, 39, 26, 40, 55, 101, 108, 109, 111, 117, 133, 156, 160, 181, 203, 213, 216–217, 238, 256, 259, 264, 286–287, 300, 328, 350

National Aeronautics and Space Administration (NASA), personnel, continued
  awards and honors, 9, 10–11, 15, 52, 73, 105, 111, 120, 128, 135, 136, 137, 186, 287, 327
  death, 157, 203, 341, 344
  dismissal, 299–300, 329
  promotion, 119, 221, 327
  resignation, 109, 203, 296, 317
  retirement, 7, 25, 117, 126, 135, 201–202, 201–202, 203, 333, 350, 351–352, 353, 354, 359
  visit to U.S.S.R., 325–326, 326–327
  press comment, 270, 276, 329, 330
  press conference, 16, 142, 143–144, 147, 181–182, 184–185, 266–267, 270, 282–283
  programs
    aeronautics, 2, 5, 10, 26, 27, 29, 40, 41, 43, 50, 56, 59, 73, 68, 72, 74–75, 80, 83, 89–90, 90–91, 92, 95, 101, 103, 107–108, 114, 118–119, 135, 139–140, 143, 160, 167, 187, 204, 206, 216, 219, 231, 232, 235, 243, 245, 247, 253, 266, 269, 270, 272, 279, 285, 291, 300, 301, 302, 307, 319, 344, 358, 359
    airborne science, 245
    astronomy, 5, 13, 22, 26, 32, 35, 36, 38–39, 46, 64–65, 73, 75, 78, 98, 104, 106–107, 110, 111–112, 119, 124, 125, 128, 131, 138–139, 140, 142, 160, 165, 172, 204, 219, 220, 222, 224, 227, 228, 230, 249, 257, 280–281, 285, 294, 303, 307, 315, 316, 317, 341, 354, 361
    automotive research, 199, 263–264
    communications, 48, 222, 250, 271
    earth resources, 25, 29, 36, 37–38, 42, 44, 46, 61–62, 70, 76, 80–81, 85, 90–91, 92–93, 95, 98, 108–109, 110, 127, 139, 140, 142, 143, 161, 167, 170, 213, 219–220, 222, 231, 238, 240, 243, 245–246, 247, 255–256, 257, 267, 271, 281, 283, 289, 303, 341, 344–345, 352, 357—359, 361
    education, 51, 246, 253, 282
    energy research and development, 154, 316, 318, 330, 332, 337, 340, 349, 362
    equal employment opportunity, 10, 15, 39, 95, 118, 299–301, 329
    lifting body, 206, 216, 220, 235, 247, 253, 284, 285, 302, 303, 318, 319, 342, 358
    lunar and planetary exploration, 2–3, 27, 32–33, 64, 66, 82, 93, 96, 128, 140, 142, 157, 166–167, 174, 183–184, 185–186, 197, 219, 279, 303, 307, 308, 331, 350
    manned space flight (see also Apollo missions; Apollo-Soyuz Test Project; Gemini; *Skylab 1* through *Skylab 4* missions; Space shuttle), 76, 82, 166–167, 174–175, 186–187, 188, 194, 196–197, 238, 239–240, 251–252, 276, 296
    achievements, 66, 213, 248, 251, 255, 274, 279
    benefits, 87, 270, 275
    long duration, 271, 280–281, 298
    policy and plans, 88
    press comment, 179
    meteorology, 162, 189, 210, 212, 245, 253, 255–256, 261, 265, 280, 293, 295, 312, 313–314, 315, 316, 329, 344, 358
    nuclear propulsion, 7–8
    sounding rocket, 224, 279, 294, 358–359, 361
    space biology and medicine, 12, 22–23, 27, 35, 41, 44, 48, 52–53, 55, 67–68, 82, 88–89, 122, 126, 133, 134, 139, 142, 177, 180, 192, 193, 195, 196–197, 199, 207, 209, 214, 217, 219, 222, 223, 225–226, 227, 228, 229, 238, 244–245, 246–247, 251, 257–262, 274, 280, 283, 289, 295, 296, 298, 303, 317, 319–320, 321, 322, 323, 341, 350, 351, 361
    space shuttle. See Space shuttle.
    space station (see also Skylab), 177, 179, 193–194, 271
    space tug, 46, 69–70, 71, 75, 111, 209, 256–257, 302–303, 358
    Spacelab, 210, 231–232, 244, 268, 280–281, 289, 297, 302–303, 350, 358, 360
    technology utilization. See Technology utilization, space.
    tracking and data acquisition (see also Tracking), 83, 91, 94, 97, 140, 219, 298
  publications
    "Foundations of Space Biology and Medicine," 244–245
    NASA Patent Abstracts Bibliography (SP-7039), 252
    1973 Payload Model: Space Opportunities 1973–1991, 303
    Nondestructive Testing, a Survey (SP-5113), 258
    The Quiet Sun (SP-303), 233
    Solar Energy Research: A Multidisciplinary Approach, 205
  test
    aircraft. See Aircraft, flight tests.
    Apollo-Soyuz Test Project. See Apollo-Soyuz Test Project.
    launch vehicle, 264
    lifting body, 206, 216, 220, 235, 247, 253, 284, 285, 291, 302, 303, 318, 319, 342, 358
    rescue vehicle, 336
    Skylab, 90, 116
    space shuttle, 327–328
    Viking lander and orbiter, 357
National Aeronautics and Space Council (NASC), 21, 24, 49, 51–52, 101, 203
National Air and Space Museum. See Air and Space Museum.

National Applications of Science Administration (proposed NASA name change), 272
National Assn. for Equal Opportunity in Higher Education, 52
National Assn. of Student Councils, 181
National Aviation Club, 183
National Aviation Facilities Experimental Center, FAA, 72
National Bureau of Standards (NBS), 189, 293, 330
National Center for Atmospheric Research (NCAR), 59, 161, 261
National Center for Space Studies (Centre National d'Etudes Spatiales, CNES), France, 24
National Civil Service League, 136
National Commission for Outer Space (Comision Nacional del Espacio Exterior, CNEE), Mexico, 87
National Cosmonaut Day, 115
National Energy Information System (proposed), 338–339
National Energy Office (NEO, President's), 121, 155
National Environmental Satellite Service (NESS), 62, 312, 316
National Football League, 166
National Geographic Society, 128, 129
National Heart and Lung Institute, 351
National Institutes of Health (NIH), 130
National Medal of Science, 284–285, 290, 292, 301
National Military Command System, 56
National Museum of History and Technology, 124
National Oceanic and Atmospheric Administration (NOAA, (see also *Essa 1*; *Itos 1*; National Operational Meteorological Satellite System; *Noaa 1*; *Noaa 2*; *Noaa 3*)
  budget, 28–29
  conference, 260
  cooperation, 48, 51, 61, 64, 70–71, 110, 154, 210, 237, 261, 293, 312, 313–314, 316, 352
  Gilmore Creek Command Data and Acquisition Center, 352
  National Environmental Satellite Service (NESS), 62, 312, 316
  National Severe Storm Laboratories, 118, 220
  research, 24, 32, 59, 175, 183, 258, 298
  satellite, 24, 48, 212, 214, 312, 316, 343, 358
  sounding rocket experiment, 16
  Space Environmental Laboratory, 132
  World Weather Program for 1974, 210
National Operational Meteorological Satellite System (NOMSS), 312, 316
National Park Service, 5, 13, 162–163
National Patterns of R&D Resources: Funds & Manpower in the United States 1953–1973 (NSF 73–303), 188
National Register of Historic Places, 162–163

National Research Council (NRC), 94
  Assembly of Behavioral and Social Sciences, 79
  Astronomy Survey Committee, 115
  Committee on Atmospheric Sciences, 194
  International Magnetospheric Study: Guidelines for United States Participation, 202
  Space Science Board (SSB), 341
    Committee on Space Biology and Medicine
      Radiobiological Advisory Panel, 133
    HZE-Particle Effects in Manned Spaceflight, 133
    United States Space Science Program, 162
National Research Council of Canada, 32
National Satellite Services, Inc. See Hughes Aircraft Co.
National Science and Policy Priorities Act of 1973, 4–5
National Science Foundation (NSF), 17.4, 65.5, 129.3, 174.2, 199.17, 230.6, 259.4, 263.5, 269.3, 281.6, 305.3, 318.1, 382.2, 392.5, 436.4, 450.8
  An Analysis of Federal R&D Funding by Function: FY 1963–73 (NSF 72–313), 38
  budget and funding, 4, 28, 231, 237, 243–244, 299
  Continued Growth Planned for Federal Civilian R&D Programs (NSF 73–314), 293
  cooperation, 89, 97, 154, 205–206, 240, 277, 293, 330, 332
  Energy Research Div., 181
  Federal Funds for Research, Development and Other Scientific Activities, Fiscal Years 1971, 1972, and 1973 (NSF 72–317), 39
  Federal Scientific, Technical, and Health Personnel, 1971. (NSF 73–309), 278
  Federal Support to Universities, Colleges, and Selected Nonprofit Institutions, Fiscal Year 1971 (NSF 73–300), 129–130
  Graduate Science Education: Student Support and Postdoctorals, Fall 1972 (NSF 73–315), 305
  Graduate Student Support and Manpower Resources in Graduate Science Education (NSF 73–304), 232
  Immigrant Scientists and Engineers Decline in FY 1972: Physicians Increase Sharply (NSF 72–311), 248
  Immigrant Scientists and Engineers in the United States (NSF 73–302), 241
  National Science Board
    Science Indicators, 1972, 256
  NSF Forecasts Rise in Company-Funded Research and Development and R&D Employment (NSF 73–301), 58–59

personnel, 9, 49, 51–52, 153, 215, 260, 286–287, 331
Research and Development in Industry 1971: Scientists and Engineers, January 1972 (NSF 73–305), 317
Resources for Scientific Activities at Universities and Colleges (NSF 72–315), 102
Scientists, Engineers, and Physicians from Abroad: Trends Through Fiscal Year 1970 (NSF 72–312), 34
Selected Characteristics of Five Engineering and Scientific Occupational Groups, 1972 (NSF 73–306), 218
Solar Energy Research: A Multidisciplinary Approach, 205
Twenty-Second Annual Report for Fiscal Year 1972, 80
Wind Energy Conversion Systems Workshop, 181
World Weather Program Plan for 1974, 210
National Scientific Balloon Facility, 26, 90
National security, 258–259, 293
National Security Agency (NSA), 236
National Security Industrial Assn., 17
National Space Club (NSC), 21, 52, 76, 194, 279
National Space Council, 261
National Space Day (proposed), 54–55
National Space Development Agency, Japan, 315
National Space Week (proposed), 161, 264, 266, 270, 282
National Technology Resources Council (proposed), 272
National Weather Service, 29, 62
National Wildlife Assn., 85
NATO. See North Atlantic Treaty Organization.
Natural Resources, Dept. of (proposed), 361
Naugle, Dr. John E., 16, 78, 96, 213, 269, 309
Naval Research Laboratory (NRL), 232, 254, 261
  personnel, 126, 213
  sounding rocket experiment, 104, 111–112, 121, 182–183, 254, 256, 317
Naval Shipyard, Long Beach, Calif., 43
Naval Torpedo Station Keyport, 49
Navigation (see also Air traffic control), 280
  aircraft, 135–136, 261
  spacecraft, 242, 276, 347
Navigation satellite, 5, 28, 86, 253, 294, 304, 358, 359, 361
Navy Navigation Satellite System (NNSS; see also *Nnss 0–20)*), 301
NBC. See National Broadcasting Co.
NBS. See National Bureau of Standards.
NC–135 (aircraft), 205
NCAR. See National Center for Atmospheric Research.
Nebula. See Gum Nebula.

Nectaris Basin (moon), 164
Neilsen Engineering and Research, Inc., 233
Nelson, M/G Douglas T. (USAF), 210
Nelson, Sen. Gaylord, 338–339
Nelson, Richard H., 105
NEO.. See National Energy Office.
Neon, 66
Neptune (planet), 310, 350
NERVA. See Nuclear engine for rocket vehicle application.
NESS. See National Environmental Satellite Service.
Netherlands, 24, 210, 268–269, 361
Neutral Buoyancy Simulator, 146
Neutron, 77
New Carrollton, Md., 66
New Hampshire, Univ. of, 85
New Mexico, 130, 149, 155
New Mexico, Univ. of, 13, 23
New York, 73, 342, 349
New York Bight, 110
New York City Applications Project, 291
New York, N.Y., 3, 22, 49, 158, 201, 217, 223, 360
New Zealand, 159–160, 219
Newark, N. J., 220
Newell, Dr. Homer E., 160, 280, 351–352, 354, 359
Newfoundland, 133, 261
Newton, Sir Isaac, 4
Nicaragua, 198
Nicholson, David, 135
Nickel, 247–248, 323–324
Niger, 200
NIH. See National Institutes of Health.
Nike-Apache (sounding rocket), 15–16, 63, 65, 201, 236, 237, 239, 240–241, 243, 245
Nike-Cajun (sounding rocket), 214, 237, 240–241, 275
Nike-Tomahawk (sounding rocket), 15–16, 26, 40, 42–43, 85, 89, 90, 95, 98, 228
Nikolayev, Andrian G., 26
Nikolayeva-Tereshkova, Valentina, 26
Nikon Instrument Div. See Ehrenreich Photo-Optical Industries.
Nimbus 2 (meteorological satellite), 62
Nimbus 4, 46, 48, 51, 174, 247, 265, 295, 329
Nimbus 5, 61–62, 104, 214, 255–256, 255–256
Nimbus–G, 27–28, 29–30
1973 Payload Model: Space Opportunities 1973–1991 (NASA report), Nippon Electric Co., Tokyo, Japan, 287, 315
Nitric acid, 315
Nitric oxide, 59, 63, 68, 288
Nitrogen, 139, 321
Nixon, Mrs. Richard M., 189–190
Nixon, President Richard M.
  appointments and nominations by, 5, 40, 48–49, 55, 109, 112, 117, 134, 141, 153, 168, 174, 200, 203, 204–205, 208–209, 216–217, 223, 239, 246, 248, 258, 330, 331, 336, 337

astronaut visit, 63–64
astronauts, reception for, 193
Aviation Advisory Commission recommendations, 31
awards presented by, 73, 231, 265–266, 284–285, 290, 292, 301
bills signed,
  appropriations, 219, 299
  daylight savings time, 346
  MSC renaming, 26, 40, 49–50, 251–252
bills submitted to, 3, 119, 209, 216, 335
criticism, 25-2, 49
Department of Natural Resources (proposed), 361–362
energy crisis, 314, 318–319, 329–330, 335, 361–362
Energy Research and Development (proposed), 361–362
Executive Order, 121, 153, 200, 337
inaugural parade, 19
international cooperation, 89, 264
messages, 154–155
  anniversary, manned space flight, 323–324
  anniversary, NASA, 285
  Apollo 16 mission, 21
  budget, 5, 26–27, 29–30, 42, 56–57, 97, 259, 290, 359
  congressional, 24, 26, 80, 86–87, 121, 134, 210, 258–259, 336
  Pioneer 10 mission, 336
  Skylab 2, 150, 151, 162, 191
  Skylab 3, 270
  Skylab 4, 323–324
NASA minorities program, 15
National Space Day, 54–55
National Space Week, 264
Nicolaus Copernicus Week, 112
Nuclear Energy Commission (proposed), 361–362
nuclear testing, 140
presidential aircraft, 41
press comment, 25-2, 49, 301
Project Independence–1980, 329–330
reports submitted to, 335
reports transmitted to Congress, 256, 257
resignations accepted by, 2, 5, 19, 55, 98, 153–154, 173, 174, 261, 336
resolutions, 237
Rickenbacker, Capt. Edward V., death, 220
science and technology, 41
space program, national, 24–25, 323–324
speech, 178
supersonic transport (SST), 42
Trident submarine program funds, 236
U.S.S.R. agreements
  Agreement on the Prevention of Nuclear War, 192
  Basic Principles of Negotiations on the Further Limitation of Strategic Offensive Arms, 189–190
  summit accord, 176, 184, 186, 187–188, 194
  U.S.S.R. space program, 50
  Watergate controversy, 154–155
  Wright Brothers Day, 346
Nnss 0-20 (Navy navigation satellite), 301, 358
NOA: new obligational authority.
NOAA. See National Oceanic and Atmospheric Administration.
Noaa 1 (ITOS-A); Improved Tiros Operational Satellite), 212, 312
Noaa 2 (ITOS-D), 24–25, 61–62, 110, 212, 214, 312, 316, 352
Noaa 3 (ITOS-E), 1, 209–210, 212, 312, 352, 358
Nobel Prize (1973)
  chemistry, 296
  medicine, 290
  physics, 296
Noise, aircraft
  abatement, 280, 358
    acoustic nacelles, 68–69
    ATT program, 72
    award, 9–11
    contract, 95, 204
    EPA control requirements, 219
    FAA program, 129
    funds, 28–29, 140
    landing systems, 182, 231, 358
    refan engine, 89–90
    sonic boom research, 13, 95
    Supreme Court decision, 153
    symposium, 97, 333
    technology utilization, space, 178
    testing, 50, 138, 232
  Noise Pollution Resource Compendium, 13
  QCSEE, 68–69, 90, 300
  QUESTOL, 5
  STOL, 20
  supersonic transport (SST), 42, 101, 175
Noise Pollution Resource Compendium, 13
NOMSS. See National Operational Meteorological Satellite System.
Nondestructive Testing, a Survey (NASA SP-5113), 258
Noordwijk, Netherlands, 122
NORAD. See North American Air Defense Command.
Nordberg, Dr. William, 219–220
Norman Engineering Co., 338
North America, 149–150, 170, 340, 342
North American Air Defense Command (NORAD), 185–186
North American Rockwell Corp. (see also Rockwell International Corp.), 17, 47
North Atlantic Treaty Organization (NATO), 57, 175–176, 185–186, 202, 204–205
North Carolina, 130, 227
North Dakota, 227
North Pole, 255–256
North Ray Crater (moon), 7
North Sea, 261
Northampton County, Va., 129

Northern Lights. See Aurora Borealis.
Northrop Corp., 40, 132
  Northrop Services, Inc., 54
Norway, 361
Norwegian Defense Research Establishment, 26, 40
Nottingham Univ., Eng., 221
Nova. See Supernova.
Noyes, Dr. Robert, 184
NP-3A (NASA research aircraft), 87
NRC. See National Research Council.
NRL. See Naval Research Laboratory.
NSA. See National Security Agency.
NSC. See National Space Club.
NSF. See National Science Foundation.
NSF Forecasts Rise in Company-Funded Research and Development and R&D Employment (NSF 73-301), 58-59
Nuclear energy, 5, 13, 14, 137, 159, 165, 187, 194, 213, 232, 240, 257-258, 265, 268, 271, 292, 302, 313, 314, 318-319, 341, 359
Nuclear Energy Commission (proposed), 258-259, 361-362
Nuclear engine for rocket vehicle application (NERVA), 7-8, 359-360
Nuclear reactor, 165
Nuclear Rocket Development Station, 5
Nuclear test, 140, 157, 161, 197, 211, 218, 219, 221-222, 232, 311
Nuclear Test Ban Treaty, 211, 294
Nuclear weapons, 51, 97, 112-113, 159-160, 189-190, 192, 353-354
Numeroff, William J., 223
Nurses, 261-262, 296
Nutrition, 189, 289

O

OA. See NASA Office of Applications.
Oak Ridge National Laboratory (ORNL), 116
Oakey Air Field, Queensland, Australia, 26, 90
Oao 1 (Orbiting Astronomical Observatory), 279-280
Oao 2, 45
Oao 3 (*Copernicus*), 13, 124, 191, 224, 249, 280, 329
Oasis 2 (U.S.S.R. experiment), 348
OAST. See NASA Office of Aeronautics and Space Technology.
Oberg, James E., 116
Oberpfaffenhofen, West Germany, 138-139
O'Brian, Hugh, Youth Foundation, 181
O'Brien, John E., 26, 40
O'Bryant, Capt. William T. (USN, Ret.), 64, 93, 354
Observatory (see also individual observatories), 132, 268
Ocean of Storms (Oceanus Procellarum; moon), 53, 82
Oceanography, 45
  budget, 28-29, 62
  expedition, 38, 48, 62
  international cooperation, 22, 74, 87, 187-188, 237
  meteorological data, 206, 255-256
  ocean dynamics, 75, 79, 82, 261
  remote sensing, 41, 42, 180, 219-220, 227, 240, 255-256, 328
O'Connor, L/G Edmund F. (USAF), 282
O'Connor, Joseph H., 53
Odyssey (Apollo 13 command module). See Command module.
Oertel, Dr. Goetz K., 142
Office of Aeronautics and Space Technology. See NASA Office of Aeronautics and Space Technology.
Office of Applications. See NASA Office of Applications.
Office of Assistant Secretary of Transportation for Congressional and Intergovernmental Affairs, 39
Office of Coal Research (DOI), 337-338
Office of Emergency Preparedness, 134-135
Office of Management and Budget (OMB), 50, 51-52, 259, 330, 342
Office of Science and Technology (OST, President's), 2, 3, 19, 24, 49, 51-52, 132, 154-155, 203, 208-209, 286-287, 354
Office of the Surgeon General (USAF), 350
Office of Technology Assessment (OTA), 63, 254, 307
Office of Technology Assessment: Background and Status (congressional report), 254
Office of Telecommunications Policy (President's), 239
Office of Transportation Energy Policy (DOT), 341-342
OFO. See Orbiting Frog Otolith.
Ogo 5 (Orbiting Geophysical Observatory), 32
O'Hare International Airport, 201
Ohio, 245
Ohio Development Center (proposed), 69-70
Oil (see also Energy crisis), 206, 240, 246, 330, 361-362
Oklahoma, 130
O'Leary, Dr. Brian T., 174-175, 188
Olindo, Perez, 201
O'Melia, Richard J., 258
OMB. See Office of Management and Budget.
OMSF. See NASA Office of Manned Space Flight.
O'Neill, Rep. Thomas P., Jr., 41
ONR. See Office of Naval Research.
Operation Line Backer II, 183
Optimal Data Corp., 300-301
Orbita network (U.S.S.R. communications satellite system), 38, 107, 209, 253, 294, 318, 333, 351
Orbital Workshop (OWS). See *Skylab 1.*
Orbiter. See Space shuttle.

Orbiting Astronomical Observatory. See *Oao 1*, *Oao 2*, and *Oao 3*.
Orbiting Frog Otolith (OFO), 131
Orbiting laboratory (see also Salyut and *Skylab 1*), 194, 271
Orbiting Solar Observatory (OSO; see also *Oso 7*), 169
Oregon, 149, 243, 354
Oreol (Soviet-French satellite), 359
Oreol 1, 351
Oreol 2, 351
Orgueil C1 (meteorite), 137
Orion (antisubmarine patrol aircraft). See P–3, P–3C antisubmarine patrol aircraft.
Orion (constellation), 4
Orkand Corp., 156
Orlando, Fla., 178
Orly Field, Paris, France, 271
ORNL. See Oak Ridge National Laboratory.
Oscillograph, 270
OSO. See Orbiting Solar Observatory.
Oso 7, (OSO–H), 78–79, 191, 224
OSO–I, 78–79
OSS. See NASA Office of Space Science.
OSSA. See NASA Office of Space Science and Applications.
OST. See Office of Science and Technology (President's).
OTA. See Office of Technology Assessment.
Otago (New Zealand frigate), 218
OTDA. See NASA Office of Tracking and Data Acquisition.
Ousley, Gilbert W., 139
Outer planet orbiter, automated (proposed), 281
Outer Space Treaty, 252
Outstanding Handicapped Federal Employee of the Year Award, 111
Outstanding Leadership Medal (NASA), 135, 327
OV–1 (aircraft), 172
Overland, Mo., 261
Overmyer, Maj. Robert F. (USMC), 31, 326
Owl, 232
OWS (Orbital Workshop). See *Skylab 1*.
Oxygen, 48, 75, 139, 302
Ozone, 46, 51, 53, 59, 174, 223, 247, 265, 288, 295, 329

P

P–3 (Orion; antisubmarine patrol aircraft), 115, 116–118, 128
P–3C, 131
Pacific Ocean, 49, 74, 113, 123–124, 136, 159–160, 175, 189, 197, 203, 210, 217, 218, 219, 224–225, 251, 313–314, 319, 320
Paine, Dr. Thomas O., 42, 203
Pakistan, 186
Pan (Jovian moon). See Jupiter.

Pan American World Airways, Inc., 5, 33, 193
Papeete, Tahiti, 311
PAR (long-range perimeter acquisition radar), 277
Parachute, 174, 273, 291, 327–328
Paris Air Show, 163, 173, 175, 176–177, 179
Paris, France, 17, 111, 159, 184, 186, 217, 244, 264, 271, 294–295
Parker, Calvin, 291
Parker, Dr. Robert A., 142, 223
Parks, Robert J., 108
Parris, Rep. Stanford E., 23
Pascagoula, Miss., 291, 295
Passive reflection satellite, 273
Pastore, Sen. John O., 18
Patent, 156, 241, 256
  AEC regulations, 257–258
  aluminum alloy, 136–137
  automated visual sensitivity tester, 298
  burglar-intrusion detection system, 310
  ergometer, 209
  fluoride-metal composite material, 247–248
  multispectral aerial photography, 46
  NASA Patents Abstracts Bibliography (NASA SP–7039), 252
  NASA policy, 360
  Patent Licensing Conference, 156
  planetary exploration vehicle, 131–132
  satellite solar power station, 351
  solar engine, 347
  tobaccoless cigarette, 207
Patent Law Assn., Los Angeles, Calif., 48
Patman, Rep. Wright, 31
Patrick AFB, Fla., 142–143, 144, 222, 318
Patrol hydrofoil missile ship (PHMS), 140, 185
Pearson, Sen. James B., 31
Pegasus 15 (V/STOL aircraft engine), 117
Peking, P.R.C., 5, 67, 135, 217
Pendray, G. Edward, Award, 11
Pennsylvania State University, 16
Pennsylvania, Univ. of, 292
Pentagon, 1, 214
People-to-People Health Foundation, 59
People's Republic of China. See China, People's Republic of.
Perimeter acquisition radar. See PAR.
Perkin-Elmer Corp., 242
Perseus (constellation), 329
Pesticide, 170
Petroleum, 210, 314
Petrone, Dr. Rocco A., 8–9, 21, 25, 82, 124, 133, 328, 350
Petrov, Dr. Boris N., 193, 282, 325
Petrov, Dr. Georgy I., 36
Petrukhin, Valentin I., 159
Petty, Richard, 199
Phantom. See F–4 aircraft.
Phelps, Dr. Richard, 167–168
Philadelphia, Pa., 182, 219, 313
Philco-Ford Corp., 57, 160
Phillips, Gen. Samuel C. (USAF), 221, 236

Phillips Univ., Okla., 319
PHMS. See Patrol hydrofoil missile ship.
Phoenix (missile), 176, 328
Phoenix Islands, 354
Phosphorus, 7
Photochemistry, 245
Photography
  aircraft accident, 177
  "Early Photography from Rockets" (symposium paper), 288
  earth environment and resources, 240, 246, 247, 253, 255, 261, 344
  Erts 1, 25, 219–220, 358
  Kohoutek (comet), 268, 299, 321, 322, 323
  Mariner 9 mission, 16, 219, 279–280
  meteorological, 13–14, 166, 255, 288
  multispectral aerial photography system, 46
  Pioneer 10 mission, 357
  reconnaissance aircraft, 144
  Skylab missions, 143, 149, 155, 185, 245, 256, 275, 299, 344–345
  solar, 132, 149, 205, 266, 321–323
  stellar, 150, 347
  television, 338
  x-ray, 266, 348
Photometer, 19
Phototransistor, 327
Physicans, 217, 248, 273, 339, 351
Physics, 296, 303
Physics and astronomy program (NASA), 219
Physiology, 209, 289, 290
Pickering, Dr. William H., 264
Pickle, Rep. J. J., 23
Pilgrims (U.K. organization), 342
Pilot, 210, 339
  accident. See Accident, aircraft.
  Apollo missions, 121–122
  Apollo-Soyuz Test Project, 326
  awards and honors, 118, 135, 220, 265–266
  Concorde 001 final flight, 294–295
  death, 220
  first South Pole flight, 294
  flight test, 265–266, 284, 285, 303, 304, 318, 319
  foreign, 338
  Friendship 7 mission, 341
  RPRV, 291, 357
  Soyuz missions, 162, 273
  training, 305
  UFO reports, 293, 296
Pioneer (program), 33–34, 106–107, 311, 359
Pioneer (spacecraft), 185
Pioneer 6 (interplanetary probe), 33–34, 107, 311
Pioneer 7, 33–34, 107, 311
Pioneer 8, 33–34, 107, 311
Pioneer 9, 33–34, 107, 311
Pioneer 10 (Pioneer–F)
  achievements and progress, 96, 106–107, 138–139, 174, 185–186, 267, 308, 339–340, 357
  asteroid encounter, 54, 248
  data transmission, 310, 337, 338, 340, 357
  experiments, 267
  flight path, 197, 241, 267, 357
  flight profile, 36, 45, 47, 274, 267, 308
  Jupiter observations, 78, 83, 96, 280, 308–309, 336, 337, 338, 339–340
  presidential messages and statements, 336
  press comment, 54, 252, 340
  press conference, 47, 54, 309, 310
  solar observations, 47
Pioneer 11 (Pioneer–G)
  achievements and progress, 138–139, 160, 174, 197
  earth observations, 197
  flight path, 197, 241, 357
  flight profile, 1, 45, 78, 96, 97, 106–107, 174, 357
  international cooperation, 361
  Jupiter observations, 47, 248, 280
Pioneer Venus mission (interplanetary probe), 185
Piper (twin-engine Seneca aircraft), 285
Pittsburgh, Univ. of, 63, 214, 245
Plagemann, Stephen, 136
Planetarium, 116
Planetary exploration. See individual planets.
Plans for U.S. Clear-Air Turbulence Research in the Global Atmospheric Research Program (NAS report), 59
Plasma, 106, 283
Plasma physics, 307
Plastic, 256
Plato Systems, Inc., 95
Plesetsk, U.S.S.R., launch
  Cosmos. See individual Cosmos satellites.
  Intercosmos 10, 302
  Meteor 14, 88
  Meteor 15, 166
  Molniya II–6, 209
  Oreol 2, 351
Plowshare program, 157, 161
Plum Brook Station, Ohio, 5, 54, 70, 126, 359
Plummer, James W., 331
Pluto (planet), 267, 310, 350
pndb: perceived noise in decibels.
Podgorny, President Nikolay V. (U.S.S.R.), 236
Pogue, L/C William R. (USAF)
  Skylab 4 mission
    achievements, 244, 320
    medical aspects, 295, 351
    mission preparations, 290, 299, 311, 313, 318
    press conference, 283
Point Mugu, Calif., 328
Poker Flats, Alaska, 85, 89, 90, 95, 98, 321, 330–331, 338
Polaire (French satellite), 159
Poland, 122
Polar icecaps, 255

Polar lights, 351
Polar night, 253
Polarimetry, 308
Polaris (missile), 124, 336
Polish Council of State, 213
Pollack, Dr. James B., 332
Pollution control (see also Air pollution; Noise, aircraft; Water pollution)
  agreement, 87
  aircraft, 59, 67, 72, 191, 263–264, 358
  automobile, 265
  electric power generation, 349
  geothermal energy, 260
  light, 98, 352
  monitoring, 62, 70, 73, 85, 231, 245
  presidential policy, 41
  press comment, 170
  space shuttle, 61
Pollux (French satellite), 159
Pompidou, President Georges (France), 264
Ponce de Leon, Juan, 289
Ponnampuruma, Dr. Cyril A., 14–15
Poor Richard Club, 313
Popkov, Valery I., 170
Population explosion, 224
"Popup" missile launching technique, 268
Port Canaveral, Fla., 97
Porter, Katherine Anne, 3
Portugal, 24
Poseidon (Jovian moon). See Jupiter.
Poseidon (missile), 28, 124, 336
Positron, 269
Postal Service. See U.S. Postal Service.
Potassium, 7, 53
Potassium-argon dating, 49
Potate, John A., 126
Potter, David S., 248
POW. See Prisoner of war.
"Powder Puff Derby." See All-Women's Transcontinental Air Race.
Power Generation Group. See General Electric Co.
Powley, Mallison, 80
Pownall, Thomas G., 326
Pratt, Perry W., 10
Pratt & Whitney Div. See United Aircraft Corp.
P.R.C. See China, People's Republic of.
Presidential Management Improvement Certificate (USAF), 122
Presidential Medal of Freedom Group Achievement Award, 120
Presidential Reorganization Plan Number One, 52
Press, Dr. Frank, 299
Press comment
  aeronautics, 289
  aerospace industry, 1, 166
  Agnew, Vice President Spiro T., 275
  Air and Space Museum, 289
  anniversary, 231, 245, 270
  Apollo (program), 166
  Apollo 13 mission, 153, 158
  Apollo-Soyuz Test Project, 276–277
  astronaut, 171–172, 174, 180, 193, 194, 274, 275, 276–277
  Aviation Advisory Commission report, 31
  comet, 342, 344
  communications, 271
  Copernicus, Nicolaus, 50
  cosmonaut, 276–277
  David, Dr. Edward E. Jr., 19
  earth resources, 271
  education, 292
  energy crisis, 224, 237–238, 325, 330, 335
  energy research and development, 206, 268, 292, 330, 340
  equal employment opportunity, 329
  Erts 1, 37–38, 231, 245
  FCC comsat policy, 14
  Federal employee reduction, 109–110
  Federal science policy, 335
  Florida, 326
  Harris, Mrs. Ruth Bates, 329
  international cooperation, 276–277
  Johnson, Lyndon B., Space Center, 54
  Jupiter (planet), 337, 338, 340
  Kohoutek, Dr. Lubos, 344
  Lunar Science Conference, Fourth, 89
  manned space flight, 238, 242, 251, 275, 276
  MIRV, 249
  NASA, 276, 297, 329, 330
  Nixon, President Richard M., 41, 49
  nuclear testing, 136, 221–222
  Office of Management and Budget, 259
  Office of Science and Technology, 286–287
  Office of Technology Assessment, 63
  orbiting laboratories, 193–194, 271
  Pioneer 10 probe, 54, 337, 338, 340
  pollution control, 237–238, 325
  presidential appointment, 286–287
  Salyut 2, 114, 141
  science, 286–287, 292, 301
  Skylab program, 114, 144, 156, 157, 163, 168, 190–191, 193–194, 276, 326
  Skylab 1 mission
    benefits, 197, 199
    damage and repair, 156–157, 158–159, 163, 166, 168, 169, 179, 180
  Skylab 2 mission
    astronaut, 158–159, 166, 168, 169, 171–172, 174, 180, 193
    benefits, 164, 197
  Skylab 3 mission
    achievements, 251, 271, 274
    astronaut, 275
    medical aspects, 283
    mission profile, 238, 239–240
    solar observations, 268, 274
  Skylab 4 mission, 323–324, 326
  solar observation, 268, 271
  solar system, 323–324
  Soyuz (spacecraft), 276–277
  Soyuz 12 mission, 276–277
  space program, national, 6, 144, 178, 190–191, 289, 326, 344

space shuttle, 157, 256, 289
space station, 179, 194, 271
storm prediction, 271
supersonic transport, 42, 120–121, 175
technology, 259, 270, 271
technology utilization, space, 259, 270, 271
Tu-144 accident, 175, 179
Tupolev, Andrey N., 9
UFO, 295–296, 297
U.S.–Canada comsat agreement, 14
U.S.–U.S.S.R. detente, 221–222
U.S.–U.S.S.R. Joint Commission on Scientific and Technical Cooperation, 101
Venus (planet), 242
Watergate controversy, 154–155, 275
Press conference
   Apollo 17 mission, 5
   Apollo-Soyuz Test Project, 35, 84, 326
   Apollo Telescope Mount, 184–185, 266–267
   astronaut, 5, 222, 223, 274, 282–283
   A–X (tactical fighter aircraft), 40
   B–1 (bomber), 210–211
   Convair 990, 116–117
   cosmonaut, 276
   Cousteau, Jacques-Yves, 62
   EREP, 143, 181–182
   Erts 1, 167–168
   fuel management, 321, 349
   Grumman Corp. contract dispute, 4
   ICBM, 247
   ILLIAC IV (computer), 250
   Jupiter (planet), 309
   Kohoutek (comet), 352
   Mariner 9 mission, 16
   military records, 261
   NASA budget, 29–30
   Pioneer 10 probe, 47, 54, 309
   Saturn IB (booster), 313, 315–318
   Skylab (program), 142
   Skylab (spacecraft), 282–283
   Skylab 1 mission, 143–144, 145, 147, 169, 191–192, 319
   Skylab 2 mission, 142, 143–144, 148, 191–192, 199, 222
   Skylab 3 mission, 171, 222, 223–224, 226–227, 228, 245, 258, 266–267, 270, 274, 282–283
   Skylab 4 mission, 283, 313, 315–318, 319, 321
   solar system exploration, 309
   space biology and medicine, 282–283
   SS–17 missile, 90
   USAF secret satellite system, 70–71
   U.S.S.R. space program, 187, 326–327
   Viking Mars mission, 138
Press, Dr. Frank, 299
Preston, G. Merritt, 135
Priboy (U.S.S.R. weather ship), 47, 49, 74
Price, Rep. Charles M., 41
Priem, Dr. Richard, 11
Princeton Univ., 165, 249
Prisoner of war (POW), 153, 183

Probe (see also individual probes such as *Luna 19*, *Mariner 10*, *Mars 1*, *Pioneer 10*), 184, 208, 280, 304–305
Professor Vize (U.S.S.R. research ship), 167, 200
Profilometer, 182, 261
Prognoz (U.S.S.R. satellite), 359
Prognoz 3, 47
Project Boomerang, 9, 26, 90–91
Project Independence–1980, 329–330
Project Mercury. See Mercury (program).
Project Wamflex (wave-induced momentum flux experiment), 59
Propellant, 206, 269–270
Propulsion, 68–69, 71, 211, 288, 300, 302, 345
Protein, 348
Proton accelerator, 77
Proton (U.S.S.R. booster), 124
Proxmire, Sen. William, 18, 39–40
Pryor, Harold E., 213
Public Law 92–484, 254
Puerto Rican Trench, 198, 222
Puerto Rico, 73
Puget, Jean-Loup, 128
Pulling, Ronald W., 123
Pulsar, 12, 115, 329, 333
Purdue Univ.
   Laboratory for the Applications of Remote Sensing, 66

Q

QCSEE. See Quiet, clean, short-haul experimental engine.
Quarantine, 145, 196–197, 229, 274, 295, 323
Quarles, John R., Jr., 246
Quasar (quasi-stellar object), 12, 38, 109, 110
Queen Elizabeth 2 (U.K. ocean liner), 342
Queensland, Australia, 26, 90
Question Mark (trimotor aircraft), 122
QUESTOL. See Quiet, experimental, short takeoff and landing aircraft.
Quiet, clean, short-haul experimental engine (QCSEE), 68–69, 90, 300
Quiet, experimental, short takeoff and landing (QUESTOL) aircraft, 5, 50, 90, 140, 187
Quiet Sun (NASA SP–303), 233

R

Rabinowitch, Dr. Eugene, 155
Radar, 339
   air-route-surveillance radar (ARSR-3), 41
   aircraft, 118, 134, 172, 183, 261
   astronomy, 64–65, 110, 169, 239, 294, 358
   Automated Radar Terminal System, ARTS, 46, 255

earth resources, 183
meteorological applications, 118, 293
perimeter acquisition radar (PAR), 277–278
reconnaissance, 50
Radiation (see also Cosmic ray, Gamma ray, Infrared radiation, Solar radiation, Ultraviolet radiation and X-ray)
  belts, 33, 182, 287, 309
  corpuscular, 47, 54
  effects, 133, 257, 303
  electromagnetic, 106
  fields, 96, 106
  helium-fluorescent, 257
  interstellar, 115, 341
  nuclear, 140, 232, 265
  planetary, 183–184, 309–310
Radiation/Meteoroid Satellite. See RM.
Radio
  astronomy, 28, 77, 179, 329, 341, 358
  frequency, 71, 203
  galaxy, 12
  photon, 329
  transmission, 132, 313, 332
    U.S.S.R. communications satellite, 7, 253, 294, 318, 333, 351
  transmitter pill, 251, 360
  waves, 339
Radio Astronomy Explorer (RAE), See Explorer 38 and Explorer 49.
Radioactivity, 53, 302
Radioelectronics, 251
Radioisotope thermoelectric generator (RTG), 5, 80, 106
Radiometer (see also individual radiometers, such as Electrically scanning microwave radiometer, Infrared radiometer, Infrared temperature profile radiometer, K–band radiometer, Ka–band radiometer, Microwave radiometer, S–band radiometer, Scanning radiometer, Vertical temperature profile radiometer, Very-high-resolution radiometer), 246, 352
Radiosonde, 245
Radiotelescope, 294, 342, 358
RAE 1 (Radio Astronomy Explorer). See Explorer 38.
RAE 2 (RAE–B). See Explorer 49.
Rall, L/C Frederick A. (USAF), 204
RAM (Research and applications module). See Spacelab.
Ramey, James T., 217
Rancho Los Amigos Hospital, Downey, Calif., 53
R&D. See Research and development.
R&PM. See Research and program management.
RAND Corp., 109
Randomdec (computer-and-analysis method), 233
Ranger (program), 350
RANN. See Research Applied to National Needs.
Rasmussen, Dr. Norman C., 271

Rasool, Dr. S. Ichtiaque, 36
Rat, 303
Ray, Dr. Dixy Lee, 40, 157, 261, 318–319, 335
Raytheon Corp., 176
RB–57 (reconnaissance aircraft), 59, 174
RBV. See Return-beam-vidicon camera.
RCA Corp.
  RCA Alaska Communications, Inc., 79, 215, 260, 262
  RCA Global Communications, Inc., 5, 14, 79, 118, 215, 252, 260
RCS. See Reaction control system.
RDC. See NASA Regional Dissemination Centers.
RDT&E. See Research, development, test, and evaluation.
Reaction control system (RCS), 225, 226, 227, 237, 244
Rechtin, Dr. Eberhardt F., 261
Reconnaissance satellite, 45, 67, 122, 287, 290, 292–293, 295, 301
Recovery ship, 186, 192, 229, 270, 274, 319, 323, 351
Red Square Parade (U.S.S.R.), 314
Redhawk (Cessna Cardinal aircraft), 285
Redstone (missile), 36, 248
Redstone Arsenal Airstrip, Ala., 86, 344
Reed, Nathaniel P., 13
Reed, Sylvanus Albert, Award, 10
Rees, Dr. Eberhard F. M., 25, 108
Reeves, Dr. Edward M., 222
Refan program, 219, 280, 358
Regolith (lunar surface sample), 323–324
Regula, Rep. Ralph S., 342
Reinecke, Lt. Gov. Edwin (Calif.), 245–246
Relativity, theory of, 208
Reliable Engineering Associates, Inc., 95
Remote access computer system, 250
Remote manipulator systems, 209
Remote sensing (see also Earth resources experiment package; Erts 1; Sensor)
  agricultural applications, 129
  aircraft, 63, 87, 95, 130
  cooperation, 95, 108–109
  environment, 95, 209–210, 344
  meteorology, 62, 128, 214, 253, 255–256, 288, 293, 312, 313–314, 316, 329, 358
  oceanography, 41, 42, 180, 219–220, 227, 240, 255–256, 328
  Skylab program, 344–345
  task site, 205
Remotely piloted research vehicle (RPRV), 291, 357
Remotely piloted vehicle (RPV), 160, 305
Rendezvous (see also Docking)
  ASTP. See Apollo-Soyuz Test Project.
  Skylab 2 mission, 144–145, 147
  Skylab 3 mission, 211–212, 219, 224, 225, 259–260
  Skylab 4 mission, 244, 296, 311, 320
  Soyuz-Salyut, 326
  space shuttle, 97
Reorganization Plan No. 1 of 1973 (President's), 24, 52

Rescue. See Space rescue.
Research and applications module (RAM). See Spacelab.
Research and development (R&D)
  aeronautics, 42, 86–87, 280
  aircraft industry, 189
  Apollo program contribution, 17
  budget, 2, 7, 26, 139–140, 167, 198, 205–206
  civil aviation, 3
  communications, 293
  Continued Growth Planned for Federal Civilian R&D Programs (NSF 73-314), 293
  employment, 102, 189, 241, 278
  energy. See Energy research and development.
  Federal program, 24–25, 133–134, 205–206, 233, 236, 240, 243–244, 256, 259
  fuel. See Energy research and development.
  funding, 38, 58–59, 104, 130, 187, 188, 219
  international cooperation, 13, 101, 205, 277
  Science Indicators, 1972 (NSF report), 256
  significance, 170
  Watergate controversy impact, 154–155
Research and Development Collaboration with the U.S.S.R. and Japan (congressional report), 277
Research and Development in Industry 1971: Funds, 1971; Scientists and Engineers, January 1972 (NSF 73-305), 317
Research and program management (R&PM), 187, 219
Research Applied to National Needs (RANN), 28, 154, 219
Research, development, test, and evaluation (RDT&E), 28
Researcher (NOAA research ship), 237
Resources for Scientific Activities at Universities and Colleges, 1971 (NSF 72-315), 102
Return-beam-vidicon (RBV) camera, 219–220
Reuel, Norman C., 313
Reykjavik, Iceland, 264
Reynolds Metals Co., 136–137
RF: radio frequency.
RFP: request for proposals.
"R. H. Goddard: Accomplishments of the Roswell Years (1930–1941)" (symposium paper), 288
Rhone Valley, Switzerland, 287
Ribicoff, Sen. Abraham A., 51
Richardson, Secretary of Health, Education, and Welfare Elliot L., 30, 51–52, 96, 122, 141
Richmond, Va., 136
Rickenbacker, Capt. Edward V., 220
Rickover, Adm. Hyman G. (USN, Ret.), 336
RIF: reduction in force.
Rifma radiation analyzer, 139

Riley, J. Patrick, 115
Rio Blanco, Colo., 161
Rio Grande Valley, 149
Ritchie, Capt. Richard S. (USAF), 135
Riverside, Calif., 245
RM (Radiation/Meteoroid Satellite), 136
RNA (ribonucleic acid), 15
Robillard, Geoffrey, 20
Rochester, N.Y., 295
Rochester, Univ. of, 341
Rockefeller Univ., 284
Rocketdyne Div. See Rockwell International.
Rockwell, Dr. Don A., 158, 197
Rockwell International Corp. (formerly North American Rockwell Corp.), 105
  Aerospace Group, 47, 118
  Atomics International Div., 5
  aircraft, 176
  contract, 97, 132, 135–136, 210–211, 270, 295, 349
  merge, 47, 296–297
  Rocketdyne Div., 314
  Skylab support, 17
  Space Div., 119–120, 186, 271
  space shuttle, 69–70, 289, 311
  wage increase order, 342–343
Roe, Rep. Robert A., 23
Rogers, Rep. Paul G., 191
Rogers, Secretary of State William P., 187
Rogerson, Dr. John B., 249
Rolls-Royce Ltd., 117
Romanenky, Yuri, 162
Romanov, Aleksander, 276
Rome, Italy, 252
Roschin, Aleksey A., 51
Rosen, Dr. Harold A., 48
Ross Sea, 87
Rotor systems research aircraft (RSRA), 92, 247
Rotorcraft, 80
Rover. See Lunar roving vehicle.
Rowley, Dr. Peter D., 104
Royal Aero Club of the United Kingdom, 220
Royal Air Force, U.K., 169
Royal Astronomical Society, 15
Royal Greenwich Observatory, U.K., 297
RPRV. See Remotely piloted research vehicle.
RPV. See Remotely piloted vehicle.
RSRA. See Rotor systems research aircraft.
RTG. See Radioisotope thermoelectric generator.
Ruckelshaus, William D., 223
Rukavishnikov, Nikolay N., 162
Runnels, Rep. Harold L., 217, 291
Ryan, Gen. John D., 205, 231
Ryan, Michael P., Jr., 263
Rykalin, Vladimir I., 159

S

S-009 (NRL experiment), 232
S-43 (amphibious aircraft), 206
S-II. See Saturn IB and V, 2nd stage.
S-IVB. See Saturn IB, 2nd stage.
S-band radiometer, 261
Sabreliner 75A (business transport aircraft), 135–136
SAC. See Strategic Air Command.
SAE. See Society of Automotive Engineers.
Safeguard Ballistic Missile Defense Facility, Grand Forks, N.D., 277-278
Safety (see also NASA Aerospace Safety Advisory Panel and Space rescue)
  aircraft, 72, 107–108, 255, 280, 332, 358
  astronaut, 238
  civil aviation, 252
  engineering, 194
  fire protection, 360
  lightning prediction, 293
  manned space flight, 239–240
  missile demonstration, 353–354
  public, 271, 291, 360
Sagan, Dr. Carl E., 186, 262
Sagdeyev, Roald, 267
Sahara desert, 200, 205
St. Johns, Newfoundland, 83
St. Louis, Mo., 261, 275
Saint Matthew Island, 47
SALT. See Strategic Arms Limitation Talks.
Salt Lake City, Utah, 80, 210
"The Salvage of *Skylab 1*" (film), 223
Salyut (U.S.S.R. scientific space station), 55, 133, 156, 174, 359
Salyut 1, 30–31, 102–103, 196, 273
Salyut 2, 102–103, 114, 123, 133, 141–142, 159, 182, 196
Sampson, George P., 53–54
Sams, David, 4
SAMSO. See Space and Missile Systems Organization.
San Andreas fault, 46, 64
San Clemente, Calif., 41, 151, 184, 193
San Diego, Calif., 53, 152, 193, 217, 229, 274, 319, 322, 323, 345
San Francisco Bay, Calif., 233, 245
San Francisco, Calif., 43, 113, 176, 245, 319
San Francisco International Airport, 201
San Francisco Univ., 243
San Jose State Univ., 243, 245
San Marco (Italian launch site), Indian Ocean, 199
San Marco C-2 (Italian satellite), 361
San Marco Equatorial Range, 35
Sandage, Dr. Alan R., 110, 125
Sandler, Dr. Harold, 296
Sanford, Dr. Peter, 329
Sansom, Robert L., 199
Santa Ana, Calif., 201
Santiago, Chile, 172

SAO. See Smithsonian Astrophysical Observatory.
Sao Paulo, Brazil, 95, 263
Sapp, Capt. Earle W. (USN), 213
Saralov, U.S.S.R., 159
Sarnoff, Robert M., 263
SAS. See Scandinavian Airlines System; Small Astronomy Satellite; Solar array system.
SAS–A (Small Astronomy Satellite). See *Explorer 42.*
SAS–B. See *Explorer 48*
Satellite Data System (SDS), 113, 312
Saturn (booster), 263, 265
  Saturn IB, 55
    Apollo spacecraft 211
    contract, 353
    damage, 296, 299, 312, 315, 316, 317–318
    2nd Stage (S–II), 353
    2nd stage (S–IV B), 320
    Skylab 2 mission, 10, 144, 147, 164
    Skylab 3 mission, 180, 211, 217, 224–225
    Skylab 4 mission, 296, 299, 311, 318, 320, 336
    testing, 98, 116, 127
  Saturn V, 56, 57, 127, 144, 145, 152, 163, 352
    2nd stage (S–II), 235
Saturn (planet; see also Mariner Jupiter-Saturn 1977 mission)
  atmosphere, 11, 119, 125, 317
  exploration, 106, 280, 357
  flyby mission, 45, 269, 310
  moon, 350
  rings, 64–65
Saturn Workshop. See *Skylab 1.*
Savage. See SS–11 missile.
Savage, Swede, 168
Sayler, Sen. Henry, 92
SCAN. See Silent Communication Alarm Network.
Scandinavian Airlines System (SAS), 294
Scanning radiometer, 316
Scatterometer, 143
Schaeffer, Dr. Oliver A., 49, 66
Schardt, Dr. Alois W., 203
Scherer, Capt. Lee R. (USN, Ret.), 127
Schilling, David C., Trophy, 238
Schlesinger, Dr. James R., Secretary of Defense, 23, 40, 117, 141, 203, 247, 281, 333, 353
Schmitt, Dr. Harrison H., 18, 32, 49, 325, 339
  ASTP backup crew, 163
  awards and honors, 20–21, 52, 63, 76
  ceremony, 105, 107
  congressional testimony, 81–82, 112
  good will tour, 186, 195
  KSC visit, 50–51
  press conference, 6
  public appearance, 13
Schmitt, Col. John J., (USAF), 122
Schneider, William C.
  awards ceremony, 300

NASA budget testimony, 57–58
  press conference, 143–144, 146, 147, 313, 316, 317–318, 319, 321, 322
  Skylab 1, 21–22
  Skylab 2, 21–22
  Skylab 3, 217, 221, 223–224, 226–227, 228, 244, 270
  Skylab 4, 299, 313, 316–317, 317–318, 319, 321, 322
Schull, Wilfred, E., 76
Schulman, Dr. Fred, 201–202
Schurmeier, Harris M., 10
Schweickart, Russell L., 126, 149–150, 223
Schweiker, Sen. Richard S., 63
SCI Systems, Inc., 294
Science, 50, 178, 290
  education, 232, 237, 243–244, 305
  Federal support, 2, 4–5, 24–25, 112, 133–134, 215, 237, 243–244, 335, 354–355
  funds, 102
  international cooperation, 13, 187, 274, 280, 331
  press comment, 155, 170, 171–172, 292, 301, 335, 354–355
Science Adviser (President's), 49
Science Advisory Committee (President's), 19
Science and Man in the Americas (AAAS meeting), 189
Science and Technology (Presidential message), 154–155
Science Applications, Inc., 12
Science, Dept. of (proposed), 19
Science Indicators, 1972 (NSF report), 256
Scientist-astronaut, 81–82, 163, 174–175, 223, 224, 225, 227, 229, 241–242, 282–283, 319, 325, 339
Scientists, 276, 282, 287, 301, 328
  Apollo program, 195
  Apollo-Soyuz Test Project, 84, 207, 257, 261
  awards, 292, 299
  Earth Resources Technology Satellite Symposium, Third, 341
  employment, 7, 58, 102, 109–110, 143, 188, 189, 218, 232, 256, 278, 286, 319
  energy research and development, 165, 340, 362
  Federal Scientific, Technical, and Health Personnel, 1971 (NSF 73–309), 278
  foreign, 34, 182, 204, 205, 221, 261, 267–268, 292, 302, 312
  Immigrant Scientists and Engineers Decline in FY 1972; Physicians Increase Sharply (NSF 72–311), 248
  Immigrant Scientists and Engineers in the United States (NSF 73–302), 241
  international cooperation, 37–38, 205, 237, 315, 329, 331, 361
  Kohoutek comet study, 358
  Large Space Telescope, 181
  lunar science, 32, 49, 89, 93, 110

  meetings, 53, 66, 126, 181, 243, 260, 341
  meteorology, 175, 220
  Nixon, messages and statements, 336
  planetary research, 53, 183–184, 308–309, 310, 332, 336
  press comment, 180, 206
  Research and Development in Industry 1971: Funds, 1971; Scientists and Engineers, January 1972 (NSF 73–305), 317
  Research and Technology Review, 52
  satellite programs, 86–87, 346
  solar research, 200, 205–206, 222, 228, 258, 268
  space shuttle, 61
Scientists, Engineers, and Physicians from Abroad: Trends Through Fiscal Year 1970 (NSF 72–312), 34
Scotland, U.K., 339
Scott, Col. David R. (USAF), 238
Scout (booster), 300, 301
Scout Project Team (LeRC), 299
SDS. See Satellite Data System.
Sea of Rains (Mare Imbrium; moon), 53
Sea of Serenity (Mare Serenitatis; moon), 8, 110–111, 346–347
Seamans, Secretary of the Air Force, Dr. Robert C., Jr., 18, 58, 63, 86, 128, 129, 153–154, 205
Seattle, Wash., 32, 140
Second World Broadcasting Conference, 337
Secret Service. See U.S. Secret Service.
Section Special Event Award (AIAA), 10
Seismology, 250
Seismometer, 82
Seitz, Dr. Frederick, 284
Sekanina, Dr. Sdeneka, 352
Seki, Tsuomi, 268
Selb Manufacturing Co., 204
Selected Characteristics of Five Engineering and Scientific Occupational Groups, 1972 (NSF 73–306), 218
Semya Air Force Station, Aleutian Islands, 176
Seneca aircraft. See Piper.
Sensors, 143, 205
Sergievsky, Boris, 206
Serpukhov, U.S.S.R., 159
Service module (SM). See Command and service module.
SES. See Surface effects ship.
SEVA (stand-up extravehicular activity). See Extravehicular activity.
Sevastyanov, Vitaly I., 287
Sever-2 (U.S.S.R. bathyscaphe), 38
Severny, Andrey, 19
Seyfert (galaxy), 12, 329
Shaffer, John H., 83
Shaffer, Philip C., 321
Shanghai, P.R.C., 5, 32
Shapiro, Dr. Maurice M., 22, 232
Shapley, Willis H., 109, 163, 236

Shatalov, Maj. Gen. Vladimir A. (U.S.S.R.), 84, 102–103, 108, 115, 187, 196, 207, 287, 326
Shcherbina, Vladimir, 84
Sheldon, Dr. Charles S., II, 31
Shell Oil Co., 317
Shepard, R/A Alan B., Jr. (USN), 137, 248, 279, 320, 339
Sherrod, Robert, 125
Shevchenko, U.S.S.R., 213
Shikarin, Andrey, 277
Short-Haul Air Transportation-STOL Symposium, 114
Short-range-attack missile (SRAM). See Missile, short-range-attack.
Short takeoff and landing aircraft. See STOL.
Shorty Crater (moon), 6, 32, 45
Shoup, Rep. Richard G., 231
Shteinberg, Valentin A., 170
Shuttle training aircraft (STA), 269, 343
Siberia, U.S.S.R., 212, 263, 269, 287
Sigma Xi (honor society), 15
Sikorsky Aircraft Div. See United Aircraft Corp.
Silent Communication Alarm Network (SCAN), 291
Silent Spring (environmental publication), 170
Silver Plaque (NASA), 195
Silvo, Joseph N., 18
Simon, Deputy Secretary of the Treasury William E., 337
Simpler, Guy W., 248
Sinclair, Gordon, 344
Sirius (star), 98
Sjogren, William L., 15
Skaramanga Naval Base, Greece, 325
Skyjacking. See Hijacking of aircraft.
Skylab (program), 181, 217, 335
   achievements, 276, 357, 359
   astronaut, 153, 207, 313
   awards, 300
   benefits, 178, 213
   CALROC support, 22, 104
   contract, 289
   cost, 207
   employment, 58, 92, 359–360
   exhibit, 265
   experiments, 87, 142, 344–345
   flight simulation, 17
   funds, 5, 39, 44, 57–58, 140, 167, 359
   instrumentation, 25, 171, 340
   international cooperation, 81, 361
   JSC reorganization plan, 339
   medical aspects, 73, 336
   meeting, 119
   Nixon praise, 178
   plans, 55, 84, 224, 279, 330
   press comment, 114, 144, 163, 166, 174, 178, 179, 180, 197, 207, 276, 326, 339–340
   press conference, 185
   problems, 171
   records, 357
   safety considerations, 194, 353
   solar research, 200, 313
   sounding rocket support, 340
   status, 75–76, 275
   technology utilization, 190
   tracking support, 83, 97
Skylab (spacecraft; see also Apollo Telescope mount), 10, 22, 349
   capabilities, 36
   cost, 173
   design, 194, 230–231
   instrumentation, 139, 318, 344–345
   mission preparation, 136, 283
   press comment, 156
   press conference, 283
Skylab 1 (Orbital Workshop)
   backup Workshop, 246
   benefits, 144, 156, 199
   components and equipment
      airlock module (AM), 145, 146, 152, 193
      batteries, 146, 149, 151, 171
      computer, 244
      coolant system, 227, 240, 320
      coronagraph, 224
      ergometer, 209
      gyroscope, 228
      instrument unit (IU), 145
      meteoroid shield, 145, 147, 157, 160–161, 211, 230–231, 235
      Model H microscope, 224
      multiple docking adapter (MDA), 127, 145, 148, 152
      power systems, 174, 223, 227
      solar array system (SAS), 145, 147, 177, 211
      solar panels, 145–147, 149–151, 160–161
      solar parasol, 146–147, 148, 152, 227, 228, 239
      solar telescope, 318, 321
      solar wing, 173
   congressional investigation, 155, 213, 235
   damage to, 144–145, 160–161, 211, 235
   deactivation, 323
   experiments, 145, 146
      Apollo Telescope Mount (ATM), 145, 148, 151, 169, 173, 182–183, 185, 192, 222, 227, 230, 359
      astronomy, 149, 169
      cosmic ray detection, 232
      earth resources experiment package (EREP), 149, 181–182, 197, 222, 230, 245
      materials science, 146–147
      medicine, 142, 145, 147, 149, 230
      press conference, 142
      student scientific experiments, 152
      temperature data, 145–146, 148, 149, 228
   funds, 173
   ground management and observation, 132–133, 142–143, 155, 244, 328
   habitability status, 319, 323–324
   human factor, 244

inflight repair, 145, 146, 147, 148, 150, 157, 160–161, 165–166, 168, 169, 174, 177–178, 179, 286, 320–321, 323, 357
Kohoutek observation, 224
launch, 145
launch vehicle, 10, 116, 118, 121
mission preparation, 21–22, 50, 98, 104, 112, 116, 118, 121, 123, 127, 134, 144, 320, 323
   countdown demonstration test (CDDT), 136
orbit, 261
orbital lifetime, 323–324
press comment, 156–157, 158–159, 163, 164, 165–166, 168, 169, 173, 178, 179, 180, 199
press conference, 142, 143–144, 145, 147, 169, 185, 198, 222, 319
Skylab 2, 3, 4 visits. See *Skylab 2, 3, 4.*
solar eclipse observation, 200
Space Object 1973-027A, 185
systems malfunction, 145, 153, 155, 160–161, 169
unmanned operation, 152, 230
Skylab 1 Investigation Board, 160, 235
Skylab 1 Investigation Report, 230–231, 235–236
Skylab 2 (first manned mission), 223
  achievements, 171–172, 180, 191–192, 193, 207, 213, 286
  astronaut
    awards and honors, 193, 286, 300
    congratulations, 162, 177–178, 186–187, 191
    extravehicular activity (EVA), 145–147 149–150, 151, 152, 161, 164, 227
    medical aspects, 134, 191–192
      inflight examination, 148, 197, 214
      postflight condition, 192, 193, 197, 214
      preflight examination, 136, 142
      quarantine, 126, 145
      temperature effects, 142, 145, 146–147, 148, 149, 150
      weightlessness effects, 151, 177, 180, 191–192, 195, 199, 222, 225–226
    Nixon, President Richard M., messages, 162, 191
    press comment, 158–159, 166, 168, 169, 171–172, 173, 174, 178, 179, 180, 190–191, 193–194, 207
    press conference, 143–144, 148, 150, 181–182, 199
    private communications, 134
    retirement, 333
    spacecraft repair, 160–161, 180, 211, 223, 324, 357
    testimony, 213
    training, 112, 127, 136, 139, 148,
  benefits, 199
  command and service module (CSM), 116, 145–147, 150–152
  command module (CM), 147, 148, 152
  components and equipment
    computer, 244
    coolant system, 227
    gyroscope, 228
    life support system, 139, 150
    mobile service structure, 147
  cuisine, 141
  docking procedure, 147
  experiments
    Apollo Telescope Mount (ATM), 121, 148, 151, 184–185
    astronomy, 145, 149–150, 151, 152, 222, 228
    earth resources, 127, 140, 142, 149, 151, 152, 213, 228
    earth resources experiment package (EREP), 143, 149–150, 181–182 198, 222
    life sciences, 142, 151
    materials sciences, 142, 146–147, 151, 216, 222
    medicine, 142, 145, 147, 149, 152, 222, 228
  ground support and observation, 132–133, 142–143, 244
  human factor, 244
  inflight repair, 145, 146, 147, 148, 150
  launch, 145, 147, 164
  mission preparation, 21–22, 90, 98, 123, 127, 143–144, 146–147
    countdown demonstration testing (CDDT), 136
  mission salvage, 171–172, 174, 179, 180, 186–187, 191–194, 197, 207, 211, 223, 230–231
  NASA budget impact, 197
  Nixon congratulations, 191
  objectives, 142, 152
  press comment, 158–159, 164, 165–166, 168, 169, 171–172, 173, 174, 178, 180, 190–191, 193–194, 197, 199, 207, 251
  press conference, 143–144, 147, 148, 181–182, 185, 191–192, 193, 197, 199, 222
  record, 150, 151, 152, 251, 357
  splashdown and recovery, 144, 152, 176, 191
Skylab 3 (second manned mission)
  achievements, 228, 251, 270, 271, 274, 281
  astronaut
    activities, 194, 225, 226, 228, 241–242, 258
    congressional comment, 213
    experimentation, 121, 192–193, 224, 227, 229, 237, 245, 256, 274, 324
    extravehicular activity (EVA), 225, 226, 227, 228–229, 239–240
    hazards and safety, 238, 239–240
    Johnson, President Lyndon B., 251–252
    medical aspects
      fluid intake and output, 227
      inflight examination, 214, 283
      lower body pressure, 274
      mental stress, 240
      motion sickness, 225–226, 228

postflight examination, 214, 217, 227, 228, 229, 274
preflight examination, 219
press conference, 274, 282–283
quarantine, 229, 274
vestibular disturbance, 226
weight loss, 228
Nixon, President Richard M., message, 270
Podgorny, message, 236
press comment, 256, 271, 274, 275
promotion, 328
scientist-astronaut, 283, 319
spacecraft repairs, 227
spacesuit, 320
training, 127, 221
command and service module (CSM), 206, 224–225, 226, 244
command module (CM), 180, 226–227, 229, 237, 244
completion, 246
experiments

Apollo Telescope Mount (ATM), 121, 222, 227, 228, 243, 258, 266
astronaut movement tests, 223
astronomy, 200, 224, 258, 274, 324
celestial sources of x-rays, 223
cellular mechanisms and metabolism in zero g, 223
circadian rhythm, 223
crystal growth, 223, 271, 275
desert mice, 223
earth resources, 224, 324
earth resources experiment package (EREP), 222, 227, 228, 245
human eye spectral capability, 223
materials science, 266–267, 271, 275
medicine, 224, 228, 274, 323–324
micrometeoroid collection, 223
minnows, 225, 241
ozone and airglow observations, 223
rocket chair test, 228
spiders, 161, 229, 241, 275
vehicle disturbance, 223
vinegar gnats, 223
ground support and observations, 224–225, 227
launch, 224, 228
launch vehicle, 22
mission preparations, 171, 180, 203, 206, 211–212, 217, 222, 226, 228
press comment, 238, 239–240, 251, 256, 268, 271, 274, 275
press conference, 223–224, 225–226, 226–227, 245, 258, 266–267, 270, 274, 283
record, 229, 228, 271, 322, 357
rendezvous and docking, 180, 211–212, 219, 224, 225, 229, 259–260
rescue readiness, 226–227, 238, 239–240, 244, 259–260
splashdown and recovery, 217, 229, 270, 275
system malfunction, 224, 226, 227, 228, 237, 238, 239–240
Skylab 4 (third manned mission)

astronaut, 299, 347
achievements, 319–320
backup crew, 290
Christmas celebration, 350–351
countdown participation, 317–318
extravehicular activity (EVA), 290, 320, 350–351, 357
medical aspects
echo-cardiography, 351
inflight examination, 320, 321, 322
motion sickness, 320
physiological changes, 322, 323
postflight examination, 323
quarantine, 295, 323
Nixon, President Richard M., message, 324
preflight briefing, 318
training, 127, 290, 311
completion, 280
command and service module (CSM), 284, 311, 320, 323
experiments
Apollo Telescope Mount (ATM), 220, 224, 246, 283, 320–321, 321–322, 322–323, 349, 358
barium cloud release observation, 330–331
earth resources, 320
earth resources experiment package (EREP), 283, 321, 322, 323
engineering, 323
laser communications, 321
light-scattering measurement, 322
magnetospheric particle collector, 322
materials science, 323
medicine, 323
solar physics, 321, 322, 323, 349
ground support and observation, 338
launch, 319–320
launch vehicle, 22, 121, 311, 312, 315–316, 316–317, 317–318, 320
mission preparation, 152, 227, 228, 230, 259–260, 283, 290, 294, 295, 296, 299, 301, 311, 317–318, 319, 326
multiple docking adapter (MDA), 320
press comment, 324, 326
press conference, 283, 313, 315–316, 316, 317–318, 319, 321
record, 320, 322, 323, 357
rendezvous and docking, 244, 319–320
rescue readiness, 226, 228, 336
sounding rocket support, 330–331
splashdown and recovery, 319, 323, 336
system malfunction, 296, 299, 319–320, 322
Skylab 5, 330, 340
Skylab Payload Integration, 289
Skylab Program Office, 283
Skylab Student Project, 25, 127, 265
Skylark (U.K. sounding rocket), 328
Skynet II-A (U.K. comsat), 1
SL-1. See *Skylab 1* (Orbital Workshop).
SL-2. See *Skylab 2*.
SL-3. See *Skylab 3*.
SL-4. See *Skylab 4*.

Slayton, Maj. Donald K. (USAF, Ret.) 31, 35, 326
SLM (submarine-launched missile). See Missile, submarine-launched.
Sloan Fellowship, 126
Sloop, John L., 288
SM (service module). See Command and service module.
Small Astronomy Satellite (SAS; see also *Explorer 42* and *Explorer 48*), 73
Smart, Jacob E., 201
Smirnov, Leonid, 176, 184
Smith, Dr. Bradford A., 16
Smith, Gerard C., 5, 109
Smith, Rep. H. Allen, 41, 336
Smith, Dr. Malcolm C., 141
Smith, Dr. Richard K., 10
Smith, Theodore D., 105
Smith, Wendell S., 344
Smithsonian Astrophysical Observatory (SAO), 218, 294, 342, 352
Smithsonian Institution (see also Air and Space Museum), 124, 155, 275, 294, 347
   Center for Short Lived Phenomena, 127
   Central Telegram Bureau, 98
   "Experimental Experimentarium" (planetarium prototype), 116
Smog, 178
SMS. See Synchronous Meteorological Satellite.
Smylie, Robert E., 300
SNAP-50 (systems for auxiliary nuclear power), 117
SNSO. See AEC–NASA Space Nuclear Systems Office.
Société Nationale Industrielle Aérospatiale. See Aérospatiale.
Society of Automotive Engineers (SAE), 107, 208
Soderman, Paul, 232
Solar array system (SAS), 145, 147, 211
   Skylab 1 problem, 177–178
   Skylab 2 repair, 178, 179
   solar panel, 103, 160–161
Solar cell, 8, 43
Solar corona, 132, 138–139, 149, 169, 185, 192–193, 222, 228, 266–267, 313
Solar eclipse, 114, 167, 192–193, 199–200, 205, 283
Solar energy
   research and development
     budget, FY 1974, 28
     congressional consideration, 177, 219, 231, 240, 293, 330, 332, 246, 347
     energy crisis application, 137, 205, 246, 362
     international aspects, 205, 239, 332, 361
     NASA role, 279
     NSF program, 154
     Office of Solar Energy Research (proposed), 347
     press comment, 330
     proposed legislation, 293, 330, 246
     U.S. industry, 206
   technology utilization
     heating and cooling, 177, 205–206, 213, 292, 318, 246, 349, 353, 360
     power plants, 355
     satellite power station, 351
     solar-powered engine, 347
     solar-powered laser, 158
Solar Energy Research: A Multidisciplinary Approach (NASA–NSF report), 205
Solar flares, 78–79, 132, 142, 151, 185, 258, 266, 274, 313, 322
Solar flux data, 121, 182
Solar gravity, 315
Solar Heating and Cooling Demonstration Act of 1973, 318, 349
Solar Heating and Cooling Information Data Bank (HUD; proposed), 349
Solar magnetic field, 185, 311
Solar neutrons, 325
Solar Observing Optical Network (SOON, USAF), 169
Solar physics (see also Apollo Telescope Mount), 19, 22, 65, 75, 138–139, 142, 171, 172, 181, 233, 243, 256, 261, 268, 274, 298, 321, 323, 329, 340, 349
Solar plasma, 47, 239, 267, 311
Solar radiation, 47, 90, 106, 115–116, 122, 139, 167, 169, 174, 182, 239, 267, 280–281, 298, 325, 340, 342, 346, 351
Solar radio emissions, 179, 239, 267
Solar storm, 132, 136
Solar system, 47, 106, 126, 128, 130, 137, 208, 224, 232, 279, 281, 294, 309, 310, 317, 331, 337, 338, 340, 341, 345, 357, 361
Solar wind, 32, 34, 66, 47, 174, 185, 222, 283, 298, 309
Solid-fueled rocket booster (SRB). See Space shuttle.
Solid-fueled rocket motor (SRM). See Space shuttle.
Solid rocket propellant, 269–270
Solid waste management. See Modular integrated utility system.
Solomon, Dr. Philip M., 328
Somali Republic, 200
Sonar, 94
Sonic boom, 13, 95, 126–127
SOON. See Solar observing optical network.
SOREL (Sun-Orbiting Relativity Experiment Satellite), 315
Sortie lab. See Spacelab.
Sortie Lab Task Force (OMSF), 61, 133, 269, 297
Sortie Lab Task Team (OMSF), 269
Sortie Laboratory. See Spacelab.
Sounding rocket (see also individual sounding rockets, such as Aerobee, Arcas, Black Brant, Centaur, INTA 300, Javelin, Nike-Apache, Skylark, Super Loki, Viper Dart)
   contract, 178–179
   experiments

aeronomy, 42–43, 214, 228, 330–331, 338
astronomy, 103, 111–112, 165, 204, 307, 315, 316, 317
astrophysics, 250
atmospheric, 46, 51, 236, 240–241, 247, 298, 358–359
auroral studies, 15–16, 32, 37, 38, 41, 63, 85, 358
barium cloud, 321, 326, 330–331, 338
corpuscular radiation study, 54
electromagnetic spectrum study, 358
electron study, 65
geomagnetic storms, 358
heat pipe, 221
ionospheric studies, 15–16, 201, 237, 239, 240–241, 243, 275, 328, 358
magnetospheric physics, 26, 40, 89, 90, 95, 98
mesospheric study, 245
meteorology, 56, 206, 214, 329
multiwire proportional counters, 92
ozone, 51, 174, 247, 265, 329
solar physics, 19, 75, 121, 172, 181, 243, 256, 261, 267, 329, 340, 349
stratospheric studies, 247, 265
x-ray astronomy, 98, 203, 254, 358
facility rehabilitation, 94
instrumentation, 181, 182–183, 191, 214
international programs, 77, 279, 361
NASA-
-Australia, 307, 315, 316, 317, 361
-Brazil, 326, 328, 361
-Canada, 32
-Germany, West, 361
-Netherlands, 361
-Norway, 361
-Sweden, 361
-U.S.S.R., 328
U.K.-Norway, 328
Kohoutek observation, 224, 294, 358
launch failure, 329
Skylab missions support, 78–79, 338, 340
South Africa, 109, 161, 208
South America, 41, 123, 245, 321, 337
South Atlantic Anomaly, 131
South Dakota, 103
South Pacific. See Pacific Ocean.
South Pole, 75, 294
South Ray Crater (moon), 7
Southeast Asia, 26, 28
"Southern California Inventor of the Year," 48
Southern California, Univ. of, 156
Southwest Research Institute, 81
Souza, Kenneth A., 317
Soviet Academy of Sciences, 24, 165, 287
Annual General Meeting, 65
Institute for Medical Biological Problems, 93
Institute of Radio Engineering and Electronics, 283
Intercosmos Council, 282
international cooperation

Joint Working Group on the Natural Environment, 45, 164
Joint Working Group on Satellite Meteorology, 22
scientist exchange, 170
international cooperation, space
Apollo-Soyuz Test Project, 93, 205, 257
Joint Editorial Board on Space Biology and Medicine, 177, 244–245, 341
Joint Working Group on Guidance, 207
Joint Working Group on Interplanetary Exploration, 36
Joint Working Group on Mission Model and Operational Plans, 207
Joint Working Group on Near-Earth Space, the Moon, and the Planets, 269
Joint Working Group on Space Biology and Medicine, 55, 122, 335–336, 361
lunar science, 84
science and applications, 188
space engineering (proposed), 165
Space Research Institute, 267
Soviet Committee for Science and Technology, 216, 275
Soviet Communist Party, 176, 184, 187, 189, 193, 194
Soviet cosmonaut training center, 282
Soviet Council for International Cooperation in the Exploration and Use of Outer Space, 269
Soviet Exhibition of Economic Achievement, Moscow, 216
Soviet Institute of High Energy Physics, 159
Soviet Ministry of Civil Aviation, 216
Soviet Red Square Parade, 314
Soviet State Research Institute of Machine Building, 261
Soyuz (U.S.S.R. spacecraft; see also Apollo-Soyuz Test Project), 1, 8–9, 30, 35, 84, 93, 133, 159, 163, 184, 196, 207–208, 217, 273, 276–277, 282, 325–326, 332, 347, 353, 359, 360
Soyuz 2 mission, 196
Soyuz 6 mission, 162
Soyuz 7 mission, 162
Soyuz 10 mission, 103, 273
Soyuz 11 mission, 103, 153, 273, 282, 326–327, 348
Soyuz 12 mission
cosmonaut, 273, 276, 335–336
flight control team, 276
press comment, 276–277
success, 359
tests and experiments, 273, 275, 276, 282, 326–327, 348
Soyuz 13 mission, 347, 359
SPAC. See NASA Space Program Advisory Council.
Space Act of 1958. See National Aeronautics and Space Act of 1958.

Space and Missile Systems Organization (SAMSO, USAF), 71, 206, 221
   Air Force Satellite Facility (AFSF), 122
   Defense Navigation Satellite System (DNSS), 304
Space biology and medicine
   animal experiments, 22-23, 241-242, 275
   Antarctic expedition, 12, 341
   Apollo-Soyuz Test Project, 207
   astronaut, 19, 214, 219, 222, 227, 228, 229, 282-283, 298, 320, 321, 322, 323-324
   award, 256
   cardiology, 82, 88-89
   Biosatellite 3, 195
   budget, FY 1974, 27
   contaminated oxygen, 48
   echo-cardiography, 351
   electric connector, 52-53
   ergometer, 197, 209
   extraterrestrial life, 317, 331-332
   "Foundations of Space Biology and Medicine" (U.S.-U.S.S.R. publication), 244-245
   growth hormone theory, 289
   handicapped assistance devices, 251
   HZE-Particle Effects in Manned Spaceflight (SSB report), 133
   immunity, 257
   International Academy of Aviation and Space Medicine, 264
   international cooperation, 55, 122, 177, 187, 335-336, 361
   Joint Editorial Board on Space Biology and Medicine, 177, 244-245, 341
   long-duration space flight, 44
   microbial exchange, 257
   NASA program
      Apollo 8 mission, 238
      Apollo 11 mission, 153
      Apollo 12 mission, 82, 238
      Apollo 13 mission, 153
      Apollo 14 mission, 238
      Apollo 16 mission, 238
      Apollo 17 mission, 55
      Mercury missions, 350
      Skylab 1 mission, 142, 145, 146-147, 149
      Skylab 2 mission, 126, 134, 136, 139, 142, 145, 146-147, 148, 149-150, 151, 152, 180, 192, 193, 195, 197, 199, 217, 230
      Skylab 3 mission, 223, 224, 225-226, 228, 274, 323
      Skylab 4 mission, 295, 319-320, 323, 351
      space shuttle, 71, 262, 303
   physiological stress research, 289
   press comment, 153
   press conference, 142, 148, 149-150, 274, 282-283
   quarantine, 145, 196-197, 229, 274, 295, 323
   radiation effects, 59, 133, 238
   radio transmitter pill, 251
   Summer Institute for Biomedical Engineering (GSFC), 246-247
   technology utilization. See Technology utilization, space, medicine.
   U.S.S.R. program, 153, 273, 287, 303, 347-348
   weightlessness, effects, 151, 153, 180, 222, 223, 224, 257, 323, 347-348
"Space Biology and Medicine," Volume One (U.S.-U.S.S.R. report), 341
Space debris, 92, 102, 122, 259
Space Detection and Traffic System, 185-186
Space Div., Rockwell International. See Rockwell International.
Space Flight Award (AAS), 351-352
Space Hall of Fame. See Albert Thomas Space Hall of Fame and International Space Hall of Fame.
Space manufacturing, 36, 53, 62, 67, 79, 142, 150, 232, 271
Space, military use of, 63, 304-305
   budget, FY 1974, 27-28
   communications, 57, 343
   meteorology, 314, 343
   press comment, 271
   U.S., 30, 129
   U.S.S.R., 30
Space Nuclear Systems Office (SNSO). See AEC-NASA Space Nuclear Systems Office.
Space, peaceful use of, 55, 63, 136, 186, 282
Space Plasma High Voltage Interaction Experiment spacecraft. See SPHINX.
Space program, national (see also individual programs, such as Apollo; Apollo-Soyuz Test Project; National Aeronautics and Space Administration; Skylab)
   achievements, 2, 21, 42, 44, 46, 49-50, 54, 55, 76, 156, 174, 178, 191-192, 197, 217-218, 228, 281, 354-355
   Aeronautics and Space Report of the President: 1972 Activities, 86-87
   anniversary, 324
   benefits. See Space results.
   budget (see also National Aeronautics and Space Administration, budget), 5, 9, 26-31, 42, 56-57, 70, 78, 88-89, 173, 188, 293
   Compendium of Meteorological Satellites and Instrumentation (NSSDC 73-02), 233
   employment, 189, 286, 329, 335, 352, 353, 359
   exhibit, 265
   goals and priorities, 6, 29-30, 36, 68, 84, 86-87, 93, 208-209, 279, 280, 282, 354-355
   international aspects (see also International cooperation, space and Space race), 186, 205, 231-232, 268, 274, 276-277, 285, 330
   Johnson, President Lyndon B., 20, 21, 25, 31, 251-252

National Space Week, 264, 266, 270, 281
Nixon, President Richard M., 24, 54, 286, 290, 324, 336
opinion, 130, 174–175, 188, 194
outlook, 5, 117, 275, 330
press comment, 6, 144, 166–167, 178, 190–191, 193, 239–240, 259, 289, 326, 344
press conference, 330
public contributions, 152–153
significance, 1, 294
U.S. technology development, 17, 272–273, 281, 330
Space race, 12–13, 30, 33, 114, 187
Space rescue, 91, 226–227, 228, 244, 259–260, 336, 353
press comment, 238, 239–240
Space Research Institute. See Soviet Academy of Sciences.
Space Research Performed in the U.S.S.R. in 1972 (U.S.S.R. report), 162
Space results (see also individual planets; probes; satellites; sounding rockets; Research Applied to National Needs; Technology utilization, space)
geology, 104
international aspects, 64, 41, 332
lunar science, 112
materials technology, 213, 216
meteorology, 104
Nixon, President Richard M., 178
press comment, 178, 197, 270, 271, 289, 297
public opinion, 233
quasar discovery, 110
science and technology, 17, 271, 280
solar astronomy, 169
U.S.S.R. program, 162
Space Science Award (AIAA), 11
Space Science Board (SSB; NAS-NRC), 341
Committee on Space Biology and Medicine, Radiobiological Advisory Panel, 133
HZE-Particle Effects in Manned Spaceflight, 133
United States Space Science Program, 162
Space Science Seminar, 22
Space shuttle, 84
AIAA evaluation, 11
arc jet, 80
astronaut, 268, 296, 339
benefits, 359
contract
actuator, 83
data acquisition system, 239, 285
data dissemination, 133
earth resources survey systems, 139
external tank (ET), 61, 101, 169, 246, 302–303
main engine (SSME), 61
orbital maneuvering system, 97
orbiter, 61, 97, 119–120, 270, 271, 358
propellant and vapor disposal, 262
runway, 340
solid-fueled rocket booster (SRB), 61, 69–70, 71, 302–303
solid-fueled rocket motor (SRM), 212, 253, 302–303, 327
technical support, 49
test stand, 338
training aircraft, 269, 343
cooperation, 103, 129
cost, 75, 172, 195–196
design, 46, 194, 358
development, 9, 208–209, 302–303
employment, 91–92, 359
environmental impact, 61
EOS compatibility, 80–81
exhibit, 265
external tank (ET) development, 69, 71, 118
facilities, 69–70, 71, 76–77, 80, 187, 198, 219, 271, 289, 295
fuel cell, 349
funds
budget cut, FY 1974, 29–30, 39–40, 161, 259
budget restrictions, 5, 359
congressional consideration, 56–57, 192
NASA request, 26–27, 42, 167
press comment, 156–157
R&D, 139–140, 219
international cooperation, 11
NASA-
-ESRO, 17–18, 61, 76–77, 136, 221, 231–232
-Europe, 114, 119, 244, 268, 280, 297
KSC student tour, 181
Large Space Telescope (LST), 125, 181, 242–243
launch equipment modification, 352
management, 105, 111, 339
NAS Summer Study, 204
1973 Payload Model: Space Opportunities 1973–1991 (NASA report), 303
orbiter
advanced applications, 12, 268, 297, 360
avionics software system, 270
development, 69, 88, 121–122
engine, 195
ferry aircraft, 270, 286
payload range, 311
simulation, 269, 343
thermal protection system, 186
postponement, 174–175
press comment, 156–157, 256, 289
remote manipulator systems, 209
rescue capabilities, 240, 273

Skylab 1 visit (proposed), 323
solid-fueled rocket booster (SRB)
cost, 75–76
recovery system test, 327–328
water impact and towing test, 43
solid-fueled rocket motor (SRM), 253
space tug impact, 45, 256–257

TDRSS support, 298
television demonstration, 207
U.S.S.R., 30–31, 277
von Kármán lecture, 9
wind-tunnel test, 80, 133
women simulation experiment, 262, 296
Space Shuttle–Skylab, 1973: Status report (congressional report), 75–76
Space station (see also Salyut and Skylab), 330
"Space Station West Virginia" (pilot housing project), 215–216
Space Task Group, 24
Space tug, 46, 69–70, 71, 75, 111, 209, 256–257, 302–303, 358
Space Week. See National Space Week.
Spacecraft (see also individual spacecraft), 178–179, 207, 224, 241, 288, 318
Spacecraft Design Award (AIAA), 11
Spaceflight Tracking and Data Network (STDN), 83, 180, 208, 298
Spacelab
  astronaut, 136
  benefits, 88
  components, 268
  cost, 268–269
  design and development, 55, 61, 136, 302
  funding, 113, 136, 268–269, 360
  international participation, 61, 76–77, 113, 119, 122, 136, 174, 210, 231–232, 244, 268, 280, 289, 297, 358, 360
  management, 111
  NASA Summer Study, 204
  NASA Spacelab Program Office, 297, 350
  1973 Payload Model: Space Opportunities 1973–1991 (NASA report), 297, 350, 303
  Preliminary Design and Definition Study, MSFC, 289
  Research and Applications Module (RAM), 17–18
  TDRSS support, 298
Spacelab Task Force (OMSF), 269
Spacelab Task Team (OMSF), 269
Spacesuit, 93, 251, 255, 273, 320, 327
Spacewalk. See Extravehicular activity.
Spain, 17–18, 60, 116, 208, 268–269, 276, 288–289, 361
Sparkman, Sen. John J., 54
Sparrow (air-to-air missile), 189
Special Committee on Energy (President's), 121, 154–155
Spectrograph, 144, 165, 273, 276
Spectrometer, 53, 131, 143, 245, 208
Speer, Dr. F. A., 38
Sperry, Elmer A., Award, 10
Sperry, Lawrence, Award, 11
SPHINX (Space Plasma High Voltage Interaction Experiment spacecraft), 10–11, 253–254, 263
Spider, 225, 229, 241, 275
Spier, Raymond A., 209
Spirit of St. Louis Award, 135

Spirit of '76, The (presidential aircraft), 41
Spoiler. See Wing, aircraft.
Sputnik (U.S.S.R. satellite), 109, 155, 335, 342
SR–71 (reconnaissance aircraft), 265
SRAM (short-range-attack missile). See Missile, short-range-attack.
SRB (solid-fueled rocket booster). See Space shuttle.
SRM (solid-fueled rocket motor). See Space shuttle.
SS–11 (Savage; U.S.S.R. ICBM), 90
SS–17, 90
SS–18, 247
SSB. See Space Science Board.
S.S. *Hope* (hospital ship), 59
SSME. See Space shuttle, main engine.
S.S. *Statendam*, 3
SST. See Supersonic transport.
ST–124M (stabilized platform), 338
STA. See Shuttle training aircraft.
Stafford, B/G Thomas P. (USAF), 31, 35, 56, 276, 287, 326
Stand-up extravehicular activity (SEVA). See Extravehicular activity.
Stanford Univ., 97, 284
  Biomedical Applications Team, 81
Star (see also individual stars, such as Aldebaran and Gamma Velorium), 39, 232, 254, 267, 280, 329, 333, 341, 350
STAR (Satellites for Telecommunications, Applications, and Research; consortium), 304
Star City. See Zvezdny Gorodok.
Starr, David O., 52
State Committee for Science and Technology (U.S.S.R.), 93
State Committee on the Utilization of Atomic Energy (U.S.S.R.), 265
State, Dept. of, 4, 70, 93, 176, 187, 210, 332
Staten Island (USCG icebreaker), 49, 74
Staten Island, N.Y., 110
STDN. See Spaceflight Tracking and Data Network.
Stecker, Dr. Floyd W., 128
Steinberg, Robert, 128
Stennis, Sen. John C., 4, 18
Stereo (U.S.S.R–France experiment), 239
Stevens, Sen. Ted, 18, 31
Stevenson, Sen. Adlai E., III, 31
Stever, Dr. H. Guyford, 24, 49, 51, 89, 155, 203, 208, 215, 286–287, 331
STIF. See NASA Scientific and Technical Information Facility.
STIO (Scientific and Technical Information Office). See NASA Office of Industry Affairs and Technology Utilization.
Stockholm, Sweden, 290, 296
Stockholm, Univ. of, 240–241
STOL (short takeoff and landing) aircraft, 20, 90, 114, 186, 333
Strategic Air Command (SAC, USAF), 56, 244, 259

Strategic Arms Limitation Talks (SALT), 5, 139, 189–190, 214, 313
Stratford, Conn., 206
Stratofortress. See B–52.
Stratosphere, 46, 51, 74, 120–121, 174, 247, 264, 288, 295, 329
Stuhlinger, Dr. Ernst, 222, 288
Submarine (see also U-boat), 49, 94, 112–113, 136, 221, 236, 305
Submarine-launched missile (SLM). See Missile, submarine-launched.
Sud-Aviation Caravelle. See Caravelle.
Sudan, 200, 256
Suharto, President T. N. J. (Indonesia), 195
Sulfuric acid, 332
Summit meeting, U.S.–U.S.S.R., 176, 184, 187, 189–190, 192, 194
Sun (see also Solar headings), 4, 6, 12, 32, 44, 47, 59, 150, 205, 220, 222, 224, 345
    satellite data, 8, 47, 138–139, 185
"Sun in the Service of Mankind" (conference), 205
Sunol, Calif., 161
Sun-Orbiting Relativity Experiment Satellite. See SOREL.
Sunshine Project (Japan), 277
Super Arcas. See Arcas.
Super Bowl (football game), 13
Super Loki (sounding rocket), 214–215
Superconductivity, 154, 296
Supercritical wing. See Wing, aircraft.
Supernova, 11–12, 39, 232, 341
Supersonic transport (SST; see also Concorde and Tu-144)
    accident, 163, 173, 176, 175, 176, 179, 189
    CIAP research, 74
    congressional consideration, 101
    contract, 83
    criticism, 175
    environmental impact, 42, 57, 59, 201, 288
    European status, 113
    foreign airline services, 135
    funding, 42, 187
    materials patent, 136–137
    press comment, 120–121
    sonic boom, 95, 126–127, 175
Supreme Court. See U.S. Supreme Court.
Surface effect ship (SES), 231, 331
    Arctic surface effect vehicle (ASEV), 207
Surinam, 200
Surveyor (program), 350
Sutherland, Dr. Earl W., Jr., 284
Swaziland, 212
Sweden, 42–43, 77, 78–79, 210, 211, 219, 236, 237, 240–241, 290, 296, 361
Swedish Space Corp., 42–43
Swigert, John L., Jr., 78, 117, 153
Switzerland, 220
    disarmament, 211, 253, 313
    European Nuclear Research Center, 77, 302
    international cooperation, space, 244, 268–269, 287, 360
SWS (Saturn Workshop). See *Skylab 1*.
Symington, Rep. James W., 23, 240
Symington, Sen. Stuart, 4
Symphonie (French-German comsat), 361
Symposium on Earth Resources Technology Satellite, Third, 341
Synchronous communications satellite. See *Syncom 1*.
Synchronous Meteorological Satellite (SMS), 210, 313–314
Syncom 1 (synchronous communications satellite), 48, 222
Syncom 2, 222
Synthetic Fuels Pioneer program (proposed), 335
Syromyatikov, Vladimir S., 261
Systems Engineering Design Program. See MSFC Systems Engineering Design Program.
Systems Engineering Laboratories, 239
Systrain (translation computer), 99

T

T–2 (Fokker aircraft), 135
T–29 (training aircraft), 135–136
T–38, 144
Table Mountain Observatory, Calif.,164
TAC. See New Mexico, Univ. of, Technical Application Center.
TACT. See Transonic aircraft technology program.
Tahiti, 311
Taiwan. See China, Republic of.
Tape, Gerald F., 141
Tape recorder, 98, 190, 220, 345–346
Taranik, James V., 130
Tarasov, Lev, 84
Target-recognition attack (TRAM), 261–262
Taurus (constellation), 267, 310
Taurus-Littrow region (moon), 6, 32, 84, 112, 325, 346–347
Taurus Massif (moon), 139
Taurus Mountains (moon), 43, 346–347
Taylor, Dr. Edward S., 11
TCV. See Terminal-configured vehicle.
TDRSS. See Tracking and Data Relay Satellite System.
Teague, Rep. Olin E., 23, 26, 40–41, 63, 117, 161, 293
Technical Application Center (TAC). See New Mexico, Univ. of.
Technical Univ., Munich, W. Germany, 296
Technology, 232, 256, 282
    Federal support, 2, 133–134, 233, 272–273
    international cooperation, 13, 187–188, 255, 275, 277, 331
    press comment, 170, 289, 335
Technology, Inc., 39

"Technology in the Service of Mankind" (exhibit), 265
Technology Resources Survey and Applications Act (proposed), 272
Technology utilization, space (see also Research Applied to National Needs), 31, 279, 280, 304–305
  aeronautics, 13, 74, 92, 178, 196, 280
  agriculture, 102–103, 280
  automotive, 58, 263–264
  burglar alarm systems, 243, 291
  civil engineering, 280, 291
  commerce, 280
  communications, 48, 67–68, 91, 102–103, 108, 123, 273, 280
  computer systems, 360
  earth resources, 61–62, 63, 66, 70, 95, 210, 238
  earthquake prediction, 360
  education, 67–68
  electronics, 190
  employment, 105–106
  energy research and development, 42, 154, 160, 293, 330, 332, 360
  environment (see also Earth Resources Technology Satellite), 25, 88–89, 92, 102–103, 162, 170, 178, 215, 219
  fire safety, 88–89, 178, 216, 261, 291
  forestry, 212
  funding, 42, 139–140, 167, 219
  geology, 81–82, 102–103, 108
  geothermal research, 267
  heating and cooling, 177, 205–206, 213, 292, 318, 246, 349, 353, 360
  housing, 216
  industry, 105–106, 156
  law enforcement, 88–89, 280, 291, 295
  materials construction technology, 216, 257
  medicine, 52–53, 67, 81, 88–89, 97, 116, 178, 241, 246–247, 251, 280, 294, 298, 327, 360
  meteorology, 61–62, 102–103, 108
  mine safety, 88–89, 242
  money identification device, 327
  navigation, 242, 253, 280
  Nixon, President Richard M., 86–87
  nuclear power safety, 271
  oceanography, 108
  photography, 266
  policy and opinion, 233
  pollution control, 73, 92, 110
  postal service, 88–89
  press comment, 178, 197, 270, 271, 297
  tobaccoless cigarettes, 207
  transportation, 88–89, 102–103, 114, 280
  urban problems, 88–89, 133–134, 206, 279
Teir, William, 46
Telecare, 294
Teledyne Industries, Inc., 120
Teledyne Ryan Aeronautical Corp., 132
Telegraph, 209, 253, 294, 333, 351
Telephone, 67, 123, 154, 209, 222, 250, 253, 260, 273, 281, 294, 318, 333, 351

Telesat Canada, 79, 123, 360
Telesat–A (Canadian comsat). See *Anik 1*.
Telesat–B. See *Anik 2*.
Telescope (see also Apollo Telescope Mount, Large Space Telescope, Radiotelescope), 181, 344
  astrophotometer, 19
  comet observations, 4, 98, 294, 344, 358
  Mariner 10 mission, 144, 208
  Oao 3 observation, 249
  observatories. See individual observatories.
  Pioneer 10 mission, 47
  Pioneer 11 mission, 106
  solar, 12, 39, 132, 169, 318, 321
  Soyuz 13 mission, 347–348
  Spacelab, 61, 268
  ultralow-energy, 131
Teletype, 154, 326–327
Television
  Apollo 8 mission coverage, 350–351
  Apollo-Soyuz Test Project, 84
  balloon relay, 9
  camera, 218, 220, 291, 208
  educational, 18
  foreign censorship and control, 337
  laser transmission, 158
  lunar photography, 174
  satellite transmission, 38, 85, 107, 123, 154, 222, 250, 253, 273
  Second World Broadcasting Conference, 337
  Skylab 2 photographs, 145, 147, 148, 149–150, 184–185
  Skylab 3 mission coverage, 224–225
  Skylab 4 mission coverage, 322
  space shuttle, 207
  stellar UV observations, 165
  U.S.S.R., 8, 38, 107, 187–188, 209, 253, 294, 314, 318, 326–327, 333, 351
Teller, Dr. Edward H., 126, 165
Temper foam, 81
Tennessee, 46, 343
Tennessee Valley Authority (TVA), 86
Terhune, Gen. Charles H., 20
TERLS. See Thumba Equatorial Rocket Launching Station.
Terminal-configured vehicles (TCV), 72
Texas, 63, 157, 217, 229, 265–266, 271, 275
  accident, 250
  balloon launch, 103
  exhibit, 120
  Houston Chamber of Commerce, 251–252
  Skylab 3 postflight operations, 274
  WESTAR earth station, 184
Texas Instruments, Inc., 86
Texas, Univ. of, 24, 66, 263, 284–285
  Health Science Center, 350
Textron, Inc.
  Bell Helicopter Co. Div., 118–119, 192
  Dalmo Victor Div., 172
TF–33–P–7 (turbofan engine), 195

ASTRONAUTICS AND AERONAUTICS, 1973

TF-34, 63
Thailand, 245, 251
Thayer, W. Paul, 326
Thermal data channel, 143
Thermosphere (earth), 345
Thiokol Corp., 212, 253, 327, 328
13: The Flight That Failed (book), 77–78
Thomas, Albert, 251–252
Thompson, William I., 1
Thor–Delta (booster), 70, 191, 215, 257, 269, 304
    Burner II stage, 247
    long-tank, thrust-augmented (LTTAT), 122, 179, 212, 298, 312, 345
    Thorad, 45
Thornton, Rep. Ray, 23
Thornton, Dr. William P., 339
Thumba Equatorial Rocket Launching Station (TERLS), 38
Ticonderoga. See *U.S.S. Ticonderoga.*
Tidal wave, 328
Tidbinbilla Deep Space Communications Complex, Australia, 116, 308
Tidewater, Va., 206
Timm, Robert D., 55
Tinbergen, Dr. Nikolaas, 290
Tindall, Howard W., Jr., 339
Tiros-N, 29
Tiros Operational Satellite (*TOS*; see also *Itos 1, Noaa 1, Noaa 2, Noaa 3*), 312
Tischler, Adelbert O., 354
Tissandier, Paul, Diploma, 255
Titan (booster), 119
    Titan IIIB, 75
    Titan IIIB–Agena, 45, 155, 274
    Titan IIIB–Centaur, 69–70
    Titan IIIC, 181, 343
    Titan IIID, 316
    Titan–Centaur, 263, 333
    3rd stage, 138–139
Titan (Saturn moon). See Saturn (planet).
Titanium, 296
Titov, Gherman S., 277
Tokamak fusion reactor, 165
Tokyo, Japan, 137, 176, 268, 277, 281, 287
Tomcat. See F-14 aircraft.
Torell, Bruce N., 349
Tornado, 118, 189, 220
Toronto, Canada, 344
TOS. See Tiros Operational Satellite.
Toulmin, Dr. Stephen, 125
Toulouse, France, 73, 294–295, 338
Townsend, Mrs. Marjorie R., 52, 73
Tracking, (see also Earth station), 28, 94, 141, 208
    aircraft, 305
    capabilities, 83, 208
    funding, 28, 139–140, 298
    IMP program, 298
    satellite, 91, 277–278, 312
    ship, 97, 102, 133, 159, 209
    Skylab 2 mission, 142–143, 147, 152, 155
    sounding rocket experiment, 330–331

Tracking and Data Relay Satellite System (TDRSS), 298
Trade and Industry, Dept. of (U.K.), 17
Trade Union Congress (U.K.), 197
Train, Russell R., 223, 246, 259
TRAM. See Target-recognition attack.
Trans World Airlines, Inc. (TWA), 33
Transducer, 59
Transistor, 271
Transit navigation satellite system, 5, 104
Transonic aircraft technology program (TACT), 307
Transponder, 71
Transportation, 102–103, 114, 129, 272, 292, 293, 359
Transportation, Dept. of (DOT; see also Climatic Impact Assessment Program)
    air-route-control-system automation program, 46
    aviation, 31, 95
    contract, 86
    cooperation, 62, 97, 288
    executive reorganization, 39, 83
    funding, 28
    noise abatement, 97
    Office of Transportation Energy Policy, 341–342
    personnel, 63–64
    Under Secretary for Civil Aviation (USCA; proposed), 3
    World Weather Program for Fiscal Year 1974, 210
Transtage (growth-stage booster), 257, 302–303
Trapeznikov, V. A., 89
Treaty for the Peaceful Uses of Outer Space, 251–252
Trent, Darrell M., 134–135
Trident (undersea long-range missile system: ULMS), 28, 49, 112–113, 124, 236, 258–259, 336
Tritium, 159
Tritonis Lacus (Mars landing site), 138
Troitsky, Vsevolod, 182, 293
Tropics, 51, 255–256, 265, 329
Trott, Jack, 133
Trubshaw, Brian, 265
Truly, L/Cdr Richard H. (USN), 31–32
Truszynski, Gerald M., 83, 91, 298
TRW, Inc., 92, 146, 292
    Energy Systems Operations Div., 313
    Systems Group, 206
Tu-144 (U.S.S.R. supersonic transport), 7, 159, 163, 172–173, 175, 176–177, 179, 184, 191
Tucker, Gardiner L., 112
Tukey, Dr. John W., 284–285
Tunguska meteorite, 263
Tunney, Sen. John V., 31, 272
Tupolev, Gen. Aleksey A., U.S.S.R., 163, 172–173, 184
Tupolev, Andrey N., 9, 172–173
Turbogenerator, 80
Turcat, André, 265, 294–295, 338

Turner, L/C Harold E. (USAF, Ret.), 344
Turner, M/G Vernon R. (USAF), 170
Turtle, 303
TVA. See Tennessee Valley Authority.
TWA. See Trans World Airlines, Inc.
"2001" (film), 148, 149

U

U-2 (reconnaissance aircraft), 50
UAL. See United Air Lines.
U.A.R. See United Arab Republic.
U-boat (German submarine), 339
Uchinoura Space Center, Japan, 26
Udall, Rep. Morris K., 63
UFO. See Unidentified flying object.
Uganda, 200
UHF: ultrahigh frequency.
Uhuru (Small Astronomy Satellite). See *Explorer 42*.
U.K. See United Kingdom.
UK-5 (stellar x-ray satellite), 361
Ulmer, Ralph E., 201–202
ULMS (undersea long-range missile system). See Trident.
Ultrasonics, 295
Ultraviolet (UV)
  astronomy and physics
    comet study, 224, 294
    interplanetary, 106, 257
    Lunokhod 2, 19
    satellite, 13, 106, 146–147, 149, 169, 190, 224, 257, 279–280, 294, 341, 346, 347
    solar study, 146–147, 149, 169, 190, 346
    sounding rocket, 46, 51, 165, 174, 247, 254, 265, 279–280, 295, 329, 358–359
    stellar, 13, 165, 254, 279–280, 341, 347
  extreme (EUV), 169, 257, 346
  radiation effects, 59, 239
  spectrometer, 53
U.N. See United Nations.
Undersea long-range missile system (ULMS). See Trident.
Underwater detection system, 339
UNESCO. See United Nations Educational, Scientific, and Cultural Organization.
Unidentified flying object (UFO), 259, 291, 293, 295–296, 297
Unidentified satellite, 67, 75, 155, 181, 211, 247, 274
Union Mineral Alloys Corp., 186
Union of Soviet Socialist Republics. See U.S.S.R.
Union of Soviet Societies for Friendship and Cultural Relations with Foreign Countries, 133
United Action for Animals, 22–23, 195
United Air Lines, Inc. (UAL), 9–10, 182, 302
United Aircraft Corp., 10, 120, 204
  Hamilton Standard Div., 188–189
  Pratt & Whitney Div., 117, 127, 195, 197, 204, 291, 349
  Sikorsky Aircraft Div., 247
United Arab Republic (U.A.R.), 284, 287, 290, 292, 301, 314
United Kingdom (U.K.), 221, 288, 339
  aircraft (see also Concorde), 263, 264, 278
  airlines, 134–135
  award, 80, 266, 296
  Hovertrain, 47
  international cooperation, 114, 117, 135, 200, 288, 333, 342
  international cooperation, space, 17, 24, 78–79, 210, 244, 328, 329, 360, 361
  laser weapons, 169
  missile purchase, 175–176
  observatory, 297
  research and development expenditures, 256
  satellite, 1, 78–79, 210
  space debris, 185–186
  space program, 233
  Trade Union Congress, 197
  World Helicopter Championships, 261
United Nations (U.N.), 156, 193–194, 223, 361
  Convention on International Liability for Damage Caused by Space Objects, 257
  Intergovernmental Maritime Consultative Organization Communications Subcommittee, 67
  Working Group on Remote Sensing of Earth by Satellites, 42
United Nations Educational, Scientific, and Cultural Organization (UNESCO), 205
United States (U.S.; see also appropriate agencies and Congress), 286
  aerospace industry. See Aerospace industry.
  airline industry. See Airlines.
  astronaut tour, 13
  budget, 26, 214, 233, 256, 258–259, 290, 292, 359
  communications, 4, 67, 217, 273
  defense, 41, 64, 90, 94–95, 187, 214, 233, 236, 249, 256, 258–259, 261–262, 272, 292, 333, 353–354
  disarmament, 97, 139, 189–190, 192, 214, 249, 258–259, 313
  economy, 134–135, 156, 224, 256, 259, 345
  energy crisis, 42, 237–238, 239, 240, 246, 314, 325, 329–330, 335, 337, 342, 346, 347, 348, 355, 361–362
  environmental program, 41, 42, 92–93, 178, 272
  foreign policy, 134, 224, 263, 301
  immigration, 241, 248
  international cooperation, 181, 202, 219–220
    Australia, 333
    Brazil, 108–109
    Canada, 333

Europe, 175–176, 278, 342
France, 41, 114, 264, 333
Germany, West, 333
India, 165
Japan, 277, 342
Mexico, 237
P.R.C., 33
U.K., 264, 333
U.S.S.R.
  air services, 193
  Direct Communications Link, 361
  environmental protection, 45
  exhibit, 163
  joint working groups. See Joint Working Groups.
  lunar sample presentation, 236
  medicine and health, 55
  nuclear testing, 221–222
  oceanography, 47, 116–117, 237
  science and technology, 187–188, 205, 277, 331
  scientist exchange, 170
  summit meeting, 134, 176, 184, 186, 187–188, 189–190, 192, 194, 244–245
World Helicopter Championships, 220
World Weather Program Plan for Fiscal Year 1974, 210
international cooperation, space. See International cooperation, space.
national security, 17, 258–259, 293
  intelligence, 17
patent policy, 256
press comment, 224, 237–238, 249, 259, 271, 286–287, 292, 301, 325, 335, 344
prisoner of war return, 183
research and development, 7, 188, 205–206, 233, 236, 240, 241, 243–244, 256, 258–259, 267, 272–273, 277, 278, 290, 293, 335, 338–339, 342
satellite monitoring, 180, 227
science and technology, 41, 48, 205, 255, 277, 285, 286–287, 290, 292, 301
space program. See Space program, national; Space race; Space results.
trade, 83, 205–206, 224, 233, 255, 256, 362
transportation. See Transportation.
Vietnam war. See Vietnam war.
United States and Soviet Progress in Space: Summary Data Through 1972 and a Forward Look (congressional report), 30–31
United States Foreign Policy for the 1970's: Shaping a Durable Peace (presidential report), 134
United States Military Posture for FY 1974, 94–95
United States Space Science Program (SSB report), 162
United Technology Center, 212, 253
Universal Oil Products Corp., 284–285
Universe, 115–116, 249, 324, 325, 331–332, 337, 342, 344

Universities (see also individual universities), 232, 271
  contract, 63
  cooperation, 52, 97, 250, 261, 298, 315
  Federal support, 102, 129–130, 237, 243–244
Universities Scientific Research Assn., 232
University College, London, Eng., 329
University Corp. for Atmospheric Research, 315
Uranium, 53
Uranus (planet), 310, 317, 350, 357
Urban affairs, 79, 272, 344–345
Urinalysis, 291
U.S. See United States.
U.S.–U.S.S.R. Aeronautical Technology Symposium and Display, 27–28
U.S.–U.S.S.R. Apollo-Soyuz Test Project. See Apollo-Soyuz Test Project.
U.S.–U.S.S.R. Direct Communications Link, 361
U.S.–U.S.S.R. Joint Commission on Cooperation in the Field of Environmental Protection, 87, 134
U.S.–U.S.S.R. Joint Commission on Scientific and Technical Cooperation, 24, 87, 89, 101, 134, 331
U.S.–U.S.S.R. Joint Committee on Cooperation in the Peaceful Uses of Atomic Energy, 189–190
U.S.–U.S.S.R. Joint Editorial Board on Space Biology and Medicine, 244–245, 341
U.S.–U.S.S.R. Joint Working Groups. See Joint Working Groups.
U.S. Air Force (USAF; see also individual bases, centers, and commands; Air Force Avionics Laboratory; Air Force Cambridge Research Laboratory; Air Force Space and Missile Test Center; Air Force Weather Service; Space and Missile Systems Command; Vandenberg AFB)
  aircraft (see also individual aircraft, such as C-5A, F-15, F-111, YF-12), 13, 24–25, 200, 343
    contract, 11, 13, 40, 56, 63, 119, 134, 143, 160, 172, 183, 210–211, 217, 218, 291, 343
  award, 122, 135, 183, 231
  balloon, 9
  communications system, 71, 104, 114, 132, 158, 343
  computer use, 99
  contract, 13, 38, 57, 65, 73, 92, 96, 108, 114, 120, 155, 172, 176, 175, 197, 204, 314, 349
  cooperation
    NASA, 22–23, 71, 90, 103, 174, 195, 206, 209, 214–215, 216, 235, 266, 284, 285, 291, 301, 307, 319, 358
    NOAA, 70–71
    USN, 301
  Defense Satellite Communication System, Phase 2, 104, 343

hypersonic-aerospacecraft research program. See X-24B.
launch, 67, 75, 155, 181, 211, 247, 274, 316, 343
lifting body, 216, 235, 284, 285, 304, 303, 318, 319, 342, 358
medical research, 19
missile program, 28, 38, 63, 73, 112-113, 165, 353-354
Office of the Surgeon General, 350
personnel
   appointment, 204-205, 248
   astronaut. See individual astronauts.
   award, 128, 129, 265-266
   promotion, 109, 170, 172, 221, 236, 248, 282, 331
   rehire, 265
   resignation, 153-154, 173
satellite, 9, 22-23, 57, 70-71, 104, 114, 288, 292, 312
Satellite Data System (planned), 114, 312
Skylab recovery force, 223-224, 319
space program (see also Defense, Dept. of), 104, 113, 128, 142-143, 155, 169
space shuttle, 71, 103, 262, 296
Veto project, 9
U.S. Arms Control and Disarmament Agency (USACDA), 5, 98, 134
U.S. Army (USA; see also individual centers and commands, such as Army Air Corps; Army Air Defense Command; Army Air Mobility Research and Development Laboratory; Army Ballistic Missile Agency; Army Engineering Development Center; Fort Myer)
   aircraft (see also individual aircraft), 32, 344
   contract, 13, 26, 40, 92, 118-119, 192
   cooperation, 209, 247
   missile, 248
   personnel, 134, 248
   radio communications, 71
   space program, 142-143
   Women's Auxiliary Corps, 172
U.S. Capitol, 19, 51
U.S. Coast Guard (USCG), 13, 22, 47, 49, 74, 108, 110, 176
U.S. Committee for the Global Atmospheric Research Program, 59
U.S. Geological Survey (USGS), 16, 64, 66, 86, 104, 110
U.S. Junior Chambers of Commerce (Jaycees), 181
U.S. Marine Corps (USMC), 328, 341
U.S. Navy (USN; see also individual centers and stations, such as Naval Air Test Center; Naval Observatory; Naval Research Laboratory; Naval Shipyard; Naval Torpedo Station Keyport; Naval Weapons Center)
   aircraft (see also individual aircraft, such as F-14, P-3), 50
   awards, 183, 286
   communications, 65, 104
   conference, 260
   contract, 4, 13, 50, 63, 104, 207
   cooperation, 110, 116-117, 127-128, 131, 209, 253, 288, 301
   Fleet Satellite Communications System. See Fleet Satellite Communications System.
   Fleet Weather Facility. See Fleet Weather Facility.
   missile program, 112-113, 328
   Navigation Satellite System. See Navy Navigation Satellite System.
   patrol hydrofoil missile ship (PHMS), 140, 185
   personnel (see also individual astronauts), 213, 218, 248, 333
   satellite, 5, 104, 115
   space program, 104, 142-143
   space shuttle, 43
   surface effect ship (SES), 231, 331
   Surface Ship Acquisition Div., 231
U.S. Participation in the U.N.: Report by the President to the Congress for the Year 1972, 257
U.S. Postal Service, 88-89, 124, 362
U.S. Secret Service, 176
U.S. Supreme Court, 153
U.S. Weather Service. See National Weather Service.
USA. See U.S. Army.
USACDA. See U.S. Arms Control and Disarmament Agency.
USAF. See U.S. Air Force.
USC. See Southern California, Univ. of.
USCG. See U.S. Coast Guard.
USGS. See U.S. Geological Survey.
USMC. See U.S. Marine Corps.
USN. See U.S. Navy.
USNS *Vanguard*, 97
U.S.S. *New Orleans*, 229, 270, 274, 319, 323
U.S.S. *Ticonderoga*, 152, 192
U.S.S. *Wasp*, 186
U.S.S.R. (Union of Soviet Socialist Republics; see also Soviet headings)
   aerospace industry, 9
   agreement. See individual agreements; International cooperation; International cooperation, space.
   aircraft, 7, 23, 80, 101, 111, 159, 191, 203, 329
      accident, 163, 173, 175-177, 179, 184, 191
      reconnaissance, 144
   Arab-Israeli war monitoring, 284, 287, 290, 292, 295, 301
   astronaut visit to, 133
   astronomy, 167, 182, 200
   award, 186, 329
   commercial aviation, 172-173, 184, 193, 277, 278, 289
   Committee for Science and Technology, 216, 275
   communications, 209, 253, 284, 294, 314, 318, 333, 348-349, 351, 361
   computer technology, 90
   cosmonaut. See Cosmonaut.

Council of Ministers, 176-177, 184, 187-188, 287
disarmament, 97, 214, 249
   SALT, 139, 189-190, 258-259, 313
Embassy, U.S., 190, 236
employment, 256
exhibit, 216
hijacking of aircraft policy, 9
hovercraft, 331
Institute of Geochemistry, 282
Institute of Space Research, 282
international cooperation
   aeronautics, 135, 216
   air services, 193
   environment, 45, 87
   exhibit, 163, 263, 277
   joint working groups. See Joint Working Groups.
   medicine and health, 55
   meteorology, 24, 71, 210
   nuclear testing, 51, 221-222
   oceanography, 22, 47, 49, 74, 87, 116-117, 237
   peaceful uses of atomic energy, 189-190, 265
   research and development, 101
   science and technology, 13, 25, 89, 187-188, 264, 331
   scientist exchange, 170
   summit talks, 134, 176, 184, 187-188, 189-190, 192, 194
   tourism, 326
   visit to U.S., 301
international cooperation, space
   Czechoslovakia, 302
   France, 24, 65, 71, 78-79, 96, 239, 351, 359
   Germany, 78-79
   Germany, East, 302
   Poland, 122
   Sweden, 78-79
   U.K., 78-79
   U.S., 65, 78-79, 162, 164, 186, 269, 344, 361
     Apollo-Soyuz Test Project. See Apollo-Soyuz Test Project.
     lunar data exchange, 64, 84, 96, 335-336
     manned Mars mission, 267, 330
     press comment, 141
Krasnoyarsk Institute of Physics, 287
launch
   Arktika (icebreaker), 292
   cancellation, 159
   probe
     Luna 19, 283
     Luna 21, 8, 174, 346-347
     Mars 4, 218
     Mars 5, 221
     Mars 6, 239
     Mars 7, 242
   Salyut 1 (space station), 273
   Salyut 2, 102-103, 114
   satellite
     Copernicus 500, 122

Cosmos. See individual Cosmos satellites.
   Intercosmos 10, 302
   Intercosmos Copernicus 500, 122
   Meteor 14, 88
   Meteor 15, 166
   Molniya I-23, 38
   Molniya I-24, 253
   Molniya II-5, 107
   Molniya II-6, 209
   Molniya II-7, 294
   Molniya II-8, 351
   Oreol 2, 351
   Prognoz 3, 47
sounding rocket, 54, 56, 71, 206
Soyuz spacecraft, 159
Soyuz 10, 198, 273
Soyuz 11, 273, 282, 348, 359
Soyuz 12, 273, 276, 348, 359
Soyuz 13, 347, 359
licensing agency, 80
military strength, 214, 221, 249, 354
Ministry of Civil Aviation, 216
Ministry of Defense, 354
missile and rocket program, 50, 90, 94-95, 181, 187, 247, 249, 268, 303, 333
missile threat to, 17, 61
Mission Control Center, 326-327
Moscow Center of Control of Manned Space Flight, 207-208
nuclear capability, 90
nuclear energy facilities, 194, 213, 265
probe
   Luna 19, 284
   Luna 21, 8, 174, 346-347
   Lunokhod 1, 8
   Lunokhod 2, 8, 24, 30, 43, 65, 77, 96, 110-111, 139, 164, 173, 346-347, 359
   Mars 1, 218
   Mars 3, 64, 65, 218
   Mars 4, 218, 221, 239, 242
   Mars 5, 221, 239, 242
   Mars 6, 239, 242
   Mars 7, 242
   Venus 8, 36
reconnaissance satellite, 45, 284, 287, 290, 292, 293-295, 301
research and development, 256, 277
science and technology, 11, 38, 54, 65, 87, 89, 126, 128-129, 134, 159, 205, 231, 233, 256, 277, 287, 293, 312, 333
space debris, 185-186
Space Flight Control Center, 282
space program, 116, 156, 179, 195-196
   achievements, 32, 42, 65, 102-103, 108, 114-115, 162, 209, 276, 287, 348, 359
   biology, 287, 303
   competition with U.S., 17
   launch record, 92, 123-124, 154, 159, 233, 357, 359
   lunar exploration (see also Luna probes), 12-13, 43

plans, 30–31, 36, 276
press comment, 179, 196
press conference, 326
secrecy, 141
space rescue capability, 240
spacesuit, 93, 255
training, 26, 54, 207
space shuttle, 277
space station, 102–103, 108, 159, 174–175, 182, 195–196, 330, 359
spacecraft. See U.S.S.R., launch; and individual spacecraft, such as *Luna 19*, Salyut, Soyuz, *Vostok 6*.
state commission, 276
State Committee for Science and Technology, 331
State Committee on the Utilization of Atomic Energy, 265
supersonic transport (SST), 7, 9, 135, 172–173, 176, 175, 176–177, 179, 184, 191
tracking ship, 209
trade, 292
weaponry, 94–95, 120, 249
World War II, 354
Usery, Willie J., Jr., 48–49
Ushuaia, Argentina, 48, 50–51, 62
Utah, 210
Utah, Univ. of, 65, 80, 110
UV. See Ultraviolet.

V

V–2 (long-range ballistic missile), 275
VAB (Vehicle Assembly Building). See Kennedy Space Center.
Vaiana, Dr. Guiseppe, 185
Valais, Switzerland, 287
Van Allen radiation belts, 33, 309
Van Staden, George, 91–92
Vanadium, 296
Vandenberg AFB, Calif., 69, 104, 155, 211, 247, 274, 301, 316
Vanguard (satellite), 33
Vanguard 1, 185–186
Vanguard Project, 351–352, 359
Vanik, Rep. Charles A., 95, 236
Vatican Astronomical Observatory, Italy, 333
Vecchietti, George J., 327
Vega (star), 308
Vegetation, 45, 247, 287, 344–345
Vehicle Assembly Building (VAB). See Kennedy Space Center.
Venezuela, 24, 37–38, 197, 217, 265–266
Venus (planet; see also *Mariner 10* and Venus probes)
atmosphere, 64, 77, 144, 183–184, 304–305, 332
equator, 350
LST observation, 125
surface, 64, 239, 242
Venus 8 (U.S.S.R. interplanetary probe), 36, 64, 65
Vermont, 73

Vernon Township, N.J., 141, 184
Vernov, Sergey, 287
Vershinin, Konstantin A., 354
Vertical/short takeoff and landing aircraft. See V/STOL.
Vertical takeoff and landing aircraft. See VTOL.
Vertical-temperature-profile radiometer (VTPR), 312
Very-high-resolution radiometer (VHRR), 24, 312
Vescelus, Glenn E., 20
Veterans, 186, 198
Veto project, 9
Vette, Col. Alan R. (USAF), 143–144, 223–224, 319
VFR. See Visual flight rules.
VFX (fighter aircraft), 281
VHF: very high frequency.
VHRR. See Very-high-resolution radiometer.
VIC (Visitor Information Center). See Kennedy Space Center.
Vienna, Austria, 139, 261
Vietnam war, 9, 26, 135, 137, 214, 258–259
Viking (program), 5, 12, 78, 251–252, 263, 341, 357, 359
Viking (spacecraft), 36, 64, 70, 78, 138, 263
Viking Biology Team, 341
VIMS. See Virginia Institute of Marine Science.
Viper Dart (sounding rocket), 214–215
Virgin Islands, 237
Virginia, 114, 129, 138, 217, 240
Virginia Institute of Marine Science (VIMS), 206, 240
Virginia Polytechnic Institute, Agronomy Dept., 129
Virgo (constellation), 329
Vishniac, Dr. Wolf V., 341
Visitor Information Center (VIC). See Kennedy Space Center.
Visual flight rules (VFR), 131
VLF: very low frequency.
Vogt, Sen. John, 92
Volcano, 85, 328
Volpe, Secretary of Transportation John A., 5
von Braun, Dr. Wernher, 21, 248, 272
von Frisch, Dr. Karl, 290
von Kármán Lecture, 9, 11
Voskhod 2 (U.S.S.R. spacecraft), 162
Vostok 1 mission, 114–115, 116, 273
Vostok 6, 26
Vought Missiles & Space Co. See LTV Aerospace Corp.
"Voyage beyond Apollo" (cruise), 3
V/STOL (vertical/short takeoff and landing) aircraft, 117, 263
VTOL (verticle takeoff and landing) aircraft, 305
VTPR. See Vertical-temperature-profile radiometer.

W

Wabash River, 197
Wac Corporal (rocket), 275
Waite, Jack H., 266-267
Walker, Gov. Daniel (Ill.), 282
Wallops Island, Va., 239, 243
Wallops Station (NASA), 261
   aircraft research, 72, 231
   conference, 260
   contract, 94, 178-179, 300-301
   cooperation, 129, 240
   employment, 30, 359-360
   launch
      balloonsondes, 214-215
      sounding rocket
         Astrobee D, 201
         Black Brant IVB, 243
         Javelin, 214
         Nike-Apache, 15-16, 201, 239, 243
         Nike-Cajun, 201, 214
         Super Arcas, 15-16
         Super Loki, 214-215
         Viper Dart, 214-215
   personnel, 201-202
Wamflex (wave-induced momentum flux experiment) Project, 59
Ward, Julian E., Award, 256
Warhead, 247, 333, 353-354
Warner, Secretary of the Navy John W., 49, 75, 117, 286
Warsaw, Poland, 54, 213
Washington, 140, 243, 353-354
Washington, D.C., 41, 67, 73, 271, 289, 335, 347
   air services, 193
   Apollo celebration, 21
   astronaut tour, 130
   awards presented in, 52, 76, 83, 122, 150, 183, 299, 304, 345
   earth resources photography, 344-345
   Intelsat secretariat, 14
   meetings, 9, 11-12, 14-15, 44, 85, 87, 101, 114, 116, 119, 126, 135, 136, 137-138, 155, 184, 186, 213, 244, 268, 331, 333, 344-345
   news media, 292
   press conference, 40, 70-71, 153, 187, 210-211, 247
   Workshop on Wind Energy Conversion Systems, 181
Washington, Univ. of, 16, 40
Wasserburg, Dr. Gerald J., 65-66, 128
Water pollution, 110, 178, 188-189, 253
Water resources management, 62, 89, 92-93, 183, 219-220, 245, 344-345
Watergate controversy, 132, 154-155, 168, 190-191, 275
Watson-Watt, Sir Robert, 339
Wave-induced momentum flux experiment. See Wamflex Project.
Weapon systems (see also Disarmament; Nuclear weapons), 210, 236, 261-262
   Giant Patriot Project, 353-354
   U.S.S.R., 249

Weather. See Meteorology.
Weather and Climate Modification: Problems and Progress (NAS-NRC report), 238
Weather Service. See National Weather Service.
Webb, James E., 21
Webster, James B., 167-168
Webster, Robert H., 190
Weicker, Sen. Lowell P., Jr., 9, 13, 30, 40, 84
Weightlessness effects, 177, 209, 241-242, 257, 273, 323
   animal experiments, 22-23, 195, 223, 225, 229, 275, 303
   astronaut response, 192, 193, 194, 199, 222, 223, 224, 323, 347-348, 351
   space processing, 79, 222, 223, 271
Weitz, Capt. Paul J. (USN), 90, 112, 126, 139, 142, 144, 145, 164, 177, 181-182, 185, 192, 193, 195, 197, 199, 218, 222, 286, 300
West German Hydrographic Institute, Hamburg, 261
West Germany. See Germany, West.
West Indies, 46, 51, 133, 174, 247, 265, 295, 329
West Palm Beach, Fla., 291
West Virginia, Univ. of, 216
WESTAR (Western Union communications satellite), 135, 141, 184
Western Assn. of State Game and Fish Commissioners, 210
Western European Union, 264
Western Hemisphere, 206
Western Test Range (WTR), 1, 17, 212, 219-220, 312, 345-346
Western Union communications satellite system. See WESTAR.
Western Union International, Inc., 5, 44, 186, 217, 252, 260
   American Satellite Corp., 260
Western Union Telegraph Co., 4, 135, 141, 184
Westinghouse Electric Corp., 5, 177, 190, 292
   Aerospace and Electronic Systems Center, 41
Westland Aircraft Ltd., 80
Wetmore, William H., 205-206
Wheeler, Dr. John A., 125
"When Earth Became a Planet" (exhibit), 4
Whipple, Dr. Fred L., 294
Whitcomb, Dr. Richard T., 143, 284-285, 307
White, George C., 31
White, Dr. Paul Dudley, 217
White, Dr. Sarah, 191
White, Gen. Thomas D., Space Trophy, 128, 129
White dwarf (star), 39
White House, 9, 154-155, 214, 301
   announcement, 141
   ceremonies, 73, 231, 265-266, 290, 336
   meeting, 89, 186, 189-190

staff, 2, 19, 117, 248
White House Fellows Program, 256
White Sands Missile Range (WSMR), N. Mex.
  "Great White Sands Missile Range Lost Gold Treasure Affair," 291
  launch
    Aerobee 150A, 161
    Aerobee 170
      astronomy experiment, 111–112, 165, 204
      solar physics experiment, 75, 261
      x-ray astronomy, 203
    Aerobee 170A, 92, 98, 172, 285
    Aerobee 200, 19, 221, 267, 298
    Aerobee 200A, 181, 250, 329, 349
    Black Brant VC
      astronomy experiment, 42–43
      CALROC subsystem test, 22, 158
      instrument evaluation, 104
      solar physics experiment, 173, 182–183, 243, 256, 340
    CALROC, 121
    Nike-Apache, 65, 245
    Nike-Cajun, 214–215
    Nike-Tomahawk, 228
White Sands Naval Ordnance Missile Test Facility, 22, 121
Whitlam, Prime Minister E. G. (Australia), 116
Whitney, John A., 26, 40
Whittle, Dr. Michael W., 207
Wichita, Kans., 107–108
Widnall, Sheila E., 11
Wienfeld, Edward, 195
Wiesner, Dr. Jerome B., 170
Wilkinson, Geoffrey, 296
Williams, Dr. Charles, 161
Williams, Dr. Donald D., 48
Wilmarth, Dr. Verl R., 143, 181–182, 197, 222, 245
Wilson, Sen. Lori, 92
Wilson, Dr. Robert Rathbun, 284–285
Winblade, Roger L., 138
Wind, 90–91, 118, 181, 185, 220, 261
Wind River Mountains, 46
Wind tunnel, 68–69, 72, 80, 107–108, 133, 143, 232, 243, 270, 286, 307, 349
Windler, Dr. Gerd, 352
Wing, aircraft, 83, 101–102
  advanced technology, 117, 285
  extendable, 159
  fixed, 32, 305, 319
  retractable, 7, 305
  serrated, 232
  supercritical, 68–69, 72, 143, 280, 307
  swing, 33
  variable-sweep, 280, 307
  wake vortex, 302
  yawed, 72
Winn, Rep. Larry, Jr., 23, 189
Winter, Frank H., 288
Wisconsin, Univ. of, 130, 307, 317
Wolfe, Dr. John H., 47, 310
Women, 54, 172, 218, 220, 255, 262, 278, 296, 299–300, 329

Wood, H. William, 53–54
Woods Hole, Mass., 204
Woomera, Australia, 307, 315, 316, 317
Working Group on Control and Guidance (ASTP), 207–208
Working Group on Mission Model and Operational Plans (ASTP), 207–208
Workshop on Wind Energy Conversion Systems (NSF), 181
World Helicopter Championships (II), 220
World War I, 220
World War II, 26, 118, 137, 186, 214, 272, 291–292, 301, 314, 339, 354, 361–362
The World Weather Program Plan for Fiscal Year 1974, 210
Wright, Orville, 345
Wright Brothers Day, 346
Wright Brothers Memorial Trophy (1973), 304, 345
Wright-Patterson AFB, Ohio, 59, 118
  Aerospace Medical Research Laboratory, 19
  Avionics Laboratory, 158
  Foreign Translation Div., 99
WSMR. See White Sands Missile Range.
WTR. See Western Test Range.
WUI. See Western Union International, Inc.
Wydler, Rep. John W., 23
Wyld, James H., Propulsion Award (AIAA, 1973), 313
Wyle Laboratories, 239

X

X-4 (U.K. technology spacecraft), 17, 361
X-15 (experimental hypersonic aircraft), 280
X-24A (lifting body), 206, 216
X-24B, 206, 216, 220, 235, 247, 253, 285, 303, 304, 318, 319, 342, 358
X2048 (aluminum alloy), 136–137
XB-70 (experimental supersonic aircraft), 59
XLR-11 (rocket engine), 216, 319
X-ray
  astronomy, 11–12, 98, 142, 204, 254, 358–359
  equipment, 228, 266
  mapping, 92, 224–225, 257
  medical applications, 88–89
  photography, 52, 266, 322–323, 348
  solar, 35, 38–39, 47, 139, 223, 329
  source, 11–12, 35, 38–39, 223, 329
  stellar, 279–280, 361
X-ray Astronomy Symposium, 128

Y

Yak (U.S.S.R. jet transport), 163
Yeliseyev, Dr. Aleksey S., 84, 163, 208
YF-12 (triplesonic research aircraft), 59

YF-12 Thermal Loads Calibration Team (FRC), 299
YF-16 (lightweight fighter aircraft), 266
York, Dr. Donald G., 249
Yost, Dr. Edward F., 46
Young, Col. Clarence M. (USA, Ret.), 113
Young, Capt. John W. (USN), 255
Young, Stanley G., 295
Yucatan Peninsula, 228
Yugoslavia, 56
Yukon River, 66

Z

Zeiglschmid, Dr. John, 70
Zero gravity. See Weightlessness effects.
Zeigler, Henri, 102, 217
Zinc, 315, 324–325
Zodiacal light, 174
Zurich, Switzerland, 220
Zvezdny Gorodok (Star City), U.S.S.R. 133, 207, 276, 326–327

www.ingramcontent.com/pod-product-compliance
Lightning Source LLC
Chambersburg PA
CBHW081714170526
45167CB00009B/3578